COMMONLY USED FUNCTIONS

	$-A$	$90 \pm A$	$180 \pm A$	$270 \pm A$	$360\,k \pm A$
sin	$-\sin A$	$\cos A$	$\mp \sin A$	$-\cos A$	$\pm \sin A$
cos	$\cos A$	$\mp \sin A$	$-\cos A$	$\pm \sin A$	$\cos A$

$$\sin (A \pm B) = \sin A \cos B \pm \cos A \sin B$$

$$\cos (A \pm B) = \cos A \cos B \mp \sin A \sin B$$

$$\sin 2A = 2 \sin A \cos A$$

$$\cos 2A = 1 - 2 \sin^2 A = 2 \cos^2 A - 1$$

$$\sin A + \sin B = 2 \sin \frac{A + B}{2} \cos \frac{A - B}{2}$$

$$\sin A - \sin B = 2 \cos \frac{A + B}{2} \sin \frac{A - B}{2}$$

$$\cos A + \cos B = 2 \cos \frac{A + B}{2} \cos \frac{A - B}{2}$$

$$\cos A - \cos B = 2 \sin \frac{A + B}{2} \sin \frac{B - A}{2}$$

$$\sin A \sin B = \frac{1}{2}[\cos (A - B) - \cos (A + B)]$$

$$\cos A \cos B = \frac{1}{2}[\cos (A - B) + \cos (A + B)]$$

$$\sin A \cos B = \frac{1}{2}[\sin (A - B) + \sin (A + B)]$$

$$\int \sin nx \, dx = -\frac{\cos nx}{n}$$

$$\int \sin^2 nx \, dx = \frac{x}{2} - \frac{\sin 2nx}{4n}$$

COMMONLY USED FUNCTIONS

$$\int \sin mx \sin nx \, dx = \frac{\sin (m - n)x}{2(m - n)} - \frac{\sin (m + n)x}{2(m + n)} \qquad \text{for } m \neq n$$

$$\int \cos nx \, dx = \frac{\sin nx}{n}$$

$$\int \cos^2 nx \, dx = \frac{x}{2} + \frac{\sin 2nx}{4n}$$

$$\int \cos mx \cos nx \, dx = \frac{\sin (m - n)x}{2(m - n)} + \frac{\sin (m + n)x}{2(m + n)} \qquad \text{for } m \neq n$$

$$\int \sin nx \cos nx \, dx = \frac{\sin^2 nx}{2n}$$

$$\int \sin mx \cos nx \, dx = \frac{\cos (m - n)x}{2(m - n)} - \frac{\cos (m + n)x}{2(m + n)} \qquad \text{for } m \neq n$$

SOME UNITS AND CONSTANTS

Quantity	Units	Equivalent
Length	1 meter (m)	3.281 feet (ft)
		39.36 inches (in)
Mass	1 kilogram (kg)	2.205 pounds (lb)
		35.27 ounces (oz)
Force	1 newton (N)	0.2248 force pounds (lbf)
Torque	1 newton-meter (N.m.)	0.738 pound-feet (lbf.ft)
Moment of inertia	1 kilogram-meter^2 (kg.m^2)	23.7 pound-feet^2 (lb.ft^2)
Power	1 watt (W)	0.7376 foot-pounds/second
		1.341×10^{-3} horsepower (hp)
Energy	1 joules (J)	1 watt-second
		0.7376 foot-pounds
		2.778×10^{-7} kilowatt-hours (kWh)
Horsepower	1 hp	746 watts
Magnetic flux	1 weber (Wb)	10^8 maxwells or lines
Magnetic flux density	1 tesla (T)	1 weber/meter^2 (Wb/m^2)
		10^4 gauss
Magnetic field intensity	1 ampere-turn/meter (At/m)	1.257×10^2 oersted
Permeability of free space	$\mu_0 = 4\pi \times 10^7$ H/m	

POWER ELECTRONICS CIRCUITS, DEVICES, AND APPLICATIONS

Third Edition

POWER ELECTRONICS CIRCUITS, DEVICES, AND APPLICATIONS

Third Edition

Muhammad H. Rashid

Electrical and Computer Engineering
University of West Florida

PEARSON

Prentice
Hall

Pearson Education International

Vice President and Editorial Director, ECS: *Marcia J. Horton*
Publisher: *Tom Robbins*
Associate Editor: *Alice Dworkin*
Vice President and Director of Production and Manufacturing, ESM: *David W. Riccardi*
Executive Managing Editor: *Vince O'Brien*
Managing Editor: *David George*
Production Editor: *Donna King*
Director of Creative Services: *Paul Belfanti*
Creative Director: *Carole Anson*
Art Director: *Jayne Conte*
Cover Designer: *Bruce Kenselaar*
Art Editor: *Greg Dulles*
Manufacturing Manager: *Trudy Pisciotti*
Manufacturing Buyer: *Lynda Castillo*
Marketing Manager: *Holly Stark*

© 2004 by Pearson Education, Inc.
Pearson Prentice Hall
Pearson Education, Inc.
Upper Saddle River, NJ 07458

ORCAD is a registered trademark of the Cadence Design Systems, Inc.
Mathcad is a registered trademark of the MathSoft, Inc.
IBM®PC is a registered trademark of International Business Machines Corporation.
PSpice® is a registered trademark of MicroSim Corporation.

Printed in the United States of America

10 9 8 7 6 5

ISBN 0-13-122815-3

Pearson Education Ltd., *London*
Pearson Education Australia Pty., Ltd., *Sydney*
Pearson Education Singapore, Pte. Ltd.
Pearson Education North Asia Ltd., *Hong Kong*
Pearson Education Canada, Inc., *Toronto*
Pearson Educación de Mexico, S.A. de C.V.
Pearson Education—Japan, *Tokyo*
Pearson Education Malaysia, Pte. Ltd.
Pearson Education, Inc., *Upper Saddle River, New Jersey*

To my parents, my wife Fatema, and
my family: Faeza, Farzana, Hasan and Hussain

Contents

Preface

The third edition of *Power Electronics* is intended as a textbook for a course on power electronics/static power converters for junior or senior undergraduate students in electrical and electronic engineering. It can also be used as a textbook for graduate students and as a reference book for practicing engineers involved in the design and applications of power electronics. The prerequisites are courses on basic electronics and basic electrical circuits. The content of *Power Electronics* is beyond the scope of a one-semester course. The time allocated to a course on power electronics in a typical undergraduate curriculum is normally only one semester. Power electronics has already advanced to the point where it is difficult to cover the entire subject in a one-semester course. For an undergraduate course, Chapters 1 to 11 should be adequate to provide a good background on power electronics. Chapters 12 to 16 could be left for other courses or included in a graduate course. Table P.1 shows suggested topics for a one-semester course on "Power Electronics" and Table P.2 for one semester course on "Power Electronics and Motor Drives."

TABLE P.1   Suggested Topics for One Semester Course on Power Electronics

Chapter	Topics	Sections	Lectures
1	Introduction	1.1 to 1.12	2
2	Power semiconductor diodes and circuits	2.1 to 2.4, 2.7, 2.10 to 2.13	2
3	Diode rectifiers	3.1 to 3.9	5
4	Power transistors	4.2, 4.10, 4.11	2
5	DC–DC converters	5.1 to 5.7	5
6	PWM inverters	6.1 to 6.6, 6.8 to 6.11	7
7	Thyristors	7.1 to 7.5, 7.9, 7.10	2
8	Resonant pulse inverters	8.1 to 8.5	3
10	Controlled rectifiers	10.1 to 10.6	6
11	AC voltage controllers	11.1 to 11.5	3
12	Static switches	12.1 to 12.8	2
	Mid-term exams and quizzes		3
	Final exam		3
	Total lectures in a 15-week semester		45

TABLE P.2   Suggested Topics for One Semester Course on Power Electronics and Motor Drives

Chapter	Topics	Sections	Lectures
1	Introduction	1.1 to 1.12	2
2	Power semiconductor diodes and circuits	2.1 to 2.4, 2.7, 2.10 to 2.13	2
3	Diode rectifiers	3.1 to 3.8	4
4	Power transistors	4.2, 4.10, 4.11	1
5	DC–DC converters	5.1 to 5.7	4
6	PWM inverters	6.1 to 6.6, 6.8 to 6.11	5
7	Thyristors	7.1 to 7.5, 7.9, 7.10	1
10	Controlled rectifiers	10.1 to 10.7	5
11	AC voltage controllers	11.1 to 11.5	2
Appendix	Magnetic circuits	B 1.6 to 16.6	1
15	DC drives	15.1 to 15.7	5
Appendix	Three-phase circuits	A 1.6 to 16.6	1
14	AC drives	16.1 to 16.6	6
	Mid-term exams and quizzes		3
	Final exam		3
	Total lectures in a 15-week semester		45

The fundamentals of power electronics are well established and they do not change rapidly. However, the device characteristics are continuously being improved and new devices are added. *Power Electronics,* which employs the bottom-up approach, covers device characteristics conversion techniques first and then applications. It emphasizes the fundamental principles of power conversions. This third edition of *Power Electronics* is a complete revision of the second edition, and (i) features bottom-up approach rather than top-down approach; (ii) introduces the state-of-the-art advanced Modulation Techniques; (iii) presents three new chapters on "Multilevel Inverters" (Chapter 9), "Flexible AC Transmission Systems" (Chapter 13), and "Gate Drive Circuits" (Chapter 17) and covers state-of-the-art techniques; (iv) integrates the industry standard software, SPICE, and design examples that are verified by SPICE simulation; (v) examines converters with RL-loads under both continuous and discontinuous current conduction; and (vi) has expanded sections and/or paragraphs to add explanations. The book is divided into five parts:

1. Introduction—Chapter 1
2. Devices and gate-drive circuits—Chapters 2, 4, 7, and 17
3. Power conversion techniques—Chapters 3, 5, 6, 8, 9, 10, and 11
4. Applications—Chapters 12, 13, 14, 15, and 16
5. Protection and thermal modeling—Chapter 18

Topics like three-phase circuits, magnetic circuits, switching functions of converters, DC transient analysis, and Fourier analysis are reviewed in the Appendices.

Power electronics deals with the applications of solid-state electronics for the control and conversion of electric power. Conversion techniques require the switching on and off of power semiconductor devices. Low-level electronics circuits, which normally consist of integrated circuits and discrete components, generate the required gating signals for the power devices. Integrated circuits and discrete components are being replaced by microprocessors and signal processing ICs.

An ideal power device should have no switching-on and -off limitations in terms of turn-on time, turn-off time, current, and voltage handling capabilities. Power semiconductor technology is rapidly developing fast switching power devices with increasing voltage and current limits. Power switching devices such as power BJTs, power MOSFETs, SITs, IGBTs, MCTs, SITHs, SCRs, TRIACs, GTOs, MTOs, ETOs, IGCTs, and other semiconductor devices are finding increasing applications in a wide range of products. With the availability of faster switching devices, the applications of modern microprocessors and digital signal processing in synthesizing the control strategy for gating power devices to meet the conversion specifications are widening the scope of power electronics. The power electronics revolution has gained momentum, since the early 1990s. Within the next 20 years, power electronics will shape and condition the electricity somewhere between its generation and all its users. The potential applications of power electronics are yet to be fully explored but we've made every effort to cover as many applications as possible in this book.

Any comments and suggestions regarding this book are welcomed and should be sent to the author.

Dr. Muhammad H. Rashid
Professor and Director
Electrical and Computer Engineering
University of West Florida
11000 University Parkway
Pensacola, FL 32514–5754

E-mail: mrashid@uwf.edu

PSPICE SOFTWARE AND PROGRAM FILES

The student version PSpice schematics and/or Orcad capture software can be obtained or downloaded from

Cadence Design Systems, Inc.
2655 Seely Avenue
San Jose, CA 95134

Websites: http://www.cadence.com
          http://www.orcad.com
          http://www.pspice.com

The website http://uwf.edu/mrashid contains all PSpice circuits, PSpice schematics, Orcad capture, and Mathcad files for use with this book.

Important Note: The PSpice circuit files (with an extension .CIR) are self-contained and each file contains any necessary device or component models. However, the PSpice schematic files (with an extension .SCH) need the user-defined model library file ***Rashid_PE3_MODEL.LIB***, which is included with the schematic files, and ***must be included*** from the Analysis menu of PSpice Schematics. Similarly, the Orcad schematic files (with extensions .OPJ and .DSN) need the user-defined model library file **Rashid_PE3_MODEL.LIB**, which is included with the Orcad schematic files, ***must be included*** from the PSpice Simulation settings menu of Orcad Capture. Without these files being included while running the simulation, it will not run and will give errors.

ACKNOWLEDGMENTS

Many people have contributed to this edition and made suggestions based on their classroom experience as a professor or a student. I would like to thank the following persons for their comments and suggestions:

Mazen Abdel-Salam, *King Fahd University of Petroleum and Minerals, Saudi Arabia*
Johnson Asumadu, *Western Michigan University*
Ashoka K. S. Bhat, *University of Victoria, Canada*
Fred Brockhurst, *Rose-Hulman Institution of Technology*
Jan C. Cochrane, *The University of Melbourne, Australia*
Ovidiu Crisan, *University of Houston*
Joseph M. Crowley, *University of Illinois, Urbana-Champaign*
Mehrad Ehsani, *Texas A&M University*
Alexander E. Emanuel, *Worcester Polytechnic Institute*
George Gela, *Ohio State University*
Herman W. Hill, *Ohio University*
Constantine J. Hatziadoniu, *Southern Illinois University, Carbondale*
Wahid Hubbi, *New Jersey Institute of Technology*
Marrija Ilic-Spong, *University of Illinois, Urbana-Champaign*
Shahidul I. Khan, *Concordia University, Canada*
Hussein M. Kojabadi, *Sahand University of Technology , Iran*
Peter Lauritzen, *University of Washington*
Jack Lawler, *University of Tennessee*
Arthur R. Miles, *North Dakota State University*
Medhat M. Morcos, *Kansas State University*
Hassan Moghbelli, *Purdue University Calumet*
H. Rarnezani-Ferdowsi, *University of Mashhad, Iran*
Prasad Enjeti, *Texas A&M University*
Saburo Mastsusaki, *TDK Corporation, Japan*
Vedula V. Sastry, *Iowa State University*
Elias G. Strangas, *Michigan State University*
Selwyn Wright, *The University of Huddersfield, Queensgate, UK*
S. Yuvarajan, *North Dakota State University*

It has been a great pleasure working with the editor, Alice Dworkin and the production editor, Donna King. Finally, I would thank my family for their love, patience, and understanding.

MUHAMMAD H. RASHID
Pensacola, Florida

About the Author

Muhammad H. Rashid received the B.Sc. degree in electrical engineering from the Bangladesh University of Engineering and Technology and the M.Sc. and Ph.D. degrees from the University of Birmingham, UK.

Currently, he is a Professor of electrical engineering with the University of Florida and the Director of the UF/UWF Joint Program in Electrical and Computer Engineering. Previously, he was a Professor of electrical engineering and the Chair of the Engineering Department at Indiana University–Purdue University at Fort Wayne. In addition, he was a Visiting Assistant Professor of electrical engineering at the University of Connecticut, Associate Professor of electrical engineering at Concordia University (Montreal, Canada), Professor of electrical engineering at Purdue University, Calumet, and Visiting Professor of electrical engineering at King Fahd University of Petroleum and Minerals, Saudi Arabia. He has also been employed as a design and development engineer with Brush Electrical Machines Ltd. (UK), as a Research Engineer with Lucas Group Research Centre (UK), and as a Lecturer and Head of Control Engineering Department at the Higher Institute of Electronics (Malta). He is actively involved in teaching, researching, and lecturing in power electronics. He has published 14 books and more than 100 technical papers. His books have been adopted as textbooks all over the world. His book *Power Electronics* has been translated into Spanish, Portuguese, Indonesian, Korean and Persian. His book *Microelectronics* has been translated into Spanish in Mexico and Spain. He has had many invitations from foreign governments and agencies to be a keynote lecturer and consultant, from foreign universities to serve as an external Ph.D. examiner, and from funding agencies to serve as a research proposal reviewer. His contributions in education have been recognized by foreign governments and agencies. He has previously lectured and consulted for NATO for Turkey in 1994, UNDP for Bangladesh in 1989 and 1994, Saudi Arabia in 1993, Pakistan in 1993, Malaysia in 1995 and 2002, and Bangkok in 2002, and has been invited by foreign universities in Australia, Canada, Hong Kong, India, Malaysia, Singapore to serve as an external examiner for undergraduate, master's and Ph.D. degree examinations, by funding agencies in Australia, Canada, United States, and Hong Kong to review research proposals, and by U.S. and foreign universities to evaluate promotion cases for professorship. He has previously authored seven books published by Prentice Hall: *Power Electronics—Circuits, Devices, and Applications* (1988, 2/e 1993), *SPICE For Power Electronics* (1993), *SPICE for Circuits and Electronics Using PSpice*

(1990, 2/e 1995), *Electromechanical and Electrical Machinery* (1986), and *Engineering Design for Electrical Engineers* (1990). He has authored five IEEE self-study guides: *Self-Study Guide on Fundamentals of Power Electronics, Power Electronics Laboratory Using PSpice, Selected Readings on SPICE Simulation of Power Electronics*, and *Selected Readings on Power Electronics* (IEEE Press, 1996) and *Microelectronics Laboratory Using Electronics Workbench* (IEEE Press, 2000). He also wrote two books: *Electronic Circuit Design using Electronics Workbench* (January 1998), and *Microelectronic Circuits—Analysis and Design* (April 1999) by PWS Publishing). He is editor of *Power Electronics Handbook* published by Academic Press, 2001.

Dr. Rashid is a registered Professional Engineer in the Province of Ontario (Canada), a registered Chartered Engineer (UK), a Fellow of the Institution of Electrical Engineers (IEE, UK) and a Fellow of the Institute of Electrical and Electronics Engineers (IEEE, USA). He was elected as an IEEE Fellow with the citation "*Leadership in power electronics education and contributions to the analysis and design methodologies of solid-state power converters.*" He was the recipient of the *1991 Outstanding Engineer Award* from The Institute of Electrical and Electronics Engineers (IEEE). He received the 2002 IEEE Educational Activity Award (EAB) Meritorious Achievement Award in Continuing Education with the citation "*for contributions to the design and delivery of continuing education in power electronics and computer-aided-simulation*". He was also an ABET program evaluator for electrical engineering from 1995 to 2000 and he is currently an engineering evaluator for the Southern Association of Colleges and Schools (SACS, USA). He has been elected as an IEEE-Industry Applications Society (IAS) Distinguished Lecturer. He is the Editor-in-Chief of the *Power Electronics and Applications Series,* published by CRC Press.

CHAPTER 1

Introduction

The learning objectives of this chapter are as follows:

- To get an overview of power electronics and its history of development
- To get an overview of different types of power semiconductor devices and their switching characteristics
- To learn about the types of power converters
- To know about resources for finding manufacturers of power semiconductors
- To know about resources for finding published articles of power electronics and applications

1.1   APPLICATIONS OF POWER ELECTRONICS

The demand for control of electric power for electric motor drive systems and industrial controls existed for many years, and this led to early development of the Ward–Leonard system to obtain a variable dc voltage for the control of dc motor drives. Power electronics have revolutionized the concept of power control for power conversion and for control of electrical motor drives.

Power electronics combine power, electronics, and control. Control deals with the steady-state and dynamic characteristics of closed-loop systems. Power deals with the static and rotating power equipment for the generation, transmission, and distribution of electric energy. Electronics deal with the solid-state devices and circuits for signal processing to meet the desired control objectives. *Power electronics* may be defined as the applications of solid-state electronics for the control and conversion of electric power. The interrelationship of power electronics with power, electronics, and control is shown in Figure 1.1.

Power electronics are based primarily on the switching of the power semiconductor devices. With the development of power semiconductor technology, the power-handling capabilities and the switching speed of the power devices have improved tremendously. The development of microprocessors and microcomputer technology has a great impact on the control and synthesizing the control strategy for the power semiconductor devices. Modern power electronics equipment uses (1) power semiconductors that can be regarded as the muscle, and (2) microelectronics that have the power and intelligence of a brain.

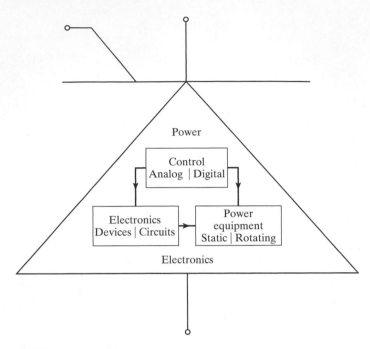

FIGURE 1.1

Relationship of power electronics to power, electronics, and control.

Power electronics have already found an important place in modern technology and are now used in a great variety of high-power products, including heat controls, light controls, motor controls, power supplies, vehicle propulsion systems, and high-voltage direct-current (HVDC) systems. It is difficult to draw the flexible ac transmissions (FACTs) boundaries for the applications of power electronics, especially with the present trends in the development of power devices and microprocessors. Table 1.1 shows some applications of power electronics [5].

1.1.1   History of Power Electronics

The history of power electronics began with the introduction of the mercury arc rectifier in 1900. Then the metal tank rectifier, grid-controlled vacuum-tube rectifier, ignitron, phanotron, and thyratron were introduced gradually. These devices were applied for power control until the 1950s.

The first electronics revolution began in 1948 with the invention of the silicon transistor at Bell Telephone Laboratories by Bardeen, Brattain, and Schockley. Most of today's advanced electronic technologies are traceable to that invention. Modern microelectronics evolved over the years from silicon semiconductors. The next breakthrough, in 1956, was also from Bell Laboratories: the invention of the *PNPN* triggering transistor, which was defined as a thyristor or silicon-controlled rectifier (SCR).

The second electronics revolution began in 1958 with the development of the commercial thyristor by the General Electric Company. That was the beginning of a

TABLE 1.1   Some Applications of Power Electronics

Advertising	Magnets
Air-conditioning	Mass transits
Aircraft power supplies	Mercury-arc lamp ballasts
Alarms	Mining
Appliances	Model trains
Audio amplifiers	Motor controls
Battery charger	Motor drives
Blenders	Movie projectors
Blowers	Nuclear reactor control rod
Boilers	Oil well drilling
Burglar alarms	Oven controls
Cement kiln	Paper mills
Chemical processing	Particle accelerators
Clothes dryers	People movers
Computers	Phonographs
Conveyors	Photocopies
Cranes and hoists	Photographic supplies
Dimmers	Power supplies
Displays	Printing press
Electric blankets	Pumps and compressors
Electric door openers	Radar/sonar power supplies
Electric dryers	Range surface unit
Electric fans	Refrigerators
Electric vehicles	Regulators
Electromagnets	RF amplifiers
Electromechanical electroplating	Security systems
Electronic ignition	Servo systems
Electrostatic precipitators	Sewing machines
Elevators	Solar power supplies
Fans	Solid-state contactors
Flashers	Solid-state relays
Food mixers	Space power supplies
Food warmer trays	Static circuit breakers
Forklift trucks	Static relays
Furnaces	Steel mills
Games	Synchronous machine starting
Garage door openers	Synthetic fibers
Gas turbine starting	Television circuits
Generator exciters	Temperature controls
Grinders	Timers
Hand power tools	Toys
Heat controls	Traffic signal controls
High-frequency lighting	Trains
High-voltage dc (HVDC)	TV deflections
Induction heating	Ultrasonic generators
Laser power supplies	Uninterruptible power supplies
Latching relays	Vacuum cleaners
Light dimmers	Volt-ampere reactive (VAR) compensation
Light flashers	Vending machines
Linear induction motor controls	Very low frequency (VLF) transmitters
Locomotives	Voltage regulators
Machine tools	Washing machines
Magnetic recordings	Welding

Source: Ref. 5.

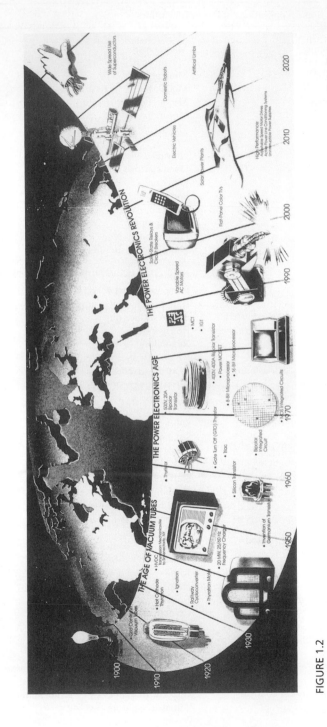

FIGURE 1.2

History of power electronics. (Courtesy of Tennessee Center for Research and Development.)

new era of power electronics. Since then, many different types of power semiconductor devices and conversion techniques have been introduced. The microelectronics revolution gave us the ability to process a huge amount of information at incredible speed. The power electronics revolution is giving us the ability to shape and control large amounts of power with ever-increasing efficiency. Due to the marriage of power electronics, the muscle, with microelectronics, the brain, many potential applications of power electronics are now emerging, and this trend will continue. Within the next 30 years, power electronics will shape and condition the electricity somewhere in the transmission network between its generation and all its users. The power electronics revolution has gained momentum since the late 1980s and early 1990s [1]. A chronological history of power electronics is shown in Figure 1.2.

1.2   POWER SEMICONDUCTOR DEVICES

Since the first thyristor SCR was developed in late 1957, there have been tremendous advances in the power semiconductor devices. Until 1970, the conventional thyristors had been exclusively used for power control in industrial applications. Since 1970, various types of power semiconductor devices were developed and became commercially available. Figure 1.3 shows the classification of the power semiconductors, which are made of either silicon or silicon carbide. Silicon carbide devices are, however, under development. A majority of the devices are made of silicon. These devices can be divided broadly into three types: (1) power diodes, (2) transistors, and (3) thyristors. These can be divided broadly into five types: (1) power diodes, (2) thyristors, (3) power bipolar junction transistors (BJTs), (4) power metal oxide semiconductor field-effect transistors (MOSFETs), and (5) insulated-gate bipolar transistors (IGBTs) and static induction transistors (SITs).

1.2.1   Power Diodes

A diode has two terminals: a cathode and an anode. Power diodes are of three types: general purpose, high speed (or fast recovery), and Schottky. General-purpose diodes are available up to 6000 V, 4500 A, and the rating of fast-recovery diodes can go up to 6000 V, 1100 A. The reverse recovery time varies between 0.1 and 5 μs. The fast-recovery diodes are essential for high-frequency switching of power converters. Schottky diodes have low on-state voltage and very small recovery time, typically nanoseconds. The leakage current increases with the voltage rating and their ratings are limited to 100 V, 300 A. A diode conducts when its anode voltage is higher than that of the cathode; and the forward voltage drop of a power diode is very low, typically 0.5 and 1.2 V. If the cathode voltage is higher than its anode voltage, a diode is said to be in a *blocking mode*. Figure 1.4 shows various configurations of general-purpose diodes, which basically fall into two types. One is called a *stud*, or *stud-mounted* type, and the other is called a *disk, press pak*, or *hockey puck* type. In a stud-mounted type, either the anode or the cathode could be the stud.

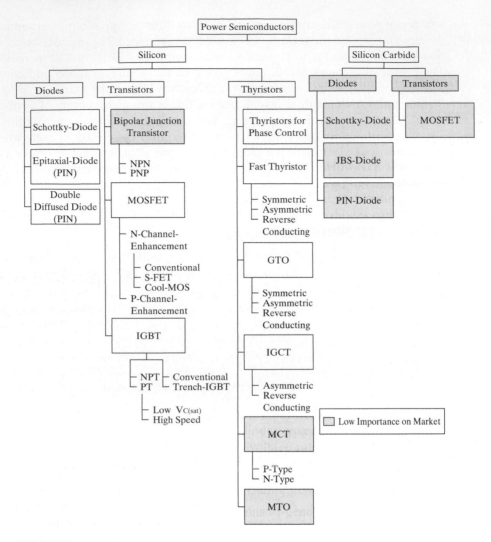

FIGURE 1.3

Classification of the power semiconductors. [Ref. 2, S. Bernet]

1.2.2 Thyristors

A thyristor has three terminals: an anode, a cathode, and a gate. When a small current is passed through the gate terminal to cathode, the thyristor conducts, provided that the anode terminal is at a higher potential than the cathode. The thyristors can be subdivided into eleven types: (a) forced-commutated thyristor, (b) line-commutated thyristor, (c) gate-turn-off thyristor (GTO), (d) reverse-conducting thyristor (RCT), (e) static induction thyristor (SITH), (f) gate-assisted turn-off thyristor (GATT), (g) light-activated silicon-controlled rectifier (LASCR), (h) MOS turn-off (MTO)

FIGURE 1.4

Various general-purpose diode configurations. (Courtesy of Powerex, Inc.)

thyristor, (i) emitter turn-off (ETO) thyristor, (j) integrated gate-commutated thyristor (IGCT), and (k) MOS-controlled thyristors (MCTs). Once a thyristor is in a conduction mode, the gate circuit has no control and the thyristor continues to conduct. When a thyristor is in a conduction mode, the forward voltage drop is very small, typically 0.5 to 2 V. A conducting thyristor can be turned off by making the potential of the anode equal to or less than the cathode potential. The line-commutated thyristors are turned off due to the sinusoidal nature of the input voltage, and forced-commutated thyristors are turned off by an extra circuit called *commutation circuitry*. Figure 1.5 shows various configurations of phase control (or line-commutated) thyristors: stud, hockey puck, flat, and pin types.

Natural or line-commutated thyristors are available with ratings up to 6000 V, 4500 A. The *turn-off time* of high-speed reverse-blocking thyristors has been improved substantially and it is possible to have 10 to 20 µs in a 3000-V, 3600-A thyristor. The turn-off time is defined as the time interval between the instant when the principal current has decreased to zero after external switching of the principal voltage circuit, and the instant when the thyristor is capable of supporting a specified principal voltage without

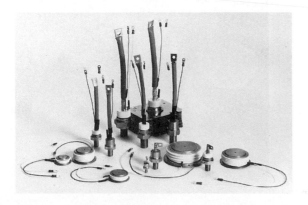

FIGURE 1.5

Various thyristor configurations. (Courtesy of Powerex, Inc.)

turning on. RCTs and GATTs are widely used for high-speed switching, especially in traction applications. An RCT can be considered as a thyristor with an inverse-parallel diode. RCTs are available up to 4000 V, 2000 A (and 800 A in reverse conduction) with a switching time of 40 μs. GATTs are available up to 1200 V, 400 A with a switching speed of 8 μs. LASCRs, which are available up to 6000 V, 1500 A, with a switching speed of 200 to 400 μs, are suitable for high-voltage power systems, especially in HVDC. For low-power ac applications, TRIACs are widely used in all types of simple heat controls, light controls, motor controls, and ac switches. The characteristics of TRIACs are similar to two thyristors connected in inverse parallel and having only one gate terminal. The current flow through a TRIAC can be controlled in either direction.

GTOs and SITHs are self-turned-off thyristors. GTOs and SITHs are turned on by applying a short positive pulse to the gates and are turned off by the applications of short negative pulse to the gates. They do not require any commutation circuit. GTOs are very attractive for forced commutation of converters and are available up to 6000 V, 6000 A. SITHs, whose ratings can go as high as 1200 V, 300 A, are expected to be applied for medium-power converters with a frequency of several hundred kilohertz and beyond the frequency range of GTOs. Figure 1.6 shows various configurations of GTOs. An MTO [3] is a combination of a GTO and a MOSFET, which together overcome the limitations of the GTO turn-off ability. Its structure is similar to that of a GTO and retains the GTO advantages of high voltage (up to 10 kV) and high current (up to 4000 A). MTOs can be used in high power applications ranging from 1 to 20 MVA. An ETO is a MOS-GTO hybrid device that combines the advantages of both the GTO and MOSFET. ETO has two gates: one normal gate for turn-on and one with a series MOSFET for turn-off. ETOs with a current rating of up to 4 kA and a voltage rating of up to 6 kV have been demonstrated.

An IGCT [4] integrates a gate-commutated thyristor (GCT) with a multilayered printed circuit board gate drive. The GCT is a hard-switched GTO with a very fast and large gate current pulse, as large as the full-rated current, that draws out all the current from the cathode into the gate in about 1 μs to ensure a fast turn-off. Similar to a GTO,

FIGURE 1.6

Gate-turn-off thyristors. (Courtesy of International Rectifiers.)

the IGCT is turned on by applying the turn-on current to its gate. The IGCT is turned off by a multilayered gate-driver circuit board that can supply a fast-rising turn-off pulse (i.e., a gate current of 4 kA/μs with gate-cathode voltage of 20 V only). An MCT can be turned "on" by a small negative voltage pulse on the MOS gate (with respect to its anode), and turned "off" by a small positive voltage pulse. It is like a GTO, except that the turn-off gain is very high. MCTs are available up to 4500 V, 250 A.

1.2.3   Power Transistors

Power transistors are of four types: (1) BJTs, (2) power MOSFETs, (3) IGBTs, and (4) SITs. A bipolar transistor has three terminals: base, emitter, and collector. It is normally operated as a switch in the common-emitter configuration. As long as the base of an *NPN*-transistor is at a higher potential than the emitter and the base current is sufficiently large to drive the transistor in the saturation region, the transistor remains on, provided that the collector-to-emitter junction is properly biased. High-power bipolar transistors are commonly used in power converters at a frequency below 10 kHz and are effectively applied in the power ratings up to 1200 V, 400 A. The various configurations of bipolar power transistors are shown in Figure 4.2. The forward drop of a conducting transistor is in the range 0.5 to 1.5 V. If the base drive voltage is withdrawn, the transistor remains in the nonconduction (or off) mode.

Power MOSFETs are used in high-speed power converters and are available at a relatively low power rating in the range of 1000 V, 100 A at a frequency range of several tens of kilohertz. The various power MOSFETs of different sizes are shown in Figure 4.24. IGBTs are voltage-controlled power transistors. They are inherently faster than BJTs, but still not quite as fast as MOSFETs. However, they offer far superior drive and output characteristics to those of BJTs. IGBTs are suitable for high voltage, high current, and frequencies up to 20 kHz. IGBTs are available up to 1700 V, 2400 A.

COOLMOS [8] is a new technology for high-voltage power MOSFETs, and it implements a compensation structure in the vertical drift region of a MOSFET to improve the on-state resistance. It has a lower on-state resistance for the same package compared with that of other MOSFETs. The conduction losses are at least 5 times less as compared with those of the conventional MOSFET technology. COOLMOS is capable of handling two to three times more output power as compared to the conventional MOSFET in the same package. The active chip area of COOLMOS is approximately 5 times smaller than that of a standard MOSFET. The on-state resistance of a 600 V, 47 A COOLMOS is 70 mΩ.

A SIT is a high-power, high-frequency device. It is essentially the solid-state version of the triode vacuum tube, and is similar to a junction field-effect transistor (JFET). It has a low-noise, low-distortion, high-audio frequency power capability. The turn-on and turn-off times are very short, typically 0.25 μs. The normally on-characteristic and the high on-state drop limit its applications for general power conversions. The current rating of SITs can be up to 1200 V, 300 A, and the switching speed can be as high as 100 kHz. SITs are most suitable for high-power, high-frequency applications (e.g., audio, VHF/ultrahigh frequency [UHF], and microwave amplifiers).

Figure 1.7 shows the power range of commercially available power semiconductors. The ratings of commercially available power semiconductor devices are shown in Table 1.2, where the on-voltage is the on-state voltage drop of the device at

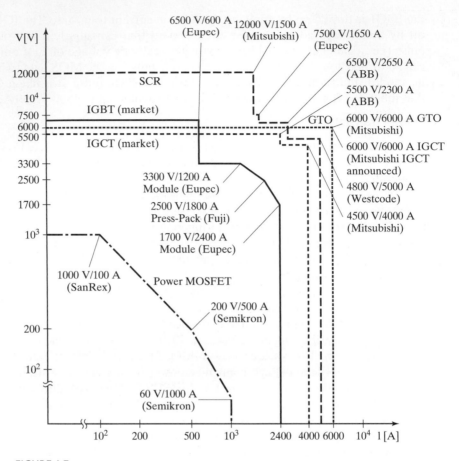

FIGURE 1.7

Power ranges of commercially available power semiconductors. [Ref. 2, S. Bernet]

the specified current. Table 1.3 shows the symbols and the $v-i$ characteristics of commonly used power semiconductor devices.

Figure 1.8 shows the applications and frequency range of power devices. A super-power device should (1) have a zero on-state voltage, (2) withstand an infinite off-state voltage, (3) handle an infinite current, and (4) turn on and off in zero time, thereby having infinite switching speed.

1.3   CONTROL CHARACTERISTICS OF POWER DEVICES

The power semiconductor devices can be operated as switches by applying control signals to the gate terminal of thyristors (and to the base of bipolar transistors). The required output is obtained by varying the conduction time of these switching devices. Figure 1.9 shows the output voltages and control characteristics of commonly used power switching devices. Once a thyristor is in a conduction mode, the gate signal of either positive or negative magnitude has no effect and this is shown in Figure 1.9a.

TABLE 1.2   Ratings of Power Semiconductor Devices

Device Type	Devices		Voltage/Current Rating	Upper Frequency (Hz)	Switching Time (μs)	On-State Resistance (Ω)
Power Diodes	Power diodes	General purpose	4000 V/4500 A	1 k	50–100	0.32 m
			6000 V/3500 A	1 k	50–100	0.6 m
			600 V/9570 A	1 k	50–100	0.1 m
			2800 V/1700 A	20 k	5–10	0.4 m
		High speed	4500 V/1950 A	20 k	5–10	1.2 m
			6000 V/1100 A	20 k	5–10	1.96 m
			600 V/17 A	30 k	0.2	0.14
		Schottky	150 V/80 A	30 k	0.2	8.63 m
Power Transistors	Bipolar transistors	Single	400 V/250 A	25 k	9	4 m
			400 V/40 A	30 k	6	31 m
			630 V/50 A	35 k	2	15 m
		Darlington	1200 V/400 A	20 k	30	10 m
	MOSFETs	Single	800 V/7.5 A	100 k	1.6	1
	COOLMOS	Single	800 V/7.8 A	125 k	2	1.2 m
			600 V/40 A	125 k	1	0.12 m
			1000 V/6.1 A	125 k	1.5	2 Ω
	IGBTs	Single	2500 V/2400 A	100 k	5–10	2.3 m
			1200 V/52 A	100 k	5–10	0.13
			1200 V/25 A	100 k	5–10	0.14
			1200 V/80 A	100 k	5–10	44 m
			1800 V/2200 A	100 k	5–10	1.76 m
	SITs		1200 V/300 A	100 k	0.5	1.2
Thyristors (Silicon-Controlled Rectifiers)	Phase control thyristors	Line-commutated low speed	6500 V/4200 A	60	100–400	0.58 m
			2800 V/1500 A	60	100–400	0.72 m
			5000 V/4600 A	60	100–400	0.48 m
			5000 V/3600 A	60	100–400	0.50 m
			5000 V/5000 A	60	100–400	0.45 m
	Forced-turned-off thyristors	Reverse blocking high speed	2800 V/1850 A	20 k	20–100	0.87 m
			1800 V/2100 A	20 k	20–100	0.78 m
			4500 V/3000 A	20 k	20–100	0.5 m
			6000 V/2300 A	20 k	20–100	0.52 m
			4500 V/3700 A	20 k	20–100	0.53 m
		Bidirectional	4200 V/1920 A	20 k	20–100	0.77 m
		RCT	2500 V/1000 A	20 k	20–100	2.1 m
		GATT	1200 V/400 A	20 k	10–50	2.2 m
		Light triggered	6000 V/1500 A	400	200–400	0.53 m
	Self-turned-off thyristors	GTO	4500 V/4000 A	10 k	50–110	1.07 m
		HD-GTO	4500 V/3000 A	10 k	50–110	1.07 m
		Pulse GTO	5000 V/4600 A	10 k	50–110	0.48 m
		SITH	4000 V/2200 A	20 k	5–10	5.6 m
		MTO	4500 V/500 A	5 k	80–110	10.2 m
		ETO	4500 V/4000 A	5 k	80–110	0.5 m
		IGCT	4500 V/3000 A	5 k	80–110	0.8 m
	TRIACs	Bidirectional	1200 V/300 A	400	200–400	3.6 m
	MCTs	Single	4500 V/250 A	5 k	50–110	10.4 m
			1400 V/65 A	5 k	50–110	28 m

TABLE 1.3 Characteristics and Symbols of Some Power Devices

Devices	Symbols	Characteristics
Diode	A I_D ⊳⊢ K + V_{AK} −	I_D / 0 / V_{AK}
Thyristor	I_A ⊳⊢ ∘G A + V_{AK} − K	I_A / Gate triggered / 0 / V_{AK}
SITH	G A ∘⊳⊢ ∘K	
GTO	I_A G A + V_{AK} − K	I_A / Gate triggered / 0 / V_{AK}
MCT	A∘⌐ ⌐∘K G	
MTO	Cathode Turn-on gate Turn-off gate Anode	
ETO	Cathode Turn-off gate Turn-on gate Anode	
IGCT	Cathode Gate (turn-on & turn-off) Anode	
TRIAC	∘ I_A B A ∘G	I_A / Gate triggered / 0 / V_{AB} / Gate triggered
LASCR	A K I_A ∘G	I_A / Gate triggered / 0 / V_{AK}
NPN BJT	I_B I_C C B∘ E I_E	I_C / I_{Bn} $I_{Bn} > I_{B1}$ / I_{B1} / 0 / V_{CE}
IGBT	C I_C G∘ E I_E	I_C / V_{GSn} $V_{GSn} > V_{GS1}$ / V_{GS1} / V_T / 0 / V_{CE}
N-Channel MOSFET	D I_D G∘ S	I_D / V_{GS0} $V_{GS1} > V_{GSn}$ / V_{GS1} / V_{GSn} / 0 / V_{DS}
SIT	D ∘ S	I_D / $V_{GS1} = 0$ V $V_{GS1} < V_{GSn}$ / V_{GSn} / 0 / V_{DS}

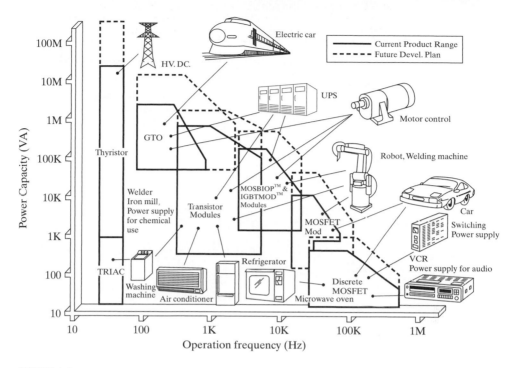

FIGURE 1.8

Applications of power devices. (Courtesy of Powerex, Inc.)

When a power semiconductor device is in a normal conduction mode, there is a small voltage drop across the device. In the output voltage waveforms in Figure 1.9, these voltage drops are considered negligible, and unless specified this assumption is made throughout the following chapters.

The power semiconductor switching devices can be classified on the basis of:

1. Uncontrolled turn on and off (e.g., diode);
2. Controlled turn on and uncontrolled turn off (e.g., SCR);
3. Controlled turn-on and -off characteristics (e.g., BJT, MOSFET, GTO, SITH, IGBT, SIT, MCT);
4. Continuous gate signal requirement (BJT, MOSFET, IGBT, SIT);
5. Pulse gate requirement (e.g., SCR, GTO, MCT);
6. Bipolar voltage-withstanding capability (SCR, GTO);
7. Unipolar voltage-withstanding capability (BJT, MOSFET, GTO, IGBT, MCT);
8. Bidirectional current capability (TRIAC, RCT);
9. Unidirectional current capability (SCR, GTO, BJT, MOSFET, MCT, IGBT, SITH, SIT, diode).

Table 1.4 shows the switching characteristics in terms of its voltage, current, and gate signals.

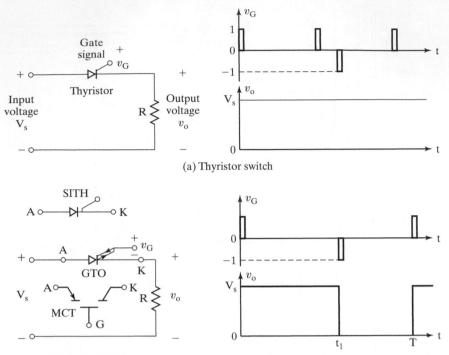

(a) Thyristor switch

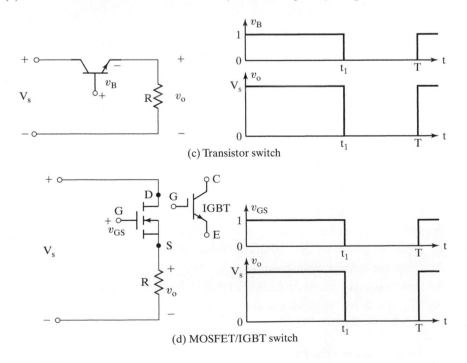

(b) GTO/MTO/ETO/IGCT/MCT/SITH switch (For MCT, the polarity of V_G is reversed as shown)

(c) Transistor switch

(d) MOSFET/IGBT switch

FIGURE 1.9

Control characteristics of power switching devices.

TABLE 1.4 Switching Characteristics of Power Semiconductors

Device Type	Device	Continuous Gate	Pulse Gate	Controlled Turn-On	Controlled Turn-Off	Unipolar Voltage	Bipolar Voltage	Unidirectional Current	Bidirectional Current
Diodes	Power diode							x	
Transistors	BJT	x		x	x	x		x	
	MOSFET	x		x	x	x		x	x
	COOLMOS	x		x	x	x			x
	IGBT	x		x	x	x		x	
	SIT	x		x	x	x		x	
Thyristors	SCR		x	x			x	x	
	RCT		x	x			x		x
	TRIAC		x	x			x		x
	GTO		x	x	x		x	x	
	MTO		x	x	x		x	x	
	ETO		x	x	x		x	x	
	IGCT		x	x	x		x	x	
	SITH		x	x	x		x	x	
	MCT	x		x	x		x	x	

1.4 CHARACTERISTICS AND SPECIFICATIONS OF SWITCHES

There are many types of power switching devices. Each device, however, has its advantages and disadvantages and is suitable to specific applications. The motivation behind the development of any new device is to achieve the characteristics of a "super device." Therefore, the characteristics of any real device can be compared and evaluated with reference to the ideal characteristics of a super device.

1.4.1 Ideal Characteristics

The characteristics of an ideal switch are as follows:

1. In the on-state when the switch is on, it must have (a) the ability to carry a high forward current I_F, tending to infinity; (b) a low on-state forward voltage drop V_{ON}, tending to zero; and (c) a low on-state resistance R_{ON}, tending to zero. Low R_{ON} causes low on-state power loss P_{ON}. These symbols are normally referred to under dc steady-state conditions.

2. In the off-state when the switch is off, it must have (a) the ability to withstand a high forward or reverse voltage V_{BR}, tending to infinity; (b) a low off-state leakage current I_{OFF}, tending to zero; and (c) a high off-state resistance R_{OFF}, tending to infinity. High R_{OFF} cause low off-state power loss P_{OFF}. These symbols are normally referred to under dc steady-state conditions.

3. During the turn-on and turn-off process, it must be completely turned on and off instantaneously so that the device can be operated at high frequencies. Thus, it must have (a) a low delay time t_d, tending to zero; (b) a low rise time t_r, tending to zero; (c) a low storage time t_s, tending to zero; and (d) a low fall time t_f, tending to zero.

4. For turn-on and turn-off, it must require (a) a low gate-drive power P_G, tending to zero; (b) a low gate-drive voltage V_G, tending to zero; and (c) a low gate-drive current I_G, tending to zero.

5. Both turn-on and turn-off must be controllable. Thus, it must turn on with a gate signal (e.g., positive) and must turn off with another gate signal (e.g., zero or negative).

6. For turning on and off, it should require a pulse signal only, that is, a small pulse with a very small width t_W, tending to zero.

7. It must have a high dv/dt, tending to infinity. That is, the switch must be capable of handling rapid changes of the voltage across it.

8. It must have a high di/dt, tending to infinity. That is, the switch must be capable of handling a rapid rise of the current through it.

9. It requires very low thermal impedance from the internal junction to the ambient R_{JA}, tending to zero so that it can transmit heat to the ambient easily.

10. The ability to sustain any fault current for a long time is needed; that is, it must have a high value of i^2t, tending to infinity.

11. Negative temperature coefficient on the conducted current is required to result in an equal current sharing when the devices are operated in parallel.

12. Low price is a very important consideration for reduced cost of the power electronics equipment.

1.4.2   Characteristics of Practical Devices

During the turn-on and -off process, a practical switching device, shown in Figure 1.10a, requires a finite delay time (t_d), rise time (t_r), storage time (t_s), and fall time (t_f). As the device current i_{sw} rises during turn-on, the voltage across the device v_{sw} falls. As the device current falls during turn-off, the voltage across the device rises. The typical waveforms of device voltages v_{sw} and currents i_{sw} are shown in Figure 1.10b. The turn-on

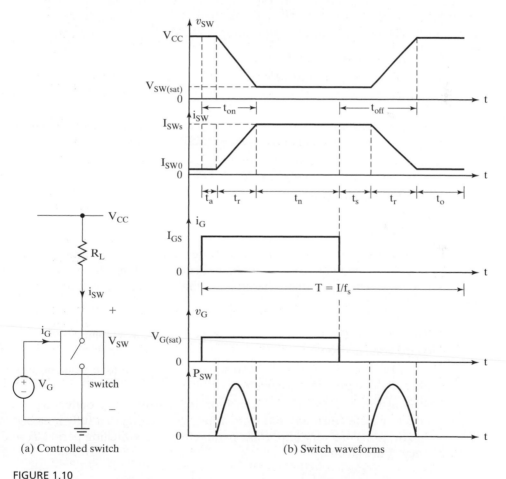

(a) Controlled switch   (b) Switch waveforms

FIGURE 1.10

Typical waveforms of device voltages and currents.

time (t_{on}) of a device is the sum of the delay time and the rise time, whereas the turn-off time (t_{off}) of a device is the sum of the storage time and the fall time. In contrast to an ideal, lossless switch, a practical switching device dissipates some energy when conducting and switching. Voltage drop across a conducting power device is at least on the order of 1 V, but can often be higher, up to several volts. The goal of any new device is to improve the limitations imposed by the switching parameters.

The average conduction power loss, P_{ON} is given by

$$P_{ON} = \frac{1}{T_S} \int_0^{t_{ON}} p \, dt \tag{1.1}$$

where T_S denotes the conduction period and p is the instantaneous power loss (i.e., product of the voltage drop v_{sw} across the switch and the conducted current i_{sw}). Power losses increase during turn-on and turn-off of the switch because during the transition from one conduction state to another state both the voltage and current have significant values. The resultant switching power loss P_{SW} during the turn-on and turn-off periods, is given by

$$P_{SW} = f_s \left(\int_0^{t_r} p \, dt + \int_0^{t_s} p \, dt + \int_0^{t_f} p \, dt \right) \tag{1.2}$$

$f_S = 1/T_S$ is the switching frequency; t_r, t_s, and f_f are the rise time, storage time, and fall time respectively. Therefore, the power dissipation of a switching device is given by:

$$P_D = P_{ON} + P_{SW} + P_G \tag{1.3}$$

where P_G is the gate driver power.

1.4.3   Switch Specifications

The characteristics of practical semiconductor devices differ from those of an ideal device. The device manufacturers supply data sheets describing the device parameters and their ratings. There are many parameters that are important to the devices. The most important among these are:

Voltage ratings: Forward and reverse repetitive peak voltages, and an on-state forward voltage drop.

Current ratings: Average, root-mean-square (rms), repetitive peak, nonrepetitive peak, and off-state leakage currents.

Switching speed or frequency: Transition from a fully nonconducting to a fully conducting state (turn-on) and from a fully conducting to a fully nonconducting state (turn-off) are very important parameters. The switching period T_S and frequency f_S are given by

$$f_S = \frac{1}{T_S} = \frac{1}{t_d + t_r + t_{on} + t_s + t_f + t_{off}} \tag{1.4}$$

where t_{off} is the off time during which the switch remains off.

di/dt rating: The device needs a minimum amount of time before its whole conducting surface comes into play in carrying the full current. If the current rises rapidly, the current flow may be concentrated to a certain area and the device may be damaged. The *di/dt* of the current through the device is normally limited by connecting a small inductor in series with the device, known as a *series snubber.*

dv/dt rating: A semiconductor device has an internal junction capacitance C_J. If the voltage across the switch changes rapidly during turn-on, turn-off and also while connecting the main supply the initial current, the current $C_J \, dv/dt$ flowing through C_J may be too high, thereby causing damage to the device. The *dv/dt* of the voltage across the device is limited by connecting an *RC* circuit across the device, known as a *shunt snubber*, or simply *snubber.*

Switching losses: During turn-on the forward current rises before the forward voltage falls, and during turn-off the forward voltage rises before the current falls. Simultaneous existence of high voltage and current in the device represents power losses as shown in Figure 1.10b. Because of their repetitiveness, they represent a significant part of the losses, and often exceed the on-state conduction losses.

Gate drive requirements: The gate-drive voltage and current are important parameters to turn-on and -off a device. The gate-driver power and the energy requirement are very important parts of the losses and total equipment cost. With large and long current pulse requirements for turn-on and turn-off, the gate drive losses can be significant in relation to the total losses and the cost of the driver circuit can be higher than the device itself.

Safe operating area (SOA): The amount of heat generated in the device is proportional to the power loss, that is, the voltage-current product. For this product to be constant $P = vi$ and equal to the maximum allowable value, the current must be inverse proportional to the voltage. This yields the SOA limit on the allowable steady-state operating points in the voltage-current coordinates.

$I^{2l}t$ *for fusing:* This parameter is needed for fuse selection. The $I^2 t$ of the device must be less than that of the fuse so that the device is protected under fault current conditions.

Temperatures: Maximum allowable junction, case and storage temperatures, usually between 150°C and 200°C for junction and case, and between −50°C and 175°C for storage.

Thermal resistance: Junction-to-case thermal resistance, Q_{JC}; case-to-sink thermal resistance, Q_{CS}; and sink-ambient thermal resistance, Q_{SA}. Power dissipation must be rapidly removed from the internal wafer through the package and ultimately to the cooling medium. The size of semiconductor power switches is small, not exceeding 150 mm, and the thermal capacity of a bare device is too low to safely remove the heat generated by internal losses. Power devices are generally mounted on heat sinks. Thus, removing heat represents a high cost of equipment.

1.4.4 Device Choices

Although, there are many power semiconductor devices, none of them have the ideal characteristics. Continuous improvements are made to the existing devices and new devices are under development. For high power applications from the ac 50- to 60-Hz main supply, the phase control and bidirectional thyristors are the most economical choices. COOLMOSs and IGBTs are the potential replacements for MOSFETS and BJTs, respectively, in low and medium power applications. GTOs and IGCTs are most suited for high-power applications requiring forced commutation. With the continuous advancement in technology, IGBTs are increasingly employed in high-power applications and MCTs may find potential applications that require bidirectional blocking voltages.

1.5   TYPES OF POWER ELECTRONIC CIRCUITS

For the control of electric power or power conditioning, the conversion of electric power from one form to another is necessary and the switching characteristics of the power devices permit these conversions. The static power converters perform these functions of power conversions. A converter may be considered as a switching matrix. The power electronics circuits can be classified into six types:

1. Diode rectifiers
2. Ac–dc converters (controlled rectifiers)
3. Ac–ac converters (ac voltage controllers)
4. Dc–dc converters (dc choppers)
5. Dc–ac converters (inverters)
6. Static switches

The devices in the following converters are used to illustrate the basic principles only. The switching action of a converter can be performed by more than one device. The choice of a particular device depends on the voltage, current, and speed requirements of the converter.

Diode Rectifiers.   A diode rectifier circuit converts ac voltage into a fixed dc voltage and is shown in Figure 1.11. The input voltage to the rectifier v_i could be either single phase or three phase.

Ac–dc converters.   A single-phase converter with two natural commutated thyristors is shown in Figure 1.12. The average value of the output voltage v_0 can be controlled by varying the conduction time of thyristors or firing delay angle, α. The input could be a single- or three-phase source. These converters are also known as *controlled rectifiers*.

Ac–ac converters.   These converters are used to obtain a variable ac output voltage v_0 from a fixed ac source and a single-phase converter with a TRIAC is shown in Figure 1.13. The output voltage is controlled by varying the conduction time of a TRIAC or firing delay angle, α. These types of converters are also known as *ac voltage controllers*.

Dc–dc converters.   A dc–dc converter is also known as a *chopper*, or *switching regulator*, and a transistor chopper is shown in Figure 1.14. The average output voltage v_o is controlled by varying the conduction time t, of transistor Q_1. If T is the chopping period, then $t_1 = \delta T$. δ is called as the *duty cycle* of the chopper.

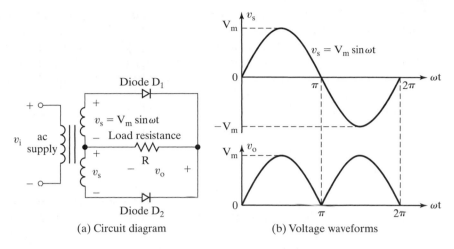

FIGURE 1.11

Single-phase diode rectifier circuit.

Dc–ac converters.   A dc–ac converter is also known as an *inverter*. A single-phase transistor inverter is shown in Figure 1.15. If transistors M_1 and M_2 conduct for one half of a period and M_3 and M_4 conduct for the other half, the output voltage is of the alternating form. The output voltage can be controlled by varying the conduction time of transistors.

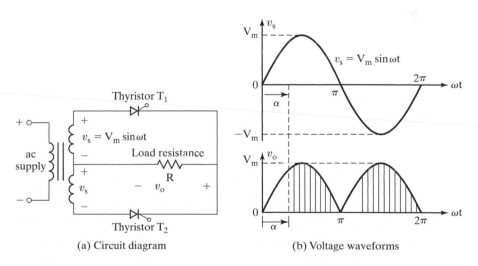

FIGURE 1.12

Single-phase ac–dc converter.

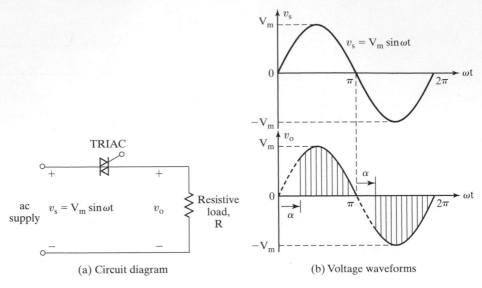

(a) Circuit diagram   (b) Voltage waveforms

FIGURE 1.13

Single-phase ac–ac converter.

Static switches.   Because the power devices can be operated as static switches or contactors, the supply to these switches could be either ac or dc and the switches are called as *ac static switches* or *dc switches*.

A number of conversion stages are often cascaded to produce the desired output as shown in Figure 1.16. Mains 1 supplies the normal ac supply to the load through the static bypass. The ac–dc converter charges the standby battery from mains 2. The dc–ac

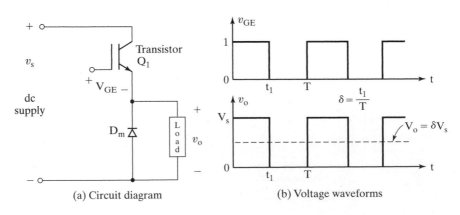

(a) Circuit diagram   (b) Voltage waveforms

FIGURE 1.14

Dc–dc converter.

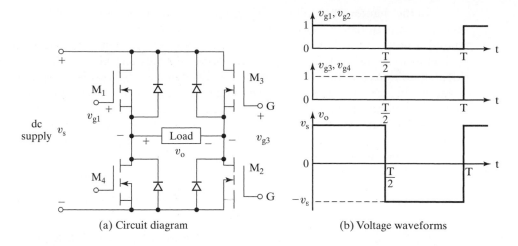

(a) Circuit diagram   (b) Voltage waveforms

FIGURE 1.15

Single-phase dc–ac converter.

converter supplies the emergency power to the load through an isolating transformer. Mains 1 and mains 2 are normally connected to the same ac supply.

1.6   DESIGN OF POWER ELECTRONICS EQUIPMENT

The design of a power electronics equipment can be divided into four parts:

1. Design of power circuits
2. Protection of power devices
3. Determination of control strategy
4. Design of logic and gating circuits

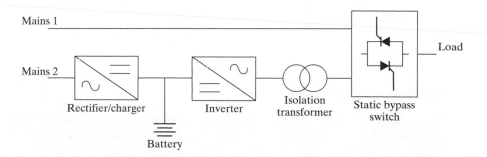

FIGURE 1.16

Block diagram of an uninterruptible power supply (UPS).

In the chapters that follow, various types of power electronic circuits are described and analyzed. In the analysis, the power devices are assumed to be ideal switches unless stated otherwise; and effects of circuit stray inductance, circuit resistances, and source inductance are neglected. The practical power devices and circuits differ from these ideal conditions and the designs of the circuits are also affected. However, in the early stage of the design, the simplified analysis of a circuit is very useful to understand the operation of the circuit and to establish the characteristics and control strategy.

Before a prototype is built, the designer should investigate the effects of the circuit parameters (and devices imperfections) and should modify the design if necessary. Only after the prototype is built and tested, the designer can be confident about the validity of the design and can estimate more accurately some of the circuit parameters (e.g., stray inductance).

1.7   DETERMINING THE ROOT-MEAN-SQUARE VALUES OF WAVEFORMS

To accurately determine the conduction losses in a device and the current ratings of the device and components, the rms values of the current waveforms must be known. The current waveforms are rarely simple sinusoids or rectangles, and this can pose some problems in determining the rms values. The rms value of a waveform $i(t)$ can be calculated as

$$I_{\text{rms}} = \sqrt{\frac{1}{T} \int_0^T i^2 \, dt} \tag{1.5}$$

where T is the time period. If a waveform can be broken up into harmonics whose rms values can be calculated individually, the rms values of the actual waveform can be approximated satisfactorily by combining the rms values of the harmonics. That is, the rms value of the waveform can be calculated from

$$I_{\text{rms}} = \sqrt{I_{\text{dc}}^2 + I_{\text{rms}(1)}^2 + I_{\text{rms}(2)}^2 + \cdots + I_{\text{rms}(n)}^2} \tag{1.6}$$

where I_{dc} = the dc component. $I_{\text{rms}(1)}$ and $I_{\text{rms}(n)}$ are the rms values of the fundamental and nth harmonic components, respectively.

Figure 1.17 shows the rms values of different waveforms that are commonly encountered in power electronics.

1.8   PERIPHERAL EFFECTS

The operations of the power converters are based mainly on the switching of power semiconductor devices; and as a result the converters introduce current and voltage harmonics into the supply system and on the output of the converters. These can cause problems of distortion of the output voltage, harmonic generation into the supply system, and interference with the communication and signaling circuits. It is normally necessary to introduce filters on the input and output of a converter system to reduce the

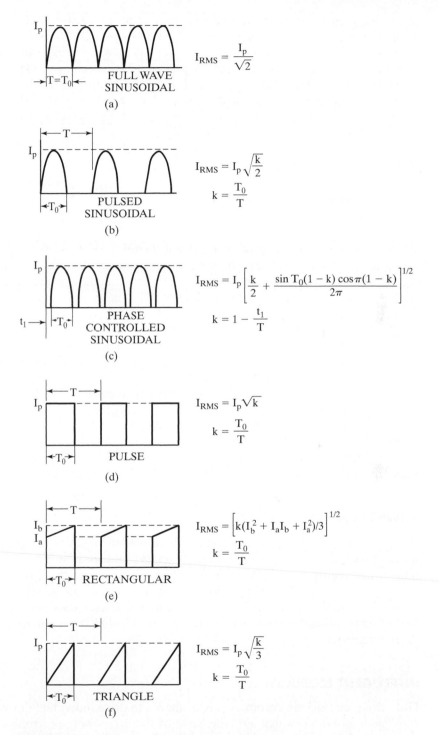

FIGURE 1.17

The rms values of commonly encountered waveforms.

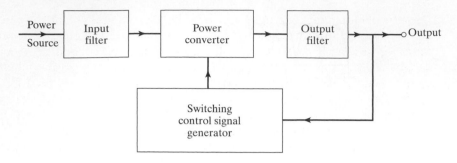

FIGURE 1.18

Generalized power converter system.

harmonic level to an acceptable magnitude. Figure 1.18 shows the block diagram of a generalized power converter. The application of power electronics to supply the sensitive electronic loads poses a challenge on the power quality issues and raises problems and concerns to be resolved by researchers. The input and output quantities of converters could be either ac or dc. Factors such as total harmonic distortion (THD), displacement factor (DF), and input power factor (IPF) are measures of the quality of a waveform. To determine these factors, finding the harmonic content of the waveforms is required. To evaluate the performance of a converter, the input and output voltages and currents of a converter are expressed in a Fourier series. The quality of a power converter is judged by the quality of its voltage and current waveforms.

The control strategy for the power converters plays an important part on the harmonic generation and output waveform distortion, and can be aimed to minimize or reduce these problems. The power converters can cause radio-frequency interference due to electromagnetic radiation, and the gating circuits may generate erroneous signals. This interference can be avoided by *grounded shielding*.

1.9 POWER MODULES

Power devices are available as a single unit or in a module. A power converter often requires two, four, or six devices, depending on its topology. Power modules with dual (in half-bridge configuration) or quad (in full bridge) or six (in three phase) are available for almost all types of power devices. The modules offer the advantages of lower on-state losses, high voltage and current switching characteristics, and higher speed than that of conventional devices. Some modules even include transient protection and gate drive circuitry.

1.10 INTELLIGENT MODULES

Gate drive circuits are commercially available to drive individual devices or modules. *Intelligent modules*, which are the state-of-the-art power electronics, integrate the power module and the peripheral circuit. The peripheral circuit consists of input or output isolation from, and interface with, the signal and high-voltage system, a drive circuit, a protection and diagnostic circuit (against excess current, short circuit, an open load, overheating, and an excess voltage), microcomputer control, and a control power supply. The users need only to connect external (floating) power supplies. An intelligent module is also known as *smart power*. These modules are used increasingly in power electronics [6]. Smart power technology can be viewed as a box that interfaces power source to any load. The box interface function is realized with high-density complementary metal oxide semiconductor (CMOS) logic circuits, its sensing and protection function with bipolar analog and detection circuits, and its power control function with power devices and their associated drive circuits. The functional block diagram of a smart power system [7] is shown in Figure 1.19.

The analog circuits are used for creating the sensors necessary for self-protection and for providing a rapid feedback loop, which can terminate chip operation harmlessly when the system conditions exceed the normal operating conditions. For example, smart power chips must be designed to shut down without damage when a short circuit occurs across a load such as a motor winding. With smart power technology, the load current is monitored, and whenever this current exceeds a preset limit, the drive voltage to the power switches is shut off. In addition to this over-current protection features such as

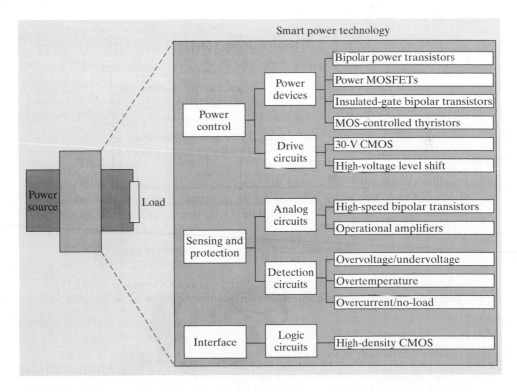

FIGURE 1.19

Functional block diagram of a smart power. [Ref. 7, J. Baliga]

overvoltage and overtemperature protection are commonly included to prevent destructive failures. Some manufacturers of devices and modules and their Web sites are as follows:

Advanced Power Technology, Inc.	www.advancedpower.com/
ABB Semiconductors	www.abbsem.com/
Eupec	www.eupec.com/p/index.htm
Fuji Electric	www.fujielectric.co.jp/eng/denshi/scd/index.htm
Collmer Semiconductor, Inc.	www.collmer.com
Dynex Semiconductor	www.dynexsemi.com
Harris Corp.	www.harris.com/
Hitachi, Ltd. Power Devices	www.hitachi.co.jp/pse
Infineon Technology	www.infineon.com/
International Rectifier	www.irf.com
Marconi Electronic Devices, Inc.	www.marconi.com/
Mitsubishi Semiconductors	www.mitsubishielectric.com/
Mitel Semiconductors	www.mitelsemi.com
Motorola, Inc.	www.motorola.com
National Semiconductors, Inc.	www.national.com/
Nihon International Electronics Corp.	www.abbsem.com/english/salesb.htm
On Semiconductor	www.onsemi.com
Philips Semiconductors	www.semiconductors.philips.com/catalog/
Power Integrations, Inc.	www.powerint.com/
Powerex, Inc.	www.pwrx.com/
PowerTech, Inc.	www.power-tech.com/
RCA Corp.	www.rca.com/
Rockwell Inc.	www.rockwell.com
Reliance Electric	www.reliance.com
Siemens	www.siemens.com
Silicon Power Corp.	www.siliconpower.com/
Semikron International	www.semikron.com/
Siliconix, Inc.	www.siliconix.com
Tokin, Inc.	www.tokin.com/
Toshiba America Electronic Components, Inc.	www.toshiba.com/taec/
Unitrode Integrated Circuits Corp.	www.unitrode.com/
Westcode Semiconductors Ltd.	www.westcode.com/ws-prod.html

1.11   POWER ELECTRONICS JOURNALS AND CONFERENCES

There are many professional journals and conferences in which the new developments are published. The Institute of Electrical and Electronics Engineers (IEEE) e-library *Explore* is an excellent tool in finding articles published in the IEE journals and magazines, and in the IEEE journals, magazines, and sponsored conferences. Some of them are:

IEEE e_Library	ieeexplore.ieee.org/
IEEE Transactions on Aerospace and Systems	www.ieee.org/
IEEE Transactions on Industrial Electronics	www.ieee.org/
IEEE Transactions on Industry Applications	www.ieee.org/
IEEE Transactions on Power Delivery	www.ieee.org/
IEEE Transactions on Power Electronics	www.ieee.org/
IEE *Proceedings on Electric Power*	www.iee.org/Publish/
Journal of Electrical Machinery and Power Systems	

Applied Power Electronics Conference (APEC)
European Power Electronics Conference (EPEC)
IEEE Industrial Electronics Conference (IECON)
IEEE Industry Applications Society (IAS) Annual Meeting
International Conference on Electrical Machines (ICEM)
International Power Electronics Conference (IPEC)
International Power Electronics Congress (CIEP)
International Telecommunications Energy Conference (INTELEC)
Power Conversion Intelligent Motion (PCIM)
Power Electronics Specialist Conference (PESC)

SUMMARY

As the technology for the power semiconductor devices and integrated circuits develops, the potential for the applications of power electronics becomes wider. There are already many power semiconductor devices that are commercially available; however, the development in this direction is continuing. The power converters fall generally into six categories: (1) rectifiers, (2) ac–dc converters, (3) ac–ac converters, (4) dc–dc converters, (5) dc–ac converters, and (6) static switches. The design of power electronics circuits requires designing the power and control circuits. The voltage and current harmonics that are generated by the power converters can be reduced (or minimized) with a proper choice of the control strategy.

REFERENCES

[1]   E. I. Carroll, "Power electronics: where next?" *Power Engineering Journal*, December 1996, pp. 242–243.

[2]   S. Bernet, "Recent developments of high power converters for industry and traction applications," *IEEE Transactions on Power Electronics*, Vol. 15, No. 6, November 2000, pp. 1102–1117.

[3]   E. I. Carroll, "Power electronics for very high power applications," *Power Engineering Journal*, April 1999, pp. 81–87.

[4]   P. K. Steimer, H. E. Gruning, J. Werninger, E. Carroll, S. Klada, and S. Linder, "IGCT—a new emerging for high power, low cost inverters," *IEEE Industry Applications Magazine*, July/August 1999, pp. 12–18.

[5]   R. G. Hoft, *Semiconductor Power Electronics*. New York: Van Nostrand Reinhold. 1986.

[6]   K. Gadi, "Power electronics in action," *IEEE Spectrum*, July 1995, p. 33.

[7]   J. Baliga, "Power ICs in the daddle," *IEEE Spectrum*, July 1995, pp. 34–49.

[8]   I. Zverev and J. Hancock, "CoolMOS Selection Guide," Application note: AN-CoolMOS-02, *Infineon Technologies*, June 2000.

[9]   "Power Electronics Books," SMPS Technology Knowledge Base, March 1, 1999. www.smpstech.com/books/booklist.htm

[10]   "Power Communities," Darnell.Com Inc., March 1, 2002. www.darnell.com/

REVIEW QUESTIONS

1.1   What are power electronics?

1.2   What are the various types of thyristors?

1.3   What is a commutation circuit?

1.4   What are the conditions for a thyristor to conduct?

1.5   How can a conducting thyristor be turned off?

1.6   What is a line commutation?

1.7   What is a forced commutation?

1.8   What is the difference between a thyristor and a TRIAC?

1.9   What is the gating characteristic of a GTO?

1.10   What is the gating characteristic of an MTO?

1.11   What is the gating characteristic of an ETO?

1.12   What is the gating characteristic of an IGCT?

1.13   What is turn-off time of a thyristor?

1.14   What is a converter?

1.15   What is the principle of ac–dc conversion?

1.16   What is the principle of ac–ac conversion?

1.17   What is the principle of dc–dc conversion?

1.18   What is the principle of dc–ac conversion?

1.19   What are the steps involved in designing power electronics equipment?

1.20   What are the peripheral effects of power electronics equipment?

1.21   What are the differences in the gating characteristics of GTOs and thyristors?

1.22   What are the differences in the gating characteristics of thyristors and transistors?

1.23   What are the differences in the gating characteristics of BJTs and MOSFETs?

1.24   What is the gating characteristic of an IGBT?

1.25   What is the gating characteristic of an MCT?

1.26   What is the gating characteristic of an SIT?

1.27   What are the differences between BJTs and IGBTs?

1.28   What are the differences between MCTs and GTOs?

1.29   What are the differences between SITHs and GTOs?

C H A P T E R   2

Power Semiconductor Diodes and Circuits

The learning objectives of this chapter are as follows:

- To understand the diode characteristics and its models
- To learn the types of diodes
- To learn the series and parallel operation of diodes
- To learn the SPICE diode model
- To study the effects of a unidirectional device like a diode on RLC circuits
- To study the applications of diodes in freewheeling and stored-energy recovery

2.1 INTRODUCTION

Many applications have been found for diodes in electronics and electrical engineering circuits. Power diodes play a significant role in power electronics circuits for conversion of electric power. Some diode circuits that are commonly encountered in power electronics for power processing are reviewed in this chapter.

A diode acts as a switch to perform various functions, such as switches in rectifiers, freewheeling in switching regulators, charge reversal of capacitor and energy transfer between components, voltage isolation, energy feedback from the load to the power source, and trapped energy recovery.

Power diodes can be assumed as ideal switches for most applications but practical diodes differ from the ideal characteristics and have certain limitations. The power diodes are similar to *pn*-junction signal diodes. However, the power diodes have larger power-, voltage-, and current-handling capabilities than those of ordinary signal diodes. The frequency response (or switching speed) is low compared with that of signal diodes.

2.2 SEMICONDUCTOR BASICS

Power semiconductor devices are based on high-purity, single-crystal silicon. Single crystals of several meters long and with the required diameter (up to 150 mm) are grown in the so-called *float zone* furnaces. Each huge crystal is sliced into thin wafers, which then go through numerous process steps to turn into power devices.

Silicon, is a member of Group IV of the periodic table of elements, that is, having four electrons per atom in its outer orbit. A pure silicon material is known as an *intrinsic semiconductor* with resistivity that is too low to be an insulator and too high to be a conductor. It has high resistivity and very high dielectric strength (over 200 kV/cm). The resistivity of an intrinsic semiconductor and its charge carriers that are available for conduction can be changed, shaped in layers, and *graded* by implantation of specific impurities. The process of adding impurities is called *doping*, which involves a single atom of the added impurity per over a million silicon atoms. With different impurities, levels and shapes of doping, high technology of photolithography, laser cutting, etching, insulation, and packaging, the finished power devices are produced from various structures of *n*-type and *p*-type semiconductor layers.

> *n-Type material*: If pure silicon is doped with a small amount of a Group V element, such as phosphorus, arsenic, or antimony, each atom of the *dopant* forms a covalent bond within the silicon lattice, leaving a loose electron. These loose electrons greatly increase the conductivity of the material. When the silicon is lightly doped with an impurity such as phosphorus, the doping is denoted as *n doping* and the resultant material is referred to as *n-type semiconductor*. When it is heavily doped, it is denoted as *n+* doping and the material is referred to as *n+-type semiconductor*.
>
> *p-Type material*: If pure silicon is doped with a small amount of a Group III element, such as boron, gallium, or indium, a vacant location called a *hole* is introduced into the silicon lattice. Analogous to an electron, a hole can be considered a mobile charge carrier as it can be filled by an adjacent electron, which in this way leaves a hole behind. These holes greatly increase the conductivity of the material. When the silicon is lightly doped with an impurity such as boron, the doping is denoted as *p-doping* and the resultant material is referred to as *p-type semiconductor*. When it is heavily doped, it is denoted as *p+* doping and the material is referred to as *p+-type semiconductor*.

Therefore, there are free electrons available in an *n*-type material and free holes available in a *p*-type material. In a *p*-type material the holes are called the majority carriers and electrons are called the minority carriers. In the *n*-type material, the electrons are called the majority carriers, and holes are called the minority carriers. These carriers are continuously generated by thermal agitations, they combine and recombine in accordance to their lifetime, and they achieve an equilibrium density of carriers from about 10^{10} to $10^{13}/cm^3$ over a range of about 0 °C to 1000 °C. Thus, an applied electric field can cause a current flow in an *n*-type or *p*-type material.

Key Points of Section 2.2

- Free electrons or holes are made available by adding impurities to the pure silicon or germanium through a doping process. The electrons are the majority carriers in the *n*-type material whereas the holes are the majority carriers in a *p*-type material. Thus, the application of electric field can cause a current flow in an *n*-type or a *p*-type material.

2.3   DIODE CHARACTERISTICS

A power diode is a two-terminal *pn*-junction device [1, 2] and a *pn*-junction is normally formed by alloying, diffusion, and epitaxial growth. The modern control techniques in diffusion and epitaxial processes permit the desired device characteristics. Figure 2.1 shows the sectional view of a *pn*-junction and diode symbol.

When the anode potential is positive with respect to the cathode, the diode is said to be forward biased and the diode conducts. A conducting diode has a relatively small forward voltage drop across it; and the magnitude of this drop depends on the manufacturing process and junction temperature. When the cathode potential is positive with respect to the anode, the diode is said to be reverse biased. Under reverse-biased conditions, a small reverse current (also known as *leakage current*) in the range of micro- or milliampere, flows and this leakage current increases slowly in magnitude with the reverse voltage until the avalanche or zener voltage is reached. Figure 2.2a shows the steady-state *v−i* characteristics of a diode. For most practical purposes, a diode can be regarded as an ideal switch, whose characteristics are shown in Figure 2.2b.

The *v−i* characteristics shown in Figure 2.2a can be expressed by an equation known as *Schockley diode equation*, and it is given under dc steady-state operation by

$$I_D = I_s(e^{V_D/nV_T} - 1) \tag{2.1}$$

where I_D = current through the diode, A;
V_D = diode voltage with anode positive with respect to cathode, V;
I_s = leakage (or reverse saturation) current, typically in the range 10^{-6} to 10^{-15} A;
n = empirical constant known as *emission coefficient*, or *ideality factor*, whose value varies from 1 to 2.

The emission coefficient *n* depends on the material and the physical construction of the diode. For germanium diodes, *n* is considered to be 1. For silicon diodes, the predicted value of *n* is 2, but for most practical silicon diodes, the value of *n* falls in the range 1.1 to 1.8.

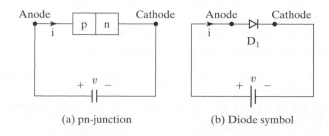

(a) pn-junction   (b) Diode symbol

FIGURE 2.1
pn-Junction and diode symbol.

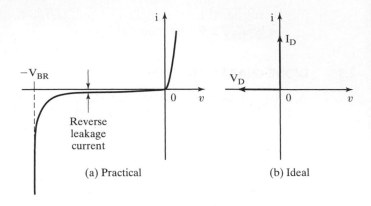

FIGURE 2.2

$v-i$ Characteristics of diode.

(a) Practical

(b) Ideal

V_T in Eq. (2.1) is a constant called *thermal voltage* and it is given by

$$V_T = \frac{kT}{q} \qquad (2.2)$$

where q = electron charge: 1.6022×10^{-19} coulomb (C);
$\quad\ T$ = absolute temperature in Kelvin (K = 273 + °C);
$\quad\ k$ = Boltzmann's constant: 1.3806×10^{-23} J/K.

At a junction temperature of 25 °C, Eq. (2.2) gives

$$V_T = \frac{kT}{q} = \frac{1.3806 \times 10^{-23} \times (273 + 25)}{1.6022 \times 10^{-19}} \approx 25.7 \text{ mV}$$

At a specified temperature, the leakage current I_s is a constant for a given diode. The diode characteristic of Figure 2.2a can be divided into three regions:

Forward-biased region, where $V_D > 0$
Reverse-biased region, where $V_D < 0$
Breakdown region, where $V_D < -V_{BR}$

Forward-biased region. In the forward-biased region, $V_D > 0$. The diode current I_D is very small if the diode voltage V_D is less than a specific value V_{TD} (typically 0.7 V). The diode conducts fully if V_D is higher than this value V_{TD}, which is referred to as the *threshold voltage, cut-in voltage*, or *turn-on voltage*. Thus, the threshold voltage is a voltage at which the diode conducts fully.

Let us consider a small diode voltage $V_D = 0.1$ V, $n = 1$, and $V_T = 25.7$ mV. From Eq. (2.1) we can find the corresponding diode current I_D as

$$I_D = I_s(e^{V_D/nV_T} - 1) = I_s[e^{0.1/(1 \times 0.0257)} - 1] = I_s(48.96 - 1) = 47.96 \, I_s$$

which can be approximated to $I_D \approx I_s e^{V_D/nV_T} = 48.96 \, I_s$, that is, with an error of 2.1%. As v_D increases, the error decreases rapidly.

Therefore, for $V_D > 0.1$ V, which is usually the case, $I_D \gg I_s$, and Eq. (2.1) can be approximated within 2.1% error to

$$I_D = I_s(e^{V_D/nV_T} - 1) \approx I_s e^{V_D/nV_T} \tag{2.3}$$

Reverse-biased region. In the reverse-biased region, $V_D < 0$. If V_D is negative and $|V_D| \gg V_T$, which occurs for $V_D < -0.1$ V, the exponential term in Eq. (2.1) becomes negligibly small compared with unity and the diode current I_D becomes

$$I_D = I_s(e^{-|V_D|/nV_T} - 1) \approx -I_s \tag{2.4}$$

which indicates that the diode current I_D in the reverse direction is constant and equals I_s.

Breakdown region. In the breakdown region, the reverse voltage is high, usually with a magnitude greater than 1000 V. The magnitude of the reverse voltage may exceed a specified voltage known as the *breakdown voltage V_{BR}*, with a small change in reverse voltage beyond V_{BR}. The reverse current increases rapidly. The operation in the breakdown region will not be destructive, provided that the power dissipation is within a "safe level" that is specified in the manufacturer's data sheet. However, it is often necessary to limit the reverse current in the breakdown region to limit the power dissipation within a permissible value.

Example 2.1 Finding the Saturation Current

The forward voltage drop of a power diode is $V_D = 1.2$ V at $I_D = 300$ A. Assuming that $n = 2$ and $V_T = 25.7$ mV, find the reverse saturation current I_s.

Solution
Applying Eq. (2.1), we can find the leakage (or saturation) current I_s from

$$300 = I_s[e^{1.2/(2 \times 25.7 \times 10^{-3})} - 1]$$

which gives $I_s = 2.17746 \times 10^{-8}$ A.

Key Points of Section 2.3

- A diode exhibits a nonlinear $v-i$ characteristic, consisting of three regions: forward biased, reverse-biased, and breakdown. In the forward condition the diode drop is small, typically 0.7 V. If the reverse voltage exceeds the breakdown voltage, the diode may be damaged.

2.4 REVERSE RECOVERY CHARACTERISTICS

The current in a forward-biased junction diode is due to the net effect of majority and minority carriers. Once a diode is in a forward conduction mode and then its forward current is reduced to zero (due to the natural behavior of the diode circuit or application of a reverse voltage), the diode continues to conduct due to minority carriers that remain stored in the *pn*-junction and the bulk semiconductor material. The minority carriers require a certain time to recombine with opposite charges and to be neutralized. This time

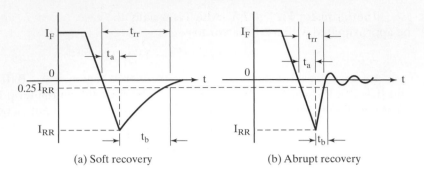

(a) Soft recovery   (b) Abrupt recovery

FIGURE 2.3

Reverse recovery characteristics.

is called the *reverse recovery time* of the diode. Figure 2.3 shows two reverse recovery characteristics of junction diodes. The soft-recovery type is more common. The reverse recovery time is denoted as t_{rr} and is measured from the initial zero crossing of the diode current to 25% of maximum (or peak) reverse current I_{RR}. The t_{rr} consists of two components, t_a and t_b. Variable t_a is due to charge storage in the depletion region of the junction and represents the time between the zero crossing and the peak reverse current I_{RR}. The t_b is due to charge storage in the bulk semiconductor material. The ratio t_b/t_a is known as the *softness factor* (SF). For practical purposes, one needs be concerned with the total recovery time t_{rr} and the peak value of the reverse current I_{RR}.

$$t_{rr} = t_a + t_b \qquad (2.5)$$

The peak reverse current can be expressed in reverse di/dt as

$$I_{RR} = t_a \frac{di}{dt} \qquad (2.6)$$

Reverse recovery time t_{rr}, may be defined as the time interval between the instant the current passes through zero during the changeover from forward conduction to reverse blocking condition and the moment the reverse current has decayed to 25% of its peak reverse value I_{RR}. Variable t_{rr} is dependent on the junction temperature, rate of fall of forward current, and forward current prior to commutation, I_F.

Reverse recovery charge Q_{RR}, is the amount of charge carriers that flows across the diode in the reverse direction due to changeover from forward conduction to reverse blocking condition. Its value is determined from the area enclosed by the path of the reverse recovery current.

The storage charge, which is the area enclosed by the path of the recovery current, is approximately

$$Q_{RR} \cong \tfrac{1}{2} I_{RR} t_a + \tfrac{1}{2} I_{RR} t_b = \tfrac{1}{2} I_{RR} t_{rr} \qquad (2.7)$$

or

$$I_{RR} \cong \frac{2Q_{RR}}{t_{rr}} \qquad (2.8)$$

Equating I_{RR} in Eq. (2.6) to I_{RR} in Eq. (2.8) gives

$$t_{rr}t_a = \frac{2Q_{RR}}{di/dt} \tag{2.9}$$

If t_b is negligible as compared to t_a, which is usually the case, $t_{rr} \approx t_a$, and Eq. (2.9) becomes

$$t_{rr} \cong \sqrt{\frac{2Q_{RR}}{di/dt}} \tag{2.10}$$

and

$$I_{RR} = \sqrt{2Q_{RR}\frac{di}{dt}} \tag{2.11}$$

It can be noticed from Eqs. (2.10) and (2.11) that the reverse recovery time t_{rr} and the peak reverse recovery current I_{RR} depend on the storage charge Q_{RR} and the reverse (or reapplied) di/dt. The storage charge is dependent on the forward diode current I_F. The peak reverse recovery current I_{RR}, reverse charge Q_{RR}, and the SF are all of interest to the circuit designer, and these parameters are commonly included in the specification sheets of diodes.

If a diode is in a reverse-biased condition, a leakage current flows due to the minority carriers. Then the application of forward voltage would force the diode to carry current in the forward direction. However, it requires a certain time known as *forward recovery (or turn-on) time* before all the majority carriers over the whole junction can contribute to the current flow. If the rate of rise of the forward current is high and the forward current is concentrated to a small area of the junction, the diode may fail. Thus, the forward recovery time limits the rate of the rise of the forward current and the switching speed.

Example 2.2   Finding the Reverse Recovery Current

The reverse recovery time of a diode is $t_{rr} = 3$ μs and the rate of fall of the diode current is $di/dt = 30$ A/μs. Determine (a) the storage charge Q_{RR}, and (b) the peak reverse current I_{RR}.

Solution

$t_{rr} = 3$ μs and $di/dt = 30$ A/μs.

   a. From Eq. (12.10),

$$Q_{RR} = \frac{1}{2}\frac{di}{dt}\, t_{rr}^2 = 0.5 \times 30 \text{ A/μs} \times (3 \times 10^{-6})^2 = 135 \text{ μC}$$

   b. From Eq. (2.11),

$$I_{RR} = \sqrt{2Q_{RR}\frac{di}{dt}} = \sqrt{2 \times 135 \times 10^{-6} \times 30 \times 10^6} = 90 \text{ A}$$

Key Points of Section 2.4

- During the reverse recovery time t_{rr}, the diode behaves effectively as a short circuit and is not capable of blocking reverse voltage, allowing reverse current flow, and then suddenly disrupting the current. Parameter t_{rr} is important for switching applications.

2.5 POWER DIODE TYPES

Ideally, a diode should have no reverse recovery time. However, the manufacturing cost of such a diode may increase. In many applications, the effects of reverse recovery time is not significant, and inexpensive diodes can be used. Depending on the recovery characteristics and manufacturing techniques, the power diodes can be classified into the following three categories:

1. Standard or general-purpose diodes
2. Fast-recovery diodes
3. Schottky diodes

The characteristics and practical limitations of these types restrict their applications.

2.5.1 General-Purpose Diodes

The general-purpose rectifier diodes have relatively high reverse recovery time, typically 25 μs; and are used in low-speed applications, where recovery time is not critical (e.g., diode rectifiers and converters for a low-input frequency up to 1-kHz applications and line-commutated converters). These diodes cover current ratings from less than 1 A to several thousands of amperes, with voltage ratings from 50 V to around 5 kV. These diodes are generally manufactured by diffusion. However, alloyed types of rectifiers that are used in welding power supplies are most cost-effective and rugged, and their ratings can go up to 1500 V, 400 A.

Figure 2.4 shows various configurations of general-purpose diodes, which basically fall into two types. One is called a *stud*, or *stud-mounted* type; the other one is called a *disk*, *press pak*, or *hockey-puck* type. In a stud-mounted type, either the anode or the cathode could be the stud.

2.5.2 Fast-Recovery Diodes

The fast-recovery diodes have low recovery time, normally less than 5 μs. They are used in dc–dc and dc–ac converter circuits, where the speed of recovery is often of critical importance. These diodes cover current ratings of voltage from 50 V to around 3 kV, and from less than 1 A to hundreds of amperes.

For voltage ratings above 400 V, fast-recovery diodes are generally made by diffusion and the recovery time is controlled by platinum or gold diffusion. For voltage ratings below 400 V, epitaxial diodes provide faster switching speeds than those of diffused diodes. The epitaxial diodes have a narrow base width, resulting in a fast recovery time of as low as 50 ns. Fast-recovery diodes of various sizes are shown in Figure 2.4.

FIGURE 2.4

Fast-recovery diodes. (Courtesy of Powerex, Inc.)

2.5.3   Schottky Diodes

The charge storage problem of a *pn*-junction can be eliminated (or minimized) in a Schottky diode. It is accomplished by setting up a "barrier potential" with a contact between a metal and a semiconductor. A layer of metal is deposited on a thin epitaxial layer of *n*-type silicon. The potential barrier simulates the behavior of a *pn*-junction. The rectifying action depends on the majority carriers only, and as a result there are no excess minority carriers to recombine. The recovery effect is due solely to the self-capacitance of the semiconductor junction.

The recovered charge of a Schottky diode is much less than that of an equivalent *pn*-junction diode. Because it is due only to the junction capacitance, it is largely independent of the reverse *di/dt*. A Schottky diode has a relatively low forward voltage drop.

The leakage current of a Schottky diode is higher than that of a *pn*-junction diode. A Schottky diode with relatively low-conduction voltage has relatively high leakage current, and vice versa. As a result, the maximum allowable voltage of this diode is generally limited to 100 V. The current ratings of Schottky diodes vary from 1 to 400 A. The Schottky diodes are ideal for high-current and low-voltage dc power supplies. However, these diodes are also used in low-current power supplies for increased efficiency. In Figure 2.5, 20- and 30-A dual Schottky rectifiers are shown.

Key Points of Section 2.5

- Depending on the switching recovery time and the on-state drop, the power diodes are of three types: general purpose, fast recovery, and Schottky.

2.6   SILICON CARBIDE DIODES

Silicon Carbide (SiC) is a new material for power electronics. Its physical properties outperform Si and GaAs by far. For example, the Schottky SiC diodes manufactured

FIGURE 2.5

Dual Schottky center rectifiers of 20 and 30 A.
(Courtesy of International Rectifier.)

by Infineon Technologies [3] have ultralow power losses and high reliability. They also
have the following features:

- No reverse recovery time;
- Ultrafast switching behavior;
- No temperature influence on the switching behavior.

The typical storage charge Q_{RR} is 21 nC for a 600-V, 6-A diode and is 23 nC for a
600-V, 10-A, device.

2.7   SPICE DIODE MODEL

The SPICE model of a diode [4–6] is shown in Figure 2.6a. The diode current I_D that depends on its voltage is represented by a current source. R_s is the series resistance, and it
is due to the resistance of the semiconductor. R_s, also known as *bulk resistance*, is dependent on the amount of doping. The small-signal and static models that are generated
by SPICE are shown in Figures 2.6b and 12.6c, respectively. C_D is a nonlinear function of
the diode voltage v_D and is equal to $C_D = dq_d/dv_D$, where q_d is the depletion-layer
charge. SPICE generates the small-signal parameters from the operating point.

The SPICE model statement of a diode has the general form

```
.MODEL DNAME D (P1=V1 P2=V2 P3=V3 ..... PN=VN)
```

DNAME is the model name and it can begin with any character; however, its word size
is normally limited to 8. D is the type symbol for diodes. P1, P2, ... and V1, V2, ... are
the model parameters and their values, respectively.

Among many diode parameters, the important parameters [5] for power switching are:

IS	Saturation current
BV	Reverse breakdown voltage
IBV	Reverse breakdown current
TT	Transit time
CJO	Zero-bias *pn* capacitance

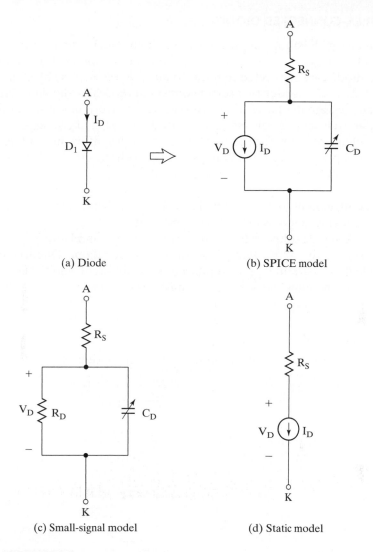

FIGURE 2.6

SPICE diode model with reverse-biased diode.

Because the SiC diodes use a completely new type of technology, the use of SPICE models for silicon diodes may introduce a significant amount of errors. The manufacturers [3] are, however, providing the SPICE models of SiC diodes.

Key Points of Section 2.7

- The SPICE parameters, which can be derived from the data sheet, may significantly affect the transient behavior of a switching circuit.

2.8 SERIES-CONNECTED DIODES

In many high-voltage applications (e.g., high-voltage direct current [HVDC] transmission lines), one commercially available diode cannot meet the required voltage rating, and diodes are connected in series to increase the reverse blocking capabilities.

Let us consider two series-connected diodes as shown in Figure 2.7a. Variables i_D and v_D are the current and voltage, respectively, in the forward direction; v_{D1} and v_{D2} are the sharing reverse voltages of diodes D_1 and D_2, respectively. In practice, the $v-i$ characteristics for the same type of diodes differ due to tolerances in their production process. Figure 2.7b shows two $v-i$ characteristics for such diodes. In the forward-biased condition, both diodes conduct the same amount of current, and the forward voltage drop of each diode would be almost equal. However, in the reverse blocking condition, each diode has to carry the same leakage current, and as a result the blocking voltages may differ significantly.

A simple solution to this problem, as shown in Figure 2.8a, is to force equal voltage sharing by connecting a resistor across each diode. Due to equal voltage sharing, the leakage current of each diode would be different, and this is shown in Figure 2.8b. Because the total leakage current must be shared by a diode and its resistor,

$$I_s = I_{s1} + I_{R1} = I_{s2} + I_{R2} \tag{2.12}$$

However, $I_{R1} = V_{D1}/R_1$ and $I_{R2} = V_{D2}/R_2 = V_{D1}/R_2$. Equation (2.12) gives the relationship between R_1 and R_2 for equal voltage sharing as

$$I_{s1} + \frac{V_{D1}}{R_1} = I_{s2} + \frac{V_{D1}}{R_2} \tag{2.13}$$

If the resistances are equal, then $R = R_1 = R_2$ and the two diode voltages would be slightly different depending on the dissimilarities of the two $v-i$ characteristics. The

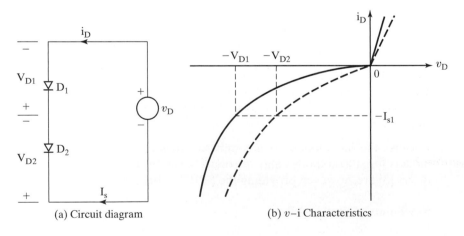

(a) Circuit diagram   (b) $v-i$ Characteristics

FIGURE 2.7

Two series-connected diodes with reverse bias.

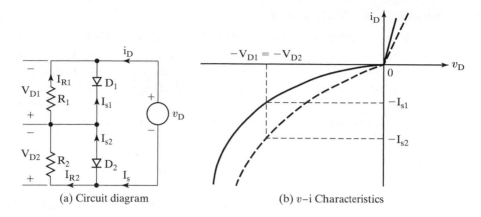

(a) Circuit diagram   (b) v–i Characteristics

FIGURE 2.8

Series-connected diodes with steady-state voltage-sharing characteristics.

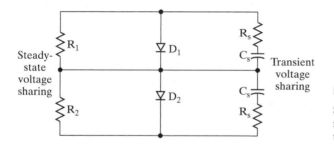

FIGURE 2.9

Series diodes with voltage-sharing networks under steady-state and transient conditions.

values of V_{D1} and V_{D2} can be determined from Eqs. (2.14) and (2.15):

$$I_{s1} + \frac{V_{D1}}{R} = I_{s2} + \frac{V_{D2}}{R} \tag{2.14}$$

$$V_{D1} + V_{D2} = V_s \tag{2.15}$$

The voltage sharings under transient conditions (e.g., due to switching loads, the initial applications of the input voltage) are accomplished by connecting capacitors across each diode, which is shown in Figure 2.9. R_s limits the rate of rise of the blocking voltage.

Example 2.3   Finding the Voltage Sharing Resistors

Two diodes are connected in series, shown in Figure 2.8a to share a total dc reverse voltage of $V_D = 5\,\text{kV}$. The reverse leakage currents of the two diodes are $I_{s1} = 30\,\text{mA}$ and $I_{s2} = 35\,\text{mA}$. (a) Find the diode voltages if the voltage-sharing resistances are equal, $R_1 = R_2 = R = 100\,\text{k}\Omega$. (b) Find the voltage-sharing resistances R_1 and R_2 if the diode voltages are equal, $V_{D1} = V_{D2} = V_D/2$. (c) Use PSpice to check your results of part (a). PSpice model parameters of the diodes are BV = 3 kV and IS = 30 mA for diode D_1, and IS = 35 mA for diode D_2.

Solution

a. $I_{s1} = 30\,\text{mA}$, $I_{s2} = 35\,\text{mA}$, and $R_1 = R_2 = R = 100\,\text{k}\Omega$. $-V_D = -V_{D1} - V_{D2}$ or $V_{D2} = V_D - V_{D1}$. From Eq. (2.14),

$$I_{s1} + \frac{V_{D1}}{R} = I_{s2} + \frac{V_{D2}}{R}$$

Substituting $V_{D2} = V_D - V_{D1}$ and solving for the diode voltage D_1, we get

$$V_{D1} = \frac{V_D}{2} + \frac{R}{2}(I_{s2} - I_{s1})$$

$$= \frac{5\,\text{kV}}{2} + \frac{100\,\text{k}\Omega}{2}(35 \times 10^{-3} - 30 \times 10^{-3}) = 2750\,\text{V} \tag{2.16}$$

and $V_{D2} = V_D - V_{D1} = 5\,\text{kV} - 2750 = 2250\,\text{V}$.

b. $I_{s1} = 30\,\text{mA}$, $I_{s2} = 35\,\text{mA}$, and $V_{D1} = V_{D2} = V_D/2 = 2.5\,\text{kV}$. From Eq. (2.13),

$$I_{s1} + \frac{V_{D1}}{R_1} = I_{s2} + \frac{V_{D2}}{R_2}$$

which gives the resistance R_2 for a known value of R_1 as

$$R_2 = \frac{V_{D2}R_1}{V_{D1} - R_1(I_{s2} - I_{s1})} \tag{2.17}$$

Assuming that $R_1 = 100\,\text{k}\Omega$, we get

$$R_2 = \frac{2.5\,\text{kV} \times 100\,\text{k}\Omega}{2.5\,\text{kV} - 100\,\text{k}\Omega \times (35 \times 10^{-3} - 30 \times 10^{-3})} = 125\,\text{k}\Omega$$

c. The diode circuit for PSpice simulation is shown in Figure 2.10. The list of the circuit file is as follows:

```
Example 2.3              Diode Voltage-Sharing Circuit
VS     1    0      DC      5KV
R      1    2      0.01
R1     2    3      100K
R2     3    0      100K
D1     3    2      MOD1
D2     0    3      MOD2
.MODEL MOD1 D (IS=30MA BV=3KV)   ; Diode model parameters
.MODEL MOD2 D (IS=35MA BV=3KV)   ; Diode model parameters
.OP                              ; Dc operating point analysis
.END
```

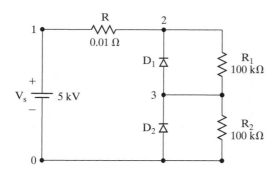

FIGURE 2.10

Diode circuit for PSpice simulation
for Example 2.3.

The results of PSpice simulation are

NAME	D1			D2	
ID	-3.00E-02	I_{D1}=-30 mA		-3.50E-02	I_{D2}=-35 mA
VD	-2.75E+03	V_{D1}=-2750 V	expected -2750 V	-2.25E+03	V_{D2}=-2250 V
					expected -2250 V
REQ	1.00E+12	R_{D1}=1 GΩ		1.00E+12	R_{D2}=1 GΩ

Note: The SPICE gives the same voltages as expected. A small resistance of $R = 10\,\text{m}\Omega$ is inserted to avoid SPICE error due to a zero-resistance voltage loop.

Key Points of Section 2.8

- When diodes of the same type are connected in series, they do not share the same reverse voltage due to mismatches in their reverse v–i characteristics. Voltage-sharing networks are needed to equalize the voltage sharing.

2.9   PARALLEL-CONNECTED DIODES

In high-power applications, diodes are connected in parallel to increase the current-carrying capability to meet the desired current requirements. The current sharings of diodes would be in accord with their respective forward voltage drops. Uniform current sharing can be achieved by providing equal inductances (e.g., in the leads) or by connecting current-sharing resistors (which may not be practical due to power losses); and this is depicted in Figure 2.11. It is possible to minimize this problem by selecting diodes with equal forward voltage drops or diodes of the same type. Because the diodes are connected in parallel, the reverse blocking voltages of each diode would be the same.

The resistors of Figure 2.11a help current sharing under steady-state conditions. Current sharing under dynamic conditions can be accomplished by connecting coupled inductors as shown in Figure 2.11b. If the current through D_1 rises, the $L\,di/dt$ across L_1 increases, and a corresponding voltage of opposite polarity is induced across inductor L_2. The result is a low-impedance path through diode D_2 and the current is shifted to D_2. The inductors may generate voltage spikes and they may be expensive and bulky, especially at high currents.

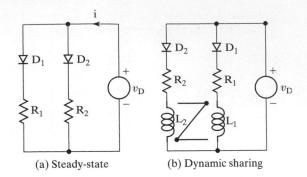

FIGURE 2.11

Parallel-connected diodes.

(a) Steady-state   (b) Dynamic sharing

Key Points of Section 2.9

- When diodes of the same type are connected in parallel, they do not share the same on-state current due to mismatches in their forward v–i characteristics. Current sharing networks are needed to equalize the current sharing.

2.10   DIODES WITH *RC* AND *RL* LOADS

Figure 2.12a shows a diode circuit with an RC load. For the sake of simplicity the diodes are considered to be ideal. By "ideal" we mean that the reverse recovery time t_{rr} and the forward voltage drop V_D are negligible. That is, $t_{rr} = 0$ and $V_D = 0$. The source voltage V_S is a dc constant voltage. When the switch S_1 is closed at $t = 0$, the charging current i that flows through the capacitor can be found from

$$V_s = v_R + v_c = v_R + \frac{1}{C}\int_{t_0}^{t} i\, dt + v_c(t = 0) \tag{2.18}$$

$$v_R = Ri \tag{2.19}$$

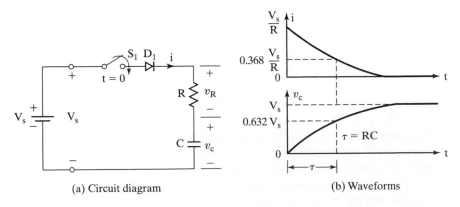

(a) Circuit diagram   (b) Waveforms

FIGURE 2.12

Diode circuit with an RC load.

With initial condition $v_c(t = 0) = 0$, the solution of Eq. (2.18) (which is derived in Appendix D, Eq. D.1) gives the charging current i as

$$i(t) = \frac{V_s}{R} e^{-t/RC} \tag{2.20}$$

The capacitor voltage v_c is

$$v_c(t) = \frac{1}{C} \int_0^t i \, dt = V_s(1 - e^{-t/RC}) = V_s(1 - e^{-t/\tau}) \tag{2.21}$$

where $\tau = RC$ is the time constant of an *RC* load. The rate of change of the capacitor voltage is

$$\frac{dv_c}{dt} = \frac{V_s}{RC} e^{-t/RC} \tag{2.22}$$

and the initial rate of change of the capacitor voltage (at $t = 0$) is obtained from Eq. (2.22)

$$\left. \frac{dv_c}{dt} \right|_{t=0} = \frac{V_s}{RC} \tag{2.23}$$

A diode circuit with an *RL* load is shown in Figure 2.13a. When switch S_1 is closed at $t = 0$, the current i through the inductor increases and is expressed as

$$V_s = v_L + v_R = L \frac{di}{dt} + Ri \tag{2.24}$$

With initial condition $i(t = 0) = 0$, the solution of Eq. (2.24) (which is derived in Appendix D, Eq. D.2) yields

$$i(t) = \frac{V_s}{R}(1 - e^{-tR/L}) \tag{2.25}$$

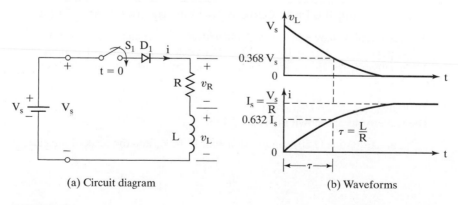

(a) Circuit diagram   (b) Waveforms

FIGURE 2.13

Diode circuit with an *RL* load.

The rate of change of this current can be obtained from Eq. (2.25) as

$$\frac{di}{dt} = \frac{V_s}{L}e^{-tR/L} \tag{2.26}$$

and the initial rate of rise of the current (at $t = 0$) is obtained from Eq. (2.26):

$$\left.\frac{di}{dt}\right|_{t=0} = \frac{V_s}{L} \tag{2.27}$$

The voltage v_L across the inductor is

$$v_L(t) = L\frac{di}{dt} = V_s e^{-tR/L} \tag{2.28}$$

where $L/R = \tau$ is the time constant of an RL load.

The waveforms for voltage v_L and current are shown in Figure 2.13b. If $t \gg L/R$, the voltage across the inductor tends to be zero and its current reaches a steady-state value of $I_s = V_s/R$. If an attempt is then made to open switch S_1, the energy stored in the inductor $(= 0.5Li^2)$ will be transformed into a high reverse voltage across the switch and diode. This energy dissipates in the form of sparks across the switch; and diode D_1 is likely to be damaged in this process. To overcome such a situation, a diode commonly known as a *freewheeling diode* is connected across an inductive load as shown in Figure 2.21a.

Note: Because the current i in Figures 2.12a and 2.13a is unidirectional and does not tend to change its polarity, the diodes have no effect on circuit operation.

Key Points of Section 2.10

- The current of an RC or RL circuit that rises or falls exponentially with a circuit time constant does not reverse its polarity. The initial dv/dt of a charging capacitor in an RC circuit is V_s/RC, and the initial di/dt in an RL circuit is V_s/L.

Example 2.4   Finding the Peak Current and Energy Loss in an RC Circuit

A diode circuit is shown in Figure 2.14a with $R = 44 \ \Omega$ and $C = 0.1 \ \mu F$. The capacitor has an initial voltage, $V_{c0} = V_c(t = 0) = 220 \ V$. If switch S_1 is closed at $t = 0$, determine (a) the peak diode current, (b) the energy dissipated in the resistor R, and (c) the capacitor voltage at $t = 2 \ \mu s$.

Solution

The waveforms are shown in Figure 2.14b.

a. Equation (2.20) can be used with $V_s = V_{c0}$ and the peak diode current I_p is

$$I_p = \frac{V_{c0}}{R} = \frac{220}{44} = 5 \ A$$

b. The energy W dissipated is

$$W = 0.5CV_{c0}^2 = 0.5 \times 0.1 \times 10^{-6} \times 220^2 = 0.00242 \ J = 2.42 \ mJ$$

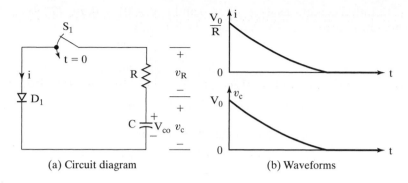

(a) Circuit diagram   (b) Waveforms

FIGURE 2.14

Diode circuit with an *RC* load.

---

c. For $RC = 44 \times 0.1\ \mu = 4.4\ \mu s$ and $t = t_1 = 2\ \mu s$, the capacitor voltage is

$$v_c(t = 2\ \mu s) = V_{c0}e^{-t/RC} = 220 \times e^{-2/4.4} = 139.64\ V$$

Note: Because the current is unidirectional, the diode does not affect circuit operation.

2.11   DIODES WITH *LC* AND *RLC* LOADS

A diode circuit with an *LC* load is shown in Figure 2.15a. The source voltage V_s is a dc constant voltage. When switch S_1 is closed at $t = 0$, the charging current i of the capacitor is expressed as

$$V_s = L\frac{di}{dt} + \frac{1}{C}\int_{t_0}^{t} i\,dt + v_c(t = 0) \tag{2.29}$$

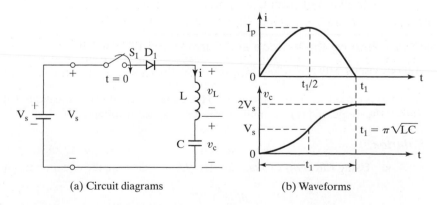

(a) Circuit diagrams   (b) Waveforms

FIGURE 2.15

Diode circuit with an *LC* load.

With initial conditions $i(t = 0) = 0$ and $v_c(t = 0) = 0$, Eq. (2.29) can be solved for the capacitor current i as (in Appendix D, Eq. D.3)

$$i(t) = V_s \sqrt{\frac{C}{L}} \sin \omega_0 t \tag{2.30}$$

$$= I_p \sin \omega_0 t \tag{2.31}$$

where $\omega_0 = 1/\sqrt{LC}$ and the peak current I_p is

$$I_p = V_s \sqrt{\frac{C}{L}} \tag{2.32}$$

The rate of rise of the current is obtained from Eq. (2.30) as

$$\frac{di}{dt} = \frac{V_s}{L} \cos \omega_0 t \tag{2.33}$$

and Eq. (2.33) gives the initial rate of rise of the current (at $t = 0$) as

$$\left. \frac{di}{dt} \right|_{t=0} = \frac{V_s}{L} \tag{2.34}$$

The voltage v_c across the capacitor can be derived as

$$v_c(t) = \frac{1}{C} \int_0^t i \, dt = V_s(1 - \cos \omega_0 t) \tag{2.35}$$

At a time $t = t_1 = \pi\sqrt{LC}$, the diode current i falls to zero and the capacitor is charged to $2V_s$. The waveforms for the voltage v_L and current i are shown in Figure 2.15b.

Note: Because there is no resistance in the circuit, there can be no energy loss. Thus, in the absence of any resistance, the current of an LC circuit oscillates and the energy is transferred from C to L and vice versa.

Example 2.5 Finding the Voltage and Current in an *LC* Circuit

A diode circuit with an LC load is shown in Figure 2.16a with the capacitor having an initial voltage; $V_c(t = 0) = -V_{c0} = V_0 - 220$ V, capacitance, $C = 20$ µF; and inductance, $L = 80$ µH. If switch S_1 is closed at $t = 0$, determine (a) the peak current through the diode, (b) the conduction time of the diode, and (c) the final steady-state capacitor voltage.

Solution

a. Using *Kirchhoff's voltage law* (KVL), we can write the equation for the current i as

$$L\frac{di}{dt} + \frac{1}{C} \int_{t_0}^t i \, dt + v_c(t = 0) = 0$$

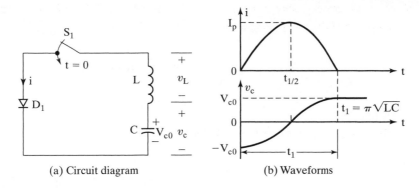

(a) Circuit diagram   (b) Waveforms

FIGURE 2.16

Diode circuit with an *LC* load.

and the current i with initial conditions of $i(t = 0) = 0$ and $v_c(t = 0) = -V_{c0}$ is solved as

$$i(t) = V_{c0}\sqrt{\frac{C}{L}}\,\sin\omega_0 t$$

where $\omega_0 = 1/\sqrt{LC} = 10^6/\sqrt{20 \times 80} = 25{,}000$ rad/s. The peak current I_p is

$$I_p = V_{c0}\sqrt{\frac{C}{L}} = 220\sqrt{\frac{20}{80}} = 110 \text{ A}$$

b. At $t = t_1 = \pi\sqrt{LC}$, the diode current becomes zero and the conduction time t_1 of diode is

$$t_1 = \pi\sqrt{LC} = \pi\sqrt{20 \times 80} = 125.66 \text{ μs}$$

c. The capacitor voltage can easily be shown to be

$$v_c(t) = \frac{1}{C}\int_0^t i\,dt - V_{c0} = -V_{c0}\cos\omega_0 t$$

For $t = t_1 = 125.66$ μs, $v_c(t = t_1) = -220\cos\pi = 220$ V.

A diode circuit with an *RLC* load is shown in Figure 2.17. If switch S_1 is closed at $t = 0$, we can use the KVL to write the equation for the load current i as

$$L\frac{di}{dt} + Ri + \frac{1}{C}\int^t i\,dt + v_c(t = 0) = V_s \qquad (2.36)$$

with initial conditions $i(t = 0)$ and $v_c(t = 0) = V_{c0}$. Differentiating Eq. (2.36) and dividing both sides by L gives the characteristic equation

$$\frac{d^2i}{dt^2} + \frac{R}{L}\frac{di}{dt} + \frac{i}{LC} = 0 \qquad (2.37)$$

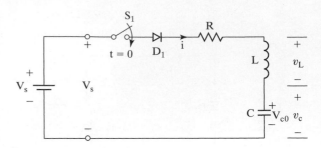

FIGURE 2.17

Diode circuit with an *RLC* load.

Under final steady-state conditions, the capacitor is charged to the source voltage V_s and the steady-state current is zero. The forced component of the current in Eq. (2.37) is also zero. The current is due to the natural component.

The characteristic equation in Laplace's domain of s is

$$s^2 + \frac{R}{L}s + \frac{1}{LC} = 0 \tag{2.38}$$

and the roots of quadratic equation (2.38) are given by

$$s_{1,2} = -\frac{R}{2L} \pm \sqrt{\left(\frac{R}{2L}\right)^2 - \frac{1}{LC}} \tag{2.39}$$

Let us define two important properties of a second-order circuit: the *damping factor*,

$$\alpha = \frac{R}{2L} \tag{2.40}$$

and the *resonant frequency*,

$$\omega_0 = \frac{1}{\sqrt{LC}} \tag{2.41}$$

Substituting these into Eq. (2.39) yields

$$s_{1,2} = -\alpha \pm \sqrt{\alpha^2 - \omega_0^2} \tag{2.42}$$

The solution for the current, which depends on the values of α and ω_0, would follow one of the three possible cases.

Case 1. If $\alpha = \omega_0$, the roots are equal, $s_1 = s_2$, and the circuit is called *critically damped*. The solution takes the form

$$i(t) = (A_1 + A_2 t)e^{s_1 t} \tag{2.43}$$

Case 2. If $\alpha > \omega_0$, the roots are real and the circuit is said to be *over-damped*. The solution takes the form

$$i(t) = A_1 e^{s_1 t} + A_2 e^{s_2 t} \tag{2.44}$$

Case 3. If $\alpha < \omega_0$, the roots are complex and the circuit is said to be *underdamped*. The roots are

$$s_{1,2} = -\alpha \pm j\omega_r \tag{2.45}$$

where ω_r is called the *ringing frequency* (or damped resonant frequency) and $\omega_r = \sqrt{\omega_0^2 - \alpha^2}$. The solution takes the form

$$i(t) = e^{-\alpha t}(A_1 \cos \omega_r t + A_2 \sin \omega_r t) \tag{2.46}$$

which is a *damped* or *decaying sinusoidal*.

Note: The constants A_1 and A_2 can be determined from the initial conditions of the circuit. The ratio of α/ω_0 is commonly known as the *damping ratio*, $\delta = R/2\sqrt{C/L}$. Power electronic circuits are generally underdamped such that the circuit current becomes near sinusoidal, to cause a nearly sinusoidal ac output or to turn off a power semiconductor device.

Example 2.6   Finding the Current in an *RLC* Circuit

The second-order *RLC* circuit of Figure 2.17 has the dc source voltage $V_s = 220$ V, inductance $L = 2$ mH, capacitance $C = 0.05$ μF, and resistance $R = 160$ Ω. The initial value of the capacitor voltage is $v_c(t = 0) = V_{c0} = 0$ and conductor current $i(t = 0) = 0$. If switch S_1 is closed at $t = 0$, determine (a) an expression for the current $i(t)$, and (b) the conduction time of diode. (c) Draw a sketch of $i(t)$. (d) Use PSpice to plot the instantaneous current i for $R = 50$ Ω, 160 Ω, and 320 Ω.

Solution

a. From Eq. (2.40), $\alpha = R/2L = 160 \times 10^3/(2 \times 2) = 40{,}000$ rad/s, and from Eq. (2.41), $\omega_0 = 1/\sqrt{LC} = 10^5$ rad/s. The ringing frequency becomes

$$\omega_r = \sqrt{10^{10} - 16 \times 10^8} = 91{,}652 \text{ rad/s}$$

Because $\alpha < \omega_0$, it is an underdamped circuit and the solution is of the form

$$i(t) = e^{-\alpha t}(A_1 \cos \omega_r t + A_2 \sin \omega_r t)$$

At $t = 0$, $i(t = 0) = 0$ and this gives $A_1 = 0$. The solution becomes

$$i(t) = e^{-\alpha t} A_2 \sin \omega_r t$$

The derivative of $i(t)$ becomes

$$\frac{di}{dt} = \omega_r \cos \omega_r t A_2 e^{-\alpha t} - \alpha \sin \omega_r t A_2 e^{-\alpha}$$

When the switch is closed at $t = 0$, the capacitor offers a low impedance and the inductor offers a high impedance. The initial rate of rise of the current is limited only by the inductor L. Thus at $t = 0$, the circuit di/dt is V_s/L. Therefore,

$$\left.\frac{di}{dt}\right|_{t=0} = \omega_r A_2 = \frac{V_s}{L}$$

which gives the constant as

$$A_2 = \frac{V_s}{\omega_r L} = \frac{220 \times 1{,}000}{91{,}652 \times 2} = 1.2 \text{ A}$$

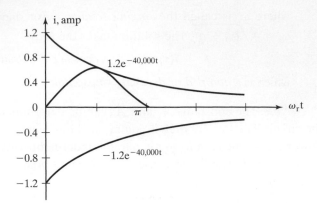

FIGURE 2.18

Current waveform for Example 2.6.

The final expression for the current $i(t)$ is

$$i(t) = 1.2 \sin(91,652t)e^{-40,000t} \text{ A}$$

b. The conduction time t_1 of the diode is obtained when $i = 0$. That is,

$$\omega_r t_1 = \pi \quad \text{or} \quad t_1 = \frac{\pi}{91,652} = 34.27 \text{ μs}$$

c. The sketch for the current waveform is shown in Figure 2.18.

d. The circuit for PSpice simulation [4] is shown in Figure 2.19. The list of the circuit file is as follows:

```
Example 2.6        RLC Circuit with Diode
.PARAM   VALU = 160                            ;Define parameter VALU
.STEP    PARAM    VALU LIST 50 160 320         ; Vary parameter VALU
VS       1   0     PWL (0   0   INS 220V  1MS 220V) ; Piecewise linear
R        2   3     {VALU}                      ; Variable resistance
L        3   4     2MH
C        4   0     0.05UF
D1       1   2     DMOD                         ; Diode with model DMOD
.MODEL DMOD   D(IS=2.22E-15 BV=1800V)          ; Diode model parameters
.TRAN  0.1US   60US                            ; Transient analysis
.PROBE                                         ; Graphics postprocessor
.END
```

The PSpice plot of the current $I(R)$ through resistance R is shown in Figure 2.20. The current response depends on the resistance R. With a higher value of R, the current becomes more damped; and with a lower value, it tends more toward sinusoidal. For $R = 0$, the peak current becomes $V_s(C/L) = 220 \times (0.05 \text{ μ}/2\text{m}) = 1.56$ A.

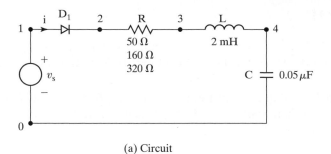

(a) Circuit

(b) Input voltage

FIGURE 2.19

RLC circuit for PSpice simulation.

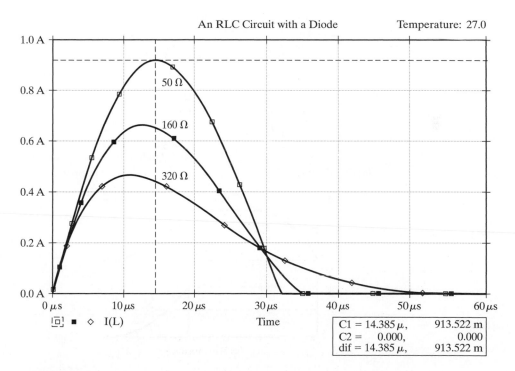

FIGURE 2.20

Plots for Example 2.6.

Key Points of Section 2.11

- The current of an LC circuit goes through resonant oscillation with a peak value of $V_s \ (C/L)$. The diode D_1 stops the reverse current flow and the capacitor is charged to $2V_s$.
- The current of an RLC depends on the damping ratio $\delta = (R/2)(C/L)$. Power electronics circuits are generally underdamped such that the circuit current becomes near sinusoidal.

2.12 FREEWHEELING DIODES

If switch S_1 in Figure 2.21a is closed for time t_1, a current is established through the load; and then if the switch is opened, a path must be provided for the current in the inductive load. Otherwise, the inductive energy induces a very high voltage and this energy is dissipated as heat across the switch as sparks. This is normally done by connecting a diode D_m as shown in Figure 2.21a, and this diode is usually called a *freewheeling diode*. The circuit operation can be divided into two modes. Mode 1 begins when the switch is closed at $t = 0$, and mode 2 begins when the switch is then opened. The equivalent circuits for the modes are shown in Figure 2.21b. Variables i_1

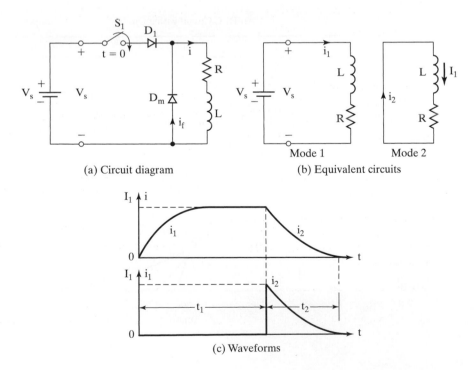

(a) Circuit diagram   (b) Equivalent circuits

(c) Waveforms

FIGURE 2.21

Circuit with a freewheeling diode.

and i_2 are defined as the instantaneous currents for mode 1 and mode 2, respectively, t_1 and t_2 are the corresponding durations of these modes.

Mode 1. During this mode, the diode current i_1, which is similar to Eq. (2.25), is

$$i_1(t) = \frac{V_s}{R}(1 - e^{-tR/L}) \tag{2.47}$$

When the switch is opened at $t = t_1$ (at the end of this mode), the current at that time becomes

$$I_1 = i_1(t = t_1) = \frac{V_s}{R}(1 - e^{t_1 R/L}) \tag{2.48}$$

If the time t_1 is sufficiently long, the current practically reaches a steady-state current of $I_s = V_s/R$ flows through the load.

Mode 2.   This mode begins when the switch is opened and the load current starts to flow through the freewheeling diode D_m. Redefining the time origin at the beginning of this mode, the current through the freewheeling diode is found from

$$0 = L\frac{di_2}{dt} + Ri_2 \tag{2.49}$$

with initial condition $i_2(t = 0) = I_1$. The solution of Eq. (2.49) gives the freewheeling current $i_f = i_2$ as

$$i_2(t) = I_1 e^{-tR/L} \tag{2.50}$$

and at $t = t_2$ this current decays exponentially to practically zero provided that $t_2 \gg L/R$. The waveforms for the currents are shown in Figure 2.21c.

Note: Figure 2.21c shows that at t_1 and t_2, the currents have reached the steady-state conditions. These are the extreme cases. A circuit normally operates under conditions such that the current remains continuous.

Example 2.7   Finding the Stored Energy in an Inductor with a Freewheeling Diode

In Figure 2.21a, the resistance is negligible ($R = 0$), the source voltage is $V_s = 220$ V (constant time), and the load inductance is $L = 220$ μH. (a) Draw the waveform for the load current if the switch is closed for a time $t_1 = 100$ μs and is then opened. (b) Determine the final energy stored in the load inductor.

Solution

a.   The circuit diagram is shown in Figure 2.22a with a zero initial current. When the switch is closed at $t = 0$, the load current rises linearly and is expressed as

$$i(t) = \frac{V_s}{L}t$$

and at $t = t_1$, $I_0 = V_s t_1/L = 220 \times 100/220 = 100$ A.

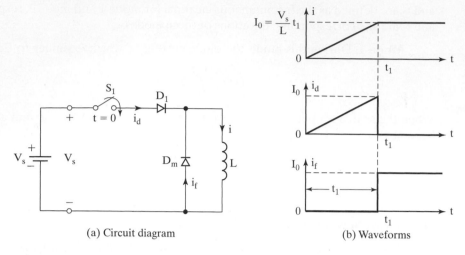

(a) Circuit diagram   (b) Waveforms

FIGURE 2.22

Diode circuit with an L load.

b. When switch S_1 is opened at a time $t = t_1$, the load current starts to flow through diode D_m. Because there is no dissipative (resistive) element in the circuit, the load current remains constant at $I_0 = 100$ A and the energy stored in the inductor is $0.5LI_0^2 -$ 1.1 J. The current waveforms are shown in Figure 2.22b.

Key Points of Section 2.12

- If the load is inductive, an antiparallel diode known as the freewheeling diode must be connected across the load to provide a path for the inductive current to flow. Otherwise, energy may be trapped into an inductive load.

2.13   RECOVERY OF TRAPPED ENERGY WITH A DIODE

In the ideal lossless circuit [7] of Figure 2.22a, the energy stored in the inductor is trapped there because no resistance exists in the circuit. In a practical circuit it is desirable to improve the *efficiency* by returning the stored energy into the supply source. This can be achieved by adding to the inductor a second winding and connecting a diode D_1 as shown in Figure 2.23a. The inductor and the secondary winding behave as a transformer. The transformer secondary is connected such that if v_1 is positive, v_2 is negative with respect to v_1, and vice versa. The secondary winding that facilitates returning the stored energy to the source via diode D_1 is known as a *feedback winding*. Assuming a transformer with a magnetizing inductance of L_m, the equivalent circuit is as shown in Figure 2.23b.

   If the diode and secondary voltage (source voltage) are referred to the primary side of the transformer, the equivalent circuit is as shown in Figure 2.23c. Parameters i_1 and i_2 define the primary and secondary currents of the transformer, respectively.

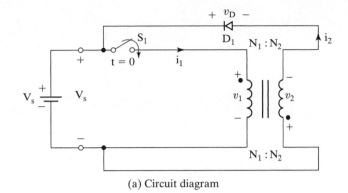

(a) Circuit diagram

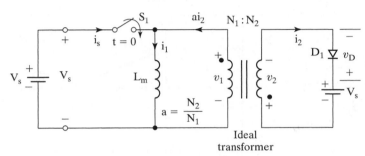

(b) Equivalent circuit

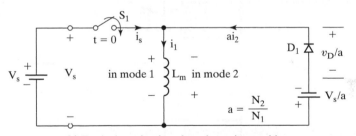

(c) Equivalent circuit, referred to primary side

FIGURE 2.23

Circuit with an energy recovery diode. [Ref. 7, S. Dewan]

The *turns ratio* of an ideal transformer is defined as

$$a = \frac{N_2}{N_1} \tag{2.51}$$

The circuit operation can be divided into two modes. Mode 1 begins when switch S_1 is closed at $t = 0$ and mode 2 begins when the switch is opened. The equivalent circuits for the modes are shown in Figure 2.24a, with t_1 and t_2 the durations of mode 1 and mode 2, respectively.

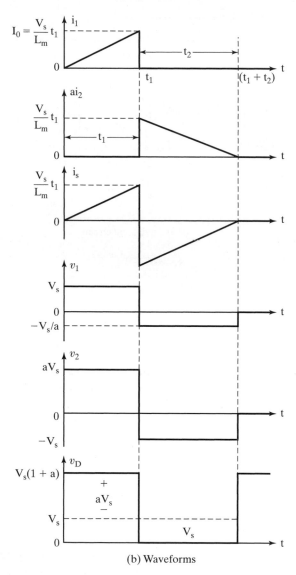

(a) Equivalent circuit

(b) Waveforms

FIGURE 2.24

Equivalent circuits and waveforms.

Mode 1.   During this mode switch S_1 is closed at $t = 0$. Diode D_1 is reverse biased and the current through the diode (secondary current) is $ai_2 = 0$ or $i_2 = 0$. Using the KVL in Figure 2.24a for mode 1, $V_s = (v_D - V_s)/a$, and this gives the reverse diode voltage as

$$v_D = V_s(1 + a) \tag{2.52}$$

Assuming that there is no initial current in the circuit, the primary current is the same as the switch current i_s and is expressed as

$$V_s = L_m \frac{di_1}{dt} \tag{2.53}$$

which gives

$$i_1(t) = i_s(t) = \frac{V_s}{L_m} t \quad \text{for} \quad 0 \le t \le t_1 \tag{2.54}$$

This mode is valid for $0 \le t \le t_1$ and ends when the switch is opened at $t = t_1$. At the end of this mode the primary current becomes

$$I_0 = \frac{V_s}{L_m} t_1 \tag{2.55}$$

Mode 2.   During this mode the switch is opened, the voltage across the inductor is reversed, and the diode D_1 is forward biased. A current flows through the transformer secondary and the energy stored in the inductor is returned to the source. Using the KVL and redefining the time origin at the beginning of this mode, the primary current is expressed as

$$L_m \frac{di_1}{dt} + \frac{V_s}{a} = 0 \tag{2.56}$$

with initial condition $i_1(t = 0) = I_0$, and we can solve the current as

$$i_1(t) = -\frac{V_s}{aL_m} t + I_0 \quad \text{for} \quad 0 \le t \le t_2 \tag{2.57}$$

The conduction time of diode D_1 is found from the condition $i_1(t = t_2) = 0$ of Eq. (2.57) and is

$$t_2 = \frac{aL_m I_0}{V_s} = at_1 \tag{2.58}$$

Mode 2 is valid for $0 \le t \le t_2$. At the end of this mode at $t = t_2$, all the energy stored in the inductor L_m is returned to the source. The various waveforms for the currents and voltage are shown in Figure 2.24b for $a = 10/6$.

Example 2.8   Finding the Recovery Energy in an Inductor with a Feedback Diode

For the energy recovery circuit of Figure 2.23a, the magnetizing inductance of the transformer is $L_m = 250$ µH, $N_1 = 10$, and $N_2 = 100$. The leakage inductances and resistances of the transformer are negligible. The source voltage is $V_s = 220$ V and there is no initial current in the circuit. If switch S_1 is closed for a time $t_1 = 50$ µs and is then opened, (a) determine the reverse voltage of diode D_1, (b) calculate the peak value of primary current, (c) calculate the peak value of secondary current, (d) determine the conduction time of diode D_1, and (e) determine the energy supplied by the source.

Solution

The turns ratio is $a = N_2/N_1 = 100/10 = 10$.

a. From Eq. (2.52) the reverse voltage of the diode,

$$v_D = V_s(1 + a) = 220 \times (1 + 10) = 2420 \text{ V}$$

b. From Eq. (2.55) the peak value of the primary current,

$$I_0 = \frac{V_s}{L_m} t_1 = 220 \times \frac{50}{250} = 44 \text{ A}$$

c. The peak value of the secondary current $I_0' = I_0/a = 44/10 = 4.4$ A.

d. From Eq. (2.58) the conduction time of the diode

$$t_2 = \frac{aL_mI_0}{V_s} = 250 \times 44 \times \frac{10}{220} = 500 \text{ µs}$$

e. The source energy,

$$W = \int_0^{t_1} vi \, dt = \int_0^{t_1} V_s \frac{V_s}{L_m} t \, dt = \frac{1}{2} \frac{V_s^2}{L_m} t_1^2$$

Using I_0 from Eq. (2.55) yields

$$W = 0.5 L_m I_0^2 = 0.5 \times 250 \times 10^{-6} \times 44^2 = 0.242 \text{ J} = 242 \text{ mJ}$$

Key Points of Section 2.13

- The trapped energy of an inductive load can be fed back to the input supply through a diode known as the feedback diode.

SUMMARY

The characteristics of practical diodes differ from those of ideal diodes. The reverse recovery time plays a significant role, especially at high-speed switching applications. Diodes can be classified into three types: (1) general-purpose diodes, (2) fast-recovery diodes, and (3) Schottky diodes. Although a Schottky diode behaves as a *pn*-junction diode, there is no physical junction; and as a result a Schottky diode is a majority carrier device. On the other hand, a *pn*-junction diode is both a majority and a minority carrier diode.

If diodes are connected in series to increase the blocking voltage capability, voltage-sharing networks under steady-state and transient conditions are required. When diodes are connected in parallel to increase the current-carrying ability, current-sharing elements are also necessary.

In this chapter we have seen the applications of power diodes in voltage reversal of a capacitor, charging a capacitor more than the dc input voltage, freewheeling action, and energy recovery from an inductive load.

REFERENCES

[1]   M. H. Rashid, *Microelectronic Circuits: Analysis and Design*. Boston: PWS Publishing. 1999, Chapter 2.

[2]   P. R. Gray and R. G. Meyer, *Analysis and Design of Analog Integrated Circuits*. New York: John Wiley & Sons. 1993, Chapter 1.

[3]   Infineon Technologies: *Power Semiconductors*. Germany: Siemens, 2001. www.infineon.com/

[4]   M. H. Rashid, *SPICE for Circuits and Electronics Using PSpice*. Englewood Cliffs, NJ: Prentice-Hall Inc. 1995.

[5]   M. H. Rashid, *SPICE for Power Electronics and Electric Power*. Englewood Cliffs, NJ: Prentice-Hall. 1993.

[6]   P. W. Tuinenga, *SPICE: A guide to Circuit Simulation and Analysis Using PSpice*. Englewood Cliffs, NJ: Prentice-Hall. 1995.

[7]   S. B. Dewan and A. Straughen, *Power Semiconductor Circuits*. New York: John Wiley & Sons. 1975, Chapter 2.

REVIEW QUESTIONS

2.1   What are the types of power diodes?
2.2   What is a leakage current of diodes?
2.3   What is a reverse recovery time of diodes?
2.4   What is a reverse recovery current of diodes?
2.5   What is a softness factor of diodes?
2.6   What are the recovery types of diodes?
2.7   What is the cause of reverse recovery time in a *pn*-junction diode?
2.8   What is the effect of reverse recovery time?
2.9   Why is it necessary to use fast-recovery diodes for high-speed switching?
2.10   What is a forward recovery time?
2.11   What are the main differences between *pn*-junction diodes and Schottky diodes?
2.12   What are the limitations of Schottky diodes?
2.13   What is the typical reverse recovery time of general-purpose diodes?
2.14   What is the typical reverse recovery time of fast-recovery diodes?
2.15   What are the problems of series-connected diodes, and what are the possible solutions?
2.16   What are the problems of parallel-connected diodes, and what are the possible solutions?
2.17   If two diodes are connected in series with equal-voltage sharings, why do the diode leakage currents differ?

2.18 What is the time constant of an RL circuit?

2.19 What is the time constant of an RC circuit?

2.20 What is the resonant frequency of an LC circuit?

2.21 What is the damping factor of an RLC circuit?

2.22 What is the difference between the resonant frequency and the ringing frequency of an RLC circuit?

2.23 What is a freewheeling diode, and what is its purpose?

2.24 What is the trapped energy of an inductor?

2.25 How is the trapped energy recovered by a diode?

PROBLEMS

2.1 The reverse recovery time of a diode is $t_{rr} = 5$ μs, and the rate of fall of the diode current is $di/dt = 80$ A/μs. If the softness factor is SF = 0.5, determine **(a)** the storage charge Q_{RR}, and **(b)** the peak reverse current I_{RR}.

2.2 The measured values of a diode at a temperature of 25 °C are

$$V_D = 1.0 \text{ V at } I_D = 50 \text{ A}$$
$$= 1.5 \text{ V at } I_D = 600 \text{ A}$$

Determine **(a)** the emission coefficient n, and **(b)** the leakage current I_s.

2.3 Two diodes are connected in series and the voltage across each diode is maintained the same by connecting a voltage-sharing resistor, such that $V_{D1} = V_{D2} = 2000$ V and $R_1 = 100$ kΩ. The $v-i$ characteristics of the diodes are shown in Figure P2.3. Determine the leakage currents of each diode and the resistance R_2 across diode D_2.

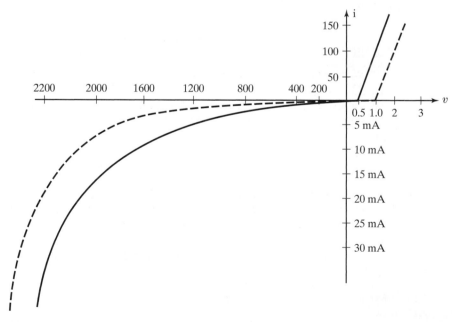

FIGURE P2.3

2.4 Two diodes are connected in parallel and the forward voltage drop across each diode is 1.5 V. The $v-i$ characteristics of diodes are shown in Figure P2.3. Determine the forward currents through each diode.

2.5 Two diodes are connected in parallel as shown in Figure 2.11a, with current-sharing resistances. The $v-i$ characteristics are shown in Figure P2.3. The total current is $I_T = 200$ A. The voltage across a diode and its resistance is $v = 2.5$ V. Determine the values of resistances R_1 and R_2 if the current is shared equally by the diodes.

2.6 Two diodes are connected in series as shown in Figure 2.8a. The resistance across the diodes is $R_1 = R_2 = 10$ kΩ. The input dc voltage is 5 kV. The leakage currents are $I_{s1} = 25$ mA and $I_{s2} = 40$ mA. Determine the voltage across the diodes.

2.7 The current waveforms of a capacitor are shown in Figure P2.7. Determine the average, root mean square (rms), and peak current ratings of the capacitor.

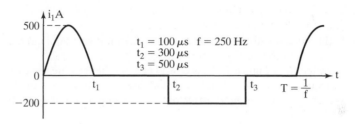

FIGURE P2.7

2.8 The waveforms of the current flowing through a diode are shown in Figure P2.8. Determine the average, rms, and peak current ratings of the diode.

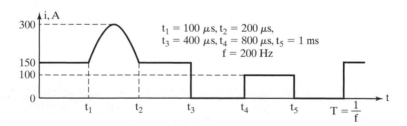

FIGURE P2.8

2.9 A diode circuit is shown in Figure P2.9 with $R = 22\ \Omega$ and $C = 10\ \mu$F. If switch S_1 is closed at $t = 0$, determine the expression for the voltage across the capacitor and the energy lost in the circuit.

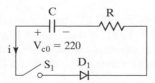

FIGURE P2.9

2.10 A diode circuit is shown in Figure P2.10 with $R = 10\ \Omega$, $L = 5$ mH, and $V_s = 220$ V. If a load current of 10 A is flowing through freewheeling diode D_m and switch S_1 is closed at $t = 0$, determine the expression for the current i through the switch.

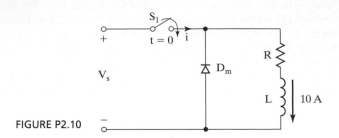

FIGURE P2.10

2.11 If the inductor of the circuit in Figure 2.15 has an initial current of I_0, determine the expression for the voltage across the capacitor.

2.12 If switch S_1 of Figure P2.12 is closed at $t = 0$, determine the expression for **(a)** the current flowing through the switch $i(t)$, and **(b)** the rate of rise of the current di/dt. **(c)** Draw sketches of $i(t)$ and di/dt. **(d)** What is the value of initial di/dt? For Figure P2.12e, find the initial di/dt only.

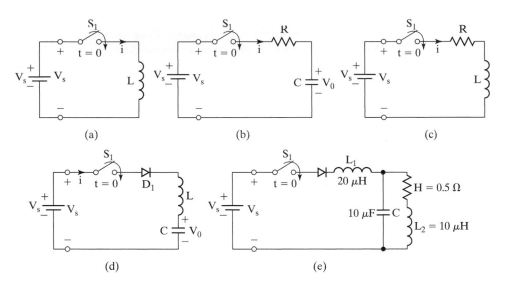

FIGURE P2.12

2.13 The second-order circuit of Figure 2.17 has the source voltage $V_s = 220$ V, inductance $L = 5$ mH, capacitance $C = 10\ \mu$F, and resistance $R = 22\ \Omega$. The initial voltage of the capacitor is $V_{c0} = 50$ V. If the switch is closed at $t = 0$, determine **(a)** an expression for the current, and **(b)** the conduction time of the diode. **(c)** Draw a sketch of $i(t)$.

2.14 For the energy recovery circuit of Figure 2.23a, the magnetizing inductance of the transformer is $L_m = 150\ \mu H$, $N_1 = 10$, and $N_2 = 200$. The leakage inductances and resistances of the transformer are negligible. The source voltage is $V_s = 200$ V and there is no initial current in the circuit. If switch S_1 is closed for a time $t_1 = 100\ \mu s$ and is then opened, **(a)** determine the reverse voltage of diode D_1, **(b)** calculate the peak primary current, **(c)** calculate the peak secondary current, **(d)** determine the time for which diode D_1 conducts, and **(e)** determine the energy supplied by the source.

2.15 A diode circuit is shown in Figure P2.15 where the load current is flowing through diode D_m. If switch S_1 is closed at a time $t = 0$, determine **(a)** expressions for $v_c(t)$, $i_c(t)$, and $i_d(t)$; **(b)** time t_1 when the diode D_1 stops conducting; **(c)** time t_q when the voltage across the capacitor becomes zero; and **(d)** the time required for capacitor to recharge to the supply voltage V_s.

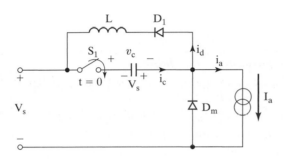

FIGURE P2.15

CHAPTER 3

Diode Rectifiers

The learning objectives of this chapter are as follows:

- To understand the operation and characteristics of diode rectifiers
- To learn the types of diode rectifiers
- To understand the performance parameters of diode rectifiers
- To learn the techniques for analyzing and design of diode rectifier circuits
- To learn the techniques for simulating diode rectifiers by using SPICE
- To study the effects of load inductance on the load current

3.1 INTRODUCTION

Diodes are extensively used in rectifiers. A *rectifier* is a circuit that converts an ac signal into a unidirectional signal. A rectifier is a type of dc–ac converter. Depending on the type of input supply, the rectifiers are classified into two types: (1) single phase and (2) three phase. For the sake of simplicity the diodes are considered to be ideal. By "ideal" we mean that the reverse recovery time t_{rr} and the forward voltage drop V_D are negligible. That is, $t_{rr} = 0$ and $V_D = 0$.

3.2 SINGLE-PHASE HALF-WAVE RECTIFIERS

A single-phase half-wave rectifier is the simplest type, but it is not normally used in industrial applications. However, it is useful in understanding the principle of rectifier operation. The circuit diagram with a resistive load is shown in Figure 3.1a. During the positive half-cycle of the input voltage, diode D_1 conducts and the input voltage appears across the load. During the negative half-cycle of the input voltage, the diode is in a *blocking condition* and the output voltage is zero. The waveforms for the input voltage and output voltage are shown in Figure 3.1b.

Key Points of Section 3.2

- The half-wave rectifier is the simplest power electronics circuit that is used for low-cost power supplies for electronics like radios.

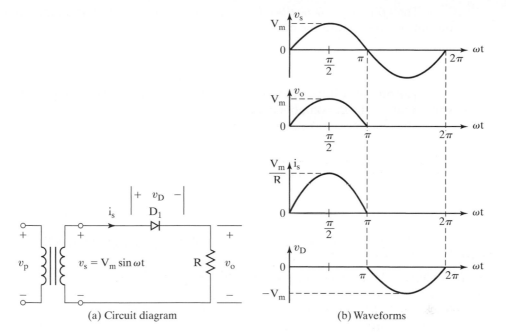

(a) Circuit diagram   (b) Waveforms

FIGURE 3.1

Single-phase half-wave rectifier.

3.3   PERFORMANCE PARAMETERS

Although the output voltage as shown in Figure 3.1b is dc, it is discontinuous and contains harmonics. A rectifier is a power processor that should give a dc output voltage with a minimum amount of harmonic contents. At the same time, it should maintain the input current as sinusoidal as possible and in phase with the input voltage so that the power factor is near unity. The power-processing quality of a rectifier requires the determination of harmonic contents of the input current, the output voltage, and the output current. We can use Fourier series expansions to find the harmonic contents of voltages and currents. There are different types of rectifier circuits and the performances of a rectifier are normally evaluated in terms of the following parameters:

The *average* value of the output (load) voltage, V_{dc}
The average value of the output (load) current, I_{dc}
The output dc power,

$$P_{dc} = V_{dc}I_{dc} \tag{3.1}$$

The root-mean-square (rms) value of the output voltage, V_{rms}
The rms value of the output current, I_{rms}

The output ac power

$$P_{ac} = V_{rms}I_{rms} \tag{3.2}$$

The *efficiency* (or *rectification ratio*) of a rectifier, which is a figure of merit and permits us to compare the effectiveness, is defined as

$$\eta = \frac{P_{dc}}{P_{ac}} \tag{3.3}$$

The output voltage can be considered as composed of two components: (1) the dc value, and (2) the ac component or ripple.

The *effective* (rms) value of the ac component of output voltage is

$$V_{ac} = \sqrt{V_{rms}^2 - V_{dc}^2} \tag{3.4}$$

The *form factor*, which is a measure of the shape of output voltage, is

$$FF = \frac{V_{rms}}{V_{dc}} \tag{3.5}$$

The *ripple factor*, which is a measure of the ripple content, is defined as

$$RF = \frac{V_{ac}}{V_{dc}} \tag{3.6}$$

Substituting Eq. (3.4) in Eq. (3.6), the ripple factor can be expressed as

$$RF = \sqrt{\left(\frac{V_{rms}}{V_{dc}}\right)^2 - 1} = \sqrt{FF^2 - 1} \tag{3.7}$$

The *transformer utilization factor* is defined as

$$TUF = \frac{P_{dc}}{V_s I_s} \tag{3.8}$$

where V_s and I_s are the rms voltage and rms current of the transformer secondary, respectively. Let us consider the waveforms of Figure 3.2, where v_s is the sinusoidal input voltage, i_s is the instantaneous input current, and i_{s1} is its fundamental component.

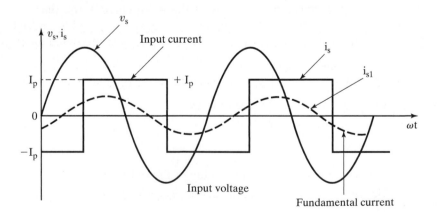

FIGURE 3.2

Waveforms for input voltage and current.

If ϕ is the angle between the fundamental components of the input current and voltage, ϕ is called the *displacement angle*. The *displacement factor* is defined as

$$DF = \cos \phi \tag{3.9}$$

The *harmonic factor* (HF) of the input current is defined as

$$HF = \left(\frac{I_s^2 - I_{s1}^2}{I_{s1}^2} \right)^{1/2} = \left[\left(\frac{I_s}{I_{s1}} \right)^2 - 1 \right]^{1/2} \tag{3.10}$$

where I_{s1} is the fundamental component of the input current I_s. Both I_{s1} and I_s are expressed here in rms. The input *power factor* (PF) is defined as

$$PF = \frac{V_s I_{s1}}{V_s I_s} \cos \phi = \frac{I_{s1}}{I_s} \cos \phi \tag{3.11}$$

Crest factor (CF), which is a measure of the peak input current $I_{s(\text{peak})}$ as compared with its rms value I_s, is often of interest to specify the peak current ratings of devices and components. CF of the input current is defined by

$$CF = \frac{I_{s(\text{peak})}}{I_s} \tag{3.12}$$

Notes

1. HF is a measure of the distortion of a waveform and is also known as *total harmonic distortion* (THD).
2. If the input current i_s is purely sinusoidal, $I_{s1} = I_s$ and the power factor PF equals the displacement factor DF. The displacement angle ϕ becomes the impedance angle $\theta = \tan^{-1}(\omega L/R)$ for an *RL* load.
3. Displacement factor DF is often known as *displacement power factor* (DPF).
4. An ideal rectifier should have $\eta = 100\%$, $V_{\text{ac}} = 0$, RF $= 0$, TUF $= 1$, HF $=$ THD $= 0$, and PF $=$ DPF $= 1$.

Example 3.1   Finding the Performance Parameters of a Half-Wave Rectifier

The rectifier in Figure 3.1a has a purely resistive load of R. Determine (a) the efficiency, (b) the FF, (c) the RF, (d) the TUF, (e) the PIV of diode D_1, (f) the CF of the input current, and (g) input PF.

Solution

The average output voltage V_{dc} is defined as

$$V_{\text{dc}} = \frac{1}{T} \int_0^T v_L(t) \, dt$$

We can notice from Figure 3.1b that $v_L(t) = 0$ for $T/2 \leq t \leq T$. Hence, we have

$$V_{\text{dc}} = \frac{1}{T} \int_0^{T/2} V_m \sin \omega t \, dt = \frac{-V_m}{\omega T} \left(\cos \frac{\omega T}{2} - 1 \right)$$

However, the frequency of the source is $f = 1/T$ and $\omega = 2\pi f$. Thus

$$V_{dc} = \frac{V_m}{\pi} = 0.318V_m$$

$$I_{dc} = \frac{V_{dc}}{R} = \frac{0.318V_m}{R} \tag{3.13}$$

The rms value of a periodic waveform is defined as

$$V_{rms} = \left[\frac{1}{T} \int_0^T v_L^2(t)\, dt \right]^{1/2}$$

For a sinusoidal voltage of $v_0(t) = V_m \sin \omega t$ for $0 \leq t \leq T/2$, the rms value of the output voltage is

$$V_{rms} = \left[\frac{1}{T} \int_0^{T/2} (V_m \sin \omega t)^2\, dt \right]^{1/2} = \frac{V_m}{2} = 0.5V_m$$

$$I_{rms} = \frac{V_{rms}}{R} = \frac{0.5V_m}{R} \tag{3.14}$$

From Eq. (3.1), $P_{dc} = (0.318V_m)^2/R$, and from Eq. (3.2), $P_{ac} = (0.5V_m)^2/R$.

a. From Eq. (3.3), the efficiency $\eta = (0.318V_m)^2/(0.5V_m)^2 = 40.5\%$.

b. From Eq. (3.5), the FF $= 0.5V_m/0.318V_m = 1.57$ or 157%.

c. From Eq. (3.7), the RF $= \sqrt{1.57^2 - 1} = 1.21$ or 121%.

d. The rms voltage of the transformer secondary is

$$V_s = \left[\frac{1}{T} \int_0^T (V_m \sin \omega t)^2\, dt \right]^{1/2} = \frac{V_m}{\sqrt{2}} = 0.707V_m \tag{3.15}$$

The rms value of the transformer secondary current is the same as that of the load:

$$I_s = \frac{0.5V_m}{R}$$

The *volt-ampere* rating (VA) of the transformer, VA $= V_s I_s = 0.707V_m \times 0.5V_m/R$. From Eq. (3.8) TUF $= P_{ac}/(V_s I_s) = 0.318^2/(0.707 \times 0.5) = 0.286$.

e. The peak reverse (or inverse) blocking voltage PIV $= V_m$.

f. $I_{s(peak)} = V_m/R$ and $I_s = 0.5V_m/R$. The CF of the input current is CF $= I_{s(peak)}/I_s = 1/0.5 = 2$.

g. The input PF for a resistive load can be found from

$$PF = \frac{P_{ac}}{VA} = \frac{0.5^2}{0.707 \times 0.5} = 0.707$$

Note: $1/TUF = 1/0.286 = 3.496$ signifies that the transformer must be 3.496 times larger than that when it is used to deliver power from a pure ac voltage. This rectifier has a high ripple factor, 121%; a low efficiency, 40.5%; and a poor TUF, 0.286. In addition, the transformer has to carry a dc current, and this results in a dc saturation problem of the transformer core.

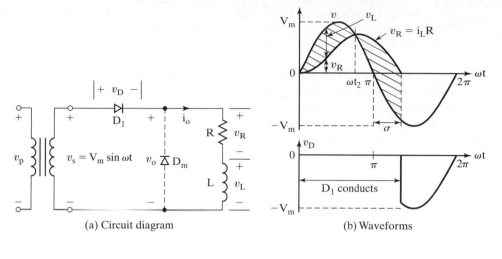

(a) Circuit diagram   (b) Waveforms

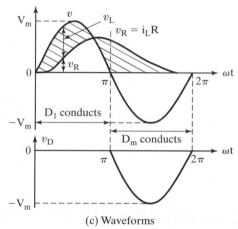

(c) Waveforms

FIGURE 3.3

Half-wave rectifier with *RL* load.

Let us consider the circuit of Figure 3.1a with an *RL* load as shown in Figure 3.3a. Due to inductive load, the conduction period of diode D_1 will extend beyond 180° until the current becomes zero at $\omega t = \pi + \sigma$. The waveforms for the current and voltage are shown in Figure 3.3b. It should be noted that the average v_L of the inductor is zero. The average output voltage is

$$V_{dc} = \frac{V_m}{2\pi} \int_0^{\pi+\sigma} \sin \omega t \, d(\omega t) = \frac{V_m}{2\pi} [-\cos \omega t]_0^{\pi+\sigma}$$

$$= \frac{V_m}{2\pi} [1 - \cos(\pi + \sigma)] \tag{3.16}$$

The average load current is $I_{dc} = V_{dc}/R$.

It can be noted from Eq. (3.16) that the average voltage (and current) can be increased by making $\sigma = 0$, which is possible by adding a freewheeling diode D_m as shown in Figure 3.3a with dashed lines. The effect of this diode is to prevent a negative voltage appearing across the load; and as a result, the magnetic stored energy is increased. At $t = t_1 = \pi/\omega$, the current from D_1 is transferred to D_m and this process is called *commutation* of diodes and the waveforms are shown in Figure 3.3c. Depending on the load time constant, the load current may be discontinuous. Load current i_0 is discontinuous with a resistive load and continuous with a very high inductive load. The continuity of the load current depends on its time constant $\tau = \omega L/R$.

If the output is connected to a battery, the rectifier can be used as a battery charger. This is shown in Figure 3.4a. For $v_s > E$, diode D_1 conducts. The angle α when the diode starts conducting can be found from the condition

$$V_m \sin \alpha = E$$

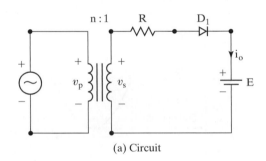

(a) Circuit

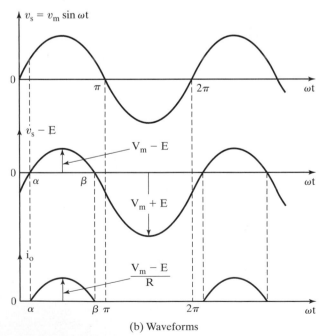

(b) Waveforms

FIGURE 3.4

Battery charger.

which gives

$$\alpha = \sin^{-1} \frac{E}{V_m} \tag{3.17}$$

Diode D_1 is turned off when $v_s < E$ at

$$\beta = \pi - \alpha$$

The charging current i_L, which is shown in Figure 3.4b, can be found from

$$i_0 = \frac{v_s - E}{R} = \frac{V_m \sin \omega t - E}{R} \quad \text{for } \alpha < \omega t < \beta$$

Example 3.2   Finding the Performance Parameters of a Battery Charger

The battery voltage in Figure 3.4a is $E = 12$ V and its capacity is 100 Wh. The average charging current should be $I_{dc} = 5$ A. The primary input voltage is $V_p = 120$ V, 60 Hz, and the transformer has a turn ratio of $n = 2{:}1$. Calculate (a) the conduction angle δ of the diode, (b) the current-limiting resistance R, (c) the power rating P_R of R, (d) the charging time h_o in hours, (e) the rectifier efficiency η, and (f) the PIV of the diode.

Solution

$E = 12$ V, $V_p = 120$ V, $V_s = V_p/n = 120/2 = 60$ V, and $V_m = \sqrt{2} V_s = \sqrt{2} \times 60 = 84.85$ V.

a. From Eq. (3.17), $\alpha = \sin^{-1}(12/84.85) = 8.13°$ or 0.1419 rad. $\beta = 180 - 8.13 = 171.87°$. The conduction angle is $\delta = \beta - \alpha = 171.87 - 8.13 = 163.74°$.

b. The average charging current I_{dc} is

$$I_{dc} = \frac{1}{2\pi} \int_\alpha^\beta \frac{V_m \sin \omega t - E}{R} \, d(\omega t)$$

$$= \frac{1}{2\pi R} (2V_m \cos \alpha + 2E\alpha - \pi E), \quad \text{for } \beta = \pi - \alpha \tag{3.18}$$

which gives

$$R = \frac{1}{2\pi I_{dc}} (2V_m \cos \alpha + 2E\alpha - \pi E)$$

$$= \frac{1}{2\pi \times 5} (2 \times 84.85 \times \cos 8.13° + 2 \times 12 \times 0.1419 - \pi \times 12) = 4.26 \ \Omega$$

c. The rms battery current I_{rms} is

$$I_{rms}^2 = \frac{1}{2\pi} \int_\alpha^\beta \frac{(V_m \sin \omega t - E)^2}{R^2} \, d(\omega t)$$

$$= \frac{1}{2\pi R^2} \left[\left(\frac{V_m^2}{2} + E^2 \right)(\pi - 2\alpha) + \frac{V_m^2}{2} \sin 2\alpha - 4V_m E \cos \alpha \right]$$

$$= 67.4 \tag{3.19}$$

or $I_{rms} = \sqrt{67.4} = 8.2$ A. The power rating of R is $P_R = 8.2^2 \times 4.26 = 286.4$ W.

d. The power delivered P_{dc} to the battery is

$$P_{dc} = EI_{dc} = 12 \times 5 = 60 \text{ W}$$

$$h_o P_{dc} = 100 \quad \text{or} \quad h_o = \frac{100}{P_{dc}} = \frac{100}{60} = 1.667 \text{ h}$$

e. The rectifier efficiency η is

$$\eta = \frac{\text{power delivered to the battery}}{\text{total input power}} = \frac{P_{dc}}{P_{dc} + P_R} = \frac{60}{60 + 286.4} = 17.32\%$$

f. The peak inverse voltage PIV of the diode is

$$\text{PIV} = V_m + E$$
$$= 84.85 + 12 = 96.85 \text{ V}$$

Example 3.3   Finding the Fourier Series of the Output Voltage for a Half-Wave Rectifier

The single-phase half-wave rectifier of Figure 3.1a is connected to a source of $V_s = 120$ V, 60 Hz. Express the instantaneous output voltage $v_0(t)$ in Fourier series.

Solution

The rectifier output voltage v_0 may be described by a Fourier series as

$$v_0(t) = V_{dc} + \sum_{n=1,2,\ldots}^{\infty} (a_n \sin n\omega t + b_n \cos n\omega t)$$

$$V_{dc} = \frac{1}{2\pi} \int_0^{2\pi} v_0 \, d(\omega t) = \frac{1}{2\pi} \int_0^{\pi} V_m \sin \omega t \, d(\omega t) = \frac{V_m}{\pi}$$

$$a_n = \frac{1}{\pi} \int_0^{2\pi} v_0 \sin n\omega t \, d(\omega t) = \frac{1}{\pi} \int_0^{\pi} V_m \sin \omega t \sin n\omega t \, d(\omega t)$$

$$= \frac{V_m}{2} \qquad \text{for } n = 1$$

$$= 0 \qquad \text{for } n = 2, 3, 4, 5, 6, \ldots$$

$$b_n = \frac{1}{\pi} \int_0^{2\pi} v_0 \cos n\omega t \, d(\omega t) = \frac{1}{\pi} \int_0^{\pi} V_m \sin \omega t \cos n\omega t \, d(\omega t)$$

$$= \frac{V_m}{\pi} \frac{1 + (-1)^n}{1 - n^2} \qquad \text{for } n = 2, 4, 6, \ldots$$

$$= 0 \qquad \text{for } n = 1, 3, 5, \ldots$$

Substituting a_n and b_n, the instantaneous output voltage becomes

$$v_0(t) = \frac{V_m}{\pi} + \frac{V_m}{2} \sin \omega t - \frac{2V_m}{3\pi} \cos 2\omega t - \frac{2V_m}{15\pi} \cos 4\omega t - \frac{2V_m}{35\pi} \cos 6\omega t - \cdots \qquad (3.20)$$

where $V_m = \sqrt{2} \times 120 = 169.7$ V and $\omega = 2\pi \times 60 = 377$ rad/s.

Key Points of Section 3.3

- The performance of a half-wave rectifier that is measured by certain parameters is poor. The load current can be made continuous by adding an inductor and a freewheeling diode. The output voltage is discontinuous and contains harmonics at multiples of the supply frequency.

3.4   SINGLE-PHASE FULL-WAVE RECTIFIERS

A full-wave rectifier circuit with a center-tapped transformer is shown in Figure 3.5a. Each half of the transformer with its associated diode acts as a half-wave rectifier and the output of a full-wave rectifier is shown in Figure 3.5b. Because there is no dc current flowing through the transformer, there is no dc saturation problem of transformer core. The average output voltage is

$$V_{dc} = \frac{2}{T} \int_0^{T/2} V_m \sin \omega t \, dt = \frac{2V_m}{\pi} = 0.6366 V_m \qquad (3.21)$$

Instead of using a center-tapped transformer, we could use four diodes, as shown in Figure 3.6a. During the positive half-cycle of the input voltage, the power is supplied

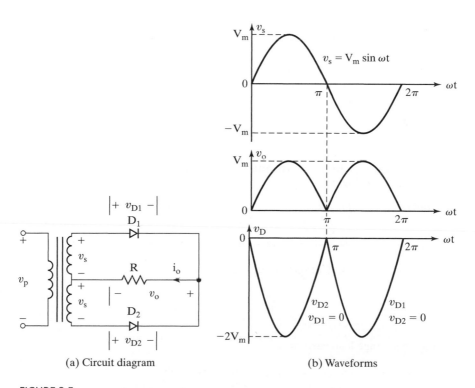

(a) Circuit diagram   (b) Waveforms

FIGURE 3.5

Full-wave rectifier with center-tapped transformer.

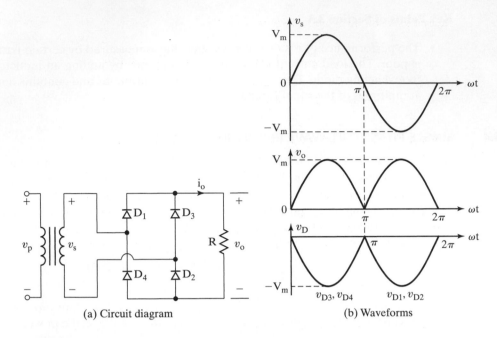

(a) Circuit diagram   (b) Waveforms

FIGURE 3.6

Full-wave bridge rectifier.

to the load through diodes D_1 and D_2. During the negative cycle, diodes D_3 and D_4 conduct. The waveform for the output voltage is shown in Figure 3.6b and is similar to that of Figure 3.5b. The peak-inverse voltage of a diode is only V_m. This circuit is known as a *bridge rectifier*, and it is commonly used in industrial applications [1, 2].

Example 3.4   Finding the Performance Parameters of a Full-Wave Rectifier with Center-Tapped Transformer

If the rectifier in Figure 3.5a has a purely resistive load of R, determine (a) the efficiency, (b) the FF, (c) the RF, (d) the TUF, (e) the PIV of diode D_1, and (f) the CF of the input current.

Solution

From Eq. (3.21), the average output voltage is

$$V_{dc} = \frac{2V_m}{\pi} = 0.6366V_m$$

and the average load current is

$$I_{dc} = \frac{V_{dc}}{R} = \frac{0.6366V_m}{R}$$

The rms value of the output voltage is

$$V_{rms} = \left[\frac{2}{T}\int_0^{T/2}(V_m\sin\omega t)^2\,dt\right]^{1/2} = \frac{V_m}{\sqrt{2}} = 0.707V_m$$

$$I_{rms} = \frac{V_{rms}}{R} = \frac{0.707V_m}{R}$$

From Eq. (3.1) $P_{dc} = (0.6366V_m)^2/R$, and from Eq. (3.2) $P_{ac} = (0.707V_m)^2/R$.

a. From Eq. (3.3), the efficiency $\eta = (0.6366V_m)^2/(0.707V_m)^2 = 81\%$.

b. From Eq. (3.5), the form factor FF $= 0.707V_m/0.6366V_m = 1.11$.

c. From Eq. (3.7), the ripple factor RF $= \sqrt{1.11^2 - 1} = 0.482$ or 48.2%.

d. The rms voltage of the transformer secondary $V_s = V_m/\sqrt{2} = 0.707V_m$. The rms value of transformer secondary current $I_s = 0.5V_m/R$. The volt-ampere rating (VA) of the transformer, VA $= 2V_sI_s = 2 \times 0.707V_m \times 0.5V_m/R$. From Eq. (3.8),

$$\text{TUF} = \frac{0.6366^2}{2 \times 0.707 \times 0.5} = 0.5732 = 57.32\%$$

e. The peak reverse blocking voltage, PIV $= 2V_m$.

f. $I_{s(peak)} = V_m/R$ and $I_s = 0.707V_m/R$. The CF of the input current is CF $= I_{s(peak)}/I_s = 1/0.707 = \sqrt{2}$.

g. The input PF for a resistive load can be found from

$$\text{PF} = \frac{P_{ac}}{\text{VA}} = \frac{0.707^2}{2 \times 0.707 \times 0.5} = 0.707$$

Note: 1/TUF $= 1/0.5732 = 1.75$ signifies that the input transformer, if present, must be 1.75 times larger than that when it is used to deliver power from a pure ac sinusoidal voltage. The rectifier has an RF of 48.2% and a rectification efficiency of 81%.

Note: The performance of a full-wave rectifier is significantly improved compared with that of a half-wave rectifier.

Example 3.5   Finding the Fourier Series of the Output Voltage for a Full-Wave Rectifier

The rectifier in Figure 3.5a has an RL load. Use the method of Fourier series to obtain expressions for output voltage $v_0(t)$.

Solution

The rectifier output voltage may be described by a Fourier series (which is reviewed in Appendix E) as

$$v_0(t) = V_{dc} + \sum_{n=2,4,\ldots}^{\infty}(a_n\cos n\omega t + b_n\sin n\omega t)$$

where

$$V_{dc} = \frac{1}{2\pi} \int_0^{2\pi} v_0(t)\, d(\omega t) = \frac{2}{2\pi} \int_0^{\pi} V_m \sin \omega t\, d(\omega t) = \frac{2V_m}{\pi}$$

$$a_n = \frac{1}{\pi} \int_0^{2\pi} v_0 \cos n\omega t\, d(\omega t) = \frac{2}{\pi} \int_0^{\pi} V_m \sin \omega t \cos n\omega t\, d(\omega t)$$

$$= \frac{4V_m}{\pi} \sum_{n=2,4,\ldots}^{\infty} \frac{-1}{(n-1)(n+1)} \qquad \text{for } n = 2, 4, 6, \ldots$$

$$= 0 \qquad \text{for } n = 1, 3, 5, \ldots$$

$$b_n = \frac{1}{\pi} \int_0^{2\pi} v_0 \sin n\omega t\, d(\omega t) = \frac{2}{\pi} \int_0^{\pi} V_m \sin \omega t \sin n\omega t\, d(\omega t) - 0$$

Substituting the values of a_n and b_n, the expression for the output voltage is

$$v_0(t) = \frac{2V_m}{\pi} - \frac{4V_m}{3\pi} \cos 2\omega t - \frac{4V_m}{15\pi} \cos 4\omega t - \frac{4V_m}{35\pi} \cos 6\omega t - \cdots \qquad (3.22)$$

Note: The output of a full-wave rectifier contains only even harmonics and the second harmonic is the most dominant one and its frequency is $2f (= 120 \text{ Hz})$. The output voltage in Eq. (3.22) can be derived by spectrum multiplication of switching function, and this is explained in Appendix C.

Example 3.6 Finding the Input Power Factor of a Full-Wave Rectifier

A single-phase bridge rectifier that supplies a very high inductive load such as a dc motor is shown in Figure 3.7a. The turns ratio of the transformer is unity. The load is such that the motor draws a ripple-free armature current of I_a as shown in Figure 3.7b. Determine (a) the HF of input current, and (b) the input PF of the rectifier.

Solution

Normally, a dc motor is highly inductive and acts like a filter in reducing the ripple current of the load.

a. The waveforms for the input current and input voltage of the rectifier are shown in Figure 3.7b. The input current can be expressed in a Fourier series as

$$i_s(t) = I_{dc} + \sum_{n=1,3,\ldots}^{\infty} (a_n \cos n\omega t + b_n \sin n\omega t)$$

where

$$I_{dc} = \frac{1}{2\pi} \int_0^{2\pi} i_s(t)\, d(\omega t) = \frac{1}{2\pi} \int_0^{2\pi} I_a\, d(\omega t) = 0$$

$$a_n = \frac{1}{\pi} \int_0^{2\pi} i_s(t) \cos n\omega t\, d(\omega t) = \frac{2}{\pi} \int_0^{\pi} I_a \cos n\omega t\, d(\omega t) = 0$$

$$b_n = \frac{1}{\pi} \int_0^{2\pi} i_s(t) \sin n\omega t\, d(\omega t) = \frac{2}{\pi} \int_0^{\pi} I_a \sin n\omega t\, d(\omega t) = \frac{4I_a}{n\pi}$$

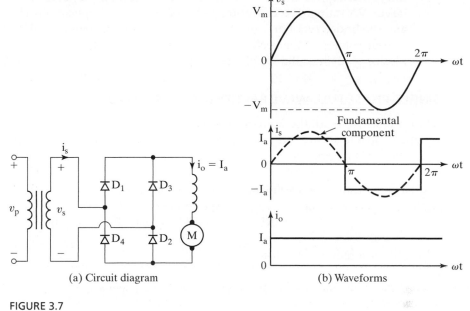

(a) Circuit diagram   (b) Waveforms

FIGURE 3.7

Full-wave bridge rectifier with dc motor load.

Substituting the values of a_n and b_n, the expression for the input current is

$$i_s(t) = \frac{4I_a}{\pi}\left(\frac{\sin \omega t}{1} + \frac{\sin 3\omega t}{3} + \frac{\sin 5\omega t}{5} + \cdots\right) \tag{3.23}$$

The rms value of the fundamental component of input current is

$$I_{s1} = \frac{4I_a}{\pi\sqrt{2}} = 0.90I_a$$

The rms value of the input current is

$$I_s = \frac{4}{\pi\sqrt{2}}I_a\left[1 + \left(\frac{1}{3}\right)^2 + \left(\frac{1}{5}\right)^2 + \left(\frac{1}{7}\right)^2 + \left(\frac{1}{9}\right)^2 + \cdots\right]^{1/2} = I_a$$

From Eq. (3.10),

$$\text{HF} = \text{THD} = \left[\left(\frac{1}{0.90}\right)^2 - 1\right]^{1/2} = 0.4843 \quad \text{or} \quad 48.43\%$$

b. The displacement angle $\phi = 0$ and $\text{DF} = \cos \phi = 1$. From Eq. (3.11), the PF = $(I_{s1}/I_s) \cos \phi = 0.90$ (lagging).

Key Points of Section 3.4

- There are two types of single-phase rectifiers: center-tapped transformer and bridge. Their performances are almost identical, except the secondary current of the center-tapped transformer carries unidirectional (dc) current and it requires larger VA rating. The center-tapped type is used in applications less than 100 W and the bridge rectifier is used in applications ranging from 100 W to 100 kW. The output voltage of the rectifiers contains harmonics whose frequencies are multiples of $2f$ (2 times the supply frequency).

3.5   SINGLE-PHASE FULL-WAVE RECTIFIER WITH *RL* LOAD

With a resistive load, the load current is identical in shape to the output voltage. In practice, most loads are inductive to a certain extent and the load current depends on the values of load resistance R and load inductance L. This is shown in Figure 3.8a. A battery of voltage E is added to develop generalized equations. If $v_s = V_m \sin \omega t = \sqrt{2}\, V_s \sin \omega t$ is the input voltage, the load current i_0 can be found from

$$L\frac{di_0}{dt} + Ri_0 + E = \sqrt{2}\, V_s \sin \omega t \quad \text{for} \quad i_0 \geq 0$$

which has a solution of the form

$$i_0 = \frac{\sqrt{2}V_s}{Z}\sin(\omega t - \theta) + A_1 e^{-(R/L)t} - \frac{E}{R} \tag{3.24}$$

where load impedance $Z = [R^2 + (\omega L)^2]^{1/2}$, load impedance angle $\theta = \tan^{-1}(\omega L/R)$, and V_s is the rms value of the input voltage.

Case 1: continuous load current.   This is shown in Figure 3.8b. The constant A_1 in Eq. (3.24) can be determined from the condition: at $\omega t = \pi$, $i_0 = I_0$.

$$A_1 = \left(I_0 + \frac{E}{R} - \frac{\sqrt{2}V_s}{Z}\sin\theta\right)e^{(R/L)(\pi/\omega)}$$

Substitution of A_1 in Eq. (3.24) yields

$$i_0 = \frac{\sqrt{2}V_s}{Z}\sin(\omega t - \theta) + \left(I_0 + \frac{E}{R} - \frac{\sqrt{2}V_s}{Z}\sin\theta\right)e^{(R/L)(\pi/\omega - t)} - \frac{E}{R} \tag{3.25}$$

Under a steady-state condition, $i_0(\omega t = 0) = i_0(\omega t = \pi)$. That is, $i_0(\omega t = 0) = I_0$. Applying this condition, we get the value of I_0 as

$$I_0 = \frac{\sqrt{2}V_s}{Z}\sin\theta\,\frac{1 + e^{-(R/L)(\pi/\omega)}}{1 - e^{-(R/L)(\pi/\omega)}} - \frac{E}{R} \quad \text{for } I_0 \geq 0 \tag{3.26}$$

which, after substituting I_0 in Eq. (3.25) and simplification, gives

$$i_0 = \frac{\sqrt{2}V_s}{Z}\left[\sin(\omega t - \theta) + \frac{2}{1 - e^{-(R/L)(\pi/\omega)}}\sin\theta\, e^{-(R/L)t}\right] - \frac{E}{R}$$

$$\text{for } 0 \leq (\omega t - \theta) \leq \pi \text{ and } i_0 \geq 0 \tag{3.27}$$

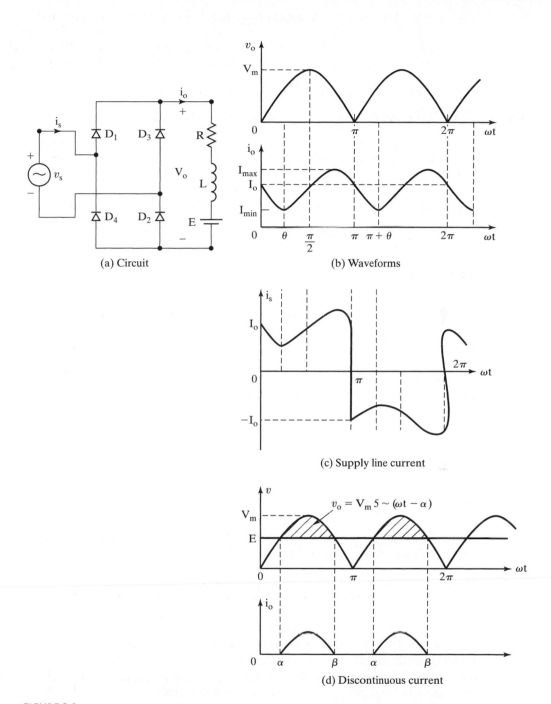

(a) Circuit

(b) Waveforms

(c) Supply line current

(d) Discontinuous current

FIGURE 3.8

Full-bridge rectifier with *RL* load.

The rms diode current can be found from Eq. (3.27) as

$$I_r = \left[\frac{1}{2\pi} \int_0^\pi i_0^2 \, d(\omega t) \right]^{1/2}$$

and the rms output current can then be determined by combining the rms current of each diode as

$$I_{\text{rms}} = (I_r^2 + I_r^2)^{1/2} = \sqrt{2} I_r$$

The average diode current can also be found from Eq. (3.27) as

$$I_d = \frac{1}{2\pi} \int_0^\pi i_0 \, d(\omega t)$$

Case 2: discontinuous load current.   This is shown in Figure 3.8d. The load current flows only during the period $\alpha \leq \omega t \leq \beta$. Let us define $x = E/V_m = E/\sqrt{2}V_s$ as the load battery (*emf*) constant, called the *voltage ratio*. The diodes start to conduct at $\omega t = \alpha$ given by

$$\alpha = \sin^{-1} \frac{E}{V_m} = \sin^{-1}(x)$$

At $\omega t = \alpha$, $i_0(\omega t) = 0$ and Eq. (3.24) gives

$$A_1 = \left[\frac{E}{R} - \frac{\sqrt{2}V_s}{Z} \sin(\alpha - \theta) \right] e^{(R/L)(\alpha/\omega)}$$

which, after substituting in Eq. (3.24), yields the load current

$$i_0 = \frac{\sqrt{2}V_s}{Z} \sin(\omega t - \theta) + \left[\frac{E}{R} - \frac{\sqrt{2}V_s}{Z} \sin(\alpha - \theta) \right] e^{(R/L)(\alpha/\omega - t)} - \frac{E}{R} \quad (3.28)$$

At $\omega t = \beta$, the current falls to zero, and $i_0(\omega t = \beta) = 0$. That is,

$$\frac{\sqrt{2}V_s}{Z} \sin(\beta - \theta) + \left[\frac{E}{R} - \frac{\sqrt{2}V_s}{Z} \sin(\alpha - \theta) \right] e^{(R/L)(\alpha - \beta)/\omega} - \frac{E}{R} = 0 \quad (3.29)$$

Dividing Eq. (3.29) by $\sqrt{2}V_s/Z$, and substituting $R/Z = \cos\theta$ and $\omega L/R = \tan\theta$, we get

$$\sin(\beta - \theta) + \left(\frac{x}{\cos(\theta)} - \sin(\alpha - \theta) \right) e^{\frac{(\alpha - \beta)}{\tan(\theta)}} - \frac{x}{\cos(\theta)} = 0 \quad (3.30)$$

β can be determined from this transcendental equation by an iterative (trial and error) method of solution. Start with $\beta = 0$, and increase its value by a very small amount until the left-hand side of this equation becomes zero.

As an example, Mathcad was used to find the value of β for $\theta = 30°, 60°$, and $x = 0$ to 1. The results are shown in Table 3.1. As k increases, β decreases. At $k = 1.0$, the diodes do not conduct and no current flows.

TABLE 3.1 Variations of Angle β with the Voltage Ratio, *x*

Voltage Ratio, *x*	0	0.1	0.2	0.3	0.4	0.5	0.6	0.7	0.8	0.9	1.0
β for θ = 30°	210	203	197	190	183	175	167	158	147	132	90
β for θ = 60°	244	234	225	215	205	194	183	171	157	138	90

The rms diode current can be found from Eq. (3.28) as

$$I_r = \left[\frac{1}{2\pi} \int_\alpha^\beta i_0^2 \, d(\omega t) \right]^{1/2}$$

The average diode current can also be found from Eq. (3.28) as

$$I_d = \frac{1}{2\pi} \int_\alpha^\beta i_0 \, d(\omega t)$$

Boundary conditions: The condition for the discontinuous current can be found by setting I_0 in Eq. (3.26) to zero.

$$0 = \frac{V_s \sqrt{2}}{Z} \sin(\theta) \left[\frac{1 + e^{-(\frac{R}{L})(\frac{\pi}{\omega})}}{1 - e^{-(\frac{R}{L})(\frac{\pi}{\omega})}} \right] - \frac{E}{R}$$

which can be solved for the voltage ratio $x = E/(\sqrt{2}V_s)$ as

$$x(\theta): = \left[\frac{1 + e^{-(\frac{\pi}{\tan(\theta)})}}{1 - e^{-(\frac{\pi}{\tan(\theta)})}} \right] \sin(\theta) \cos(\theta) \tag{3.31}$$

The plot of the voltage ratio *x* against the load impedance angle θ is shown in Figure 3.9. The load angle θ cannot exceed π/2. The value of *x* is 63.67% at θ = 1.5567 rad, 43.65% at θ = 0.52308 rad (30°) and 0% at θ = 0.

Example 3.7 Finding the Performance Parameters of a Full-Wave Rectifier with an *RL* Load

The single-phase full-wave rectifier of Figure 3.8a has $L = 6.5$ mH, $R = 2.5 \, \Omega$, and $E = 10$ V. The input voltage is $V_s = 120$ V at 60 Hz. (a) Determine (1) the steady-state load current I_0 at $\omega t = 0$, (2) the average diode current I_d, (3) the rms diode current I_r, and (4) the rms output current I_{rms}. (b) Use PSpice to plot the instantaneous output current i_0. Assume diode parameters IS = 2.22E − 15, BV = 1800 V.

Solution

It is not known whether the load current is continuous or discontinuous. Assume that the load current is continuous and proceed with the solution. If the assumption is not correct, the load current is zero and then moves to the case for a discontinuous current.

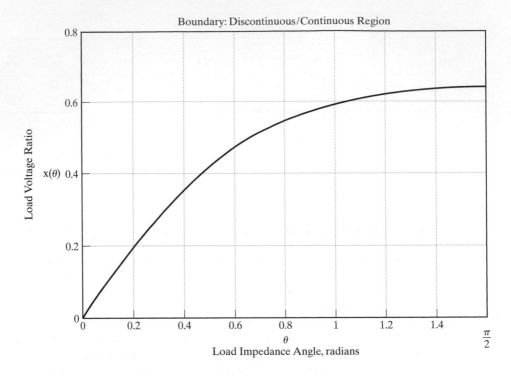

FIGURE 3.9

Boundary of continuous and discontinuous regions for single-phase rectifier.

a. $R = 2.5\ \Omega$, $L = 6.5\ \text{mH}$, $f = 60\ \text{Hz}$, $\omega = 2\pi \times 60 = 377\ \text{rad/s}$, $V_s = 120\ \text{V}$, $Z = [R^2 + (\omega L)^2]^{1/2} = 3.5\ \Omega$, and $\theta = \tan^{-1}(\omega L/R) = 44.43°$.

(1) The steady-state load current at $\omega t = 0$, $I_0 = 32.8\ \text{A}$. Because $I_0 > 0$, the load current is continuous and the assumption is correct.

(2) The numerical integration of i_0 in Eq. (3.27) yields the average diode current as $I_d = 19.61\ \text{A}$.

(3) By numerical integration of i_0^2 between the limits $\omega t = 0$ and π, we get the rms diode current as $I_r = 28.5\ \text{A}$.

(4) The rms output current $I_{\text{rms}} = \sqrt{2}I_r = \sqrt{2} \times 28.50 = 40.3\ \text{A}$.

Notes

1. i_0 has a minimum value of 25.2 A at $\omega t = 25.5°$ and a maximum value of 51.46 A at $\omega t = 125.25°$. i_0 becomes 27.41 A at $\omega t = \theta$ and 48.2 A at $\omega t = \theta + \pi$. Therefore, the minimum value of i_0 occurs approximately at $\omega t = \theta$.

2. The switching action of diodes makes the equations for currents nonlinear. A numerical method of solution for the diode currents is more efficient than the classical techniques. A Mathcad program is used to solve for I_0, I_d, and I_r by using numerical integration. Students are encouraged to verify the results of this example and to appreciate the usefulness of numerical solution, especially in solving nonlinear equations of diode circuits.

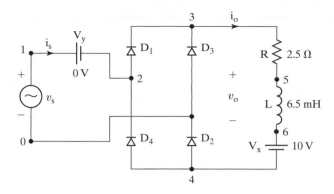

FIGURE 3.10

Single-phase bridge rectifier for PSpice simulation.

b. The single-phase bridge rectifier for PSpice simulation is shown in Figure 3.10. The list of the circuit file is as follows:

```
Example 3.7 Single-Phase Bridge Rectifier with RL load
VS     1    0    SIN (0    169.7V      60HZ)
L      5    6    6.5MH
R      3    5    2.5
VX     6    4    DC   10V ; Voltage source to measure the output current
D1     2    3    DMOD                      ; Diode model
D2     4    0    DMOD
D3     0    3    DMOD
D4     4    2    DMOD
VY     1         2    0DC
.MODEL      DMOD   D(IS=2.22E-15    BV=1800V)  ; Diode model parameters
.TRAN    1US    32MS    16.667MS             ; Transient analysis
.PROBE                                       ; Graphics postprocessor
.END
```

The PSpice plot of instantaneous output current i_0 is shown in Figure 3.11, which gives $I_0 = 31.83$ A, compared with the expected value of 32.8 A. A Dbreak diode was used in PSpice simulation to specify the diode parameters.

Key Points of Section 3.5

- An inductive load can make the load current continuous. There is a critical value of the load impedance angle θ for a given value of the load *emf* constant *x* to keep the load current continuous.

3.6   MULTIPHASE STAR RECTIFIERS

We have seen in Eq. (3.21) the average output voltage that could be obtained from single-phase full-wave rectifiers is $0.6366V_m$ and these rectifiers are used in applications up to a power level of 15 kW. For larger power output, *three-phase* and *multiphase* rectifiers are

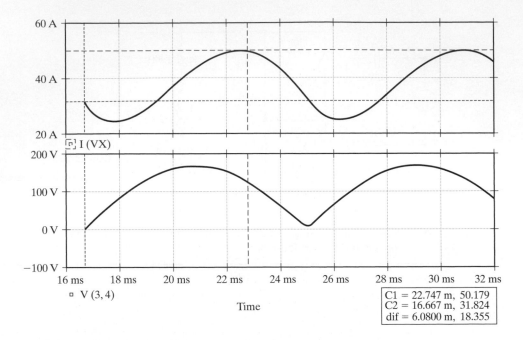

FIGURE 3.11

PSpice plot for Example 3.7.

used. The Fourier series of the output voltage given by Eq. (3.22) indicates that the output contains harmonics and the frequency of the *fundamental component* is two times the source frequency (2*f*). In practice, a filter is normally used to reduce the level of harmonics in the load; and the size of the filter decreases with the increase in frequency of the harmonics. In addition to the larger power output of multiphase rectifiers, the fundamental frequency of the harmonics is also increased and is *q* times the source frequency (*qf*). This rectifier is also known as a star rectifier.

The rectifier circuit of Figure 3.5a can be extended to multiple phases by having multiphase windings on the transformer secondary as shown in Figure 3.12a. This circuit may be considered as *q* single-phase half-wave rectifiers and can be considered as a half-wave type. The *k*th diode conducts during the period when the voltage of *k*th phase is higher than that of other phases. The waveforms for the voltages and currents are shown in Figure 3.12b. The conduction period of each diode is 2π/*q*.

It can be noticed from Figure 3.12b that the current flowing through the secondary winding is unidirectional and contains a dc component. Only one secondary winding carries current at a particular time, and as a result the primary must be connected in delta to eliminate the dc component in the input side of the transformer. This minimizes the harmonic content of the primary line current.

Assuming a cosine wave from π/*q* to 2π/*q*, the average output voltage for a *q*-phase rectifier is given by

$$V_{dc} = \frac{2}{2\pi/q} \int_0^{\pi/q} V_m \cos \omega t \, d(\omega t) = V_m \frac{q}{\pi} \sin \frac{\pi}{q} \tag{3.32}$$

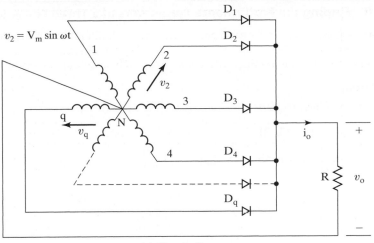

(a) Circuit diagram

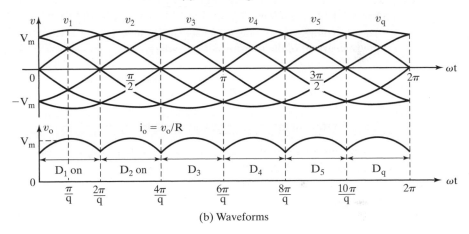

(b) Waveforms

FIGURE 3.12

Multiphase rectifiers.

$$V_{\text{rms}} = \left[\frac{2}{2\pi/q} \int_0^{\pi/q} V_m^2 \cos^2 \omega t \, d(\omega t) \right]^{1/2}$$

$$= V_m \left[\frac{q}{2\pi} \left(\frac{\pi}{q} + \frac{1}{2} \sin \frac{2\pi}{q} \right) \right]^{1/2} \tag{3.33}$$

If the load is purely resistive, the peak current through a diode is $I_m = V_m/R$ and we can find the rms value of a diode current (or transformer secondary current) as

$$I_s = \left[\frac{2}{2\pi} \int_0^{\pi/q} I_m^2 \cos^2 \omega t \, d(\omega t) \right]^{1/2}$$

$$= I_m \left[\frac{1}{2\pi} \left(\frac{\pi}{q} + \frac{1}{2} \sin \frac{2\pi}{q} \right) \right]^{1/2} = \frac{V_{\text{rms}}}{R} \tag{3.34}$$

Example 3.8   Finding the Performance Parameters of a Three-Phase Star Rectifier

A three-phase star rectifier has a purely resistive load with R ohms. Determine (a) the efficiency, (b) the FF, (c) the RF, (d) the TUF factor, (e) the PIV of each diode, and (f) the peak current through a diode if the rectifier delivers $I_{dc} = 30$ A at an output voltage of $V_{dc} = 140$ V.

Solution

For a three-phase rectifier $q = 3$ in Eqs. (3.32) to (3.34)

a. From Eq. (3.32), $V_{dc} = 0.827V_m$ and $I_{dc} = 0.827V_m/R$. From Eq. (3.33), $V_{rms} = 0.84068V_m$ and $I_{rms} = 0.84068V_m/R$. From Eq. (3.1), $P_{dc} = (0.827V_m)^2/R$; from Eq. (3.2), $P_{ac} = (0.84068V_m)^2/R$; and from Eq. (3.3), the efficiency

$$\eta = \frac{(0.827V_m)^2}{(0.84068V_m)^2} = 96.77\%$$

b. From Eq. (3.5), the FF = $0.84068/0.827 = 1.0165$ or 101.65%.
c. From Eq. (3.7), the RF = $\sqrt{1.0165^2 - 1} = 0.1824 = 18.24\%$.
d. The rms voltage of the transformer secondary, $V_s = V_m/\sqrt{2} = 0.707V_m$. From Eq. (3.34), the rms current of the transformer secondary,

$$I_s = 0.4854I_m = \frac{0.4854V_m}{R}$$

The VA rating of the transformer for $q = 3$ is

$$VA = 3V_sI_s = 3 \times 0.707V_m \times \frac{0.4854V_m}{R}$$

From Eq. (3.8),

$$TUF = \frac{0.827^2}{3 \times 0.707 \times 0.4854} = 0.6643$$

$$PF = \frac{0.84068^2}{3 \times 0.707 \times 0.4854} = 0.6844$$

e. The peak inverse voltage of each diode is equal to the peak value of the secondary line-to-line voltage. Three-phase circuits are reviewed in Appendix A. The line-to-line voltage is $\sqrt{3}$ times the phase voltage and thus PIV = $\sqrt{3}\, V_m$.
f. The average current through each diode is

$$I_d = \frac{2}{2\pi} \int_0^{\pi/q} I_m \cos \omega t \; d(\omega t) = I_m \frac{1}{\pi} \sin \frac{\pi}{q} \qquad (3.35)$$

For $q = 3$, $I_d = 0.2757I_m$. The average current through each diode is $I_d = 30/3 = 10$ A and this gives the peak current as $I_m = 10/0.2757 = 36.27$ A.

Example 3.9   Finding the Fourier Series of a q-Phase Rectifier

a. Express the output voltage of a q-phase rectifier in Figure 3.12a in Fourier series.

b. If $q = 6$, $V_m = 170$ V, and the supply frequency is $f = 60$ Hz, determine the rms value of the dominant harmonic and its frequency.

Solution

a. The waveforms for q-pulses are shown in Figure 3.12b and the frequency of the output is q times the fundamental component (qf). To find the constants of the Fourier series, we integrate from $-\pi/q$ to π/q and the constants are

$$b_n = 0$$

$$a_n = \frac{1}{\pi/q} \int_{-\pi/q}^{\pi/q} V_m \cos \omega t \cos n\omega t \, d(\omega t)$$

$$= \frac{qV_m}{\pi} \left\{ \frac{\sin[(n-1)\pi/q]}{n-1} + \frac{\sin[(n+1)\pi/q]}{n+1} \right\}$$

$$= \frac{qV_m}{\pi} \frac{(n+1)\sin[(n-1)\pi/q] + (n-1)\sin[(n+1)\pi/q]}{n^2 - 1}$$

After simplification and then using trigonometric relationships, we get

$$\sin(A + B) = \sin A \cos B + \cos A \sin B$$

and

$$\sin(A - B) = \sin A \cos B - \cos A \sin B$$

we get

$$a_n = \frac{2qV_m}{\pi(n^2 - 1)} \left(n \sin \frac{n\pi}{q} \cos \frac{\pi}{q} - \cos \frac{n\pi}{q} \sin \frac{\pi}{q} \right) \tag{3.36}$$

For a rectifier with q pulses per cycle, the harmonics of the output voltage are: qth, $2q$th, $3q$th, and $4q$th, and Eq. (3.36) is valid for $n = 0$, $1q$, $2q$, $3q$. The term $\sin(n\pi/q) = \sin \pi = 0$ and Eq. (3.36) becomes

$$a_n = \frac{-2qV_m}{\pi(n^2 - 1)} \left(\cos \frac{n\pi}{q} \sin \frac{\pi}{q} \right)$$

The dc component is found by letting $n = 0$ and is

$$V_{dc} = \frac{a_0}{2} = V_m \frac{q}{\pi} \sin \frac{\pi}{q} \tag{3.37}$$

which is the same as Eq. (3.32). The Fourier series of the output voltage v_0 is expressed as

$$v_0(t) = \frac{a_0}{2} + \sum_{n=q,2q,\ldots}^{\infty} a_n \cos n\omega t$$

Substituting the value of a_n, we obtain

$$v_0 = V_m \frac{q}{\pi} \sin \frac{\pi}{q} \left(1 - \sum_{n=q,2q,\ldots}^{\infty} \frac{2}{n^2 - 1} \cos \frac{n\pi}{q} \cos n\omega t \right) \tag{3.38}$$

b. For $q = 6$, the output voltage is expressed as

$$v_0(t) = 0.9549 V_m \left(1 + \frac{2}{35} \cos 6\omega t - \frac{2}{143} \cos 12\omega t + \cdots \right) \tag{3.39}$$

The sixth harmonic is the dominant one. The rms value of a sinusoidal voltage is $1/\sqrt{2}$ times its peak magnitude, and the rms of the sixth harmonic is $V_{6h} = 0.9549 V_m \times 2/(35 \times \sqrt{2}) = 6.56$ V and its frequency is $f_6 = 6f = 360$ Hz.

Key Points of Section 3.6

- A multiphase rectifier increases the amount of dc component and lowers the amount of the harmonic components. The output voltage of a p-phase rectifier contains harmonics whose frequencies are multiples of p (p times the supply frequency), pf.

3.7   THREE-PHASE BRIDGE RECTIFIERS

A three-phase bridge rectifier is commonly used in high-power applications and it is shown in Figure 3.13. This is a *full-wave rectifier*. It can operate with or without a transformer and gives six-pulse ripples on the output voltage. The diodes are numbered in order of conduction sequences and each one conducts for 120°. The conduction sequence for diodes is $D_1 - D_2$, $D_3 - D_2$, $D_3 - D_4$, $D_5 - D_4$, $D_5 - D_6$, and $D_1 - D_6$. The pair of diodes which are connected between that pair of supply lines having the highest amount of instantaneous line-to-line voltage will conduct. The line-to-line voltage is $\sqrt{3}$ times the phase voltage of a three-phase Y-connected source. The waveforms and conduction times of diodes are shown in Figure 3.14 [4].

If V_m is the peak value of the phase voltage, then the instantaneous phase voltages can be described by

$$v_{an} = V_m \sin(\omega t) \quad v_{bn} = V_m \sin(\omega t - 120°) \quad v_{cn} = V_m \sin(\omega t - 240°)$$

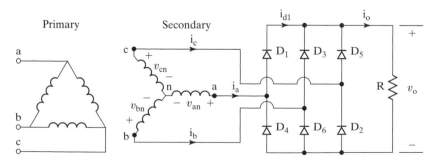

FIGURE 3.13

Three-phase bridge rectifier.

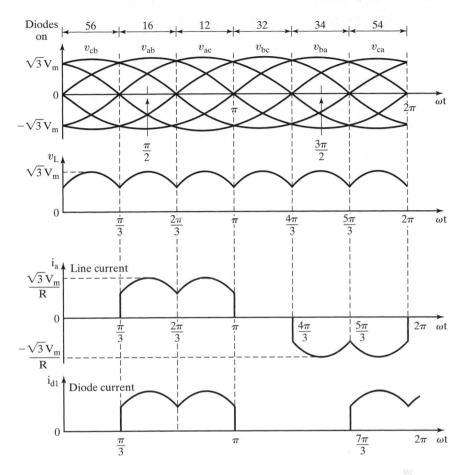

FIGURE 3.14

Waveforms and conduction times of diodes.

Because the line–line voltage leads the phase voltage by 30°, the instantaneous line–line voltages can be described by

$$v_{ab} = \sqrt{3}\, V_m \sin(\omega t + 30°) \quad v_{bc} = \sqrt{3}\, V_m \sin(\omega t - 90°)$$
$$v_{ca} = \sqrt{3}\, V_m \sin(\omega t - 210°)$$

The average output voltage is found from

$$V_{dc} = \frac{2}{2\pi/6} \int_0^{\pi/6} \sqrt{3}\, V_m \cos \omega t \; d(\omega t)$$

$$= \frac{3\sqrt{3}}{\pi} V_m = 1.654 V_m \tag{3.40}$$

where V_m is the peak phase voltage. The rms output voltage is

$$V_{\text{rms}} = \left[\frac{2}{2\pi/6} \int_0^{\pi/6} 3V_m^2 \cos^2 \omega t \, d(\omega t) \right]^{1/2}$$

$$= \left(\frac{3}{2} + \frac{9\sqrt{3}}{4\pi} \right)^{1/2} V_m = 1.6554V_m \tag{3.41}$$

If the load is purely resistive, the peak current through a diode is $I_m = \sqrt{3}\, V_m/R$ and the rms value of the diode current is

$$I_r = \left[\frac{4}{2\pi} \int_0^{\pi/6} I_m^2 \cos^2 \omega t \, d(\omega t) \right]^{1/2}$$

$$= I_m \left[\frac{1}{\pi} \left(\frac{\pi}{6} + \frac{1}{2} \sin \frac{2\pi}{6} \right) \right]^{1/2}$$

$$= 0.5518 I_m \tag{3.42}$$

and the rms value of the transformer secondary current,

$$I_s = \left[\frac{8}{2\pi} \int_0^{\pi/6} I_m^2 \cos^2 \omega t \, d(\omega t) \right]^{1/2}$$

$$= I_m \left[\frac{2}{\pi} \left(\frac{\pi}{6} + \frac{1}{2} \sin \frac{2\pi}{6} \right) \right]^{1/2}$$

$$= 0.7804 I_m \tag{3.43}$$

where I_m is the peak secondary line current.

For a three-phase rectifier $q = 6$, Eq. (3.38) gives the instantaneous output voltage as

$$v_0(t) = 0.9549V_m \left(1 + \frac{2}{35} \cos(6\omega t) - \frac{2}{143} \cos(12\omega t) + \cdots \right) \tag{3.44}$$

Example 3.10 Finding the Performance Parameters of a Three-Phase Bridge Rectifier

A three-phase bridge rectifier has a purely resistive load of R. Determine (a) the efficiency, (b) the FF, (c) the RF, (d) the TUF, (e) the peak inverse (or reverse) voltage (PIV) of each diode, and (f) the peak current through a diode. The rectifier delivers $I_{\text{dc}} = 60$ A at an output voltage of $V_{\text{dc}} = 280.7$ V and the source frequency is 60 Hz.

Solution

a. From Eq. (3.40), $V_{\text{dc}} = 1.654V_m$ and $I_{\text{dc}} = 1.654V_m/R$. From Eq. (3.41), $V_{\text{rms}} = 1.6554V_m$ and $I_{\text{rms}} = 1.6554V_m/R$. From Eq. (3.1), $P_{\text{dc}} = (1.654V_m)^2/R$, from Eq. (3.2), $P_{\text{ac}} = (1.6554V_m)^2/R$, and from Eq. (3.3) the efficiency

$$\eta = \frac{(1.654V_m)^2}{(1.6554V_m)^2} = 99.83\%$$

b. From Eq. (3.5), the FF = 1.6554/1.654 = 1.0008 = 100.08%.

c. From Eq. (3.6), the RF = $\sqrt{1.0008^2 - 1}$ = 0.04 = 4%.

d. From Eq. (3.15), the rms voltage of the transformer secondary, $V_s = 0.707V_m$.
From Eq. (3.43), the rms current of the transformer secondary,

$$I_s = 0.7804I_m = 0.7804 \times \sqrt{3}\,\frac{V_m}{R}$$

The VA rating of the transformer,

$$VA = 3V_sI_s = 3 \times 0.707V_m \times 0.7804 \times \sqrt{3}\,\frac{V_m}{R}$$

From Eq. (3.8),

$$TUF = \frac{1.654^2}{3 \times \sqrt{3} \times 0.707 \times 0.7804} = 0.9542$$

e. From Eq. (3.40), the peak line-to-neutral voltage is $V_m = 280.7/1.654 = 169.7$ V. The peak inverse voltage of each diode is equal to the peak value of the secondary line-to-line voltage, PIV = $\sqrt{3}\,V_m = \sqrt{3} \times 169.7 = 293.9$ V.

f. The average current through each diode is

$$I_d = \frac{4}{2\pi}\int_0^{\pi/6} I_m \cos \omega t \; d(\omega t) = I_m \frac{2}{\pi}\sin\frac{\pi}{6} = 0.3183I_m$$

The average current through each diode is $I_d = 60/3 = 20$ A; therefore, the peak current is $I_m = 20/0.3183 = 62.83$ A.

Note: This rectifier has considerably improved performances compared with those of the multiphase rectifier in Figure 3.12 with six pulses.

Key Points of Section 3.6

- A three-phase bridge rectifier has considerably improved performances compared with those of single-phase rectifiers.

3.8   THREE-PHASE BRIDGE RECTIFIER WITH *RL* LOAD

Equations that are derived in Section 3.5 can be applied to determine the load current of a three-phase rectifier with an *RL* load (similar to Figure 3.15). It can be noted from Figure 3.14 that the output voltage becomes

$$v_{ab} = \sqrt{2}\,V_{ab} \sin \omega t \quad \text{for} \quad \frac{\pi}{3} \leq \omega t \leq \frac{2\pi}{3}$$

where V_{ab} is the line-to-line rms input voltage. The load current i_0 can be found from

$$L\frac{di_0}{dt} + Ri_0 + E = \sqrt{2}\,V_{ab} \sin \omega t$$

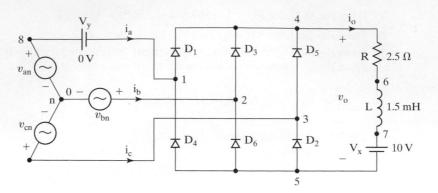

FIGURE 3.15

Three-phase bridge rectifier for PSpice simulation.

which has a solution of the form

$$i_0 = \frac{\sqrt{2}V_{ab}}{Z} \sin(\omega t - \theta) + A_1 e^{-(R/L)t} - \frac{E}{R} \tag{3.45}$$

where load impedance $Z = [R^2 + (\omega L)^2]^{1/2}$ and load impedance angle $\theta = \tan^{-1} (\omega L/R)$. The constant A_1 in Eq. (3.45) can be determined from the condition: at $\omega t = \pi/3$, $i_0 = I_0$.

$$A_1 = \left[I_0 + \frac{E}{R} - \frac{\sqrt{2}V_{ab}}{Z} \sin\left(\frac{\pi}{3} - \theta\right) \right] e^{(R/L)(\pi/3\omega)}$$

Substitution of A_1 in Eq. (3.45) yields

$$i_0 = \frac{\sqrt{2}V_{ab}}{Z} \sin(\omega t - \theta) + \left[I_0 + \frac{E}{R} - \frac{\sqrt{2}V_{ab}}{Z} \sin\left(\frac{\pi}{3} - \theta\right) \right] e^{(R/L)(\pi/3\omega - t)} - \frac{E}{R}$$

$$\tag{3.46}$$

Under a steady-state condition, $i_0(\omega t = 2\pi/3) = i_0(\omega t = \pi/3)$. That is, $i_0(\omega t = 2\pi/3) = I_0$. Applying this condition, we get the value of I_0 as

$$I_0 = \frac{\sqrt{2}V_{ab}}{Z} \frac{\sin(2\pi/3 - \theta) - \sin(\pi/3 - \theta)e^{-(R/L)(\pi/3\omega)}}{1 - e^{-(R/L)(\pi/3\omega)}} - \frac{E}{R} \quad \text{for } I_0 \geq 0 \tag{3.47}$$

which, after substitution in Eq. (3.46) and simplification, gives

$$i_0 = \frac{\sqrt{2}V_{ab}}{Z} \left[\sin(\omega t - \theta) + \frac{\sin(2\pi/3 - \theta) - \sin(\pi/3 - \theta)}{1 - e^{-(R/L)(\pi/3\omega - t)}} e^{(R/L)(\pi/3\omega - t)} \right]$$

$$- \frac{E}{R} \quad \text{for } \pi/3 \leq \omega t \leq 2\pi/3 \text{ and } i_0 \geq 0 \tag{3.48}$$

The rms diode current can be found from Eq. (3.48) as

$$I_r = \left[\frac{2}{2\pi} \int_{\pi/3}^{2\pi/3} i_0^2\, d(\omega t) \right]^{1/2}$$

and the rms output current can then be determined by combining the rms current of each diode as

$$I_{rms} = (I_r^2 + I_r^2 + I_r^2)^{1/2} = \sqrt{3}\, I_r$$

The average diode current can also be found from Eq. (3.47) as

$$I_d = \frac{2}{2\pi} \int_{\pi/3}^{2\pi/3} i_0\, d(\omega t)$$

Boundary conditions: The condition for the discontinuous current can be found by setting I_0 in Eq. (3.47) to zero.

$$\frac{\sqrt{2}V_{AB}}{Z} \cdot \left[\frac{\sin\left(\dfrac{2\pi}{3} - \theta\right) - \sin\left(\dfrac{\pi}{3} - \theta\right) e^{-\left(\frac{R}{L}\right)\left(\frac{\pi}{3\omega}\right)}}{1 - e^{-\left(\frac{R}{L}\right)\left(\frac{\pi}{3\omega}\right)}} \right] - \frac{E}{R} = 0$$

which can be solved for the voltage ratio $x = E/(\sqrt{2}V_{AB})$ as

$$x(\theta) := \left[\frac{\sin\left(\dfrac{2\pi}{3} - \theta\right) - \sin\left(\dfrac{\pi}{3} - \theta\right) e^{\left(\frac{\pi}{3\tan(\theta)}\right)}}{1 - e^{-\left(\frac{\pi}{3\tan(\theta)}\right)}} \right] \cos(\theta) \qquad (3.49)$$

The plot of the voltage ratio x against the load impedance angle θ is shown in Figure 3.16. The load angle θ cannot exceed $\pi/2$. The value of x is 95.49% at $\theta = 1.5598$ rad, 95.03% at $\theta = 0.52308$ (30°), and 86.68% at $\theta = 0$.

Example 3.11 Finding the Performance Parameters of a Three-Phase Bridge Rectifier with an *RL* Load

The three-phase full-wave rectifier of Figure 3.15 has a load of $L = 1.5$ mH, $R = 2.5\ \Omega$, and $E = 10$ V. The line-to-line input voltage is $V_{ab} = 208$ V, 60 Hz. (a) Determine (1) the steady-state load current I_0 at $\omega_t = \pi/3$, (2) the average diode current I_0, (3) the rms diode current I_r, and (4) the rms output current I_{rms}. (b) Use PSpice to plot the instantaneous output current i_0. Assume diode parameters IS = 2.22E − 15, BV = 1800 V.

Solution

a. $R = 2.5\ \Omega$, $L = 1.5$ mH, $f = 60$ Hz, $\omega = 2\pi \times 60 = 377$ rad/s, $V_{ab} = 208$ V, $Z = [R^2 + (\omega L)^2]^{1/2} = 2.56\ \Omega$, and $\theta = \tan^{-1}(\omega L/R) = 12.74°$.

   1. The steady-state load current at $\omega t = \pi/3$, $I_0 = 105.77$ A.

   2. The numerical integration of i_0 in Eq. (3.48) yields the average diode current as $I_d = 36.09$ A. Because $I_0 > 0$, the load current is continuous.

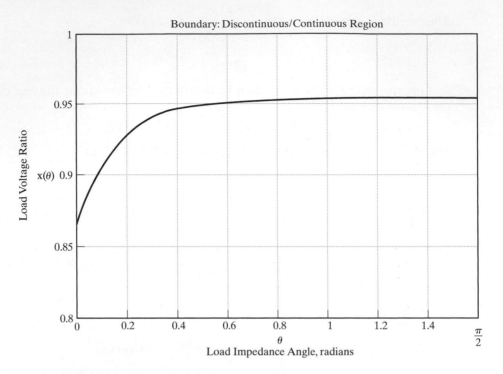

FIGURE 3.16

Boundary of continuous and discontinuous regions for three-phase rectifier.

3. By numerical integration of i_0^2 between the limits $\omega t = \pi/3$ and $2\pi/3$, we get the rms diode current as $I_r = 62.733$ A.

4. The rms output current $I_{rms} = \sqrt{3}I_r = \sqrt{3} \times 62.53 = 108.31$ A.

b. The three-phase bridge rectifier for PSpice simulation is shown in Figure 3.15. The list of the circuit file is as follows:

```
Example 3.11      Three-Phase Bridge Rectifier with RL load
VAN   8   0    SIN (0 169.7V 60HZ)
VBN   2   0    SIN (0 169.7V 60HZ 0 0 120DEG)
VCN   3   0    SIN (0 169.7V 60HZ 0 0 240DEG)
L     6   7    1.5MH
R     4   6    2.5
VX    7   5    DC 10V ; Voltage source to measure the output current
VY    8   1    DC  0V ; Voltage source to measure the input current
D1    1   4    DMOD                     ; Diode model
D3    2   4    DMOD
D5    3   4    DMOD
D2    5   3    DMOD
D4    5   1    DMOD
```

```
D6      5     2      DMOD
.MODEL    DMOD   D (IS=2.22E-15 BV=1800V)  ; Diode model parameters
.TRAN     1OUS   25MS    16.667MS  1OUS   ; Transient analysis
.PROBE                                    ; Graphics postprocessor
.options ITL5=0 abstol = 1.000n reltol = .01 vntol = 1.000m
.END
```

The PSpice plot of instantaneous output current i_0 is shown in Figure 3.17, which gives $I_0 = 104.89$ A, compared with the expected value of 105.77 A. A Dbreak diode was used in PSpice simulation to include the specified diode parameters.

Example 3.12 Finding the Input Power Factor of a Three-Phase Rectifier with a Highly Inductive Load

The load current of a three-phase rectifier in Figure 3.13 is continuous, with a negligible ripple content. Express the input current in Fourier series, and determine the HF of the input current, the DF, and the input PF.

Solution

The waveform of the line current is shown in Figure 3.14. The line current is symmetric at the angle $(q = p/6)$ when the phase voltage becomes zero, not when the line–line voltage v_{ab}

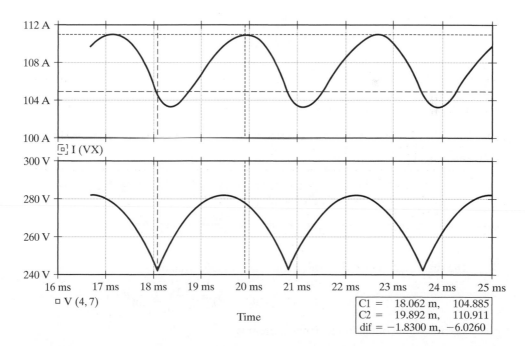

FIGURE 3.17

PSpice plot for Example 3.11.

becomes zero. Thus, for satisfying the condition of $f(x + 2\pi) = f(x)$, the input current can be described by

$$i_s(t) = I_a \quad \text{for} \quad \frac{\pi}{6} \le \omega t \le \frac{5\pi}{6}$$

$$i_s(t) = -I_a \quad \text{for} \quad \frac{7\pi}{6} \le \omega t \le \frac{11\pi}{6}$$

which can be expressed in a Fourier series as

$$i_s(t) = I_{dc} + \sum_{n=1}^{\infty} (a_n \cos(n\omega t) + b_n \sin(n\omega t)) = \sum_{n=1}^{\infty} c_n \sin(n\omega t + \phi_n)$$

where the coefficients are

$$I_{dc} = \frac{1}{2\pi} \int_0^{2\pi} i_s(t) \cdot d(\omega t) = \frac{1}{2\pi} \int_0^{2\pi} I_a \cdot d(\omega t) = 0$$

$$a_n = \frac{1}{\pi} \int_0^{2\pi} i_s(t) \cos(n\omega t)\, d(\omega t) = \frac{1}{\pi} \left[\int_{\frac{\pi}{6}}^{\frac{5\pi}{6}} I_a \cos(n\omega t)\, d(\omega t) - \int_{\frac{7\pi}{6}}^{\frac{11\pi}{6}} I_a \cos(n\omega t)\, d(\omega t) \right] = 0$$

$$b_n = \frac{1}{\pi} \int_0^{2\pi} i_s(t) \cos(n\omega t)\, d(\omega t) = \frac{1}{\pi} \left[\int_{\frac{\pi}{6}}^{\frac{5\pi}{6}} I_a \sin(n\omega t)\, d(\omega t) - \int_{\frac{7\pi}{6}}^{\frac{11\pi}{6}} I_a \sin(n\omega t)\, d(\omega t) \right]$$

which, after integration and simplification, gives b_n as

$$b_n = \frac{-4I_a}{n\pi} \cos(n\pi)\sin\left(\frac{n\pi}{2}\right) \sin\left(\frac{n\pi}{3}\right) \quad \text{for} \quad n = 1, 5, 7, 11, 13, \ldots$$

$$b_n = 0 \quad \text{for} \quad n = 2, 3, 4, 6, 8, 9, \ldots$$

$$c_n = \sqrt{(a_n)^2 + (b_n)^2} = \frac{-4I_a}{n\pi} \cos(n\pi)\sin\left(\frac{n\pi}{2}\right) \sin\left(\frac{n\pi}{3}\right)$$

$$\phi_n = \arctan\left(\frac{a_n}{b_n}\right) = 0$$

Thus, the Fourier series of the input current is given by

$$i_s = \sum_{n=1}^{\infty} \frac{4 \cdot \sqrt{3} I_a}{2\pi} \left(\frac{\sin(\omega t)}{1} - \frac{\sin(5\omega t)}{5} - \frac{\sin(7\omega t)}{7} \right.$$
$$\left. + \frac{\sin(11\omega t)}{11} + \frac{\sin(13\omega t)}{13} - \frac{\sin(17\omega t)}{17} - \cdots \right) \tag{3.50}$$

The rms value of the nth harmonic input current is given by

$$I_{sn} = \frac{1}{\sqrt{2}} (a_n^2 + b_n^2)^{1/2} = \frac{2\sqrt{2} I_a}{n\pi} \sin\frac{n\pi}{3} \tag{3.51}$$

The rms value of the fundamental current is

$$I_{s1} = \frac{\sqrt{6}}{\pi} I_a = 0.7797 I_a$$

The rms input current

$$I_s = \left[\frac{2}{2\pi} \int_{\pi/6}^{5\pi/6} I_a^2 \, d(\omega t) \right]^{1/2} = I_a \sqrt{\frac{2}{3}} = 0.8165 I_a$$

$$\text{HF} = \left[\left(\frac{I_s}{I_{s1}} \right)^2 - 1 \right]^{1/2} = \left[\left(\frac{\pi}{3} \right)^2 - 1 \right]^{1/2} = 0.3108 \quad \text{or} \quad 31.08\%$$

$$\text{DF} = \cos \phi_1 = \cos(0) = 1$$

$$\text{PF} = \frac{I_{s1}}{I_s} \cos(0) = \frac{0.7797}{0.8165} = 0.9549$$

Note: If we compare the PF with that of Example 3.10, where the load is purely resistive, we can notice that the input PF depends on the load angle. For a purely resistive load, PF = 0.8166.

Key Points of Section 3.8

- An inductive load can make the load current continuous. The critical value of the load electromotive force (emf) constant x for a given load impedance angle θ is higher than that of a single-phase rectifier; that is, $x = 86.68\%$ at $\theta = 0$.
- With a highly inductive load, the input current of a rectifier becomes an ac square wave. The input power factor of a three-phase rectifier is 0.955, which is higher than 0.9 for a single-phase rectifier.

3.9   COMPARISONS OF DIODE RECTIFIERS

The goal of a rectifier is to yield a dc output voltage at a given dc output power. Therefore, it is more convenient to express the performance parameters in terms of V_{dc} and P_{dc}. For example, the rating and turns ratio of the transformer in a rectifier circuit can easily be determined if the rms input voltage to the rectifier is in terms of the required output voltage V_{dc}. The important parameters are summarized in Table 3.2 [3]. Due to their relative merits, the single-phase and three-phase bridge rectifiers are commonly used.

Key Points of Section 3.9

- The single-phase and three-phase bridge rectifiers, which have relative merits, are commonly used for dc–ac conversion.

3.10   RECTIFIER CIRCUIT DESIGN

The design of a rectifier involves determining the ratings of semiconductor diodes. The ratings of diodes are normally specified in terms of average current, rms current, peak current, and peak inverse voltage. There are no standard procedures for the design, but it is required to determine the shapes of the diode currents and voltages.

We have noticed in Eqs. (3.20), (3.22), and (3.39) that the output of the rectifiers contain harmonics. Filters can be used to smooth out the dc output voltage of

TABLE 3.2   Performance Parameters of Diode Rectifiers with a Resistive Load

Performance Parameters	Single-Phase Rectifier with Center-Tapped Transformer	Single-Phase Bridge Rectifier	Six-Phase Star Rectifier	Three-Phase Bridge Rectifier
Peak repetitive reverse voltage, V_{RRM}	$3.14V_{dc}$	$1.57V_{dc}$	$2.09V_{dc}$	$1.05V_{dc}$
Rms input voltage per transformer leg, V_s	$1.11V_{dc}$	$1.11V_{dc}$	$0.74V_{dc}$	$0.428V_{dc}$
Diode average current, $I_{F(AV)}$	$0.50I_{dc}$	$0.50I_{dc}$	$0.167I_{dc}$	$0.333I_{dc}$
Peak repetitive forward current, I_{FRM}	$1.57I_{dc}$	$1.57I_{dc}$	$6.28I_{dc}$	$3.14I_{dc}$
Diode rms current, $I_{F(RMS)}$	$0.785I_{dc}$	$0.785I_{dc}$	$0.409I_{dc}$	$0.579I_{dc}$
Form factor of diode current, $I_{F(RMS)}/I_{F(AV)}$	1.57	1.57	2.45	1.74
Rectification ratio, η	0.81	0.81	0.998	0.998
Form factor, FF	1.11	1.11	1.0009	1.0009
Ripple factor, RF	0.482	0.482	0.042	0.042
Transformer rating primary, VA	$1.23P_{dc}$	$1.23P_{dc}$	$1.28P_{dc}$	$1.05P_{dc}$
Transformer rating secondary, VA	$1.75P_{dc}$	$1.23P_{dc}$	$1.81P_{dc}$	$1.05P_{dc}$
Output ripple frequency, f_r	$2f_s$	$2f_s$	$6f_s$	$6f_s$

the rectifier and these are known as *dc filters*. The dc filters are usually of *L*, *C*, and *LC* type, as shown in Figure 3.18. Due to rectification action, the input current of the rectifier contains harmonics also and an *ac filter* is used to filter out some of the harmonics from the supply system. The ac filter is normally of *LC* type, as shown in Figure 3.19.

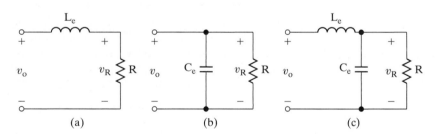

(a)   (b)   (c)

FIGURE 3.18

Dc filters.

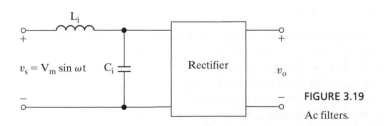

FIGURE 3.19

Ac filters.

Normally, the filter design requires determining the magnitudes and frequencies of the harmonics. The steps involved in designing rectifiers and filters are explained by examples.

Example 3.13 Finding the Diode Ratings from the Diode Currents

A three-phase bridge rectifier supplies a highly inductive load such that the average load current is $I_{dc} = 60$ A and the ripple content is negligible. Determine the ratings of the diodes if the line-to-neutral voltage of the Y-connected supply is 120 V at 60 Hz.

Solution

The currents through the diodes are shown in Figure 3.20. The average current of a diode $I_d = 60/3 = 20$ A. The rms current is

$$I_r = \left[\frac{1}{2\pi} \int_{\pi/3}^{\pi} I_{dc}^2 \, d(\omega t) \right]^{1/2} = \frac{I_{dc}}{\sqrt{3}} = 34.64 \text{ A}$$

The PIV $= \sqrt{3} \, V_m = \sqrt{3} \times \sqrt{2} \times 120 = 294$ V.

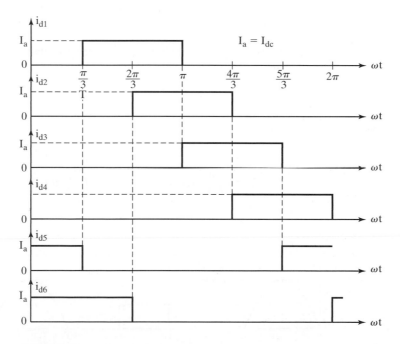

FIGURE 3.20

Current through diodes.

Note: The factor of $\sqrt{2}$ is used to convert rms to peak value.

Example 3.14 Finding the Diode Average and rms Currents from the Waveforms

The current through a diode is shown in Figure 3.21. Determine (a) the rms current, and (b) the average diode current if $t_1 = 100$ μs, $t_2 = 350$ μs, $t_3 = 500$ μs, $f = 250$ Hz, $f_s = 5$ kHz, $I_m = 450$ A, and $I_a = 150$ A.

Solution

a. The rms value is defined as

$$I_r = \left[\frac{1}{T} \int_0^{t_1} (I_m \sin \omega_s t)^2 \, dt + \frac{1}{T} \int_{t_2}^{t_3} I_a^2 \, dt \right]^{1/2}$$

$$= (I_{r1}^2 + I_{r2}^2)^{1/2} \tag{3.52}$$

where $\omega_s = 2\pi f_s = 31{,}415.93$ rad/s, $t_1 = \pi/\omega_s = 100$ μs, and $T = 1/f$.

$$I_{r1} = \left[\frac{1}{T} \int_0^{t_1} (I_m \sin \omega_s t)^2 \, dt \right]^{1/2} = I_m \sqrt{\frac{ft_1}{2}} \tag{3.53}$$

$$= 50.31 \text{ A}$$

and

$$I_{r2} = \left(\frac{1}{T} \int_{t_2}^{t_3} I_a \, dt \right)^2 = I_a \sqrt{f(t_3 - t_2)} \tag{3.54}$$

$$= 29.05 \text{ A}$$

Substituting Eqs. (3.53) and (3.54) in Eq. (3.52), the rms value is

$$I_r = \left[\frac{I_m^2 f t_1}{2} + I_a^2 f (t_3 - t_2) \right]^{1/2} \tag{3.55}$$

$$= (50.31^2 + 29.05^2)^{1/2} = 58.09 \text{ A}$$

b. The average current is found from

$$I_d = \left[\frac{1}{T} \int_0^{t_1} (I_m \sin \omega_s t) \, dt + \frac{1}{T} \int_{t_2}^{t_3} I_a \, dt \right]$$

$$= I_{d1} + I_{d2}$$

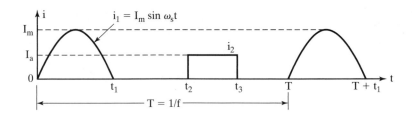

FIGURE 3.21

Current waveform.

where

$$I_{d1} = \frac{1}{T} \int_0^{t_1} (I_m \sin \omega_s t) \, dt = \frac{I_m f}{\pi f_s} \tag{3.56}$$

$$I_{d2} = \frac{1}{T} \int_{t_2}^{t_3} I_a \, dt = I_a f (t_3 - t_2) \tag{3.57}$$

Therefore, the average current becomes

$$I_{dc} = \frac{I_m f}{\pi f_s} + I_a f (t_3 - t_2) = 7.16 + 5.63 = 12.79 \text{ A}$$

Example 3.15 Finding the Load Inductance to Limit the Amount of Ripple Current

The single-phase bridge rectifier is supplied from a 120-V, 60-Hz source. The load resistance is $R = 500 \ \Omega$. Calculate the value of a series inductor L that limits the rms ripple current I_{ac} to less than 5% of I_{dc}.

Solution

The load impedance

$$Z = R + j(n\omega L) = \sqrt{R^2 + (n\omega L)^2} \ \underline{/\theta_n} \tag{3.58}$$

and

$$\theta_n = \tan^{-1} \frac{n\omega L}{R} \tag{3.59}$$

and the instantaneous current is

$$i_0(t) = I_{dc} - \frac{4V_m}{\pi \sqrt{R^2 + (n\omega L)^2}} \left[\frac{1}{3} \cos(2\omega t - \theta_2) + \frac{1}{15} \cos(4\omega t - \theta_4) \dots \right] \tag{3.60}$$

where

$$I_{dc} = \frac{V_{dc}}{R} = \frac{2V_m}{\pi R}$$

Equation (3.60) gives the rms value of the ripple current as

$$I_{ac}^2 = \frac{(4V_m)^2}{2\pi^2 [R^2 + (2\omega L)^2]} \left(\frac{1}{3} \right)^2 + \frac{(4V_m)^2}{2\pi^2 [R^2 + (4\omega L)^2]} \left(\frac{1}{15} \right)^2 + \cdots$$

Considering only the lowest order harmonic ($n = 2$), we have

$$I_{ac} = \frac{4V_m}{\sqrt{2}\pi \sqrt{R^2 + (2\omega L)^2}} \left(\frac{1}{3} \right)$$

Using the value of I_{dc} and after simplification, the ripple factor is

$$RF = \frac{I_{ac}}{I_{dc}} = \frac{0.4714}{\sqrt{1 + (2\omega L/R)^2}} = 0.05$$

For $R = 500\ \Omega$ and $f = 60$ Hz, the inductance value is obtained as $0.4714^2 = 0.05^2\,[1 + (4 \times 60 \times \pi L/500)^2]$ and this gives $L = 6.22$ H.

We can notice from Eq. (3.60) that an inductance in the load offers a high impedance for the harmonic currents and acts like a filter in reducing the harmonics. However, this inductance introduces a time delay of the load current with respect to the input voltage; and in the case of the single-phase half-wave rectifier, a freewheeling diode is required to provide a path for this inductive current.

Example 3.16 Finding the Filter Capacitance to Limit the Amount of Output Ripple Voltage

A single-phase bridge-rectifier is supplied from a 120-V, 60-Hz source. The load resistance is $R = 500\ \Omega$. (a) Design a C filter so that the ripple factor of the output voltage is less than 5%. (b) With the value of capacitor C in part (a), calculate the average load voltage V_{dc}.

Solution

a. When the instantaneous voltage v_s in Figure 3.22a is higher than the instantaneous capacitor voltage v_c, the diodes (D_1 and D_2 or D_3 and D_4) conduct; and the capacitor is then charged from the supply. If the instantaneous supply voltage v_s falls below the instantaneous capacitor voltage v_c, the diodes (D_1 and D_2 or D_3 and D_4) are reverse biased and the capacitor C_e discharges through the load resistance R. The capacitor voltage v_c varies between a minimum $V_{c(min)}$ and maximum value $V_{c(max)}$. This is shown in Figure 3.22b.

Let us assume that t_1 is the charging time and that t_2 is the discharging time of capacitor C_e. The equivalent circuit during charging is shown in Figure 3.22c. The capacitor charges almost instantaneously to the supply voltage v_s. The capacitor C_e will be charged to the peak supply voltage V_m, so that $v_c(t = t_1) = V_m$. Figure 3.22d shows the equivalent circuit during discharging. The capacitor discharges exponentially through R.

$$\frac{1}{C_e} \int i_0\, dt + v_c(t = 0) + Ri_0 = 0$$

which, with an initial condition of $v_c(t = 0) = V_m$, gives the discharging current as

$$i_0 = \frac{V_m}{R} e^{-t/RC_e}$$

The output (or capacitor) voltage v_0 during the discharging period can be found from

$$v_0(t) = Ri_0 = V_m e^{-t/RC_e}$$

The peak-to-peak ripple voltage $V_{r(pp)}$ can be found from

$$V_{r(pp)} = v_0(t = t_1) - v_0(t = t_2) = V_m - V_m e^{-t_2/RC_e} = V_m(1 - e^{-t_2/RC_e}) \tag{3.61}$$

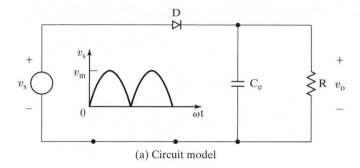

(a) Circuit model

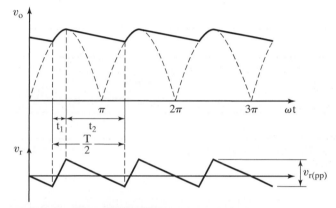

(b) Waveforms for full-wave rectifier

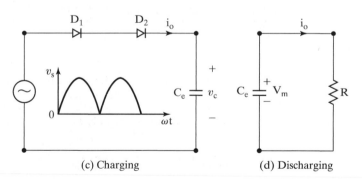

(c) Charging   (d) Discharging

FIGURE 3.22

Single-phase bridge rectifier with C filter.

Since, $e^{-x} \approx 1 - x$, Eq. (3.61) can be simplified to

$$V_{r(pp)} = V_m \left(1 - 1 + \frac{t_2}{RC_e} \right) = \frac{V_m t_2}{RC_e} = \frac{V_m}{2fRC_e}$$

Therefore, the average load voltage V_{dc} is given by (assuming $t_2 = 1/2 f$)

$$V_{dc} = V_m - \frac{V_{r(pp)}}{2} = V_m - \frac{V_m}{4fRC_e} \qquad (3.62)$$

Thus, the rms output ripple voltage V_{ac} can be found approximately from

$$V_{ac} = \frac{V_{r(pp)}}{2\sqrt{2}} = \frac{V_m}{4\sqrt{2}\,fRC_e}$$

The RF can be found from

$$RF = \frac{V_{ac}}{V_{dc}} = \frac{V_m}{4\sqrt{2}\,fRC_e} \frac{4fRC_e}{V_m(4fRC_e - 1)} = \frac{1}{\sqrt{2}\,(4fRC_e - 1)} \tag{3.63}$$

which can be solved for C_e:

$$C_e = \frac{1}{4fR}\left(1 + \frac{1}{\sqrt{2}\,RF}\right) = \frac{1}{4 \times 60 \times 500}\left(1 + \frac{1}{\sqrt{2} \times 0.05}\right) = 126.2\ \mu F$$

b. From Eq. (3.62), the average load voltage V_{dc} is

$$V_{dc} = 169.7 - \frac{169.7}{4 \times 60 \times 500 \times 126.2 \times 10^{-6}} = 169.7 - 11.21 = 158.49\ V$$

Example 3.17 Finding the Values of an *LC* Output Filter to Limit the Amount of Output Ripple Voltage

An *LC* filter as shown in Figure 3.18c is used to reduce the ripple content of the output voltage for a single-phase full-wave rectifier. The load resistance is $R = 40\ \Omega$, load inductance is $L = 10$ mH, and source frequency is 60 Hz (or 377 rad/s). (a) Determine the values of L_e and C_e so that the RF of the output voltage is 10%. (b) Use PSpice to calculate Fourier components of the output voltage v_0. Assume diode parameters IS = 2-22E − 15, BV = 1800 V.

Solution

a. The equivalent circuit for the harmonics is shown in Figure 3.23. To make it easier for the *n*th harmonic ripple current to pass through the filter capacitor, the load impedance must be much greater than that of the capacitor. That is,

$$\sqrt{R^2 + (n\omega L)^2} \gg \frac{1}{n\omega C_e}$$

This condition is generally satisfied by the relation

$$\sqrt{R^2 + (n\omega L)^2} = \frac{10}{n\omega C_e} \tag{3.64}$$

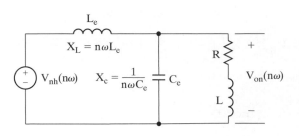

FIGURE 3.23

Equivalent circuit for harmonics.

and under this condition, the effect of the load is negligible. The rms value of the nth harmonic component appearing on the output can be found by using the voltage-divider rule and is expressed as

$$V_{\text{o}n} = \left| \frac{-1/(n\omega C_e)}{(n\omega L_e) - 1/(n\omega C_e)} \right| V_{nh} = \left| \frac{-1}{(n\omega)^2 L_e C_e - 1} \right| V_{nh} \tag{3.65}$$

The total amount of ripple voltage due to all harmonics is

$$V_{\text{ac}} = \left(\sum_{n=2,4,6,\dots}^{\infty} V_{\text{o}n}^2 \right)^{1/2} \tag{3.66}$$

For a specified value of V_{ac} and with the value of C_e from Eq. (3.64), the value of L_e can be computed. We can simplify the computation by considering only the dominant harmonic. From Eq. (3.22) we find that the second harmonic is the dominant one and its rms value is $V_{2h} = 4V_m/(3\sqrt{2}\pi)$ and the dc value, $V_{\text{dc}} = 2V_m/\pi$.

For $n = 2$, Eqs. (3.65) and (3.66) give

$$V_{\text{ac}} = V_{o2} = \left| \frac{-1}{(2\omega)^2 L_e C_e - 1} \right| V_{2h}$$

The value of the filter capacitor C_e is calculated from

$$\sqrt{R^2 + (2\omega L)^2} = \frac{10}{2\omega C_e}$$

or

$$C_e = \frac{10}{4\pi f \sqrt{R^2 + (4\pi f L)^2}} = 326 \ \mu\text{F}$$

From Eq. (3.6) the RF is defined as

$$\text{RF} = \frac{V_{\text{ac}}}{V_{\text{dc}}} = \frac{V_{o2}}{V_{\text{dc}}} = \frac{V_{2h}}{V_{\text{dc}}} \frac{1}{(4\pi f)^2 L_e C_e - 1} = \frac{\sqrt{2}}{3} \left| \frac{1}{[(4\pi f)^2 L_e C_e - 1]} \right| = 0.1$$

or $(4\pi f)^2 L_e C_e - 1 = 4.714$ and $L_e = 30.83$ mH.

b. The single-phase bridge rectifier for PSpice simulation is shown in Figure 3.24. The list of the circuit file is as follows:

```
Example 3.17.  Single-Phase Bridge Rectifier with LC Filter
VS      1       0       SIN (0 169.7V 60HZ)
LE      3       8       30.83MH
CE      7       4       326UF
RX      8       7       80M                ; Used to converge the solution
L       5       6       10MH
R       7       5       40
VX      6       4       DC 0V ; Voltage source to measure the output current
VY      1       2       DC 0V ; Voltage source to measure the input current
D1      2       3       DMOD               ; Diode models
```

```
D2        4        0        DMOD
D3        0        3        DMOD
D4        4        2        DMOD
.MODEL DMOD    D  (IS=2.22E-15 BV=1800V)   ; Diode model parameters
.TRAN    10US    50MS 33MS 50US               ; Transient analysis
.FOUR    120HZ   V(6,5)                    ; Fourier analysis of output voltage
.options  ITL5=0    abstol = 1.000u reltol = .05 vntol = 0.01m
.END
```

The results of PSpice simulation for the output voltage $V(6, 5)$ are as follows:

```
FOURIER COMPONENTS OF TRANSIENT RESPONSE V(6,5)
DC COMPONENT = 1.140973E+02
```

HARMONIC NO	FREQUENCY (HZ)	FOURIER COMPONENT	NORMALIZED COMPONENT	PHASE (DEG)	NORMALIZED PHASE (DEG)
1	1.200E+02	1.304E+01	1.000E+00	1.038E+02	0.000E+00
2	2.400E+02	6.496E-01	4.981E-02	1.236E+02	1.988E+01
3	3.600E+02	2.277E-01	1.746E-02	9.226E+01	-1.150E+01
4	4.800E+02	1.566E-01	1.201E-02	4.875E+01	-5.501E+01
5	6.000E+02	1.274E-01	9.767E-03	2.232E+01	-8.144E+01
6	7.200E+02	1.020E-01	7.822E-03	8.358E+00	-9.540E+01
7	8.400E+02	8.272E-02	6.343E-03	1.997E+00	-1.018E+02
8	9.600E+02	6.982E-02	5.354E-03	-1.061E+00	-1.048E+02
9	1.080E+03	6.015E-02	4.612E-03	-3.436E+00	-1.072E+02

```
TOTAL HARMONIC DISTORTION = 5.636070E+00 PERCENT
```

which verifies the design.

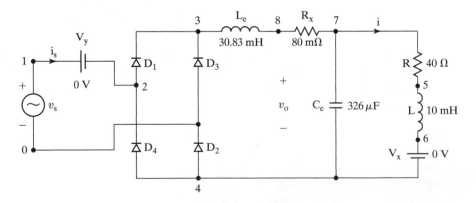

FIGURE 3.24

Single-phase bridge rectifier for PSpice simulation.

Example 3.18 Finding the Values of an *LC* Input Filter to Limit the Amount of Input Ripple Current

An *LC* input filter as shown in Figure 3.19 is used to reduce the input current harmonics in a single-phase full-wave rectifier of Figure 3.7a. The load current is ripple free and its average value is I_a. If the supply frequency is $f = 60$ Hz (or 377 rad/s), determine the resonant frequency of the filter so that the total input harmonic current is reduced to 1% of the fundamental component.

Solution

The equivalent circuit for the *n*th harmonic component is shown in Figure 3.25. The rms value of the *n*th harmonic current appearing in the supply is obtained by using the current-divider rule,

$$I_{sn} = \left| \frac{1/(n\omega C_i)}{(n\omega L_i - 1/(n\omega C_i))} \right| I_{nh} = \left| \frac{1}{(n\omega)^2 L_i C_i - 1} \right| I_{nh} \tag{3.67}$$

where I_{nh} is the rms value of the *n*th harmonic current. The total amount of harmonic current in the supply line is

$$I_h = \left(\sum_{n=2,3,\dots}^{\infty} I_{sn}^2 \right)^{1/2}$$

and the harmonic factor of input current (with the filter) is

$$r = \frac{I_h}{I_{s1}} = \left[\sum_{n=2,3,\dots}^{\infty} \left(\frac{I_{sn}}{I_{s1}} \right)^2 \right]^{1/2} \tag{3.68}$$

From Eq. (3.23), $I_{1h} = 4I_a/\sqrt{2}\,\pi$ and $I_{nh} = 4I_a/(\sqrt{2}\,n\pi)$ for $n = 3, 5, 7, \dots$ From Eqs. (3.67) and (3.68) we get

$$r^2 = \sum_{n=3,5,7,\dots}^{\infty} \left(\frac{I_{sn}}{I_{s1}} \right)^2 = \sum_{n=3,5,7,\dots}^{\infty} \left| \frac{(w^2 L_i C_i - 1)^2}{n^2 [(n\omega)^2 L_i C_i - 1]^2} \right| \tag{3.69}$$

This can be solved for the value of $L_i C_i$. To simplify the calculations, if we consider only the third harmonic, $3[(3 \times 2 \times \pi \times 60)^2 L_i C_i - 1]/(w^2 L_i C_i - 1) = 1/0.01 = 100$ or $L_i C_i = 9.349 \times 10^{-6}$ and the filter frequency is $1/\sqrt{L_i C_i} = 327.04$ rad/s, or 52.05 Hz. Assuming that $C_i = 1000$ μF, we obtain $L_i = 9.349$ mH.

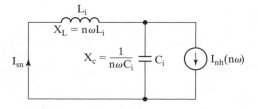

FIGURE 3.25

Equivalent circuit for harmonic current.

Note: The ac filter is generally tuned to the harmonic frequency involved, but it requires a careful design to avoid the possibility of resonance with the power system. The resonant frequency of the third-harmonic current is $377 \times 3 = 1131$ rad/s.

Key Points of Section 3.10

- The design of a rectifier requires determining the diode ratings and the ratings of filter components at the input and output side. Filters are used to smooth out the output voltage by a dc filter and to reduce the amount of harmonic injection to the input supply by an ac filter.

3.11   OUTPUT VOLTAGE WITH *LC* FILTER

The equivalent circuit of a full-wave rectifier with an *LC* filter is shown in Figure 3.26a. Assume that the value of C_e is very large, so that its voltage is ripple free with an average value of $V_{o(dc)}$. L_e is the total inductance, including the source or line inductance, and is generally placed at the input side to act as an ac inductance instead of a dc choke.

If V_{dc} is less than V_m, the current i_0 begins to flow at α, which is given by

$$V_{dc} = V_m \sin \alpha$$

This in turn gives

$$\alpha = \sin^{-1} \frac{V_{dc}}{V_m} = \sin^{-1} x$$

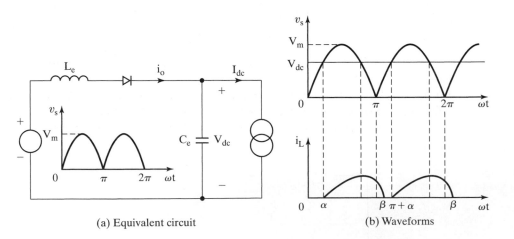

(a) Equivalent circuit   (b) Waveforms

FIGURE 3.26

Output voltage with *LC* filter.

where $x = V_{dc}/V_m$. The output current i_0 is given by

$$L_e \frac{di_L}{dt} = V_m \sin \omega t - V_{dc}$$

which can be solved for i_0.

$$i_0 = \frac{1}{\omega L_e} \int_\alpha^{\omega_t} (V_m \sin \omega t - V_{dc}) \, d(\omega t)$$

$$= \frac{V_m}{\omega L_e} (\cos \alpha - \cos \omega t) - \frac{V_{dc}}{\omega L_e} (\omega t - \alpha) \qquad \text{for } \omega t \geq \alpha \qquad (3.70)$$

The critical value of $\omega t = \beta = \pi + \alpha$ at which the current i_0 falls to zero can be found from the condition $i_0(\omega t = \beta) = \pi + \alpha = 0$.

The average current I_{dc} can be found from

$$I_{dc} = \frac{1}{\pi} \int_\alpha^{\pi + \alpha} i_0(t) \, d(\omega t)$$

which, after integration and simplification gives

$$I_{dc} = \frac{V_m}{\omega L_e} \left[\sqrt{1 - x^2} + x \left(\frac{2}{\pi} - \frac{\pi}{2} \right) \right] \qquad (3.71)$$

For $V_{dc} = 0$, the peak current that can flow through the rectifier is $I_{pk} = V_m/\omega L_e$. Normalizing I_{dc} with respect to I_{pk}, we get

$$k(x) = \frac{I_{dc}}{I_{pk}} = \sqrt{1 - x^2} + x \left(\frac{2}{\pi} - \frac{\pi}{2} \right) \qquad (3.72)$$

Normalizing the rms value I_{rms} with respect to I_{pk}, we get

$$k_r(x) = \frac{I_{rms}}{I_{pk}} = \sqrt{\frac{1}{\pi} \int_\alpha^{\pi + \alpha} i_0(t)^2 \, d(\omega \cdot t)} \qquad (3.73)$$

Because α depends on the voltage ratio x, Eqs. (3.71) and (3.72) are dependent on x. Table 3.3 shows the values of $k(x)$ and $k_r(x)$ against the voltage ratio x.

Because the average voltage of the rectifier is $V_{dc} = 2 V_m/\pi$, the average current equal to

$$I_{dc} = \frac{2 V_m}{\pi R}$$

Thus,

$$\frac{2 V_m}{\pi R} = I_{dc} = I_{pk} k(x) = \frac{V_m}{\omega L_e} \left[\sqrt{1 - x^2} + x \left(\frac{2}{\pi} - \frac{\pi}{2} \right) \right]$$

which gives the critical value of the inductance $L_{cr}(= L_e)$ for a continuous current as

$$L_{cr} = \frac{\pi R}{2\omega}\left[\sqrt{1 - x^2} + x\left(\frac{2}{\pi} - \frac{\pi}{2}\right)\right] \tag{3.74}$$

Thus, for a continuous current through the inductor, the value of L_e must be larger than the value of L_{cr}. That is,

$$L_e > L_{cr} = \frac{\pi R}{2\omega}\left[\sqrt{1 - x^2} + x\left(\frac{2}{\pi} - \frac{\pi}{2}\right)\right] \tag{3.75}$$

Discontinuous case. The current is discontinuous if $\omega t = \beta \leq (\pi + \alpha)$. The angle β at which the current is zero can be found by setting in Eq. (3.70) to zero. That is,

$$\cos(\alpha) - \cos(\beta) - x(\beta - \alpha) = 0$$

which in terms of x becomes

$$\sqrt{1 - x^2} - x(\beta - \arcsin(x)) = 0 \tag{3.76}$$

Example 3.19   Finding the Critical Value of Inductor for Continuous Load Current

The rms input voltage to the circuit in Figure 3.26a is 220 V, 60 Hz. (a) If the dc output voltage is $V_{dc} = 100$ V at $I_{dc} = 10$ A, determine the values of critical inductance L_e, α, and I_{rms}. (b) If $I_{dc} = 15$ A and $L_e = 6.5$ mH, use Table 3.3 to determine the values of V_{dc}, α, β, and I_{rms}.

TABLE 3.3   Normalized Load Current

x %	I_{dc}/I_{pk} %	I_{rms}/I_{pk} %	α Degrees	β Degrees
0	100.0	122.47	0	180
5	95.2	115.92	2.87	182.97
10	90.16	109.1	5.74	185.74
15	84.86	102.01	8.63	188.63
20	79.30	94.66	11.54	191.54
25	73.47	87.04	14.48	194.48
30	67.37	79.18	17.46	197.46
35	60.98	71.1	20.49	200.49
40	54.28	62.82	23.58	203.58
45	47.26	54.43	26.74	206.74
50	39.89	46.06	30.00	210.00
55	32.14	38.03	33.37	213.37
60	23.95	31.05	36.87	216.87
65	15.27	26.58	40.54	220.54
70	6.02	26.75	44.27	224.43
72	2.14	28.38	46.05	226.05
72.5	1.15	28.92	46.47	226.47
73	0.15	29.51	46.89	226.89
73.07	0	29.60	46.95	226.95

Solution

$\omega = 2\pi \times 60 = 377$ rad/s, $V_s = 120\,V$, $V_m = \sqrt{2} \times 120 = 169.7\,V$.

a. Voltage ratio $x = V_{dc}/V_m = 100/169.7 = 58.93\%$; $\alpha = \sin^{-1}(x) = 36.87\%$. Equation (3.72) gives the average current ratio $k = I_{dc}/I_{pk} = 25.75\%$. Thus, $I_{pk} = I_{dc}/k = 10/0.2575 = 38.84$ A. The critical value of inductance is

$$L_{cr} = \frac{V_m}{\omega I_{pk}} = \frac{169.7}{377 \times 38.84} = 11.59 \text{ mH}$$

Equation (3.73) gives the rms current ratio $k_r = I_{rms}/I_{pk} = 32.4\%$. Thus,

$$I_{rms} = k_r I_{pk} = 0.324 \times 38.84 = 12.58 \text{ A}.$$

b. $L_e = 6.5$ mH, $I_{pk} = V_m/(\omega L_e) = 169.7/(377 \times 6.5 \text{ mH}) = 69.25$ A.

$$k = \frac{I_{dc}}{I_{pk}} = \frac{15}{69.25} = 21.66\%$$

Using linear interpolation, we get

$$x = x_n + \frac{(x_{n+1} - x_n)(k - k_n)}{k_{n+1} - k_n}$$

$$= 60 + \frac{(65 - 60)(21.66 - 23.95)}{15.27 - 23.95} = 61.32\%$$

$$V_{dc} = xV_m = 0.6132 \times 169.7 = 104.06 \text{ V}$$

$$\alpha = \alpha_n + \frac{(\alpha_{n+1} - \alpha_n)(k - k_n)}{k_{n+1} - k_n}$$

$$= 36.87 + \frac{(40.54 - 36.87)(21.66 - 23.95)}{15.27 - 23.95} = 37.84°$$

$$\beta = \beta_n + \frac{(\beta_{n+1} - \beta_n)(k - k_n)}{k_{n+1} - k_n}$$

$$= 216.87 + \frac{(220.54 - 216.87)(21.66 - 23.95)}{15.27 - 23.95} = 217.85°$$

$$k_r = \frac{I_{rms}}{I_{pk}} = k_{r(n)} + \frac{(k_{r(n+1)} - k_{r(n)})(k - k_n)}{k_{n+1} - k_n}$$

$$= 31.05 + \frac{(26.58 - 31.05)(21.66 - 23.95)}{15.27 - 23.95} = 29.87\%$$

Thus, $I_{rms} = 0.2987 \times I_{pk} = 0.2987 \times 69.25 = 20.68$ A.

Key Points of Section 3.11

- With a high value of output filter capacitance C_e the output voltage remains almost constant. A minimum value of the filter inductance L_e is required to maintain a continuous current. The inductor L_e is generally placed at the input side to act as an ac inductor instead of a dc choke.

3.12   EFFECTS OF SOURCE AND LOAD INDUCTANCES

In the derivations of the output voltages and the performance criteria of rectifiers, it was assumed that the source has no inductances and resistances. However, in a practical transformer and supply, these are always present and the performances of rectifiers are slightly changed. The effect of the source inductance, which is more significant than that of resistance, can be explained with reference to Figure 3.27a.

The diode with the most positive voltage conducts. Let us consider the point $\omega t = \pi$ where voltages v_{ac} and v_{bc} are equal as shown in Figure 3.27b. The current I_{dc} is still flowing through diode D_1. Due to the inductance L_1, the current cannot fall to zero immediately and the transfer of current cannot be on an instantaneous basis. The current i_{d1} decreases, resulting in an induced voltage across L_1 of $+v_{01}$ and the output voltage becomes $v_0 = v_{ac} + v_{01}$. At the same time the current through D_3, i_{d3} increases from zero, inducing an equal voltage across L_2 of $-v_{02}$ and the output voltage becomes $v_{02} = v_{bc} - v_{02}$. The result is that the anode voltages of diodes D_1 and D_3 are equal;

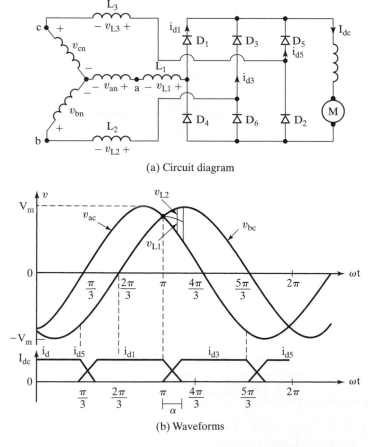

(a) Circuit diagram

(b) Waveforms

FIGURE 3.27

Three-phase bridge rectifier with source inductances.

and both diodes conduct for a certain period which is called *commutation* (or *overlap*) angle μ. This transfer of current from one diode to another is called *commutation*. The reactance corresponding to the inductance is known as *commutating reactance*.

The effect of this overlap is to reduce the average output voltage of converters. The voltage across L_2 is

$$v_{L2} = L_2 \frac{di}{dt} \qquad (3.77)$$

Assuming a linear rise of current i from 0 to I_{dc} (or a constant $di/dt = \Delta i/\Delta t$), we can write Eq. (3.77) as

$$v_{L2}\, \Delta t = L_2\, \Delta i \qquad (3.78)$$

and this is repeated six times for a three-phase bridge rectifier. Using Eq. (3.78), the average voltage reduction due to the commutating inductances is

$$V_x = \frac{1}{T} 2(v_{L1} + v_{L2} + v_{L3})\, \Delta t = 2f(L_1 + L_2 + L_3)\, \Delta i$$

$$= 2f(L_1 + L_2 + L_3)I_{dc} \qquad (3.79)$$

If all the inductances are equal and $L_c = L_1 = L_2 = L_3$, Eq. (3.79) becomes

$$V_x = 6fL_cI_{dc} \qquad (3.80)$$

where f is the supply frequency in hertz.

Example 3.20 Finding the Effect of Line Inductance on the Output Voltage of a Rectifier

A three-phase bridge rectifier is supplied from a Y-connected 208-V 60-Hz supply. The average load current is 60 A and has negligible ripple. Calculate the percentage reduction of output voltage due to commutation if the line inductance per phase is 0.5 mH.

Solution

$L_c = 0.5$ mH, $V_s = 208/\sqrt{3} = 120$ V, $f = 60$ Hz, $I_{dc} = 60$ A, and $V_m = \sqrt{2} \times 120 = 169.7$ V. From Eq. (3.40), $V_{dc} = 1.654 \times 169.7 = 280.7$ V. Equation (3.80) gives the output voltage reduction,

$$V_x = 6 \times 60 \times 0.5 \times 10^{-3} \times 60 = 10.8 \text{ V} \quad \text{or} \quad 10.8 \times \frac{100}{280.7} = 3.85\%$$

and the effective output voltage is $(280.7 - 10.8) = 269.9$ V.

Example 3.21 Finding the Effect of Diode Recovery Time on the Output Voltage of a Rectifier

The diodes in the single-phase full-wave rectifier in Figure 3.6a have a reverse recovery time of $t_{rr} = 50$ μs and the rms input voltage is $V_s = 120$ V. Determine the effect of the reverse recovery time on the average output voltage if the supply frequency is (a) $f_s = 2$ kHz, and (b) $f_s = 60$ Hz.

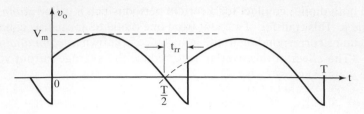

FIGURE 3.28

Effect of reverse recovery time on output voltage.

Solution

The reverse recovery time would affect the output voltage of the rectifier. In the full-wave rectifier of Figure 3.6a, the diode D_1 is not off at $\omega t = \pi$; instead, it continues to conduct until
$t = \pi/\omega + t_{rr}$. As a result of the reverse recovery time, the average output voltage is reduced and
the output voltage waveform is shown in Figure 3.28.

If the input voltage is $v = V_m \sin \omega t = \sqrt{2}\, V_s \sin \omega t$, the average output voltage reduction is

$$V_{rr} = \frac{2}{T} \int_0^{t_{rr}} V_m \sin \omega t \, dt = \frac{2V_m}{T} \left[-\frac{\cos \omega t}{\omega} \right]_0^{t_{rr}}$$

$$= \frac{V_m}{\pi}(1 - \cos \omega t_{rr})$$

$$V_m = \sqrt{2}\, V_s = \sqrt{2} \times 120 = 169.7 \text{ V} \tag{3.81}$$

Without any reverse recovery time, Eq. (3.21) gives the average output voltage $V_{dc} = 0.6366V_m = 108.03$ V.

    a. For $t_{rr} = 50$ μs and $f_s = 2000$ Hz, the reduction of the average output voltage is

$$V_{rr} = \frac{V_m}{\pi}(1 - \cos 2\pi f_s t_{rr})$$

$$= 0.061V_m = 10.3 \text{ V} \quad \text{or} \quad 9.51\% \text{ of } V_{dc}$$

    b. For $t_{rr} = 50$ μs and $f_s = 60$ Hz, the reduction of the output dc voltage

$$V_{rr} = \frac{V_m}{\pi}(1 - \cos 2\pi f_s t_{rr}) = 5.65 \times 10^{-5} V_m$$

$$= 9.6 \times 10^{-3} \text{ V} \quad \text{or} \quad 8.88 \times 10^{-3}\% \text{ of } V_{dc}$$

Note: The effect of t_{rr} is significant for high-frequency source and for the case of
normal 60-Hz source, its effect can be considered negligible.

Key Points of Section 3.12

- A practical supply has a source reactance. As a result, the transfer of current from
  one diode to another one cannot be instantaneous. There is an overlap known as

commutation angle, which lowers the effective output voltage of the rectifier. The effect of the diode reverse time may be significant for a high-frequency source.

SUMMARY

There are different types of rectifiers depending on the connections of diodes and input transformer. The performance parameters of rectifiers are defined and it has been shown that the performances of rectifiers vary with their types. The rectifiers generate harmonics into the load and the supply line; and these harmonics can be reduced by filters. The performances of the rectifiers are also influenced by the source and load inductances.

REFERENCES

[1] J. Schaefer, *Rectifier Circuits—Theory and Design,* New York: John Wiley & Sons, 1975.

[2] R. W. Lee, *Power Converter Handbook—Theory Design and Application*, Canadian General Electric, Peterborough, Ontario, 1979.

[3] Y.-S. Lee and M. H. L. Chow, *Power Electronics Handbook*, edited by M. H. Rashid. San Diego, CA: Academic Press, 2001, Chapter 10.

[4] IEEE Standard 597, *Practices and Requirements for General Purpose Thyristor Drives*, Piscataway, NJ, 1983.

REVIEW QUESTIONS

3.1 What is the turns ratio of a transformer?

3.2 What is a rectifier? What is the difference between a rectifier and a converter?

3.3 What is the blocking condition of a diode?

3.4 What are the performance parameters of a rectifier?

3.5 What is the significance of the form factor of a rectifier?

3.6 What is the significance of the ripple factor of a rectifier?

3.7 What is the efficiency of rectification?

3.8 What is the significance of the transformer utilization factor?

3.9 What is the displacement factor?

3.10 What is the input power factor?

3.11 What is the harmonic factor?

3.12 What is the difference between a half-wave and a full-wave rectifier?

3.13 What is the dc output voltage of a single-phase half-wave rectifier?

3.14 What is the dc output voltage of a single-phase full-wave rectifier?

3.15 What is the fundamental frequency of the output voltage of a single-phase full-wave rectifier?

3.16 What are the advantages of a three-phase rectifier over a single-phase rectifier?

3.17 What are the disadvantages of a multi-phase half-wave rectifier?

3.18 What are the advantages of a three-phase bridge rectifier over a six-phase star rectifier?

3.19 What are the purposes of filters in rectifier circuits?

3.20 What are the differences between ac and dc filters?

3.21 What are the effects of source inductances on the output voltage of a rectifier?

3.22 What are the effects of load inductances on the rectifier output?

3.23 What is a commutation of diodes?

3.24 What is the commutation angle of a rectifier?

PROBLEMS

3.1 A single-phase bridge rectifier has a purely resistive load $R = 10\ \Omega$, the peak supply voltage $V_m = 170$ V, and the supply frequency $f = 60$ Hz. Determine the average output voltage of the rectifier if the source inductance is negligible.

3.2 Repeat Problem 3.1 if the source inductance per phase (including transformer leakage inductance) is $L_c = 0.5$ mH.

3.3 A six-phase star rectifier has a purely resistive load of $R = 10\ \Omega$, the peak supply voltage $V_m = 170$ V, and the supply frequency $f = 60$ Hz. Determine the average output voltage of the rectifier if the source inductance is neglibile.

3.4 Repeat Problem 3.3 if the source inductance per phase (including the transformer leakage inductance) is $L_c = 0.5$ mH.

3.5 A three-phase bridge rectifier has a purely resistive load of $R = 100\ \Omega$ and is supplied from a 280-V, 60-Hz supply. The primary and secondary of the input transformer are connected in Y. Determine the average output voltage of the rectifier if the source inductances are negligible.

3.6 Repeat Problem 3.5 if the source inductance per phase (including transformer leakage inductance) is $L_c = 0.5$ mH.

3.7 The single-phase bridge rectifier of Figure 3.6a is required to supply an average voltage of $V_{dc} = 400$ V to a resistive load of $R = 10\ \Omega$. Determine the voltage and current ratings of diodes and transformer.

3.8 A three-phase bridge rectifier is required to supply an average voltage of $V_{dc} = 750$ V at a ripple-free current of $I_{dc} = 9000$ A. The primary and secondary of the transformer are connected in Y. Determine the voltage and current ratings of diodes and transformer.

3.9 The single-phase rectifier of Figure 3.5a has an RL load. If the peak input voltage is $V_m = 170$ V, the supply frequency $f = 60$ Hz, and the load resistance $R = 15\ \Omega$, determine the load inductance L to limit the load current harmonic to 4% of the average value I_{dc}.

3.10 The three-phase star rectifier of Figure 3.12a has an RL load. If the secondary peak voltage per phase is $V_m = 170$ V at 60 Hz, and the load resistance is $R = 15\ \Omega$, determine the load inductance L to limit the load current harmonics to 2% of the average value I_{dc}.

3.11 The battery voltage in Figure 3.4a is $E = 20$ V and its capacity is 200 Wh. The average charging current should be $I_{dc} = 10$ A. The primary input voltage is $V_p = 120$ V, 60 Hz, and the transformer has a turns ratio of $h = 2$:1. *Calculate* **(a)** the conduction angle δ of the diode, **(b)** the current-limiting resistance R, **(c)** the power rating P_R of R, **(d)** the charging time h in hours, **(e)** the rectifier efficiency η, and **(f)** the peak inverse voltage PIV of the diode.

3.12 The single-phase full-wave rectifier of Figure 3.8a has $L = 4.5$ mH, $R = 5\ \Omega$, and $E = 20$ V. The input voltage is $V_s = 120$ V at 60 Hz. **(a)** Determine (1) the steady-state load current I_0 at $\omega t = 0$, (2) the average diode current I_d, (3) the rms diode current I_r, and (4) the rms output current I_{rms}. **(b)** Use PSpice to plot the instantaneous output current i_0. Assume diode parameters IS $= 2.22\text{E} - 15$, BV $= 1800$ V.

3.13 The three-phase full-wave rectifier of Figure 3.13a has a load of $L = 2.5$ mH, $R = 5\ \Omega$, and $E = 20$ V. The line-to-line input voltage is $V_{ab} = 208$ V, 60 Hz. **(a)** Determine (1) the steady-state load current I_0 at $\omega t = \pi/3$, (2) the average diode current I_d, (3) the rms diode current I_r, and (4) the rms output current I_{rms}. **(b)** Use PSpice to plot the instantaneous output current i_0. Assume diode parameters IS $= 2.22\text{E} - 15$, BV $= 1800$ V.

3.14 A single-phase bridge rectifier is supplied from a 120-V, 60-Hz source. The load resistance is $R = 200\ \Omega$. **(a)** Design a C-filter so that the ripple factor of the output voltage is less than 5%. **(b)** With the value of capacitor C in part (a), calculate the average load voltage V_{dc}.

3.15 Repeat Problem 3.14 for the single-phase half-wave rectifier.

3.16 The rms input voltage to the circuit in Figure 3.22a is 120 V, 60 Hz. **(a)** If the dc output voltage is $V_{dc} = 48$ V at $I_{dc} = 25$ A, determine the values of inductance L_e, α, and I_{rms}. **(b)** If $I_{dc} = 15$ A and $L_e = 6.5$ mH, use Table 3.3 to calculate the values of V_{dc}, α, β, and I_{rms}.

3.17 The single-phase rectifier of Figure 3.15a has a resistive load of R, and a capacitor C is connected across the load. The average load current is I_{dc}. Assuming that the charging time of the capacitor is negligible compared with the discharging time, determine the rms output voltage harmonics, V_{ac}.

3.18 The LC filter shown in Figure 3.18c is used to reduce the ripple content of the output voltage for a six-phase star rectifier. The load resistance is $R = 20$ Ω, load inductance is $L = 5$ mH, and source frequency is 60 Hz. Determine the filter parameters L_e and C_e so that the ripple factor of the output voltage is 5%.

3.19 The three-phase bridge rectifier of Figure 3.3a has an RL load and is supplied from a Y connected supply. **(a)** Use the method of Fourier series to obtain expressions for the output voltage $v_0(t)$ and load current $i_0(t)$. **(b)** If peak phase voltage is $V_m = 170$ V at 60 Hz and the load resistance is $R = 200$ Ω, determine the load inductance L to limit the ripple current to 2% of the average value I_{dc}.

3.20 The single-phase half-wave rectifier of Figure 3.3a has a freewheeling diode and a ripple-free average load current of I_a. **(a)** Draw the waveforms for the currents in D_1, D_m, and the transformer primary; **(b)** express the primary current in Fourier series; and **(c)** determine the input PF and HF of the input current at the rectifier input. Assume a transformer turns ratio of unity.

3.21 The single-phase full-wave rectifier of Figure 3.5a has a ripple-free average load current of I_a. **(a)** Draw the waveforms for currents in D_1, D_2, and transformer primary; **(b)** express the primary current in Fourier series; and **(c)** determine the input PF and HF of the input current at the rectifier input. Assume a transformer turns ratio of unity.

3.22 The multiphase star rectifier of Figure 3.12a has three pulses and supplies a ripple-free average load current of I_a. The primary and secondary of the transformer are connected in Y. Assume a transformer turns ratio of unity. **(a)** Draw the waveforms for currents in D_1, D_2, D_3, and transformer primary; **(b)** express the primary current in Fourier series; and **(c)** determine the input PF and HF of input current.

3.23 Repeat Problem 3.22 if the primary of the transformer is connected in delta and secondary in Y.

3.24 The multiphase star rectifier of Figure 3.12a has six pulses and supplies a ripple-free average load current of I_a. The primary of the transformer is connected in delta and secondary in Y. Assume a transformer turns ratio of unity. **(a)** Draw the waveforms for currents in D_1, D_2, D_3, and transformer primary; **(b)** express the primary current in Fourier series; and **(c)** determine the input PF and HF of the input current.

3.25 The three-phase bridge rectifier of Figure 3.13a supplies a ripple-free load current of I_a. The primary and secondary of the transformer are connected in Y. Assume a transformer turns ratio of unity. **(a)** Draw the waveforms for currents in D_1, D_3, D_5 and the secondary phase current of the transformer; **(b)** express the secondary phase current in Fourier series; and **(c)** determine the input PF and HF of the input current.

3.26 Repeat Problem 3.25 if the primary of the transformer is connected in delta and secondary in Y.

3.27 Repeat Problem 3.25 if the primary and secondary of the transformer are connected in delta.

CHAPTER 4

Power Transistors

The learning objectives of this chapter are as follows:

- To learn the characteristics of an ideal switch
- To learn about different power transistors such as BJTs, MOSFETs, SITs, IGBTs, and COOLMOS.
- To learn the limitations of transistors as switches
- To understand the characteristics, gate control requirements, and models of power transistors

4.1 INTRODUCTION

Power transistors have controlled turn-on and turn-off characteristics. The transistors, which are used as switching elements, are operated in the saturation region, resulting in a low on-state voltage drop. The switching speed of modern transistors is much higher than that of thyristors and they are extensively employed in dc–dc and dc–ac converters, with inverse parallel-connected diodes to provide bidirectional current flow. However, their voltage and current ratings are lower than those of thyristors and transistors are normally used in low- to medium-power applications. The power transistors can be classified broadly into five categories:

1. Bipolar junction transistors (BJTs)
2. Metal oxide semiconductor field-effect transistors (MOSFETs)
3. Static induction transistors (SITs)
4. Insulated-gate bipolar transistors (IGBTs)
5. COOLMOS

BJTs, MOSFETs, SITs, IGBTs, or COOLMOS can be assumed as ideal switches to explain the power conversion techniques. A transistor can be operated as a switch. However, the choice between a BJT and an MOSFET in the converter circuits is not obvious, but each of them can replace a switch, provided that its voltage and current ratings meet the output requirements of the converter. Practical transistors differ from ideal devices. The transistors have certain limitations and are restricted to some applications. The characteristics and ratings of each type should be examined to determine its suitability to a particular application.

4.2 BIPOLAR JUNCTION TRANSISTORS

A bipolar transistor is formed by adding a second *p*- or *n*-region to a *pn*-junction diode. With two *n*-regions and one *p*-region, two junctions are formed and it is known as an *NPN-transistor*, as shown in Figure 4.1a. With two *p*-regions and one *n*-region, it is called as a *PNP-transistor*, as shown in Figure 4.1b. The three terminals are named as *collector*, *emitter*, and *base*. A bipolar transistor has two junctions, collector–base junction (CBJ) and base–emitter junction (BEJ) [1–5] *NPN*-transistors of various sizes are shown in Figure 4.2.

There are two n^+-regions for the emitter of *NPN*-type transistor shown in Figure 4.3a and two p^+-regions for the emitter of the *PNP*-type transistor shown in Figure 4.3b. For an *NPN*-type, the emitter side *n*-layer is made wide, the *p*-base is narrow, and the collector side *n*-layer is narrow and heavily doped. For a *PNP*-type,

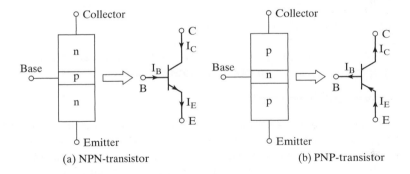

(a) NPN-transistor   (b) PNP-transistor

FIGURE 4.1

Bipolar transistors.

FIGURE 4.2

NPN-transistors. (Courtesy of Powerex, Inc.)

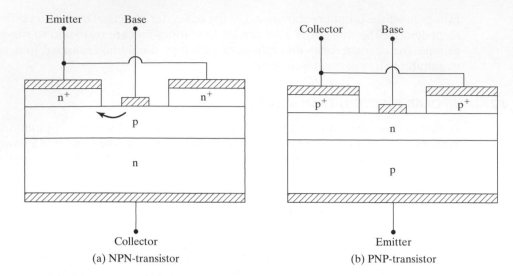

FIGURE 4.3

Cross sections of BJTs.

the emitter side p-layer is made wide, the n-base is narrow, and the collector side p-layer is narrow and heavily doped. The base and collector currents flow through two parallel paths, resulting in a low on-state collector–emitter resistance, $R_{CE(ON)}$.

4.2.1 Steady-State Characteristics

Although there are three possible configurations—common collector, common base, and common emitter, the common-emitter configuration, which is shown in Figure 4.4a for an *NPN*-transistor, is generally used in switching applications. The typical input characteristics of base current I_B, against base–emitter voltage V_{BE}, are shown in Figure 4.4b. Figure 4.4c shows the typical output characteristics of collector current I_C, against collector–emitter voltage V_{CE}. For a *PNP*-transistor, the polarities of all currents and voltages are reversed.

There are three operating regions of a transistor: cutoff, active, and saturation. In the cutoff region, the transistor is off or the base current is not enough to turn it on and both junctions are reverse biased. In the active region, the transistor acts as an amplifier, where the base current is amplified by a gain and the collector–emitter voltage decreases with the base current. The CBJ is reverse biased, and the BEJ is forward biased. In the saturation region, the base current is sufficiently high so that the collector–emitter voltage is low, and the transistor acts as a switch. Both junctions (CBJ and BEJ) are forward biased. The transfer characteristic, which is a plot of V_{CE} against I_B, is shown in Figure 4.5.

The model of an *NPN*-transistor is shown in Figure 4.6 under large-signal dc operation. The equation relating the currents is

$$I_E = I_C + I_B \tag{4.1}$$

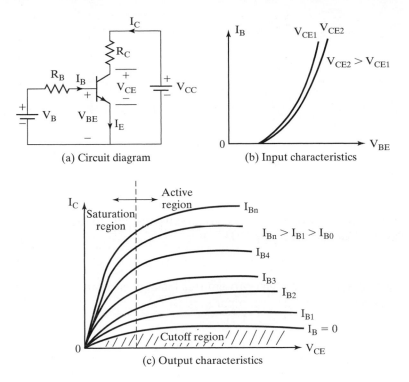

(a) Circuit diagram

(b) Input characteristics

(c) Output characteristics

FIGURE 4.4

Characteristics of *NPN*-transistors.

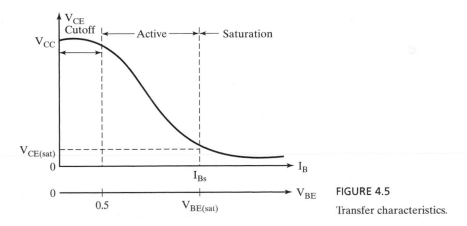

FIGURE 4.5

Transfer characteristics.

The base current is effectively the input current and the collector current is the output current. The ratio of the collector current I_C, to base current I_B, is known as the forward *current gain*, β_F:

$$\beta_F = h_{FE} = \frac{I_C}{I_B} \tag{4.2}$$

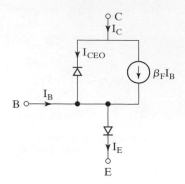

FIGURE 4.6

Model of *NPN*-transistors.

The collector current has two components: one due to the base current and the other is the leakage current of the CBJ.

$$I_C = \beta_F I_B + I_{CEO} \tag{4.3}$$

where I_{CEO} is the collector-to-emitter leakage current with base open circuit and can be considered negligible compared to $\beta_F I_B$.

From Eqs. (4.1) and (4.3),

$$I_E = I_B(1 + \beta_F) + I_{CEO} \tag{4.4}$$

$$\approx I_B(1 + \beta_F) \tag{4.4a}$$

$$I_E \approx I_C\left(1 + \frac{1}{\beta_F}\right) = I_C \frac{\beta_F + 1}{\beta_F} \tag{4.5}$$

Because $\beta_F \gg 1$, the collector current can be expressed as

$$I_C \approx \alpha_F I_E \tag{4.6}$$

where the constant α_F is related to β by

$$\alpha_F = \frac{\beta_F}{\beta_F + 1} \tag{4.7}$$

or

$$\beta_F = \frac{\alpha_F}{1 - \alpha_F} \tag{4.8}$$

Let us consider the circuit of Figure 4.7, where the transistor is operated as a switch.

$$I_B = \frac{V_B - V_{BE}}{R_B} \tag{4.9}$$

$$V_C = V_{CE} = V_{CC} - I_C R_C = V_{CC} - \frac{\beta_F R_C}{R_B}(V_B - V_{BE})$$

$$V_{CE} = V_{CB} + V_{BE} \tag{4.10}$$

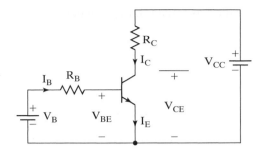

FIGURE 4.7

Transistor switch.

or

$$V_{CB} = V_{CE} - V_{BE} \tag{4.11}$$

Equation (4.11) indicates that as long as $V_{CE} \geq V_{BE}$, the CBJ is reverse biased and the transistor is in the active region. The maximum collector current in the active region, which can be obtained by setting $V_{CB} = 0$ and $V_{BE} = V_{CE}$, is

$$I_{CM} = \frac{V_{CC} - V_{CE}}{R_C} = \frac{V_{CC} - V_{BE}}{R_C} \tag{4.12}$$

and the corresponding value of base current

$$I_{BM} = \frac{I_{CM}}{\beta_F} \tag{4.13}$$

If the base current is increased above I_{BM}, V_{BE} increases, the collector current increases, and the V_{CE} falls below V_{BE}. This continues until the CBJ is forward biased with V_{BC} of about 0.4 to 0.5 V. The transistor then goes into saturation. The *transistor saturation* may be defined as the point above which any increase in the base current does not increase the collector current significantly.

In the saturation, the collector current remains almost constant. If the collector–emitter saturation voltage is $V_{CE(\text{sat})}$, the collector current is

$$I_{CS} = \frac{V_{CC} - V_{CE(\text{sat})}}{R_C} \tag{4.14}$$

and the corresponding value of base current is

$$I_{BS} = \frac{I_{CS}}{\beta_F} \tag{4.15}$$

Normally, the circuit is designed so that I_B is higher than I_{BS}. The ratio of I_B to I_{BS} is called the *overdrive factor* (ODF):

$$\text{ODF} = \frac{I_B}{I_{BS}} \tag{4.16}$$

a

T

A
e
tl

Example

T
ta
cl
sa

S
V
V
1
o

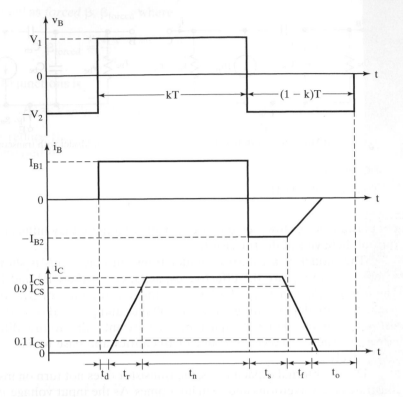

FIGURE 4.9

Switching times of bipolar transistors.

to the change in the polarity of v_B from V_1 to $-V_2$. The reverse current, $-I_{B2}$, helps to discharge the base and remove the extra charge from the base. Without $-I_{B2}$, the saturating charge has to be removed entirely by recombination and the storage time would be longer.

Once the extra charge is removed, the BEJ capacitance charges to the input voltage $-V_2$, and the base current falls to zero. The fall time t_f depends on the time constant, which is determined by the capacitance of the reverse-biased BEJ.

Figure 4.10a shows the extra storage charge in the base of a saturated transistor. During turn-off, this extra charge is removed first in time t_s and the charge profile is changed from a to c as shown in Figure 4.10b. During fall time, the charge profile decreases from profile c until all charges are removed.

The turn-on time t_{on} is the sum of delay time t_d and rise time t_r:

$$t_{on} = t_d + t_r$$

and the turn-off time t_{off} is the sum of storage time t_s and fall time t_f:

$$t_{off} = t_s + t_f$$

2
is
cl
a
th

4.2.2   S

A
p
h

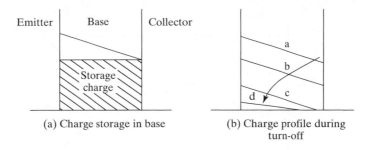

(a) Charge storage in base   (b) Charge profile during turn-off

FIGURE 4.10

Charge storage in saturated bipolar transistors.

Example 4.2   Finding the Switching Loss of a BJT

The waveforms of the transistor switch in Figure 4.7 are shown in Figure 4.11. The parameters are $V_{CC} = 250$ V, $V_{BE(\text{sat})} = 3$ V, $I_B = 8$ A, $V_{CS(\text{sat})} = 2$ V, $I_{CS} = 100$ A, $t_d = 0.5$ μs, $t_r = 1$ μs, $t_s = 5$ μs, $t_f = 3$ μs, and $f_s = 10$ kHz. The duty cycle is $k = 50\%$. The collector-to-emitter leakage current is $I_{CEO} = 3$ mA. Determine the power loss due to collector current (a) during turn-on $t_{\text{on}} = t_d + t_r$, (b) during conduction period t_n, (c) during turn-off $t_{\text{off}} = t_s + t_f$, (d) during off-time t_o, and (e) total average power losses P_T. (f) Plot the instantaneous power due to collector current $P_c(t)$.

Solution
$T = 1/f_s = 100$ μs, $k = 0.5$, $kT = t_d + t_r + t_n = 50$ μs, $t_n = 50 - 0.5 - 1 = 48.5$ μs, $(1 - k)T = t_s + t_f + t_o = 50$ μs, and $t_o = 50 - 5 - 3 = 42$ μs.

a.   During delay time, $0 \le t \le t_d$:

$$i_c(t) = I_{CEO}$$
$$v_{CE}(t) = V_{CC}$$

The instantaneous power due to the collector current is

$$P_c(t) = i_c v_{CE} = I_{CEO} V_{CC}$$
$$= 3 \times 10^{-3} \times 250 = 0.75 \text{ W}$$

The average power loss during the delay time is

$$P_d = \frac{1}{T} \int_0^{t_d} P_c(t)\,dt = I_{CEO} V_{CC} t_d f_s \tag{4.21}$$

$$= 3 \times 10^{-3} \times 250 \times 0.5 \times 10^{-6} \times 10 \times 10^3 = 3.75 \text{ mW}$$

During rise time, $0 \le t \le t_r$:

$$i_c(t) = \frac{I_{CS}}{t_r} t$$

$$v_{CE}(t) = V_{CC} + (V_{CE(\text{sat})} - V_{CC}) \frac{t}{t_r}$$

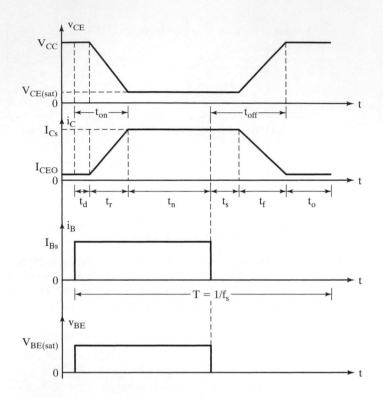

FIGURE 4.11

Waveforms of transistor switch.

$$P_c(t) = i_c v_{CE} = I_{CS} \frac{t}{t_r} \left[V_{CC} + (V_{CE\,(\text{sat})} - V_{CC}) \frac{t}{t_r} \right] \tag{4.22}$$

The power $P_c(t)$ is maximum when $t = t_m$, where

$$t_m = \frac{t_r V_{CC}}{2[V_{CC} - V_{CE\,(\text{sat})}]} \tag{4.23}$$

$$= 1 \times \frac{250}{2(250 - 2)} = 0.504 \ \mu s$$

and Eq. (4.22) yields the peak power

$$P_p = \frac{V_{CC}^2 I_{CS}}{4[V_{CC} - V_{CE(\text{sat})}]} \tag{4.24}$$

$$= 250^2 \times \frac{100}{4(250 - 2)} = 6300 \ \text{W}$$

$$P_r = \frac{1}{T} \int_0^{t_r} P_c(t)dt = f_s I_{CS} t_r \left[\frac{V_{CC}}{2} + \frac{V_{CE(\text{sat})} - V_{CC}}{3} \right]$$

$$= 10 \times 10^3 \times 100 \times 1 \times 10^{-6} \left[\frac{250}{2} + \frac{2 - 250}{3} \right] = 42.33 \text{ W}$$

(4.25)

The total power loss during the turn-on is

$$\begin{aligned} P_{\text{on}} &= P_d + P_r \\ &= 0.00375 + 42.33 = 42.33 \text{ W} \end{aligned}$$

(4.26)

b. The conduction period, $0 \le t \le t_n$:

$$i_c(t) = I_{CS}$$

$$v_{CE}(t) = V_{CE(\text{sat})}$$

$$P_c(t) = i_c v_{CE} = V_{CE(\text{sat})} I_{CS}$$

$$= 2 \times 100 = 200 \text{ W}$$

$$P_n = \frac{1}{T} \int_0^{t_n} P_c(t)dt = V_{CE(\text{sat})} I_{CS} t_n f_s$$

$$= 2 \times 100 \times 48.5 \times 10^{-6} \times 10 \times 10^3 = 97 \text{ W}$$

(4.27)

c. The storage period, $0 \le t \le t_s$:

$$i_c(t) = I_{CS}$$

$$v_{CE}(t) = V_{CE(\text{sat})}$$

$$P_c(t) = i_c v_{CE} = V_{CE(\text{sat})} I_{CS}$$

$$= 2 \times 100 = 200 \text{ W}$$

$$P_s = \frac{1}{T} \int_0^{t_s} P_c(t)dt = V_{CE(\text{sat})} I_{CS} t_s f_s$$

$$= 2 \times 100 \times 5 \times 10^{-6} \times 10 \times 10^3 = 10 \text{ W}$$

(4.28)

The fall time, $0 \le t \le t_f$:

$$i_c(t) = I_{CS}\left(1 - \frac{t}{t_f}\right), \text{ neglecting } I_{CEO}$$

$$v_{CE}(t) = \frac{V_{CC}}{t_f} t, \text{ neglecting } I_{CEO}$$

(4.29)

$$P_c(t) = i_c v_{CE} = V_{CC} I_{CS} \left[\left(1 - \frac{t}{t_f}\right)\frac{t}{t_f} \right]$$

This power loss during fall time is maximum when $t = t_f/2 = 1.5 \text{ μs}$ and Eq. (4.29) gives the peak power,

$$P_m = \frac{V_{CC} I_{CS}}{4}$$

$$= 250 \times \frac{100}{4} = 6250 \text{ W}$$

(4.30)

$$P_f = \frac{1}{T} \int_0^{t_f} P_c(t)dt = \frac{V_{CC} I_{CS} t_f f_s}{6}$$

$$= \frac{250 \times 100 \times 3 \times 10^{-6} \times 10 \times 10^3}{6} = 125 \text{ W}$$

(4.31)

The power loss during turn-off is

$$P_{\text{off}} = P_s + P_f = I_{CS} f_s \left(t_s V_{CE(\text{sat})} + \frac{V_{CC} t_f}{6} \right)$$

$$= 10 + 125 = 135 \text{ W}$$

(4.32)

d. Off-period, $0 \le t \le t_o$:

$$i_c(t) = I_{CEO}$$
$$v_{CE}(t) = V_{CC}$$
$$P_c(t) = i_c v_{CE} = I_{CEO} V_{CC}$$

(4.33)

$$= 3 \times 10^{-3} \times 250 = 0.75 \text{ W}$$

$$P_0 = \frac{1}{T} \int_0^{t_o} P_c(t)dt = I_{CEO} V_{CC} t_o f_s$$

$$= 3 \times 10^{-3} \times 250 \times 42 \times 10^{-6} \times 10 \times 10^3 = 0.315 \text{ W}$$

e. The total power loss in the transistor due to collector current is

$$P_T = P_{\text{on}} + P_n + P_{\text{off}} + P_0$$

$$= 42.33 + 97 + 135 + 0.315 = 274.65 \text{ W}$$

(4.34)

f. The plot of the instantaneous power is shown in Figure 4.12

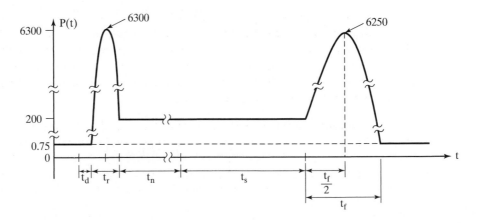

FIGURE 4.12

Plot of instantaneous power for Example 4.2.

Example 4.3 Finding the Base Drive Loss of a BJT

For the parameters in Example 4.2, calculate the average power loss due to the base current.

Solution

$V_{BE(sat)} = 3$ V, $I_B = 8$ A, $T = 1/f_s = 100$ μs, $k = 0.5$, $kT = 50$ μs, $t_d = 0.5$ μs, $t_r = 1$ μs, $t_n = 50 - 1.5 = 48.5$ μs, $t_s = 5$ μs, $t_f = 3$ μs, $t_{on} = t_d + t_r = 1.5$ μs, and $t_{off} = t_s + t_f = 5 + 3 = 8$ μs.

During the period, $0 \leq t \leq (t_{on} + t_n)$:

$$i_b(t) = I_{BS}$$
$$v_{BE}(t) = V_{BE(sat)}$$

The instantaneous power due to the base current is

$$P_b(t) = i_b v_{BE} = I_{BS} V_{BS(sat)}$$
$$= 8 \times 3 = 24 \text{ W}$$

During the period, $0 \leq t \leq t_o = (T - t_{on} - t_n - t_s - t_f)$: $P_b(t) = 0$. The average power loss is

$$P_B = I_{BS} V_{BE(sat)} (t_{on} + t_n + t_s + t_f) f_s$$
$$= 8 \times 3 \times (1.5 + 48.5 + 5 + 3) \times 10^{-6} \times 10 \times 10^3 = 13.92 \text{ W}$$

(4.35)

4.2.3 Switching Limits

Second breakdown (SB). The SB, which is a destructive phenomenon, results from the current flow to a small portion of the base, producing localized hot spots. If the energy in these hot spots is sufficient, the excessive localized heating may damage the transistor. Thus, secondary breakdown is caused by a localized thermal runaway, resulting from high current concentrations. The current concentration may be caused by defects in the transistor structure. The SB occurs at certain combinations of voltage, current, and time. Because the time is involved, the secondary breakdown is basically an energy dependent phenomenon.

Forward-biased safe operating area (FBSOA). During turn-on and on-state conditions, the average junction temperature and second breakdown limit the power-handling capability of a transistor. The manufacturers usually provide the FBSOA curves under specified test conditions. FBSOA indicates the i_c–v_{CE} limits of the transistor; and for reliable operation the transistor must not be subjected to greater power dissipation than that shown by the FBSOA curve.

Reverse-biased safe operating area (RBSOA). During turn-off, a high current and high voltage must be sustained by the transistor, in most cases with the base-to-emitter junction reverse biased. The collector–emitter voltage must be held to a safe level at, or below, a specified value of collector current. The manufacturers provide the I_C–V_{CE} limits during reverse-biased turn-off as RBSOA.

Power derating. The thermal equivalent circuit is shown in Figure 4.13. If the total average power loss is P_T, the case temperature is

$$T_C = T_J - P_T R_{JC}$$

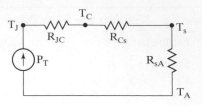

FIGURE 4.13

Thermal equivalent circuit of a transistor.

The sink temperature is

$$T_S = T_C - P_T R_{CS}$$

The ambient temperature is

$$T_A = T_S - P_T R_{SA}$$

and

$$T_J - T_A = P_T(R_{JC} + R_{CS} + R_{SA}) \tag{4.36}$$

where R_{JC} = thermal resistance from junction to case, °C/W;
R_{CS} = thermal resistance from case to sink, °C/W;
R_{SA} = thermal resistance from sink to ambient, °C/W.

The maximum power dissipation P_T is normally specified at $T_C = 25°C$. If the ambient temperature is increased to $T_A = T_{J(\text{max})} = 150°C$, the transistor can dissipate zero power. On the other hand, if the junction temperature is $T_C = 0°C$, the device can dissipate maximum power and this is not practical. Therefore, the ambient temperature and thermal resistances must be considered when interpreting the ratings of devices. Manufacturers show the derating curves for the thermal derating and second breakdown derating.

Breakdown voltages. A *breakdown voltage* is defined as the absolute maximum voltage between two terminals with the third terminal open, shorted, or biased in either forward or reverse direction. At breakdown the voltage remains relatively constant, where the current rises rapidly. The following breakdown voltages are quoted by the manufacturers:

V_{EBO}: the maximum voltage between the emitter terminal and base terminal with collector terminal open circuited.

V_{CEV} or V_{CEX}: the maximum voltage between the collector terminal and emitter terminal at a specified negative voltage applied between base and emitter.

$V_{CEO(\text{SUS})}$: the maximum sustaining voltage between the collector terminal and emitter terminal with the base open circuited. This rating is specified at the maximum collector current and voltage, appearing simultaneously across the device with a specified value of load inductance.

Let us consider the circuit in Figure 4.14a. When the switch SW is closed, the collector current increases, and after a transient, the steady-state collector current is

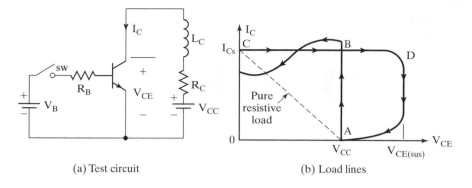

(a) Test circuit

(b) Load lines

FIGURE 4.14

Turn-on and turn-off load lines.

$I_{CS} = (V_{CC} - V_{CE(\text{sat})})/R_C$. For an inductive load, the load line would be the path ABC shown in Figure 4.14b. If the switch is opened to remove the base current, the collector current begins to fall and a voltage of $L(di/dt)$ is induced across the inductor to oppose the current reduction. The transistor is subjected to a transient voltage. If this voltage reaches the sustaining voltage level, the collector voltage remains approximately constant and the collector current falls. After a short time, the transistor is in the off-state and the turn-off load line is shown in Figure 4.14b by the path CDA.

Example 4.4   Finding the Case Temperature of a BJT

The maximum junction temperature of a transistor is $T_J = 150°C$ and the ambient temperature is $T_A = 25°C$. If the thermal impedances are $R_{JC} = 0.4°C/W$, $R_{CS} = 0.1°C/W$, and $R_{SA} = 0.5°C/W$, calculate (a) the maximum power dissipation, and (b) the case temperature.

Solution

a. $T_J - T_A = P_T(R_{JC} + R_{CS} + R_{SA}) = P_T R_{JA}$, $R_{JA} = 0.4 + 0.1 + 0.5 = 1.0$, and $150 - 25 = 1.0 P_T$, which gives the maximum power dissipation as $P_T = 125$ W.

b. $T_C = T_J - P_T R_{JC} = 150 - 125 \times 0.4 = 100°C$.

4.3   POWER MOSFETs

A BJT is a current-controlled device and requires base current for current flow in the collector. Because the collector current is dependent on the input (or base) current, the current gain is highly dependent on the junction temperature.

A power MOSFET is a voltage-controlled device and requires only a small input current. The switching speed is very high and the switching times are of the order of nanoseconds. Power MOSFETs find increasing applications in low-power high-frequency converters. MOSFETs do not have the problems of second breakdown phenomena as do BJTs. However, MOSFETs have the problems of electrostatic discharge and require special care in handling. In addition, it is relatively difficult to protect them under short-circuited fault conditions.

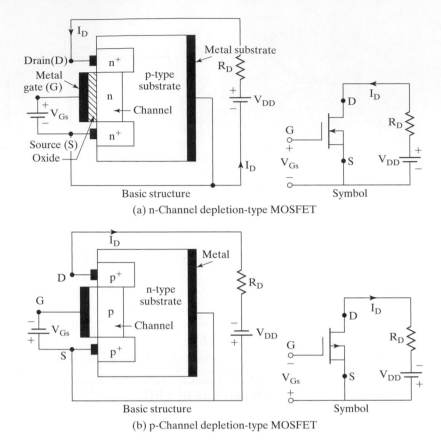

FIGURE 4.15

Depletion-type MOSFETs.

   The two types of MOSFETs are (1) depletion MOSFETs, and (2) enhancement MOSFETs [6–8]. An *n*-channel depletion-type MOSFET is formed on a *p*-type silicon substrate as shown in Figure 4.15a, with two heavily doped n^+ silicon for low-resistance connections. The gate is isolated from the channel by a thin oxide layer. The three terminals are called *gate, drain*, and *source*. The substrate is normally connected to the source. The gate-to-source voltage V_{GS} could be either positive or negative. If V_{GS} is negative, some of the electrons in the *n*-channel area are repelled and a depletion region is created below the oxide layer, resulting in a narrower effective channel and a high resistance from the drain to source R_{DS}. If V_{GS} is made negative enough, the channel becomes completely depleted, offering a high value of R_{DS}, and no current flows from the drain to source, $I_{DS} = 0$. The value of V_{GS} when this happens is called *pinch-off voltage* V_P. On the other hand, V_{GS} is made positive, the channel becomes wider, and I_{DS} increases due to reduction in R_{DS}. With a *p*-channel depletion-type MOSFET, the polarities of V_{DS}, I_{DS}, and V_{GS} are reversed as shown in Figure 4.15b.
   An *n*-channel enhancement-type MOSFET has no physical channel, as shown in Figure 4.16a. If V_{GS} is positive, an induced voltage attracts the electrons from the

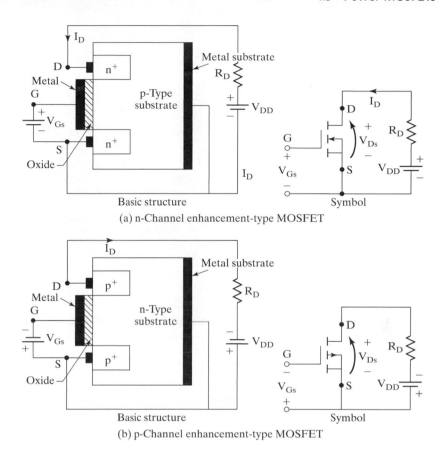

FIGURE 4.16

Enhancement-type MOSFETs.

p-substrate and accumulate them at the surface beneath the oxide layer. If V_{GS} is greater than or equal to a value known as *threshold voltage V_T*, a sufficient number of electrons are accumulated to form a virtual n-channel and the current flows from the drain to source. The polarities of V_{DS}, I_{DS}, and V_{GS} are reversed for a p-channel enhancement-type MOSFET as shown in Figure 4.16b. Power MOSFETs of various sizes are shown in Figure 4.17.

Because a depletion MOSFET remains on at zero gate voltage whereas an enhancement type MOSFET remains off at zero gate voltage, the enhancement type MOSFETS are generally used as switching devices in power electronics. The cross section of a power MOSFET known as a vertical (V) MOSFET is shown in Figure 4.18a.

When the gate has a sufficiently positive voltage with respect to the source, the effect of its electric field pulls electrons from the $n+$ layer into the p layer. This opens a channel closest to the gate, which in turn allows the current to flow from the drain to the source. There is a silicon oxide (SiO) dielectric layer between the gate metal and the $n+$ and p junction. MOSFET is heavily doped on the drain side to create an $n+$ buffer below the n-drift layer. This buffer prevents the depletion layer from reaching

FIGURE 4.17

Power MOSFETs. (Courtesy of
International Rectifier.)

the metal, evens out the voltage stress across the n layer, and also reduces the forward
voltage drop during conduction. The buffer layer also makes it an asymmetric device
with rather low reverse voltage capability.

MOSFETs require low gate energy, and have a very fast switching speed and low
switching losses. The input resistance is very high, 10^9 to 10^{11} Ω. MOSFETs, however,
suffer from the disadvantage of high forward on-state resistance as shown in Figure 4.18b,
and hence high on-state losses, which makes them less attractive as power devices, but
they are excellent as gate amplifying devices for thyristors (see Chapter 7).

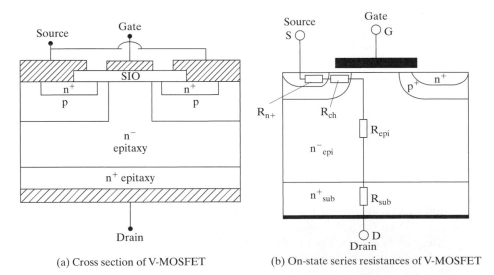

(a) Cross section of V-MOSFET   (b) On-state series resistances of V-MOSFET

FIGURE 4.18

Cross sections of MOSFETs. [Ref. 10, G. Deboy]

4.3.1   Steady-State Characteristics

The MOSFETs are voltage-controlled devices and have a very high input impedance. The gate draws a very small leakage current, on the order of nanoamperes. The current gain, which is the ratio of drain current I_D, to input gate current I_G, is typically on the order of 10^9. However, the current gain is not an important parameter. The *transconductance*, which is the ratio of drain current to gate voltage, defines the transfer characteristics and is a very important parameter.

The transfer characteristics of *n*-channel and *p*-channel MOSFETs are shown in Figure 4.19. Figure 4.20 shows the output characteristics of an *n*-channel enhancement MOSFET. There are three regions of operation: (1) cutoff region, where $V_{GS} \leq V_T$; (2) pinch-off or saturation region, where $V_{DS} \geq V_{GS} - V_T$; and (3) linear region, where $V_{DS} \leq V_{GS} - V_T$. The pinch-off occurs at $V_{DS} = V_{GS} - V_T$. In the linear region, the drain current I_D varies in proportion to the drain–source voltage V_{DS}. Due to high drain current and low drain voltage, the power MOSFETs are operated in the linear region for switching actions. In the saturation region, the drain current remains almost constant for any increase in the value of V_{DS} and the transistors are used in this region for voltage amplification. It should be noted that saturation has the opposite meaning to that for bipolar transistors.

The steady-state model, which is the same for both depletion-type and enhancement-type MOSFETs, is shown in Figure 4.21. The transconductance g_m is defined as

$$g_m = \frac{\Delta I_D}{\Delta V_{GS}}\bigg|_{V_{DS}=\text{constant}} \tag{4.37}$$

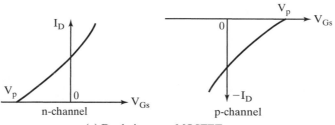

(a) Depletion-type MOSFET

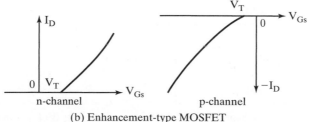

(b) Enhancement-type MOSFET

FIGURE 4.19

Transfer characteristics of MOSFETs.

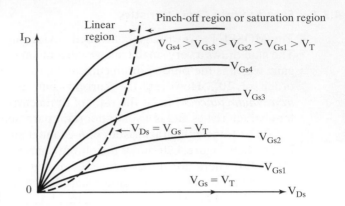

FIGURE 4.20

Output characteristics of enhancement-type MOSFET.

The output resistance, $r_o = R_{DS}$, which is defined as

$$R_{DS} = \frac{\Delta V_{DS}}{\Delta I_D} \tag{4.38}$$

is normally very high in the pinch-off region, typically on the order of megohms and is very small in the linear region, typically on the order of milliohms.

For the depletion-type MOSFETs, the gate (or input) voltage could be either positive or negative. However, the enhancement-type MOSFETs respond to a positive gate voltage only. The power MOSFETs are generally of the enhancement type. However, depletion-type MOSFETs would be advantageous and simplify the logic design in some applications that require some form of logic-compatible dc or ac switch that would remain on when the logic supply falls and V_{GS} becomes zero. The characteristics of depletion-type MOSFETs are not discussed further.

4.3.2 Switching Characteristics

Without any gate signal, an enhancement-type MOSFET may be considered as two diodes connected back to back or as an *NPN*-transistor. The gate structure has parasitic

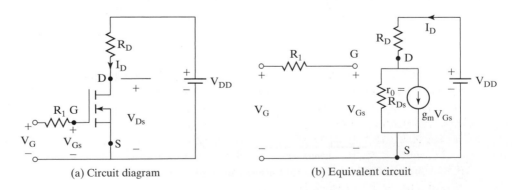

(a) Circuit diagram   (b) Equivalent circuit

FIGURE 4.21

Steady-state switching model of MOSFETs.

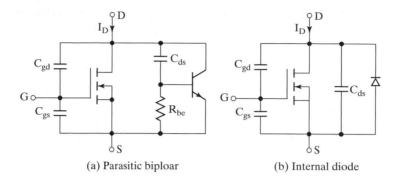

(a) Parasitic biploar

(b) Internal diode

FIGURE 4.22

Parasitic model of enhancement of MOSFETs.

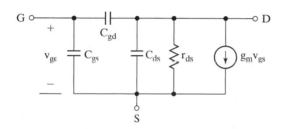

FIGURE 4.23

Switching model of MOSFETs.

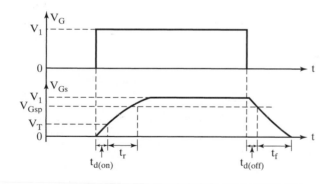

FIGURE 4.24

Switching waveforms and times.

capacitances to the source, C_{gs}, and to the drain, C_{gd}. The *npn*-transistor has a reverse-bias junction from the drain to the source and offers a capacitance, C_{ds}. Figure 4.22a shows the equivalent circuit of a parasitic bipolar transistor in parallel with a MOSFET. The base-to-emitter region of an *NPN*-transistor is shorted at the chip by metalizing the source terminal and the resistance from the base to emitter due to bulk resistance of *n*- and *p*-regions, R_{be}, is small. Hence, as MOSFET may be considered as having an internal diode and the equivalent circuit is shown in Figure 4.22b. The parasitic capacitances are dependent on their respective voltages.

The switching model of MOSFETs is shown in Figure 4.23. The typical switching waveforms and times are shown in Figure 4.24. The *turn-on delay* $t_{d(\text{on})}$ is the time that is required to charge the input capacitance to threshold voltage level. The *rise time* t_r is the gate-charging time from the threshold level to the full-gate voltage V_{GSP}, which is required to drive the transistor into the linear region. The *turn-off delay time* $t_{d(\text{off})}$ is the time required for the input capacitance to discharge from the overdrive gate voltage V_1 to the pinch-off region. V_{GS} must decrease significantly before V_{DS} begins to rise. The *fall time* t_f is the time that is required for the input capacitance to discharge from the pinch-off region to threshold voltage. If $V_{GS} \le V_T$, the transistor turns off.

4.4   COOLMOS

COOLMOS [9–11], which is a new technology for high voltage power MOSFETs, implements a compensation structure in the vertical drift region of a MOSFET to improve the on-state resistance. It has a lower on-state resistance for the same package compared with that of other MOSFETs. The conduction losses are at least five times less as compared with those of the conventional MOSFET technology. It is capable of handling two to three times more output power as compared with that of the conventional MOSFET in the same package. The active chip area of COOLMOS is approximately five times smaller than that of a standard MOSFET.

Figure 4.25 shows the cross section of a COOLMOS. The device enhances the doping of the current conducting *n*-doped layer by roughly one order of magnitude without altering the blocking capability of the device. A high-blocking voltage V_{BR} of

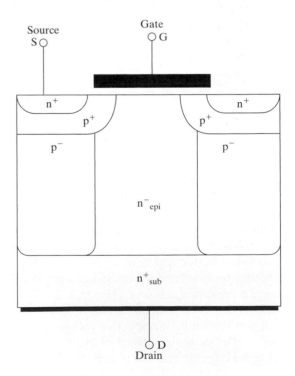

FIGURE 4.25

Cross section of COOLMOS.

the transistor requires a relative thick and low-doped epitaxial layer leading to the well-known law [12] that relates the drain to source resistance to V_{BR} by

$$R_{D(on)} = V_{BR}^{k_c} \tag{4.39}$$

where k_c is a constant between 2.4 and 2.6.

This limitation is overcome by adding columns of the opposite doping type that are implemented into the drift region in a way that the doping integral along a line perpendicular to the current flow remains smaller than the material specific breakthrough charge, which for silicon is about 2×10^{12} cm^{-2}. This concept requires a compensation of the additional charge in the n region by adjacently situated p-doped regions. These charges create a lateral electric field, which does not contribute to the vertical field profile. In other words, the doping concentration is integrated along a line perpendicular to the interface created by the p- and n-regions.

Majority carriers provide the electrical conductivity only. Because there is no bipolar current contribution, the switching losses are equal to that of conventional MOSFETs. The doping of the voltage sustaining the layer is raised by roughly one order of magnitude; additional vertical p-stripes, which are inserted into the structure compensate for the surplus current conducting n-charge. The electric field inside the structure is fixed by the net charge of the two opposite doped columns. Thus, a nearly horizontal field distribution can be achieved if both regions counterbalance each other perfectly. The fabrication of adjacent pairs of p- and n-doped regions with practically zero net charge requires a precision manufacturing. Any charge imbalance impacts the blocking voltage of the device. For higher blocking voltages only the depth of the columns has to be increased without the necessity to alter the doping. This leads to a linear relationship [10] between blocking voltage and on-resistance as shown in Figure 4.26. The on-state resistance of a 600-V, 47-A COOLMOS is 70 mΩ. The COOLMOS has a linear v–i characteristic with a low-threshold voltage [10].

The COOLMOS devices can be used in applications up to power range of 2 kVA such as power supplies for workstations and server, uninterruptible power supplies (UPS), high-voltage converters for microwave and medical systems, induction ovens, and welding equipment. These devices can replace conventional power MOSFETs in all applications in most cases without any circuit adaptation. At switching frequencies above 100 kHz, COOLMOS devices offer a superior current-handling capability such as smallest required chip area at a given current. The devices have the advantage of an intrinsic inverse diode. Any parasitic oscillations, which could cause negative undershoots of the drain-source voltage, are clamped by the diode to a defined value.

4.5 SITs

An SIT is a high-power, high-frequency device. Since the invention of the static induction devices in Japan by J. Nishizawa [17], the number of devices in this family is growing [19]. It is essentially the solid-state version of the triode vacuum tube. The silicon cross section of an SIT [15] and its symbol are shown in Figure 4.27. It is a vertical structure device with short multichannels. Thus, it is not subject to area limitation and is suitable for high-speed, high-power operation. The gate electrodes are buried within the drain and source n-epi layers. An SIT is identical to a JFET except for vertical and buried gate construction,

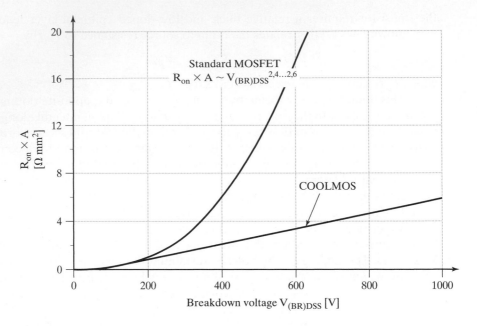

FIGURE 4.26

The linear relationship between blocking voltage and on-resistance. [Ref. 10, G. Deboy]

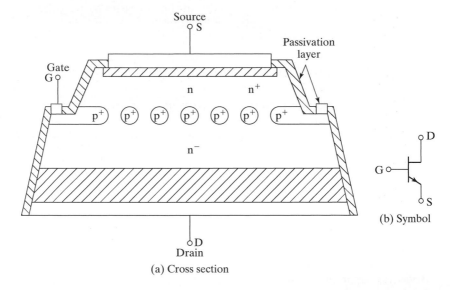

(a) Cross section

(b) Symbol

FIGURE 4.27

Cross section and symbol for SITs.

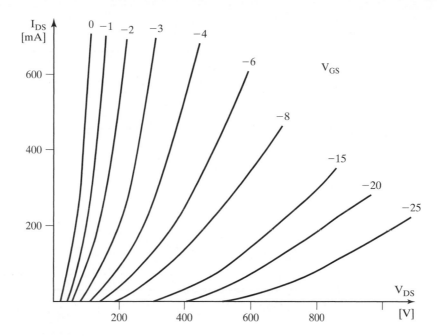

FIGURE 4.28

Typical characteristics of SITs. [Ref. 18, 19]

which gives a lower channel resistance, causing a lower drop. An SIT has a short channel length, low gate series resistance, low gate–source capacitance, and small thermal resistance. It has a low noise, low distortion, and high audiofrequency power capability. The turn-on and turn-off times are very small, typically 0.25 μs.

The on-state drop is high, typically 90 V for a 180-A device and 18 V for an 18-A device. An SIT normally is an on device, and a negative gate voltage holds it off. The normally on-characteristic and the high on-state drop limit its applications for general power conversions. The typical characteristics of SITs are shown in Figure 4.28 [18]. An electrostatically induced potential barrier controls the current in static induction devices. The SITs can operate with the power of 100 KVA at 100 kHz or 10 VA at 10 GHz. The current rating of SITs can be up to 1200 V, 300 A, and the switching speed can be as high as 100 kHz. It is most suitable for high-power, high-frequency applications (e.g., audio, VHF/UHF, and microwave amplifiers).

4.6   IGBTs

An IGBT combines the advantages of BJTs and MOSFETs. An IGBT has high input impedance, like MOSFETs, and low on-state conduction losses, like BJTs. However, there is no second breakdown problem, as with BJTs. By chip design and structure, the equivalent drain-to-source resistance R_{DS} is controlled to behave like that of a BJT [13–14].

The silicon cross section of an IGBT is shown in Figure 4.29a, which is identical to that of an MOSFET except the p^+ substrate. However, the performance of an IGBT is

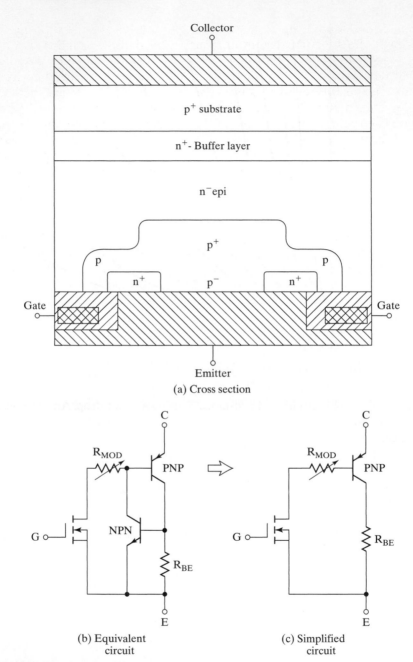

FIGURE 4.29

Cross section and equivalent circuit for IGBTs.

closer to that of a BJT than an MOSFET. This is due to the p^+ substrate, which is responsible for the minority carrier injection into the n-region. The equivalent circuit is shown in Figure 4.29b, which can be simplified to Figure 4.29c. An IGBT is made of four alternate *PNPN* layers, and could latch like a thyristor given the necessary condition: $(\alpha_{npn} + \alpha_{pnp}) > 1$. The n^+-buffer layer and the wide epi base reduce the gain of the *NPN*-terminal by internal design, thereby avoiding latching. IGBTs have two structures of IGBTs: punch-through (PT) and nonpunch through (NPT). In the PT IGBT structure, the switching time is reduced by use of a heavily doped *n*-buffer layer in the drift region near the collector. In the NPT structure, carrier lifetime is kept more than that of a PT structure, which causes conductivity modulation of the drift region and reduces the on-state voltage drop. An IGBT is a voltage-controlled device similar to a power MOSFET. Like an MOSFET, when the gate is made positive with respect to the emitter for turn-on, *n* carriers are drawn into the *p*-channel near the gate region; this results in a forward bias of the base of the *npn*-transistor, which thereby turns on. An IGBT is turned on by just applying a positive gate voltage to open the channel for *n* carriers and is turned off by removing the gate voltage to close the channel. It requires a very simple driver circuit. It has lower switching and conducting losses while sharing many of the appealing features of power MOSFETS, such as ease of gate drive, peak current, capability, and ruggedness. An IGBT is inherently faster than a BJT. However, the switching speed of IGBTs is inferior to that of MOSFETs.

The symbol and circuit of an IGBT switch are shown in Figure 4.30. The three terminals are gate, collector, and emitter instead of gate, drain, and source for an MOSFET. The typical output characteristics of i_C versus v_{CE} are shown in Figure 4.31a for various gate–emitter voltage v_{GE}. The typical transfer characteristic of i_C versus v_{GE} is

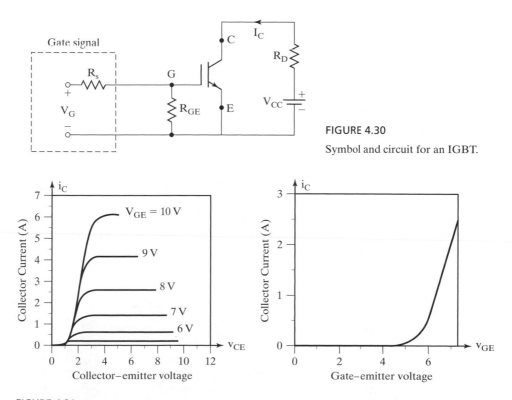

FIGURE 4.30

Symbol and circuit for an IGBT.

FIGURE 4.31

Typical output and transfer characteristics of IGBTs.

shown in Figure 4.31b. The parameters and their symbols are similar to that of MOSFETs, except that the subscripts for source and drain are changed to emitter and collector, respectively. The current rating of a single IGBT can be up to 1200 V, 400 A, and the switching frequency can be up to 20 kHz. IBGTs are finding increasing applications in medium-power applications such as dc and ac motor drives, power supplies, solid-state relays, and contractors.

As the upper limits of commercially available IGBT ratings are increasing (e.g., as high as 6500 V and 2400 A), IGBTs are finding and replacing applications where BJTs and conventional MOSFETs were predominantly used as switches.

4.7   SERIES AND PARALLEL OPERATION

Transistors may be operated in series to increase their voltage-handling capability. It is very important that the series-connected transistors are turned on and off simultaneously. Otherwise, the slowest device at turn-on and the fastest device at turn-off may be subjected to the full voltage of the collector–emitter (or drain–source) circuit and that particular device may be destroyed due to a high voltage. The devices should be matched for gain, transconductance, threshold voltage, on-state voltage, turn-on time, and turn-off time. Even the gate or base drive characteristics should be identical. Voltage-sharing networks similar to diodes could be used.

Transistors are connected in parallel if one device cannot handle the load current demand. For equal current sharings, the transistors should be matched for gain, transconductance, saturation voltage, and turn-on time and turn-off time. In practice, it is not always possible to meet these requirements. A reasonable amount of current sharing (45 to 55% with two transistors) can be obtained by connecting resistors in series with the emitter (or source) terminals, as shown in Figure 4.32.

The resistors in Figure 4.32 help current sharing under steady-state conditions. Current sharing under dynamic conditions can be accomplished by connecting coupled inductors as shown in Figure 4.33. If the current through Q_1 rises, the $L(di/dt)$ across L_1 increases, and a corresponding voltage of opposite polarity is induced across inductor L_2. The result is a low-impedance path, and the current is shifted to Q_2. The inductors would generate voltage spikes and they may be expensive and bulky, especially at high currents.

BJTs have a negative temperature coefficient. During current sharing, if one BJT carries more current, its on-state resistance decreases and its current increases further, whereas MOSFETs have a positive temperature coefficient and parallel operation is

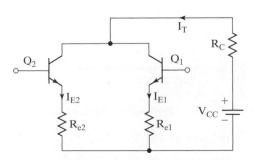

FIGURE 4.32

Parallel connection of transistors.

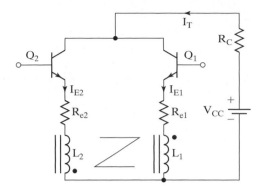

FIGURE 4.33

Dynamic current sharing.

relatively easy. The MOSFET that initially draws higher current heats up faster and its on-state resistance increases, resulting in current shifting to the other devices. IGBTs require special care to match the characteristics due to the variations of the tempera-ture coefficients with the collector current.

Example 4.5   Finding the Current Sharing by Two Parallel MOSFETs

Two MOSFETs that are connected in parallel similar to Figure 4.32 carry a total current of $I_T = 20$ A. The drain-to-source voltage of MOSFET M_1 is $V_{DS1} = 2.5$ V and that of MOSFET M_2 is $V_{DS2} = 3$ V. Determine the drain current of each transistor and difference in current shar-ing if the current sharing series resistances are (a) $R_{s1} = 0.3\ \Omega$ and $R_{s2} = 0.2\ \Omega$, and (b) $R_{s1} = R_{s2} = 0.5\ \Omega$.

Solution

a. $I_{D1} + I_{D2} = I_T$ and $V_{DS1} + I_{D1}R_{S1} = V_{DS2} + I_{D2}R_{S2} = V_{DS2} = R_{S2}(I_T - I_{D1})$.

$$SI_{D1} = \frac{V_{DS2} - V_{DS1} + I_T R_{s2}}{R_{s1} + R_{s2}}$$

$$= \frac{3 - 2.5 + 20 \times 0.2}{0.3 + 0.2} = 9\ \text{A} \quad \text{or} \quad 45\% \tag{4.40}$$

$$I_{D2} = 20 - 9 = 11\ \text{A} \quad \text{or} \quad 55\%$$

$$\Delta I = 55 - 45 = 10\%$$

b. $I_{D1} = \dfrac{3 - 2.5 + 20 \times 0.5}{0.5 + 0.5} = 10.5\ \text{A} \quad \text{or} \quad 52.5\%$

$$I_{D2} = 20 - 10.5 = 9.5\ \text{A} \quad \text{or} \quad 47.5\%$$

$$\Delta I = 52.5 - 47.5 = 5\%$$

4.8   *di/dt* AND *dv/dt* LIMITATIONS

Transistors require certain turn-on and turn-off times. Neglecting the delay time t_d and the storage time t_s, the typical voltage and current waveforms of a BJT switch are

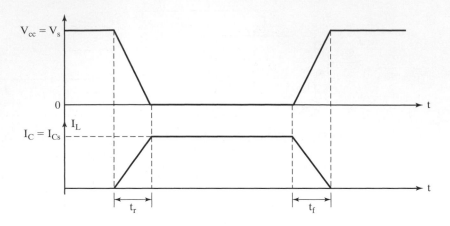

FIGURE 4.34

Voltage and current waveforms.

shown in Figure 4.34. During turn-on, the collector current rises and the *di/dt* is

$$\frac{di}{dt} = \frac{I_L}{t_r} = \frac{I_{cs}}{t_r} \tag{4.41}$$

During turn-off, the collector–emitter voltage must rise in relation to the fall of the collector current, and *dv/dt* is

$$\frac{dv}{dt} = \frac{V_s}{t_f} = \frac{V_{cs}}{t_f} \tag{4.42}$$

The conditions *di/dt* and *dv/dt* in Eqs. (4.41) and (4.42) are set by the transistor switching characteristics and must be satisfied during turn-on and turn-off. Protection circuits are normally required to keep the operating *di/dt* and *dv/dt* within the allowable limits of the transistor. A typical transistor switch with *di/dt* and *dv/dt* protection is shown in Figure 4.35a, with the operating waveforms in Figure 4.35b. The *RC* network across the transistor is known as the *snubber circuit*, or *snubber*, and limits the *dv/dt*. The inductor L_s, which limits the *di/dt*, is sometimes called a *series snubber*.

Let us assume that under steady-state conditions the load current I_L is freewheeling through diode D_m, which has negligible reverse recovery time. When transistor Q_1 is turned on, the collector current rises and current of diode D_m falls, because D_m behaves as short-circuited. The equivalent circuit during turn-on is shown in Figure 4.36a and turn-on *di/dt* is

$$\frac{di}{dt} = \frac{V_s}{L_s} \tag{4.43}$$

Equating Eq. (4.41) to Eq. (4.43) gives the value of L_s,

$$L_s = \frac{V_s t_r}{I_L} \tag{4.44}$$

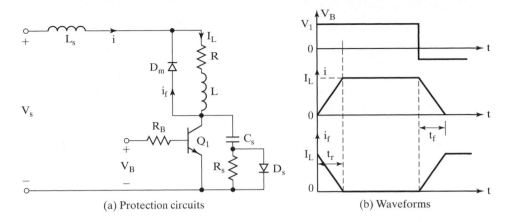

(a) Protection circuits   (b) Waveforms

FIGURE 4.35

Transistor switch with *di/dt* and *dv/dt* protection.

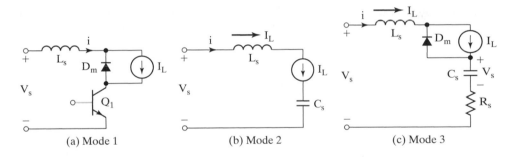

(a) Mode 1   (b) Mode 2   (c) Mode 3

FIGURE 4.36

Equivalent circuits.

During turn-off, the capacitor C_s charges by the load current and the equivalent circuit is shown in Figure 4.36b. The capacitor voltage appears across the transistor and the *dv/dt* is

$$\frac{dv}{dt} = \frac{I_L}{C_s} \tag{4.45}$$

Equating Eq. (4.42) to Eq. (4.45) gives the required value of capacitance,

$$C_s = \frac{I_L t_f}{V_s} \tag{4.46}$$

Once the capacitor is charged to V_s, the freewheeling diode turns on. Due to the energy stored in L_s, there is a damped resonant circuit as shown in Figure 4.36c. The transient analysis of *RLC* circuit is discussed in Section 16.4. The *RLC* circuit is normally

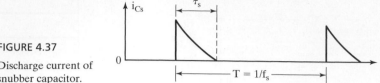

FIGURE 4.37

Discharge current of
snubber capacitor.

made critically damped to avoid oscillations. For unity critical damping, $\delta = 1$, and Eq. (18.11) yields

$$R_s = 2\sqrt{\frac{L_s}{C_s}} \qquad (4.47)$$

The capacitor C_s has to discharge through the transistor and this increases the peak current rating of the transistor. The discharge through the transistor can be avoided by placing resistor R_s across C_s instead of placing R_s across D_s.

The discharge current is shown in Figure 4.37. When choosing the value of R_s, the discharge time, $R_sC_s = \tau_s$ should also be considered. A discharge time of one-third the switching period T_s is usually adequate.

$$3R_sC_s = T_s = \frac{1}{f_s}$$

or

$$R_s = \frac{1}{3f_sC_s} \qquad (4.48)$$

Example 4.6   Finding the Snubber Values for Limiting *dv/dt* and *di/dt* Values of a BJT Switch

A bipolar transistor is operated as a chopper switch at a frequency of $f_s = 10$ kHz. The circuit arrangement is shown in Figure 4.35a. The dc voltage of the chopper is $V_s = 220$ V and the load current is $I_L = 100$ A. $V_{CE(\text{sat})} = 0$ V. The switching times are $t_d = 0, t_r = 3$ μs, and $t_f = 1.2$ μs. Determine the values of (a) L_s; (b) C_s; (c) R_s for critically damped condition; (d) R_s, if the discharge time is limited to one-third of switching period; (e) R_s, if the peak discharge current is limited to 10% of load current; and (f) power loss due to RC snubber P_s, neglecting the effect of inductor L_s on the voltage of the snubber capacitor C_s.

Solution
$I_L = 100$ A, $V_s = 220$ V, $f_s = 10$ kHz, $t_r = 3$ μs, and $t_f = 1.2$ μs.

a. From Eq. (4.44), $L_s = V_s t_r / I_L = 220 \times 3/100 = 6.6$ μH.

b. From Eq. (4.46), $C_s = I_L t_f / V_s = 100 \times 1.2/220 = 0.55$ μF.

c. From Eq. (4.47), $R_s = 2\sqrt{L_s/C_s} = 2\sqrt{6.6/0.55} = 6.93$ Ω.

d. From Eq. (4.48), $R_s = 1/(3f_sC_s) = 10^3/(3 \times 10 \times 0.55) = 60.6$ Ω.

e. $V_s/R_s = 0.1 \times I_L$ or $220/R_s = 0.1 \times 100$ or $R_s = 22$ Ω.

f. The snubber loss, neglecting the loss in diode D_s, is

$$P_s \cong 0.5 C_s V_s^2 f_s \tag{4.49}$$
$$= 0.5 \times 0.55 \times 10^{-6} \times 220^2 \times 10 \times 10^3 = 133.1 \text{ W}$$

4.9 SPICE MODELS

Due to the nonlinear behavior of power electronics circuits, the computer-aided simulation plays an important role in the design and analysis of power electronics circuits and systems. Device manufacturers often provide SPICE models for power devices.

4.9.1 BJT SPICE Model

The PSpice model, which is based on the integral charge-control model of Gummel and Poon [16], is shown in Figure 4.38a. The static (dc) model that is generated by PSpice is shown in Figure 4.38b. If certain parameters are not specified, PSpice assumes the simple model of Ebers–Moll as shown in Figure 4.38c.

The model statement for *NPN*-transistors has the general form

```
.MODEL QNAME NPN (P1=V1 P2=V2 P3=V3 . . . PN=VN)
```

and the general form for *PNP*-transistors is

```
.MODEL QNAME PNP (P1=V1 P2=V2 P3=V3 . . . PN=VN)
```

where QNAME is the name of the BJT model. *NPN* and *PNP* are the type symbols for *NPN*- and *PNP*-transistors, respectively. P1, P2, ... and V1, V2, ... are the parameters and their values, respectively. The parameters that affect the switching behavior of a BJT in power electronics are IS, BF, CJE, CJC, TR, TF. The symbol for a BJT is Q, and its name must start with Q. The general form is

```
Q <name> NC NB NE NS QNAME [(area) value]
```

where NC, NB, NE, and NS are the collector, base, emitter, and substrate nodes, respectively. The substrate node is optional: If not specified, it defaults to ground. Positive current is the current that flows into a terminal. That is, the current flows from the collector node, through the device, to the emitter node for an *NPN*-BJT.

The parameters that significantly influence the switching behavior of a BJT are:

IS	P-N saturation current
BF	Ideal maximum forward beta
CJE	Base-emitter zero-bias *pn* capacitance
CJC	Base-collector zero-bias *pn* capacitance
TR	Ideal reverse transit time
TF	Ideal forward transit time

4.9.2 MOSFET SPICE Model

The PSpice model [16] of an *n*-channel MOSFET is shown in Figure 4.39a. The static (dc) model that is generated by PSpice is shown in Figure 4.39b. The model statement

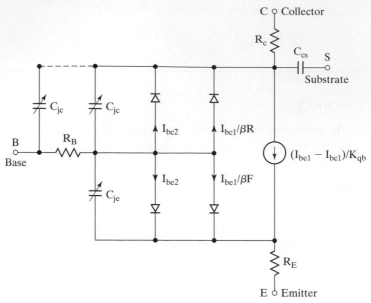

(a) Gummel–Poon model

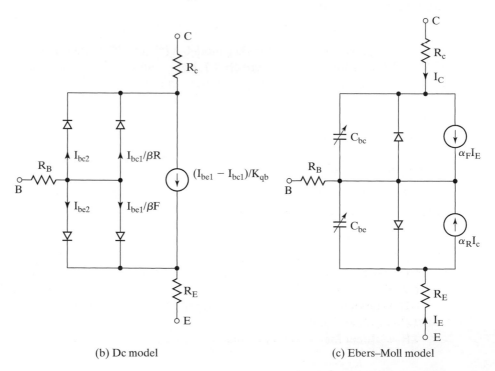

(b) Dc model   (c) Ebers–Moll model

FIGURE 4.38

PSpice BJT model.

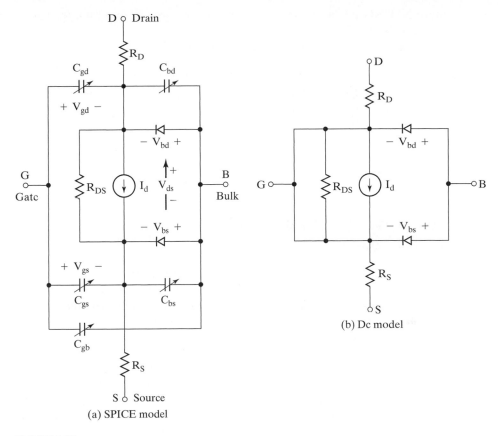

(a) SPICE model

(b) Dc model

FIGURE 4.39

PSpice *n*-channel MOSFET model.

of *n*-channel MOSFETs has the general form

```
.MODEL MNAME NMOS (P1=V1 P2=V2 P3=V3 ... PN=VN)
```

and the statement for *p*-channel MOSFETs has the form

```
.MODEL MNAME PMOS (P1=V1 P2=V2 P3=V3 ... PN=VN)
```

where MNAME is the model name. NMOS and PMOS are the type symbols of *n*-channel and *p*-channel MOSFETs, respectively. The parameters that affect the switching behavior of an MOSFET in power electronics are L, W, VTO, KP, IS, CGSO, and CGDO.

The symbol for an MOSFET is M. The name of MOSFETs must start with M and it takes the general form

```
M<name>   ND    NG    NS    NB    MNAME
+              [L=<value]   [W=<value>]
+              [AD=<value>] [AS=<value>]
+              [PD=<value>] [PS=<value>]
+              [NRD=<value>] [NRS=<value>]
+              [NRG=<value>] [NRB=<value>]
```

where ND, NG, NS, and NB are the drain, gate, source, and bulk (or substrate) nodes, respectively.

The parameters that significantly influence the switching behavior of an MOSFET are:

L	Channel length
W	Channel width
VTO	Zero-bias threshold voltage
IS	Bulk *pn*-saturation current
CGSO	Gate–source overlap capacitance and channel width
CGDO	Gate–drain overlap capacitance and channel width

For COOLMOS, SPICE does not support any models. However, the manufacturers provide models for COOLMOS [11].

4.9.3   IGBT SPICE Model

The *n*-channel IGBT consists of a *pnp*-bipolar transistor that is driven by an *n*-channel MOSFET. Therefore, the IGBT behavior is determined by physics of the bipolar and MOSFET devices. Several effects dominate the static and dynamic device characteristics. The internal circuit of an IGBT is shown in Figure 4.40a.

An IGBT circuit model [16], which relates the currents between terminal nodes as a nonlinear function of component variables and their rate of change, is shown in Figure 4.40b. The capacitance of the emitter–base junction C_{eb} is implicitly defined by the emitter–base voltage as a function of base charge. I_{ceb} is the emitter–base capacitor current that defines the rate of change of the base charge. The current through the collector–emitter redistribution capacitance I_{ccer} is part of the collector current, which in contrast to I_{css} depends on the rate of change of the base-emitter voltage. I_{bss} is part of the base current that does not flow through C_{eb} and does not depend on rate of change of base–collector voltage.

There are two main ways to model IGBT in SPICE: (1) composite model and (2) equation model. The composite model connects the existing SPICE *pnp*-BJT and *n*-channel MOSFET models. The equivalent circuit of the composite model is shown in Figure 4.41a. It connects the existing BIT and MOSFET models of PSpice in a Darlington configuration and uses the built-in equations of the two. The model computes quickly and reliably, but it does not model the behavior of the IGBT accurately.

The equation model [22, 23] implements the physics-based equations and models the internal carrier and charge to simulate the circuit behavior of the IGBT accurately. This model is complicated, often unreliable, and computationally slow because the equations are derived from the complex semiconductor physics theory. Simulation times can be over 10 times longer than those for the composite model.

There are numerous papers of SPICE modeling of IGBTs and Sheng [24] compares the merits and limitations of various models. Figure 4.41b shows the equivalent circuit of Sheng's model [21] that adds a current source from the drain to the gate. It has been found that the major inaccuracy in dynamic electrical properties is associated with the modeling of the drain to gate capacitance of the *n*-channel MOSFET. During high-voltage switching, the drain-to-gate capacitance C_{dg} changes by two orders of

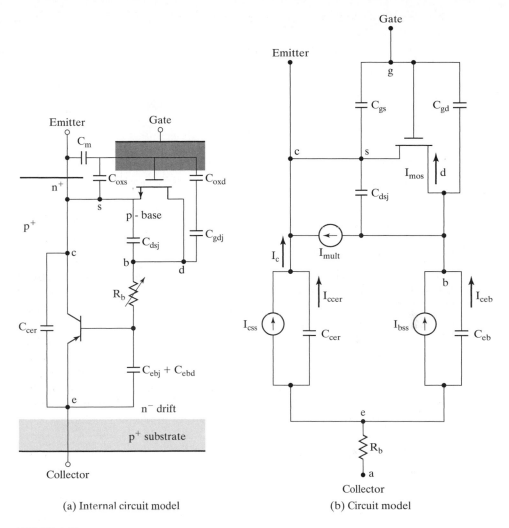

(a) Internal circuit model (b) Circuit model

FIGURE 4.40

IGBT model. [Ref. 16, K. Shenai]

magnitude due to any changes in drain-to-gate voltage V_{dg}. This is, C_{dg} is expressed by

$$C_{dg} = \frac{\epsilon_{si} C_{oxd}}{\sqrt{\dfrac{2\epsilon_{si} V_{dg}}{q N_B}}\, C_{oxd} + A_{dg}\epsilon_{si}}$$

where A_{dg} is the area of the gate over the base;
ϵ_{si} is the dielectric constant of silicon;
C_{oxd} is the gate–drain overlap oxide capacitance;
q is the electron charge;
N_B is the base doping density.

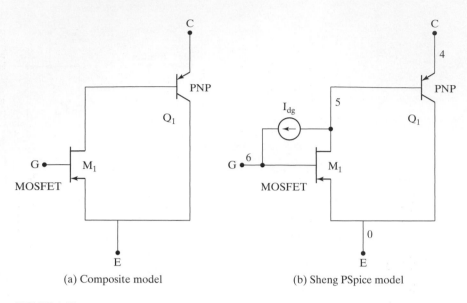

FIGURE 4.41

Equivalent circuits of IGBT SPICE models. [Ref. 21, K. Sheng]

PSpice does not incorporate a capacitance model involving the square root, which models the space charge layer variation for a step junction. PSpice model can implement the equations describing the highly nonlinear gate–drain capacitance into the composite model by using the analog behavioral modeling function of PSpice.

4.10   COMPARISONS OF TRANSISTORS

Table 4.1 shows the comparisons of BJTs, MOSFETs, and IGBTs.

SUMMARY

Power transistors are generally of five types: BJTs, MOSFETs, SITs, IGBTs, and COOLMOS. BJTs are current-controlled devices and their parameters are sensitive to junction temperature. BJTs suffer from second breakdown and require reverse base current during turn-off to reduce the storage time, but they have low on-state or saturation voltage.

MOSFETs are voltage-controlled devices and require very low gating power and their parameters are less sensitive to junction temperature. There is no second breakdown problem and no need for negative gate voltage during turn-off. The conduction losses of COOLMOS devices is reduced by a factor of five as compared with those of the conventional technology. It is capable to handle two to three times more output power as compared with that of a standard MOSFET of the same package. IGBTs, which combine the advantages of BJTs and MOSFETs, are voltage-controlled devices and have low on-state voltage similar to BJTs. COOLMOS, which has very low on-state loss, is used in high-efficiency, low-power applications. IGBTs have no second

TABLE 4.1   Comparisons of Power Transistors

Switch Type	Base/Gate Control Variable	Control Characteristic	Switching Frequency	On-State Voltage Drop	Max. Voltage Rating V_s	Max. Current Rating I_s	Advantages	Limitations
BJT	Current	Continuous	Medium 20 kHz	Low	1.5 kV $S_s = V_s I_s$ = 1.5 MVA	1 kA $S_s = V_s I_s$ = 1.5 MVA	Simple switch Low on-state drop Higher off-state voltage capability High switching loss	Current controlled device and requires a higher base current to turn-on and sustain on-state current Base drive power loss Charge recovery time and slower switching speed Secondary breakdown region High switching losses Unipolar voltage device
MOSFET	Voltage	Continuous	Very high	High	1 kV $S_s = V_s I_s$ = 0.1 MVA	150 A $S_s = V_s I_s$ = 0.1 MVA	Higher switching speed Low switching loss Simple gate drive circuit Little gate power. Negative temperature coefficient on rain current and facilitates parallel operation	High on-state drop as high as 10 V Lower off-state voltage capability Unipolar voltage device
COOLMOS	Voltage	Continuous	Very high	Low	1 kV	100 A	Low gate drive requirement and low on-state power drop	Low-power device Low voltage and current ratings
IGBT	Voltage	Continuous	High	Medium	3.5 kV $S_s = V_s I_s$ = 1.5 MVA	2 kA $S_s = V_s I_s$ = 1.5 MVA	Low on-state voltage Little gate power	Lower off-state voltage capability Unipolar voltage device
SIT	Voltage	Continuous	Very high	High	$S_s = V_s I_s$ = 1.5 MVA		High-voltage rating	Higher on-state voltage drop Lower current ratings

breakdown phenomena. SITs are high-power, high-frequency devices. They are most suitable for audio, VHF/UHF, and microwave amplifiers. They have a normally on-characteristic and a high on-state drop.

Transistors can be connected in series or parallel. Parallel operation usually requires current-sharing elements. Series operation requires matching of parameters, especially during turn-on and turn-off. To maintain the voltage and current relationship of transistors during turn-on and turn-off, it is generally necessary to use snubber circuits to limit the *di/dt* and *dv/dt*.

The gate signals can be isolated from the power circuit by pulse transformers or optocouplers. The pulse transformers are simple, but the leakage inductance should be very small. The transformers may be saturated at a low frequency and a long pulse. Optocouplers require separate power supply.

REFERENCES

[1]   B. J. Baliga, *Power Semiconductor Devices*. Boston, MA: PWS Publishing. 1996.

[2]   S. K. Ghandi, *Semiconductor Power Devices*. New York: John Wiley & Sons. 1977.

[3]   S. M. Sze, *Modern Semiconductor Device Physics*. New York: John Wiley & Sons. 1998.

[4]   B. I. Baliga and D. Y. Chen, *Power Transistors: Device Design and Applications*. New York: IEEE Press. 1984.

[5]   Westinghouse Electric, *Silicon Power Transistor Handbook*. Pittsburgh: Westinghouse Electric Corp. 1967.

[6]   R. Severns and J. Armijos, *MOSPOWER Application Handbook*. Santa Clara, CA: Siliconix Corp. 1984.

[7]   S. Clemente and B. R. Pelly, "Understanding power MOSFET switching performance," *Solid-State Electronics,* Vol. 12, No. 12, 1982, pp. 1133–1141.

[8]   D. A. Grant and I. Gower, *Power MOSFETs: Theory and Applications*. New York: John Wiley & Sons. 1988.

[9]   L. Lorenz, G. Deboy, A. Knapp, and M. Marz, "COOLMOS™—a new milestone in high voltage power MOS," *Proc. ISPSD 99*, Toronto, 1999, pp. 3–10.

[10] G. Deboy, M. Marz, J. P. Stengl, H. Strack, J. Tilhanyi, and H. Weber, "A new generation of high voltage MOSFETs breaks the limit of silicon," *Proc. IEDM 98*, San Francisco, 1998, pp. 683–685.

[11] Infineon Technologies, *CoolMOS™: Power Semiconductors*. Germany: Siemens. 2001. www.infineon.co

[12] C. Hu, "Optimum doping profile for minimum ohmic resistance and high breakdown voltage," *IEEE Transactions on Electronic Devices*, Vol. ED-26, No. 3, 1979.

[13] B. J. Baliga, M. Cheng, P. Shafer, and M. W. Smith, "The insulated gate transistor (IGT): a new power switching device," IEEE Industry Applications Society Conference Record, 1983, pp. 354–363.

[14] B. J. Baliga, M. S. Adler, R. P. Love, P. V. Gray, and N. Zommer, "The insulated gate transistor: a new three-terminal MOS controlled bipolar power device," *IEEE Transactions Electron Devices*, ED-31, 1984, pp. 821–828.

[15] *IGBT Designer's Manual*. El Segundo, CA. International Rectifier, 1991.

[16] K. Shenai, *Power Electronics Handbook*, edited by M. H. Rashid. Los Angeles, CA: Academic Press. 2001, Chapter 7.

[17] I. Nishizawa and K. Yamamoto, "High-frequency high-power static induction transistor," *IEEE Transactions on Electron Devices,* Vol. ED25, No. 3, 1978, pp. 314–322.

[18] J. Nishizawa, T. Terasaki, and J. Shibata, "Field-effect transistor versus analog transistor (static induction transistor)," *IEEE Transactions on Electron Devices*, Vol. 22, No. 4, April 1975, pp. 185–197.

[19] B. M. Wilamowski, *Power Electronics Handbook*, edited by M. H. Rashid. Los Angeles, CA: Academic Press. 2001, Chapter 9.

[20] M. H. Rashid, *SPICE for Power Electronics and Electric Power*. Englewood Cliffs, NJ: Prentice-Hall. 1993.

[21] K. Sheng, S. J. Finney, and B. W. Williams, "Fast and accurate IGBT model for PSpice," *Electronics Letters*, Vol. 32, No. 25, December 5, 1996, pp. 2294–2295.

[22] A. G. M. Strollo, "A new IGBT circuit model for SPICE simulation," *Power Electronics Specialists Conference*, June 1997, Vol. 1, pp. 133–138.

[23] K. Sheng, S. J. Finney, and B. W. Williams, "A new analytical IGBT model with improved electrical characteristics," *IEEE Transactions on Power Electronics*, Vol. 14, No. 1, January 1999, pp. 98–107.

[24] K. Sheng, B. W. Williams, and S. J. Finney, "A review of IGBT models," *IEEE Transactions on Power Electronics*, Vol. 15, No. 6, November 2000, pp. 1250–1266.

[25] A. R. Hefner, "An investigation of the drive circuit requirements for the power insulated gate bipolar transistor (IGBT)," *IEEE Transactions on Power Electronics*, Vol. 6, 1991, pp. 208–219.

[26] C. Licitra, S. Musumeci, A. Raciti, A. U. Galluzzo, and R. Letor, "A new driving circuit for IGBT devices," *IEEE Transactions Power Electronics*, Vol. 10, 1995, pp. 373–378.

[27] H. G. Lee, Y. H. Lee, B. S. Suh, and J. W. Lee, "A new intelligent gate control scheme to drive and protect high power IGBTs," *European Power Electronics Conference Records*, 1997, pp. 1.400–1.405.

[28] S. Bernet, "Recent developments of high power converters for industry and traction applications," *IEEE Transactions on Power Electronics*, Vol. 15, No. 6, November 2000, pp. 1102–1117.

REVIEW QUESTIONS

4.1 What is a bipolar transistor (BJT)?

4.2 What are the types of BJTs?

4.3 What are the differences between *NPN*-transistors and *PNP*-transistors?

4.4 What are the input characteristics of *NPN*-transistors?

4.5 What are the output characteristics of *NPN*-transistors?

4.6 What are the three regions of operation for BJTs?

4.7 What is a beta (β) of BJTs?

4.8 What is the difference between beta, β, and forced beta, β_F of BJTs?

4.9 What is a transductance of BJTs?

4.10 What is an overdrive factor of BJTs?

4.11 What is the switching model of BJTs?

4.12 What is the cause of delay time in BJTs?

4.13 What is the cause of storage time in BJTs?

4.14 What is the cause of rise time in BJTs?

4.15 What is the cause of fall time in BJTs?

4.16 What is a saturation mode of BJTs?

4.17 What is a turn-on time of BJTs?

4.18 What is a turn-off time of BJTs?

4.19 What is a FBSOA of BJTs?

4.20 What is a RBSOA of BJTs?

4.21 Why is it necessary to reverse bias BJTs during turn-off?

4.22 What is a second breakdown of BJTs?

4.23 What are the advantages and disadvantages of BJTs?

4.24 What is an MOSFET?

4.25 What are the types of MOSFETs?

4.26 What are the differences between enhancement-type MOSFETs and depletion-type MOSFETs?

4.27 What is a pinch-off voltage of MOSFETs?

4.28 What is a threshold voltage of MOSFETs?

4.29 What is a transconductance of MOSFETs?

4.30 What is the switching model of n-channel MOSFETs?

4.31 What are the transfer characteristics of MOSFETs?

4.32 What are the output characteristics of MOSFETs?

4.33 What are the advantages and disadvantages of MOSFETs?

4.34 Why do the MOSFETs not require negative gate voltage during turn-off?

4.35 Why does the concept of saturation differ in BJTs and MOSFETs?

4.36 What is a turn-on time of MOSFETs?

4.37 What is a turn-off time of MOSFETs?

4.38 What is an SIT?

4.39 What are the advantages of SITs?

4.40 What are the disadvantages of SITs?

4.41 What is an IGBT?

4.42 What are the transfer characteristics of IGBTs?

4.43 What are the output characteristics of IGBTs?

4.44 What are the advantages and disadvantages of IGBTs?

4.45 What are the main differences between MOSFETs and BJTs?

4.46 What are the problems of parallel operation of BJTs?

4.47 What are the problems of parallel operation of MOSFETs?

4.48 What are the problems of parallel operation of IGBTs?

4.49 What are the problems of series operation of BJTs?

4.50 What are the problems of series operations of MOSFETs?

4.51 What are the problems of series operations of IGBTs?

4.52 What are the purposes of shunt snubber in transistors?

4.53 What is the purpose of series snubber in transistors?

PROBLEMS

4.1 The beta (β) of bipolar transistor in Figure 4.7 varies from 10 to 60. The load resistance is $R_C = 5\ \Omega$. The dc supply voltage is $V_{CC} = 100$ V and the input voltage to the base circuit is $V_B = 8$ V. If $V_{CE(\text{sat})} = 2.5$ V and $V_{BE(\text{sat})} = 1.75$ V, find **(a)** the value of R_B that will result in saturation with an overdrive factor of 20; **(b)** the forced β, and **(c)** the power loss in the transistor P_T.

4.2 The beta (β) of bipolar transistor in Figure 4.7 varies from 12 to 75. The load resistance is $R_C = 1.5\ \Omega$. The dc supply voltage is $V_{CC} = 40$ V and the input voltage to the base circuit is $V_B = 6$ V. If $V_{CE(\text{sat})} = 1.2$ V, $V_{BE(\text{sat})} = 1.6$ V, and $R_B = 0.7\ \Omega$, determine **(a)** the ODF, **(b)** the forced β, and **(c)** the power loss in the transistor P_T.

4.3 A transistor is used as a switch and the waveforms are shown in Figure 4.11. The parameters are $V_{CC} = 200$ V, $V_{BE(\text{sat})} = 3$ V, $I_B = 8$ A, $V_{CE(\text{sat})} = 2$ V, $I_{CS} = 100$ A, $t_d = 0.5$ μs, $t_r = 1$ μs, $t_s = 5$ μs, $t_f = 3$ μs, and $f_s = 10$ kHz. The duty cycle is $k = 50\%$. The collector–emitter leakage current is $I_{CEO} = 3$ mA. Determine the power loss due to the collector current **(a)** during turn-on $t_{\text{on}} = t_d + t_r$; **(b)** during conduction period t_n, **(c)** during turn-off $t_{\text{off}} = t_s + t_f$, **(d)** during off-time t_o, and **(e)** the total average power losses P_T. **(f)** Plot the instantaneous power due to the collector current $P_c(t)$.

4.4 The maximum junction temperature of the bipolar transistor in Problem 4.3 is $T_j = 150°C$ and the ambient temperature is $T_A = 25°C$. If the thermal resistances are $R_{JC} = 0.4°C/W$ and $R_{CS} = 0.05°C/W$, calculate the thermal resistance of heat sink R_{SA}. (*Hint:* Neglect the power loss due to base drive.)

4.5 For the parameters in Problem 4.3, calculate the average power loss due to the base current P_B.

4.6 Repeat Problem 4.3 if $V_{BE(\text{sat})} = 2.3$ V, $I_B = 8$ A, $V_{CE(\text{sat})} = 1.4$ V, $t_d = 0.1$ μs, $t_r = 0.45$ μs, $t_s = 3.2$ μs, and $t_f = 1.1$ μs.

4.7 An MOSFET is used as a switch. The parameters are $V_{DD} = 40$ V, $I_D = 35$ A, $R_{DS} = 28$ mΩ, $V_{GS} = 10$ V, $t_{d(\text{on})} = 25$ ns, $t_r = 60$ ns, $t_{d(\text{off})} = 70$ ns, $t_f = 25$ ns, and $f_s = 20$ kHz. The drain source leakage current is $I_{DSS} = 250$ μA. The duty cycle is $k = 60\%$. Determine the power loss due to the drain current **(a)** during turn-on $t_{\text{on}} = t_{d(n)} + t_r$; **(b)** during conduction period t_n, **(c)** during turn-off $t_{\text{off}} = t_{d(\text{off})} + t_f$, **(d)** during off-time t_o, and **(e)** the total average power losses P_T.

4.8 The maximum junction temperature of the MOSFET in Problem 4.7 is $T_j = 150°C$ and the ambient temperature is $T_A = 30°C$. If the thermal resistances are $R_{JC} = 1$ K/W and $R_{CS} = 1$ K/W, calculate the thermal resistance of the heat sink R_{SA}. (*Note:* K = °C + 273.)

4.9 Two BJTs are connected in parallel similar to Figure 4.32. The total load current of $I_T = 200$ A. The collector–emitter voltage of transistor Q_1 is $V_{CE1} = 1.5$ V and that of transistor Q_2 is $V_{CE2} = 1.1$ V. Determine the collector current of each transistor and difference in current sharing if the current sharing series resistances are **(a)** $R_{e1} = 10$ mΩ and $R_{e2} = 20$ mΩ, and **(b)** $R_{e1} = R_{e2} = 20$ mΩ.

4.10 A bipolar transistor is operated as a chopper switch at a frequency of $f_s = 20$ kHz. The circuit arrangement is shown in Figure 4.35a. The dc input voltage of the chopper is $V_s = 400$ V and the load current is $I_L = 100$ A. The switching times are $t_r = 1$ μs and $t_f = 3$ μs. Determine the values of **(a)** L_s; **(b)** C_s; **(c)** R_s for critically damped condition; **(d)** R_s if the discharge time is limited to one-third of switching period; **(e)** R_s if peak discharge current is limited to 5% of load current; and **(f)** power loss due to RC snubber P_s, neglecting the effect of inductor L_s on the voltage of snubber capacitor C_s. Assume that $V_{CE(\text{sat})} = 0$.

4.11 An MOSFET is operated as a chopper switch at a frequency of $f_s = 50$ kHz. The circuit arrangement is shown in Figure 4.35a. The dc input voltage of the chopper is $V_s = 30$ V and the load current is $I_L = 40$ A. The switching times are $t_r = 60$ ns and $t_f = 25$ ns. Determine the values of **(a)** L_s; **(b)** C_s; **(c)** R_s for critically damped condition; **(d)** R_s if the discharge time is limited to one-third of switching period; **(e)** R_s if peak discharge current is limited to 5% of load current; and **(f)** power loss due to RC snubber P_s, neglecting the effect of inductor L_s on the voltage of snubber capacitor C_s. Assume that $V_{CE(\text{sat})} = 0$.

Dc–Dc Converters

The learning objectives of this chapter are as follows:

- To learn the switching technique for dc–dc conversion and the types of dc–dc converters
- To study the operation of dc–dc converters
- To understand the performance parameters of dc converters
- To learn the techniques for the analysis and design of dc converters
- To learn the techniques for simulating dc converters by using SPICE
- To study effects of load inductance on the load current and the conditions for continuous current

5.1   INTRODUCTION

In many industrial applications, it is required to convert a fixed-voltage dc source into a variable-voltage dc source. A dc–dc converter converts directly from dc to dc and is simply known as a dc converter. A dc converter can be considered as dc equivalent to an ac transformer with a continuously variable turns ratio. Like a transformer, it can be used to step down or step up a dc voltage source.

Dc converters are widely used for traction motor control in electric automobiles, trolley cars, marine hoists, forklift trucks, and mine haulers. They provide smooth acceleration control, high efficiency, and fast dynamic response. Dc converters can be used in regenerative braking of dc motors to return energy back into the supply, and this feature results in energy savings for transportation systems with frequent stops. Dc converters are used in dc voltage regulators; and also are used, in conjunction with an inductor, to generate a dc current source, especially for the current source inverter.

5.2   PRINCIPLE OF STEP-DOWN OPERATION

The principle of operation can be explained by Figure 5.1a. When switch SW, known as the chopper, is closed for a time t_1, the input voltage V_s appears across the load. If the switch remains off for a time t_2, the voltage across the load is zero. The waveforms for the output voltage and load current are also shown in Figure 5.1b. The converter switch can be implemented by using a (1) power bipolar junction transistor (BJT), (2) power metal oxide semiconductor field-effect transistor (MOSFET), (3) gate-turn-off thyristor (GTO), or (4) insulated-gate bipolar transistor (IGBT). The

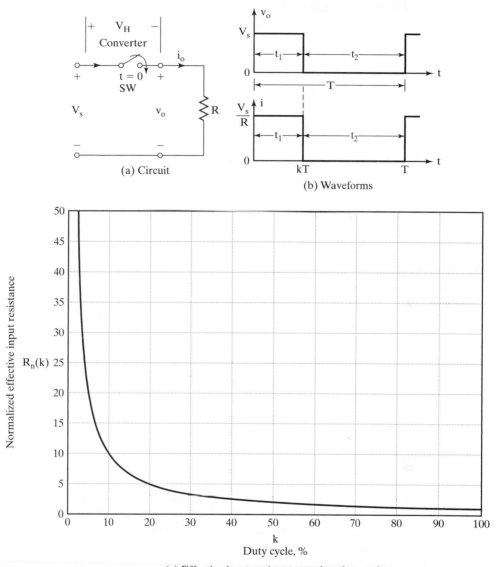

(a) Circuit

(b) Waveforms

(c) Effective input resistance against duty cycle

FIGURE 5.1

Step-down converter with resistive load.

practical devices have a finite voltage drop ranging from 0.5 to 2 V, and for the sake of simplicity we shall neglect the voltage drops of these power semiconductor devices.

The average output voltage is given by

$$V_a = \frac{1}{T} \int_0^{t_1} v_0 \, dt = \frac{t_1}{T} V_s = f t_1 V_s = k V_s \tag{5.1}$$

and the average load current, $I_a = V_a/R = kV_s/R$,

where T is the chopping period;
$k = t_1/T$ is the duty cycle of chopper;
f is the chopping frequency.

The rms value of output voltage is found from

$$V_o = \left(\frac{1}{T} \int_0^{kT} v_0^2 \, dt \right)^{1/2} = \sqrt{k} \, V_s \tag{5.2}$$

Assuming a lossless converter, the input power to the converter is the same as the output power and is given by

$$P_i = \frac{1}{T} \int_0^{kT} v_0 i \, dt = \frac{1}{T} \int_0^{kT} \frac{v_0^2}{R} \, dt = k \frac{V_s^2}{R} \tag{5.3}$$

The effective input resistance seen by the source is

$$R_i = \frac{V_s}{I_a} = \frac{V_s}{kV_s/R} = \frac{R}{k} \tag{5.4}$$

which indicates that the converter makes the input resistance R_i as a variable resistance of R/k. The variation of the normalized input resistance against the duty cycle is shown in Figure 5.1c. It should be noted that the switch in Figure 5.1 could be implemented by a BJT, an MOSFET, an IGBT, or a GTO.

The duty cycle k can be varied from 0 to 1 by varying t_1, T, or f. Therefore, the output voltage V_o can be varied from 0 to V_s by controlling k, and the power flow can be controlled.

1. *Constant-frequency operation*: The converter, or switching, frequency f (or chopping period T) is kept constant and the on-time t_1 is varied. The width of the pulse is varied and this type of control is known as *pulse-width-modulation* (PWM) control.

2. *Variable-frequency operation*: The chopping, or switching, frequency f is varied. Either on-time t_1 or off-time t_2 is kept constant. This is called *frequency modulation*. The frequency has to be varied over a wide range to obtain the full output voltage range. This type of control would generate harmonics at unpredictable frequencies and the filter design would be difficult.

Example 5.1   Finding the Performances of a Dc–Dc Converter

The dc converter in Figure 5.1a has a resistive load of $R = 10 \, \Omega$ and the input voltage is $V_s = 220$ V. When the converter switch remains on, its voltage drop is $v_{ch} = 2$ V and the chopping frequency is $f = 1$ kHz. If the duty cycle is 50%, determine (a) the average output voltage V_a, (b) the rms output voltage V_o, (c) the converter efficiency, (d) the effective input resistance R_i of the converter, and (e) the rms value of the fundamental component of output harmonic voltage.

Solution
$V_s = 220$ V, $k = 0.5$, $R = 10 \, \Omega$, and $v_{ch} = 2$ V.

    a. From Eq. (5.1), $V_a = 0.5 \times (220 - 2) = 109$ V.
    b. From Eq. (5.2), $V_o = \sqrt{0.5} \times (220 - 2) = 154.15$ V.

c.   The output power can be found from

$$P_o = \frac{1}{T}\int_0^{kT}\frac{v_o^2}{R}\,dt = \frac{1}{T}\int_0^{kT}\frac{(V_s - v_{ch})^2}{R}\,dt = k\frac{(V_s - v_{ch})^2}{R}$$

(5.5)

$$= 0.5 \times \frac{(220 - 2)^2}{10} = 2376.2 \text{ W}$$

The input power to the converter can be found from

$$P_i = \frac{1}{T}\int_0^{kT}V_s i\,dt = \frac{1}{T}\int_0^{kT}\frac{V_s(V_s - v_{ch})}{R}\,dt = k\frac{V_s(V_s - v_{ch})}{R}$$

(5.6)

$$= 0.5 \times 220 \times \frac{220 - 2}{10} = 2398 \text{ W}$$

The converter efficiency is

$$\frac{P_o}{P_i} = \frac{2376.2}{2398} = 99.09\%$$

d.   From Eq. (5.4), $R_i = 10/0.5 = 20 \ \Omega$.

e.   The output voltage as shown in Figure 5.1b can be expressed in a Fourier series as

$$v_o(t) = kV_s + \frac{V_s}{n\pi}\sum_{n=1}^{\infty}\sin 2n\pi k \cos 2n\pi ft$$

$$+ \frac{V_s}{n\pi}\sum_{n=1}^{\infty}(1 - \cos 2n\pi k)\sin 2n\pi ft$$

(5.7)

The fundamental component (for $n = 1$) of output voltage harmonic can be determined from Eq. (5.7) as

$$v_1(t) = \frac{V_s}{\pi}[\sin 2\pi k \cos 2\pi ft + (1 - \cos 2\pi k)\sin 2\pi ft]$$

(5.8)

$$= \frac{220 \times 2}{\pi}\sin(2\pi \times 1000t) = 140.06 \sin(6283.2t)$$

and its root-mean-square (rms) value is $V_1 = 140.06/\sqrt{2} = 99.04$ V.

Note: The efficiency calculation, which includes the conduction loss of the converter, does not take into account the switching loss due to turn-on and turn-off of practical converters. The efficiency of a practical converter varies between 92 and 99%.

Key Points of Section 5.2

- A step-down chopper, or dc converter, that acts as a variable resistance load can produce an output voltage from 0 to V_S.

- Although a dc converter can be operated either at a fixed or variable frequency, it is usually operated at a fixed frequency with a variable duty cycle.
- The output voltage contains harmonics and a dc filter is needed to smooth out the ripples.

5.2.1 Generation of Duty Cycle

The duty cycle k can be generated by comparing a dc reference signal v_r with a saw-tooth carrier signal v_{cr}. This is shown in Figure 5.2, where V_r is the peak value of v_r, and V_{cr} is the peak value of v_{cr}. The reference signal v_r is given by

$$v_r = \frac{V_r}{T}t \tag{5.9}$$

which must equal to the carrier signal $v_{cr} = V_{cr} = kT$. That is,

$$V_{cr} = \frac{V_r}{T}kT$$

which gives the duty cycle duty k as

$$k = \frac{V_{cr}}{V_r} = M \tag{5.10}$$

where M is called the *modulation index*. By varying the carrier signal v_{cr} from 0 to V_{cr}, the duty cycle k can be varied from 0 to 1.

The algorithm to generate the gating signal is as follows:

1. Generate a triangular waveform of period T as the reference signal v_r and a dc carrier signal v_{cr}.
2. Compare these signals by a comparator to generate the difference $v_c - v_{cr}$ and then a hard limiter to obtain a square-wave gate pulse of width kT, which must be applied to the switching device through an isolating circuit.
3. Any variation in v_{cr} varies linearly with the duty cycle k.

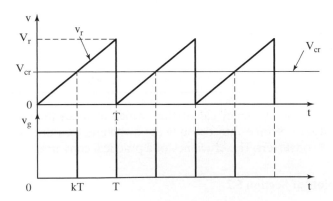

FIGURE 5.2

Comparing a reference signal with a carrier signal.

5.3   STEP-DOWN CONVERTER WITH *RL* LOAD

A converter [1] with an *RL* load is shown in Figure 5.3. The operation of the converter can be divided into two modes. During mode 1, the converter is switched on and the current flows from the supply to the load. During mode 2, the converter is switched off and the load current continues to flow through freewheeling diode D_m. The equivalent circuits for these modes are shown in Figure 5.4a. The load current and output voltage waveforms are shown in Figure 5.4b with the assumption that the load current rises linearly. However, the current flowing through an *RL* load rises or falls exponentially with a time constant. The load time constant ($\tau = L/R$) is generally much higher than the switching period *T*. Thus, the linear approximation is valid for many circuit conditions and simplified expressions can be derived within reasonable accuracies.

The load current for mode 1 can be found from

$$V_s = Ri_1 + L\frac{di_1}{dt} + E$$

which with initial current $i_1(t = 0) = I_1$ gives the load current as

$$i_1(t) = I_1 e^{-tR/L} + \frac{V_s - E}{R}(1 - e^{-tR/L}) \tag{5.11}$$

This mode is valid $0 \le t \le t_1 \, (= kT)$; and at the end of this mode, the load current becomes

$$i_1(t = t_1 = kT) = I_2 \tag{5.12}$$

The load current for mode 2 can be found from

$$0 = Ri_2 + L\frac{di_2}{dt} + E$$

With initial current $i_2(t = 0) = I_2$ and redefining the time origin (i.e., $t = 0$) at the beginning of mode 2, we have

$$i_2(t) = I_2 e^{-tR/L} - \frac{E}{R}(1 - e^{-tR/L}) \tag{5.13}$$

This mode is valid for $0 \le t \le t_2 \, [= (1 - k)T]$. At the end of this mode, the load current becomes

$$i_2(t = t_2) = I_3 \tag{5.14}$$

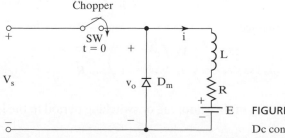

FIGURE 5.3

Dc converter with *RL* loads.

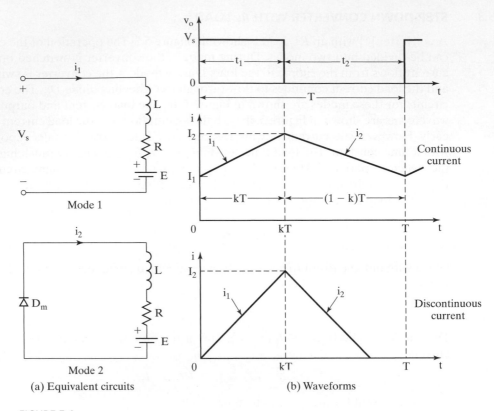

FIGURE 5.4

Equivalent circuits and waveforms for *RL* loads.

At the end of mode 2, the converter is turned on again in the next cycle after time, $T = 1/f = t_1 + t_2$.

Under steady-state conditions, $I_1 = I_3$. The peak-to-peak load ripple current can be determined from Eqs. (5.11) to (5.14). From Eqs. (5.11) and (5.12), I_2 is given by

$$I_2 = I_1 e^{-kTR/L} + \frac{V_s - E}{R}(1 - e^{-kTR/L}) \qquad (5.15)$$

From Eqs. (5.13) and (5.14), I_3 is given by

$$I_3 = I_1 = I_2 e^{-(1-k)TR/L} - \frac{E}{R}(1 - e^{-(1-k)TR/L}) \qquad (5.16)$$

Solving for I_1 and I_2, we get

$$I_1 = \frac{V_S}{R}\left(\frac{e^{kz} - 1}{e^z - 1}\right) - \frac{E}{R} \qquad (5.17)$$

where $z = \dfrac{TR}{L}$ is the ratio of the chopping or switching period to the load time constant.

$$I_2 = \frac{V_s}{R}\left(\frac{e^{-kz}-1}{e^{-z}-1}\right) - \frac{E}{R} \tag{5.18}$$

The peak-to-peak ripple current is

$$\Delta I = I_2 - I_1$$

which after simplifications becomes

$$\Delta I = \frac{V_s}{R}\frac{1 - e^{-kz} + e^{-z} - e^{-(1-k)z}}{1 - e^{-z}} \tag{5.19}$$

The condition for maximum ripple,

$$\frac{d(\Delta I)}{dk} = 0 \tag{5.20}$$

gives $e^{-kz} - e^{-(1-k)z} = 0$ or $-k = -(1-k)$ or $k = 0.5$. The maximum peak-to-peak ripple current (at $k = 0.5$) is

$$\Delta I_{max} = \frac{V_s}{R}\tanh\frac{R}{4fL} \tag{5.21}$$

For $4fL \gg R$, $\tanh\theta \approx \theta$ and the maximum ripple current can be approximated to

$$\Delta I_{max} = \frac{V_s}{4fL} \tag{5.22}$$

Note: Equations (5.11) to (5.22) are valid only for continuous current flow. For a large off-time, particularly at low-frequency and low-output voltage, the load current may be discontinuous. The load current would be continuous if $L/R \gg T$ or $Lf \gg R$. In case of discontinuous load current, $I_1 = 0$ and Eq. (5.11) becomes

$$i_1(t) = \frac{V_s - E}{R}(1 - e^{-tR/L})$$

and Eq. (5.13) is valid for $0 \leq t \leq t_2$ such that $i_2(t = t_2) = I_3 = I_1 = 0$, which gives

$$t_2 = \frac{L}{R}\ln\left(1 + \frac{RI_2}{E}\right)$$

Because $t = kT$, we get

$$i_1(t) = I_2 = \frac{V_s - E}{R}\left(1 - e^{-kz}\right)$$

which after substituting for I_2 becomes

$$t_2 = \frac{L}{R}\ln\left[1 + \left(\frac{V_s - E}{E}\right)\left(1 - e^{-z}\right)\right]$$

Condition for continuous current: For $I_1 \geq 0$, Eq. (5.17) gives

$$\left(\frac{e^{kz} - 1}{e^z - 1} - \frac{E}{V_s} \right) \geq 0$$

which gives the value of the load electromotive force (emf) ratio $x = E/V_s$ as

$$x = \frac{E}{V_s} \leq \frac{e^{kz} - 1}{e^z - 1} \tag{5.23}$$

Example 5.2   Finding the Currents of a Dc Converter with an *RL* Load

A converter is feeding an *RL* load as shown in Figure 5.3 with $V_s = 220$ V, $R = 5\ \Omega$, $L = 7.5$ mH, $f = 1$ kHz, $k = 0.5$, and $E = 0$ V. Calculate (a) the minimum instantaneous load current I_1, (b) the peak instantaneous load current I_2, (c) the maximum peak-to-peak load ripple current, (d) the average value of load current I_a, (e) the rms load current I_o, (f) the effective input resistance R_i seen by the source, (g) the rms chopper current I_R, and (h) the critical value of the load inductance for continuous load current. Use PSpice to plot the load current, the supply current, and the freewheeling diode current.

Solution
$V_s = 220$ V, $R = 5\ \Omega$, $L = 7.5$ mH, $E = 0$ V, $k = 0.5$, and $f = 1000$ Hz. From Eq. (5.15), $I_2 = 0.7165 I_1 + 12.473$ and from Eq. (5.16), $I_1 = 0.7165 I_2 + 0$.

    a. Solving these two equations yields $I_1 = 18.37$ A.

    b. $I_2 = 25.63$ A.

    c. $\Delta I = I_2 - I_1 = 25.63 - 18.37 = 7.26$ A. From Eq. (5.21), $\Delta I_{max} = 7.26$ A and Eq. (5.22) gives the approximate value, $\Delta I_{max} = 7.33$ A.

    d. The average load current is, approximately,

$$I_a = \frac{I_2 + I_1}{2} = \frac{25.63 + 18.37}{2} = 22\ \text{A}$$

    e. Assuming that the load current rises linearly from I_1 to I_2, the instantaneous load current can be expressed as

$$i_1 = I_1 + \frac{\Delta I t}{kT} \qquad \text{for } 0 < t < kT$$

The rms value of load current can be found from

$$I_o = \left(\frac{1}{kT} \int_0^{kT} i_1^2 \, dt \right)^{1/2} = \left[I_1^2 + \frac{(I_2 - I_1)^2}{3} + I_1(I_2 - I_1) \right]^{1/2} \tag{5.24}$$

$$= 22.1\ \text{A}$$

    f. The average source current

$$I_s = kI_a = 0.5 \times 22 = 11\ \text{A}$$

and the effective input resistance $R_i = V_s/I_s = 220/11 = 20\ \Omega$.

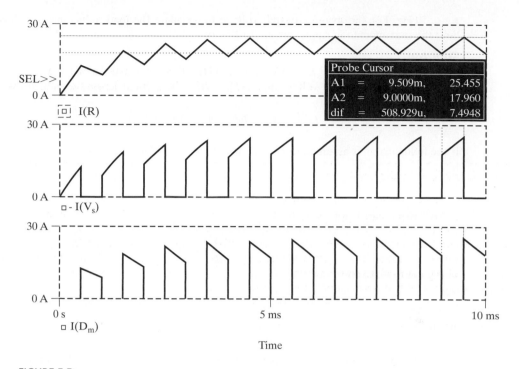

FIGURE 5.5

SPICE plots of load, input, and diode currents for Example 5.2.

g. The rms converter current can be found from

$$I_R = \left(\frac{1}{T}\int_0^{kT} i_1^2 \, dt\right)^{1/2} = \sqrt{k}\left[I_1^2 + \frac{(I_2 - I_1)^2}{3} + I_1(I_2 - I_1)\right]^{1/2}$$ (5.25)

$$= \sqrt{k}I_o = \sqrt{0.5} \times 22.1 = 15.63 \text{ A}$$

h. We can rewrite Eq. (5.23) as

$$V_S\left(\frac{e^{kz} - 1}{e^z - 1}\right) = E$$

which, after iteration, gives, $z = TR/L = 52.5$ and $L = 1 \text{ ms} \times 5/52.5 = 0.096 \text{ mH}$. The SPICE simulation results [32] are shown in Figure 5.5, which shows the load current $I(E)$, the supply current $-I(V_s)$, and the diode current $I(D_m)$. We get $I_1 = 17.96$ A and $I_2 = 25.46$ A.

Example 5.3 Finding the Load Inductance to Limit the Load Ripple Current

The converter in Figure 5.3 has a load resistance $R = 0.25 \ \Omega$, input voltage $V_s = 550$ V, and battery voltage $E = 0$ V. The average load current $I_a = 200$ A, and chopping frequency

$f = 250$ Hz. Use the average output voltage to calculate the load inductance L, which would limit the maximum load ripple current to 10% of I_a.

Solution
$V_s = 550$ V, $R = 0.25$ Ω, $E = 0$ V, $f = 250$ Hz, $T = 1/f = 0.004$ s, and $\Delta i = 200 \times 0.1 = 20$ A. The average output voltage $V_a = kV_s = RI_a$. The voltage across the inductor is given by

$$L\frac{di}{dt} = V_s - RI_a = V_s - kV_s = V_s(1 - k)$$

If the load current is assumed to rise linearly, $dt = t_1 = kT$ and $di = \Delta i$:

$$\Delta i = \frac{V_s(1 - k)}{L} kT$$

For the worst-case ripple conditions,

$$\frac{d(\Delta i)}{dk} = 0$$

This gives $k = 0.5$ and

$$\Delta i \, L = 20 \times L = 550(1 - 0.5) \times 0.5 \times 0.004$$

and the required value of inductance is $L = 27.5$ mH.

Note: For $\Delta I = 20$ A, Eq. (5.19) gives $z = 0.036$ and $L = 27.194$ mH.

Key Points of Section 5.3

- An inductive load can make the load current continuous. However, the critical value of inductance, which is required for continuous current, is influenced by the load emf ratio. The peak-to-peak load current ripple becomes maximum at $k = 0.5$.

5.4   PRINCIPLE OF STEP-UP OPERATION

A converter can be used to step up a dc voltage and an arrangement for step-up operation is shown in Figure 5.6a. When switch SW is closed for time t_1, the inductor current rises and energy is stored in the inductor L. If the switch is opened for time t_2, the energy stored in the inductor is transferred to load through diode D_1 and the inductor current falls. Assuming a continuous current flow, the waveform for the inductor current is shown in Figure 5.6b.

When the converter is turned on, the voltage across the inductor is

$$v_L = L\frac{di}{dt}$$

and this gives the peak-to-peak ripple current in the inductor as

$$\Delta I = \frac{V_s}{L} t_1 \tag{5.26}$$

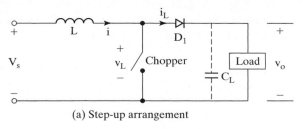

(a) Step-up arrangement

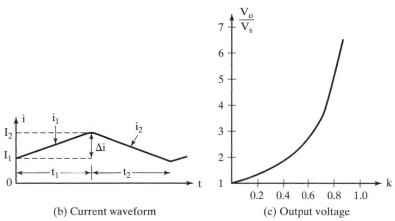

(b) Current waveform   (c) Output voltage

FIGURE 5.6

Arrangement for step-up operation.

The average output voltage is

$$v_o = V_s + L\frac{\Delta I}{t_2} = V_s\left(1 + \frac{t_1}{t_2}\right) = V_s\frac{1}{1 - k} \tag{5.27}$$

If a large capacitor C_L is connected across the load as shown by dashed lines in Figure 5.6a, the output voltage is continuous and v_o becomes the average value V_a. We can notice from Eq. (5.27) that the voltage across the load can be stepped up by varying the duty cycle k and the minimum output voltage is V_s when $k = 0$. However, the converter cannot be switched on continuously such that $k = 1$. For values of k tending to unity, the output voltage becomes very large and is very sensitive to changes in k, as shown in Figure 5.6c.

This principle can be applied to transfer energy from one voltage source to another as shown in Figure 5.7a. The equivalent circuits for the modes of operation are shown in Figure 5.7b and the current waveforms in Figure 5.7c. The inductor current for mode 1 is given by

$$V_s = L\frac{di_1}{dt}$$

and is expressed as

$$i_1(t) = \frac{V_s}{L}t + I_1 \tag{5.28}$$

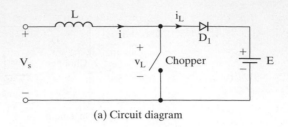

(a) Circuit diagram

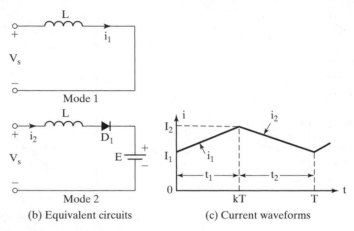

(b) Equivalent circuits   (c) Current waveforms

FIGURE 5.7

Arrangement for transfer of energy.

where I_1 is the initial current for mode 1. During mode 1, the current must rise and the necessary condition,

$$\frac{di_1}{dt} > 0 \quad \text{or} \quad V_s > 0$$

The current for mode 2 is given by

$$V_s = L\frac{di_2}{dt} + E$$

and is solved as

$$i_2(t) = \frac{V_s - E}{L}t + I_2 \tag{5.29}$$

where I_2 is initial current for mode 2. For a stable system, the current must fall and the condition is

$$\frac{di_2}{dt} < 0 \quad \text{or} \quad V_s < E$$

If this condition is not satisfied, the inductor current continues to rise and an unstable situation occurs. Therefore, the conditions for controllable power transfer are

$$0 < V_s < E \tag{5.30}$$

Equation (5.30) indicates that the source voltage V_s must be less than the voltage E to permit transfer of power from a fixed (or variable) source to a fixed dc voltage. In electric braking of dc motors, where the motors operate as dc generators, terminal voltage falls as the machine speed decreases. The converter permits transfer of power to a fixed dc source or a rheostat.

When the converter is turned on, the energy is transferred from the source V_s to inductor L. If the converter is then turned off, a magnitude of the energy stored in the inductor is forced to battery E.

Note: Without the chopping action, v_s must be greater than E for transferring power from V_s to E.

Key Points of Section 5.4

- A step-up dc converter can produce an output voltage that is higher than the input. The input current can be transferred to a voltage source higher than the input voltage.

5.5   STEP-UP CONVERTER WITH A RESISTIVE LOAD

A step-up converter with a resistive load is shown in Figure 5.8a. When switch S_1 is closed, the current rises through L and the switch. The equivalent circuit during mode 1 is shown in Figure 5.8b and the current is described by

$$V_s = L \frac{d}{dt} i_1$$

which for an initial current of I_1 gives

$$i_1(t) = \frac{V_s}{L} t + I_1 \tag{5.31}$$

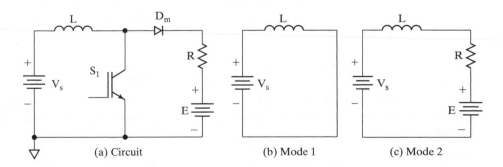

(a) Circuit   (b) Mode 1   (c) Mode 2

FIGURE 5.8

Step-up converter with a resistive load.

which is valid for $0 \leq t \leq kT$. At the end of mode 1 at $t = kT$,

$$I_2 = i_1(t = kT) = \frac{V_s}{L} kT + I_1 \qquad (5.32)$$

When switch S_1 is opened, the inductor current flows through the RL load.

The equivalent current is shown in Figure 5.8c and the current during mode 2 is described by

$$V_s = Ri_2 + \frac{d}{dt} i_2 + E$$

which for an initial current of I_2 gives

$$i_2(t) = \frac{V_s - E}{L} \left(1 - e^{\frac{-tR}{L}} \right) + I_2 e^{\frac{-tR}{L}} \qquad (5.33)$$

which is valid for $0 \leq t \leq (1 - k)T$. At the end of mode 2 at $t = (1 - k)T$.

$$I_1 = i_2[t = (1 - k)T] = \frac{V_s - E}{L} \left[1 - e^{-(1-k)z} \right] + I_2 e^{-(1-k)z} \qquad (5.34)$$

where $z = TR/L$. Solving for I_1 and I_2 from Eqs. (5.32) and (5.34), we get

$$I_1 = \frac{V_s kz}{R} \frac{e^{-(1-k)z}}{1 - e^{-(1-k)z}} + \frac{V_s - E}{R} \qquad (5.35)$$

$$I_2 = \frac{V_s kz}{R} \frac{1}{1 - e^{-(1-k)z}} + \frac{V_s - E}{R} \qquad (5.36)$$

The ripple current is given by

$$\Delta I = I_2 - I_1 = \frac{V_s}{L} kT \qquad (5.37)$$

These equations are valid for $E \leq V_s$. If $E \geq V_s$ and the converter switch S_1 is opened, the inductor transfers its stored energy through R to the source and the inductor current is discontinuous.

Example 5.4 Finding the Currents of a Step-up Dc Converter

The step-up converter in Figure 5.8a has $V_s = 10 \text{ V}, f = 1 \text{ kHz}, R = 5 \text{ }\Omega, L = 6.5 \text{ mH}, E = 0 \text{ V}$, and $k = 0.5$. Find I_1, I_2, and ΔI. Use SPICE to find these values and plot the load, diode, and switch current.

Solution
Equations (5.35) and (5.36) give $I_1 = 3.64 \text{ A}$ (3.36 A from SPICE) and $I_2 = 4.4 \text{ A}$ (4.15 A from SPICE). The plots of the load current $I(L)$, the diode current $I(D_m)$ and the switch current $IC(Q_1)$ are shown in Figure 5.9.

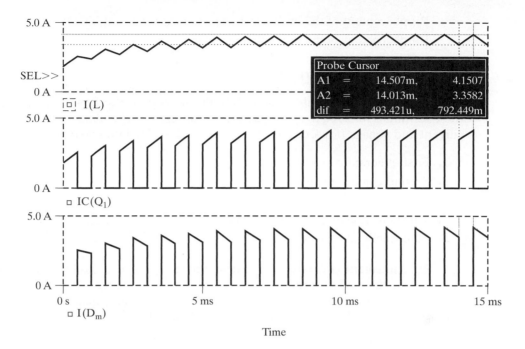

FIGURE 5.9

SPICE plots of load, input, and diode current for Example 5.4.

Key Points of Section 5.5

- With a resistive load, the load current and the voltage are pulsating. An output filter is required to smooth the output voltage.

5.6 PERFORMANCE PARAMETERS

The power semiconductor devices require a minimum time to turn on and turn off. Therefore, the duty cycle k can only be controlled between a minimum value k_{min} and a maximum value k_{max}, thereby limiting the minimum and maximum value of output voltage. The switching frequency of the converter is also limited. It can be noticed from Eq. (5.22) that the load ripple current depends inversely on the chopping frequency f. The frequency should be as high as possible to reduce the load ripple current and to minimize the size of any additional series inductor in the load circuit.

The performance parameters of the step-up and step-down converters are as follows:

Ripple current of the inductor, ΔI_L;

Maximum switching frequency, f_{max};

Condition for continuous or discontinuous inductor current;

Minimum value of inductor to maintain continuous inductor current;

Ripple content of the output voltage and output current, THD;

Ripple content of the input current, THD.

5.7 CONVERTER CLASSIFICATION

The step-down converter in Figure 5.1a only allows power to flow from the supply to the load, and is referred to as first quadrant converter. Depending on the directions of current and voltage flows, dc converters can be classified into five types:

1. First quadrant converter
2. Second quadrant converter
3. First and second quadrant converter
4. Third and fourth quadrant converter
5. Four-quadrant converter

First quadrant converter. The load current flows into the load. Both the load voltage and the load current are positive, as shown in Figure 5.10a. This is a single-quadrant converter and is said to be operated as a rectifier. Equations in Sections 5.2 and 5.3 can be applied to evaluate the performance of a first quadrant converter.

Second quadrant converter. The load current flows out of the load. The load voltage is positive, but the load current is negative, as shown in Figure 5.10b. This is also

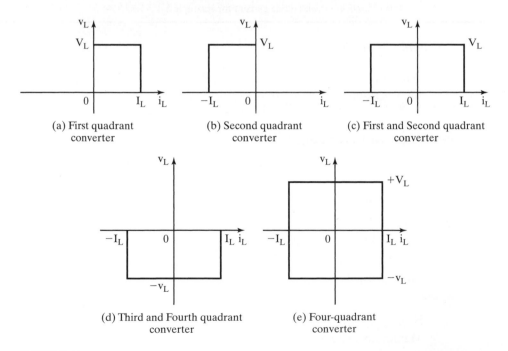

(a) First quadrant converter

(b) Second quadrant converter

(c) First and Second quadrant converter

(d) Third and Fourth quadrant converter

(e) Four-quadrant converter

FIGURE 5.10

Dc converter classification.

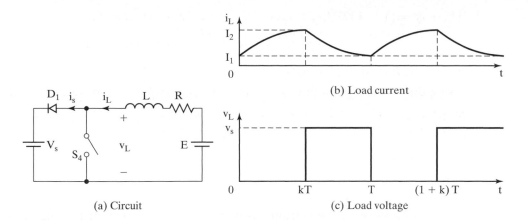

(a) Circuit

(b) Load current

(c) Load voltage

FIGURE 5.11

Second quadrant converter.

a single-quadrant converter, but operates in the second quadrant and is said to be operated as an inverter. A second quadrant converter is shown in Figure 5.11a, where the battery E is a part of the load and may be the back emf of a dc motor.

When switch S_4 is turned on, the voltage E drives current through inductor L and load voltage v_L becomes zero. The instantaneous load voltage v_L and load current i_L are shown in Figure 5.11b and 5.11c, respectively. The current i_L, which rises, is described by

$$0 = L\frac{di_L}{dt} + Ri_L + E$$

which, with initial condition $i_L(t = 0) = I_1$, gives

$$i_L = I_1 e^{-(R/L)t} - \frac{E}{R}(1 - e^{-(R/L)t}) \qquad \text{for } 0 \le t \le kT \qquad (5.38)$$

At $t = t_1$,

$$i_L(t = t_1 = kT) = I_2 \qquad (5.39)$$

When switch S_4 is turned off, a magnitude of the energy stored in inductor L is returned to the supply V_s via diode D_1. The load current i_L falls. Redefining the time origin $t - 0$, the load current i_L is described by

$$V_s = L\frac{di_L}{dt} + Ri_L + E$$

which, with initial condition $i(t = t_2) = I_2$, gives

$$i_L = I_2 e^{-(R/L)t} + \frac{V_s - E}{R}(1 - e^{-(R/L)t}) \qquad \text{for } 0 \le t \le t_2 \qquad (5.40)$$

where $t_2 = (1 - k)T$. At $t = t_2$,

$$i_L(t = t_2) = I_1 \quad \text{for steady-state continuous current}$$
$$= 0 \quad \text{for steady-state discontinuous current} \tag{5.41}$$

Using the boundary conditions in Eqs. (5.39) and (5.41), we can solve for I_1 and I_2 as

$$I_1 = \frac{V_S}{R}\left[\frac{1 - e^{-(1-k)z}}{1 - e^{-x}}\right] - \frac{E}{R} \tag{5.42}$$

$$I_2 = \frac{V_S}{R}\left(\frac{e^{-kz} - e^{-z}}{1 - e^{-z}}\right) - \frac{E}{R} \tag{5.43}$$

First and second quadrant converter.   The load current is either positive or negative, as shown in Figure 5.10c. The load voltage is always positive. This is known as a *two-quadrant converter*. The first and second quadrant converters can be combined to form this converter, as shown in Figure 5.12. S_1 and D_4 operate as a first quadrant converter. S_2 and D_4 D_1 operate as a second quadrant converter. Care must be taken to ensure that the two switches are not fired together; otherwise, the supply V_s becomes short-circuited. This type of converter can operate either as a rectifier or as an inverter.

Third and fourth quadrant converter.   The circuit is shown in Figure 5.13. The load voltage is always negative. The load current is either positive or negative, as

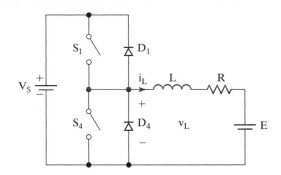

FIGURE 5.12

First and second quadrant converter.

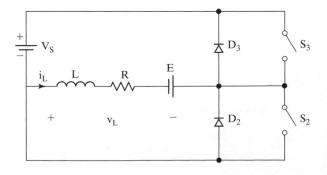

FIGURE 5.13

Third and fourth quadrant converter.

shown in Figure 5.10d. S_3 and D_2 operate to yield both a negative voltage and a load current. When S_3 is closed, a negative current flows through the load. When S_3 is opened, the load current freewheels through diode D_2. S_2 and D_3 operate to yield a negative voltage and a positive load current. When S_2 is closed, a positive load current flows. When S_2 is opened, the load current free wheels through diode D_3. It is important to note that the polarity of E must be reversed for this circuit to yield a negative voltage and a positive current. This is a negative two-quadrant converter. This converter can also operate as a rectifier or as an inverter.

Four-quadrant converter [2]. The load current is either positive or negative, as shown in Figure 5.10e. The load voltage is also either positive or negative. One first and second quadrant converter and one third and fourth quadrant converter can be combined to form the four-quadrant converter, as shown in Figure 5.14a. The polarities of the load voltage and load currents are shown in Figure 5.14b. The devices that are operative in different quadrants are shown in Figure 5.14c. For operation in the fourth quadrant, the direction of the battery E must be reversed. This converter forms the basis for the single-phase full-bridge inverter in Section 6.4.

For an inductive load with an emf (E) such as a dc motor, the four-quadrant converter can control the power flow and the motor speed in the forward direction (v_L positive and i_L positive), forward regenerative braking (v_L positive and i_L reverse), reverse direction (v_L negative and i_L reversing) and reverse regenerative braking (v_L negative and i_L negative).

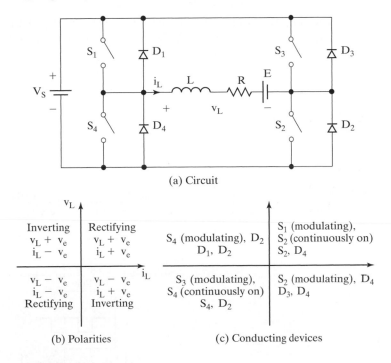

(a) Circuit

v_L			
Inverting $v_L + v_e$ $i_L - v_e$	Rectifying $v_L + v_e$ $i_L + v_e$	S_4 (modulating), D_2 D_1, D_2	S_1 (modulating), S_2 (continuously on) S_2, D_4
$v_L - v_e$ $i_L - v_e$ Rectifying	$v_L - v_e$ $i_L + v_e$ Inverting	S_3 (modulating), S_4 (continuously on) S_4, D_2	S_2 (modulating), D_4 D_3, D_4

(b) Polarities

(c) Conducting devices

FIGURE 5.14

Four-quadrant converter.

Key Points of Section 5.7

- With proper switch control, the four-quadrant converter can operate and control flow in any of the four quadrants. For operation in the third and fourth quadrants, the direction of the load emf E must be reversed internally.

5.8   SWITCHING-MODE REGULATORS

Dc converters can be used as switching-mode regulators to convert a dc voltage, normally unregulated, to a regulated dc output voltage. The regulation is normally achieved by PWM at a fixed frequency and the switching device is normally BJT, MOSFET, or IGBT. The elements of switching-mode regulators are shown in Figure 5.15. We can notice from Figure 5.1b that the output of dc converters with resistive load is discontinuous and contains harmonics. The ripple content is normally reduced by an LC filter.

Switching regulators are commercially available as integrated circuits. The designer can select the switching frequency by choosing the values of R and C of frequency oscillator. As a rule of thumb, to maximize efficiency, the minimum oscillator period should be about 100 times longer than the transistor switching time; for example, if a transistor has a switching time of 0.5 μs, the oscillator period would be 50 μs, which gives the maximum oscillator frequency of 20 kHz. This limitation is due to a switching loss in the transistor. The transistor switching loss increases with the switching frequency and as a result the efficiency decreases. In addition, the core loss of inductors limits the high-frequency operation. Control voltage v_c is obtained by comparing the output voltage with its desired value. The v_{cr} can be compared with a sawtooth voltage v_r to generate the PWM control signal for the dc converter. There are four basic topologies of switching regulators [33, 34]:

1. Buck regulators
2. Boost regulators
3. Buck–boost regulators
4. Cúk regulators

5.8.1   Buck Regulators

In a buck regulator, the average output voltage V_a, is less than the input voltage, V_s—hence the name "buck," a very popular regulator [6, 7]. The circuit diagram of a buck

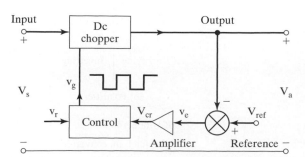

FIGURE 5.15

Elements of switching-mode regulators.

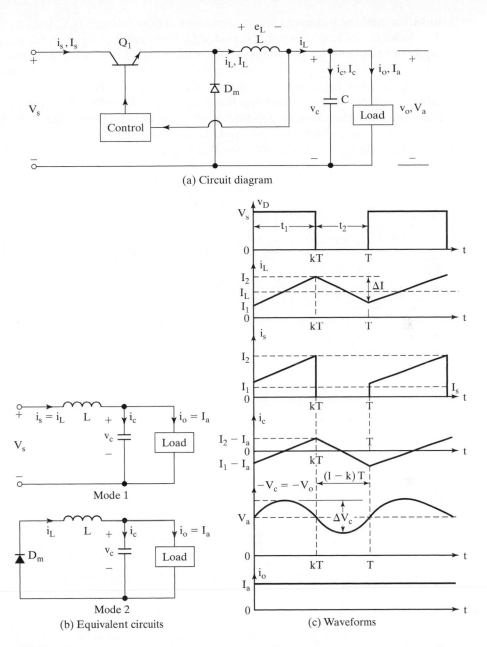

FIGURE 5.16

Buck regulator with continuous i_L.

regulator using a power BJT is shown in Figure 5.16a, and this is like a step-down converter. The circuit operation can be divided into two modes. Mode 1 begins when transistor Q_1 is switched on at $t = 0$. The input current, which rises, flows through filter inductor L, filter capacitor C, and load resistor R. Mode 2 begins when transistor Q_1 is switched off at $t = t_1$. The freewheeling diode D_m conducts due to energy stored in the inductor; and the inductor current continues to flow through L, C, load, and diode D_m. The inductor current falls until transistor Q_1 is switched on again in the next cycle. The equivalent circuits for the modes of operation are shown in Figure 5.16b. The waveforms for the voltages and currents are shown in Figure 5.16c for a continuous current flow in the inductor L. It is assumed that the current rises and falls linearly. In practical circuits, the switch has a finite, nonlinear resistance. Its effect can generally be negligible in most applications. Depending on the switching frequency, filter inductance, and capacitance, the inductor current could be discontinuous.

The voltage across the inductor L is, in general,

$$e_L = L\frac{di}{dt}$$

Assuming that the inductor current rises linearly from I_1 to I_2 in time t_1,

$$V_s - V_a = L\frac{I_2 - I_1}{t_1} = L\frac{\Delta I}{t_1} \tag{5.44}$$

or

$$t_1 = \frac{\Delta I\, L}{V_s - V_a} \tag{5.45}$$

and the inductor current falls linearly from I_2 to I_1 in time t_2,

$$-V_a = -L\frac{\Delta I}{t_2} \tag{5.46}$$

or

$$t_2 = \frac{\Delta I\, L}{V_a} \tag{5.47}$$

where $\Delta I = I_2 - I_1$ is the peak-to-peak ripple current of the inductor L. Equating the value of ΔI in Eqs. (5.44) and (5.46) gives

$$\Delta I = \frac{(V_s - V_a)t_1}{L} = \frac{V_a t_2}{L}$$

Substituting $t_1 = kT$ and $t_2 = (1 - k)T$ yields the average output voltage as

$$V_a = V_s\frac{t_1}{T} = kV_s \tag{5.48}$$

Assuming a lossless circuit, $V_s I_s = V_a I_a = kV_s I_a$ and the average input current

$$I_s = kI_a \tag{5.49}$$

The switching period T can be expressed as

$$T = \frac{1}{f} = t_1 + t_2 = \frac{\Delta I\, L}{V_s - V_a} + \frac{\Delta I\, L}{V_a} = \frac{\Delta I\, L V_s}{V_a(V_s - V_a)} \tag{5.50}$$

which gives the peak-to-peak ripple current as

$$\Delta I = \frac{V_a(V_s - V_a)}{fLV_s} \tag{5.51}$$

or

$$\Delta I = \frac{V_s k(1 - k)}{fL} \tag{5.52}$$

Using Kirchhoff's current law, we can write the inductor current i_L as

$$i_L = i_c + i_o$$

If we assume that the load ripple current Δi_o is very small and negligible, $\Delta i_L = \Delta i_c$. The average capacitor current, which flows into for $t_1/2 + t_2/2 = T/2$, is

$$I_c = \frac{\Delta I}{4}$$

The capacitor voltage is expressed as

$$v_c = \frac{1}{C} \int i_c \, dt + v_c(t = 0)$$

and the peak-to-peak ripple voltage of the capacitor is

$$\Delta V_c = v_c - v_c(t = 0) = \frac{1}{C} \int_0^{T/2} \frac{\Delta I}{4} \, dt = \frac{\Delta I\, T}{8C} = \frac{\Delta I}{8fC} \tag{5.53}$$

Substituting the value of ΔI from Eq. (5.51) or (5.52) in Eq. (5.53) yields

$$\Delta V_c = \frac{V_a(V_s - V_a)}{8LCf^2 V_s} \tag{5.54}$$

or

$$\Delta V_c = \frac{V_s k(1 - k)}{8LCf^2} \tag{5.55}$$

Condition for continuous inductor current and capacitor voltage.   If I_L is the average inductor current, the inductor ripple current $\Delta I = 2I_L$.
Using Eqs. (5.48) and (5.52), we get

$$\frac{V_S(1 - k)k}{fL} = 2I_L = 2I_a = \frac{2kV_S}{R}$$

which gives the critical value of the inductor L_c as

$$L_c = L = \frac{(1 - k)R}{2f} \tag{5.56}$$

If V_c is the average capacitor voltage, the capacitor ripple voltage $\Delta V_c = 2V_a$. Using Eqs. (5.48) and (5.55), we get

$$\frac{V_S(1-k)k}{8LCf^2} = 2V_a = 2kV_S$$

which gives the critical value of the capacitor C_c as

$$C_c = C = \frac{1-k}{16Lf^2} \tag{5.57}$$

The buck regulator requires only one transistor, is simple, and has high efficiency greater than 90%. The di/dt of the load current is limited by inductor L. However, the input current is discontinuous and a smoothing input filter is normally required. It provides one polarity of output voltage and unidirectional output current. It requires a protection circuit in case of possible short circuit across the diode path.

Example 5.5   Finding the Values of *LC* Filter for the Buck Regulator

The buck regulator in Figure 5.16a has an input voltage of $V_s = 12$ V. The required average output voltage is $V_a = 5$ V at $R = 500\ \Omega$ and the peak-to-peak output ripple voltage is 20 mV. The switching frequency is 25 kHz. If the peak-to-peak ripple current of inductor is limited to 0.8 A, determine (a) the duty cycle k, (b) the filter inductance L, and (c) the filter capacitor C, and (d) the critical values of L and C.

Solution
$V_s = 12$ V, $\Delta V_c = 20$ mV, $\Delta I - 0.8$ A, $f = 25$ kHz, and $V_a = 5$ V.

   a. From Eq. (5.48), $V_a = kV_s$ and $k = V_a/V_s = 5/12 = 0.4167 = 41.67\%$.
   b. From Eq. (5.51),

$$L = \frac{5(12-5)}{0.8 \times 25{,}000 \times 12} = 145.83\ \mu\text{H}$$

   c. From Eq. (5.53),

$$C = \frac{0.8}{8 \times 20 \times 10^{-3} \times 25{,}000} = 200\ \mu\text{F}$$

   d. From Eq. (5.56), we get $L_c = \dfrac{(1-k)R}{2f} = \dfrac{(1-0.4167) \times 500}{2 \times 25 \times 10^3} = 5.83$ mH

      From Eq. (5.57), we get $C_c = \dfrac{1-k}{16Lf^2} = \dfrac{1-0.4167}{16 \times 5.83 \times 10^2 \times (25 \times 10^3)^2} = 0.4\ \mu\text{F}$

5.8.2   Boost Regulators

In a boost regulator [8, 9] the output voltage is greater than the input voltage—hence the name "boost." A boost regulator using a power MOSFET is shown in Figure 5.17a. The circuit operation can be divided into two modes. Mode 1 begins when transistor

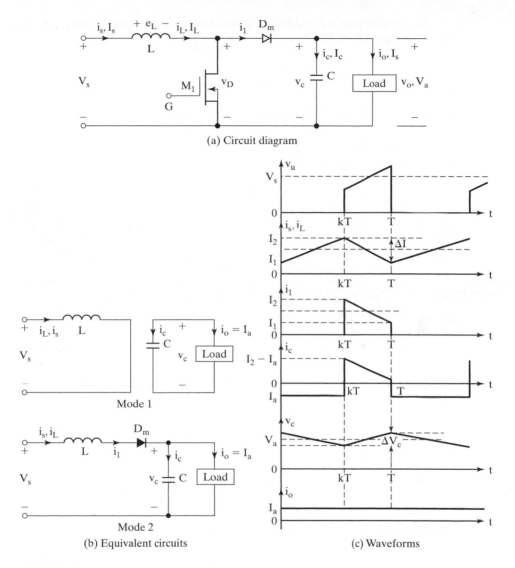

FIGURE 5.17

Boost regulator with continuous i_L.

M_1 is switched on at $t = 0$. The input current, which rises, flows through inductor L and transistor Q_1. Mode 2 begins when transistor M_1 is switched off at $t = t_1$. The current that was flowing through the transistor would now flow through L, C, load, and diode D_m. The inductor current falls until transistor M_1 is turned on again in the next cycle. The energy stored in inductor L is transferred to the load. The equivalent circuits for the modes of operation are shown in Figure 5.17b. The waveforms for voltages and currents are shown in Figure 5.17c for continuous load current, assuming that the current rises or falls linearly.

Assuming that the inductor current rises linearly from I_1 to I_2 in time t_1,

$$V_s = L\frac{I_2 - I_1}{t_1} = L\frac{\Delta I}{t_1} \tag{5.58}$$

or

$$t_1 = \frac{\Delta I L}{V_s} \tag{5.59}$$

and the inductor current falls linearly from I_2 to I_1 in time t_2,

$$V_s - V_a = -L\frac{\Delta I}{t_2} \tag{5.60}$$

or

$$t_2 = \frac{\Delta I L}{V_a - V_s} \tag{5.61}$$

where $\Delta I = I_2 - I_1$ is the peak-to-peak ripple current of inductor L. From Eqs. (5.58) and (5.60),

$$\Delta I = \frac{V_s t_1}{L} = \frac{(V_a - V_s)t_2}{L}$$

Substituting $t_1 = kT$ and $t_2 = (1 - k)T$ yields the average output voltage,

$$V_a = V_s\frac{T}{t_2} = \frac{V_s}{1 - k} \tag{5.62}$$

which gives

$$(1 - k) = \frac{V_S}{V_a} \tag{5.63}$$

Substituting $k = t_1/T = t_1 f$ into Eq. (5.63) yields

$$t_1 = \frac{V_a - V_S}{V_a f} \tag{5.64}$$

Assuming a lossless circuit, $V_s I_s = V_a I_a = V_s I_a/(1 - k)$ and the average input current is

$$I_s = \frac{I_a}{1 - k} \tag{5.65}$$

The switching period T can be found from

$$T = \frac{1}{f} = t_1 + t_2 = \frac{\Delta I L}{V_s} + \frac{\Delta I L}{V_a - V_s} = \frac{\Delta I L V_a}{V_s(V_a - V_s)} \tag{5.66}$$

and this gives the peak-to-peak ripple current:

$$\Delta I = \frac{V_s(V_a - V_s)}{fLV_a} \tag{5.67}$$

or

$$\Delta I = \frac{V_s k}{fL} \tag{5.68}$$

When the transistor is on, the capacitor supplies the load current for $t = t_1$. The average capacitor current during time t_1 is $I_c = I_a$ and the peak-to-peak ripple voltage of the capacitor is

$$\Delta V_c = v_c - v_c(t = 0) = \frac{1}{C}\int_0^{t_1} I_c \, dt = \frac{1}{C}\int_0^{t_1} I_a = \frac{I_a t_1}{C} \tag{5.69}$$

Substituting $t_1 = (V_a - V_s)/(V_a f)$ from Eq. (5.64) gives

$$\Delta V_c = \frac{I_a(V_a - V_s)}{V_a f C} \tag{5.70}$$

or

$$\Delta V_c = \frac{I_a k}{fC} \tag{5.71}$$

Condition for continuous inductor current and capacitor voltage. If I_L is the average inductor current, the inductor ripple current $\Delta I = 2I_L$.
Using Eqs. (5.62) and (5.68), we get

$$\frac{kV_S}{fL} = 2I_L = 2I_a = \frac{2V_S}{(1 - k)R}$$

which gives the critical value of the inductor L_c as

$$L_c = L = \frac{k(1 - k)R}{2f} \tag{5.72}$$

If V_c is the average capacitor voltage, the capacitor ripple voltage $\Delta V_c = 2V_a$. Using Eq. (5.71), we get

$$\frac{I_a k}{Cf} = 2V_a = 2I_a R$$

which gives the critical value of the capacitor C_c as

$$C_c = C = \frac{k}{2fR} \tag{5.73}$$

A boost regulator can step up the output voltage without a transformer. Due to a single transistor, it has a high efficiency. The input current is continuous. However, a high-peak current has to flow through the power transistor. The output voltage is very sensitive to changes in duty cycle k and it might be difficult to stabilize the regulator. The average output current is less than the average inductor current by a factor of $(1 - k)$, and a much higher rms current would flow through the filter capacitor, resulting in the use of a larger filter capacitor and a larger inductor than those of a buck regulator.

Example 5.6   Finding the Currents and Voltage in the Boost Regulator

A boost regulator in Figure 5.17a has an input voltage of $V_s = 5$ V. The average output voltage $V_a = 15$ V and the average load current $I_a = 0.5$ A. The switching frequency is 25 kHz. If $L = 150$ μH and $C = 220$ μF, determine (a) the duty cycle k, (b) the ripple current of inductor ΔI, (c) the peak current of inductor I_2, (d) the ripple voltage of filter capacitor ΔV_c, and (e) the critical values of L and C.

Solution
$V_s = 5$ V, $V_a = 15$ V, $f = 25$ kHz, $L = 150$ μH, and $C = 220$ μF.

a. From Eq. (5.62), $15 = 5/(1 - k)$ or $k = 2/3 = 0.6667 = 66.67\%$.

b. From Eq. (5.67),

$$\Delta I = \frac{5 \times (15 - 5)}{25{,}000 \times 150 \times 10^{-6} \times 15} = 0.89 \text{ A}$$

c. From Eq. (5.65), $I_s = 0.5/(1 - 0.667) = 1.5$ A and peak inductor current,

$$I_2 = I_s + \frac{\Delta I}{2} = 1.5 + \frac{0.89}{2} = 1.945 \text{ A}$$

d. From Eq. (5.71),

$$\Delta V_c = \frac{0.5 \times 0.6667}{25{,}000 \times 220 \times 10^{-6}} = 60.61 \text{ mV}$$

e. $R = \dfrac{V_a}{I_a} = \dfrac{15}{0.5} = 30 \ \Omega$

From Eq. (5.72), we get $L_c = \dfrac{(1 - k)kR}{2f} = \dfrac{(1 - 0.6667) \times 0.6667 \times 30}{2 \times 25 \times 10^3} = 133 \ \mu$H

From Eq. (5.73), we get $C_c = \dfrac{k}{2fR} = \dfrac{0.6667}{2 \times 25 \times 10^3 \times 30} = 0.44 \ \mu$F

5.8.3   Buck–Boost Regulators

A buck–boost regulator provides an output voltage that may be less than or greater than the input voltage—hence the name "buck–boost"; the output voltage polarity is opposite to that of the input voltage. This regulator is also known as an *inverting regulator*. The circuit arrangement of a buck–boost regulator is shown in Figure 5.18a.

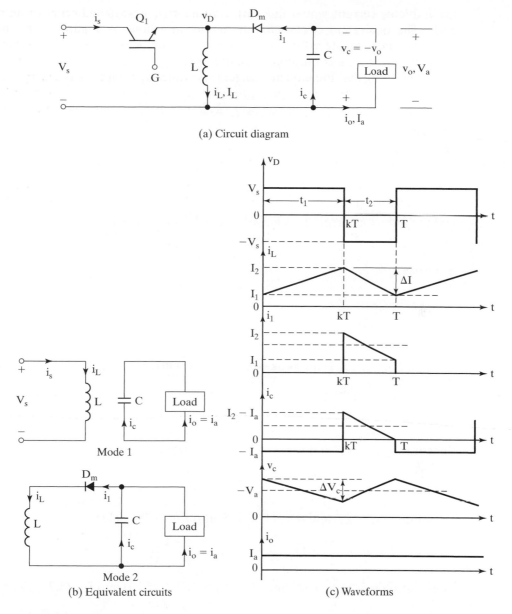

FIGURE 5.18

Buck–boost regulator with continuous i_L.

The circuit operation can be divided into two modes. During mode 1, transistor Q_1 is turned on and diode D_m is reversed biased. The input current, which rises, flows through inductor L and transistor Q_1. During mode 2, transistor Q_1 is switched off and the current, which was flowing through inductor L, would flow through L, C, D_m, and the load. The energy stored in inductor L would be transferred to the load and the inductor current would fall until transistor Q_1 is switched on again in the next cycle. The equivalent circuits for the modes are shown in Figure 5.18b. The waveforms for steady-state voltages and currents of the buck–boost regulator are shown in Figure 5.18c for a continuous load current.

Assuming that the inductor current rises linearly from I_1 to I_2 in time t_1,

$$V_s = L\frac{I_2 - I_1}{t_1} = L\frac{\Delta I}{t_1} \tag{5.74}$$

or

$$t_1 = \frac{\Delta I L}{V_s} \tag{5.75}$$

and the inductor current falls linearly from I_2 to I_1 in time t_2,

$$V_a = -L\frac{\Delta I}{t_2} \tag{5.76}$$

or

$$t_2 = \frac{-\Delta I L}{V_a} \tag{5.77}$$

where $\Delta I = I_2 - I_1$ is the peak-to-peak ripple current of inductor L. From Eqs. (5.74) and (5.76),

$$\Delta I = \frac{V_s t_1}{L} = \frac{-V_a t_2}{L}$$

Substituting $t_1 = kT$ and $t_2 = (1 - k)T$, the average output voltage is

$$V_a = -\frac{V_s k}{1 - k} \tag{5.78}$$

Substituting $t_1 = kT$ and $t_2 = (1 - k)T$ into Eq. (5.78) yields

$$(1 - k) = \frac{-V_S}{V_a - V_S} \tag{5.79}$$

Substituting $t_2 = (1 - k)T$, and $(1 - k)$ from Eq. (5.79) into Eq. (5.78) yields

$$t_1 = \frac{V_a}{(V_a - V_S)f} \tag{5.80}$$

Assuming a lossless circuit, $V_s I_s = -V_a I_a = V_s I_a k/(1 - k)$ and the average input current I_s is related to the average output current I_a by

$$I_s = \frac{I_a k}{1 - k} \tag{5.81}$$

The switching period T can be found from

$$T = \frac{1}{f} = t_1 + t_2 = \frac{\Delta IL}{V_s} + \frac{\Delta IL}{V_a} = \frac{\Delta IL(V_a - V_s)}{V_s V_a} \tag{5.82}$$

and this gives the peak-to-peak ripple current,

$$\Delta I = \frac{V_s V_a}{f L(V_a - V_s)} \tag{5.83}$$

or

$$\Delta I = \frac{V_s k}{f L} \tag{5.84}$$

When transistor Q_1 is on, the filter capacitor supplies the load current for $t = t_1$. The average discharging current of the capacitor is $I_c = I_a$ and the peak-to-peak ripple voltage of the capacitor is

$$\Delta V_c = \frac{1}{C} \int_0^{t_1} I_c \, dt = \frac{1}{C} \int_0^{t_1} I_a \, dt = \frac{I_a t_1}{C} \tag{5.85}$$

Substituting $t_1 = V_a/[(V_a - V_s)f]$ from Eq. (5.80) becomes

$$\Delta V_c = \frac{I_a V_a}{(V_a - V_s)fC} \tag{5.86}$$

or

$$\Delta V_c = \frac{I_a k}{fC} \tag{5.87}$$

Condition for continuous inductor current and capacitor voltage.   If I_L is the average inductor current, the inductor ripple current $\Delta I = 2I_L$. Using Eqs. (5.78) and (5.84), we get

$$\frac{kV_s}{fL} = 2I_L = 2I_a = \frac{2kV_s}{(1-k)R}$$

which gives the critical value of the inductor L_c as

$$L_c = L = \frac{(1-k)R}{2f} \tag{5.88}$$

If V_c is the average capacitor voltage, the capacitor ripple voltage $\Delta V_c = 2V_a$. Using Eq. (5.87), we get

$$\frac{I_a k}{Cf} = 2V_a = 2I_a R$$

which gives the critical value of the capacitor C_c as

$$C_c = C = \frac{k}{2fR} \tag{5.89}$$

A buck–boost regulator provides output voltage polarity reversal without a transformer. It has high efficiency. Under a fault condition of the transistor, the *di/dt* of the fault current is limited by the inductor L and will be V_s/L. Output short-circuit protection would be easy to implement. However, the input current is discontinuous and a high peak current flows through transistor Q_1.

Example 5.7   Finding the Currents and Voltage in the Buck–Boost Regulator

The buck–boost regulator in Figure 5.18a has an input voltage of $V_s = 12$ V. The duty cycle $k = 0.25$ and the switching frequency is 25 kHz. The inductance $L = 150$ μH and filter capacitance $C = 220$ μF. The average load current $I_a = 1.25$ A. Determine (a) the average output voltage, V_a; (b) the peak-to-peak output voltage ripple, ΔV_c; (c) the peak-to-peak ripple current of inductor, ΔI; (d) the peak current of the transistor, I_p; and (e) the critical values of L and C.

Solution

$V_s = 12$ V, $k = 0.25$, $I_a = 1.25$ A, $f = 25$ kHz, $L = 150$ μH, and $C = 220$ μF.

a. From Eq. (5.78), $V_a = -12 \times 0.25/(1 - 0.25) = -4$ V.

b. From Eq. (5.87), the peak-to-peak output ripple voltage is

$$\Delta V_c = \frac{1.25 \times 0.25}{25{,}000 \times 220 \times 10^{-6}} = 56.8 \text{ mV}$$

c. From Eq. (5.84), the peak-to-peak inductor ripple is

$$\Delta I = \frac{12 \times 0.25}{25{,}000 \times 150 \times 10^{-6}} = 0.8 \text{ A}$$

d. From Eq. (5.81), $I_s = 1.25 \times 0.25/(1 - 0.25) = 0.4167$ A. Because I_s is the average of duration kT, the peak-to-peak current of the transistor,

$$I_p = \frac{I_s}{k} + \frac{\Delta I}{2} = \frac{0.4167}{0.25} + \frac{0.8}{2} = 2.067 \text{ A}$$

e. $R = \dfrac{-V_a}{I_a} = \dfrac{4}{1.25} = 3.2 \ \Omega$

From Eq. (5.88), we get $L_c = \dfrac{(1 - k)R}{2f} = \dfrac{(1 - 0.25) \times 3.2}{2 \times 25 \times 10^3} = 450$ μH.

From Eq. (5.89), we get $C_c = \dfrac{k}{2fR} = \dfrac{0.25}{2 \times 25 \times 10^3 \times 3.2} = 1.56$ μF.

5.8.4   Cúk Regulators

The circuit arrangement of the Cúk regulator [10] using a power bipolar junction transistor (BJT) is shown in Figure 5.19a. Similar to the buck–boost regulator, the Cúk regulator provides an output voltage that is less than or greater than the input voltage, but the output voltage polarity is opposite to that of the input voltage. It is named after its inventor [1]. When the input voltage is turned on and transistor Q_1 is switched off,

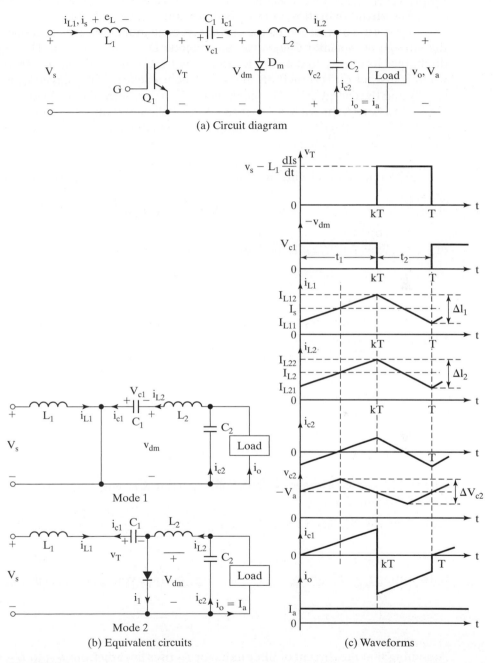

FIGURE 5.19

Cúk regulator.

diode D_m is forward biased and capacitor C_1 is charged through L_1, D_m, and the input supply V_s.

The circuit operation can be divided into two modes. Mode 1 begins when transistor Q_1 is turned on at $t = 0$. The current through inductor L_1 rises. At the same time, the voltage of capacitor C_1 reverse biases diode D_m and turns it off. The capacitor C_1 discharges its energy to the circuit formed by C_1, C_2, the load, and L_2. Mode 2 begins when transistor Q_1 is turned off at $t = t_1$. The capacitor C_1 is charged from the input supply and the energy stored in the inductor L_2 is transferred to the load. The diode D_m and transistor Q_1 provide a synchronous switching action. The capacitor C_1 is the medium for transferring energy from the source to the load. The equivalent circuits for the modes are shown in Figure 5.19b and the waveforms for steady-state voltages and currents are shown in Figure 5.19c for a continuous load current.

Assuming that the current of inductor L_1 rises linearly from I_{L11} to I_{L12} in time t_1,

$$V_s = L_1 \frac{I_{L12} - I_{L11}}{t_1} = L_1 \frac{\Delta I_1}{t_1} \tag{5.90}$$

or

$$t_1 = \frac{\Delta I_1 L_1}{V_s} \tag{5.91}$$

and due to the charged capacitor C_1, the current of inductor L_1 falls linearly from I_{L12} to I_{L11} in time t_2,

$$V_s - V_{c1} = -L_1 \frac{\Delta I_1}{t_2} \tag{5.92}$$

or

$$t_2 = \frac{-\Delta I_1 L_1}{V_s - V_{c1}} \tag{5.93}$$

where V_{c1} is the average voltage of capacitor C_1, and $\Delta I_1 = I_{L12} - I_{L11}$. From Eqs. (5.90) and (5.92).

$$\Delta I_1 = \frac{V_s t_1}{L_1} = \frac{-(V_s - V_{c1})t_2}{L_1}$$

Substituting $t_1 = kT$ and $t_2 = (1 - k)T$, the average voltage of capacitor C_1 is

$$V_{c1} = \frac{V_s}{1 - k} \tag{5.94}$$

Assuming that the current of filter inductor L_2 rises linearly from I_{L21} to I_{L22} in time t_1,

$$V_{c1} + V_a = L_2 \frac{I_{L22} - I_{L21}}{t_1} = L_2 \frac{\Delta I_2}{t_1} \tag{5.95}$$

or

$$t_1 = \frac{\Delta I_2 L_2}{V_{c1} + V_a} \tag{5.96}$$

and the current of inductor L_2 falls linearly from I_{L22} to I_{L21} in time t_2,

$$V_a = -L_2 \frac{\Delta I_2}{t_2} \tag{5.97}$$

or

$$t_2 = -\frac{\Delta I_2 L_2}{V_a} \tag{5.98}$$

where $\Delta I_2 = I_{L22} - I_{L21}$. From Eqs. (5.95) and (5.97),

$$\Delta I_2 = \frac{(V_{c1} + V_a)t_1}{L_2} = -\frac{V_a t_2}{L_2}$$

Substituting $t_1 = kT$ and $t_2 = (1 - k)T$, the average voltage of capacitor C_1 is

$$V_{c1} = -\frac{V_a}{k} \tag{5.99}$$

Equating Eq. (5.94) to Eq. (5.99), we can find the average output voltage as

$$V_a = -\frac{kV_s}{1 - k} \tag{5.100}$$

which gives

$$k = \frac{V_a}{V_a - V_S} \tag{5.101}$$

$$1 - k = \frac{V_S}{V_S - V_a} \tag{5.102}$$

Assuming a lossless circuit, $V_s I_s = -V_a I_a = V_s I_a k / (1 - k)$ and the average input current,

$$I_s = \frac{kI_a}{1 - k} \tag{5.103}$$

The switching period T can be found from Eqs. (5.91) and (5.93):

$$T = \frac{1}{f} = t_1 + t_2 = \frac{\Delta I_1 L_1}{V_s} - \frac{\Delta I_1 L_1}{V_s - V_{c1}} = \frac{-\Delta I_1 L_1 V_{c1}}{V_s(V_s - V_{c1})} \tag{5.104}$$

which gives the peak-to-peak ripple current of inductor L_1 as

$$\Delta I_1 = \frac{-V_s(V_s - V_{c1})}{fL_1 V_{c1}} \tag{5.105}$$

or

$$\Delta I_1 = \frac{V_s k}{f L_1} \tag{5.106}$$

The switching period T can also be found from Eqs. (5.96) and (5.98):

$$T = \frac{1}{f} = t_1 + t_2 = \frac{\Delta I_2 L_2}{V_{c1} + V_a} - \frac{\Delta I_2 L_2}{V_a} = \frac{-\Delta I_2 L_2 V_{c1}}{V_a (V_{c1} + V_a)} \tag{5.107}$$

and this gives the peak-to-peak ripple current of inductor L_2 as

$$\Delta I_2 = \frac{-V_a (V_{c1} + V_a)}{f L_2 V_{c1}} \tag{5.108}$$

or

$$\Delta I_2 = -\frac{V_a (1 - k)}{f L_2} = \frac{k V_s}{f L_2} \tag{5.109}$$

When transistor Q_1 is off, the energy transfer capacitor C_1 is charged by the input current for time $t = t_2$. The average charging current for C_1 is $I_{c1} = I_s$ and the peak-to-peak ripple voltage of the capacitor C_1 is

$$\Delta V_{c1} = \frac{1}{C_1} \int_0^{t_2} I_{c1} \, dt = \frac{1}{C_1} \int_0^{t_2} I_s \, dt = \frac{I_s t_2}{C_1} \tag{5.110}$$

Equation (5.102) gives $t_2 = V_s/[(V_s - V_a)f]$ and Eq. (5.110) becomes

$$\Delta V_{c1} = \frac{I_s V_s}{(V_s - V_a) f C_1} \tag{5.111}$$

or

$$\Delta V_{c1} = \frac{I_s (1 - k)}{f C_1} \tag{5.112}$$

If we assume that the load current ripple Δi_o is negligible, $\Delta i_{L2} = \Delta i_{c2}$. The average charging current of C_2, which flows for time $T/2$, is $I_{c2} = \Delta I_2/4$ and the peak-to-peak ripple voltage of capacitor C_2 is

$$\Delta V_{c2} = \frac{1}{C_2} \int_0^{T/2} I_{c2} \, dt = \frac{1}{C_2} \int_0^{T/2} \frac{\Delta I_2}{4} \, dt = \frac{\Delta I_2}{8 f C_2} \tag{5.113}$$

or

$$\Delta V_{c2} = -\frac{V_a (1 - k)}{8 C_2 L_2 f^2} = \frac{k V_s}{8 C_2 L_2 f^2} \tag{5.114}$$

Condition for continuous inductor current and capacitor voltage. If I_{L1} is the average current of inductor L_1, the inductor ripple current $\Delta I_1 = 2I_{L1}$. Using Eqs. (5.103) and (5.106), we get

$$\frac{kV_S}{fL_1} = 2I_{L1} = 2I_S = \frac{2kI_a}{1-k} = 2\left(\frac{k}{1-k}\right)^2 \frac{V_S}{R}$$

which gives the critical value of the inductor L_{c1} as

$$L_{c1} = L_1 = \frac{(1-k)^2 R}{2kf} \tag{5.115}$$

If I_{L2} is the average current of inductor L_2, the inductor ripple current $\Delta I_1 = 2I_{L2}$. Using Eqs. (5.100) and (5.109), we get

$$\frac{kV_S}{fL_2} = 2I_{L2} = 2I_a = \frac{2V_a}{R} = \frac{2kV_S}{(1-k)R}$$

which gives the critical value of the inductor L_{c2} as

$$L_{c2} = L_2 = \frac{(1-k)R}{2f} \tag{5.116}$$

If V_{c1} is the average capacitor voltage, the capacitor ripple voltage $\Delta V_{c1} = 2V_a$. Using $\Delta V_{c1} = 2\,V_a$ into Eq. (5.112), we get

$$\frac{I_S(1-k)}{fC_1} = 2V_a = 2I_a R$$

which, after substituting for I_S, gives the critical value of the capacitor C_{c1} as

$$C_{c1} = C_1 = \frac{k}{2fR} \tag{5.117}$$

If V_{c2} is the average capacitor voltage, the capacitor ripple voltage $\Delta V_{c2} = 2V_a$. Using Eq. (5.100) and (5.114), we get

$$\frac{kV_S}{8C_2 L_2 f^2} = 2V_a = \frac{2kV_S}{1-k}$$

which, after substituting for L_2 from Eq. (5.116), gives the critical value of the capacitor C_{c2} as

$$C_{c2} = C_2 = \frac{1}{8fR} \tag{5.118}$$

The Cúk regulator is based on the capacitor energy transfer. As a result, the input current is continuous. The circuit has low switching losses and has high efficiency. When transistor Q_1 is turned on, it has to carry the currents of inductors L_1 and L_2. As a result a high peak current flows through transistor Q_1. Because the capacitor provides the energy transfer, the ripple current of the capacitor C_1 is also high. This circuit also requires an additional capacitor and inductor.

Example 5.8 Finding the Currents and Voltages in the Cúk Regulator

The input voltage of a Cúk converter in Figure 5.19a, $V_s = 12$ V. The duty cycle $k = 0.25$ and the switching frequency is 25 kHz. The filter inductance is $L_2 = 150$ μH and filter capacitance is $C_2 = 220$ μF. The energy transfer capacitance is $C_1 = 200$ μF and inductance $L_1 = 180$ μH. The average load current is $I_a = 1.25$ A. Determine (a) the average output voltage V_a; (b) the average input current I_s; (c) the peak-to-peak ripple current of inductor L_1, ΔI_1; (d) the peak-to-peak ripple voltage of capacitor C_1, ΔV_{c1}; (e) the peak-to-peak ripple current of inductor L_2, ΔI_2; (f) the peak-to-peak ripple voltage of capacitor C_2, ΔV_{c2}; and (g) the peak current of the transistor I_p.

Solution
$V_s = 12$ V, $k = 0.25$, $I_a = 1.25$ A, $f = 25$ kHz, $L_1 = 180$ μH, $C_1 = 200$ μF, $L_2 = 150$ μH, and $C_2 = 220$ μF.

    a. From Eq. (5.100), $V_a = -0.25 \times 12/(1 - 0.25) = -4$ V.

    b. From Eq. (5.103), $I_s = 1.25 \times 0.25/(1 - 0.25) = 0.42$ A.

    c. From Eq. (5.106), $\Delta I_1 = 12 \times 0.25/(25{,}000 \times 180 \times 10^{-6}) = 0.67$ A.

    d. From Eq. (5.112), $\Delta V_{c1} = 0.42 \times (1 - 0.25)/(25{,}000 \times 200 \times 10^{-6}) = 63$ mV.

    e. From Eq. (5.109), $\Delta I_2 = 0.25 \times 12/(25{,}000 \times 150 \times 10^{-6}) = 0.8$ A.

    f. From Eq. (5.113), $\Delta V_{c2} = 0.8/(8 \times 25{,}000 \times 220 \times 10^{-6}) = 18.18$ mV.

    g. The average voltage across the diode can be found from

$$V_{dm} = -kV_{c1} = -V_a k\, \frac{1}{-k} = V_a \tag{5.119}$$

For a lossless circuit, $I_{L2}V_{dm} = V_a I_a$ and the average value of the current in inductor L_2 is

$$I_{L2} = \frac{I_a V_a}{V_{dm}} = I_a \tag{5.120}$$
$$= 1.25 \text{ A}$$

Therefore, the peak current of transistor is

$$I_p = I_s + \frac{\Delta I_1}{2} + I_{L2} + \frac{\Delta I_2}{2} = 0.42 + \frac{0.67}{2} + 1.25 + \frac{0.8}{2} = 2.405 \text{ A}$$

5.8.5 Limitations of Single-Stage Conversion

The four regulators use only one transistor, employing only one stage conversion, and require inductors or capacitors for energy transfer. Due to the current-handling limitation of a single transistor, the output power of these regulators is small, typically tens of watts. At a higher current, the size of these components increases, with increased component losses, and the efficiency decreases. In addition, there is no isolation between the input and output voltage, which is a highly desirable criterion in most applications. For high-power applications, multistage conversions are used, where a dc voltage is converted to ac by an inverter. The ac output is isolated by a transformer and then converted to dc by rectifiers. The multistage conversions are discussed in Section 14-4.

Key Points of Section 5.8

- A dc regulator can produce a dc output voltage, which is higher or lower than the dc supply voltage. *LC* filters are used to reduce the ripple content of the output voltage. Depending on the type of the regulator, the polarity of the output voltage can be opposite of the input voltage.

5.9 COMPARISON OF REGULATORS

When a current flows through an inductor, a magnetic field is set up. Any change in this current changes this field and an emf is induced. This emf acts in such a direction as to maintain the flux at its original density. This effect is known as the *self-induction*. An inductor limits the rise and fall of its currents and tries to maintain the ripple current low.

There is no change in the position of the main switch Q_1 for the buck and buck–boost regulators. Switch Q_1 is connected to the dc supply line. Similarly, there is no change in the position of the main switch Q_1 for the boost and Cúk regulators. Switch Q_1 is connected between the two supply lines. When the switch is closed, the supply is shorted through an inductor L, which limits the rate of rise of the supply current.

In Section 5.8, we derive the voltage gain of the regulators with the assumptions that there were no resistances associated with the inductors and capacitors. However, such resistances though small may reduce the gain significantly [11, 12]. Table 5.1 summarizes the voltage gains of the regulators.

Inductors and capacitors act as energy storage elements in switched-mode regulators, and as filter elements to smooth out the current harmonics. We can notice from Eqs. (B.17) and (B.18) in Appendix B that the magnetic loss increases with the square of frequency. On the other hand, a higher frequency reduces the size of inductors for

TABLE 5.1 Summaries of Regulator Gains [Ref. 11]

Regulator	Voltage Gain, $G(k) = V_a/V_S$ with Negligible Values of r_L and r_C	Voltage Gain, $G(k) = V_a/V_S$ with Finite Values of r_L and r_C
Buck	k	$\dfrac{kR}{R + r_L}$
Boost	$\dfrac{1}{1-k}$	$\dfrac{1}{1-k}\left[\dfrac{(1-k)^2 R}{(1-k)^2 R + r_L + k(1-k)\left(\dfrac{r_C R}{r_C + R}\right)}\right]$
Buck–boost	$\dfrac{-k}{1-k}$	$\dfrac{-k}{1-k}\left[\dfrac{(1-k)^2 R}{(1-k)^2 R + r_L + k(1-k)\left(\dfrac{r_C R}{r_C + R}\right)}\right]$

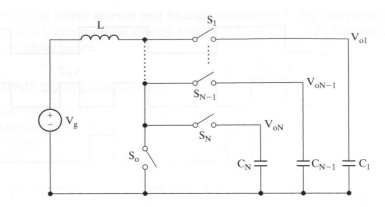

FIGURE 5.22

Topology of boost converter with N outputs.

5.11 DIODE RECTIFIER-FED BOOST CONVERTER

Diode rectifiers are the most commonly used circuits for applications where the input is the ac supply (e.g., in computers, telecommunications, fluorescent lighting, and air-conditioning). The power factor of diode rectifiers with a resistive load can be as high as 0.9, and it is lower with a reactive load. With the aid of a modern control technique, the input current of the rectifiers can be made sinusoidal and in phase with the input voltage, thereby having an input PF of approximate unity. A unity PF circuit that combines a full-bridge rectifier and a boost converter is shown in Figure 5.23a. The input current of the converter is controlled to follow the full-rectified waveform of the sinusoidal input voltage by PWM control [16–23]. The PWM control signals can be generated by using the bang–bang hysteresis (BBH) technique, similar to the delta modulation in Figure 6.26. This technique, which is shown in Figure 5.23b, has the advantage of yielding instantaneous current control, resulting in a fast response. However, the switching frequency is not constant and varies over a wide range during each half-cycle of the ac input voltage. The frequency is also sensitive to the values of the circuit components.

The switching frequency can be maintained constant by using the reference current I_{ref} and feedback current I_{fb} averaged over the per-switching period. This is shown in Figure 5.23c. I_{ref} is compared with I_{fb}. If $I_{ref} > I_{fb}$, the duty cycle is more than 50%. For $I_{ref} = I_{fb}$, the duty cycle is 50%. For $I_{ref} < I_{fb}$, the duty cycle is less than 50%. The error is forced to remain between the maximum and the minimum of the triangular waveform and the inductor current follows the reference sine wave, which is superimposed with a triangular waveform. The reference current I_{ref} is generated from the error voltage $V_e(= V_{ref} - V_o)$ and the input voltage V_{in} to the boost converter.

The boost converter can also be used for the power factor (PF) correction of three-phase diode rectifiers with capacitive output filters [19, 29] as shown in Figure 5.24. The boost converter is operated under DCM of the inductor current mode to achieve a sinusoidal input current shaping. This circuit uses only one active switch,

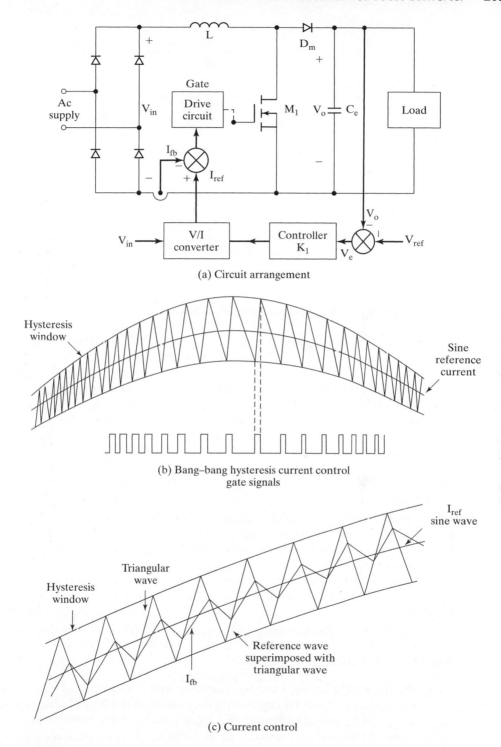

(a) Circuit arrangement

(b) Bang–bang hysteresis current control
gate signals

(c) Current control

FIGURE 5.23

Power factor conditioning of diode rectifiers.

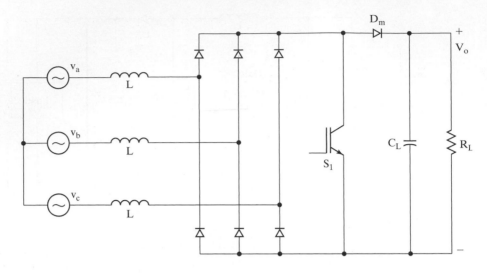

FIGURE 5.24

Three-phase rectifier-fed boost converter. [Ref. 29, C. Mufioz]

with no active control of the current. The drawbacks of the simple converter are excessive output voltage and the presence of fifth harmonics in the line current. This kind of converter is commonly used in industrial and commercial applications requiring a high input power factor because their input-current waveform automatically follows the input-voltage waveform. Also, the circuit has an extremely high efficiency.

However, if the circuit is implemented with the conventional constant-frequency, low-bandwidth, output-voltage feedback control, which keeps the duty cycle of the switch constant during a rectified line period, the rectifier input current exhibits a relatively large fifth-order harmonic. As a result, at power levels above 5 kW, the fifth-order harmonic imposes severe design, performance, and cost trade-offs to meet the maximum permissible harmonic current levels defined by the IEC555-2 document [30]. Advanced control methods such as the harmonic injection method [31] can reduce the fifth-order harmonic of the input current so that the power level at which the input current harmonic content still meets the IEC555-2 standard is extended.

Figure 5.25 shows the block diagram of the robust, harmonic injection technique introduced in [3–5]. A voltage signal that is proportional to the inverted ac component of the rectified, three-phase, line-to-line input voltages is injected into the output-voltage feedback loop. The injected signal varies the duty cycle of the rectifier within a line cycle to reduce the fifth-order harmonic and improve the THD of the rectifier input currents.

Key Points of Section 5.11

- The full-bridge rectifier can be combined with a boost converter to form a unity power factor circuit. By controlling the current of the boost inductor with the aid of feedback control technique, the input current of the rectifier can be made sinusoidal and in phase with the input voltage, thereby having an input PF of approximate unity.

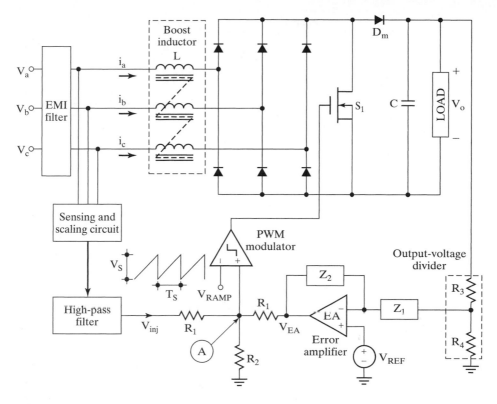

FIGURE 5.25

Three-phase DCM boost rectifier with a harmonic injection method. [Ref. 31, Y. Jang]

5.12   CHOPPER CIRCUIT DESIGN

We can notice from Eq. (5.7) that the output voltage contains harmonics. An output filter of C, LC, L type may be connected to the output to reduce the output harmonics [24, 25]. The techniques for filter design are similar to that of Examples 3.17 and 10.15.

A converter with a highly inductive load is shown in Figure 5.26a. The load current ripple is negligible ($\Delta I = 0$). If the average load current is I_a, the peak load current is $I_m = I_a + \Delta I = I_a$. The input current, which is of pulsed shape as shown in Figure 5.26b, contains harmonics and can be expressed in Fourier series as

$$i_{nh}(t) = kI_a + \frac{I_a}{n\pi} \sum_{n-1}^{\infty} \sin 2n\pi k \cos 2n\pi ft$$

$$+ \frac{I_a}{n\pi} \sum_{n=1}^{\infty} (1 - \cos 2n\pi k) \sin 2n\pi ft \qquad (5.123)$$

The fundamental component ($n = 1$) of the converter-generated harmonic current at the input side is given by

$$i_{1h}(t) = \frac{I_a}{\pi} \sin 2\pi k \cos 2\pi ft + \frac{I_a}{\pi} (1 - \cos 2\pi k) \sin 2\pi ft \qquad (5.124)$$

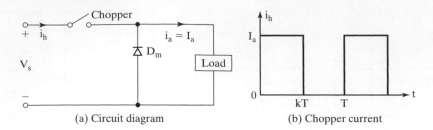

(a) Circuit diagram   (b) Chopper current

FIGURE 5.26

Input current waveform of converter.

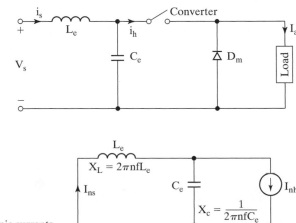

FIGURE 5.27

Converter with input filter.

FIGURE 5.28

Equivalent circuit for harmonic currents.

In practice, an input filter as shown in Figure 5.27 is normally connected to filter out the converter-generated harmonics from the supply line. The equivalent circuit for the converter-generated harmonic currents is shown in Figure 5.28, and the rms value of the nth harmonic component in the supply can be calculated from

$$I_{ns} = \frac{1}{1 + (2n\pi f)^2 L_e C_e} I_{nh} = \frac{1}{1 + (nf/f_0)^2} I_{nh} \qquad (5.125)$$

where f is the chopping frequency and $f_0 = 1/(2\pi\sqrt{L_e C_e})$ is the filter resonant frequency. If $(f/f_0) \gg 1$, which is generally the case, the nth harmonic current in the supply becomes

$$I_{ns} = I_{nh}\left(\frac{f_0}{nf}\right)^2 \qquad (5.126)$$

A high chopping frequency reduces the sizes of input filter elements, but the frequencies of converter-generated harmonics in the supply line are also increased; and this may cause interference problems with control and communication signals.

If the source has some inductances, L_s, and the converter switch as in Figure 5.1a is turned on, an amount of energy can be stored in the source inductance. If an attempt is made to turn off the converter switch, the power semiconductor devices might be damaged due to an induced voltage resulting from this stored energy. The LC input filter provides a low-impedance source for the converter action.

Example 5.9 Finding the Harmonic Input Current of a Dc Converter

A highly inductive load is supplied by a converter as shown in Figure 5.26a. The average load current is $I_a = 100$ A and the load ripple current can be considered negligible ($\Delta I = 0$). A simple LC input filter with $L_e = 0.3$ mH and $C_e = 4500$ μF is used. If the converter is operated at a frequency of 350 Hz and a duty cycle of 0.5, determine the maximum rms value of the fundamental component of converter-generated harmonic current in the supply line.

Solution
For $I_a = 100$ A, $f = 350$ Hz, $k = 0.50$, $C_e = 4500$ μF, and $L_e = 0.3$ mH, $f_0 = 1/(2\pi\sqrt{C_e L_e}) = 136.98$ Hz. Equation (5.124) can be written as

$$I_{1h}(t) = A_1 \cos 2\pi f t + B_1 \sin 2\pi f t$$

where $A_1 = (I_a/\pi) \sin 2\pi k$ and $B_1 = (I_a/\pi)(1 - \cos 2\pi k)$. The peak magnitude of this current is calculated from

$$I_{ph} = (A_1^2 + B_1^2)^{1/2} = \frac{\sqrt{2}I_a}{\pi}(1 - \cos 2\pi k)^{1/2} \tag{5.127}$$

The rms value of this current is

$$I_{1h} = \frac{I_a}{\pi}(1 - \cos 2\pi k)^{1/2} = 45.02 \text{ A}$$

and this becomes maximum at $k = 0.5$. The fundamental component of converter-generated harmonic current in the supply can be calculated from Eq. (5.125) and is given by

$$I_{1s} = \frac{1}{1 + (f/f_0)^2} I_{1h} = \frac{45.02}{1 + (350/136.98)^2} = 5.98 \text{ A}$$

If $f/f_0 \gg 1$, the harmonic current in the supply becomes approximately

$$I_{1s} = I_{1h}\left(\frac{f_0}{f}\right)^2$$

Key Points of Section 5.12

- The design of a dc–dc converter circuit requires (a) determining the converter topology, (b) finding the voltage and currents of the switching devices, (c) finding the values and ratings of passive elements such as capacitors and inductors, and (d) choosing the control strategy and the gating algorithm in order to obtain the desired output.

Example 5.10

A buck converter is shown in Figure 5.29. The input voltage is $V_s = 110$ V, the average load voltage is $V_a = 60$ V, and the average load current is $I_a = 20$ A. The chopping frequency is $f = 20$ kHz. The peak-to-peak ripples are 2.5% for load voltage, 5% for load current, and 10% for filter L_e current. (a) Determine the values of L_e, L, and C_e. Use PSpice (b) to verify the results by plotting the instantaneous capacitor voltage v_C, and instantaneous load current i_L; and (c) to calculate the Fourier coefficients and the input current i_S. The SPICE model parameters of the transistor are IS = 6.734f, BF = 416.4, BR = 0.7371, CJC = 3.638P, CJE = 4.493P, TR = 239.5N, TF = 301.2P, and that of the diode are IS = 2.2E−15, BV = 1800V, TT = 0.

Solution

$$V_s = 110 \text{ V}, V_a = 60 \text{ V}, I_a = 20 \text{ A}.$$
$$\Delta V_c = 0.025 \times V_a = 0.025 \times 60 = 1.5 \text{ V}$$
$$R = \frac{V_a}{I_a} = \frac{60}{20} = 3 \ \Omega$$

From Eq. (5.48),

$$k = \frac{V_a}{V_s} = \frac{60}{110} = 0.5455$$

From Eq. (5.49),

$$I_s = kI_a = 0.5455 \times 20 = 10.91 \text{ A}$$
$$\Delta I_L = 0.05 \times I_a = 0.05 \times 20 = 1 \text{ A}$$
$$\Delta I = 0.1 \times I_a = 0.1 \times 20 = 2 \text{ A}$$

a. From Eq. (5.51), we get the value of L_e:

$$L_e = \frac{V_a(V_s - V_a)}{\Delta I f V_s} = \frac{60 \times (110 - 60)}{2 \times 20 \text{ kHz} \times 110} = 681.82 \ \mu\text{H}$$

From Eq. (5.53) we get the value of C_e:

$$C_e = \frac{\Delta I}{\Delta V_c \times 8f} = \frac{2}{1.5 \times 8 \times 20 \text{ kHz}} = 8.33 \ \mu\text{F}$$

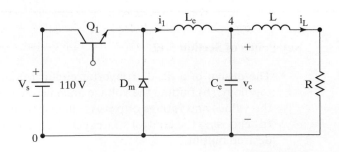

FIGURE 5.29

Buck converter.

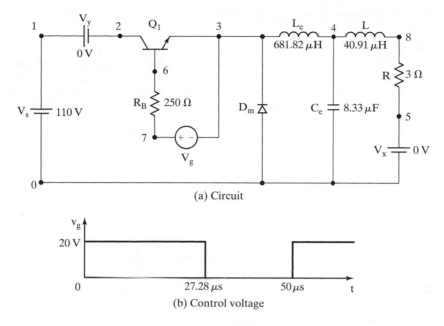

(a) Circuit

(b) Control voltage

FIGURE 5.30

Buck chopper for PSpice simulation.

Assuming a linear rise of load current i_L during the time from $t = 0$ to $t_1 = kT$, we can write approximately

$$L \frac{\Delta I_L}{t_1} = L \frac{\Delta I_L}{kT} = \Delta V_C$$

which gives the approximate value of L:

$$L = \frac{kT \Delta V_c}{\Delta I_L} = \frac{k \Delta V_c}{\Delta I_L f}$$

$$= \frac{0.5454 \times 1.5}{1 \times 20 \text{ kHz}} = 40.91 \text{ } \mu\text{H}$$

(5.128)

b. $k = 0.5455$, $f = 20 \text{ kHz}$, $T = 1/f = 50 \text{ } \mu\text{s}$, and $t_{\text{on}} - k \times T = 27.28 \text{ } \mu\text{s}$. The buck chopper for PSpice simulation is shown in Figure 5.30a. The control voltage V_g is shown in Figure 5.30b. The list of the circuit file is as follows:

```
Example 5.10    Buck Converter
VS    1   0    DC     110V
VY    1   2    DC     0V    ; Voltage source to measure input current
Vg    7   3    PULSE (0V   20V   0   0.1NS   0.1NS    27.28US 50US)
RB    7   6    250                        ; Transistor base resistance
LE    3   4    681.82UH
```

```
CE      4   0    8.33UF       IC=60V  ;  initial voltage
L       4   8    40.91UH
R       8   5    3
VX      5   0    DC    0V              ; Voltage source to measure load current
DM      0   3    DMOD                  ; Freewheeling diode
.MODEL    DMOD    D(IS=2.2E-15 BV=1800V TT=0)    ; Diode model parameters
Q1      2   6    3     QMOD            ; BJT switch
.MODEL  QMOD  NPN  (IS=6.734F BF=416.4 BR=.7371 CJC=3.638P
+ CJE=4.493P TR=239.5N TF=301.2P)              ; BJT model parameters
.TRAN  1US 1.6MS 1.5MS 1US  UIC                ; Transient analysis
.PROBE                                         ; Graphics postprocessor
.options  abstol = 1.00n reltol = 0.01 vntol = 0.1 ITL5=50000 ; convergence
.FOUR    20KHZ   I(VY)                         ; Fourier analysis
.END
```

The PSpice plots are shown in Figure 5.31, where $I(VX)$ = load current, $I(Le)$ = inductor L_e current, and $V(4)$ = capacitor voltage. Using the PSpice cursor in Figure 5.31 gives $V_a = V_c = 59.462$ V, $\Delta V_c = 1.782$ V, $\Delta I = 2.029$ A, $I_{(av)} = 19.813$ A, $\Delta I_L = 0.3278$ A, and $I_a = 19.8249$ A. This verifies the design; however, ΔI_L gives a better result than expected.

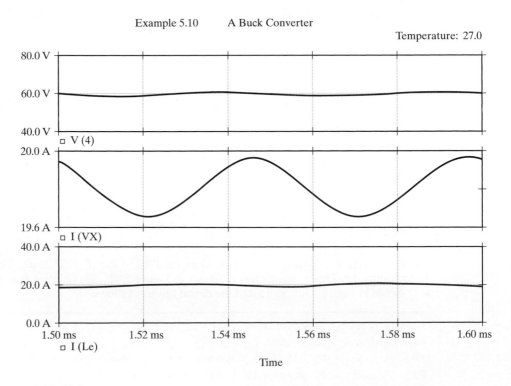

FIGURE 5.31

PSpice plots for Example 5.10.

c. The Fourier coefficients of the input current are

FOURIER COMPONENTS OF TRANSIENT RESPONSE I(VY)

DC COMPONENT = 1.079535E+01

HARMONIC NO	FREQUENCY (HZ)	FOURIER COMPONENT	NORMALIZED COMPONENT	PHASE (DEG)	NORMALIZED PHASE (DEG)
1	2.000E+04	1.251E+01	1.000E+00	−1.195E+01	0.000E+00
2	4.000E+04	1.769E+00	1.415E-01	7.969E+01	9.163E+01
3	6.000E+04	3.848E+00	3.076E-01	−3.131E+01	−1.937E+01
4	8.000E+04	1.686E+00	1.348E-01	5.500E+01	6.695E+01
5	1.000E+05	1.939E+00	1.551E-01	−5.187E+01	−3.992E+01
6	1.200E+05	1.577E+00	1.261E-01	3.347E+01	4.542E+01
7	1.400E+05	1.014E+00	8.107E-02	−7.328E+01	−6.133E+01
8	1.600E+05	1.435E+00	1.147E-01	1.271E+01	2.466E+01
9	1.800E+05	4.385E-01	3.506E-02	−9.751E+01	−8.556E+01

TOTAL HARMONIC DISTORTION =   4.401661E+01 PERCENT

Key Points of Section 5.12

- The design of a dc–dc converter circuit requires (1) determining the converter topology, (2) finding the voltage and currents of the switching devices, (3) finding the values and ratings of passive elements such as capacitors and inductors, and (4) choosing the control strategy and the gating algorithm to obtain the desired output.

5.13   STATE–SPACE ANALYSIS OF REGULATORS

Any nth order linear or nonlinear differential equation in one time-dependent variable can be written [26] as n first-order differential equation in n time-dependent variables x_1 through x_n. Let us consider, for example, the following third-order equation:

$$y''' + a_2 y'' + a_1 y' + a_0 = 0 \qquad (5.129)$$

where y' is the first derivative of y, $y' = (d/dt)\,y$. Let y be x_1. Then Eq. (5.129) can be represented by the three equations

$$x_1' = x_2 \qquad (5.130)$$

$$x_2'' = x_3 \qquad (5.131)$$

$$x_3'' = -a_0 x_1 - a_1 x_2 - a_3 x_3 \qquad (5.132)$$

In each case, n initial conditions must be known before an exact solution can be found. For any nth-order system, a set of n-independent variables is necessary and sufficient to describe that system completely. These variables $x_1, x_2, \ldots, x_n$. are called the *state variables* for the system. If the initial conditions of a linear system are known at time t_0, then we can find the states of the systems for all time $t > t_0$ and for a given set of input sources.

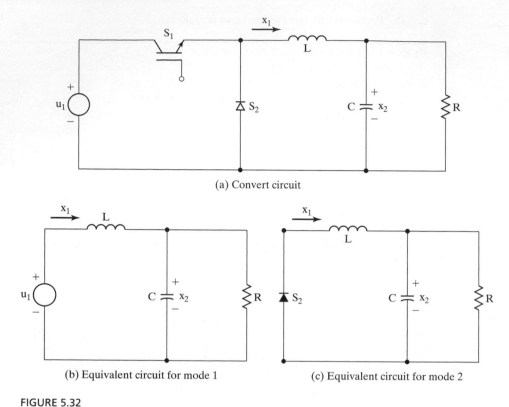

(a) Convert circuit

(b) Equivalent circuit for mode 1

(c) Equivalent circuit for mode 2

FIGURE 5.32

Buck converter with state variables.

All state variables are subscribed x's and all sources are subscribed u's. Let us consider the basic buck converter of Figure 5.16a, which is redrawn in Figure 5.32a. The dc source V_s is replaced with the more general source u_1.

Mode 1. Switch S_1 is on and switch S_2 is off. The equivalent circuit is shown in Figure 5.32b. Applying Kirchoff's voltage law (KVL), we get

$$u_1 = Lx_1' + x_2$$

$$Cx_2' = x_1 - \frac{1}{R}x_2$$

which can be rearranged to

$$x_1' = \frac{-1}{L}x_2 + \frac{1}{L}u_1 \tag{5.133}$$

$$x_2' = \frac{-1}{C}x_2 + \frac{1}{RC}x_2 \tag{5.134}$$

These equations can be written in the universal format:

$$x' = A_1 x + B_1 u_1 \qquad (5.135)$$

where $\quad \mathbf{x} = $ state vector $= \begin{pmatrix} \mathbf{x}_1 \\ \mathbf{x}_2 \end{pmatrix}$

$$A_1 = \text{state coefficient matrix} = \begin{pmatrix} 0 & \dfrac{-1}{L} \\ \dfrac{1}{C} & \dfrac{-1}{RC} \end{pmatrix}$$

$\mathbf{u}_1 = $ source vector

$$B_1 = \text{source coefficient matrix} = \begin{pmatrix} \dfrac{1}{L} \\ 0 \end{pmatrix}$$

Mode 2.   Switch S_1 is off and switch S_2 is on. The equivalent circuit is shown in Figure 5-32c. Applying KVL, we get

$$0 - Lx_1' + x_2$$

$$Cx_2' = x_1 - \frac{1}{R} x_2$$

which can be rearranged to

$$x_1' = \frac{-1}{L} x_2 \qquad (5.136)$$

$$x_2' = \frac{-1}{C} x_2 + \frac{1}{RC} x_2 \qquad (5.137)$$

These equations can be written in the universal format:

$$x' = A_2 x + B_2 u_1 \qquad (5.138)$$

where $\quad \mathbf{x} = $ state vector $= \begin{pmatrix} x_1 \\ x_2 \end{pmatrix}$

$$A_2 = \text{state coefficient matrix} = \begin{pmatrix} 0 & \dfrac{-1}{L} \\ \dfrac{1}{C} & \dfrac{-1}{RC} \end{pmatrix}$$

$\mathbf{u}_1 = $ source vector $= 0$

$$B_2 = \text{source coefficient matrix} = \begin{pmatrix} 0 \\ 0 \end{pmatrix}$$

In feedback systems, the duty cycle k is a function of $\mathbf{x}$ and may be a function of $\mathbf{u}$ as well. Thus, the total solution can be obtained by state–space averaging, that is, by summing the terms for each analysis the switched linear mode. Using the universal format, we get

$$A = A_1 k + A_2 (1 - k) \tag{5.139}$$

$$B = B_1 k + B_2 (1 - k) \tag{5.140}$$

Substituting for A_1, A_2, B_1, and B_2, we can find

$$A = \begin{pmatrix} 0 & \dfrac{-1}{L} \\ \dfrac{1}{C} & \dfrac{1}{RC} \end{pmatrix} \tag{5.141}$$

$$B = \begin{pmatrix} \dfrac{k}{L} \\ 0 \end{pmatrix} \tag{5.142}$$

which in turn lead to the following state equations:

$$x_1' = \frac{-1}{L} x_2 + \frac{k}{L} u_1 \tag{5.143}$$

$$x_2' = \frac{-1}{C} x_2 + \frac{1}{RC} x_2 \tag{5.144}$$

A continuous but nonlinear circuit that is described by Eqs. (5.143) and (5.144) is shown in Figure 5.33. It is a nonlinear circuit because k in general can be a function of x_1, x_2, and u_1.

The state–space averaging is an approximation technique that, for high enough switching frequencies, allows a continuous-time signal frequency analysis to be carried out separately from the switching frequency analysis. Although the original system is linear for any given switch condition, the resulting system (i.e., in Figure 5.33) generally is nonlinear. Therefore, small-signal approximations have to be employed to obtain the linearized small-signal behavior before other techniques [27, 28], such as Laplace transforms and Bode plots, can be applied.

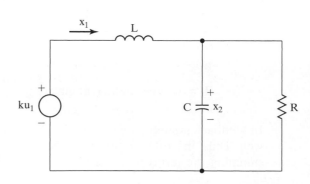

FIGURE 5.33

Continuous equivalent circuit of the
Buck converter with state variables.

Key Points of Section 5.13

- The state–space averaging is an approximate technique that can be applied to describe the input and output relations of a switching converter, having different switching modes of operation. Although the original system is linear for any given switch condition, the resulting system generally is nonlinear. Therefore, small-signal approximations have to be employed to obtain the linearized small-signal behavior before other techniques can be applied.

SUMMARY

A dc converter can be used as a dc transformer to step up or step down a fixed dc voltage. The converter can also be used for switching-mode voltage regulators and for transferring energy between two dc sources. However, harmonics are generated at the input and load side of the converter, and these harmonics can be reduced by input and output filters. A converter can operate on either fixed frequency or variable frequency. A variable-frequency converter generates harmonics of variable frequencies and a filter design becomes difficult. A fixed-frequency converter is normally used. To reduce the sizes of filters and to lower the load ripple current, the chopping frequency should be high. The state–space averaging technique can be applied to describe the input and output relations of a switching converter, having different switching modes of operation.

REFERENCES

[1]   J. A. M. Bleijs, and J. A. Gow, "Fast maximum power point control of current-fed DC–DC converter for photovoltaic arrays," *Electronics Letters*, Vol. 37, No. 1, January 2001, pp. 5–6.

[2]   A. J. Forsyth, and S. V. Mollov, "Modeling and control of DC–DC converters," *Power Engineering Journal*, Vol. 12, No. 5, 1998, pp. 229–236.

[3]   A. L. Baranovski, A. Mogel, W. Schwarz, and O. Woywode, "Chaotic control of a DC–DC-Converter," *Proc. IEEE International Symposium on Circuits and Systems*, Geneva, Switzerland, 2000, Vol. 2, pp. II-108–II-111.

[4]   H. Matsuo, F. Kurokawa, H. Etou, Y. Ishizuka, and C. Chen, Changfeng, "Design oriented analysis of the digitally controlled dc–dc converter," *Proc. IEEE Power Electronics Specialists Conference*, Galway, U.K., 2000, pp. 401–407.

[5]   J. L. Rodriguez Marrero, R. Santos Bueno, and G. C. Verghese, "Analysis and control of chaotic DC–DC switching power converters," *Proc. IEEE International Symposium on Circuits and Systems*, Orlando, FL, Vol. 5, 1999, pp. V-287–V-292.

[6]   G. Ioannidis, A. Kandianis, and S. N. Manias, "Novel control design for the buck converter," *IEE Proceedings: Electric Power Applications,* Vol. 145, No. 1, January 1998, pp. 39–47.

[7]   R. Oruganti, and M. Palaniapan, "Inductor voltage control of buck-type single-phase ac–dc converter," *IEEE Transactions on Power Electronics*, Vol. 15, No. 2, 2000, pp. 411–417.

[8]   V. J. Thottuvelil, and G. C. Verghese, "Analysis and control design of paralleled DC/DC converters with current sharing," *IEEE Transactions on Power Electronics*, Vol. 13, No. 4, 1998, pp. 635–644.

[9]   Y. Berkovich, and A. Ioinovici, "Dynamic model of PWM zero-voltage-transition DC–DC boost converter," *Proc. IEEE International Symposium on Circuits and Systems*, Orlando, FL, Vol. 5, 1999, pp. V-254–V-25.

[10] S. Cúk and R. D. Middlebrook, "Advances in switched mode power conversion," *IEEE Transactions on Industrial Electronics*, Vol. IE30, No. 1, 1983, pp. 10–29.

[11] K. Kit Sum, *Switch Mode Power Conversion—Basic Theory and Design*. New York: Marcel Dekker. 1984, Chapter 1.

[12] D. Ma, W. H. Ki, C. Y. Tsui, and P. K. T. Mok, "A 1.8-V single-inductor dual-output switching converter for power reduction techniques," Symposium on VLSI Circuits, 2001, pp. 137–140.

[13] R. D. Middlebrook and S. Cúk, "A general unified approach to modeling dc-to-dc converters in discontinuous conduction mode," *IEEE Power Electronics Specialist Conference*, 1977, pp. 36–57.

[14] H. S. H. Chung, "Design and analysis of a switched-capacitor-based step-up DC/DC converter with continuous input current," *IEEE Transactions on Circuits and Systems I: Fundamental Theory and Applications*, Vol. 46, No. 6, 1999, pp. 722–730.

[15] H. S. H. Chung, S. Y. R. Hui, and S. C. Tang, "Development of low-profile DC/DC converter using switched-capacitor circuits and coreless PCB gate drive," *Proc. IEEE Power Electronics Specialists Conference*, Charleston, SC, Vol. 1, 1999, pp. 48–53.

[16] M. Kazerani, P. D. Ziogas, and G. Ioos, "A novel active current wave shaping technique for solid-state input power factor conditioners," *IEEE Transactions on Industrial Electronics*, Vol. IE38, No. 1, 1991, pp. 72–78.

[17] B. I. Takahashi, "Power factor improvements of a diode rectifier circuit by dither signals," *Conference Proc. IEEE-IAS Annual Meeting*, Seattle, WA, October 1990, pp. 1279–1294.

[18] A. R. Prasad and P. D. Ziogas, "An active power factor correction technique for three phase diode rectifiers," *IEEE Transactions on Power Electronics*, Vol. 6, No. 1, 1991, pp. 83–92.

[19] A. R. Prasad, P. D. Ziogas, and S. Manias, "A passive current wave shaping method for three phase diode rectifiers," *Proc. IEEE APEC-91 Conference Record*, 1991, pp. 319–330.

[20] M. S. Dawande, and G. K. Dubey, "Programmable input power factor correction method for switch-mode rectifiers," *IEEE Transactions on Power Electronics*, Vol. 2, No. 4, 1996, pp. 585–591.

[21] M. S. Dawande, V. R. Kanetkar, and G. K. Dubey, "Three-phase switch mode rectifier with hysteresis current control," *IEEE Transactions on Power Electronics*, Vol. 2, No. 3, 1996, pp. 466–471.

[22] E. L. M. Mehl, and I. Barbi, "An improved high-power factor and low-cost three-phase rectifier," *IEEE Transactions on Industry Applications*, Vol. 33, No. 2, 1997, pp. 485–492.

[23] F. Daniel, R. Chaffai, and K. AI-Haddad, "Three-phase diode rectifier with low harmonic distortion to feed capacitive loads," *IEEE APEC Conference Proc.*, 1996, pp. 932–938.

[24] M. Florez-Lizarraga and A. F. Witulski, Input filter design for multiple-module DC power systems," *IEEE Transactions on Power Electronics*, Vol. 2, No. 3, 1996, pp. 472–479.

[25] M. Alfayyoumi, A. H. Nayfeh, and D. Borojevic, "Input filter interactions in DC–DC switching regulators," *Proc. IEEE Power Electronics Specialists Conference*, Charleston, SC, Vol. 2, 1999, pp. 926–932.

[26] D. M. Mitchell, *DC–DC Switching Regulator*. New York: McGraw-Hill. 1988, Chapters 2 and 4.

[27] B. Lehman and R. M. Bass, "Extensions of averaging theory for power electronic systems," *IEEE Transactions on Power Electronics*, Vol. 2, No. 4, 1996, pp. 542–553.

[28] H. Bevrani, M. Abrishamchian, and N. Safari-shad," Nonlinear and linear robust control of switching power converters," *Proc. IEEE International Conference on Control Applications*, Vol. 1, 1999, pp. 808–813.

[29] C. A. Mufioz, and I. Barbi, "A new high-power-factor three-phase ac–dc converter: analysis, design, and experimentation," *IEEE Transactions on Power Electronics*, Vol. 14, No. 1, January 1999, pp. 90–97.

[30] *IEC Publication 555:* Disturbances in supply systems caused by household appliances and similar equipment; Part 2: Harmonics.

[31] Y. Jang, and M. M. Jovanovic, "A new input-voltage feed forward harmonic-injection technique with nonlinear gain control for single-switch, three-phase, DCM boost rectifiers," *IEEE Transactions on Power Electronics*, Vol. 28, No. 1, March 2000, pp. 268–277.

[32] M. H. Rashid, *SPICE for Power Electronics Using PSpice*. Englewood Cliffs, N.J.: Prentice-Hall. 1993, Chapters 10 and 11.

[33] P. Wood, *Switching Power Converters*. New York: Van Nostrand Reinhold. 1981.

[34] R. P. Sevems and G. E. Bloom, *Modern DC-to-DC Switch Mode Power Converter Circuits*. New York: Van Nostrand Reinhold. 1983.

REVIEW QUESTIONS

5.1 What is a dc chopper, or dc–dc converter?

5.2 What is the principle of operation of a step-down converter?

5.3 What is the principle of operation of a step-up converter?

5.4 What is pulse-width-modulation control of a converter?

5.5 What is frequency-modulation control of a converter?

5.6 What are the advantages and disadvantages of a variable-frequency converter?

5.7 What is the effect of load inductance on the load ripple current?

5.8 What is the effect of chopping frequency on the load ripple current?

5.9 What are the constraints for controllable transfer of energy between two dc voltage sources?

5.10 What is the algorithm for generating the duty cycle of a converter?

5.11 What is the modulation index for a PWM control?

5.12 What is a first and second quadrant converter?

5.13 What is a third and fourth quadrant converter?

5.14 What is a four-quadrant converter?

5.15 What are the performance parameters of a converter?

5.16 What is a switching-mode regulator?

5.17 What are the four basic types of switching-mode regulators?

5.18 What are the advantages and disadvantages of a buck regulator?

5.19 What are the advantages and disadvantages of a boost regulator?

5.20 What are the advantages and disadvantages of a buck–boost regulator?

5.21 What are the advantages and disadvantages of a Cúk regulator?

5.22 At what duty cycle does the load ripple current become maximum?

5.23 What are the effects of chopping frequency on filter sizes?

5.24 What is the discontinuous mode of operation of a regulator?

5.25 What is a multioutput boost converter?

5.26 Why must the multioutput boost converter be operated with time multiplexing control?

5.27 Why must the multioutput boost converter be operated in discontinuous mode?

5.28 How can the input current of the rectifier-fed boost converter be made sinusoidal and in phase with the input voltage?

5.29 What is a state–space averaging technique?

PROBLEMS

5.1 The dc converter in Figure 5.1a has a resistive load, $R = 20\ \Omega$ and input voltage, $V_s = 220$ V. When the converter remains on, its voltage drop is $V_{ch} = 1.5$ V and chopping frequency is $f = 10$ kHz. If the duty cycle is 80%, determine **(a)** the average output voltage V_a, **(b)** the rms output voltage V_o, **(c)** the converter efficiency, **(d)** the effective input resistance R_i, and **(e)** the rms value of the fundamental component of harmonics on the output voltage.

5.2 A converter is feeding an RL load as shown in Figure 5.3 with $V_s = 220$ V, $R = 10\ \Omega$, $L = 15.5$ mH, $f = 5$ kHz, and $E = 20$ V. Calculate **(a)** the minimum instantaneous load current I_1, **(b)** the peak instantaneous load current I_2, **(c)** the maximum peak-to-peak ripple current in the load, **(d)** the average load current I_a, **(e)** the rms load current I_o, **(f)** the effective input resistance R_i, and **(g)** the rms value of converter current I_R.

5.3 The converter in Figure 5.3 has load resistance, $R = 0.2\ \Omega$; input voltage $V_s = 220$ V; and battery voltage, $E = 10$ V. The average load current, $I_a = 200$ A, and the chopping frequency is $f = 200$ Hz ($T = 5$ ms). Use the average output voltage to calculate the value of load inductance L, which would limit the maximum load ripple current to 5% of I_a.

5.4 The dc converter shown in Figure 5.7a is used to control power flow from a dc voltage, $V_s = 110$ V to a battery voltage, $E = 220$ V. The power transferred to the battery is 30 kW. The current ripple of the inductor is negligible. Determine **(a)** the duty cycle K, **(b)** the effective load resistance R_{eq}, and **(c)** the average input current I_s.

5.5 For Problem 5.4, plot the instantaneous inductor current and current through the battery E if inductor L has a finite value of $L = 7.5$ mH, $f = 250$ Hz, and $k = 0.5$.

5.6 An RL load as shown in Figure 5.3 is controlled by a converter. If load resistance $R = 0.25\ \Omega$, inductance $L = 20$ mH, supply voltage $V_s = 600$, battery voltage $E = 150$ V, and chopping frequency $f = 250$ Hz, determine the minimum and maximum load current, the peak-to-peak load ripple current, and average load current for $k = 0.1$ to 0.9 with a step of 0.1.

5.7 Determine the maximum peak-to-peak ripple current of Problem 5.6 by using Eqs. (5.21) and (5.22), and compare the results.

5.8 The step-up converter in Figure 5.8a has $R = 10\ \Omega$, $L = 6.5$ mH, $E = 5$ V, and $k = 0.5$. Find I_1, I_2, and ΔI. Use SPICE to find these values and plot the load, diode, and switch current.

5.9 The buck regulator in Figure 5.16a has an input voltage, $V_s = 15$ V. The required average output voltage $V_a = 5$ V and the peak-to-peak output ripple voltage is 10 mV. The switching frequency is 20 kHz. The peak-to-peak ripple current of inductor is limited to 0.5 A. Determine **(a)** the duty cycle k, **(b)** the filter inductance L, **(c)** the filter capacitor C, and **(d)** the critical values of L and C.

5.10 The boost regulator in Figure 5.17a has an input voltage, $V_s = 6$ V. The average output voltage, $V_a = 15$ V and average load current, $I_a = 0.5$ A. The switching frequency is 20 kHz. If $L = 250\ \mu$H and $C = 440\ \mu$F, determine **(a)** the duty cycle k **(b)** the ripple current of inductor, ΔI, **(c)** the peak current of inductor, I_2, **(d)** the ripple voltage of filter capacitor, ΔV_c, and **(e)** the critical values of L and C.

5.11 The buck–boost regulator in Figure 5.18a has an input voltage, $V_s = 12$ V. The duty cycle, $k = 0.6$, and the switching frequency is 25 kHz. For the inductance, $L = 250\ \mu$H and for filter capacitance, $C = 220\ \mu$F. For the average load current, $I_a = 1.5$ A. Determine **(a)** the average output voltage V_a, **(b)** the peak-to-peak output ripple voltage ΔV_c, **(c)** the peak-to-peak ripple current of inductor ΔI, **(d)** the peak current of the transistor I_p, and **(e)** the critical values of L and C.

5.12 The Cúk regulator in Figure 5.19a has an input voltage, $V_s = 15$ V. The duty cycle is $k = 0.4$ and the switching frequency is 25 kHz. The filter inductance is $L_2 = 350\ \mu$H and filter capacitance is $C_2 = 220$ μF. The energy transfer capacitance is $C_1 = 400$ μF and inductance is $L_1 = 250$ μH. The average load current is $I_a = 1.25$ A. Determine **(a)** the average output voltage V_a, **(b)** the average input current I_s, **(c)** the peak-to-peak ripple current of inductor L_1, ΔI_1, **(d)** the peak-to-peak ripple voltage of capacitor C_1, ΔV_{c1}, **(e)** the peak-to-peak ripple current of inductor L_2, ΔI_2, **(f)** the peak-to-peak ripple voltage of capacitor C_2, ΔV_{c2}, and **(g)** the peak current of the transistor I_p.

5.13 In Problem 5.12 for the Cúk regulator, find the critical values of L_1, C_1, L_2, and C_2.

5.14 The buck converter in Figure 5.29 has a dc input voltage $V_s = 110$ V, average load voltage $V_a = 80$ V, and average load current $I_a = 20$ A. The chopping frequency is $f = 10$ kHz. The peak-to-peak ripples are 5% for load voltage, 2.5% for load current, and 10% for filter L_e current. **(a)** Determine the values of L_e, L, and C_e. Use PSpice **(b)** to verify the results by plotting the instantaneous capacitor voltage v_C and instantaneous load current i_L, and **(c)** to calculate the Fourier coefficients of the input current i_s. Use SPICE model parameters of Example 5.10.

5.15 The boost converter in Figure 5.17a has a dc input voltage $V_S = 5$ V. The load resistance R is 100 Ω. The inductance is $L = 150$ μH, and the filter capacitance is $C = 220$ μF. The chopping frequency is $f = 20$ kHz and the duty cycle of the converter is $k = 60\%$. Use PSpice **(a)** to plot the output voltage v_C, the input current i_s, and the MOSFET voltage v_T; and **(b)** to calculate the Fourier coefficients of the input current i_s. The SPICE model parameters of the MOSFET are $L = 2U$, $W = 0.3$, $VTO = 2.831$, $KP = 20.53U$, $IS = 194E{-}18$, $CGSO = 9.027N$, $CGDO = 1.679N$.

5.16 A dc regulator is operated at a duty cycle of $k = 0.4$. The load resistance is $R = 150$ Ω, the inductor resistance is $r_L = 1$ Ω, and the resistance of the filter capacitor is $r_c = 0.2$ Ω. Determine the voltage gain for the **(a)** buck converter, **(b)** boost converter and **(c)** buck–boost converter.

C H A P T E R   6

Pulse-Width-Modulated Inverters

The learning objectives of this chapter are as follows:

- To learn the switching technique for dc–ac converters known as inverters and the types of inverters
- To study the operation of inverters
- To understand the performance parameters of inverters
- To learn the different types of modulation techniques to obtain a near sinusoidal output waveform and the techniques to eliminate certain harmonics from the output
- To learn the techniques for analyzing and designing inverters and for simulating inverters by using SPICE
- To study the effects of load impedance on the load current

6.1 INTRODUCTION

Dc-to-ac converters are known as *inverters*. The function of an inverter is to change a dc input voltage to a symmetric ac output voltage of desired magnitude and frequency [1]. The output voltage could be fixed or variable at a fixed or variable frequency. A variable output voltage can be obtained by varying the input dc voltage and maintaining the gain of the inverter constant. On the other hand, if the dc input voltage is fixed and it is not controllable, a variable output voltage can be obtained by varying the gain of the inverter, which is normally accomplished by pulse-width-modulation (PWM) control within the inverter. The *inverter gain* may be defined as the ratio of the ac output voltage to dc input voltage.

The output voltage waveforms of ideal inverters should be sinusoidal. However, the waveforms of practical inverters are nonsinusoidal and contain certain harmonics. For low- and medium-power applications, square-wave or quasi-square-wave voltages may be acceptable; and for high-power applications, low distorted sinusoidal waveforms are required. With the availability of high-speed power semiconductor devices, the harmonic contents of output voltage can be minimized or reduced significantly by switching techniques.

Inverters are widely used in industrial applications (e.g., variable-speed ac motor drives, induction heating, standby power supplies, and uninterruptible power supplies). The input may be a battery, fuel cell, solar cell, or other dc source. The typical single-phase outputs are (1) 120 V at 60 Hz, (2) 220 V at 50 Hz, and (3) 115 V at 400 Hz. For high-power three-phase systems, typical outputs are (1) 220 to 380 V at 50 Hz, (2) 120 to 208 V at 60 Hz, and (3) 115 to 200 V at 400 Hz.

Inverters can be broadly classified into two types: (1) single-phase inverters, and (2) three-phase inverters. Each type can use controlled turn-on and turn-off devices (e.g., bipolar junction transistors [BJTs], metal oxide semiconductor field-effect transistors [MOSFETs], insulated-gate bipolar transistors [IGBTs], metal oxide semiconductor-controlled thyristors [MCTs], static induction transistors, [SITs], and gate-turn-off thyristors [GTOs]). These inverters generally use PWM control signals for producing ac output voltage. An inverter is called a *voltage-fed inverter* (VFI) if the input voltage remains constant, a *current-fed inverter* (CFI) if the input current is maintained constant, and a *variable dc linked inverter* if the input voltage is controllable. If the output voltage or current of the inverter is forced to pass through zero by creating an *LC* resonant circuit, this type of inverter is called *resonant-pulse inverter* and it has wide applications in power electronics. Chapter 8 is devoted to resonant-pulse inverters only.

6.2   PRINCIPLE OF OPERATION

The principle of single-phase inverters [1] can be explained with Figure 6.1a. The inverter circuit consists of two choppers. When only transistor Q_1 is turned on for a time $T_0/2$, the instantaneous voltage across the load v_0 is $V_s/2$. If transistor Q_2 only is turned on for a time $T_0/2$, $-V_s/2$ appears across the load. The logic circuit should be designed such that Q_1 and Q_2 are not turned on at the same time. Figure 6.1b shows the waveforms for the output voltage and transistor currents with a resistive load. This inverter requires a three-wire dc source, and when a transistor is off, its reverse voltage is V_s instead of $V_s/2$. This inverter is known as a *half-bridge inverter*.

The root-mean-square (rms) output voltage can be found from

$$V_o = \left(\frac{2}{T_0} \int_0^{T_0/2} \frac{V_s^2}{4} \, dt \right)^{1/2} = \frac{V_s}{2} \tag{6.1}$$

The instantaneous output voltage can be expressed in Fourier series as

$$v_o = \frac{a_0}{2} + \sum_{n=1}^{\infty} (a_n \cos(n\omega t) + b_n \sin(n\omega t))$$

Due to the quarter-wave symmetry along the x-axis, both a_0 and a_n are zero. We get b_n as

$$b_n = \frac{1}{\pi} \left[\int_{\frac{-\pi}{2}}^{0} \frac{-V_s}{2} \, d(\omega t) + \int_0^{\frac{\pi}{2}} \frac{V_s}{2} \, d(\omega t) \right] = \frac{4V_s}{n\pi}$$

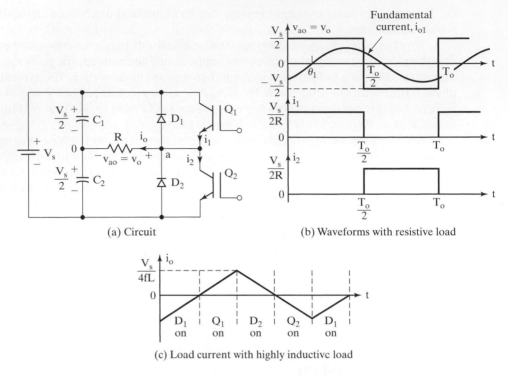

(a) Circuit

(b) Waveforms with resistive load

(c) Load current with highly inductive load

FIGURE 6.1

Single-phase half-bridge inverter.

which gives the instantaneous output voltage v_o as

$$v_0 = \sum_{n=1,3,5,\ldots}^{\infty} \frac{2V_s}{n\pi} \sin n\omega t$$

$$= 0 \quad \text{for } n = 2, 4, \ldots \tag{6.2}$$

where $\omega = 2\pi f_0$ is the frequency of output voltage in rads per second. Due to the quarter-wave symmetry of the output voltage along the x-axis, the even harmonics voltages are absent. For $n = 1$, Eq. (6.2) gives the rms value of fundamental component as

$$V_{o1} = \frac{2V_s}{\sqrt{2}\pi} = 0.45V_s \tag{6.3}$$

For an inductive load, the load current cannot change immediately with the output voltage. If Q_1 is turned off at $t = T_0/2$, the load current would continue to flow through D_2, load, and lower half of the dc source until the current falls to zero. Similary, when Q_2 is turned off at $t = T_0$, the load current flows through D_1, load, and upper half of the dc source. When diode D_1 or D_2 conducts, energy is fed back to the dc source and these diodes are known as *feedback diodes*. Figure 6.1c shows the load current and conduction intervals of devices for a purely inductive load. It can be noticed that for a purely inductive load, a transistor conducts only for $T_0/2$ (or 90°). Depending on the load impedance angle, the conduction period of a transistor would vary from 90° to 180°.

Any switching devices can replace the transistors. If t_{off} is the turn-off time of a device, there must be a minimum delay time of $t_d(= t_{off})$ between the outgoing device and triggering of the next incoming device. Otherwise, short-circuit condition would result through the two devices. Therefore, the maximum conduction time of a device would be $t_{on} = T_0/2 - t_d$. All practical devices require a certain turn-on and turn-off time. For successful operation of inverters, the logic circuit should take these into account.

For an RL load, the instantaneous load current i_0 can be found by dividing the instantaneous output voltage by the load impedance $Z = R + jn\omega L$. Thus, we get

$$i_0 = \sum_{n=1,3,5,\ldots}^{\infty} \frac{2V_s}{n\pi\sqrt{R^2 + (n\omega L)^2}} \sin(n\omega t - \theta_n) \tag{6.4}$$

where $\theta_n = \tan^{-1}(n\omega L/R)$. If I_{01} is the rms fundamental load current, the fundamental output power (for $n = 1$) is

$$P_{01} = V_{o1}I_{01}\cos\theta_1 = I_{01}^2 R \tag{6.5}$$

$$= \left[\frac{2V_s}{\sqrt{2}\pi\sqrt{R^2 + (\omega L)^2}}\right]^2 R \tag{6.5a}$$

Note: In most applications (e.g., electric motor drives) the output power due to the fundamental current is generally the useful power, and the power due to harmonic currents is dissipated as heat and increases the load temperature.

Dc supply current. Assuming a lossless inverter, the average power absorbed by the load must be equal to the average power supplied by the dc source. Thus, we can write

$$\int_0^T v_s(t)i_s(t)dt = \int_0^T v_o(t)i_o(t)dt$$

where T is the period of the ac output voltage. For an inductive load and a relatively high switching frequency, the load current i_o is nearly sinusoidal; therefore, only the fundamental component of the ac output voltage provides power to the load. Because the dc supply voltage remains constant $v_s(t) = V_s$, we can write

$$\int_0^T i_s(t)dt = \frac{1}{V_s}\int_0^T \sqrt{2}V_{o1}\sin(\omega t)\sqrt{2}I_o\sin(\omega t - \theta_1)dt = I_s$$

where V_{o1} is the fundamental rms output voltage;
  I_o is the rms load current;
  θ_1 is the load angle at the fundamental frequency.

Thus, the dc supply current I_s can be simplified to

$$I_s = \frac{V_{o1}}{V_s} I_o \cos(\theta_1) \tag{6.6}$$

Gating sequence.   The gating sequence for the switching devices is as follows:

1. Generate a square-wave gating-signal v_{g1} at an output frequency f_o and a 50% duty cycle. The gating-signal v_{g2} should be a logic invert of v_{g1}.
2. Signal v_{g1} will drive switch Q_1 through a gate-isolating circuit, and v_{g2} can drive Q_2 without any isolating circuit.

Key Points of Section 6.2

- An ac output voltage can be obtained by alternatively connecting the positive and negative terminals of the dc source across the load by turning on and off the switching devices accordingly. The rms fundamental component V_{o1} of the output voltage is 0.45 V_s.
- Feedback diodes are required to transfer the energy stored in the load inductance back to the dc source.

6.3   PERFORMANCE PARAMETERS

The output of practical inverters contain harmonics and the quality of an inverter is normally evaluated in terms of the following performance parameters.

Harmonic factor of nth harmonic (HF$_n$).   The harmonic factor (of the nth harmonic), which is a measure of individual harmonic contribution, is defined as

$$HF_n = \frac{V_{on}}{V_{o1}} \qquad \text{for } n > 1 \tag{6.7}$$

where V_1 is the rms value of the fundamental component and V_{on} is the rms value of the nth harmonic component.

Total harmonic distortion (THD).   The total harmonic distortion, which is a measure of closeness in shape between a waveform and its fundamental component, is defined as

$$\text{THD} = \frac{1}{V_{o1}} \left(\sum_{n=2,3,\ldots}^{\infty} V_{on}^2 \right)^{1/2} \tag{6.8}$$

Distortion factor (DF).   THD gives the total harmonic content, but it does not indicate the level of each harmonic component. If a filter is used at the output of inverters, the higher order harmonics would be attenuated more effectively. Therefore, a knowledge of both the frequency and magnitude of each harmonic is important. The DF indicates the amount of HD that remains in a particular waveform after the harmonics of that waveform have been subjected to a second-order attenuation (i.e., divided by n^2). Thus, DF is a measure of effectiveness in reducing unwanted harmonics without having to specify the values of a second-order load filter and is defined as

$$\text{DF} = \frac{1}{V_{o1}} \left[\sum_{n=2,3,\ldots}^{\infty} \left(\frac{V_{on}}{n^2} \right)^2 \right]^{1/2} \tag{6.9}$$

The DF of an individual (or nth) harmonic component is defined as

$$\text{DF}_n = \frac{V_{on}}{V_{o1} n^2} \qquad \text{for } n > 1 \tag{6.10}$$

Lowest order harmonic (LOH). The LOH is that harmonic component whose frequency is closest to the fundamental one, and its amplitude is greater than or equal to 3% of the fundamental component.

Example 6.1 Finding the Parameters of the Single-Phase Half-Bridge Inverter

The single-phase half-bridge inverter in Figure 6.1a has a resistive load of $R = 2.4\ \Omega$ and the dc input voltage is $V_s = 48$ V. Determine (a) the rms output voltage at the fundamental frequency V_{o1}, (b) the output power P_o, (c) the average and peak currents of each transistor, (d) the peak reverse blocking voltage V_{BR} of each transistor, (e) the THD, (f) the DF, and (g) the HF and DF of the LOH.

Solution
$V_s = 48$ V and $R - 2.4\ \Omega$.

a. From Eq. (6.3), $V_{o1} = 0.45 \times 48 = 21.6$ V.

b. From Eq. (6.1), $V_o = V_s/2 = 48/2 = 24$ V. The output power, $P_o = V_o^2/R = 24^2/2.4 = 240$ W.

c. The peak transistor current $I_p = 24/2.4 = 10$ A. Because each transistor conducts for a 50% duty cycle, the average current of each transistor is $I_Q = 0.5 \times 10 = 5$ A.

d. The peak reverse blocking voltage $V_{BR} = 2 \times 24 = 48$ V.

e. From Eq. (6.3), $V_{o1} = 0.45 V_s$ and the rms harmonic voltage V_h

$$V_h = \left(\sum_{n=3,5,7,\ldots}^{\infty} V_{on}^2 \right)^{1/2} = (V_0^2 - V_{o1}^2)^{1/2} = 0.2176 V_s$$

From Eq. (6.8), THD $= (0.2176 V_s)/(0.45 V_s) = 48.34\%$.

f. From Eq. (6.2), we can find V_{on} and then find,

$$\left[\sum_{n=3,5,\ldots}^{\infty} \left(\frac{V_{on}}{n^2} \right)^2 \right]^{1/2} = \left[\left(\frac{V_{o3}}{3^2} \right)^2 + \left(\frac{V_{o5}}{5^2} \right)^2 + \left(\frac{V_{o7}}{7^2} \right)^2 + \cdots \right]^{1/2} = 0.024 V_s$$

From Eq. (6.9), DF $= 0.024 V_s/(0.45 V_s) = 5.382\%$.

g. The LOH is the third, $V_{o3} = V_{o1}/3$. From Eq. (6.7), $\text{HF}_3 = V_{o3}/V_{o1} = 1/3 = 33.33\%$, and from Eq. (6.10), $\text{DF}_3 = (V_{o3}/3^2)/V_{o1} = 1/27 = 3.704\%$. Because $V_{o3}/V_{o1} = 33.33\%$, which is greater than 3%, LOH $= V_{o3}$.

Key Points of Section 6.3

- The performance parameters, which measure the quality of the inverter output voltage, are HF, THD, DF, and LOH.

6.4 SINGLE-PHASE BRIDGE INVERTERS

A single-phase bridge voltage source inverter (VSI) is shown in Figure 6.2a. It consists of four choppers. When transistors Q_1 and Q_2 are turned on simultaneously, the input voltage V_s appears across the load. If transistors Q_3 and Q_4 are turned on at the same time, the voltage across the load is reversed and is $-V_s$. The waveform for the output voltage is shown in Figure 6.2b.

Table 6.1 shows the five switch states. Transistors Q_1, Q_4 in Figure 6.2a act as the switching devices S_1, S_4, respectively. If two switches: one upper and one lower conduct at the same time such that the output voltage is $\pm V_s$, the switch state is 1, whereas if these switches are off at the same time, the switch state is 0.

The rms output voltage can be found from

$$V_o = \left(\frac{2}{T_0} \int_0^{T_0/2} V_s^2 \, dt \right)^{1/2} = V_s \tag{6.11}$$

Equation (6.2) can be extended to express the instantaneous output voltage in a Fourier series as

$$v_o = \sum_{n=1,3,5,\ldots}^{\infty} \frac{4V_s}{n\pi} \sin n\omega t \tag{6.12}$$

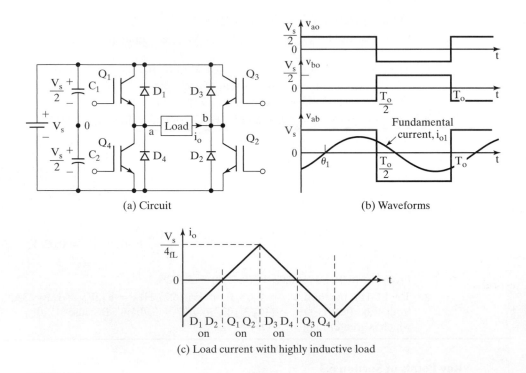

(a) Circuit

(b) Waveforms

(c) Load current with highly inductive load

FIGURE 6.2

Single-phase full-bridge inverter.

TABLE 6.1   Switch States for a Single-Phase Full-Bridge Voltage-Source Inverter (VSI)

State	State No.	Switch State*	v_{ao}	v_{bo}	v_o	Components Conducting
S_1 and S_2 are on and S_4 and S_3 are off	1	10	$V_S/2$	$-V_S/2$	V_S	S_1 and S_2 if $i_o > 0$
						D_1 and D_2 if $i_o < 0$
S_4 and S_3 are on and S_1 and S_2 are off	2	01	$-V_S/2$	$V_S/2$	$-V_S$	D_4 and D_3 if $i_o > 0$
						S_4 and S_3 if $i_o < 0$
S_1 and S_3 are on and S_4 and S_2 are off	3	11	$V_S/2$	$V_S/2$	0	S_1 and D_3 if $i_o > 0$
						D_1 and S_3 if $i_o < 0$
S_4 and S_2 are on and S_1 and S_3 are off	4	00	$-V_S/2$	$-V_S/2$	0	D_4 and S_2 if $i_o > 0$
						S_4 and D_2 if $i_o < 0$
$S_1, S_2, S_3,$ and S_4 are all off	5	off	$-V_S/2$	$V_S/2$	$-V_S$	D_4 and D_3 if $i_o > 0$
			$V_S/2$	$-V_S/2$	V_S	D_4 and D_2 if $i_o < 0$

* 1 if an upper switch is on and 0 if a lower switch is on.

and for $n = 1$, Eq. (6.12) gives the rms value of fundamental component as

$$V_1 = \frac{4V_s}{\sqrt{2}\pi} = 0.90V_s \tag{6.13}$$

Using Eq. (6.4), the instantaneous load current i_0 for an RL load becomes

$$i_0 = \sum_{n=1,3,5,\ldots}^{\infty} \frac{4V_s}{n\pi\sqrt{R^2 + (n\omega L)^2}} \sin(n\omega t - \theta_n) \tag{6.14}$$

where $\theta_n = \tan^{-1}(n\omega L/R)$.

When diodes D_1 and D_2 conduct, the energy is fed back to the dc source; thus, they are known as *feedback diodes*. Figure 6.2c shows the waveform of load current for an inductive load.

Dc supply current.   Neglecting any losses, the instantaneous power balance gives,

$$v_s(t)i_s(t) = v_o(t)i_o(t)$$

For inductive load and relatively high-switching frequencies, the load current i_o and the output voltage may be assumed sinusoidal. Because the dc supply voltage remains constant $v_s(t) = V_s$, we get

$$i_s(t) = \frac{1}{V_s}\sqrt{2}V_{o1} \sin(\omega t)\sqrt{2}I_o \sin(\omega t - \theta_1)$$

which can be simplified to find the dc supply current as

$$i_s(t) = \frac{V_{o1}}{V_s}I_o \cos(\theta_1) - \frac{V_{o1}}{V_s}I_o \cos(2\omega t - \theta_1) \tag{6.15}$$

where V_{o1} is the fundamental rms output voltage;
$\quad I_o$ is the rms load current;
$\quad \theta_1$ is the load impedance angle at the fundamental frequency.

Equation (6.15) indicates the presence of a second-order harmonic of the same order of magnitude as the dc supply current. This harmonic is injected back into the dc voltage source. Thus, the design should consider this to guarantee a nearly constant dc link voltage. A large capacitor is normally connected across the dc voltage source and such a capacitor is costly and demands space; both features are undesirable, especially in medium to high power supplies.

Example 6.2   Finding the Parameters of the Single-Phase Full-Bridge Inverter

Repeat Example 6.1 for a single-phase bridge inverter in Figure 6.2a.

Solution

V_s = 48 V and R = 2.4 Ω.

  a. From Eq. (6.13), V_1 = 0.90 × 48 = 43.2 V.
  b. From Eq. (6.11), $V_o = V_s$ = 48 V. The output power is $P_o = V_s^2/R$ = $48^2/2.4$ = 960 W.
  c. The peak transistor current is I_p = 48/2.4 = 20 A. Because each transistor conducts for a 50% duty cycle, the average current of each transistor is I_Q = 0.5 × 20 = 10 A.
  d. The peak reverse blocking voltage is V_{BR} = 48 V.
  e. From Eq. (6.13), $V_1 = 0.9V_s$. The rms harmonic voltage V_h is

$$V_h = \left(\sum_{n=3,5,7,\ldots}^{\infty} V_{on}^2 \right)^{1/2} = (V_0^2 - V_{o1}^2)^{1/2} = 0.4352V_s$$

  From Eq. (6.8), THD = $0.4359V_s/(0.9V_s)$ = 48.34%.

  f. $$\left[\sum_{n-3,5,7,\ldots}^{\infty} \left(\frac{V_{on}}{n^2} \right)^2 \right]^{1/2} = 0.048V_s$$
  From Eq. (6.9), DF = $0.048V_s/(0.9V_s)$ = 5.382%.

  g. The LOH is the third, $V_3 = V_1/3$. From Eq. (6.7), $HF_3 = V_{o3}/V_{o1}$ = 1/3 = 33.33% and from Eq. (6.10), $DF_3 = (V_{o3}/3^2)/V_{o1}$ = 1/27 = 3.704%.

Note: The peak reverse blocking voltage of each transistor and the quality of output voltage for half-bridge and full-bridge inverters are the same. However, for full-bridge inverters, the output power is four times higher and the fundamental component is twice that of half-bridge inverters.

Example 6.3   Finding the Output Voltage and Current of a Single-Phase Full-Bridge Inverter with an *RLC* Load

The bridge inverter in Figure 6.2a has an *RLC* load with R = 10 Ω, L = 31.5 mH, and C = 112 μF. The inverter frequency is f_0 = 60 Hz and dc input voltage is V_s = 220 V. (a) Express the instantaneous load current in Fourier series. Calculate (b) the rms load current at the fundamental frequency I_{o1}; (c) the THD of the load current; (d) the power absorbed by the load P_0 and the fundamental power P_{01}; (e) the average current of dc supply I_s; and (f) the rms and peak current of each transistor. (g) Draw the waveform of fundamental load current and show the conduction intervals of transistors and diodes. Calculate the conduction time of (h) the transistors, and (i) the diodes.

Solution

V_s = 220 V, f_0 = 60 Hz, R = 10 Ω, L = 31.5 mH, C = 112 μF, and $\omega = 2\pi \times 60$ = 377 rad/s. The inductive reactance for the *n*th harmonic voltage is

$$X_L = j_n\omega L = j2n\pi \times 60 \times 31.5 \times 10^{-3} = j11.87n \ \Omega$$

The capacitive reactance for the nth harmonic voltage is

$$X_c = -\frac{j}{n\omega C} = -\frac{j10^6}{2n\pi \times 60 \times 112} = \frac{-j23.68}{n} \ \Omega$$

The impedance for the nth harmonic voltage is

$$|Z_n| = \sqrt{R^2 + \left(n\omega L - \frac{1}{n\omega C}\right)^2} = [10^2 + (11.87n - 23.68/n)^2]^{1/2}$$

and the load impedance angle for the nth harmonic voltage is

$$\theta_n = \tan^{-1}\frac{11.87n - 23.68/n}{10} = \tan^{-1}\left(1.187n - \frac{2.368}{n}\right)$$

a. From Eq. (6.12), the instantaneous output voltage can be expressed as

$$v_o(t) = 280.1\sin(377t) + 93.4\sin(3 \times 377t) + 56.02\sin(5 \times 377t)$$
$$+ 40.02\sin(7 \times 377t) + 31.12\sin(9 \times 377t) + \cdots$$

Dividing the output voltage by the load impedance and considering the appropriate delay due to the load impedance angles, we can obtain the instantaneous load current as

$$i_o(t) = 18.1\sin(377t + 49.72°) + 3.17\sin(3 \times 377t - 70.17°)$$
$$+ \sin(5 \times 377t - 79.63°) + 0.5\sin(7 \times 377t - 82.85°)$$
$$+ 0.3\sin(9 \times 377t - 84.52°) + \cdots$$

b. The peak fundamental load current is $I_{m1} = 18.1$ A. The rms load current at fundamental frequency is $I_{o1} = 18.1/\sqrt{2} = 12.8$ A.

c. Considering up to the ninth harmonic, the peak load current,

$$I_m = (18.1^2 + 3.17^2 + 1.0^2 + 0.5^2 + 0.3^2)^{1/2} = 18.41 \text{ A}$$

The rms harmonic load current is

$$I_h = (I_m^2 - I_{m1}^2)^{1/2} = \frac{18.41^2 - 18.1^2}{\sqrt{2}} = 2.3789 \text{ A}$$

Using Eq. (6.8), the THD of the load current,

$$\text{THD} = \frac{(I_m^2 - I_{m1}^2)^{1/2}}{I_{m1}} = \left[\left(\frac{18.41}{18.1}\right)^2 - 1\right]^{1/2} = 18.59\%$$

d. The rms load current is $I_o \cong I_m/\sqrt{2} = 18.41/\sqrt{2} = 13.02$ A, and the load power is $P_o = 13.02^2 \times 10 = 1695$ W. Using Eq. (6.5), the fundamental output power is

$$P_{o1} = I_{o1}^2 R = 12.8^2 \times 10 = 1638.4 \text{ W}$$

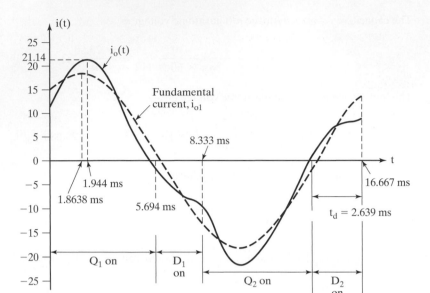

FIGURE 6.3

Waveforms for Example 6.3.

e. The average supply current $I_s = P_o/V_s = 1695/220 - 7.7$ A.

f. The peak transistor current $I_p \cong I_m = 18.41$ A. The maximum permissible rms current of each transistor is $I_{Q(max)} = I_o/\sqrt{2} = I_p/2 = 18.41/2 = 9.2$ A.

g. The waveform for fundamental load current $i_1(t)$ is shown in Figure 6.3.

h. From Fig. (6.3), the conduction time of each transistor is found approximately from $\omega t_0 = 180 - 49.72 = 130.28°$ or $t_0 = 130.28 \times \pi/(180 \times 377) = 6031$ μs.

i. The conduction time of each diode is approximately

$$t_d = (180 - 130.28) \times \frac{\pi}{180 \times 377} = 2302 \text{ μs}$$

Notes:

1. To calculate the exact values of the peak current, the conduction time of transistors and diodes, the instantaneous load current $i_o(t)$ should be plotted as shown in Figure 6.3. The conduction time of a transistor must satisfy the condition $i_o(t = t_0) = 0$, and a plot of $i_o(t)$ by a computer program gives $I_p = 21.14$ A, $t_0 = 5694$ μs, and $t_d = 2639$ μs.

2. This example can be repeated to evaluate the performance of an inverter with R, RL, or RLC load with an appropriate change in load impedance Z_L and load angle θ_n.

Gating sequence. The gating sequence for the switching devices is as follows:

1. Generate two square-wave gating-signals v_{g1} and v_{g2} at an output frequency f_o and a 50% duty cycle. The gating-signals v_{g3} and v_{g4} should be the logic invert of v_{g2} and v_{g4}, respectively.

2. Signals v_{g1} and v_{g3} drive Q_1 and Q_3, respectively, through gate isolation circuits. Signals v_{g2} and v_{g4} can drive Q_2 and Q_4, respectively, without any isolation circuits.

Key Points of Section 6.4

- The full-bridge inverter requires four switching devices and four diodes. The output voltage switches between $+V_s$ and $-V_s$. The rms fundamental component V_1 of the output voltage is $0.9V_s$.

- The design of an inverter requires the determination of the average, rms, and peak currents of the switching devices and diodes.

6.5   THREE-PHASE INVERTERS

Three-phase inverters are normally used for high-power applications. Three single-phase half (or full)-bridge inverters can be connected in parallel as shown in Figure 6.4a to form the configuration of a three-phase inverter. The gating signals of single-phase inverters should be advanced or delayed by 120° with respect to each other to obtain three-phase balanced (fundamental) voltages. The transformer primary windings must be isolated from each other, whereas the secondary windings may be connected in Y or delta. The transformer secondary is normally connected in delta to eliminate triplen harmonics ($n = 3, 6, 9, \ldots$) appearing on the output voltages and the circuit arrangement is shown in Figure 6.4b. This arrangement requires three single-phase transformers, 12 transistors, and 12 diodes. If the output voltages of single-phase inverters are not perfectly balanced in magnitudes and phases, the three-phase output voltages are unbalanced.

A three-phase output can be obtained from a configuration of six transistors and six diodes as shown in Figure 6.5a. Two types of control signals can be applied to the transistors: 180° conduction or 120° conduction. The 180° conduction has better utilization of the switches and is the preferred method.

6.5.1   180-Degree Conduction

Each transistor conducts for 180°. Three transistors remain on at any instant of time. When transistor Q_1 is switched on, terminal a is connected to the positive terminal of the dc input voltage. When transistor Q_4 is switched on, terminal a is brought to the negative terminal of the dc source. There are six modes of operation in a cycle and the duration of each mode is 60°. The transistors are numbered in the sequence of gating the transistors (e.g., 123, 234, 345, 456, 561, and 612). The gating signals shown in Figure 6.5b are shifted from each other by 60° to obtain three-phase balanced (fundamental) voltages.

The load may be connected in Y or delta as shown in Figure 6.6. The switches of any leg of the inverter (S_1 and S_4, S_3 and S_6, or S_5 and S_2) cannot be switched on simultaneously; this would result in a short circuit across the dc link voltage supply. Similarly, to avoid undefined states and thus undefined ac output line voltages, the switches of any leg of the inverter cannot be switched off simultaneously; this can result in voltages that depend on the respective line current polarity.

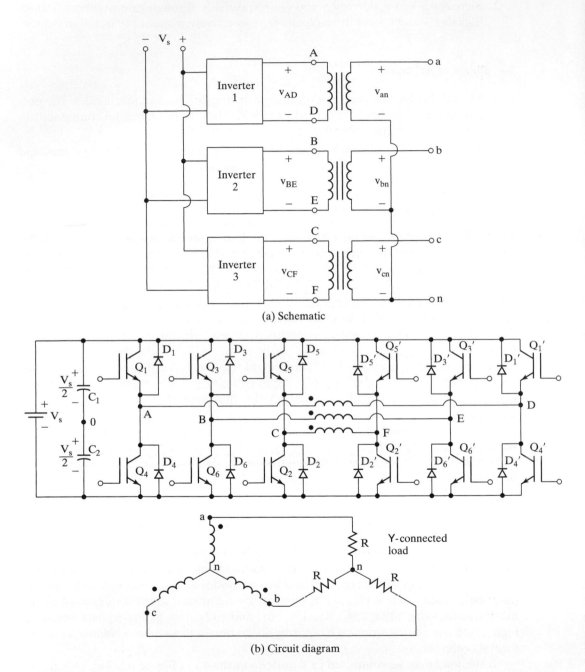

(a) Schematic

(b) Circuit diagram

FIGURE 6.4

Three-phase inverter formed by three single-phase inverters.

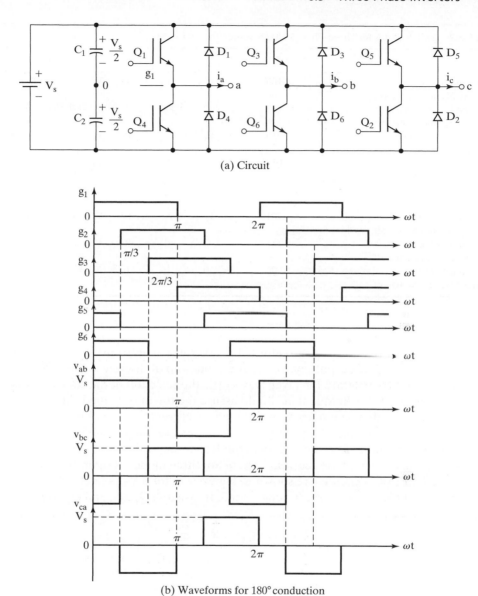

(a) Circuit

(b) Waveforms for 180° conduction

FIGURE 6.5

Three-phase bridge inverter.

Table 6.2 shows eight valid switch states. Transistors $Q1$, $Q6$ in Figure 6.6a act as the switching devices $S1$, $S6$, respectively. If two switches: one upper and one lower conduct at the same time such that the output voltage is $\pm V_s$, the switch state is 1, whereas if these switches are off at the same time, the switch state is 0. States 1 to 6 produce nonzero output voltages. States 7 and 8 produce zero line voltages and the line currents freewheel through either the upper or the lower freewheeling diodes. To

TABLE 6.2   Switch States for Three-Phase Voltage-Source Inverter (VSI)

State	State No.	Switch States	v_{ab}	v_{bc}	v_{ca}	Space Vector
S_1, S_2, and S_6 are on and S_4, S_5, and S_3 are off	1	100	V_S	0	$-V_S$	$\mathbf{V_1} = 1 + j0.577 = 2/\sqrt{3} \angle 30°$
S_2, S_3, and S_1 are on and S_5, S_6, and S_4 are off	2	110	0	V_S	$-V_S$	$\mathbf{V_2} = j1.155 = 2/\sqrt{3} \angle 90°$
S_3, S_4, and S_2 are on and S_6, S_1, and S_5 are off	3	010	$-V_S$	V_S	0	$\mathbf{V_3} = -1 + j0.577 = 2/\sqrt{3} \angle 150°$
S_4, S_5, and S_3 are on and S_1, S_2, and S_6 are off	4	011	$-V_S$	0	V_S	$\mathbf{V_4} = -1 - j0.577 = 2/\sqrt{3} \angle 210°$
S_5, S_6, and S_4 are on and S_2, S_3, and S_1 are off	5	001	0	$-V_S$	V_S	$\mathbf{V_5} = -j1.155 = 2/\sqrt{3} \angle 270°$
S_6, S_1, and S_5 are on and S_3, S_4, and S_2 are off	6	101	V_S	$-V_S$	0	$\mathbf{V_6} = 1 - j0.577 = 2/\sqrt{3} \angle 330°$
S_1, S_3, and S_5 are on and S_4, S_6, and S_2 are off	7	111	0	0	0	$\mathbf{V_7} = 0$
S_4, S_6, and S_2 are on and S_1, S_3, and S_5 are off	8	000	0	0	0	$\mathbf{V_0} = 0$

generate a given voltage waveform, the inverter moves from one state to another. Thus, the resulting ac output line voltages are built up of discrete values of voltages of V_s, 0, and $-V_s$. To generate the given waveform, the selection of the states is usually done by a modulating technique that should assure the use of only the valid states.

For a delta-connected load, the phase currents can be obtained directly from the line-to-line voltages. Once the phase currents are known, the line currents can be determined. For a Y-connected load, the line-to-neutral voltages must be determined to find the line (or phase) currents. There are three modes of operation in a half-cycle and the equivalent circuits are shown in Figure 6.7a for a Y-connected load.

During mode 1 for $0 \le \omega t < \pi/3$, transistors Q_1, Q_5, and Q_6 conduct

$$R_{eq} = R + \frac{R}{2} = \frac{3R}{2}$$

$$i_1 = \frac{V_s}{R_{eq}} = \frac{2V_s}{3R}$$

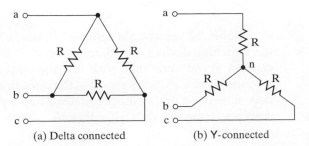

FIGURE 6.6

Delta- and Y-connected load.

(a) Delta connected   (b) Y-connected

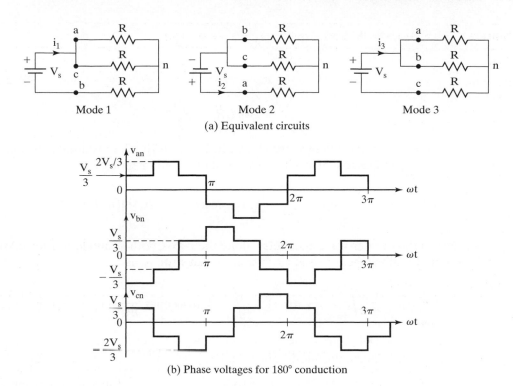

(a) Equivalent circuits

(b) Phase voltages for 180° conduction

FIGURE 6.7

Equivalent circuits for Y-connected resistive load.

$$v_{an} = v_{cn} = \frac{i_1 R}{2} = \frac{V_s}{3}$$

$$v_{bn} = -i_1 R = \frac{-2V_s}{3}$$

During mode 2 for $\pi/3 \le \omega t < 2\pi/3$, transistors Q_1, Q_2, and Q_6 conduct

$$R_{eq} = R + \frac{R}{2} = \frac{3R}{2}$$

$$i_2 = \frac{V_s}{R_{eq}} = \frac{2V_s}{3R}$$

$$v_{an} = i_2 R = \frac{2V_s}{3}$$

$$v_{bn} = v_{cn} = \frac{-i_2 R}{2} = \frac{-V_s}{3}$$

During mode 3 for $2\pi/3 \le \omega t < \pi$, transistors Q_1, Q_2, and Q_3 conduct

$$R_{eq} = R + \frac{R}{2} = \frac{3R}{2}$$

$$i_3 = \frac{V_s}{R_{eq}} = \frac{2V_s}{3R}$$

$$v_{an} = v_{bn} = \frac{i_3 R}{2} = \frac{V_s}{3}$$

$$v_{cn} = i_3 R = \frac{-2V_s}{3}$$

The line-to-neutral voltages are shown in Figure 6.7b. The instantaneous line-to-line voltage v_{ab} in Figure 6.5b can be expressed in a Fourier series,

$$v_{ab} = \frac{a_0}{2} + \sum_{n=1}^{\infty} (a_n \cos(n\omega t) + b_n \sin(n\omega t))$$

Due to the quarter-wave symmetry along the x-axis, both a_0 and a_n are zero. Assuming symmetry along the y-axis at $\omega t = \pi/6$, we can write b_n as

$$b_n = \frac{1}{\pi} \left[\int_{-5\pi/6}^{-\pi/6} -V_s d(\omega t) + \int_{\pi/6}^{5\pi/6} V_s d(\omega t) \right] = \frac{4V_s}{n\pi} \sin\left(\frac{n\pi}{2}\right) \sin\left(\frac{n\pi}{3}\right)$$

which, recognizing that v_{ab} is phase shifted by $\pi/6$ and the even harmonics are zero, gives the instantaneous line-to-line voltage v_{ab} (for a Y-connected load) as

$$v_{ab} = \sum_{n=1,3,5,...}^{\infty} \frac{4V_s}{n\pi} \sin\frac{n\pi}{3} \sin n\left(\omega t + \frac{\pi}{6}\right) \tag{6.16a}$$

Both v_{bc} and v_{ca} can be found from Eq. (6.16a) by phase shifting v_{ab} by 120° and 240°, respectively,

$$v_{bc} = \sum_{n=1,3,5,...}^{\infty} \frac{4V_s}{n\pi} \sin\frac{n\pi}{3} \sin n\left(\omega t - \frac{\pi}{2}\right) \tag{6.16b}$$

$$v_{ca} = \sum_{n=1,3,5,...}^{\infty} \frac{4V_s}{n\pi} \sin\frac{n\pi}{3} \sin n\left(\omega t - \frac{7\pi}{6}\right) \tag{6.16c}$$

We can notice from Eqs. (6.16a) to (6.16c) that the triplen harmonics ($n = 3, 9, 15, \ldots$) would be zero in the line-to-line voltages.

The line-to-line rms voltage can be found from

$$V_L = \left[\frac{2}{2\pi} \int_0^{2\pi/3} V_s^2 \, d(\omega t)\right]^{1/2} = \sqrt{\frac{2}{3}} V_s = 0.8165 V_s \tag{6.17}$$

From Eq. (6.16a), the rms nth component of the line voltage is

$$V_{Ln} = \frac{4V_s}{\sqrt{2}n\pi} \sin\frac{n\pi}{3} \tag{6.18}$$

which, for $n = 1$, gives the rms fundamental line voltage.

$$V_{L1} = \frac{4V_s \sin 60°}{\sqrt{2}\pi} = 0.7797V_s \qquad (6.19)$$

The rms value of line-to-neutral voltages can be found from the line voltage,

$$V_p = \frac{V_L}{\sqrt{3}} = \frac{\sqrt{2}\,V_s}{3} = 0.4714V_s \qquad (6.20)$$

With resistive loads, the diodes across the transistors have no functions. If the load is inductive, the current in each arm of the inverter would be delayed to its voltage as shown in Figure 6.8. When transistor Q_4 in Figure 6.5a is off, the only path for the negative line current i_a is through D_1. Hence, the load terminal a is connected to the dc source through D_1 until the load current reverses its polarity at $t = t_1$. During the period for $0 \le t \le t_1$, transistor Q_1 cannot conduct. Similarly, transistor Q_4 only starts to conduct at $t = t_2$. The transistors must be continuously gated, because the conduction time of transistors and diodes depends on the load power factor.

For a Y-connected load, the phase voltage is $v_{an} = v_{ab}/\sqrt{3}$ with a delay of 30° with respect to v_{ab}. Therefore, the instantaneous phase voltages (for a Y-connected load) are

$$v_{aN} = \sum_{n=1}^{\infty} \frac{4V_s}{\sqrt{3}n\pi} \sin\left(\frac{n\pi}{3}\right)\sin(n\omega t) \qquad \text{for } n = 1, 3, 5, \ldots \quad (6.21a)$$

$$v_{bN} = \sum_{n=1}^{\infty} \frac{4V_s}{\sqrt{3}n\pi} \sin\left(\frac{n\pi}{3}\right)\sin n\left(\omega t - \frac{2\pi}{3}\right) \qquad \text{for } n = 1, 3, 5, \ldots \quad (6.21b)$$

$$v_{bN} = \sum_{n=1}^{\infty} \frac{4V_s}{\sqrt{3}n\pi} \sin\left(\frac{n\pi}{3}\right)\sin n\left(\omega t - \frac{4\pi}{3}\right) \qquad \text{for } n = 1, 3, 5, \ldots \quad (6.21c)$$

Dividing the instantaneous phase voltage v_{aN} by the load impedance,

$$Z = R + jn\omega L$$

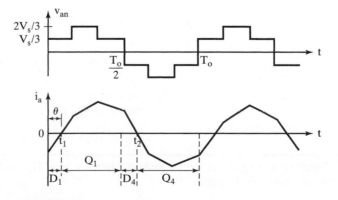

FIGURE 6.8

Three-phase inverter with RL load.

Using Eq. (6.21a), the line current i_a for an RL load is given by

$$i_a = \sum_{n=1,3,5,\ldots}^{\infty} \left[\frac{4V_s}{\sqrt{3}[n\pi\sqrt{R^2 + (n\omega L)^2}]} \sin\frac{n\pi}{3} \right] \sin(n\omega t - \theta_n) \qquad (6.22)$$

where $\theta_n = \tan^{-1}(n\omega L/R)$.

Note: For a delta-connected load, the phase voltages (v_{aN}, v_{bN}, and v_{cN}) are the same as the line-to-line voltages (v_{ab}, v_{bc}, and v_{ca}) as shown in Figure 6.6b and as described by Eq. (6.16).

Dc supply current.   Neglecting losses, the instantaneous power balance gives

$$v_s(t)i_s(t) = v_{ab}(t)i_a(t) + v_{bc}(t)i_b(t) + v_{ca}(t)i_c(t)$$

where $i_a(t)$, $i_b(t)$, and $i_c(t)$ are the phase currents in a delta-connected load. Assuming that the ac output voltages are sinusoidal and the dc supply voltage is constant $v_s(t) = V_s$, we get the dc supply current

$$i_s(t) = \frac{1}{V_s} \left\{ \begin{array}{l} \sqrt{2}V_{o1} \sin(\omega t) \times \sqrt{2}I_o \sin(\omega t - \theta_1) \\ + \sqrt{2}V_{o1} \sin(\omega t - 120°) \times \sqrt{2}I_o \sin(\omega t - 120° - \theta_1) \\ + \sqrt{2}V_{o1} \sin(\omega t - 240°) \times \sqrt{2}I_o \sin(\omega t - 240° - \theta_1) \end{array} \right\}$$

The dc supply current can be simplified to

$$I_s = 3\frac{V_{o1}}{V_s} I_o \cos(\theta_1) = \sqrt{3}\frac{V_{o1}}{V_s} I_L \cos(\theta_1) \qquad (6.23)$$

where $I_L = \sqrt{3}I_o$ is the rms load line current;
$\quad V_{o1}$ is the fundamental rms output line voltage;
$\quad I_o$ is the rms load phase current;
$\quad \theta_1$ is the load impedance angle at the fundamental frequency.

Thus, if the load voltages are harmonic free, the dc supply current becomes harmonic free. However, because the load line voltages contain harmonics, the dc supply current also contains harmonics.

Gating sequence.   The gating sequence for switching devices is as follows

1. Generate three square-wave gating-signals v_{g1}, v_{g3}, and v_{g5} at an output frequency f_0 and a 50% duty cycle. Signals v_{g4}, v_{g6}, and v_{g2} should be logic invert signals of v_{g1}, v_{g3}, and v_{g5}, respectively. Each signal is shifted from the other by 60°.
2. Signals v_{g1}, v_{g3}, and v_{g5} drive Q_1, Q_3, and Q_5, respectively, through gate-isolating circuits. Signals v_{g2}, v_{g4}, and v_{g6} can drive Q_2, Q_4, and Q_6, respectively, without any isolating circuits.

Example 6.4   Finding the Output Voltage and Current of a Three-Phase Full-Bridge Inverter with an *RL* load

The three-phase inverter in Figure 6.5a has a Y-connected load of $R = 5\ \Omega$ and $L = 23$ mH. The inverter frequency is $f_0 = 60$ Hz and the dc input voltage is $V_s = 220$ V. (a) Express the instantaneous line-to-line voltage $v_{ab}(t)$ and line current $i_a(t)$ in a Fourier series. Determine (b) the rms line voltage V_L; (c) the rms phase voltage V_p; (d) the rms line voltage V_{L1} at the fundamental frequency; (e) the rms phase voltage at the fundamental frequency V_{p1}; (f) the THD; (g) the DF; (h) the HF and DF of the LOH; (i) the load power P_o; (j) the average transistor current $I_{Q(av)}$; and (k) the rms transistor current $I_{Q(rms)}$.

Solution

$V_s = 220$ V, $R = 5$ Ω, $L = 23$ mH, $f_0 = 60$ Hz, and $\omega = 2\pi \times 60 = 377$ rad/s.

a. Using Eq. (6.16a), the instantaneous line-to-line voltage $v_{ab}(t)$ can be written as

$$v_{ab}(t) = 242.58 \sin(377t + 30°) - 48.52 \sin 5(377t + 30°)$$
$$- 34.66 \sin 7(377t + 30°) + 22.05 \sin 11(377t + 30°)$$
$$+ 18.66 \sin 13(377t + 30°) - 14.27 \sin 17(377t + 30°) + \cdots$$

$$Z_L = \sqrt{R^2 + (n\omega L)^2}\big/\tan^{-1}(n\omega L/R) = \sqrt{5^2 + (8.67n)^2}\big/\tan^{-1}(8.67n/5)$$

Using Eq. (6.22), the instantaneous line (or phase) current is given by

$$i_{a(t)} = 14 \sin(377t - 60°) - 0.64 \sin(5 \times 377t - 83.4°)$$
$$- 0.33 \sin(7 \times 377t - 85.3°) + 0.13 \sin(11 \times 377t - 87°)$$
$$+ 0.10 \sin(13 \times 377t - 87.5°) - 0.06 \sin(17 \times 377t - 88°) - \cdots$$

h. From Eq. (6.17), $V_L = 0.8165 \times 220 = 179.63$ V.

c. From Eq. (6.20), $V_p = 0.4714 \times 220 = 103.7$ V.

d. From Eq. (6.19), $V_{L1} = 0.7797 \times 220 = 171.53$ V.

e. $V_{p1} = V_{L1}/\sqrt{3} = 99.03$ V.

f. From Eq. (6.19), $V_{L1} = 0.7797 V_S$

$$\left(\sum_{n=5,7,11,\ldots}^{\infty} V_{Ln}^2\right)^{1/2} = (V_L^2 - V_{L1}^2)^{1/2} = 0.24236 V_s$$

From Eq. (6.8), THD $= 0.24236 V_s/(0.7797 V_s) = 31.08\%$. The rms harmonic line voltage is

g. $V_{Lh} = \left[\sum_{n=5,7,11,\ldots}^{\infty} \left(\dfrac{V_{Ln}}{n^2}\right)^2\right]^{1/2} = 0.00941 V_s$

From Eq. (6.9), DF $= 0.00941 V_s/(0.7797 V_s) = 1.211\%$.

h. The LOH is the fifth, $V_{L5} = V_{L1}/5$. From Eq. (6.7), $HF_5 = V_{L5}/V_{L1} = 1/5 = 20\%$, and from Eq. (6.10), $DF_5 = (V_{L5}/5^2)/V_{L1} = 1/125 = 0.8\%$.

i. For Y-connected loads, the line current is the same as the phase current and the rms line current,

$$I_L = \frac{(14^2 + 0.64^2 + 0.33^2 + 0.13^2 + 0.10^2 + 0.06^2)^{1/2}}{\sqrt{2}} = 9.91 \text{ A}$$

The load power $P_0 = 3I_L^2 R = 3 \times 9.91^2 \times 5 = 1473$ W.

j. The average supply current $I_s = P_0/220 = 1473/220 = 6.7$ A and the average transistor current $I_{Q(av)} = 6.7/3 = 2.23$ A.

k. Because the line current is shared by three transistors, the rms value of a transistor current is $I_{Q(rms)} = I_L/\sqrt{3} = 9.91/\sqrt{3} = 5.72$ A.

6.5.2   120-Degree Conduction

In this type of control, each transistor conducts for 120°. Only two transistors remain on at any instant of time. The gating signals are shown in Figure 6.9. The conduction sequence of transistors is 61, 12, 23, 34, 45, 56, 61. There are three modes of operation in one half-cycle and the equivalent circuits for a Y-connected load are shown in Figure 6.10. During mode 1 for ≤ ωt ≤ π/3, transistors 1 and 6 conduct.

$$v_{an} = \frac{V_s}{2} \qquad v_{bn} = -\frac{V_s}{2} \qquad v_{cn} = 0$$

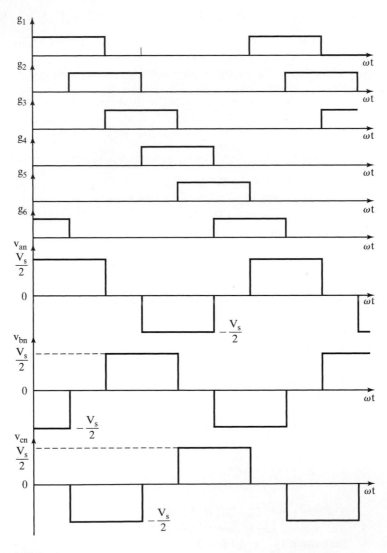

FIGURE 6.9

Gating signals for 120° conduction.

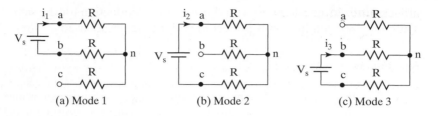

(a) Mode 1   (b) Mode 2   (c) Mode 3

FIGURE 6.10

Equivalent circuits for Y-connected resistive load.

During mode 2 for $\pi/3 \leq \omega t \leq 2\pi/3$, transistors 1 and 2 conduct.

$$v_{an} = \frac{V_s}{2} \qquad v_{bn} = 0 \qquad v_{cn} = -\frac{V_s}{2}$$

During mode 3 for $2\pi/3 \leq \omega t \leq 3\pi/3$, transistors 2 and 3 conduct.

$$v_{an} = 0 \qquad v_{bn} = \frac{V_s}{2} \qquad v_{cn} = -\frac{V_s}{2}$$

The line-to-neutral voltages that are shown in Figure 6.9 can be expressed in Fourier series as

$$v_{an} = \sum_{n=1,3,5,\ldots}^{\infty} \frac{2V_s}{n\pi} \sin \frac{n\pi}{3} \sin n\left(\omega t + \frac{\pi}{6}\right) \tag{6.24a}$$

$$v_{bn} = \sum_{n=1,3,5,\ldots}^{\infty} \frac{2V_s}{n\pi} \sin \frac{n\pi}{3} \sin n\left(\omega t - \frac{\pi}{2}\right) \tag{6.24b}$$

$$v_{cn} = \sum_{n=1,3,5,\ldots}^{\infty} \frac{2V_s}{n\pi} \sin \frac{n\pi}{3} \sin n\left(\omega t - \frac{7\pi}{6}\right) \tag{6.24c}$$

The line a-to-b voltage is $v_{ab} = \sqrt{3}\, v_{an}$ with a phase advance of 30°. Therefore, the instantaneous line-to-line voltages (for a Y-connected load) are

$$v_{ab} = \sum_{n=1}^{\infty} \frac{2\sqrt{3}V_S}{n\pi} \sin\left(\frac{n\pi}{3}\right) \sin n\left(\omega t + \frac{\pi}{3}\right) \quad \text{for } n = 1, 3, 5, \ldots \tag{6.25a}$$

$$v_{bc} = \sum_{n=1}^{\infty} \frac{2\sqrt{3}V_S}{n\pi} \sin\left(\frac{n\pi}{3}\right) \sin n\left(\omega t - \frac{\pi}{3}\right) \quad \text{for } n = 1, 3, 5, \ldots \tag{6.25b}$$

$$v_{ca} = \sum_{n=1}^{\infty} \frac{2\sqrt{3}V_S}{n\pi} \sin\left(\frac{n\pi}{3}\right) \sin n\left(\omega t - \pi\right) \quad \text{for } n = 1, 3, 5, \ldots \tag{6.25c}$$

There is a delay of $\pi/6$ between the turning off Q_1 and turning on Q_4. Thus, there should be no short circuit of the dc supply through one upper and one lower transistors. At any time, two load terminals are connected to the dc supply and the third one remains open. The potential of this open terminal depends on the load characteristics and would be unpredictable. Because one transistor conducts for 120°, the transistors are less utilized as compared with those of $180°$ conduction for the same load condition. Thus, the $180°$ conduction is preferred and it is generally used in three-phase inverters.

Key Points of Section 6.5

- The three-phase bridge inverter requires six switching devices and six diodes. The rms fundamental component V_{L1} of the output line voltage is $0.7798V_s$ and that for phase voltage is $V_{p1} = V_{L1}/\sqrt{3} = 0.45V_s$ for $180°$ conduction. For $120°$ conduction, $V_{P1} = 0.3898V_s$ and $V_{L1} = \sqrt{3}\,V_{P1} = 0.6753V_s$. The $180°$ conduction is the preferred control method.
- The design of an inverter requires the determination of the average, rms, and peak currents of the switching devices and diodes.

6.6 VOLTAGE CONTROL OF SINGLE-PHASE INVERTERS

In many industrial applications, to control of the output voltage of inverters is often necessary (1) to cope with the variations of dc input voltage, (2) to regulate voltage of inverters, and (3) to satisfy the constant volts and frequency control requirement. There are various techniques to vary the inverter gain. The most efficient method of controlling the gain (and output voltage) is to incorporate PWM control within the inverters. The commonly used techniques are:

1. Single-pulse-width modulation.
2. Multiple-pulse-width modulation.
3. Sinusoidal pulse-width modulation.
4. Modified sinusoidal pulse-width modulation.
5. Phase-displacement control.

6.6.1 Single-Pulse-Width Modulation

In single-pulse-width modulation control, there is only one pulse per half-cycle and the width of the pulse is varied to control the inverter output voltage. Figure 6.11 shows the generation of gating signals and output voltage of single-phase full-bridge inverters. The gating signals are generated by comparing a rectangular reference signal of amplitude A_r with a triangular carrier wave of amplitude A_c. The frequency of the reference signal determines the fundamental frequency of output voltage. The instantaneous output voltage is $v_o = V_s(g_1 - g_4)$. The ratio of A_r to A_c is the control variable and defined as the amplitude *modulation index*. The amplitude modulation index, or simply modulation index

$$M = \frac{A_r}{A_c} \tag{6.26}$$

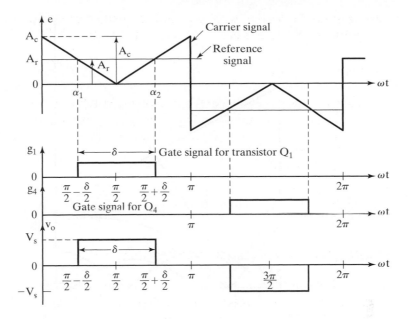

FIGURE 6.11

Single pulse-width modulation.

The rms output voltage can be found from

$$V_o = \left[\frac{2}{2\pi} \int_{(\pi-\delta)/2}^{(\pi+\delta)/2} V_s^2 \, d(\omega t) \right]^{1/2} = V_s \sqrt{\frac{\delta}{\pi}} \tag{6.27}$$

By varying A_r from 0 to A_c, the pulse width δ can be modified from $0°$ to $180°$ and the rms output voltage V_o, from 0 to V_s.

The Fourier series of output voltage yields

$$v_o(t) = \sum_{n=1,3,5,\ldots}^{\infty} \frac{4V_s}{n\pi} \sin \frac{n\delta}{2} \sin n\omega t \tag{6.28}$$

Due to the symmetry of the output voltage along the x-axis, the even harmonics (for $n = 2, 4, 6, \ldots$) are absent. A computer program is developed to evaluate the performance of single-pulse modulation for single-phase full-bridge inverters. Figure 6.12 shows the harmonic profile with the variation of modulation index M. The dominant harmonic is the third, and the DF increases significantly at a low output voltage.

The time and angles of intersections can be found from

$$t_1 = \frac{\alpha_1}{\omega} = (1 - M) \frac{T_s}{2} \tag{6.29a}$$

$$t_2 = \frac{\alpha_2}{\omega} = (1 + M) \frac{T_s}{2} \tag{6.29b}$$

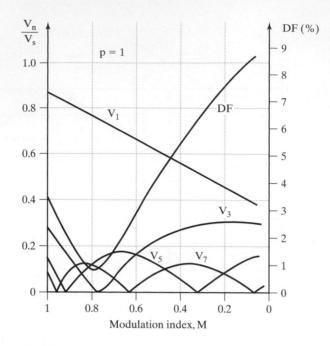

FIGURE 6.12

Harmonic profile of single-pulse-width modulation.

which gives the pulse width d (or pulse angle δ) as

$$d = \frac{\delta}{\omega} = t_2 - t_1 = MT_s \qquad (6.29c)$$

where $T_s = T/2$.

Gating sequence.   The algorithm for generating the gating signals are as follows:

1. Generate a triangular carrier signal v_{cr} of switching period $T_s = T/2$. Compare v_{cr} with a dc reference signal v_r to produce the difference $v_e = v_{cr} - v_r$, which must pass through a gain limiter to produce a square wave of width d at a switching period T_s.

2. To produce the gating signal g_1, multiply the resultant square wave by a unity signal v_z, which must be a unity pulse of 50% duty cycle at a period of T.

3. To produce the gating signal g_2, multiply the square wave by a logic-invert signal of v_z.

6.6.2   Multiple-Pulse-Width Modulation

The harmonic content can be reduced by using several pulses in each half-cycle of output voltage. The generation of gating signals (in Figure 6.13b) for turning on and off of transistors is shown in Figure 6.13a by comparing a reference signal with a triangular carrier wave. The gate signals are shown in Figure 6.13b. The frequency of reference

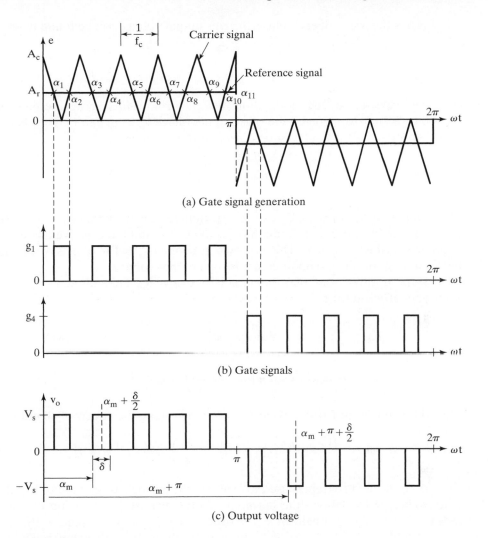

FIGURE 6.13

Multiple-pulse-width modulation.

signal sets the output frequency f_o, and the carrier frequency f_c determines the number of pulses per half-cycle p. The modulation index controls the output voltage. This type of modulation is also known as *uniform pulse-width modulation* (UPWM). The number of pulses per half-cycle is found from

$$p = \frac{f_c}{2f_o} = \frac{m_f}{2} \tag{6.30}$$

where $m_f = f_c/f_o$ is defined as the *frequency modulation ratio*.

The instantaneous output voltage is $v_o = V_s(g_1 - g_4)$. The output voltage for single-phase bridge inverters is shown in Figure 6.13c for UPWM.

If δ is the width of each pulse, the rms output voltage can be found from

$$V_o = \left[\frac{2p}{2\pi} \int_{(\pi/p-\delta)/2}^{(\pi/p+\delta)/2} V_s^2 \, d(\omega t) \right]^{1/2} = V_s \sqrt{\frac{p\delta}{\pi}} \tag{6.31}$$

The variation of the modulation index M from 0 to 1 varies the pulse width d from 0 to $T/2p$ (0 to π/p) and the rms output voltage V_o from 0 to V_s. The general form of a Fourier series for the instantaneous output voltage is

$$v_o(t) = \sum_{n=1,3,5,\dots}^{\infty} B_n \sin n\omega t \tag{6.32}$$

The coefficient B_n in Eq. (6.32) can be determined by considering a pair of pulses such that the positive pulse of duration δ starts at $\omega t = \alpha$ and the negative one of the same width starts at $\omega t = \pi + \alpha$. This is shown in Figure 6.13c. The effects of all pulses can be combined together to obtain the effective output voltage.

If the positive pulse of mth pair starts at $\omega t = \alpha_m$ and ends at $\omega t = \alpha_m + \delta$, the Fourier coefficient for a pair of pulses is

$$
\begin{aligned}
b_n &= \frac{2}{\pi} \left[\int_{\alpha_m+\delta/2}^{\alpha_m+\delta} \sin n\omega t \, d(\omega t) - \int_{\pi+\alpha_m}^{\pi+\alpha_m+\delta/2} \sin n\omega t \, d(\omega t) \right] \\
&= \frac{4V_s}{n\pi} \sin \frac{n\delta}{4} \left[\sin n \left(\alpha_m + \frac{3\delta}{4} \right) - \sin n \left(\pi + \alpha_m + \frac{\delta}{4} \right) \right]
\end{aligned} \tag{6.33}
$$

The coefficient B_n of Eq. (6.32) can be found by adding the effects of all pulses,

$$B_n = \sum_{m=1}^{2p} \frac{4V_s}{n\pi} \sin \frac{n\delta}{4} \left[\sin n \left(\alpha_m + \frac{3\delta}{4} \right) - \sin n \left(\pi + \alpha_m + \frac{\delta}{4} \right) \right] \tag{6.34}$$

A computer program is used to evaluate the performance of multiple pulse modulation. Figure 6.14 shows the harmonic profile against the variation of modulation index for five pulses per half-cycle. The order of harmonics is the same as that of single-pulse modulation. The distortion factor is reduced significantly compared with that of single-pulse modulation. However, due to larger number of switching on and off processes of power transistors, the switching losses would increase. With larger values of p, the amplitudes of LOH would be lower, but the amplitudes of some higher order harmonics would increase. However, such higher order harmonics produce negligible ripple or can easily be filtered out.

Due to the symmetry of the output voltage along the x-axis, $A_n = 0$ and the even harmonics (for $n = 2, 4, 6, \dots$) are absent.

The mth time t_m and angle α_m of intersection can be determined from

$$t_m = \frac{\alpha_m}{\omega} = (m - M) \frac{T_s}{2} \qquad \text{for} \qquad m = 1, 3, \dots, 2p \tag{6.35a}$$

$$t_m = \frac{\alpha_m}{\omega} = (m - 1 + M) \frac{T_s}{2} \qquad \text{for} \qquad m = 2, 4, \dots, 2p \tag{6.35b}$$

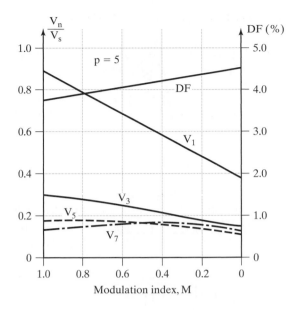

FIGURE 6.14

Harmonic profile of multiple-pulse-width modulation.

Because all widths are the same, we get the pulse width d (or pulse angle δ) as

$$d = \frac{\delta}{\omega} = t_{m+1} - t_m = MT_s \qquad (6.35c)$$

where $T_s = T/2p$.

The algorithm for generating the gating signals is the same as that for single-pulse modulation, except the switching period T_s of the triangular carrier signal v_{cr} is $T/2p$ instead of $T/2$.

6.6.3   Sinusoidal Pulse-Width Modulation

Instead of maintaining the width of all pulses the same as in the case of multiple-pulse modulation, the width of each pulse is varied in proportion to the amplitude of a sine wave evaluated at the center of the same pulse [2]. The DF and LOH are reduced significantly. The gating signals as shown in Figure 6.15a are generated by comparing a sinusoidal reference signal with a triangular carrier wave of frequency f_c. This sinusoidal pulse-width modulation (SPWM) is commonly used in industrial applications. The frequency of reference signal f_r determines the inverter output frequency f_o; and its peak amplitude A_r controls the modulation index M, and then in turn the rms output voltage V_o. Comparing the bidirectional carrier signal v_{cr} with two sinusoidal reference signals v_r and $-v_r$ shown in Figure 6.15a produces gating signals g_1 and g_4, respectively, as shown in Figure 6.15b. The output voltage is $v_o = V_s(g_1 - g_4)$. However, g_1 and g_4 cannot be released at the same time. The number of pulses per half-cycle depends on the carrier frequency. Within the constraint that two transistors of the same arm (Q_1 and

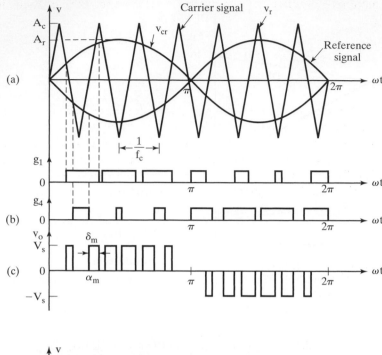

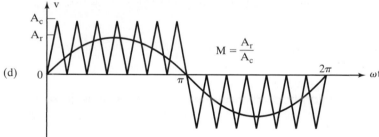

FIGURE 6.15

Sinusoidal pulse-width modulation.

Q_4) cannot conduct at the same time, the instantaneous output voltage is shown in Figure 6.15c. The same gating signals can be generated by using unidirectional triangular carrier wave as shown in Figure 6.15d. It is easier to implement this method and is preferable. The algorithm for generating the gating signals is similar to that for the uniform PWM in Section 6.6.2, except the reference signal is a sine wave $v_r = V_r \sin \omega t$, instead of a dc signal. The output voltage is $v_o = V_s(g_1 - g_4)$.

The rms output voltage can be varied by varying the modulation index M. It can be observed that the area of each pulse corresponds approximately to the area under the sine wave between the adjacent midpoints of off periods on the gating signals. If δ_m is the width of mth pulse, Eq. (6.31) can be extended to find the rms output voltage

$$V_o = V_s \left(\sum_{m=1}^{2p} \frac{\delta_m}{\pi} \right)^{1/2} \tag{6.36}$$

Equation (6.34) can also be applied to determine the Fourier coefficient of output voltage as

$$B_n = \sum_{m=1}^{2p} \frac{4V_s}{n\pi} \sin \frac{n\delta_m}{4} \left[\sin n \left(\alpha_m + \frac{3\delta_m}{4} \right) - \sin n \left(\pi + \alpha_m + \frac{\delta_m}{4} \right) \right]$$

$$\text{for } n = 1, 3, 5, \ldots \qquad (6.37)$$

A computer program is developed to determine the width of pulses and to evaluate the harmonic profile of sinusoidal modulation. The harmonic profile is shown in Figure 6.16 for five pulses per half-cycle. The DF is significantly reduced compared with that of multiple-pulse modulation. This type of modulation eliminates all harmonics less than or equal to $2p - 1$. For $p = 5$, the LOH is ninth.

The mth time t_m and angle α_m of intersection can be determined from

$$t_m = \frac{\alpha_m}{\omega} = t_x + m \frac{T_s}{2} \qquad (6.38a)$$

where t_x can be solved from

$$1 - \frac{2t}{T_s} - M \sin \left[\omega \left(t_x \mid \frac{mT_s}{2} \right) \right] \qquad \text{for } m = 1, 3, \ldots, 2p \qquad (6.38b)$$

$$\frac{2t}{T_s} = M \sin \left[\omega \left(t_x + \frac{mT_s}{2} \right) \right] \qquad \text{for } m = 2, 4, \ldots, 2p \qquad (6.38c)$$

where $T_s = T/2(p + 1)$. The width of the mth pulse d_m (or pulse angle δ_m) can be found from

$$d_m = \frac{\delta_m}{\omega} = t_{m+1} - t_m \qquad (6.38d)$$

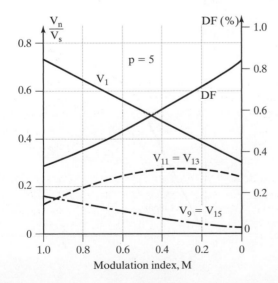

FIGURE 6.16

Harmonic profile of sinusoidal pulse-width modulation.

The output voltage of an inverter contains harmonics. The PWM pushes the harmonics into a high-frequency range around the switching frequency f_c and its multiples, that is, around harmonics m_f, $2m_f$, $3m_f$, and so on. The frequencies at which the voltage harmonics occur can be related by

$$f_n = (jm_f \pm k)f_c \tag{6.39}$$

where the nth harmonic equals the kth sideband of jth times the frequency to modulation ratio m_f.

$$n = jm_f \pm k$$
$$= 2jp \pm k \quad \text{for } j = 1, 2, 3, \ldots \text{ and } k = 1, 3, 5, \ldots \tag{6.40}$$

The peak fundamental output voltage for PWM and SPWM control can be found approximately from

$$V_{m1} = dV_s \quad \text{for } 0 \leq d \leq 1.0 \tag{6.41}$$

For $d = 1$, Eq. (6.41) gives the maximum peak amplitude of the fundamental output voltage as $V_{m1(\text{max})} = V_s$. According to Eq. (6.12), $V_{m1(\text{max})}$ could be as high as $4V_s/\pi = 1.273V_s$ for a square-wave output. To increase the fundamental output voltage, d must be increased beyond 1.0. The operation beyond $d = 1.0$ is called *overmodulation*. The value of d at which $V_{m1(\text{max})}$ equals $1.273V_s$ is dependent on the number of pulses per half-cycle p and is approximately 3 for $p = 7$, as shown in Figure 6.17. Overmodulation basically leads to a square-wave operation and adds more harmonics as compared with operation in the linear range (with $d \leq 1.0$). Overmodulation is normally avoided in applications requiring low distortion (e.g., uninterruptible power supplies [UPSs]).

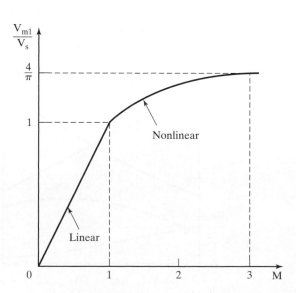

FIGURE 6.17

Peak fundamental output voltage versus modulation index M.

6.6.4   Modified Sinusoidal Pulse-Width Modulation

Figure 6.15c indicates that the widths of pulses nearer the peak of the sine wave do not change significantly with the variation of modulation index. This is due to the characteristics of a sine wave, and the SPWM technique can be modified so that the carrier wave is applied during the first and last 60° intervals per half-cycle (e.g., 0° to 60° and 120° to 180°). This modified sinusoidal pulse-width modulation (MSPWM) is shown in Figure 6.18. The fundamental component is increased and its harmonic characteristics are improved. It reduces the number of switching of power devices and also reduces switching losses.

The mth time t_m and angle α_m of intersection can be determined from

$$t_m = \frac{\alpha_m}{\omega} = t_x + m\frac{T_s}{2} \qquad \text{for} \quad m = 1, 2, 3, \ldots, p \qquad (6.42a)$$

where t_x can be solved from

$$1 - \frac{2t}{T_s} = M\sin\left[\omega\left(t_x + \frac{mT_s}{2}\right)\right] \qquad \text{for} \quad m = 1, 3, \ldots, p \qquad (6.42b)$$

$$\frac{2t}{T_s} = M\sin\left[\omega\left(t_x + \frac{mT_s}{2}\right)\right] \qquad \text{for} \quad m = 2, 4, \ldots, p \qquad (6.42c)$$

The time intersections during the last 60° intervals can be found from

$$t_{m+1} = \frac{\alpha_{m+1}}{\omega} = \frac{T}{2} - t_{2p-m} \qquad \text{for} \quad m = p, p+1\ldots, 2p-1 \qquad (6.42d)$$

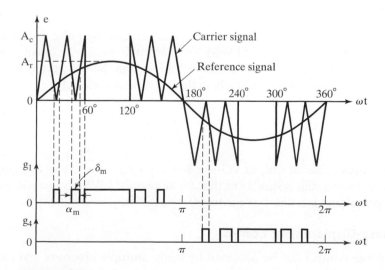

FIGURE 6.18

Modified sinusoidal pulse-width modulation.

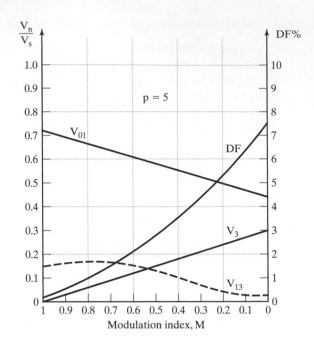

FIGURE 6.19

Harmonic profile of modified sinusoidal
pulse-width modulation.

where $T_s = T/6(p + 1)$. The width of the mth pulse d_m (or pulse angle δ_m) can be
found from

$$d_m = \frac{\delta_m}{\omega} = t_{m+1} - t_m \tag{6.42e}$$

A computer program was used to determine the pulse widths and to evaluate the
performance of modified SPWM. The harmonic profile is shown in Figure 6.19 for five
pulses per half-cycle. The number of pulses q in the 60° period is normally related to
the frequency ratio, particularly in three-phase inverters, by

$$\frac{f_c}{f_o} = 6q + 3 \tag{6.43}$$

The instantaneous output voltage is $v_o = V_s(g_1 - g_4)$. The algorithm for generating
the gating signals is similar to that for sinusoidal PWM in Section 6.6.3, except the reference signal is a sine wave from 60° to 120° only.

6.6.5 Phase-Displacement Control

Voltage control can be obtained by using multiple inverters and summing the output
voltages of individual inverters. A single-phase full-bridge inverter in Figure 6.2a can
be perceived as the sum of two half-bridge inverters in Figure 6.1a. A 180° phase

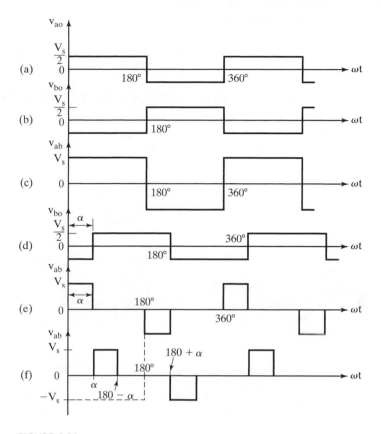

FIGURE 6.20

Phase-displacement control.

displacement produces an output voltage as shown in Figure 6.20c, whereas a delay (or displacement) angle of α produces an output as shown in Figure 6.20e.

For example, the gate signal g_1 for the half-bridge inverter can be delayed by angle α to produce the gate signal g_2.

The rms output voltage,

$$V_o = V_s \sqrt{\frac{\alpha}{\pi}} \qquad (6.44)$$

If

$$V_{ao} = \sum_{n=1,3,5,\ldots}^{\infty} \frac{2V_s}{n\pi} \sin n\omega t$$

then

$$v_{bo} = \sum_{n=1,3,5,\ldots}^{\infty} \frac{2V_s}{n\pi} \sin n(\omega t - \alpha)$$

The instantaneous output voltage,

$$v_{ab} = v_{ao} - v_{bo} = \sum_{n=1,3,5,\dots}^{\infty} \frac{2V_s}{n\pi} [\sin n\omega t - \sin n(\omega t - \alpha)]$$

which, after using $\sin A - \sin B = 2 \sin[(A - B)/2] \cos[(A + B)/2]$, can be simplified to

$$v_{ab} = \sum_{n=1,3,5,\dots}^{\infty} \frac{4V_s}{n\pi} \sin \frac{n\alpha}{2} \cos n \left(\omega t - \frac{\alpha}{2} \right) \qquad (6.45)$$

The rms value of the fundamental output voltage is

$$V_{o1} = \frac{4V_s}{\sqrt{2}} \sin \frac{\alpha}{2} \qquad (6.46)$$

Equation (6.46) indicates that the output voltage can be varied by changing the delay angle. This type of control is especially useful for high-power applications, requiring a large number of switching devices in parallel.

If the gate signals g_1 and g_2 are delayed by angles $\alpha_1 = \alpha$ and $\alpha_2 (= \pi - \alpha)$, the output voltage v_{ab} has a quarter-wave symmetry at $\pi/2$ as shown in Figure 6.20f. Thus, we get

$$v_{ao} = \sum_{n=1}^{\infty} \frac{2V_s}{n\pi} \sin(n(\omega t - \alpha)) \qquad \text{for } n = 1, 3, 5, \dots$$

$$v_{bo} = \sum_{n=1}^{\infty} \frac{2V_s}{n\pi} \sin[n(\omega t - \pi + \alpha)] \qquad \text{for } n = 1, 3, 5, \dots$$

$$v_{ab} = v_{ao} - v_{bo} = \sum_{n=1}^{\infty} \frac{4V_s}{n\pi} \cos(n\alpha) \sin(n\omega t) \qquad \text{for } n = 1, 3, 5 \qquad (6.47)$$

6.7   ADVANCED MODULATION TECHNIQUES

The SPWM, which is most commonly used, suffers from drawbacks (e.g., low fundamental output voltage). The other techniques that offer improved performances are:

    Trapezoidal modulation
    Staircase modulation
    Stepped modulation
    Harmonic injection modulation
    Delta modulation

For the sake of simplicity, we shall show the output voltage v_{ao}, for a half-bridge inverter in Figure 6.1a. For a full-bridge inverter, $v_o = v_{ao} - v_{bo}$, where v_{bo} is the inverse of v_{ao}.

Trapezoidal modulation.   The gating signals are generated by comparing a triangular carrier wave with a modulating trapezoidal wave [3] as shown in Figure 6.21.

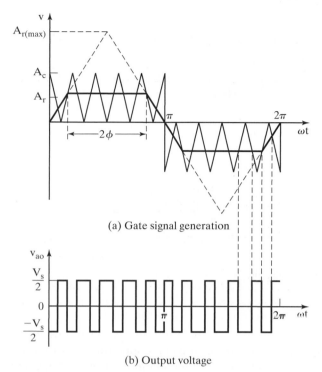

(a) Gate signal generation

(b) Output voltage

FIGURE 6.21

Trapezoidal modulation.

The trapezoidal wave can be obtained from a triangular wave by limiting its magnitude to $\pm A_r$, which is related to the peak value $A_{r(\text{max})}$ by

$$A_r = \sigma A_{r(\text{max})}$$

where σ is called the *triangular factor*, because the waveform becomes a triangular wave when $\sigma = 1$. The modulation index M is

$$M = \frac{A_r}{A_c} = \frac{\sigma A_{r(\text{max})}}{A_c} \quad \text{for } 0 \leq M \leq 1 \tag{6.48}$$

The angle of the flat portion of the trapezoidal wave is given by

$$2\phi = (1 - \sigma)\pi \tag{6.49}$$

For fixed values of $A_{r(\text{max})}$ and A_c, M that varies with the output voltage can be varied by changing the triangular factor σ. This type of modulation increases the peak fundamental output voltage up to $1.05V_s$, but the output contains LOHs.

Staircase modulation. The modulating signal is a staircase wave, as shown in Figure 6.22. The staircase is not a sampled approximation to the sine wave. The levels of the stairs are calculated to eliminate specific harmonics. The modulation frequency

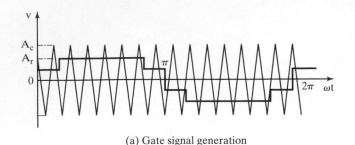

(a) Gate signal generation

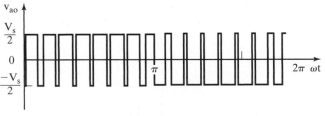

FIGURE 6.22

(b) Output voltage

Staircase modulation.

ratio m_f and the number of steps are chosen to obtain the desired quality of output voltage. This is an optimized PWM and is not recommended for fewer than 15 pulses in one cycle. It has been shown [4] that for high fundamental output voltage and low DF, the optimum number of pulses in one cycle is 15 for two levels, 21 for three levels, and 27 for four levels. This type of control provides a high-quality output voltage with a fundamental value of up to $0.94V_s$.

Stepped modulation.   The modulating signal is a stepped wave [4, 5] as shown in Figure 6.23. The stepped wave is not a sampled approximation to the sine wave. It is divided into specified intervals, say 20°, with each interval controlled individually to control the magnitude of the fundamental component and to eliminate specific harmonics. This type of control gives low distortion, but a higher fundamental amplitude compared with that of normal PWM control.

Harmonic injected modulation.   The modulating signal is generated by injecting selected harmonics to the sine wave. This results in flat-topped waveform and reduces the amount of overmodulation. It provides a higher fundamental amplitude and low distortion of the output voltage. The modulating signal [6, 7] is generally composed of

$$v_r = 1.15 \sin \omega t + 0.27 \sin 3\omega t - 0.029 \sin 9\omega t \qquad (6.50)$$

The modulating signal with third and ninth harmonic injections is shown in Figure 6.24. It should be noted that the injection of $3n$th harmonics does not affect the quality of the output voltage, because the output of a three-phase inverter does not contain triplen harmonics. If only the third harmonic is injected, v_r is given by

$$v_r = 1.15 \sin \omega t + 0.19 \sin 3\omega t \qquad (6.51)$$

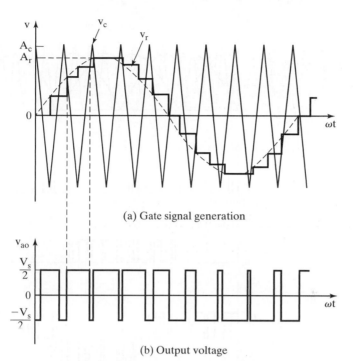

(a) Gate signal generation

(b) Output voltage

FIGURE 6.23

Stepped modulation.

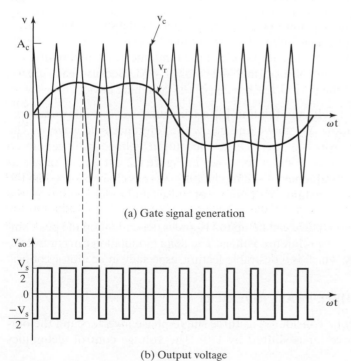

(a) Gate signal generation

(b) Output voltage

FIGURE 6.24

Selected harmonic injection modulation.

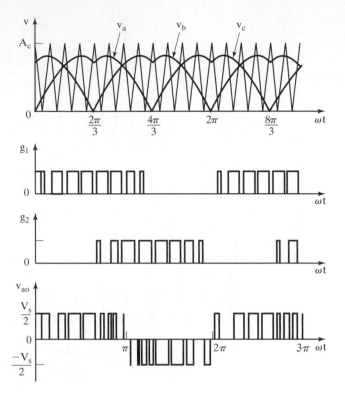

FIGURE 6.25

Harmonic injection modulation.

The modulating signal [8] can be generated from $2\pi/3$ segments of a sine wave as shown in Figure 6.25. This is the same as injecting $3n$th harmonics to a sine wave. The line-to-line voltage is sinusoidal PWM and the amplitude of the fundamental component is approximately 15% more than that of a normal sinusoidal PWM. Because each arm is switched off for one-third of the period, the heating of the switching devices is reduced.

Delta modulation.   In delta modulation [9], a triangular wave is allowed to oscillate within a defined window ΔV above and below the reference sine wave v_r. The inverter switching function, which is identical to the output voltage v_o is generated from the vertices of the triangular wave v_c as shown in Figure 6.26. It is also known as *hysteresis modulation*. If the frequency of the modulating wave is changed keeping the slope of the triangular wave constant, the number of pulses and pulses widths of the modulated wave would change.

The fundamental output voltage can be up to $1V_s$ and is dependent on the peak amplitude A_r and frequency f_r of the reference voltage. The delta modulation can control the ratio of voltage to frequency, which is a desirable feature, especially in ac motor control.

6.8   VOLTAGE CONTROL OF THREE-PHASE INVERTERS

A three-phase inverter may be considered as three single-phase inverters and the output of each single-phase inverter is shifted by 120°. The voltage control techniques

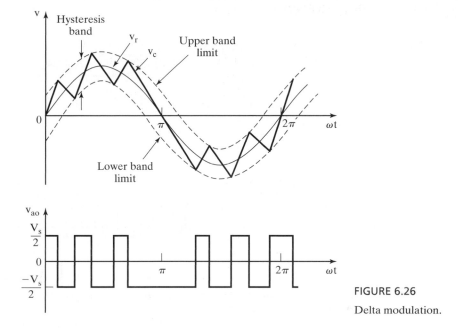

FIGURE 6.26

Delta modulation.

discussed in Section 6.6 are applicable to three-phase inverters. However, the following techniques are most commonly used for three-phase inverters.

Sinusoidal PWM

Third-harmonic PWM

60° PWM

Space vector modulation

6.8.1 Sinusoidal PWM

The generations of gating signals with sinusoidal PWM are shown in Figure 6.27a. There are three sinusoidal reference waves (v_{r_a}, v_{r_b}, and v_{r_c}) each shifted by 120°. A carrier wave is compared with the reference signal corresponding to a phase to generate the gating signals for that phase [10]. Comparing the carrier signal v_{cr} with the reference phases v_{ra}, v_{rb}, and v_{rc} produces g_1, g_3, and g_5, respectively, as shown in Figure 6.27b. The instantaneous line-to-line output voltage is $v_{ab} = V_s(g_1 - g_3)$. The output voltage as shown in Figure 6.27c, is generated by eliminating the condition that two switching devices in the same arm cannot conduct at the same time.

The normalized carrier frequency m_f should be odd multiple of three. Thus, all phase-voltage (v_{aN}, v_{bN}, and v_{cN}) are identical, but 120° out of phase without even harmonics; moreover, harmonics at frequencies multiple of three are identical in

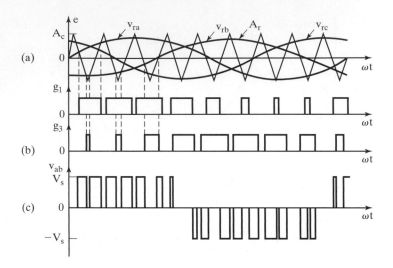

FIGURE 6.27

Sinusoidal pulse-width modulation for three-phase inverter.

amplitude and phase in all phases. For instance, if the ninth harmonic voltage in phase *a* is

$$v_{aN9}(t) = \hat{v}_9 \sin(9\omega t) \tag{6.52}$$

the corresponding ninth harmonic in phase *b* will be,

$$v_{bN9}(t) = \hat{v}_9 \sin(9(\omega t - 120°)) = \hat{v}_9 \sin(9\omega t - 1080°)) = \hat{v}_9 \sin(9\omega t) \tag{6.53}$$

Thus, the ac output line voltage $v_{ab} = v_{aN} - v_{bN}$ does not contain the ninth harmonic. Therefore, for odd multiples of three times the normalized carrier frequency m_f, the harmonics in the ac output voltage appear at normalized frequencies f_h centered around m_f and its multiples, specifically, at

$$n = jm_f \pm k \tag{6.54}$$

where $j = 1, 3, 5, \ldots$ for $k = 2, 4, 6, \ldots$; and $j = 2, 4, \ldots$ for $k = 1, 5, 7, \ldots$, such that *n* is not a multiple of three. Therefore, the harmonics are at $m_f \pm 2, m_f \pm 4 \ldots$, $2m_f \pm 1, 2m_f \pm 5, \ldots, 3m_f \pm 2, 3m_f \pm 4, \ldots, 4m_f \pm 1, 4m_f \pm 5, \ldots$. For nearly sinusoidal ac load current, the harmonics in the dc link current are at frequencies given by

$$n = jm_f \pm k \pm 1 \tag{6.55}$$

where $j = 0, 2, 4, \ldots$ for $k = 1, 5, 7, \ldots$; and $j = 1, 3, 5, \ldots$ for $k = 2, 4, 6, \ldots$, such that $n = jm_f \pm k$ is positive and not a multiple of three.

Because the maximum amplitude of the fundamental phase voltage in the linear region $(M \leq 1)$ is $V_s/2$, the maximum amplitude of the fundamental ac output line voltage is $\hat{v}_{ab1} = \sqrt{3}V_s/2$. Therefore, one can write the peak amplitude as

$$\hat{v}_{ab1} = M\sqrt{3}\,\frac{V_s}{2} \qquad \text{for } 0 < M \leq 1 \tag{6.56}$$

Overmodulation.   To further increase the amplitude of the load voltage, the amplitude of the modulating signal $\hat{v}_r$ can be made higher than the amplitude of the carrier signal $\hat{v}_{cr}$, which leads to overmodulation [11]. The relationship between the amplitude of the fundamental ac output line voltage and the dc link voltage becomes nonlinear. Thus, in the overmodulation region, the line voltages range in,

$$\sqrt{3}\,\frac{V_s}{2} < \hat{v}_{ab1} = \hat{v}_{bc1} = \hat{v}_{ca1} < \frac{4}{\pi}\sqrt{3}\,\frac{V_s}{2} \tag{6.57}$$

Large values of M in the SPWM technique lead to full overmodulation. This case is known as square-wave operation as illustrated in Figure 6.28, where the power devices are on for 180°. In this mode, the inverter cannot vary the load voltage except by

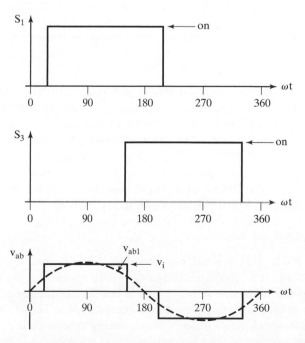

FIGURE 6.28

Square-wave operation.

varying the dc supply voltage V_s. The fundamental ac line voltage is given by

$$\hat{v}_{ab1} = \frac{4}{\pi}\sqrt{3}\,\frac{V_s}{2} \tag{6.58}$$

The ac line output voltage contains the harmonics f_n, where $n = 6k \pm 1$ ($k = 1$, $2, 3, \ldots$) and their amplitudes are inversely proportional to their harmonic order n. That is,

$$\hat{v}_{abn} = \frac{1}{n}\frac{4}{\pi}\sqrt{3}\,\frac{V_s}{2} \tag{6.59}$$

Example 6.5   Finding the Allowable Limit of the Dc Input Source

A single-phase full-bridge inverter controls the power in a resistive load. The nominal value of input dc voltage is $V_s = 220$ V and a uniform pulse-width modulation with five pulses per half-cycle is used. For the required control, the width of each pulse is 30°. (a) Determine the rms voltage of the load. (b) If the dc supply increases by 10%, determine the pulse width to maintain the same load power. If the maximum possible pulse width is 35°, determine the minimum allowable limit of the dc input source.

Solution

a. $V_s = 220$ V, $p = 5$, and $\delta = 30°$. From Eq. (6.31), $V_o = 220\sqrt{5 \times 30/180} = 200.8$ V.

b. $V_s = 1.1 \times 220 = 242$ V. By using Eq. (6.31), $242\sqrt{5\delta/180} = 200.8$ and this gives the required value of pulse width, $\delta = 24.75°$.

   To maintain the output voltage of 200.8 V at the maximum possible pulse width of $\delta = 35°$, the input voltage can be found from $200.8 = V_s\sqrt{5 \times 35/180}$, and this yields the minimum allowable input voltage, $V_s = 203.64$ V.

6.8.2   60-Degree PWM

The 60° PWM is similar to the modified PWM in Figure 6.18. The idea behind 60° PWM is to "flat top" the waveform from 60° to 120° and 240° to 300°. The power devices are held on for one-third of the cycle (when at full voltage) and have reduced switching losses. All triple harmonics (3rd, 9th, 15th, 21st, 27th, etc.) are absent in the three-phase voltages. The 60° PWM creates a larger fundamental ($2/\sqrt{3}$) and utilizes more of the available dc voltage (phased voltage $V_P = 0.57735\,V_s$ and line voltage $V_L = V_s$) than does sinusoidal PWM. The output waveform can be approximated by the fundamental and the first few terms as shown in Figure 6.29.

6.8.3   Third-Harmonic PWM

The third-harmonic PWM [12] is similar to the selected harmonic injection method shown in Figure. 6.24 and it is implemented in the same manner as sinusoidal PWM. The difference is that the reference ac waveform is not sinusoidal but consists of both a fundamental component and a third-harmonic component as shown in Figure 6.30. As a result, the peak-to-peak amplitude of the resulting reference function does not exceed the DC supply voltage V_s, but the fundamental component is higher than the available supply V_s.

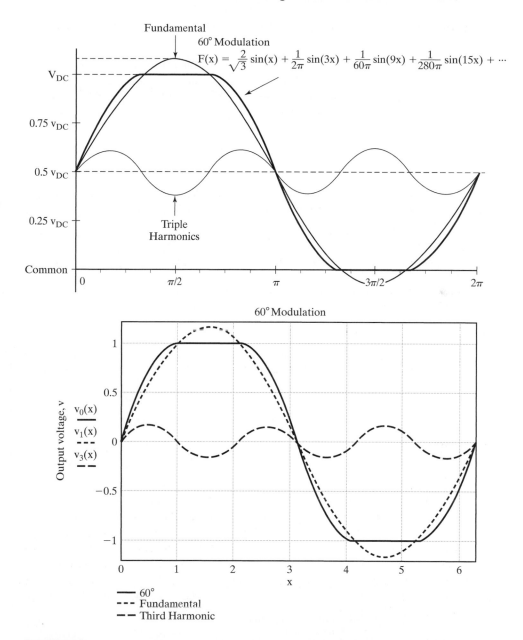

FIGURE 6.29

Output waveform for 60° PWM.

The presence of exactly the same third-harmonic component in each phase results in an effective cancellation of the third harmonic component in the neutral terminal, and the line-to-neutral phase voltages (v_{aN}, v_{bN}, and v_{cN}) are all sinusoidal with peak amplitude of $V_P = V_s/\sqrt{3} = 0.57735V_s$. The fundamental component is the same peak amplitude $V_{P1} = 0.57735V_s$ and the peak line voltage is $V_L = \sqrt{3}\,V_P = \sqrt{3} \times 0.57735\,V_s = V_s$. This

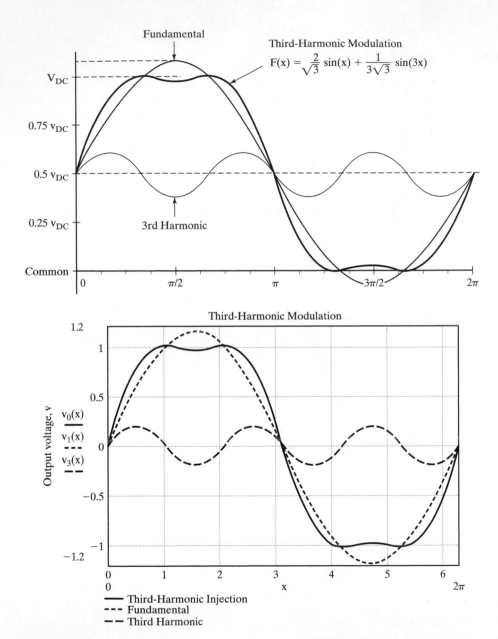

FIGURE 6.30

Output waveform for third-harmonic PWM.

is approximately 15.5% higher in amplitude than that achieved by the sinusoidal PWM. Therefore, the third-harmonic PWM provides better utilization of the dc supply voltage than the sinusoidal PWM does.

6.8.4 Space Vector Modulation

Space vector modulation (SVM) is quite different from the PWM methods. With PWMs, the inverter can be thought of as three separate push–pull driver stages, which create each phase waveform independently. SVM, however, treats the inverter as a single unit; specifically, the inverter can be driven to eight unique states, as shown in Table 6.2. Modulation is accomplished by switching the state of the inverter [13]. The control strategies are implemented in digital systems. SVM is a digital modulating technique where the objective is to generate PWM load line voltages that are in average equal to a given (or reference) load line voltages. This is done in each sampling period by properly selecting the switch states of the inverter and the calculation of the appropriate time period for each state. The selection of the states and their time periods are accomplished by the space vector (SV) transformation [25].

Space transformation.   Any three functions of time that satisfy

$$u_a(t) + u_b(t) + u_c(t) = 0 \tag{6.60}$$

can be represented in a two-dimensional space [14]. The coordinates are similar to those of three-phase voltages such that the vector $[u_a\ 0\ 0]^T$ is placed along the x-axis, the vector $[0\ u_b\ 0]^T$ is phase shifted by 120°, and the vector $[0\ 0\ u_c]^T$ is phase shifted by 240°. This is shown in Figure 6.31. The SV in complex notation is then given by

$$\mathbf{u}(t) = \frac{2}{3}\left[u_a + u_b e^{j(2/3)\pi} + u_c e^{-j(2/3)\pi}\right] \tag{6.61}$$

where 2/3 is a scaling factor. Equation (6.61) can be written in real and imaginary components in the x–y domain as

$$\mathbf{u}(t) = u_x + ju_y \tag{6.62}$$

Using Eqs. (6.61) and (6.62), we can obtain the coordinate transformation from the a–b–c-axis to the x–y axis as given by

$$\begin{pmatrix} u_x \\ u_y \end{pmatrix} = \frac{2}{3}\begin{pmatrix} 1 & \dfrac{-1}{2} & \dfrac{-1}{2} \\ 0 & \dfrac{\sqrt{3}}{2} & \dfrac{-\sqrt{3}}{2} \end{pmatrix}\begin{pmatrix} u_a \\ u_b \\ u_c \end{pmatrix} \tag{6.63}$$

which can also be written as

$$u_x = \frac{2}{3}\left[v_a - 0.5(v_b + v_c)\right] \tag{6.64a}$$

$$u_y = \frac{\sqrt{3}}{3}(v_b - v_c) \tag{6.64b}$$

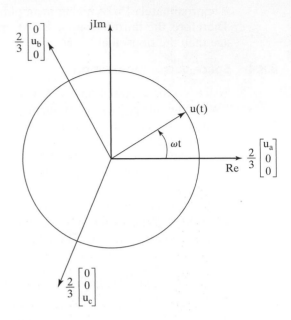

FIGURE 6.31

Three-phase coordinate vectors and space vector $\mathbf{u}(t)$.

The transformation from the x–y axis to the α–β axis, which is rotating with an angular velocity of ω, can be obtained by rotating the x–y-axis with ωt as given by

$$\begin{pmatrix} u_\alpha \\ u_\beta \end{pmatrix} = \begin{pmatrix} \cos(\omega t) & \cos\left(\dfrac{\pi}{2} + \omega t\right) \\ \sin(\omega t) & \sin\left(\dfrac{\pi}{2} + \omega t\right) \end{pmatrix} \begin{pmatrix} u_x \\ u_y \end{pmatrix} = \begin{pmatrix} \cos(\omega t) & -\sin(\omega t) \\ \sin(\omega t) & \cos(\omega t) \end{pmatrix} \begin{pmatrix} u_x \\ u_y \end{pmatrix} \qquad (6.65)$$

Using Eq. (6.61), we can find the inverse transform as

$$u_a = \mathrm{Re}(\mathbf{u}) \qquad (6.66a)$$

$$u_b = \mathrm{Re}(\mathbf{u}e^{-j(2/3)\pi}) \qquad (6.66b)$$

$$u_c = \mathrm{Re}(\mathbf{u}e^{j(2/3)\pi}) \qquad (6.66c)$$

For example, if u_a, u_b, and u_c are the three-phase voltages of a balanced supply with a peak value of V_m, we can write

$$u_a = V_m \sin(\omega t) \qquad (6.67a)$$

$$u_b = V_m \sin(\omega t - 2\pi/3) \qquad (6.67b)$$

$$u_c = V_m \sin(\omega t + 2\pi/3) \qquad (6.67c)$$

Then, using Eq. (6.61), we get the space vector representation as

$$\mathbf{u}(t) = V_m e^{j\theta} = V_m e^{j\omega t} \qquad (6.68)$$

which is a vector of magnitude V_m rotating at a constant speed ω in rads per second.

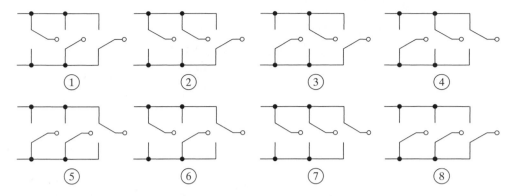

FIGURE 6.32

The on and off states of the inverter switches. [Ref. 13]

Space vector (SV).   The switching states of the inverter can be represented by binary values q_1, q_2, q_3, q_4, q_5, and q_6; that is, $q_k = 1$ when a switch is turned on and $q_k = 0$ when a switch is turned off. The pairs q_1q_4, q_3q_6, and q_5q_2 are complementary. Therefore, $q_4 = 1 - q_1$, $q_6 = 1 - q_3$, and $q_2 = 1 - q_5$. The switch on and off states are shown in Figure 6.32 [13].

Using the three-phase to two-phase transformation in Eq. (6.63) and the line voltage ($\sqrt{3}$ phase voltage) as the reference, the α–β components of the rms output voltage (peak value/$\sqrt{2}$) vectors can be expressed as functions of q_1, q_3, and q_5.

$$\begin{pmatrix} V_{L\alpha} \\ V_{L\beta} \end{pmatrix} = \frac{2}{3}\sqrt{\frac{3}{2}}\, V_s \begin{pmatrix} 1 & \dfrac{-1}{2} & \dfrac{-1}{2} \\ 0 & \dfrac{\sqrt{3}}{2} & \dfrac{-\sqrt{3}}{2} \end{pmatrix} \begin{pmatrix} q_1 \\ q_3 \\ q_5 \end{pmatrix} \tag{6.69}$$

Using the factor $\sqrt{2}$ for converting the rmns voltage to its peak value, the peak value of the line voltage is $V_{L(\text{pak})} = 2V_s/\sqrt{3}$ and that of the phase voltage is $V_{p(\text{peak})} = V_s/\sqrt{3}$. Using the phase voltage $\mathbf{V_a}$ as the reference, which is usually the case, the line voltage vector $\mathbf{V_{ab}}$ leads the phase vector by $\pi/6$. The normalized peak value of the nth line voltage vector can be found from

$$\mathbf{V_n} = \frac{\sqrt{2} \times \sqrt{2}}{\sqrt{3}}\, e^{j(2n-1)\pi/6} = \frac{2}{\sqrt{3}}\left[\cos\left(\frac{(2n-1)\pi}{6}\right) + j\sin\left(\frac{(2n-1)\pi}{6}\right)\right]$$

$$\text{for } n = 0, 1, 2, 6 \tag{6.70}$$

There are six nonzero vectors, $\mathbf{V_1}$–$\mathbf{V_6}$, and two zero vectors, $\mathbf{V_0}$ and $\mathbf{V_7}$, as shown in Figure 6.33. Let us define a performance vector $\mathbf{U}$ as the time integral function of $\mathbf{V_n}$ such that

$$\mathbf{U} = \int \mathbf{V_n}\, dt + \mathbf{U_0} \tag{6.71}$$

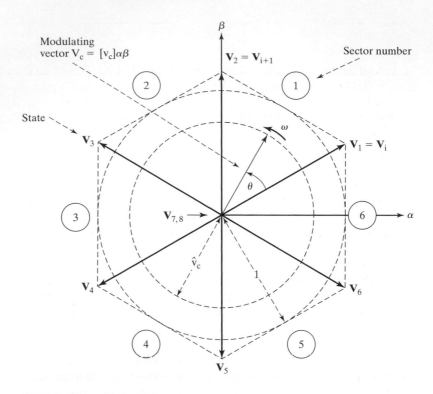

FIGURE 6.33

The space vector representation.

where $\mathbf{U_0}$ is the initial condition. According to Eq. (6.71), $\mathbf{U}$ draws a hexagon locus that is determined by the magnitude and the time period of voltage vectors. If the output voltages are purely sinusoidal, then the performance vector $\mathbf{U}$ becomes

$$\mathbf{U}^* = Me^{j\theta} = Me^{j\omega t} \tag{6.72}$$

where M is the modulation index $(0 < M < 1)$ for controlling the amplitude of the output voltage and ω is the output frequency in rads per second. $\mathbf{U}^*$ draws a pure circle locus as shown in Figure 6.33 by a dotted circle of radius $M = 1$ and it becomes the reference vector $\mathbf{Vr}$. The locus $\mathbf{U}$ can be controlled by selecting $\mathbf{V_n}$ and adjusting the time width of $\mathbf{V_n}$ to follow the $\mathbf{U}^*$ locus as closely as possible. This is called the quasi-circular locus method. The loci of $\mathbf{U}$ and $\mathbf{U}^*(=\mathbf{V_r})$ are also shown in Figure 6.33.

Modulating reference vectors. Using Eqs. (6.63) and (6.64), the vectors of three-phase line modulating signals $[v_r]_{\text{abc}} = [v_{ra} v_{rb} v_{rc}]^T$ can be represented by the complex vector $\mathbf{U}^* = \mathbf{V_r} = [v_r]_{\alpha\beta} = [v_{r\alpha} v_{r\beta}]^T$ as given by

$$v_{r\alpha} = \frac{2}{3} [v_{ra} - 0.5(v_{rb} + v_{cr})] \tag{6.73}$$

$$v_{r\beta} = \frac{\sqrt{3}}{3} (v_{rb} - v_{rc}) \tag{6.74}$$

If the line modulating signals $[v_r]_{abc}$ are three balanced sinusoidal waveforms with an amplitude of $A_c = 1$ and an angular frequency ω, the resulting modulating signals in the α–β stationary frame $\mathbf{V_c} = [v_r]_{\alpha\beta}$ becomes a vector of fixed amplitude $MA_c(=M)$ that rotates at frequency ω. This is also shown in Figure 6.33 by a dotted circle of radius M.

SV switching. The objective of the SV switching is to approximate the sinusoidal line modulating signal $\mathbf{V_r}$ with the eight space vectors ($\mathbf{V_n}$, $n = 0, 2, \ldots, 7$). However, if the modulating signal $\mathbf{V_c}$ is laying between the arbitrary vectors $\mathbf{V_n}$ and $\mathbf{V_{n+1}}$, then the two nonzero vectors ($\mathbf{V_n}$ and $\mathbf{V_{n+1}}$) and one zero SV ($\mathbf{V_z} = \mathbf{V_0}$ or $\mathbf{V_7}$) should be used to obtain the maximum load line voltage and to minimize the switching frequency. As an example, a voltage vector $\mathbf{V_r}$ in section one can be realized by the $\mathbf{V_1}$ and $\mathbf{V_2}$ vectors and one of the two null vectors ($\mathbf{V_0}$ or $\mathbf{V_7}$). In other words, $\mathbf{V_1}$ state is active for time T_1, $\mathbf{V_2}$ is active for T_2, and one of the null vectors ($\mathbf{V_0}$ or $\mathbf{V_7}$) is active for T_z. For a sufficiently high-switching frequency, the reference vector $\mathbf{V_r}$ can be assumed constant during one switching period. Because the vectors $\mathbf{V_1}$ and $\mathbf{V_2}$ are constant and $\mathbf{V_z} = 0$, we can equate the volt time of the reference vector to the SVs as

$$\mathbf{V_r} \times T_s = \mathbf{V_1} \times T_1 + \mathbf{V_2} \times T_2 + \mathbf{V_z} \times T_z \tag{6.75}$$

which is defined as the SVM. This is achieved by using two adjacent SVs with the appropriate duty cycle [15–18]. The vector diagram is shown in Figure 6.34. Expressing

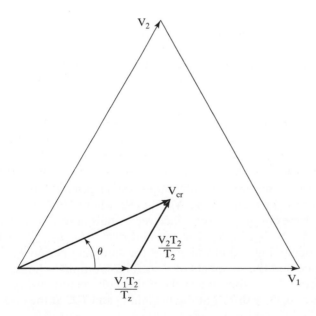

FIGURE 6.34

Determination of state times.

the SVs in rectangular coordinates, Eq. (6.75) becomes

$$T_s\text{M}\begin{pmatrix}\cos\left(\dfrac{\pi}{6}+\theta\right)\\[2mm]\sin\left(\dfrac{\pi}{6}+\theta\right)\end{pmatrix}=T_1\frac{2}{\sqrt{3}}\begin{pmatrix}\cos\left(\dfrac{\pi}{6}\right)\\[2mm]\sin\left(\dfrac{\pi}{6}\right)\end{pmatrix}+T_2\frac{2}{\sqrt{3}}\begin{pmatrix}\cos\left(\dfrac{\pi}{2}\right)\\[2mm]\sin\left(\dfrac{\pi}{2}\right)\end{pmatrix}+T_z 0 \qquad (6.76)$$

Equating the real and imaginary parts on both sides, we get

$$T_s\text{M}\cos\left(\frac{\pi}{6}+\theta\right)=T_1\frac{2}{\sqrt{3}}\cos\left(\frac{\pi}{6}\right)+T_2\frac{2}{\sqrt{3}}\cos\left(\frac{\pi}{2}\right)$$

$$T_s\text{M}\sin\left(\frac{\pi}{6}+\theta\right)=T_1\frac{2}{\sqrt{3}}\sin\left(\frac{\pi}{6}\right)+T_2\frac{2}{\sqrt{3}}\sin\left(\frac{\pi}{2}\right)$$

Solving for T_1 and T_2, we get

$$T_1=T_s\text{M}\frac{\sqrt{3}}{2}\frac{\cos\left(\dfrac{\pi}{6}+\theta\right)}{\cos\left(\dfrac{\pi}{6}\right)}=T_s\text{M}\cos\left(\frac{\pi}{6}+\theta\right)=T_s\text{M}\sin\left(\frac{\pi}{3}-\theta\right) \qquad (6.77)$$

$$T_2=T_s\text{M}\frac{\sqrt{3}}{2}\frac{\sin(\theta)}{\sin\left(\dfrac{\pi}{6}\right)}=T_s\text{M}\sin(\theta) \qquad (6.78)$$

$$T_o=T_z=T_s-T_1-T_2 \qquad (6.79)$$

where M is the modulation index;

θ is the angle between $\mathbf{V_r}$ and $\mathbf{V_n}$;

T_s is the switching or sampling period.

The same rules can be applied for calculating the time states of the vectors in sectors 2 to 6. It is assumed that the inverter operates at constant frequency and T_s remains constant.

SV sequence.   The SV sequence should assure that the load line voltages have the quarter-wave symmetry to reduce even harmonics in their spectra. To reduce the switching frequency, it is also necessary to arrange the switching sequence in such a way that the transition from one to the next is performed by switching only one inverter leg at a time. Although there is not a systematic approach to generate an SV sequence, these conditions are met by the sequence $\mathbf{V_z}$, $\mathbf{V_n}$, $\mathbf{V_{n+1}}\mathbf{V_z}$ (where $\mathbf{V_z}$ is alternately chosen between $\mathbf{V_0}$ and $\mathbf{V_7}$). If, for example, the reference vector falls in section 1, the switching sequence is $\mathbf{V_0}$, $\mathbf{V_1}$, $\mathbf{V_2}$, $\mathbf{V_7}$, $\mathbf{V_7}$, $\mathbf{V_2}$, $\mathbf{V_1}$, $\mathbf{V_0}$. The time interval $T_z(=T_0=T_7)$ can be split and distributed at the beginning and at the end of the sampling period T_s. Figure 6.35 shows both the sequence and the segments of three-phase output voltages during two samping periods. In general, the time intervals of the null vectors are equally distributed, as shown in Figure 6.35, with $T_z/2$ at the beginning and $T_z/2$ at the end.

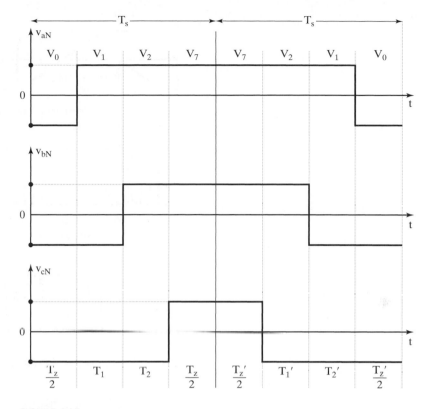

FIGURE 6.35
Pattern of SVM.

The instantaneous phase voltages can be found by time averaging of the SVs during one switching period for sector 1 as given by

$$v_{aN} = \frac{V_s}{2T_s}\left(\frac{-T_z}{2} + T_1 + T_2 + \frac{T_z}{2}\right) = \frac{V_s}{2}\sin\left(\frac{\pi}{3} + \theta\right) \tag{6.80a}$$

$$v_{bN} = \frac{V_s}{2T_s}\left(\frac{-T_z}{2} - T_1 + T_2 + \frac{T_z}{2}\right) = V_s\frac{\sqrt{3}}{2}\sin\left(\theta - \frac{\pi}{6}\right) \tag{6.80b}$$

$$v_{cN} = \frac{V_s}{T_s}\left(\frac{-T_z}{2} - T_1 - T_2 + \frac{T_z}{2}\right) = -V_{aN} \tag{6.80c}$$

To minimize uncharacteristic harmonics in SV modulation, the normalized sampling frequency f_{sn} should be an integer multiple of 6; that is, $T \geq 6nT_s$ for $n = 1, 2, 3, \ldots$. This is due to the fact that all the six sectors should be equally used in one period for producing symmetric line output voltages. As an example, Figure 6.36 shows typical waveforms of an SV modulation for $f_{sn} = 18$ and $M = 0.8$.

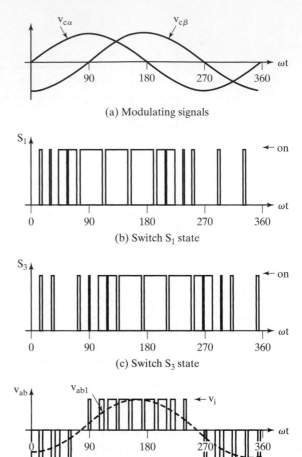

FIGURE 6.36

Three-phase waveforms for space vector modulation ($M = 0.8$, $f_{sn} = 18$).

(a) Modulating signals

(b) Switch S_1 state

(c) Switch S_3 state

(d) Ac output voltage spectrum

Overmodulation. In overmodulation, the reference vector follows a circular trajectory that extends the bounds of the hexagon [19]. The portions of the circle inside the hexagon utilize the same SVM equations for determining the state times T_n, T_{n+1}, and T_z in Eqs. (6.77) to (6.79). However, the portions of the circle outside the hexagon are limited by the boundaries of the hexagon, as shown in Figure 6.37, and the corresponding time states T_n and T_{n+1} can be found from [20]:

$$T_n = T_s \frac{\sqrt{3}\cos(\theta) - \sin(\theta)}{\sqrt{3}\cos(\theta) + \sin(\theta)} \quad (6.81a)$$

$$T_{n+1} = T_s \frac{2\sin(\theta)}{\sqrt{3}\cos(\theta) + \sin(\theta)} \quad (6.81b)$$

$$T_z = T_s - T_1 - T_2 = 0 \quad (6.81c)$$

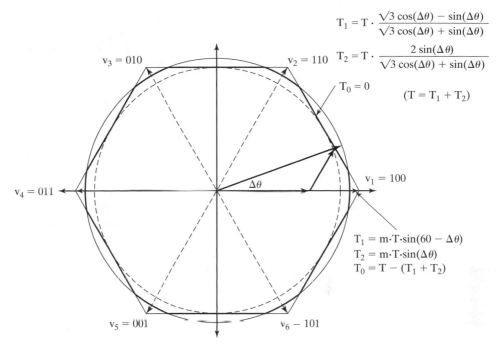

$$T_1 = T \cdot \frac{\sqrt{3}\cos(\Delta\theta) - \sin(\Delta\theta)}{\sqrt{3}\cos(\Delta\theta) + \sin(\Delta\theta)}$$

$$T_2 = T \cdot \frac{2\sin(\Delta\theta)}{\sqrt{3}\cos(\Delta\theta) + \sin(\Delta\theta)}$$

$$T_0 = 0$$

$$(T = T_1 + T_2)$$

$v_3 = 010$

$v_2 = 110$

$v_4 = 011$

$v_1 = 100$

$$T_1 = m \cdot T \cdot \sin(60 - \Delta\theta)$$
$$T_2 = m \cdot T \cdot \sin(\Delta\theta)$$
$$T_0 = T - (T_1 + T_2)$$

$v_5 = 001$

$v_6 - 101$

FIGURE 6.37

Overmodulation. [Ref. 20, R. Valentine]

The maximum modulation index M for SVM is $M_{\max} = 2/\sqrt{3}$. For $0 < M \le 1$, the inverter operates in the normal SVM, and for $M \ge 2/\sqrt{3}$, the inverter operates completely in the six-step output mode. Six-step operation switches the inverter only to the six vectors shown in Table 6.2, thereby minimizing the number of switching at one time. For $1 < M < 2/\sqrt{3}$, the inverter operates in overmodulation, which is normally used as a transitioning step from the SVM techniques into a six-step operation. Although overmodulation allows more utilization of the dc input voltage than the standard SVM techniques, it results in nonsinusoidal output voltages with a high degree of distortion, especially at a low-output frequency.

6.8.5   Comparison of PWM Techniques

Any modulation scheme can be used to create the variable-frequency, variable-voltage ac waveforms. The sinusoidal PWM compares a high frequency triangular carrier with three sinusoidal reference signals, known as the modulating signals, to generate the gating signals for the inverter switches. This is basically an analog domain technique and is commonly used in power conversion with both analog and digital implementation. Due to the cancellation of the third-harmonic components and better utilization of the dc supply, the third-harmonic PWM is preferred in three-phase applications. In contrast to sinusoidal and third-harmonic PWM techniques, the SV method does not consider each of the three modulating voltages as a separate identity. The three

TABLE 6.3   Summary of Modulation Techniques

Modulation Type	Normalized Phase Voltage, V_P/V_S	Normalized Line Voltage, V_L/V_S	Output Waveform
Sinusoidal PWM	0.5	$0.5 \times \sqrt{3} = 0.8666$	Sinusoidal
60° PWM	$1/\sqrt{3} = 0.57735$	1	Sinusoidal
Third-harmonic PWM	$1/\sqrt{3} = 0.57735$	1	Sinusoidal
SVM	$1/\sqrt{3} = 0.57735$	1	Sinusoidal
Overmodulation	Higher than the value for $M = 1$	Higher than the value for $M = 1$	Nonsinusoidal
Six-step	$\sqrt{2/3} = 0.4714$	$\sqrt{(2/3)} = 0.81645$	Nonsinusoidal

voltages are simultaneously taken into account within a two-dimensional reference frame ($\alpha-\beta$ plane) and the complex reference vector is processed as a single unit. The SVM has the advantages of lower harmonics and a higher modulation index in addition to the features of complete digital implementation by a single-chip microprocessor. Because of its flexibility of manipulation, SVM has increasing applications in power converters and motor control. Table 6.3 gives a summary of the different types of modulation schemes for three-phase inverters with $M = 1$.

Key Points of Section 6.8   Sinusoidal, harmonic injection, and SVM modulation techniques are usually used to three-phase inverters. Due to the flexibility of manipulation and digital implementation, SVM has increasing applications in power converters and motor control.

6.9   HARMONIC REDUCTIONS

We have noticed in Sections 6.6, 6.7 and 6.8 that the output voltage control of inverters requires varying both the number of pulses per half-cycle and the pulse widths that are generated by modulating techniques. The output voltage contains even harmonics over a frequency spectrum. Some applications require either a fixed or a variable output voltage, but certain harmonics are undesirable in reducing certain effects such as harmonic torque and heating in motors, interferences, and oscillations.

Phase displacement.   Equation (6.45) indicates that the nth harmonic can be eliminated by a proper choice of displacement angle α if

$$\cos n\alpha = 0$$

or

$$\alpha = \frac{90°}{n} \tag{6.82}$$

and the third harmonic is eliminated if $\alpha = 90/3 = 30°$.

Bipolar output voltage notches.   A pair of unwanted harmonics at the output of single-phase inverters can be eliminated by introducing a pair of symmetrically placed bipolar voltage *notches* [21] as shown in Figure 6.38.

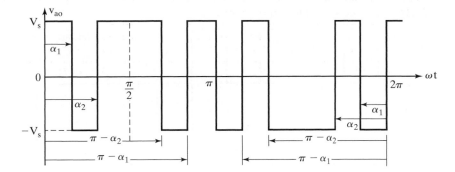

FIGURE 6.38

Output voltage with two bipolar notches per half-wave.

The Fourier series of output voltage can be expressed as

$$v_o = \sum_{n=1,3,5,\ldots}^{\infty} B_n \sin n\omega t \qquad (6.83)$$

where

$$
\begin{aligned}
B_n &= \frac{4V_s}{\pi} \left[\int_0^{\alpha_1} \sin n\omega t \, d(\omega t) - \int_{\alpha_1}^{\alpha_2} \sin n\omega t \, d(\omega t) + \int_{\alpha_2}^{\pi/2} \sin n\omega t \, d(\omega t) \right] \\
&= \frac{4V_s}{\pi} \frac{1 - 2\cos n\alpha_1 + 2\cos n\alpha_2}{n}
\end{aligned}
\qquad (6.84)
$$

Equation (6.84) can be extended to m notches per quarter-wave:

$$B_n = \frac{4V_s}{n\pi} (1 - 2\cos n\alpha_1 + 2\cos n\alpha_2 - 2\cos n\alpha_3 + 2\cos n\alpha_4 - \cdots) \qquad (6.85)$$

$$B_n = \frac{4V_s}{n\pi} \left[1 + 2\sum_{k=1}^{m} (-1)^k \cos(n\alpha_k) \right] \qquad \text{for } n = 1, 3, 5, \ldots \qquad (6.86)$$

where $\alpha_1 < \alpha_2 < \cdots < \alpha_k < \dfrac{\pi}{2}$.

The third and fifth harmonics would be eliminated if $B_3 = B_5 = 0$ and Eq. (6.84) gives the necessary equations to be solved.

$$1 - 2\cos 3\alpha_1 + 2\cos 3\alpha_2 = 0 \qquad \text{or} \qquad \alpha_2 = \tfrac{1}{3}\cos^{-1}(\cos 3\alpha_1 - 0.5)$$

$$1 - 2\cos 5\alpha_1 + 2\cos 5\alpha_2 = 0 \qquad \text{or} \qquad \alpha_1 = \tfrac{1}{5}\cos^{-1}(\cos 5\alpha_2 + 0.5)$$

These equations can be solved iteratively by initially assuming that $\alpha_1 = 0$ and repeating the calculations for α_1 and α_2. The result is $\alpha_1 = 23.62°$ and $\alpha_2 = 33.3°$.

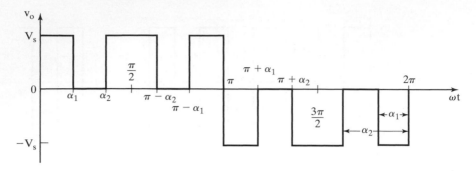

FIGURE 6.39

Unipolar output voltage with two notches per half-cycle.

Unipolar output voltage notches.   With unipolar voltage notches as shown in Figure 6.39, the coefficient B_n is given by

$$B_n = \frac{4V_s}{\pi}\left[\int_0^{\alpha_1} \sin n\omega t \ d(\omega t) + \int_{\alpha_2}^{\pi/2} \sin n\omega t \ d(\omega t)\right]$$

$$= \frac{4V_s}{\pi} \frac{1 - \cos n\alpha_1 + \cos n\alpha_2}{n} \qquad\qquad (6.87)$$

Equation (6.87) can be extended to m notches per quarter-wave as

$$B_n = \frac{4V_s}{n\pi}\left[1 + \sum_{k=1}^{m}(-1)^k \cos(n\alpha_k)\right] \qquad \text{for } n = 1, 3, 5, \ldots \qquad (6.88)$$

where $\alpha_1 < \alpha_2 < \cdots < \alpha_k < \dfrac{\pi}{2}$.

The third and fifth harmonics would be eliminated if

$$1 - \cos 3\alpha_1 + \cos 3\alpha_2 = 0$$
$$1 - \cos 5\alpha_1 + \cos 5\alpha_2 = 0$$

Solving these equations by iterations using a Mathcad program, we get $\alpha_1 = 17.83°$ and $\alpha_2 = 37.97°$.

60-Degree modulation.   The coefficient B_n is given by

$$B_n = \frac{4V_s}{n\pi}\left[\int_{\alpha_1}^{\alpha_2} \sin(n\omega t) \ d(\omega t) + \int_{\alpha_3}^{\alpha_4} \sin(n\omega t) \ d(\omega t) + \int_{\alpha_5}^{\alpha_6} \sin(n\omega t) \ d(\omega t)\right.$$

$$\left. + \int_{\pi/3}^{\pi/2} \sin(n\omega t) \ d(\omega t)\right]$$

$$B_n = \frac{4V_s}{n\pi}\left[\frac{1}{2} - \sum_{k=1}^{m}(-1)^k \cos(n\alpha_k)\right] \qquad \text{for } n = 1, 3, 5, \ldots \qquad (6.89)$$

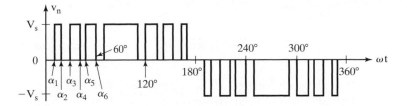

FIGURE 6.40

Output voltage for modified sinusoidal pulse-width modulation.

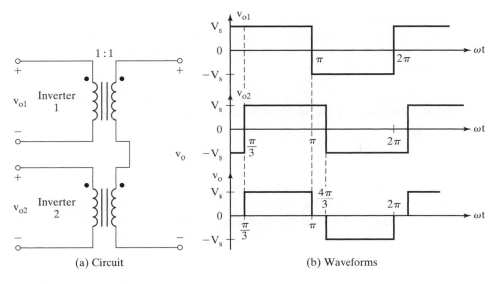

(a) Circuit   (b) Waveforms

FIGURE 6.41

Elimination of harmonics by transformer connection.

The modified sinusoidal PWM techniques can be applied to generate the notches that would eliminate certain harmonics effectively in the output voltage, as shown in Figure 6.40.

Transformer connections. The output voltages of two or more inverters may be connected in series through a transformer to reduce or eliminate certain unwanted harmonics. The arrangement for combining two inverter output voltages is shown in Figure 6.41a. The waveforms for the output of each inverter and the resultant output voltage are shown in Figure 6.41b. The second inverter is phase shifted by $\pi/3$.

From Eq. (6.12), the output of first inverter can be expressed as

$$v_{o1} = A_1 \sin \omega t + A_3 \sin 3\omega t + A_5 \sin 5\omega t + \cdots$$

Because the output of second inverter, v_{o2}, is delayed by $\pi/3$,

$$v_{o2} = A_1 \sin\left(\omega t - \frac{\pi}{3}\right) + A_3 \sin 3\left(\omega t - \frac{\pi}{3}\right) + A_5 \sin 5\left(\omega t - \frac{\pi}{3}\right) + \cdots$$

The resultant voltage v_o is obtained by vector addition.

$$v_o = v_{o1} + v_{o2} = \sqrt{3}\left[A_1 \sin\left(\omega t - \frac{\pi}{6}\right) + A_5 \sin 5\left(\omega t + \frac{\pi}{6}\right) + \cdots\right]$$

Therefore, a phase shifting of $\pi/3$ and combining voltages by transformer connection would eliminate third (and all triplen) harmonics. It should be noted that the resultant fundamental component is not twice the individual voltage, but is $\sqrt{3}/2(= 0.866)$ of that for individual output voltages and the effective output has been reduced by $(1 - 0.866 =) 13.4\%$.

The harmonic elimination techniques, which are suitable only for fixed output voltage, increase the order of harmonics and reduce the sizes of output filter. However, this advantage should be weighed against the increased switching losses of power devices and increased iron (or magnetic losses) in the transformer due to higher harmonic frequencies.

Example 6.6   Finding the Number of Notches and Their Angles

A single-phase full-wave inverter uses multiple notches to give bipolar voltage as shown in Figure 6.38, and is required to eliminate the fifth, seventh, eleventh, and thirteenth harmonics from the output wave. Determine the number of notches and their angles.

Solution
For elimination of the fifth, seventh, eleventh, and thirteenth harmonics, $A_5 = A_7 = A_{11} = A_{13} = 0$; that is, $m = 4$. Four notches per quarter-wave would be required. Equation (6.85) gives the following set of nonlinear simultaneous equations to solve for the angles.

$$1 - 2\cos 5\alpha_1 + 2\cos 5\alpha_2 - 2\cos 5\alpha_3 + 2\cos 5\alpha_4 = 0$$
$$1 - 2\cos 7\alpha_1 + 2\cos 7\alpha_2 - 2\cos 7\alpha_3 + 2\cos 7\alpha_4 = 0$$
$$1 - 2\cos 11\alpha_1 + 2\cos 11\alpha_2 - 2\cos 11\alpha_3 + 2\cos 11\alpha_4 = 0$$
$$1 - 2\cos 13\alpha_1 + 2\cos 13\alpha_2 - 2\cos 13\alpha_3 + 2\cos 13\alpha_4 = 0$$

Solution of these equations by iteration using a Mathcad program yields

$$\alpha_1 = 10.55° \qquad \alpha_2 = 16.09° \qquad \alpha_3 = 30.91° \qquad \alpha_4 = 32.87°$$

Note: It is not always necessary to eliminate the third harmonic (and triplen), which is not normally present in three-phase connections. Therefore, in three-phase inverters, it is preferable to eliminate the fifth, seventh, and eleventh harmonics of output voltages, so that the LOH is the thirteenth.

Key Points of Section 6.9

- The switching angles of the inverters can be preselected to eliminate certain harmonics on the output voltages.
- The harmonic elimination techniques that are suitable only for fixed output voltage increase the order of harmonics and reduce the sizes of output filters.

6.10   CURRENT-SOURCE INVERTERS

In the previous sections the inverters are fed from a voltage source and the load current is forced to fluctuate from positive to negative, and vice versa. To cope with inductive loads, the power switches with freewheeling diodes are required, whereas in a current-source inverter (CSI), the input behaves as a current source. The output current is maintained constant irrespective of load on the inverter and the output voltage is forced to change. The circuit diagram of a single-phase transistorized inverter is shown in Figure 6.42a. Because there must be a continuous current flow from the source, two switches must always conduct—one from the upper and one from the lower switches. The conduction sequence is 12, 23, 34, and 41 as shown in Figure 6.42b. The switch states are shown in Table 6.4. Transistors Q_1, Q_4 in Figure 6.42a act as the switching devices S_1, S_4, respectively. If two switches, one upper and one lower, conduct at the same time such that the output current is $\pm I_L$, the switch state is 1; whereas if these switches are off at the same time, the switch state is 0. The output current waveform is shown in Figure 6.42c. The diodes in series with the transistors are required to block the reverse voltages on the transistors.

When two devices in different arms conduct, the source current I_L flows through the load. When two devices in the same arm conduct, the source current is bypassed from the load. The design to the current source is similar to Example 5.10. From Eq. (6.28), the load current can be expressed as

$$i_0 = \sum_{n=1,3,5,\ldots}^{\infty} \frac{4I_L}{n\pi} \sin\frac{n\delta}{2} \sin n(\omega t) \tag{6.90}$$

Figure 6.43a shows the circuit diagram of a three-phase current-source inverter. The waveforms for gating signals and line currents for a Y-connected load are shown in Figure 6.43b. At any instant, only two thyristors conduct at the same time. Each device conducts for 120°. From Eq. (6.16a), the instantaneous current for phase a of a Y-connected load can be expressed as

$$i_a = \sum_{n=1,3,5,\ldots}^{\infty} \frac{4I_L}{n\pi} \sin\frac{n\pi}{3} \sin n\left(\omega t + \frac{\pi}{6}\right) \tag{6.91}$$

From Eq. (6.21a), the instantaneous phase current for a delta-connected load is given by

$$i_a = \sum_{n=1}^{\infty} \frac{4I_L}{\sqrt{3}n\pi} \sin\left(\frac{n\pi}{3}\right) \sin(n\omega t) \qquad \text{for } n = 1, 3, 5, \ldots \tag{6.92}$$

TABLE 6.4   Switch States for a Full-Bridge Single-Phase Current Source Inverter (CSI)

State	State No.	Switch State $S_1S_2S_3S_4$	i_o	Components Conducting
S_1 and S_2 are on and S_4 and S_3 are off	1	1100	I_L	S_1 and S_2 D_1 and D_2
S_3 and S_4 are on and S_1 and S_2 are off	2	0011	$-I_L$	S_3 and S_4 D_3 and D_4
S_1 and S_4 are on and S_3 and S_2 are off	3	1001	0	S_1 and S_4 D_1 and D_4
S_3 and S_2 are on and S_1 and S_4 are off	4	0110	0	S_3 and S_2 D_3 and D_2

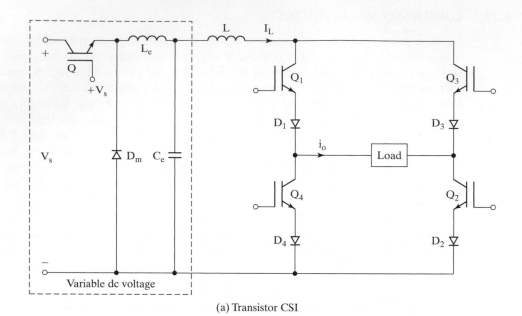

(a) Transistor CSI

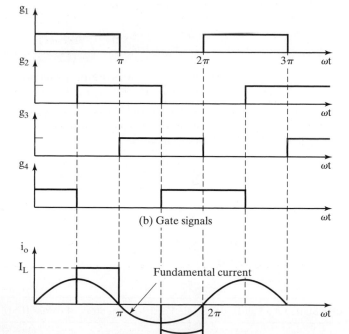

(b) Gate signals

(c) Load current

FIGURE 6.42

Single-phase current source.

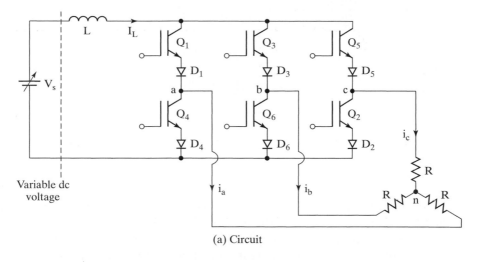

(a) Circuit

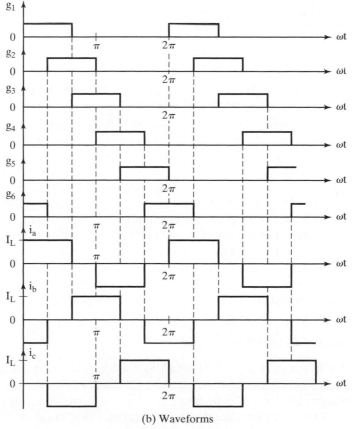

(b) Waveforms

FIGURE 6.43

Three-phase current source transistor inverter.

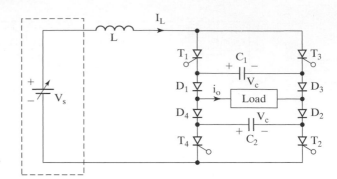

FIGURE 6.44

Single-phase thyristor current-source inverter.

The PWM, SPWM, MSPWM, MSPWN or SUM technique can be applied to vary the load current and to improve the quality of its waveform.

A current source inverter that utilizes capacitors to turn off the switching devices such as thyristors is shown in Figure 6.44. Let us assume that T_1 and T_2 are conducting, and capacitors C_1 and C_2 are charged with polarity as shown. Firing of thyristors T_3 and T_4 reverse biases thyristors T_1 and T_2. T_1 and T_2 are turned off by impulse commutation. The current now flows through $T_3 C_1 D_1$, load, and $D_2 C_2 T_4$. The capacitors C_1 and C_2 are discharged and recharged at a constant rate determined by load current, $I_m = I_L$. When C_1 and C_2 are charged to the load voltage and their currents fall to zero, the load current is transferred from diode D_1 to D_3 and D_2 to D_4. D_1 and D_2 are turned off when the load current is completely reversed. The capacitor is now ready to turn off T_3 and T_4 if thyristors T_1 and T_2 are fired in the next half-cycle. The commutation time depends on the magnitude of load current and load voltage. The diodes in Figure 6.44 isolate the capacitors from the load voltage.

The CSI is a dual of a VSI. The line-to-line voltage of a VSI is similar in shape to the line current of a CSI. The advantages of the CSI are: (1) since the input dc current is controlled and limited, misfiring of switching devices, or a short circuit, would not be serious problems; (2) the peak current of power devices is limited; (3) the commutation circuits for thyristors are simpler; and (4) it has the ability to handle reactive or regenerative load without freewheeling diodes.

A CSI requires a relatively large reactor to exhibit current-source characteristics and an extra converter stage to control the current. The dynamic response is slower. Due to current transfer from one pair of switches to another, an output filter is required to suppress the output voltage spikes.

Key Points of Section 6.10

- A CSI is a dual of the VSI. In a VSI, the load current depends on the load impedance, whereas the load voltage in a CSI depends on the load impedance. For this reason, diodes are connected in series with switching devices to protect them from transient voltages due to load current switching.

6.11   VARIABLE DC-LINK INVERTER

The output voltage of an inverter can be controlled by varying the modulation index (or pulse widths) and maintaining the dc input voltage constant; however, in this type

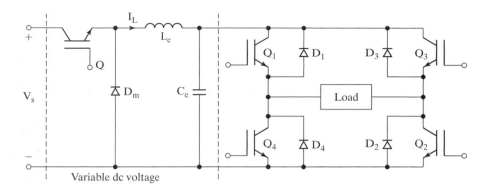

FIGURE 6.45

Variable dc-link inverter.

of voltage control, a range of harmonics would be present on the output voltage. The pulse widths can be maintained fixed to eliminate or reduce certain harmonics and the output voltage can be controlled by varying the level of dc input voltage. Such an arrangement as shown in Figure 6.45 is known as a *variable dc-link inverter*. This arrangement requires an additional converter stage; and if it is a converter, the power cannot be fed back to the dc source. To obtain the desired quality and harmonics of the output voltage, the shape of the output voltage can be predetermined, as shown in Figure 6.2b or Figure 6.42. The dc supply is varied to give variable ac output.

6.12 BOOST INVERTER

The single-phase VSI in Figure 6.2a uses the buck topology, which has the characteristic that the average output voltage is always lower than the input dc voltage. Thus, if an output voltage higher than the input one is needed, a boost dc–dc converter must be used between the dc source and inverter. Depending on the power and voltage levels, this can result in high volume, weight, cost, and reduced efficiency. The full-bridge topology can, however, be used as a boost inverter that can generate an output ac voltage higher than the input dc voltage [22, 23].

Basic principle. Let us consider two dc–dc converters feeding a resistive load R as shown in Figure 6.46a. The two converters produce a dc-biased sine wave output such that each source only produces a unipolar voltage as shown in Figure 6.46b. The modulation of each converter is 180° out of phase with the other so that the voltage excursion across the load is maximized. Thus, the output voltages of the converters are described by

$$v_a = V_{dc} + V_m \sin \omega t \tag{6.93}$$

$$v_b = V_{dc} - V_m \sin \omega t \tag{6.94}$$

Thus, the output voltage is sinusoidal as given by

$$v_o = v_a - v_b = 2V_m \sin \omega t \tag{6.95}$$

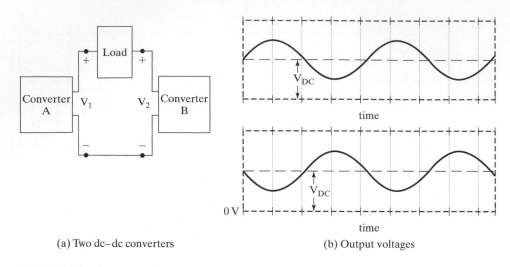

(a) Two dc–dc converters   (b) Output voltages

FIGURE 6.46

Principle of boost inverter.

Thus, a dc bias voltage appears at each end of the load with respect to ground, but the differential dc voltage across the load is zero.

Boost inverter circuit.   Each converter is a current bidirectional boost converter as shown in Figure 6.47a. The boost inverter consists of two boost converters as shown in Figure 4.47b. The output of the inverter can be controlled by one of the two methods: (1) use a duty cycle k for converter A and a duty cycle of $(1 - k)$ for converter B or (2) use a different duty cycle for each converter such that each converter produces a dc-biased sine wave output. The second method is preferred and it uses controllers A and B to make the capacitor voltages v_a and v_b follow a sinusoidal reference voltage.

Circuit operation.   The operation of the inverter can be explained by considering one converter A only as shown in Figure 6.48a, which can be simplified to Figure 6.48b. There are two modes of operation: mode 1 and mode 2.

Mode 1: When the switch S_1 is closed and S_2 is open as shown in Figure 6.49a, the inductor current i_{L1} rises quite linearly, diode D_2 is reverse biased, capacitor C_1 supplies energy to the load, and voltage V_a decreases.

Mode 2: When switch S_1 is open and S_2 is closed, as shown in Figure 6.49b, the inductor current i_{L1} flows through capacitor C_1 and the load. The current i_{L1} decreases while capacitor C_1 is recharged.

The average output voltage of converter A, which operates under the boost mode, can be found from

$$V_a = \frac{V_s}{1 - k} \tag{6.96}$$

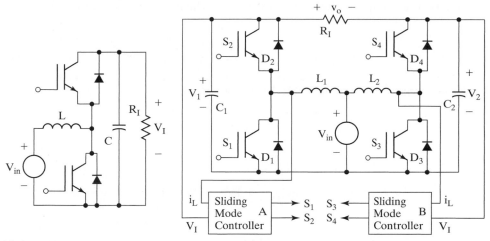

(a) One bidirectional boost converter   (b) Two bidirectional boost converters

FIGURE 6.47

Boost inverter consisting of two boost converters. [Ref. 22, R. CaCeres]

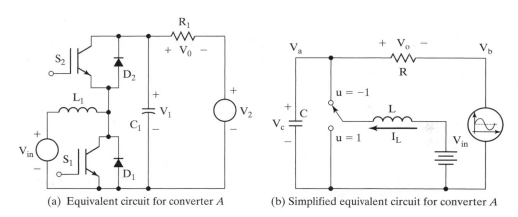

(a) Equivalent circuit for converter A (b) Simplified equivalent circuit for converter A

FIGURE 6.48

Equivalent circuit of converter A.

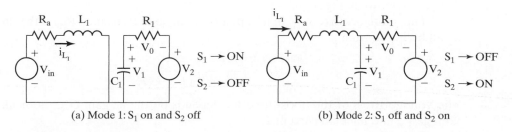

(a) Mode 1: S_1 on and S_2 off   (b) Mode 2: S_1 off and S_2 on

FIGURE 6.49

Equivalent circuits during modes of operation.

The average output voltage of converter B, which operates under the buck mode, can be found from

$$V_b = \frac{V_s}{k} \tag{6.97}$$

Therefore, the average output voltage is given by

$$V_o = V_a - V_b = \frac{V_s}{1-k} - \frac{V_s}{k}$$

which gives the dc gain of the boost inverter as

$$G_{dc} = \frac{V_o}{V_s} = \frac{2k-1}{(1-k)k} \tag{6.98}$$

where k is the duty cycle. It should be noted that V_o becomes zero at $k = 0.5$. If the duty cycle k is varied around the quiescent point of 50% duty cycle, there is an ac voltage across the load. Because the output voltage in Eq. (6.95) is twice the sinusoidal component of converter A, the peak output voltage equals to

$$V_{o(pk)} = 2V_m = 2V_a - 2V_{dc} \tag{6.99}$$

Because a boost converter cannot produce an output voltage lower than the input voltage, the dc component must satisfy the condition [24]

$$V_{dc} \geq (V_m + V_s) \tag{6.100}$$

which implies that there are many possible values of V_{dc}. However, the equal term produces the least stress on the devices. From Eqs. (6.96), (6.99), and (6.100), we get

$$V_{o(pk)} = \frac{2V_s}{1-k} - 2\left(\frac{V_{o(pk)}}{2} + V_s\right)$$

which gives the ac voltage gain as

$$G_{ac} = \frac{V_{o(pk)}}{V_s} = \frac{k}{1-k} \tag{6.101}$$

Thus, $V_{o(pk)}$ becomes V_S at $k = 0.5$. The ac and dc gain characteristics of the boost inverter are shown in Figure 6.50.

The inductor current I_L that depends on the load resistance R and the duty cycle k can be found from

$$I_L = \left[\frac{k}{1-k}\right]\frac{V_s}{(1-k)R} \tag{6.102}$$

The voltage stress of the boost inverter depends on the ac gain G_{ac}, the peak output voltage V_m and the load current I_L.

Buck–Boost inverter.   The full-bridge topology can also be operated as a buck–boost inverter [24] as shown in Figure 6.51. It has almost the same characteristic

Solution

The output voltage wavefor
$L = 31.5$ mH, and $C = 112$ μ
   The inductive reactance f

$$X_L = j2$$

The capacitive reactance for t

$$X_c =$$

The impedance for the nth ha

$$|Z_n|$$

and the PF angle for the nth h

$$\theta_n = \tan^{-1}\frac{11.}{}$$

a. Equation (6.84) giv

For $\alpha_1 = 23.62°$ an
Eq. (6.83) the instai

$$v_o(t) = 235.1 \text{ si}$$
$$+ 85.1$$

Dividing the outpu
delay due to the PF

$$i_o(t) = 15.19$$
$$+ 1.0$$

b. The nth and higher
pedance is much sm

where the filter imp
be found from

$$\left[1 \right.$$

For the seventh har

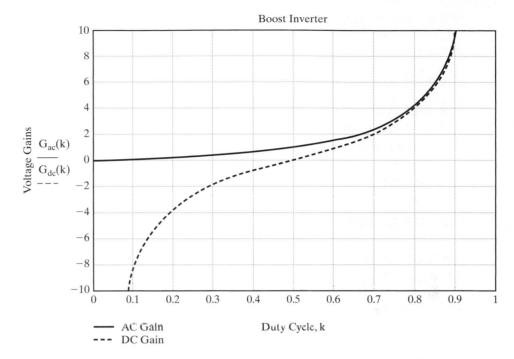

FIGURE 6.50

Gain characteristics of the boost inverter.

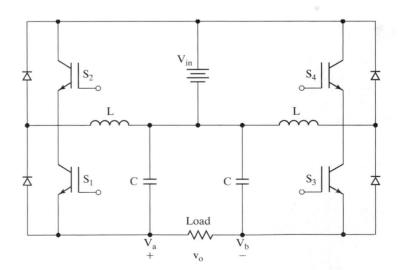

FIGURE 6.51

Buck–boost inverter. [Ref. 23, R. CaCeres]

as the boost inverter a
DC input voltage. The
as the boost inverter.

Key Points of Section

- With appropriate
  ated as a boost in
- Gating sequenc
  $(1 - k)T$. Simil
  during kT.

6.13 INVERTER CIRCUIT D

The determination of
depends on the types
The design requires (
and (2) plotting the cu
rent waveform is knov
tion of voltage ratings
    To reduce the ou
the commonly used ou
power. An LC-tuned f
erly designed CLC filt
wide bandwidth and d

Example 6.7   Finding the V

The single-phase full-bri
and $C = 112 \ \mu F$. The dc
The output voltage has t
termine the expression fo
enth and higher order ha

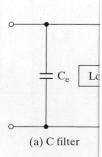

(a) C filter

FIGURE 6.52

Output filters.

```
RIN       5     0      2MEG
RF        5     3      100K
RO        6     3      75
CO        3     4      10PF
E1        6     4      0     5    2E+5     ; Voltage-controlled voltage source
.ENDS    PWM                             ; Ends subcircuit definition
.TRAN    10US 16.67MS      0     10US     ; Transient analysis
.PROBE                                    ; Graphics postprocessor
.options   abstol = 1.00n  reltol = 0.01  vntol = 0.1 ITL5=20000 ; convergence
.FOUR    60HZ      V(3, 6)                ; Fourier analysis
.END
```

The PSpice plots are shown in Figure 6.54, where V(17) = reference signal and V(3, 6) = output voltage.

b. FOURIER COMPONENTS OF TRANSIENT RESPONSE V (3, 6)
DC COMPONENT = 6.335275E-03

HARMONIC NO	FREQUENCY (HZ)	FOURIER COMPONENT	NORMALIZED COMPONENT	PHASE (DEG)	NORMALIZED PHASE (DEG)
1	6.000E+01	7.553E+01	1.000E+00	6.275E-02	0.000E+00
2	1.200E+02	1.329E-02	1.759E-04	5.651E+01	5.645E+01
3	1.800E+02	2.756E+01	3.649E-01	1.342E-01	7.141E-02
4	2.400E+02	1.216E-02	1.609E-04	6.914E+00	6.852E+00
5	3.000E+02	2.027E+01	2.683E-01	4.379E-01	3.752E-01

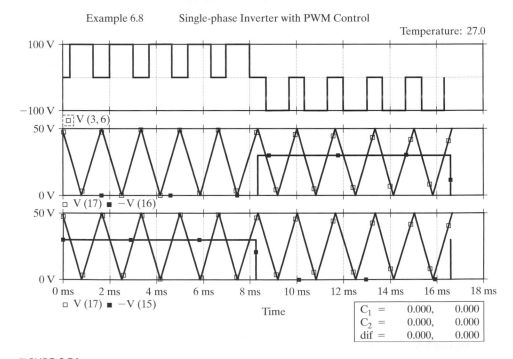

Example 6.8   Single-phase Inverter with PWM Control

FIGURE 6.54

PSpice plots for Example 6.8.

6	3.600E+02	7.502E-03	9.933E-05	−4.924E+01	−4.930E+01
7	4.200E+02	2.159E+01	2.858E-01	4.841E-01	4.213E-01
8	4.800E+02	2.435E-03	3.224E-05	−1.343E+02	−1.343E+02
9	5.400E+02	4.553E+01	6.028E-01	6.479E-01	5.852E-01

TOTAL HARMONIC DISTORTION = 8.063548E+01 PERCENT

Note: For $M = 0.6$ and $p = 5$, a Mathcad program for uniform PWM gives $V_1 = 54.59$ V (rms) and THD = 100.65% as compared with values of $V_1 = 75.53/\sqrt{2} = 53.41$ V (rms) and THD = 80.65% from PSpice. In calculating the THD, PSpice uses only up to ninth harmonics by default, instead of all harmonics. Thus, if the harmonics higher than ninth have significant values as compared with the fundamental component, PSpice gives a low and erroneous value of THD. However, the PSpice version 8.0 (or higher) allows an argument to specify the number of harmonics to be calculated. For example, the statement for calculating up to 30th harmonic is .FOUR 60HZ 30 V(3, 6). The default value is the ninth harmonic.

SUMMARY

Inverters can provide single-phase and three-phase ac voltages from a fixed or variable dc voltage. There are various voltage control techniques and they produce a range of harmonics on the output voltage. The SPWM is more effective in reducing the LOH. With a proper choice of the switching patterns for power devices, certain harmonics can be eliminated. The SV modulation is finding increasing applications in power converters and motor control. A current source inverter is a dual of a voltage source inverter. With proper gating sequence and control, the single-phase bridge inverter can be operated as a boost inverter.

REFERENCES

[1]   B. D. Bedford and R. G. Hoft, *Principle of Inverter Circuits*. New York: John Wiley & Sons. 1964.

[2]   T. Ohnishi and H. Okitsu, "A novel PWM technique for three-phase inverter/converter," *International Power Electronics Conference*, 1983, pp. 384–395.

[3]   K. Taniguchi and H. Irie, "Trapezoidal modulating signal for three-phase PWM inverter," *IEEE Transactions on Industrial Electronics*, Vol. IE3, No. 2, 1986, pp. 193–200.

[4]   K. Thorborg and A. Nystorm, "Staircase PWM: an uncomplicated and efficient modulation technique for ac motor drives," *IEEE Transactions on Power Electronics*, Vol. PE3, No. 4, 1988, pp. 391–398.

[5]   J. C. Salmon, S. Olsen, and N. Durdle, "A three-phase PWM strategy using a stepped 12 reference waveform," *IEEE Transactions on Industry Applications*, Vol. IA27, No. 5, 1991, pp. 914–920.

[6]   M. A. Boost and P. D. Ziogas, "State-of-the-art carrier PWM techniques: a critical evaluation," *IEEE Transactions on Industry Applications*, Vol. IA24, No. 2, 1988, pp. 271–279.

[7]   K. Taniguchi and H. Irie, "PWM technique for power MOSFET inverter," *IEEE Transactions on Power Electronics*, Vol. PE3, No. 3, 1988, pp. 328–334.

[8]   M. H. Ohsato, G. Kimura, and M. Shioya, "Five-stepped PWM inverter used in photovoltaic systems," *IEEE Transactions on Industrial Electronics*, Vol. 38, October, 1991, pp. 393–397.

[9]   P. D. Ziogas, "The delta modulation techniques in static PWM inverters," *IEEE Transactions on Industry Applications*, March/April 1981, pp. 199–204.

[10]   J. R. Espinoza, *Power Electronics Handbook*, edited by M. H. Rashid. San Diego, CA: Academic Press. 2001, Chapter 14–Inverters.

[11]   D.-C. Lee and G.-M. Lee, "Linear control of inverter output voltage in overmodulation," *IEEE Transactions on Industrial Electronics*, Vol. 44, No. 4, August 1997, pp. 590–592.

[12]   F. Blaabjerg, J. K. Pedersen, and P. Thoegersen, "Improved modulation techniques for PWM-V SI drives," *IEEE Transactions on Industrial Electronics*, Vol. 44, No. 1, February 1997, pp. 87–95.

[13]   H. W. Van der Broeck, H.-C. Skudelny, and G. V. Stanke "Analysis and realization of a pulse-width modulator based on voltage space vectors," *IEEE Transactions on Industry Applications*, Vol. 24, No. 1, January/February, 1988, pp. 142–150.

[14]   Y. Iwaji and S. Fukuda, "A pulse frequency modulated PWM inverter for induction motor drives," *IEEE Transactions on Power Electronics*, Vol. 7, No. 2, April 1992, pp. 404–410.

[15]   H. L. Liu and G. H. Cho, "Three-level space vector PWM in low index modulation region avoiding narrow pulse problem," *IEEE Transactions on Power Electronics*, Vol. 9, September 1994, pp. 481–486.

[16]   T.-P. Chen, Y.-S. Lai, and C.-H. Liu, "New space vector modulation technique for inverter control," *IEEE Power Electronics Specialists Conference*, Vol. 2, 1999, pp. 777–782.

[17]   S. R. Bowes and G. S. Singh, "Novel space-vector-based harmonic elimination inverter control," *IEEE Transactions on Industry Applications*, Vol. 36, No. 2, March/April 2000, pp. 549–557.

[18]   C. B. Jacobina, A. M. N. Lima, E. R. Cabral da Silva, R. N. C. Alves, and P. F. Seixas, "Digital scalar pulse-width modulation: A simple approach to introduce non-sinusoidal modulating waveforms," *IEEE Transactions on Power Electronics*, Vol. 16, No. 3, May 2001, pp. 351–359.

[19]   C. Zhan, A. Arulampalam, V. K. Ramachandaramurthy, C. Fitzer, M. Barnes, and N. Jenkins, "Novel voltage space vector PWM algorithm of 3-phase 4-wire power conditioner," *IEEE Power Engineering Society Winter Meeting*, 2001, Vol. 3, pp. 1045–1050.

[20]   R. Valentine, *Motor Control Electronics Handbook*. New York: McGraw-Hill. 1996, Chapter 8.

[21]   H. S. Patel and R. G. Hoft, "Generalized techniques of harmonic elimination and voltage control in thyristor converter," *IEEE Transactions on Industry Applications*, Vol. IA9, No. 3, 1973, pp. 310–317; Vol. IA10, No. 5, 1974, pp. 666–673.

[22]   R. O. Caceres and I. Barbi, "A boost dc–ac converter: Operation, analysis, control and experimentation," *Industrial Electronics Control and Instrumentation Conference*, November 1995, pp. 546–551.

[23]   R. O. CaCeres and I. Barbi, "A boost dc–ac converter: Analysis, design, and experimentation," *IEEE Transactions on Power Electronics*, Vol. 14, No. 1, January 1999, pp. 134–141.

[24]   J. Almazan, N. Vazquez, C. Hernandez, J. Alvarez, and J. Arau, "Comparison between the buck, boost and buck–boost inverters," *International Power Electronics Congress*, Acapulco, Mexico, October 2000, pp. 341–346.

[25]   B. H. Kwon and B. D. Min, "A fully software-controlled PWM rectifier with current link. *IEEE Transactions on Industrial Electronics*, Vol. 40, No. 3, June 1993, pp. 355–363.

REVIEW QUESTIONS

6.1 What is an inverter?

6.2 What is the principle of operation of an inverter?

6.3 What are the types of inverters?

6.4 What are the differences between half-bridge and full-bridge inverters?

6.5 What are the performance parameters of inverters?

6.6 What are the purposes of feedback diodes in inverters?

6.7 What are the arrangements for obtaining three-phase output voltages?

6.8 What are the methods for voltage control within the inverters?

6.9 What is sinusoidal PWM?

6.10 What is the purpose of overmodulation?

6.11 Why should the normalized carrier frequency m_f of a three-phase inverter be an odd multiple of 3?

6.12 What is third-harmonic PWM?

6.13 What is 60° PWM?

6.14 What is space vector modulation?

6.15 What are the advantages of SVM?

6.16 What is space vector transformation?

6.17 What are space vectors?

6.18 What are switching states of an inverter?

6.19 What are modulating reference vectors?

6.20 What is space vector switching?

6.21 What is space vector sequence?

6.22 What are null vectors?

6.23 What are the advantages and disadvantages of displacement-angle control?

6.24 What are the techniques for harmonic reductions?

6.25 What are the effects of eliminating lower order harmonics?

6.26 What is the effect of thyristor turn-off time on inverter frequency?

6.27 What are the advantages and disadvantages of current-source inverters?

6.28 What are the main differences between voltage-source and current-source inverters?

6.29 What are the main advantages and disadvantages of variable dc link inverters?

6.30 What is the basic principle of a boost inverter?

6.31 What are the two methods for voltage control of the boost inverter?

6.32 What is the dc voltage gain of the boost inverter?

6.33 What is the ac voltage gain of the boost inverter?

6.34 What are the reasons for adding a filter on the inverter output?

6.35 What are the differences between ac and dc filters?

PROBLEMS

6.1 The single-phase half-bridge inverter in Figure 6.1a has a resistive load of $R = 10 \ \Omega$ and the dc input voltage is $V_s = 220$ V. Determine **(a)** the rms output voltage at the fundamental frequency V_1; **(b)** the output power P_o; **(c)** the average, rms, and peak currents of each transistor; **(d)** the peak off-state voltage V_{BB} of each transistor; **(e)** the total harmonic distortion THD; **(f)** the distortion factor DF; and **(g)** the harmonic factor and distortion factor of the lowest order harmonic.

6.2 Repeat Problem 6.1 for the single-phase full-bridge inverter in Figure 6.2a.

6.3 The full-bridge inverter in Figure 6.2a has an RLC load with $R = 5 \ \Omega$, $L = 10$ mH, and $C = 26 \ \mu$F. The inverter frequency, $f_o = 400$ Hz, and the dc input voltage, $V_s = 220$ V. **(a)** Express the instantaneous load current in a Fourier series. Calculate **(b)** the rms load current at the fundamental frequency I_1; **(c)** the THD of the load current; **(d)** the average supply current I_s; and **(e)** the average, rms, and peak currents of each transistor.

6.4 Repeat Problem 6.3 for $f_o = 60$ Hz, $R = 4$ Ω, $L = 25$ mH, and $C = 10$ μF.

6.5 Repeat Problem 10.3 for $f_o = 60$ Hz, $R = 5$ Ω, and $L = 20$ mH.

6.6 The three-phase full-bridge inverter in Figure 6.5a has a Y-connected resistive load of $R = 5$ Ω. The inverter frequency is $f_o = 400$ Hz and the dc input voltage is $V_s = 220$ V. Express the instantaneous phase voltages and phase currents in a Fourier series.

6.7 Repeat Problem 6.6 for the line-to-line voltages and line currents.

6.8 Repeat Problem 6.6 for a delta-connected load.

6.9 Repeat Problem 6.7 for a delta-connected load.

6.10 The three-phase full-bridge inverter in Figure 6.5a has a Y-connected load and each phase consists of $R = 5$ Ω, $L = 10$ mH, and $C = 25$ μF. The inverter frequency is $f_o = 60$ Hz and the dc input voltage, $V_s = 220$ V. Determine the rms, average, and peak currents of the transistors.

6.11 The output voltage of a single-phase full-bridge inverter is controlled by pulse-width modulation with one pulse per half-cycle. Determine the required pulse width so that the fundamental rms component is 70% of dc input voltage.

6.12 A single-phase full-bridge inverter uses a uniform PWM with two pulses per half-cycle for voltage control. Plot the distortion factor, fundamental component, and lower order harmonics against modulation index.

6.13 A single-phase full-bridge inverter, which uses a uniform PWM with two pulses per half-cycle, has a load of $R = 5$ Ω, $L = 15$ mH, and $C = 25$ μF. The dc input voltage is $V_s = 220$ V. Express the instantaneous load current $i_o(t)$ in a Fourier series for $M = 0.8$, $f_o = 60$ Hz.

6.14 The single-phase full bridge inverter is operated at 1 kHz and uses a uniform PWM with four pulses per half-cycle for voltage control. Plot the fundamental component, distortion factor, and THD against the modulation index M.

6.15 A single-phase full-bridge inverter uses a uniform PWM with seven pulses per half-cycle for voltage control. Plot the distortion factor, fundamental component, and lower order harmonics against the modulation index.

6.16 The single-phase full bridge inverter is operated at 1 kHz and uses a SPWM with four pulses per half-cycle for voltage control. Plot the fundamental component, distortion factor, and THD against the modulation index M.

6.17 A single-phase full-bridge inverter uses an SPWM with seven pulses per half-cycle for voltage control. Plot the distortion factor, fundamental component, and lower order harmonics against the modulation index.

6.18 Repeat Problem 6.17 for the modified SPWM with two pulses per quarter-cycle.

6.19 The single-phase full bridge inverter is operated at 1 kHz and uses a modified SPWM as shown in Figure 6.18 with three pulses per half-cycle for voltage control. Plot the fundamental component, distortion factor, and THD against the modulation index M.

6.20 A single-phase full-bridge inverter uses a uniform PWM with five pulses per half-cycle. Determine the pulse width if the rms output voltage is 80% of the dc input voltage.

6.21 A single-phase full-bridge inverter uses displacement-angle control to vary the output voltage and has one pulse per half-cycle, as shown in Figure 6.20f. Determine the delay (or displacement) angle if the fundamental component of output voltage is 70% of dc input voltage.

6.22 The single-phase half-bridge inverter is operated at 1 kHz and uses a trapezoidal modulation shown in Figure 6.21a with five pulses per half-cycle for voltage control. Plot the fundamental component, distortion factor, and THD against the modulation index M.

6.23 The single-phase half-bridge inverter is operated at 1 kHz and uses a staircase modulation shown in Figure 6.22a with seven pulses per half-cycle for voltage control. Plot the fundamental component, distortion factor, and THD against the modulation index M.

6.24 The single-phase half-bridge inverter is operated at 1 kHz and uses a stepped modulation shown in Figure 6.23a with five pulses per half-cycle for voltage control. Plot the fundamental component, distortion factor, and THD against the modulation index M.

6.25 The single-phase half-bridge inverter is operated at 1 kHz and uses a third and ninth harmonic modulation as shown in Figure 6.24a with six pulses per half-cycle for voltage control. Plot the fundamental component, distortion factor, and THD against the modulation index M.

6.26 A single-phase full-bridge inverter uses multiple bipolar notches and it is required to eliminate third, fifth, seventh, and eleventh harmonics from the output waveform. Determine the number of notches and their angles.

6.27 Repeat Problem 6.26 to eliminate third, fifth, seventh, and ninth harmonics.

6.28 The single-phase full-bridge inverter is operated at 1 kHz and uses unipolar notches as shown in Figure 6.39. It is required to eliminate the third, fifth, seventh, and ninth harmonics. Determine the number of notches and their angles. Use PSpice to verify the elimination of those harmonics.

6.29 The single-phase full-bridge inverter is operated at 1 kHz and uses the modified SPWM as shown in Figure 6.40. It is required to eliminate the 3rd and 5th harmonics. Determine the number of pulses and their angles. Use PSpice to verify the elimination of those harmonics.

6.30 Plot the normalized state times T_1/MT_S, T_2/MT_S, and T_z/MT_S against the angle $\theta (= 0$ to $\pi/3)$ between two adjacent space vectors.

6.31 Two adjacent vectors are $\mathbf{V_1} = 1 + j0.577$ and $\mathbf{V_2} = j1.155$. If the angle between them is $\theta = \pi/6$, and the modulation index M is 0.8, find the modulating vector $\mathbf{V_{er}}$.

6.32 The parameters of the boost inverter in Figure 6.47b is operated at a duty cycle $k = 0.6$. Find (a) the dc voltage gain G_{dc}, (b) the ac voltage gain G_{ac}, and (c) the instantaneous capacitor voltages v_a and v_b.

6.33 A single-phase full-bridge inverter in Figure 6.2a supplies a load of $R = 5\ \Omega$, $L = 15$ mH, and $C = 30\ \mu$F. The dc input voltage is $V_s = 220$ V and the inverter frequency is $f_0 = 400$ Hz. The output voltage has two notches such that third and fifth harmonics are eliminated. If a tuned LC filter is used to eliminate the seventh harmonic from the output voltage, determine the suitable values of filter components.

6.34 The single-phase full-bridge inverter in Figure 6.2a supplies a load of $R = 2\ \Omega$, $L = 25$ mH, and $C = 40\ \mu$F. The dc input voltage is $V_s = 220$ V and the inverter frequency, $f_0 = 60$ Hz. The output voltage has three notches such that the third, fifth, and seventh harmonics are eliminated. If an output C filter is used to eliminate ninth and higher order harmonics, determine the value of filter capacitor C_e.

12. MOS-controlled thyristors (MCTs)

13. Static induction thyristors (SITHs)

7.6.1   Phase-Controlled Thyristors

This type of thyristors generally operates at the line frequency and is turned off by natural commutation. A thyristor starts conduction in a forward direction when a trigger current pulse is passed from gate to cathode, and rapidly latches into full conduction with a low forward voltage drop. It cannot force its current back to zero by its gate signal; instead, it relies on the natural behavior of the circuit for the current to come to zero. When the anode current comes to zero, the thyristor recovers its ability in a few tens of microseconds of reverse blocking voltage and it can block the forward voltage until the next turn-on pulse is applied. The turn-off time t_q is of the order of 50 to 100 μs. This is most suited for low-speed switching applications and is also known as a *converter thyristor*. Because a thyristor is basically a silicon-made controlled device, it is also known as *silicon-controlled rectifier* (SCR).

The on-state voltage V_T varies typically from about 1.15 V for 600 V to 2.5 V for 4000-V devices; and for a 1200-V, 5500-A thyristor it is typically 1.25 V. The modern thyristors use an amplifying gate, where an auxiliary thyristor T_A is gated on by a gate signal and then the amplified output of T_A is applied as a gate signal to the main thyristor T_M. This is shown in Figure 7.10. The amplifying gate permits high dynamic characteristics with typical dv/dt of 1000 V/μs and di/dt of 500 A/μs, and simplifies the circuit design by reducing or minimizing di/dt limiting inductor and dv/dt protection circuits.

Because of their low cost, high efficiency, ruggedness, and high voltage and current capability, these thyristors are extensively used in dc–ac converters with 50 or 60 Hz of main supply and in cost-effective applications where the turn-off capability is not an important factor. Often the turn-off capability does not offer sufficient benefits to justify higher cost and losses of the devices. They are used for almost all high-voltage dc (HVDC) transmission and a large percentage of industrial applications.

7.6.2   BCTs

The BCT [5] is a new concept for high power phase control. Its symbol is shown in Figure 7.11a. It is a unique device, combining advantages of having two thyristors in one package, enabling more compact equipment design, simplifying the cooling system, and increasing system reliability. BCTs enable designers to meet higher demands concerning size, integration, reliability, and cost for the end product. They are suitable

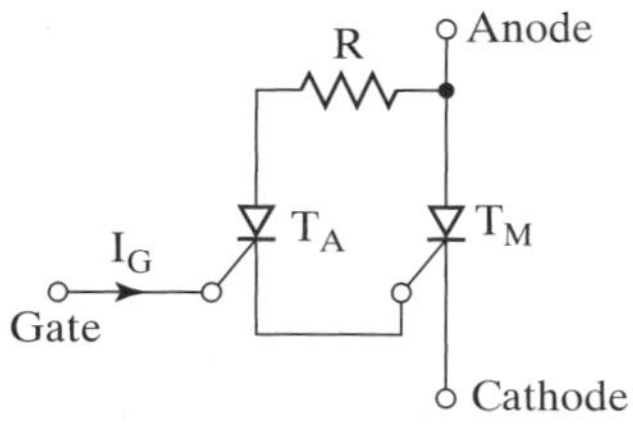

FIGURE 7.10

Amplifying gate thyristor.

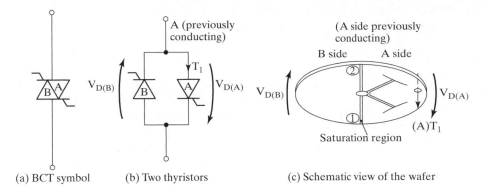

(a) BCT symbol   (b) Two thyristors   (c) Schematic view of the wafer

FIGURE 7.11

Bidirectional phase-controlled thyristor. [Ref. 5]

for applications as static volt–ampere reactive (VAR) compensators, static switches, soft starters, and motor drives. The maximum voltage rating can be as high as 6.5 kV at 1.8 kA and the maximum current rating can be as high as 3 kA at 1.8 kV.

The electrical behavior of a BCT corresponds to that of two antiparallel thyristors, integrated onto one silicon slice as shown in Figure 7.11b. Each thyristor half performs like the corresponding full-wafer thyristor with respect to its static and dynamic properties. The BCT wafer has anode and cathode regions on each face. The A and B thyristors are identified on the wafer by letters A and B, respectively.

A major challenge in the integration of two thyristor halves is to avoid harmful cross talk between the two halves under all relevant operating conditions. The device must show very high uniformity between the two halves in device parameters such as reverse recovery charge and on-state voltage drops. Regions 1 and 2 shown in Figure 7.11c are the most sensitive with respect to surge current having reapplied "reverse" voltage and the t_q capability of a BCT.

Turn-on and off.   A BCT has two gates: one for turning on the forward current and one for the reverse current. This thyristor is turned on with a pulse current to one of its gates. It is turned off if the anode current falls below the holding current due to the natural behavior of the voltage or the current.

7.6.3   Fast-Switching Thyristors

These are used in high-speed switching applications with forced commutation (e.g., resonant inverters in Chapter 8 and inverters in Chapter 6). They have fast turn-off time, generally in the range 5 to 50 μs, depending on the voltage range. The on-state forward drop varies approximately as an inverse function of the turn-off time t_q. This type of thyristor is also known as an *inverter thyristor*.

These thyristors have high dv/dt of typically 1000 V/μs and di/dt of 1000 A/μs. The fast turn-off and high di/dt are very important for reducing the size and weight of commutating or reactive circuit components. The on-state voltage of a 1800-V, 2200-A thyristor is typically 1.7 V. Inverter thyristors with a very limited reverse blocking

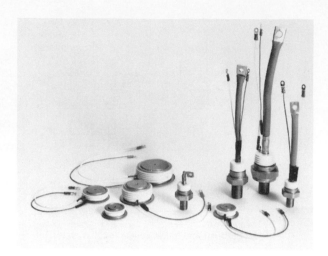

FIGURE 7.12

Fast-switching thyristors. (Courtesy of Powerex, Inc.)

capability, typically 10 V, and a very fast turn-off time between 3 and 5 μs are commonly known as *asymmetric thyristors* (ASCRs). Fast-switching thyristors of various sizes are shown in Figure 7.12.

7.6.4   LASCRs

This device is turned on by direct radiation on the silicon wafer with light. Electron–hole pairs that are created due to the radiation produce triggering current under the influence of electric field. The gate structure is designed to provide sufficient gate sensitivity for triggering from practical light sources (e.g., light-emitting diode [LED] and to accomplish high *di/dt* and *dv/dt* capabilities).

The LASCRs are used in high-voltage and high-current applications (e.g., HVDC) transmission and static reactive power or VAR compensation. An LASCR offers complete electrical isolation between the light-triggering source and the switching device of a power converter, which floats at a potential of as high as a few hundred kilovolts. The voltage rating of an LASCR could be as high as 4 kV at 1500 A with light-triggering power of less than 100 mW. The typical *di/dt* is 250 A/μs and the *dv/dt* could be as high as 2000 V/μs.

7.6.5   Bidirectional Triode Thyristors

A TRIAC can conduct in both directions and is normally used in ac-phase control (e.g., ac voltage controllers in Chapter 11). It can be considered as two SCRs connected in antiparallel with a common gate connection as shown in Figure 7.13a. The $v-i$ characteristics are shown in Figure 7.13c.

Because a TRIAC is a bidirectional device, its terminals cannot be designated as anode and cathode. If terminal MT_2 is positive with respect to terminal MT_1, the TRIAC can be turned on by applying a positive gate signal between gate G and terminal MT_1. If terminal MT_2 is negative with respect to terminal MT_1, it is turned on by applying a negative gate signal between gate G and terminal MT_1. It is not necessary to

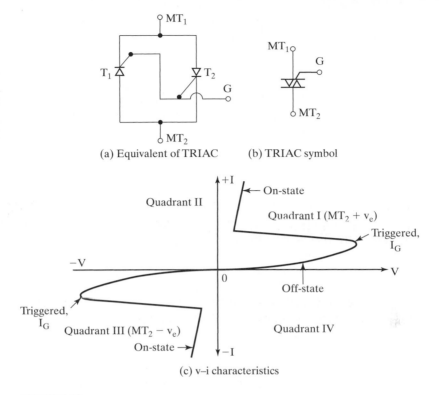

(a) Equivalent of TRIAC   (b) TRIAC symbol

(c) v–i characteristics

FIGURE 7.13

Characteristics of a TRIAC.

have both polarities of gate signals, and a TRIAC can be turned on with either a positive or a negative gate signal. In practice, the sensitivities vary from one quadrant to another, and the TRIACs are normally operated in quadrant I^+ (positive gate voltage and gate current) or quadrant III^- (negative gate voltage and gate current).

7.6.6 RCTs

In many converter and inverter circuits, an antiparallel diode is connected across an SCR to allow a reverse current flow due to inductive load and to improve the turn-off requirement of commutation circuit. The diode clamps the reverse blocking voltage of the SCR to 1 or 2 V under steady-state conditions. However, under transient conditions, the reverse voltage may rise to 30 V due to induced voltage in the circuit stray inductance within the device.

An RCT is a compromise between the device characteristics and circuit requirement; and it may be considered as a thyristor with a built-in antiparallel diode as shown in Figure 7.14. An RCT is also called as ASCR. The forward blocking voltage varies from 400 to 2000 V and the current rating goes up to 500 A. The reverse blocking voltage is typically 30 to 40 V. Because the ratio of forward current through the thyristor to the reverse current of a diode is fixed for a given device, its applications are limited to specific circuit designs.

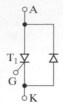

FIGURE 7.14

Reverse-conducting thyristor.

7.6.7 GTOs

A GTO like an SCR can be turned on by applying a positive gate signal. However, a GTO can be turned off by a negative gate signal. A GTO is a nonlatching device and can be built with current and voltage ratings similar to those of an SCR [8–10]. A GTO is turned on by applying a short positive pulse and turned off by a short negative pulse to its gate. The GTOs have these advantages over SCRs: (1) elimination of commutating components in forced commutation, resulting in reduction in cost, weight, and volume; (2) reduction in acoustic and electro-magnetic noise due to the elimination of commutation chokes; (3) faster turn-off, permitting high-switching frequencies; and (4) improved efficiency of converters.

In low-power applications, GTOs have the following advantages over bipolar transistors: (1) a higher blocking voltage capability; (2) a high ratio of peak controllable current to average current; (3) a high ratio of peak surge current to average current, typically 10:1; (4) a high on-state gain (anode current and gate current), typically 600; and (5) a pulsed gate signal of short duration. Under surge conditions, a GTO goes into deeper saturation due to regenerative action. On the other hand, a bipolar transistor tends to come out of saturation.

Like a thyristor, a GTO is a latch-on device, but it is also a latch-off device. The GTO symbol is shown in Figure 7.15a, and its internal cross section is shown in

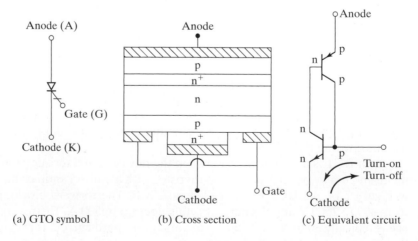

(a) GTO symbol   (b) Cross section   (c) Equivalent circuit

FIGURE 7.15

Gate turn-off (GTO) thyristor.

Figure 7.15b. Compared to a conventional thyristor, it has an additional n^+-layer near the anode that forms a turn-off circuit between the gate and the cathode in parallel with the turn-on gate. The equivalent circuit that is shown in Figure 7.15c is similar to that of a thyristor shown in Figure 7.4b, except for its internal turn-off mechanism. If a large pulse current is passed from the cathode to the gate to take away sufficient charge carriers from the cathode, that is, from the emitter of the npn transistor Q_1, the pnp transistor Q_2, can be drawn out of the regenerative action. As transistor Q_1 turns off, transistor Q_2 is left with an open base, and the GTO returns to the nonconducting state.

Turn-on. The GTO has a highly interdigited gate structure with no regenerative gate, as shown later in Figure 7.19. As a consequence, a large initial gate trigger pulse is required to turn on. A typical turn-on gate pulse and its important parameters are shown in Figure 7.16a. Minimum and maximum values of I_{GM} can be derived from the device data sheet. The value of di_g/dt is given in device characteristics of the data sheet, against turn-on time. The rate of rise of gate current di_g/dt affects the device turn-on losses. The duration of the I_{GM} pulse should not be less than half the minimum of time given in data sheet ratings. A longer period is required if the anode current di/dt is low such that I_{GM} is maintained until a sufficient level of anode current is established.

On-state. Once the GTO is turned on, forward gate current must be continued for the whole of the conduction period to ensure the device remains in conduction. Otherwise, the device cannot remain in conduction during the on-state period. The on-state gate current should be at least 1% of the turn-on pulse to ensure that the gate does not unlatch.

Turn-off. The turn-off performance of a GTO is greatly influenced by the characteristics of the gate turn-off circuit. Thus, the characteristics of the turn-off circuit must match the deice requirements. The turn-off process involves the extraction of the gate charge, the gate avalanche period, and the anode current decay. The amount of the charge extraction is a device parameter and its value is not significantly affected by the external circuit conditions. The initial peak turn-off current and turn-off time, which are important parameters of the turning-off process, depend on the external circuit components. A typical anode current versus the turn-off pulse is shown in Figure 7.16b. The device data sheet gives typical values for I_{GQ}. The GTO has a long turn-off, tail-off current at the end of the turn-off and the next turn-on must wait until the residual charge on the anode side is dissipated through the recombination process.

A turn-off circuit arrangement of a GTO is shown in Figure 7.17a. Because a GTO requires a large turn-off current, a charged capacitor C is normally used to provide the required turn-off gate current. Inductor L limits the turn-off di/dt of the gate current through the circuit formed by R_1, R_2, SW_1, and L. The gate circuit supply voltage V_{GS} should be selected to give the required value of V_{GQ}. The values of R_1 and R_2 should also be minimized.

During the off-state period, which begins after the fall of the tail current to zero, the gate should ideally remain reverse biased. This reverse bias ensures maximum blocking capability. The reverse bias can be obtained either by keeping the SW_1 closed during the whole off-state period or by using a higher impedance circuit SW_2 and R_3,

FIGURE 7.19

Junctions of 160-A GTO in Figure 7.18. (Courtesy of International Rectifier.)

by a so-called buffer layer, a heavily doped n^+-layer at the end of the n-layer. Asymmetric GTOs have lower voltage drop and higher voltage and current ratings.

Controllable peak on-state current I_{TGQ} is the peak value of on-state current that can be turned off by gate control. The off-state voltage is reapplied immediately after turn-off and the reapplied dv/dt is only limited by the snubber capacitance. Once a GTO is turned off, the load current I_L, which is diverted through and charges the snubber capacitor, determines the reapplied dv/dt.

$$\frac{dv}{dt} = \frac{I_L}{C_s}$$

where C_s is the snubber capacitance.

7.6.8 FET-CTHs

A FET-CTH device combines an MOSFET and a thyristor in parallel, as shown in Figure 7.20. If a sufficient voltage is applied to the gate of the MOSFET, typically 3 V, a triggering current for the thyristor is generated internally. It has a high switching speed, high di/dt, and high dv/dt.

This device can be turned on like conventional thyristors, but it cannot be turned off by gate control. This would find applications where optical firing is to be used for providing electrical isolation between the input or control signal and the switching device of the power converter.

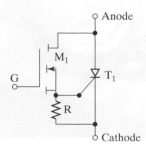

FIGURE 7.20

FET-controlled thyristor.

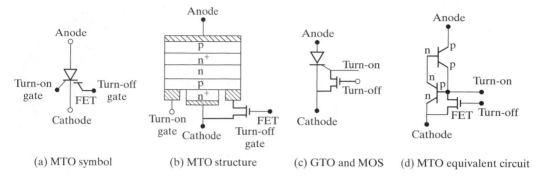

FIGURE 7.21

MOS turn-off (MTO) thyristor.

7.6.9 MTOs

The MTO was developed by Silicon Power Company (SPCO) [16]. It is a combination of a GTO and an MOSFET, which together overcome the limitations of the GTO turn-off ability. The main drawback of GTOs is that they require a high-pulse-current drive circuit for the low impedance gate. The gate circuit must provide the gate turn-off current whose typical peak amplitude is 35% of the current to be controlled. The MTO provides the same functionality as the GTO but uses a gate drive that needs to supply only the signal level voltage necessary to turn MOS transistors on and off. Figure 7.21 shows the symbol, structure, and equivalent circuit of the MTO. Its structure is similar to that of a GTO and retains the GTO's advantages of high voltage (up to 10 kV) and high current (up to 4000 A). MTOs can be used in high-power applications ranging from 1 to 20 MVA [17–20].

Turn-on. Like a GTO, the MTO is turned on by applying gate current pulse to the turn-on gate. Turn-on pulse turns on the *npn*-transistor Q_1, which then turns on the *pnp*-transistor Q_2 and latches on the MTO.

Turn-off. To turn-off the MTO, a gate pulse voltage is applied to the MOSFET gate. Turning on the MOSFETs shorts the emitter and base of the *npn*-transistor Q_1, thereby stopping the latching process. In contrast, a GTO is turned off by sweeping enough current out of the emitter base of the *npn*-transistor with a large negative pulse to stop the regenerative latching action. As a result, the MTO turns off much faster than a GTO and the losses associated with the storage time are almost eliminated. Also the MTO has a higher dv/dt and requires much smaller snubber components. Similar to a GTO, the MTO has a long turn-off tail of current at the end of the turn-off and the next turn-on must wait until the residual charge on the anode side is dissipated through the recombination process.

7.6.10 ETOs

The ETO is a MOS-GTO hybrid device [21, 22] that combines the advantages of both the GTO and the MOSFET. ETO was invented at Virginia Power Electronics Center in collaboration with SPCO [17]. The ETO symbol, its equivalent circuit, and the *pn*

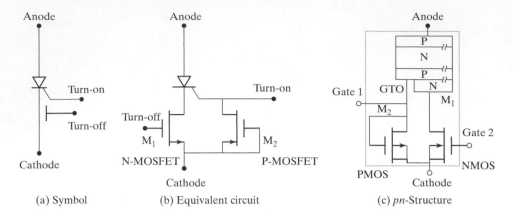

FIGURE 7.22

Emitter turn-off (ETO) thyristor. [Ref. 22, Y. Li]

structure are shown in Figure 7.22. ETO has two gates: one normal gate for turn-on and one with a series MOSFET for turn-off. High power ETOs with a current rating of up to 4 kA and a voltage rating of up to 6 kV have been demonstrated [23].

Turn-on.   An ETO is turned on by applying positive voltages to gates, gate 1 and gate 2. A positive voltage to gate 2 turns on the cathode MOSFET Q_E and turns off the gate MOSFET Q_G. An injection current into the GTO gate (through gate 1) turns on the ETO due to the existence of the GTO.

Turn-off.   When a turn-off negative voltage signal is applied to the cathode MOSFET Q_E, it turns off and transfers all the current away from the cathode (*n* emitter of the *npn*-transistor of the GTO) into the base via gate MOSFET Q_G. This stops the regenerative latching process and results in a fast turn-off.

It is important to note that both the cathode MOSFET Q_E and the gate MOSFET Q_G are not subjected to high-voltage stress, no matter how high the voltage is on the ETO. This is due to the fact that the internal structure of the GTO's gate-cathode is a *PN*-junction. The disadvantage of series MOSFET is that it has to carry the main GTO current and it increases the total voltage drop by about 0.3 to 0.5 V and corresponding losses. Similar to a GTO, the ETO has a long turn-off tail of current at the end of the turn-off and the next turn-on must wait until the residual charge on the anode side is dissipated through the recombination process.

7.6.11 IGCTs

The IGCT integrates a gate-commutated thyristor (GCT) with a multilayered printed circuit board gate drive [24, 25]. The GCT is a hard-switched GTO with a very fast and large gate current pulse, as large as the full-rated current, that draws out all the current from the cathode into the gate in about 1 μs to ensure a fast turn-off.

The internal structure and equivalent circuit of a GCT are similar to that of a GTO shown in Figure 7.14b. The cross section of an IGCT is shown in Figure 7.23. An IGCT may also have an integrated reverse diode, as shown by the n^+n^-p junction on

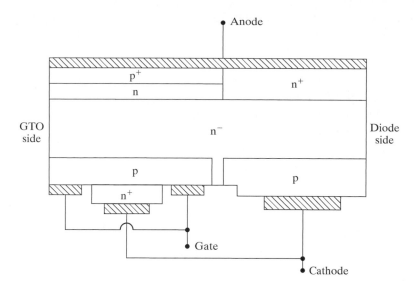

FIGURE 7.23

Cross section of IGCT with a reverse diode.

the right side of Figure 7.23. Similar to a GTO, an MTO and an ETO, the n-buffer layer evens out the voltage stress across the n^--layer, reduces the thickness of the n^--layer, decreases the on-state conduction losses, and makes the device asymmetric. The anode p-layer is made thin and lightly doped to allow faster removal of charges from the anode-side during turn-off.

Turn-on. Similar to a GTO, the IGCT is turned on by applying the turn-on current to its gate.

Turn-off. The IGCT is turned off by a multilayered gate-driver circuit board that can supply a fast rising turn-off pulse, for example, a gate current of 4 kA/μs with a gate-cathode voltage of 20 V only. With this rate of gate current, the cathode-side npn-transistor is totally turned off within about 1 μs and the anode-side pnp-transistor is effectively left with an open base and it is turned-off almost immediately. Due to a very short duration pulse, the gate–drive energy is greatly reduced and the gate–drive energy consumption is minimized. The gate–drive power requirement is decreased by a factor of five compared with that of the GTO. To apply a fast-rising and high-gate current, the IGCT incorporates a special effort to reduce the inductance of the gate circuitry as low as possible. This feature is also necessary for gate–drive circuits of the MTO and ETO.

7.6.12 MCTs

An MCT combines the features of a regenerative four-layer thyristor and a MOS-gate structure. Like the IGBT, which combines the advantages of bipolar junction and field-effect structures, an MCT is an improvement over a thyristor with a pair of MOSFETs to turn on and turn off. Although there are several devices in the MCT family with distinct combinations of channel and gate structures [26], the p-channel MCT is widely reported in literature [27, 28]. A schematic of a p-MCT cell is shown in Figure 7.24a. The equivalent circuit is shown in Figure 7.24b and the symbol in Figure 7.24c [29–36]. The $NPNP$ structure may be represented by an NPN-transistor Q_1 and a PNP-transistor Q_2. The MOS-gate structure can be represented by a p-channel MOSFET M_1, and an n-channel MOSFET M_2.

Due to an $NPNP$ structure instead of the $PNPN$ structure of a normal SCR, the anode serves as the reference terminal with respect to which all gate signals are applied. Let us assume that the MCT is in its forward blocking state and a negative voltage V_{GA} is applied. A p-channel (or an inversion layer) is formed in the n-doped material, causing holes to flow laterally from the p-emitter E_2 of Q_2 (source S_1 of p-channel MOSFET M_1) through the p-channel to the p-base B_1 of Q_1 (drain D_1 of p-channel MOSFET M_1). This hole flow is the base current for the NPN-transistor Q_1. The n^+-emitter E_1 of Q_1 then injects electrons that are collected in the n-base B_2 (and n collector C_1), which causes the p-emitter E_2 to inject holes into the n-base B_2 so that the PNP-transistor Q_2 is turned on and latches the MCT. In short, a negative gate V_{GA} turns on the p-channel MOSFET M_1, thus providing the base current for the transistor Q_2.

Let us assume that the MCT is in its conduction state, and a positive voltage V_{GA} is applied. An n-channel is formed in the p-doped material, causing electrons to flow laterally from the n-base B_2 of Q_2 (source S_2 of n-channel MOSFET M_2) through the n-channel to the heavily doped n^+-emitter E_2 of Q_2 (drain D_2 of n^+-channel MOSFET M_2). This electron flow diverts the base current of PNP-transistor Q_2 so that its base–emitter junction turns off, and holes are not available for collection by the p-base B_1 of Q_1 (and p-collector C_2 of Q_2). The elimination of this hole current at the p-base B_1 causes the NPN-transistor Q_1 to turn off, and the MCT returns to its blocking state. In short, a positive gate pulse V_{GA} diverts the current driving the base of Q_1, thereby turning off the MCT.

In the actual fabrication, each MCT is made up of a large number of cells ($\sim$100,000) each of which contains a wide-base NPN-transistor and a narrow-base PNP-transistor. Although each PNP-transistor in a cell is provided with an N-channel MOSFET across its emitter and base, only a small percentage ($\sim$4%) of PNP-transistors are provided with p-channel MOSFETs across its emitter and collector. The small percentage of PMOS cells in an MCT provides just enough current to turn on and the large number of NMOS cells provide plenty of current to turn off.

Because the gate of the p-channel MCT is preferred with respect to the anode instead of the cathode, it is sometimes referred to as complementary MCT (C-MCT). For an n-channel MCT, it is a $PNPN$ device that is represented by a PNP-transistor Q_1 and an NPN-transistor Q_2. The gate of the n-channel MCT is referenced with respect to the cathode.

Turn-on.   When a p-channel MCT is in the forward blocking state, it can be turned on by applying a negative pulse to its gate with respect to the anode. When an n-channel MCT is in the forward blocking state, it can be turned on by applying a positive pulse to its gate with respect to the cathode. An MCT remains in the on-state until the device current is reversed or a turn-off pulse is applied to its gate.

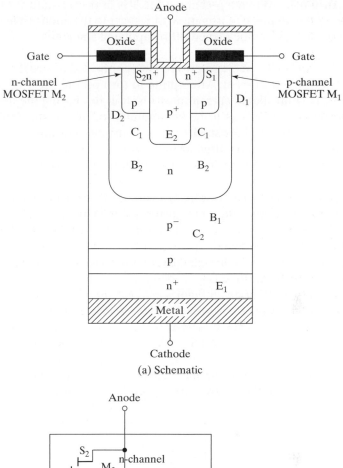

(a) Schematic

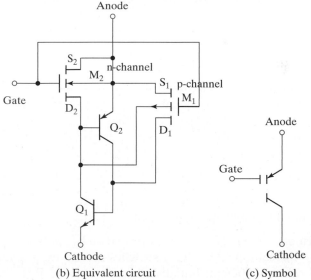

(b) Equivalent circuit

(c) Symbol

FIGURE 7.24

Schematic and equivalent circuit for *p*-channel MCTs.

Turn-off.   When a p-channel MCT is in the on-state, it can be turned off by applying a positive pulse to its gate with respect to the anode. When an n-channel MCT is in the on-state, it can be turned off by applying a negative pulse to its gate with respect to the cathode.

The MCT can be operated as a gate-controlled device if its current is less than the peak controllable current. Attempting to turn off the MCT at currents higher than its rated peak controllable current may result in destroying the device. For higher values of current, the MCT has to be commutated off like a standard SCR. The gate pulse widths are not critical for smaller device currents. For larger currents, the width of the turn-off pulse should be larger. Moreover, the gate draws a peak current during turn-off. In many applications, including inverters and converters, a continuous gate pulse over the entire on or off period is required to avoid state ambiguity.

An MCT has (1) a low forward voltage drop during conduction; (2) a fast turn-on time, typically 0.4 μs, and a fast turn-off time, typically 1.25 μs for an MCT of 500 V, 300 A; (3) low-switching losses; (4) a low reverse voltage blocking capability, and (5) a high gate input impedance, which greatly simplifies the drive circuits. It can be effectively paralleled to switch high currents with only modest deratings of the per-device current rating. It cannot easily be driven from a pulse transformer if a continuous bias is required to avoid state ambiguity.

The MOS structure is spread across the entire surface of the device that results in a fast turn-on and turn-off with low-switching losses. The power or energy required for the turn-on and turn-off is very small, and the delay time due to the charge storage is also very small. As a latching thyristor device, it has a low on-state voltage drop. Therefore, the MCT has the potential to be the near-ultimate turn-off thyristor with low on-state and switching losses, and a fast-switching speed for applications in high-power converters.

7.6.13 SITHs

The SITH, also known as the filed-controlled diode (FCD), was first introduced by Teszner in the 1960s [41]. A SITH is a minority carrier device. As a result, SITH has low on-state resistance or voltage drop and it can be made with higher voltage and current ratings. It has fast-switching speeds and high dv/dt and di/dt capabilities. The switching time is on the order of 1 to 6 μs. The voltage rating [42–46] can go up to 2500 V, and the current rating is limited to 500 A. This device is extremely process sensitive, and small perturbations in the manufacturing process would produce major changes in the device characteristics. With the advent of SiC technology, a 4H-SiC SITH has been fabricated with a forward-blocking voltage of 300 V [47]. The cross section of a half SITH cell structure is shown in Figure 7.25a, its equivalent circuit is shown in Figure 7.25b and its symbol is shown in Figure 7.25c.

Turn-on.   A SITH is normally turned on by applying a positive gate voltage with respect to the cathode. The SITH switches on rapidly, providing the gate current and voltage drive is sufficient. Initially, the gate-cathode PiN diode turns on and injects electrons from the N^+-cathode region into the base region between the P^+ gate and N^+ cathode, and into the channel, thus modulating the channel resistivity. The positive gate voltage reduces the potential barrier in the channel, which gradually becomes conductive. When electrons reach the junction J_1, the p^+ anode begins to inject holes into the

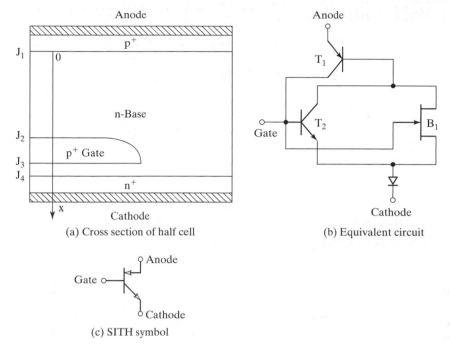

(a) Cross section of half cell

(b) Equivalent circuit

(c) SITH symbol

FIGURE 7.25

Cross section and equivalent circuit of a SITH. [Ref. 49, J. Wang]

base, providing the base current of transistor Q_2. As the base current increases, Q_2 is driven into saturation and the junction J_2 is eventually forward biased. The device is then fully turned on.

The p^+ gate and the channel region can be modeled as a junction field-effect transistor (JFET) operating in the bipolar mode. Electrons flow from the cathode into the base region under the p^+ gate through the channel, providing the base current of the p^+n-p^+-transistor. Due to the high-doping level of the p^+ gate, no electrons flow into the p^+ gate. A portion of hole current flows through the p^+ gate and the channel toward the cathode directly. The remaining hole current flows through the p^+ gate to the channel as the gate current of the bipolar mode JFET (BMFET). The short cathode and gate distance results in a uniform and large carrier concentration in this region; hence, the voltage drop is negligible.

Turn off.   An SITH is normally turned on by applying a negative gate voltage with respect to the cathode. If a sufficiently negative voltage is applied to the gate, a depletion layer forms around the p^+ gate. The depletion layer at J_2 gradually extends into the channel. A potential barrier is created in the channel, narrowing the channel and removing the excess carriers from the channel. If the gate voltage is sufficiently large, the depletion layer of adjacent gate regions merge in the channel and eventually turn off the electron current flow in the channel. Eventually the depletion layer fully cuts off the channel. In spite of no electron current, the hole current continues to flow due to the remaining slowly decaying excess carriers in the base. The removal of the channel current also stops the injection of electrons and holes into the region between the gate and cathode; then the parasitic PiN diode in this region switches off. Therefore, the negative gate voltage establishes a potential barrier in the channel that impedes the transport of electrons from the cathode to the anode. The SITH can support a high-anode voltage with a small leakage current and fully cuts off the channel.

7.6.14 Comparisons of Thyristors

Table 7.1 shows the comparisons of different thyristors in terms of their gate control, advantages, and limitations.

Example 7.2   Finding the Average On-State Current of a Thyristor

A thyristor carries a current as shown in Figure 7.26 and the current pulse is repeated at a frequency of $f_s = 50$ Hz. Determine the average on-state current I_T.

Solution
$I_p = I_{TM} = 1000$ A, $T = 1/f_s = 1/50 = 20$ ms, and $t_1 = t_2 = 5$ μs. The average on-state current is

$$I_T = \frac{1}{20,000}\,[0.5 \times 5 \times 1000 + (20{,}000 - 2 \times 5) \times 1000 + 0.5 \times 5 \times 1000]$$

$$= 999.5 \text{ A}$$

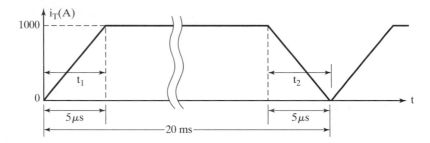

FIGURE 7.26

Thyristor current waveform.

7.7   SERIES OPERATION OF THYRISTORS

For high-voltage applications, two or more thyristors can be connected in series to provide the voltage rating. However, due to the production spread, the characteristics of thyristors of the same type are not identical. Figure 7.27 shows the off-state characteristics of two thyristors. For the same off-state current, their off-state voltages differ.

In the case of diodes, only the reverse-blocking voltages have to be shared, whereas for thyristors, the voltage-sharing networks are required for both reverse and off-state conditions. The voltage sharing is normally accomplished by connecting resistors

TABLE 7.1   Comparisons of Different Thyristors

Switch Type	Gate Control	Control Characteristic	Switching Frequency	On-State Voltage Drop	Maximum Voltage Rating	Maximum Current Rating	Advantages	Limitations
Phase-controlled SCRs	Current for turn-on No turn-off control	Turn-on with a pulse signal Turn-off with natural commutation	Low 60 Hz	Low	1.5 kV, 0.1 MVA	1 kA, 0.1 MVA	Simple turn-on Latching device Turn-on gain is very high Low-cost, high-voltage, and high-current device	Low-switching speed Most suited for line commutated applications between 50 and 60 Hz Cannot be turned off with gate control
Bidirectional thyristors	Two gates Current for turn-on No turn-off control	Turn-on with a pulse signal Turn-off with natural commutation	Low 60 Hz	Low	6.5 kV @ 1.8 kA, 0.1 MVA	3 kA @ 1.8 kV, 0.1 MVA	Same as the phase-controlled SCRs, except it has two gates and the current can flow in both directions Combines two back-to-back SCRs into one device	Similar to those of phase-controlled SCRs
Light-activated thyristors (LASCRs)	Light signal for turn-on No turn-off control	Turn-on with a pulse signal Turn-off with natural commutation	Low 60 Hz	Low			Same as the phase-controlled SCRs, except the gate is isolated and can be remotely operated	Similar to those of phase-controlled SCRs

(continued)

TABLE 7.1 (Continued)

Switch Type	Gate Control	Control Characteristic	Switching Frequency	On-State Voltage Drop	Maximum Voltage Rating	Maximum Current Rating	Advantages	Limitations
TRIAC	Current for turn-on No turn-off control	Turn-on by applying gate a pulse signal for current flow in both direction Turn-off with natural commutation	Low 60 Hz	Low			Same as the phase-controlled SCRs, except the current can flow in both directions It has one gate for turning-on in both directions Like two SCRs connected back to back	Similar to those of phase-controlled SCRs, except for low-power applications
Fast turn-off thyristors	Current for turn-on No turn-off control	Turn-on with a pulse signal Turn-off with natural commutation	Medium 5 kHz	Low			Same as the phase-controlled SCRs, except the turn-off is faster Most suited for forced commutated converters in medium- to high-power applications	Similar to those of phase-controlled SCRs
GTOs	Current for both turn-on and turn-off control	Turn-on with a positive pulse signal Turn-off with a negative pulse	Medium 5 kHz	Low			Similar to the fast turn-off thyristors, except it can be turned off with a negative gate signal	Turn-off gain is low between 5 and 8 and it requires a large gate current to turn-off a large on-state current There is a long tail current during turn-off Although a latching device, it requires a minimum gate current to sustain on-state current

TABLE 7.1 (*Continued*)

Switch Type	Gate Control	Control Characteristic	Switching Frequency	On-State Voltage Drop	Maximum Voltage Rating	Maximum Current Rating	Advantages	Limitations
MTOs	Two gates; both turn-on and turn-off control Current pulse for turn-on Voltage signal for turn-off	Turn-on with a positive pulse current to the turn-on gate Turn-off with a positive voltage to the turn-off MOS gate that unlatches the device	Medium 5 kHz	Low	10 kV@ 20 MVA, 4.5 kV @ 500 A	4 kA @ 20 MVA	Similar to those GTOs, except it can be turned on through the normal gate and turned off through the MOSFET gate Due to the MOS gate, it requires a very low turn-off current and the turn-off time is small	Similar to GTOs, it has a along tail current during turn-off
ETOs	Two gates; both turn-on and turn-off control	Turn-on with a positive pulse current to the turn-on gate and a positive pulse voltage to the turn-off MOS gate Turn-off with a negative pulse voltage to the turn-off MOS gate	Medium 5 kHz	Medium			Due to the series MOS, the transfer of current to the cathode region is rapid and fast turn-off The series MOSFET has to carry the main anode current	Similar to GTOs, it has a along tail current during turn-off The series MOSFET has to carry the main anode current and it increases the on-state voltage drop by about 0.3 to 0.5 V and the conduction losses
IGCTs	Two gates; both turn-on and turn-off control	Turned on with a positive pulse current to the turn-on gate Turned off by applying a fast rising negative current from a multilayered gate–driver circuit board	Medium 5 kHz	Low	5 kV @ 400 A		Like a hard-switched GTO Very fast turn-off due to a fast-rising and high turn-off gate current Low turn-off gate power requirement It can have a built in antiparallel diode	Similar to other GTO devices, the inductance of the gate drive and cathode loop must have a very low value

(*continued*)

TABLE 7.1 *(Continued)*

Switch Type	Gate Control	Control Characteristic	Switching Frequency	On-State Voltage Drop	Maximum Voltage Rating	Maximum Current Rating	Advantages	Limitations
MCTs	Two gates; both turn-on and turn-off control	Turned on *p*-channel MCT with a negative voltage with respect to the anode and turned off with positive voltage	Medium 5 kHz	Medium			Integrates the advantages of the GTOs and MOSFET gate into a single device. The power/energy required for the turn-on and turn-off is very small, and the delay time due to the charge storage time is also very small; as a latching thyristor device, it has a low on-state voltage drop	Has the potential to be the near-ultimate turn-off thyristor with low on-state and switching losses, and a fast-switching speed for applications in high-power converters
SITHs	One gate; both turn-on and turn-off control	Turned on by applying a positive gate drive voltage and turned off with a negative gate voltage	High 100 kHz	Low 1.5 V @ 300 A, 2.6 V @ 900 A	2500 V @		A minority carrier device. Low on-state resistance or voltage drop. It has fast switching speeds and high dv/dt and di/dt capabilities	A field-controlled device and requires a continuous gate voltage. It is extremely process sensitive, and small perturbations in the manufacturing process would produce major changes in the device characteristics

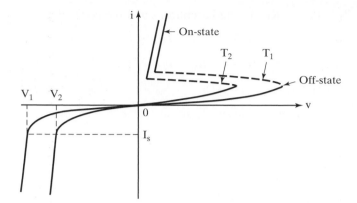

FIGURE 7.27

Off-state characteristics of two thyristors.

across each thyristor, as shown in Figure 7.28. For equal voltage sharing, the off-state currents differ, as shown in Figure 7.29. Let there be n_s thyristors in the string. The off-state current of thyristor T_1 is I_{D1} and that of other thyristors are equal such that $I_{D2} = I_{D3} = I_{Dn}$, and $I_{D1} < I_{D2}$. Because thyristor T_1 has the least off-state current, T_1 shares higher voltage.

If I_1 is the current through the resistor R across T_1 and the currents through other resistors are equal so that $I_2 = I_3 = I_n$, the off-state current spread is

$$\Delta I_D = I_{D2} - I_{D1} = I_T - I_2 - I_T + I_1 = I_1 - I_2 \quad \text{or} \quad I_2 = I_1 - \Delta I_D$$

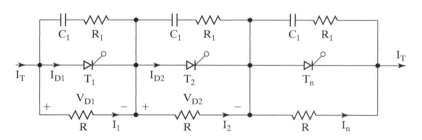

FIGURE 7.28

Three series-connected thyristors.

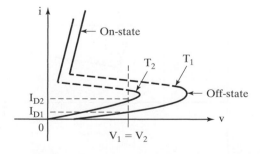

FIGURE 7.29

Forward leakage currents with equal voltage sharing.

The voltage across T_1 is $V_{D1} = RI_1$. Using Kirchhoff's voltage law yields

$$V_s = V_{D1} + (n_s - 1)I_2R = V_{D1} + (n_s - 1)(I_1 - \Delta I_D)R$$
$$= V_{D1} + (n_s - 1)I_1R - (n_s - 1)R\,\Delta I_D$$
$$= n_sV_{D1} - (n_s - 1)R\,\Delta I_D \tag{7.7}$$

Solving Eq. (7.7) for the voltage V_{D1} across T_1 gives

$$V_{D1} = \frac{V_s + (n_s - 1)R\,\Delta I_D}{n_s} \tag{7.8}$$

V_{D1} is maximum when ΔI_D is maximum. For $I_{D1} = 0$ and $\Delta I_D = I_{D2}$, Eq. (7.8) gives the worst-case steady-state voltage across T_1,

$$V_{DS(\max)} = \frac{V_s + (n_s - 1)RI_{D2}}{n_s} \tag{7.9}$$

During the turn-off, the differences in stored charge cause differences in the reverse voltage sharing, as shown in Figure 7.30. The thyristor with the least recovered charge (or reverse recovery time) faces the highest transient voltage. The junction capacitances that control the transient voltage distributions are not adequate and it is normally necessary to connect a capacitor, C_1, across each thyristor as shown in Figure 7.28. R_1 limits the discharge current. The same RC network is generally used for both transient voltage sharing and dv/dt protection.

The transient voltage across T_1 can be determined by applying the relationship of voltage difference,

$$\Delta V = R\,\Delta I_D = \frac{Q_2 - Q_1}{C_1} = \frac{\Delta Q}{C_1} \tag{7.10}$$

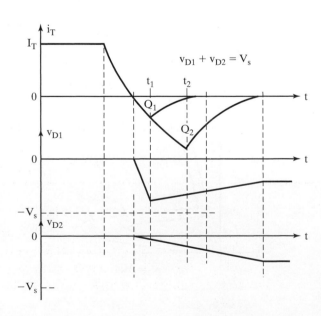

FIGURE 7.30

Reverse recovery time and voltage sharing.

where Q_1 is the stored charge of T_1 and Q_2 is the charge of other thyristors such that $Q_2 = Q_3 = Q_n$ and $Q_1 < Q_2$. Substituting Eq. (7.10) into Eq. (7.8) yields

$$V_{D1} = \frac{1}{n_s}\left[V_s + \frac{(n_s - 1)\Delta Q}{C_1}\right] \tag{7.11}$$

The worst-case transient voltage sharing that occurs when $Q_1 = 0$ and $\Delta Q = Q_2$ is

$$V_{DT(\max)} = \frac{1}{n_s}\left[V_s + \frac{(n_s - 1)Q_2}{C_1}\right] \tag{7.12}$$

A derating factor that is normally used to increase the reliability of the string is defined as

$$\text{DRF} = 1 - \frac{V_s}{n_s V_{DS(\max)}} \tag{7.13}$$

Example 7.3 Finding the Voltage Sharing of Series-Connected Thyristors

Ten thyristors are used in a string to withstand a dc voltage of $V_s = 15$ kV. The maximum leakage current and recovery charge differences of thyristors are 10 mA and 150 μC, respectively. Each thyristor has a voltage-sharing resistance of $R = 56$ kΩ and capacitance of $C_1 = 0.5$ μF. Determine (a) the maximum steady-state voltage sharing $V_{DS(\max)}$, (b) the steady-state voltage-derating factor, (c) the maximum transient voltage sharing $V_{DT(\max)}$, and (d) the transient voltage-derating factor.

Solution
$n_s = 10, V_s = 15$ kV, $\Delta I_D = I_{D2} = 10$ mA, and $\Delta Q = Q_2 = 150$ μC.

a. From Eq. (7.9), the maximum steady-state voltage sharing is

$$V_{DS(\max)} = \frac{15{,}000 + (10 - 1) \times 56 \times 10^3 \times 10 \times 10^{-3}}{10} = 2004 \text{ V}$$

b. From Eq. (7.13), the steady-state derating factor is

$$\text{DRF} = 1 - \frac{15{,}000}{10 \times 2004} = 25.15\%$$

c. From Eq. (7.12), the maximum transient voltage sharing is

$$V_{DT(\max)} = \frac{15{,}000 + (10 - 1) \times 150 \times 10^{-6}/(0.5 \times 10^{-6})}{10} = 1770 \text{ V}$$

d. From Eq. (7.13), the transient derating factor is

$$\text{DRF} = 1 - \frac{15{,}000}{10 \times 1770} = 15.25\%$$

7.8 PARALLEL OPERATION OF THYRISTORS

When thyristors are connected in parallel, the load current is not shared equally due to differences in their characteristics. If a thyristor carries more current than that of the others, its power dissipation increases, thereby increasing the junction temperature and decreasing the internal resistance. This, in turn, increases its current sharing and may

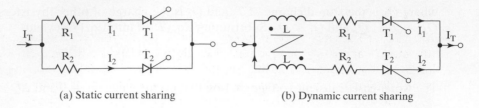

(a) Static current sharing   (b) Dynamic current sharing

FIGURE 7.31

Current sharing of thyristors.

damage the thyristor. This thermal runaway may be avoided by having a common heat sink discussed in Section 18.2, so that all units operate at the same temperature.

A small resistance, as shown in Figure 7.31a, may be connected in series with each thyristor to force equal current sharing, but there may be considerable power loss in the series resistances. A common approach for current sharing of thyristors is to use magnetically coupled inductors, as shown in Figure 7.31b. If the current through thyristor T_1 increases, a voltage of opposite polarity can be induced in the windings of thyristor T_2 and the impedance through the path of T_2 can be reduced, thereby increasing the current flow through T_2.

7.9   *di/dt* PROTECTION

A thyristor requires a minimum time to spread the current conduction uniformly throughout the junctions. If the rate of rise of anode current is very fast compared with the spreading velocity of a turn-on process, a localized "hot-spot" heating may occur due to high current density and the device may fail, as a result of excessive temperature.

The practical devices must be protected against high *di/dt*. As an example, let us consider the circuit in Figure 7.32. Under steady-state operation, D_m conducts when thyristor T_1 is off. If T_1 is fired when D_m is still conducting, *di/dt* can be very high and limited only by the stray inductance of the circuit.

In practice, the *di/dt* is limited by adding a series inductor L_s, as shown in Figure 7.32. The forward *di/dt* is

$$\frac{di}{dt} = \frac{V_s}{L_s} \tag{7.14}$$

where L_s is the series inductance, including any stray inductance.

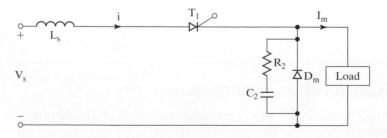

FIGURE 7.32

Thyristor switching circuit with *di/dt* limiting inductors.

7.10   *dv/dt* PROTECTION

If switch S_1 in Figure 7.33a is closed at $t = 0$, a step voltage may be applied across thyristor T_1 and *dv/dt* may be high enough to turn on the device. The *dv/dt* can be limited by connecting capacitor C_s, as shown in Figure 7.33a. When thyristor T_1 is turned on, the discharge current of capacitor is limited by resistor R_s, as shown in Figure 7.33b.

With an *RC* circuit known as a snubber circuit, the voltage across the thyristor rises exponentially as shown in Figure 7.33c and the circuit *dv/dt* can be found approximately from

$$\frac{dv}{dt} = \frac{0.632V_s}{\tau} = \frac{0.632V_s}{R_sC_s} \tag{7.15}$$

The value of snubber time constant $\tau = R_sC_s$ can be determined from Eq. (7.15) for a known value of *dv/dt*. The value of R_s is found from the discharge current I_{TD}.

$$R_s = \frac{V_s}{I_{TD}} \tag{7.16}$$

It is possible to use more than one resistor for *dv/dt* and discharging, as shown in Figure 7.33d. The *dv/dt* is limited by R_1 and C_s. $(R_1 + R_2)$ limits the discharging current such that

$$I_{TD} = \frac{V_s}{R_1 + R_2} \tag{7.17}$$

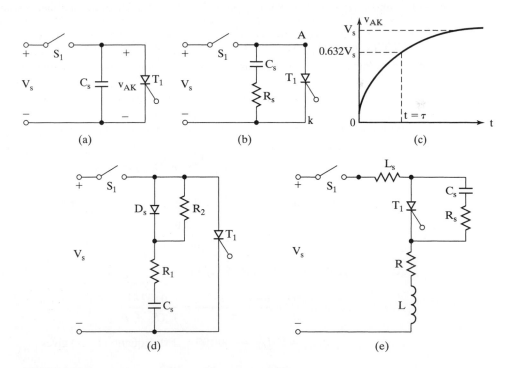

FIGURE 7.33

dv/dt protection circuits.

The load can form a series circuit with the snubber network as shown in Figure 7.33e. From Eqs. (2.40) and (2.41), the damping ratio δ of a second-order equation is

$$\delta = \frac{\alpha}{\omega_0} = \frac{R_s + R}{2}\sqrt{\frac{C_s}{L_s + L}} \tag{7.18}$$

where L_s is the stray inductance, and L and R are the load inductance and resistance, respectively.

To limit the peak voltage overshoot applied across the thyristor, the damping ratio in the range of 0.5 to 1.0 is used. If the load inductance is high, which is usually the case, R_s can be high and C_s can be small to retain the desired value of damping ratio. A high value of R_s reduces the discharge current and a low value of C_s reduces the snubber loss. The circuits of Figure 7.33 should be fully analyzed to determine the required value of damping ratio to limit the dv/dt to the desired value. Once the damping ratio is known, R_s and C_s can be found. The same RC network or snubber is normally used for both dv/dt protection and to suppress the transient voltage due to reverse recovery time. The transient voltage suppression is analyzed in Section 18.6.

Example 7.4   Finding the Values of the Snubber Circuit for a Thyristor Circuit

The input voltage of Figure 7.33e is $V_s = 200$ V with load resistance of $R = 5\ \Omega$. The load and stray inductances are negligible and the thyristor is operated at a frequency of $f_s = 2$ kHz. If the required dv/dt is 100 V/μs and the discharge current is to be limited to 100 A, determine (a) the values of R_s and C_s, (b) the snubber loss, and (c) the power rating of the snubber resistor.

Solution

$dv/dt = 100$ V/μs, $I_{TD} = 100$ A, $R = 5\ \Omega$, $L = L_s = 0$, and $V_s = 200$ V.

a. From Figure 7.33e the charging current of the snubber capacitor can be expressed as

$$V_s = (R_s + R)i + \frac{1}{C_s}\int i\,dt + v_c(t = 0)$$

With initial condition $v_c(t = 0) = 0$, the charging current is found as

$$i(t) = \frac{V_s}{R_s + R}e^{-t/\tau} \tag{7.19}$$

where $\tau = (R_s + R)C_s$. The forward voltage across the thyristor is

$$v_T(t) = V_s - \frac{RV_s}{R_s + R}e^{-t/\tau} \tag{7.20}$$

At $t = 0$, $v_T(0) = V_s - RV_s/(R_s + R)$ and at $t = \tau$, $v_T(\tau) = V_s - 0.368RV_s/(R_s + R)$:

$$\frac{dv}{dt} = \frac{v_T(\tau) - v_T(0)}{\tau} = \frac{0.632RV_s}{C_s(R_s + R)^2} \tag{7.21}$$

From Eq. (7.16), $R_s = V_s/I_{TD} = 200/100 = 2\ \Omega$. Equation (7.21) gives

$$C_s = \frac{0.632 \times 5 \times 200 \times 10^{-6}}{(2 + 5)^2 \times 100} = 0.129\ \mu\text{F}$$

b. The snubber loss

$$P_s = 0.5 C_s V_s^2 f_s$$
$$= 0.5 \times 0.129 \times 10^{-6} \times 200^2 \times 2000 = 5.2 \text{ W} \tag{7.22}$$

c. Assuming that all the energy stored in C_s is dissipated in R_s only, the power rating of the snubber resistor is 5.2 W.

7.11 SPICE THYRISTOR MODEL

As a new device is added to the thyristor family list, the question of a computer-aided model arises. Models for new devices are under development. There are published SPICE models for conventional thyristors, GTOs, MCTs, and SITHs.

7.11.1 Thyristor SPICE Model

Let us assume that the thyristor, as shown in Figure 7.34a, is operated from an ac supply. This thyristor should exhibit the following characteristics:

1. It should switch to the on-state with the application of a small positive gate voltage, provided that the anode-to-cathode voltage is positive.
2. It should remain in the on-state as long as the anode current flows.
3. It should switch to the off-state when the anode current goes through zero toward the negative direction.

The switching action of the thyristor can be modeled by a voltage-controlled switch and a polynomial current source [23]. This is shown in Figure 7.34b. The turn-on process can be explained by the following steps:

1. For a positive gate voltage V_g between nodes 3 and 2, the gate current is $I_g = I(VX) = V_g/R_G$.

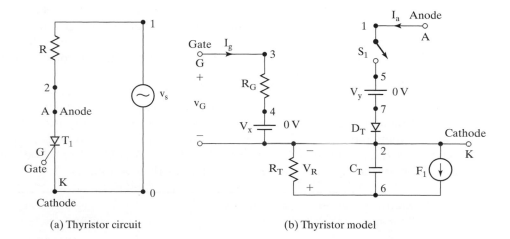

(a) Thyristor circuit   (b) Thyristor model

FIGURE 7.34

SPICE thyristor model.

2. The gate current I_g activates the current-controlled current source F_1 and produces a current of value $F_g = P_1 I_g = P_1 I(VX)$ such that $F_1 = F_g + F_a$.
3. The current source F_g produces a rapidly rising voltage V_R across resistance R_T.
4. As the voltage V_R increases above zero, the resistance R_S of the voltage-controlled switch S_1 decreases from R_{OFF} toward R_{ON}.
5. As the resistance R_S of switch S_1 decreases, the anode current $I_a = I(VY)$ increases, provided that the anode-to-cathode voltage is positive. This increasing anode current I_a produces a current $F_a = P_2 I_a = P_2 I(VY)$. This results in an increased value of voltage V_R.
6. This produces a regenerative condition with the switch rapidly driven into low resistance (on-state). The switch remains on if the gate voltage V_g is removed.
7. The anode current I_a continues to flow as long as it is positive and the switch remains in the on-state.

During the turn-off, the gate current is off and $I_g = 0$. That is, $F_g = 0$, $F_1 = F_g + F_a = F_a$. The turn-off operation can be explained by the following steps:

1. As the anode current I_a goes negative, the current F_1 reverses provided that the gate voltage V_g is no longer present.
2. With a negative F_1, the capacitor C_T discharges through the current source F_1 and the resistance R_T.
3. With the fall of voltage V_R to a low level, the resistance R_S of switch S_1 increases from low (R_{ON}) to high (R_{OFF}).
4. This is again a regenerative condition with the switch resistance driven rapidly to R_{OFF} value as the voltage V_R becomes zero.

This model works well with a converter circuit in which the thyristor current falls to zero due to natural characteristics of the current. However, for a full-wave ac–dc converter with a continuous load current discussed in Chapter 10, the current of a thyristor is diverted to another thyristor and this model may not give the true output. This problem can be remedied by adding diode D_T, as shown in Figure 7.30b. The diode prevents any reverse current flow through the thyristor resulting from the firing of another thyristor in the circuit.

This thyristor model can be used as a subcircuit. Switch S_1 is controlled by the controlling voltage V_R connected between nodes 6 and 2. The switch or diode parameters can be adjusted to yield the desired on-state drop of the thyristor. We shall use diode parameters IS = 2.2E − 15, BV = 1800V, TT = 0; and switch parameters RON = 0.0125, ROFF = 10E + 5, VON = 0.5V, VOFF = OV. The subcircuit definition for the thyristor model *SCR* can be described as follows:

```
*    Subcircuit for ac thyristor model
.SUBCKT   SCR      1        3       2
*         model   anode   control  cathode
*         name            voltage
S1   1    5    6   2       SMOD         ; Voltage-controlled switch
```

```
RG    3    4    50
VX    4    2    DC    OV
VY    5    7    DC    OV
DT    7    2    DMOD                      ; Switch diode
RT    6    2    1
CT    6    2    10UF
F1    2    6    POLY(2)    VX    VY    0    50    11
. MODEL SMOD VSWITCH (RON=0.0125 ROFF=10E+5 VON=0.5V VOFF=OV)  ; Switch model
. MODEL DMOD  D(IS=2.2E−15 BV=1800V TT=0) ; Diode model parameters
. ENDS SCR                               ; Ends subcircuit definition
```

The circuit model as shown in Figure 7.34 incorporates the switching behavior of a thyristor under dc conditions only. It does not include the second-order effects such as overvoltage, dv/dt, delay time t_d, turn-off time t_q, on-state resistance R_{on}, and threshold gate voltage or current. Gracia's model [4] shown in Figure 7.35 includes these parameters that can be extracted from the data sheets.

7.11.2 GTO SPICE Model

A GTO may be modeled with two transistors shown in Figure 7.15c. However, a GTO model [6, 11–13] consisting of two thyristors, which are connected in parallel, yields improved on-state, turn-on, and turn-off characteristics. This is shown in Figure 7.36 with four transistors.

When the anode-to-cathode voltage V_{AK} is positive and there is no gate voltage, the GTO model is in the off-state like a standard thyristor. When a small voltage is applied to the gate, then I_{B2} is nonzero; therefore, both $I_{C1} - I_{C2}$ are nonzero. There can be a current flow from the anode to the cathode. When a negative gate pulse is applied to the GTO model, the *PNP*-junction near to the cathode behaves as a diode. The diode is reverse biased because the gate voltage is negative with respect to the cathode. Therefore, the GTO stops conduction.

When the anode-to-cathode voltage is negative, that is, the anode voltage is negative with respect to the cathode, the GTO model acts like a reverse-biased diode. This is because the *PNP*-transistor sees a negative voltage at the emitter and the *NPN*-transistor sees a positive voltage at the emitter. Therefore, both transistors are in the off-state and the GTO cannot conduct. The SPICE sub-circuit description of the GTO model is as follows:

```
.SUBCIRCUIT    1       2        3           ; GTO Sub-circuit definition
*Terminal      anode   cathode  gate
Q1    5    4    1       DPNP     PNP         ; PNP transistor with model DPNP
Q3    7    6    1       DPNP     PNP
Q2    4    5    2       DNPN     NPN         ; PNP transistor with model DNPN
Q4    6    7    2       DNPN     NPN
R1    7    5    10ohms
R2    6    4    10ohms
R3    3    7    10ohms
.MODEL        DPNP      PNP      ; Model statement for an ideal PNP transistor
.MODEL        DNPN      NPN      ; Model statement for an ideal NPN transistor
.ENDS                           ; End of sub-circuit definition
```

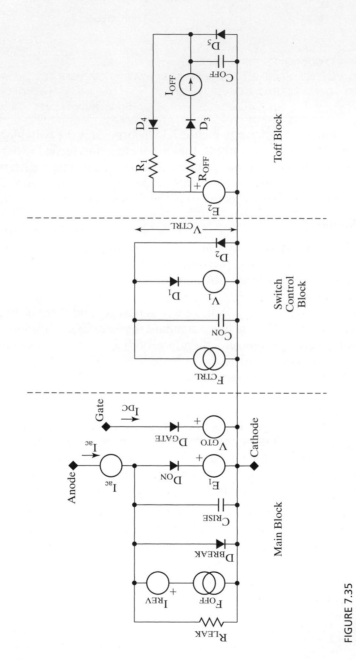

FIGURE 7.35

Complete proposed silicon-controlled rectifier model. [Ref. 4, F. Gracia]

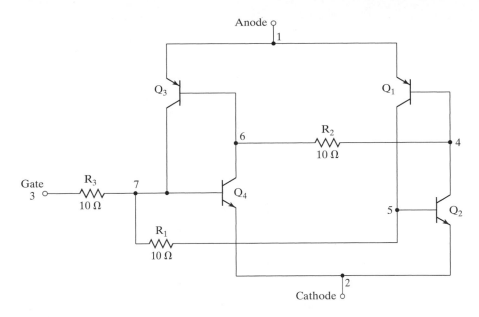

FIGURE 7.36

Four-transistor GTO model. [Ref. 12, M. El-Amia]

7.11.3 MCT SPICE Model

The MCT equivalent, as shown in Figure 7.37a, has an SCR section with integrated two MOSFET sections for turning it on and off. Because the integration of the MCT is complex, it is very difficult to obtain an exact circuit model for the device [39]. Yuvarajan's model [37], shown in Figure 7.37b, is quite simple and it is derived by expanding the SCR model [2, 3] to include the turn-on and turn-off characteristics of the MCT. The model parameters can be obtained from the manufacturer's data sheet. This model, however, does not simulate all the MCT characteristics such as breakdown and breakover voltages, high-frequency operation, and turn-on spike voltages. Arsov's model [38] is a modification of Yuvarajan's model and is derived from the transistor level equivalent circuit of the MCT by expanding the SCR model [3].

7.11.4 SITH SPICE Model

Wang's SITH model [49], which is based on device internal physical operating mechanisms of the equivalent circuit in Figure 7.25b, can predict both device static and dynamic characteristics [48, 50]. The model accounts for effects of the device structure, lifetime, and temperature. It can be implemented in circuit simulators such as PSpice as a subcircuit.

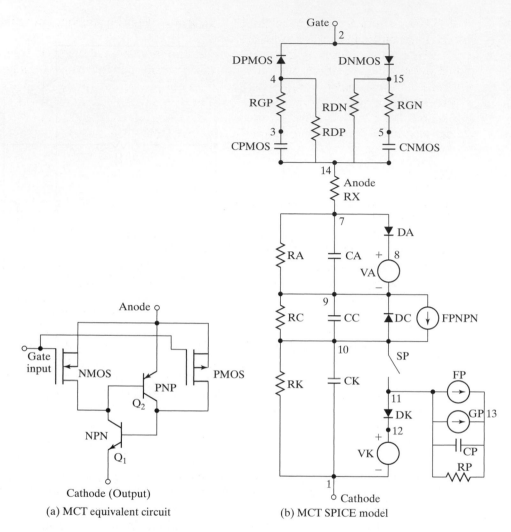

FIGURE 7.37

MCT model. [Ref. 37, S. Yuvarajan]

SUMMARY

There are 13 types of thyristors. Only the GTOs, SITHs, MTOs, ETOs, IGCTs, and MCTs are gate turn-off devices. Each type has advantages and disadvantages. The characteristics of practical thyristors differ significantly from that of ideal devices. Although there are various means of turning on thyristors, gate control is the most practical. Due to the junction capacitances and turn-on limit, thyristors must be protected from high di/dt and dv/dt failures. A snubber network is normally used to protect from high dv/dt. Due to the recovered charge, some energy is stored in the di/dt and stray inductors; and the devices must be protected from this stored energy. The switching losses of GTOs are much higher than those of normal SCRs. The snubber components of GTOs are critical to their performance.

Due to the differences in the characteristics of thyristors of the same type, the series and parallel operations of thyristors require voltage and current-sharing networks to protect them under steady-state and transient conditions. A means of isolation between the power circuit and gate circuits is necessary. A pulse transformer isolation is simple, but effective. For inductive loads, a pulse train reduces thyristor loss and is normally used for gating thyristors, instead of a continuous pulse.

REFERENCES

[1]   General Electric, D. R. Grafham and F. B. Golden, *SCR Manual,* 6th ed. Englewood Cliffs, NJ: Prentice Hall. 1982.

[2]   L. I. Giacoletto, "Simple SCR and TRIAC PSpice computer models," *IEEE Transactions on Industrial Electronics,* Vol. IE36, No. 3, 1989, pp. 451–455.

[3]   R. W. Avant and F. C. Lee, "The J3 SCR model applied to resonant converter simulation," *IEEE Transactions on Industrial Electronics,* Vol. IE-32, February 1985, pp. 1–12.

[4]   F. I. Gracia, F. Arizti, and F. I. Aranceta, "A nonideal macro-model of thyristor for transient analysis in power electronic systems," *IEEE Transactions Industrial Electronics,* Vol. 37, December 1990, pp. 514–520.

[5]   *Bi-directional control thyristor,* ABB Semiconductors, Lenzburg, Switzerland, February 1999. www.abbsemi.com

[6]   M. H. Rashid, *SPICE for Power Electronics.* Upper Saddle River, NJ: Prentice-Hall. 1995.

[7]   M. H. Rashid, *Power Electronics Handbook,* edited by M. H. Rashid. San Diego, CA: Academic Press. 2001, Chapter 4—Gate Turn-Off Thyristors (GTOs).

[8]   Westcode Semiconductor: Data GTO data-sheets. www.westcode.com/ws-gto.html

[9]   D. Grant and A. Honda, *Applying International Rectifier's Gate Turn-Off Thyristors.* El Segundo, CA: International Rectifier. Application Note AN-315A.

[10]   O. Hashimoto, H. Kirihata, M. Watanabe, A. Nishiura, and S. Tagami, "Turn-on and turn-off characteristics of a 4.5-kV 3000-A gate turn-off thyristor," *IEEE Transactions on Industrial Applications,* Vol. IA22, No. 3, 1986, pp. 478–482.

[11]   E. Y. Ho and P. C. Sen, "Effect of gate drive on GTO thyristor characteristics," *IEEE Transactions on Industrial Electronics,* Vol. IE33, No. 3, 1986, pp. 325–331.

[12]   M. A. I. El-Amin, "GTO PSpice model and its applications," *The Fourth Saudi Engineering Conference,* November 1995, Vol. III, pp. 271–277.

[13]   G. Busatto, F. Iannuzzo, and L. Fratelli, "PSpice model for GTOs," *Proceedings of Symposium on Power Electronics Electrical Drives, Advanced Machine Power Quality,* SPEEDAM Conference, June 3–5, 1998, Sorrento, Italy, Col. 1, pp. P2/5–10.

[14]   D. J. Chamund, "Characterisation of 3.3 kV asymmetrical thyristor for pulsed power application," *IEE Symposium Pulsed Power 2000* (Digest No. 00/053), May 3–4, 2000, London, pp. 35/1–4.

[15]   H. Fukui, H. Amano, and H. Miya, "Paralleling of gate turn-off thyristors," *IEEE Industrial Applications Society Conference Record,* 1982, pp. 741–746.

[16]   D. E. Piccone, R. W. DeDoncker, J. A. Barrow, and W. H. Tobin, "The MTO thyristor—A new high power bipolar MOS thyristor," *IEEE Industrial Applications Society Conference Record,* October 1996, pp. 1472–1473.

[17] "MTO data-sheets," Silicon Power Corporation (SPCO), Exton, PA. www.siliconopower.com

[18] R. Rodrigues, D. Piccone, A. Huanga, and R. De Donckerb, "MTO™ thyristor power switches," *Power Systems World '97*, Baltimore, MD, September 6–12, 1997, pp. 3–53–64.

[19] D. Piccone, J. Barrow, W. Tobin, and R. De Doncker, "MTO—A MOS turn-off disc-type thyristor for high voltage power conversion," *IEEE Industrial Applications Society Conference Record,* 1996, pp. 1472–1473.

[20] B. J. Cardoso and T. A. Lipo, "Application of MTO thyristors in current stiff converters with resonant snubbers," *IEEE Transactions on Industry Applications,* Vol. 37, No. 2, March/April 200, pp. 566–573.

[21] Y. Li, A. Q. Huang, and F. C. Lee, "Introducing the emitter turn-off thyristor," *IEEE Industrial Applications Society Conference Record,* 1998, pp. 860–864.

[22] Y. Li and A. Q. Huang, "The emitter turn-off thyristor—A new MOS-bipolar high power device," *Proc. 1997 Virginia Polytechnic Power Electronics Center Seminar,* September 28–30, 1997, pp. 179–183.

[23] L. Yuxin, A. Q. Huang, and K. Motto, "Experimental and numerical study of the emitter turn-off thyristor (ETO)," *IEEE Transactions on Power Electronics*, Vol. 15, No. 3, May 2000, pp. 561–574.

[24] P. K. Steimer, H. E. Gruning, J. Werninger, E. Carrol, S. Klaka, and S. Linder, "IGCT—A new emerging technology for high power, low cost inverters," *IEEE Industry Applications Society Conference Record,* October 5–9, 1997, New Orleans, LA, pp. 1592–1599.

[25] H. E. Gruning and B. Odegard, "High performance low cost MVA inverters realized with integrated gate commutated thyristors (IGCT)," *European Power Electronics Conference,* 1997, pp. 2060–2065.

[26] S. Lindner, S. Klaka, M. Frecker, E. Caroll, and H. Zeller, "A new range of reverse conducting gate commutated thyristors for high voltage, medium power application," *European Power Electronics Conference,* 1997, pp. 1117–1124.

[27] "Data Sheet—Reverse conducting IGCTs," ABB Semiconductors, Lenzburg, Switzerland, 1999.

[28] H. E. Gruening and A. Zuckerberger, "Hard drive of high power GTO's: Better switching capability obtained through improved gate-units," *IEEE Industry Applications Society Conference Record,* October 6–10, 1996, pp. 1474–1480.

[29] B. J. Baliga, M. S. Adler, R. P. Love, P. V. Gray, and N. D. Zommer, "The insulated gate transistor: A new three-terminal MOS-controlled bipolar power device," *IEEE Transactions on Electron Devices*, Vol. ED-31, No. 6, June 1984, pp. 821–828.

[30] V. A. K. Temple, "MOS controlled thyristors: A class of power devices," *IEEE Transactions on Electron Devices,* Vol. ED33, No. 10, 1986, pp. 1609–1618.

[31] T. M. Iahns, R. W. De Donker, I. W. A. Wilson, V. A. K. Temple, and S. L. Watrous, "Circuit utilization characteristics of MOS-controlled thyristors," *IEEE Transactions on Industry Applications*, Vol. 27, No. 3, May/June 1991, pp. 589–597.

[32] "MCT User's Guide," Harris Semiconductor Corp., Melbourne, FL, 1995.

[33] S. Yuvarajan, *Power Electronics Handbook*, edited by M. H. Rashid. San Diego, CA: Academic Press. 2001, Chapter 8—MOS-Controlled Thyristors (MCTs).

[34] P. Venkataraghavan and B. J. Baliga, "The dv/dt capability of MOS-gated thyristors," *IEEE Transactions on Power Electronics*, Vol. 13, No. 4, July 1998, pp. 660–666.

[35] S. B. Bayne, W. M. Portnoy, and A. R. Hefner, Jr., "MOS-gated thyristors (MCTs) for repetitive high power switching," *IEEE Transactions on Power Electronics*, Vol. 6, No. 1, January 2001, pp. 125–131.

[36] B. J. Cardoso and T. A. Lipo, "Application of MTO thyristors in current stiff converters with resonant snubbers," *IEEE Transactions on Industry Applications*, Vol. 37, No. 2, March/April 2001, pp. 566–573.

[37] S. Yuvarajan and D. Quek, "A PSpice model for the MOS controlled thyristor," *IEEE Transactions on Industrial Electronics*, Vol. 42, October 1995, pp. 554–558.

[38] G. L. Arsov and L. P. Panovski, "An improved PSpice model for the MOS-controlled thyristor," *IEEE Transactions on Industrial Electronics*, Vol. 46, No. 2, April 1999, pp. 473–477.

[39] Z. Hossain, K. J. Olejniczak, H. A. Mantooth, E. X. Yang, and C. L. Ma, "Physics-based MCT circuit model using the lumped-charge modeling approach," *IEEE Transactions on Power Electronics*, Vol. 16, No. 2, March 2001, pp. 264–272.

[40] S. Teszner and R. Gicquel, "Gridistor—A new field effect device," *Proc. IEEE*, Vol. 52, 1964, pp. 1502–1513.

[41] J. Nishizawa, K. Muraoka, T. Tamamushi, and Y. Kawamura, "Low-loss high-speed switching devices, 2300-V 150-A static induction thyristor," *IEEE Transactions on Electron Devices*, Vol. ED-32, No. 4, 1985, pp. 822–830.

[42] Y. Nakamura, H. Tadano, M. Takigawa, I. Igarashi, and J. Nishizawa, "Very high speed static induction thyristor," *IEEE Transactions on Industry Applications*, Vol. IA22, No. 6, 1986, pp. 1000–1006.

[43] J. Nishizawa, K. Muraoka, Y. Kawamura, and T. Tamamushi, "A low-loss high-speed switching device; Rhe 2500-V 300-A static induction thyristor," *IEEE Transactions on Electron Devices*, Vol. ED-33, No. 4, 1986, pp. 507–515.

[44] Y. Terasawa, A. Mimura, and K. Miyata, "A 2.5 kV static induction thyristor having new gate and shorted *p*-emitter structures," *IEEE Transactions on Electron Devices*, Vol. ED-33, No. 1, 1986, pp. 91–97.

[45] M. Maeda, T. Keno, Y. Suzuki, and T. Abe, "Fast-switching-speed, low-voltage-drop static induction thyristor," *Electrical Engineering in Japan*, Vol. 116, No. 3, 1996, pp. 107–115.

[46] R. Singh, K. Irvine, and J. Palmour, "4H-SiC buried gate field controlled thyristor," *Annual Device Research Conference Digest*, 1997, pp. 34–35.

[47] D. Metzner and D. Schroder, "A SITH-model for CAE in power- electronics," *International Symposium on Semiconductor Devices ICs*, Tokyo, Japan, 1990, pp. 204–210.

[48] M. A. Fukase, T. Nakamura, and J. I. Nishizawa, "A circuit simulator of the SITh," *IEEE Transactions on Power Electronics*, Vol. 7, No. 3, July 1992, pp. 581–591.

[49] J. Wang and B. W. Williams, "A new static induction thyristor (SITh) analytical model," *IEEE Transactions on Power Electronics*, Vol. 14, No. 5, September 1999, pp. 866–876.

[50] S. Yamada, Y. Morikawa, M. Kekura, T. Kawamura, S. Miyazaki, F. Ichikawa, and H. Kishibe, "A consideration on electrical characteristics of high power SITHs," *International Symposium on Power Semiconductor Devices and ICs*, ISPSD'98, Kyoto, Japan, June 3–6, 1998, pp. 241–244.

[51] S. Bernet, "Recent developments in high power converters for industry and traction applications," *IEEE Transactions on Power Electronics*, Vol. 15, No. 6, November 2000, pp. 1102–1117.

[52] Transistor Manual, *Unijunction Transistor Circuits*, 7th ed. Syracuse, NY: General Electric Company, 1964, Publication 450.37.

REVIEW QUESTIONS

7.1   What is the $v-i$ characteristic of thyristors?

7.2   What is an off-state condition of thyristors?

7.3   What is an on-state condition of thyristors?

7.4   What is a latching current of thyristors?

7.5   What is a holding current of thyristors?

7.6   What is the two-transistor model of thyristors?

7.7   What are the means of turning-on thyristors?

7.8   What is turn-on time of thyristors?

7.9   What is the purpose of di/dt protection?

7.10   What is the common method of di/dt protection?

7.11   What is the purpose of dv/dt protection?

7.12   What is the common method of dv/dt protection?

7.13   What is turn-off time of thyristors?

7.14   What are the types of thyristors?

7.15   What is an SCR?

7.16   What is the difference between an SCR and a TRIAC?

7.17   What is the turn-off characteristic of thyristors?

7.18   What are the advantages and disadvantages of GTOs?

7.19   What are the advantages and disadvantages of SITHs?

7.20   What are the advantages and disadvantages of RCTs?

7.21   What are the advantages and disadvantages of LASCRs?

7.22   What are the advantages and disadvantages of bidirectional thyristors?

7.23   What are the advantages and disadvantages of MTOs?

7.24   What are the advantages and disadvantages of ETOs?

7.25   What are the advantages and disadvantages of CGCTs?

7.26   What is a snubber network?

7.27   What are the design considerations of snubber networks?

7.28   What is the common technique for voltage sharing of series-connected thyristors?

7.29   What are the common techniques for current sharing of parallel-connected thyristors?

7.30   What is the effect of reverse recovery time on the transient voltage sharing of parallel-connected thyristors?

7.31   What is a derating factor of series-connected thyristors?

PROBLEMS

7.1   The junction capacitance of a thyristor can be assumed to be independent of off-state voltage. The limiting value of charging current to turn on the thyristor is 12 mA. If the critical value of dv/dt is 800 V/µs, determine the junction capacitance.

7.2   The junction capacitance of a thyristor is $C_{J2} = 20$ pF and can be assumed independent of off-state voltage. The limiting value of charging current to turn on the thyristor is 15 mA. If a capacitor of 0.01 µF is connected across the thyristor, determine the critical value of dv/dt.

7.3   A thyristor circuit is shown in Figure P7-3. The junction capacitance of thyristor is $C_{J2} = 15$ pF and can be assumed to be independent of the off-state voltage. The limiting value of charging current to turn on the thyristor is 5 mA and the critical value of dv/dt is 200 V/µs. Determine the value of capacitance C_s so that the thyristor cannot be turned on due to dv/dt.

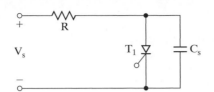

FIGURE P7.3

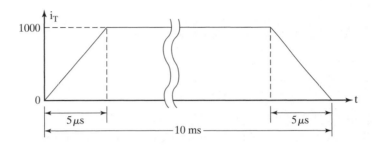

FIGURE P7.6

7.4 The input voltage in Figure 7.33e is $V_s = 200$ V with a load resistance of $R = 10\ \Omega$ and a load inductance of $L = 50\ \mu H$. If the damping ratio is 0.7 and the discharging current of the capacitor is 5 A, determine (a) the values of R_s and C_s, and (b) the maximum dv/dt.

7.5 Repeat Problem 1.4 if the input voltage is ac, $v_s = 179 \sin 377t$.

7.6 A thyristor carries a current as shown in Figure P7-6. The switching frequency is $f_s = 10$ Hz. Determine the average on-state current I_T.

7.7 A string of thyristors is connected in series to withstand a dc voltage of $V_s = 15$ kV. The maximum leakage current and recovery charge differences of thyristors are 10 mA and 150 μC, respectively. A derating factor of 20% is applied for the steady-state and transient voltage sharings of thyristors. If the maximum steady-state voltage sharing is 1000 V, determine **(a)** the steady-state voltage-sharing resistance R for each thyristor, and **(b)** the transient voltage capacitance C_1 for each thyristor.

7.8 Two thyristors are connected in parallel to share a total load current of $I_L = 600$ A. The on-state voltage drop of one thyristor is $V_{T1} = 1.0$ at 300 A and that of other thyristors is $V_{T2} = 1.5$ V at 300 A. Determine the values of series resistances to force current sharing with 10% difference. Total voltage $v = 2.5$ V.

CHAPTER 8

Resonant Pulse Inverters

The learning objectives of this chapter are as follows:

- To learn the switching technique for resonant inverters and their types
- To study the operation and frequency characteristics of resonant inverters
- To understand the performance parameters of resonant inverters
- To learn the techniques for zero-voltage and zero-current switching
- To learn the techniques for analyzing and design of resonant inverters

8.1   INTRODUCTION

The switching devices in converters with a pulse-width-modulation (PWM) control can be gated to synthesize the desired shape of the output voltage or current. However, the devices are turned on and off at the load current with a high di/dt value. The switches are subjected to a high-voltage stress, and the switching power loss of a device increases linearly with the switching frequency. The turn-on and turn-off loss could be a significant portion of the total power loss. The electromagnetic interference is also produced due to high di/dt and dv/dt in the converter waveforms.

The disadvantages of PWM control can be eliminated or minimized if the switching devices are turned "on" and "off" when the voltage across a device or its current become zero [1]. The voltage and current are forced to pass through zero crossing by creating an *LC*-resonant circuit, thereby called a *resonant pulse converter*. The resonant converters can be broadly classified into eight types:

Series-resonant inverters

Parallel-resonant inverters

Class E resonant converter

Class E resonant rectifier

Zero-voltage-switching (ZVS) resonant converters

Zero-current-switching (ZCS) resonant converters

Two-quadrant ZVS resonant converters

Resonant dc-link inverters

8.2   SERIES RESONANT INVERTERS

The series resonant inverters are based on resonant current oscillation. The resonating components and switching device are placed in series with the load to form an under-damped circuit. The current through the switching devices falls to zero due to the nat-ural characteristics of the circuit. If the switching element is a thyristor, it is said to be self-commutated. This type of inverter produces an approximately sinusoidal wave-form at a high-output frequency, ranging from 200 to 100 kHz, and is commonly used in relatively fixed output applications (e.g., induction heating, sonar transmitter, fluores-cent lighting, or ultrasonic generators). Due to the high-switching frequency, the size of resonating components is small.

There are various configurations of series resonant inverters, depending on the connections of the switching devices and load. The series inverters may be classified into two categories:

1. Series resonant inverters with unidirectional switches.
2. Series resonant inverters with bidirectional switches.

8.2.1   Series Resonant Inverters with Unidirectional Switches

Figure 8.1a shows the circuit diagram of a simple series inverter using two unidirec-tional thyristor switches. When thyristor T_1 is fired, a resonant pulse of current flows through the load and the current falls to zero at $t = t_{1m}$ and T_1 is self-commutated. Fir-ing of thyristor T_2 causes a reverse resonant current through the load and T_2 is also self-commutated. The circuit operation can be divided into three modes and the equivalent circuits are shown in Figure 8.1b. The gating signals for thyristors and the waveforms for the load current and capacitor voltage are shown in Figure 8.1c.

The series resonant circuit formed by L, C, and load (assumed resistive) must be underdamped. That is,

$$R^2 < \frac{4L}{C} \tag{8.1}$$

Mode 1.   This mode begins when T_1 is fired and a resonant pulse of current flows through T_1 and the load. The instantaneous load current for this mode is described by

$$L\frac{di_1}{dt} + Ri_1 + \frac{1}{C}\int i_1 \, dt + v_{c1}(t = 0) = V_s \tag{8.2}$$

with initial conditions $i_1(t = 0) = 0$ and $v_{c1}(t = 0) = -V_c$. Because the circuit is un-derdamped, the solution of Eq. (8.2) yields

$$i_1(t) = A_1 e^{-tR/2L} \sin \omega_r t \tag{8.3}$$

where ω_r is the resonant frequency and

$$\omega_r = \left(\frac{1}{LC} - \frac{R^2}{4L^2}\right)^{1/2} \tag{8.4}$$

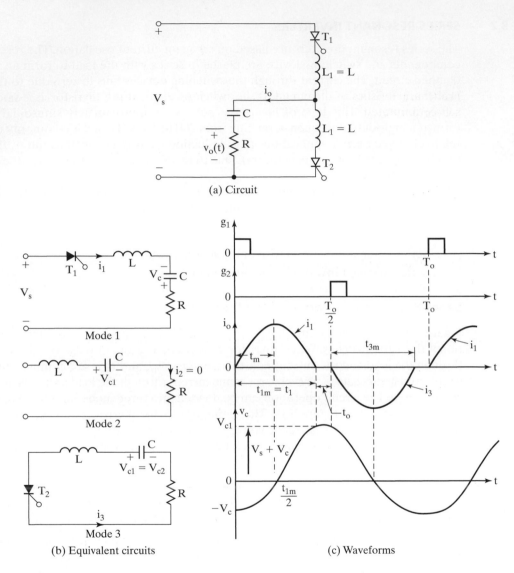

FIGURE 8.1

Basic series resonant inverter.

The constant A_1 in Eq. (8.3) can be evaluated from the initial condition:

$$\left.\frac{di_1}{dt}\right|_{t=0} = \frac{V_s + V_c}{\omega_r L} = A_1$$

and

$$i_1(t) = \frac{V_s + V_c}{\omega_r L} e^{-\alpha t} \sin \omega_r t \tag{8.5}$$

where

$$\alpha = \frac{R}{2L} \tag{8.6}$$

The time t_m when the current $i_1(t)$ in Eq. (8.5) becomes maximum can be found from the condition

$$\frac{di_1}{dt} = 0 \quad \text{or} \quad \omega_r e^{-\alpha t_m} \cos \omega_r t_m - \alpha e^{-\alpha t_m} \sin \omega_r t_m = 0$$

and this gives

$$t_m = \frac{1}{\omega_r} \tan^{-1} \frac{\omega_r}{\alpha} \tag{8.7}$$

The capacitor voltage can be found from

$$v_{c1}(t) = \frac{1}{C} \int_0^t i_1(t) \, dt - V_c$$

$$= -(V_s + V_c)e^{-\alpha t}(\alpha \sin \omega_r t + \omega_r \cos \omega_r t)/\omega_r + V_s \tag{8.8}$$

This mode is valid for $0 \le t \le t_{1m}(= \pi/\omega_r)$ and ends when $i_1(t)$ becomes zero at t_{1m}. At the end of this mode,

$$i_1(t = t_{1m}) = 0$$

and

$$v_{c1}(t = t_{1m}) = V_{c1} = (V_s + V_c)e^{-\alpha\pi/\omega_r} + V_s \tag{8.9}$$

Mode 2.   During this mode, thyristors T_1 and T_2 are off. Redefining the time origin, $t = 0$, at the beginning of this mode, this mode is valid for $0 \le t \le t_{2m}$.

$$i_2(t) = 0, \quad v_{c2}(t) = V_{c1} \quad v_{c2}(t = t_{2m}) = V_{c2} = V_{c1}$$

Mode 3.   This mode begins when T_2 is switched on and a reverse resonant current flows through the load. Let us redefine the time origin, $t = 0$, at the beginning of this mode. The load current can be found from

$$L\frac{di_3}{dt} + Ri_3 + \frac{1}{C} \int i_3 \, dt + v_{c3}(t = 0) = 0 \tag{8.10}$$

with initial conditions $i_3(t = 0) = 0$ and $v_{c3}(t = 0) = -V_{c2} = -V_{c1}$. The solution of Eq. (8.10) gives

$$i_3(t) = \frac{V_{c1}}{\omega_r L} e^{-\alpha t} \sin \omega_r t \tag{8.11}$$

The capacitor voltage can be found from

$$v_{c3}(t) = \frac{1}{C} \int_0^t i_3(t)\, dt - V_{c1}$$

$$= -V_{c1} e^{-\alpha t}(\alpha \sin \omega_r t + \omega_r \cos \omega_r t)/\omega_r \qquad (8.12)$$

This mode is valid for $0 \leq t \leq t_{3m} = \pi/\omega_r$ and ends when $i_3(t)$ becomes zero. At the end of this mode,

$$i_3(t = t_{3m}) = 0$$

and in the steady state,

$$v_{c3}(t = t_{3m}) = V_{c3} = V_c = V_{c1} e^{-\alpha \pi/\omega_r} \qquad (8.13)$$

Equations (8.9) and (8.13) yield

$$V_c = V_s \frac{1 + e^{-z}}{e^z - e^{-z}} = V_s \frac{e^z + 1}{e^{2z} - 1} = \frac{V_s}{e^z - 1} \qquad (8.14)$$

$$V_{c1} = V_s \frac{1 + e^z}{e^z - e^{-z}} = V_s \frac{e^z(1 + e^z)}{e^{2z} - 1} = \frac{V_s e^z}{e^z - 1} \qquad (8.15)$$

where $z = \alpha \pi/\omega_r$. Adding V_c from Eq. (8.14) to V_s gives

$$V_s + V_c = V_{c1} \qquad (8.16)$$

Equation (8.16) indicates that under steady-state conditions, the peak values of positive current in Eq. (8.5) and of negative current in Eq. (8.11) through the load are the same.

The load current $i_1(t)$ must be zero and T_1 must be turned off before T_2 is fired. Otherwise, a short-circuit condition results through the thyristors and dc supply. Therefore, the available off-time $t_{2m}(= t_{\text{off}})$, known as the *dead zone*, must be greater than the turn-off time of thyristors, t_q.

$$\frac{\pi}{\omega_o} - \frac{\pi}{\omega_r} = t_{\text{off}} > t_q \qquad (8.17)$$

where ω_o is the frequency of the output voltage in rads per second. Equation (8.17) indicates that the maximum possible output frequency is limited to

$$f_o \leq f_{\max} = \frac{1}{2(t_q + \pi/\omega_r)} \qquad (8.18)$$

The resonant inverter circuit in Figure 8.1a is very simple. However, it gives the basic concept and describes the characteristic equations, which can be applied to other types of resonant inverters. The power flow from the dc supply is discontinuous. The dc supply has a high peak current and would contain harmonics. An improvement of the basic inverter in Figure 8.1a can be made if inductors are closely coupled, as shown in Figure 8.2.

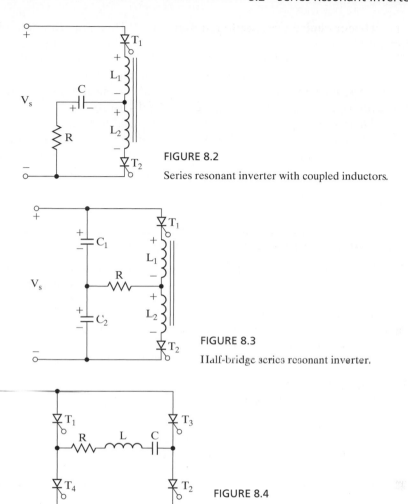

FIGURE 8.2

Series resonant inverter with coupled inductors.

FIGURE 8.3

Half-bridge series resonant inverter.

FIGURE 8.4

Full-bridge series resonant inverter.

When T_1 is fired and current $i_1(t)$ begins to rise, the voltage across L_1 is positive with polarity as shown. The induced voltage on L_2 now adds to the voltage of C in reverse biasing T_2; and T_2 can be turned off. The result is that firing of one thyristor turns off the other, even before the load current reaches zero.

The drawback of high-pulsed current from the dc supply can be overcome in a half-bridge configuration, as shown in Figure 8.3, where $L_1 = L_2$ and $C_1 = C_2$. The power is drawn from the dc source during both half-cycles of output voltage. One-half of the load current is supplied by capacitor C_1 or C_2 and the other half by the dc source.

A full-bridge inverter, which allows higher output power, is shown in Figure 8.4. When T_1 and T_2 are fired, a positive resonant current flows through the load; and when T_3 and T_4 are fired, a negative load current flows. The supply current is continuous, but pulsating.

The resonant frequency and available dead-zone depend on the load and for this reason, resonant inverters are most suitable for fixed-load applications. The inverter load (or resistor R) could also be connected in parallel with the capacitor.

Device choice and gating requirements. Thyristors can be replaced by bipolar junction transistors (BJTs), metal oxide semiconductor field-effect transistors (MOS-FETs), insulated-gate bipolar transistors (IGBTs), or gate-turn-off thyristors (GTOs). However, the device choice depends on the output power and frequency requirements. Thyristors, in general, have higher voltage and current ratings than transistors that can, however, operate at higher frequencies than thyristors.

Thyristors require only a pulse-gating signal to turn on and they are turned off naturally at the end of the half-cycle oscillation at $t = t_{1m}$. Thus, if thyristor T_1 in Figure 8.1a is triggered at $t = 0$, then thyristor T_2 is triggered at $t = T_o/2(> t_{1m})$. Transistors, however, require a continuous gate pulse. The pulse width t_{pw} of the first transistor Q_1 must satisfy the condition $t_{1m} < t_{pw} < T_o/2$ so that the resonant oscillation can complete its half-cycle before the next transistor Q_2 is turned on at $t = T_o/2(> t_{1m})$.

Example 8.1 Analysis of the Basic Resonant Inverter

The series resonant inverter in Figure 8.2 has $L_1 = L_2 = L = 50\ \mu H$, $C = 6\ \mu F$, and $R = 2\ \Omega$. The dc input voltage is $V_s = 220$ V and the frequency of output voltage is $f_o = 7$ kHz. The turn-off time of thyristors is $t_q = 10\ \mu s$. Determine (a) the available (or circuit) turn-off time t_{off}, (b) the maximum permissible frequency f_{max}, (c) the peak-to-peak capacitor voltage V_{pp}, and (d) the peak load current I_p. (e) Sketch the instantaneous load current $i_o(t)$, capacitor voltage $v_c(t)$, and dc supply current $i_s(t)$. Calculate (f) the rms load current I_o; (g) the output power P_o; (h) the average supply current I_s; and (i) the average, peak, and rms thyristor currents.

Solution

$V_s = 220$ V, $C = 6\ \mu F$, $L = 50\ \mu H$, $R = 2\ \Omega$, $f_o = 7$ kHz, $t_q = 10\ \mu s$, and $\omega_o = 2\pi \times 7000 = 43,982$ rad/s. From Eq. (8.4),

$$\omega_r = \left(\frac{1}{LC} - \frac{R^2}{4L^2}\right)^{1/2} = \left(\frac{10^{12}}{50 \times 6} - \frac{2^2 \times 10^{12}}{4 \times 50^2}\right)^{1/2} = 54{,}160 \text{ rad/s}$$

The resonant frequency is $f_r = \omega_r/2\pi = 8619.8$ Hz, $T_r = 1/f_r = 116\ \mu s$. From Eq. (8.6), $\alpha = 2/(2 \times 50 \times 10^{-6}) = 20{,}000$.

a. From Eq. (8.17),

$$t_{off} = \frac{\pi}{43{,}982} - \frac{\pi}{54{,}160} = 13.42\ \mu s$$

b. From Eq. (8.18), the maximum possible frequency is

$$f_{max} = \frac{1}{2(10 \times 10^{-6} + \pi/54{,}160)} = 7352 \text{ Hz}$$

c. From Eq. (8.14),

$$V_c = \frac{V_s}{e^{\alpha\pi/\omega r} - 1} = \frac{220}{e^{20\pi/54.16} - 1} = 100.4 \text{ V}$$

From Eq. (8.16), $V_{c1} = 220 + 100.4 = 320.4$ V. The peak-to-peak capacitor voltage is $V_{pp} = 100.4 + 320.4 = 420.8$ V.

d. From Eq. (8.7), the peak load current, which is the same as the peak supply current, occurs at

$$t_m = \frac{1}{\omega_r} \tan^{-1} \frac{\omega_r}{\alpha} = \frac{1}{54,160} \tan^{-1} \frac{54.16}{20} - 22.47 \ \mu s$$

and Eq. (8.5) gives the peak load current as

$$i_1(t = t_m) = I_p = \frac{320.4}{0.05416 \times 50} \ e^{-0.02 \times 22.47} \sin(54,160 \times 22.47 \times 10^{-6}) = 70.82 \ A$$

e. The sketches for $i(t)$, $v_c(t)$, and $i_s(t)$ are shown in Figure 8.5.

f. The rms load current is found from Eqs. (8.5) and (8.11) by a numerical method and the result is

$$I_o = \left[2f_o \int_0^{T_r/2} i_0^2(t) \ dt \right]^{1/2} = 44.1 \ A$$

g. The output power $P_o = 44.1^2 \times 2 = 3889 \ W$.

h. The average supply current $I_s = 3889/220 = 17.68 \ A$.

i. The average thyristor current

$$I_A = f_o \int_0^{T_r/2} i_0(t) \ dt = 17.68 \ A$$

The peak thyristor current $I_{pk} = I_p = 70.82 \ A$, and the root-mean-square (rms) thyristor current $I_R = I_o/\sqrt{2} = 44.1/\sqrt{2} = 31.18 \ A$.

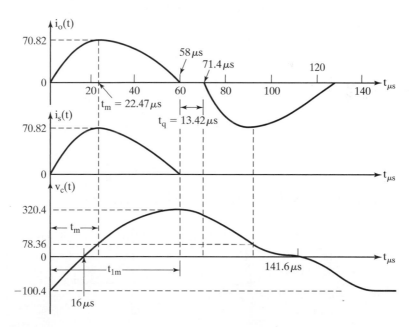

FIGURE 8.5

Waveforms for Example 8.1.

Example 8.2 Analysis of the Half-Bridge Resonant Inverter

The half-bridge resonant inverter in Figure 8.3 is operated at an output frequency, $f_o = 7$ kHz. If $C_1 = C_2 = C = 3$ μF, $L_1 = L_2 = L = 50$ μH, $R = 2$ Ω, and $V_s = 220$ V, determine (a) the peak supply current, (b) the average thyristor current I_A, and (c) the rms thyristor current I_R.

Solution

$V_s = 220$ V, $C = 3$ μF, $L = 50$ μH, $R = 2$ Ω, and $f_o = 7$ kHz. Figure 8.6a shows the equivalent circuit when thyristor T_1 is conducting and T_2 is off. The capacitors C_1 and C_2 would initially be charged to $V_{c1}(= V_s + V_c)$ and V_c, respectively, with the polarities as shown, under steady-state conditions. Because $C_1 = C_2$, the load current would be shared equally by C_1 and the dc supply, as shown in Figure 8.6b.

Figure 8.6c shows the equivalent circuit when thyristor T_2 is conducting and T_1 is off. The capacitors C_1 and C_2 are initially charged to V_{c1} and $V_S - V_{c1}$, respectively, with the polarities as shown. The load current is shared equally by C_1 and C_2, as shown in Figure 8.6d, which can be simplified to Figure 8.6e.

Considering the loop formed by C_2, dc source, L, and load, the instantaneous load current can be described (from Figure 8.6b) by

$$L \frac{di_o}{dt} + Ri_0 + \frac{1}{2C_2} \int i_o \, dt + v_{c2}(t = 0) - V_s = 0 \qquad (8.19)$$

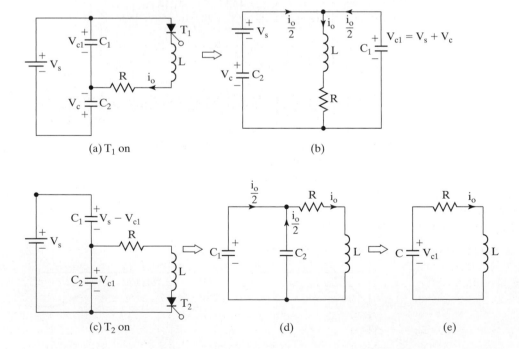

(a) T_1 on   (b)

(c) T_2 on   (d)   (e)

FIGURE 8.6

Equivalent circuit for Example 8.2.

with initial conditions $i_0(t = 0) = 0$ and $v_{c2}(t = 0) = -V_c$. For an underdamped condition and $C_1 = C_2 = C$, Eq. (8.5) is applicable:

$$i_0(t) = \frac{V_s + V_c}{\omega_r L} e^{-\alpha t} \sin \omega_r t \tag{8.20}$$

where the effective capacitance is $C_e = C_1 + C_2 = 2C$ and

$$\omega_r = \left(\frac{1}{2LC_2} - \frac{R^2}{4L^2} \right)^{1/2} = \left(\frac{10^{12}}{2 \times 50 \times 3} - \frac{2^2 \times 10^{12}}{4 \times 50^2} \right)^{1/2} = 54{,}160 \text{ rad/s} \tag{8.21}$$

The voltage across capacitor C_2 can be expressed as

$$v_{c2}(t) = \frac{1}{2C_2} \int_0^t i_0(t) \, dt - V_c$$
$$= -(V_s + V_c)e^{-\alpha t}(\alpha \sin \omega_r t + \omega_r \cos \omega_r t)/\omega_r + V_s \tag{8.22}$$

a. Because the resonant frequency is the same as that of Example 8.1, the results of Example 8.1 are valid, provided that the equivalent capacitance is $C_e = C_1 + C_2 = 6 \, \mu\text{F}$. From Example 8.1, $V_c = 100.4 \text{ V}$, $t_m = 22.47 \, \mu\text{s}$, and $I_o = 44.1 \text{ A}$. From Eq. (8.20), the peak load current is $I_p = 70.82 \text{ A}$. The peak supply current, which is half of the peak load current, is $I_{ps} = 70.82/2 = 35.41 \text{ A}$.

b. The average thyristor current $I_A = 17.68 \text{ A}$.

c. The rms thyristor current $I_R = I_o/\sqrt{2} = 31.18 \text{ A}$.

Note: For the same power output and resonant frequency, the capacitances of C_1 and C_2 in Figure 8.3 should be half that in Figures 8.1 and 8.2. The peak supply current becomes half. The analysis of full-bridge series inverters is similar to that of the basic series inverter in Figure 8.1a. That is, $i_3(t) = i_1(t) = (V_s + V_c)/(\omega_r L)e^{-\alpha t} \sin (\omega_r t)$ in the steady-state conditions.

8.2.2   Series Resonant Inverters with Bidirectional Switches

For the resonant inverters with unidirectional switches, the power devices have to be turned on in every half-cycle of output voltage. This limits the inverter frequency and the amount of energy transfer from the source to the load. In addition, the devices are subjected to high peak reverse voltage.

The performance of series inverters can be significantly improved by connecting an antiparallel diode across a device, as shown in Figure 8.7a. When device Q_1 is fired, a resonant pulse of current flows and Q_2 is self-commutated at $t = t_1$. However, the resonant oscillation continues through diode D_1 until the current falls again to zero at the end of a cycle. The waveform for the load current and the conduction intervals of the power devices are shown in Figure 8.7b.

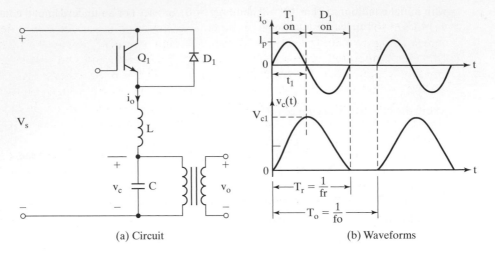

(a) Circuit   (b) Waveforms

FIGURE 8.7

Basic series resonant inverter with bidirectional switches.

If the conduction time of the diode is greater than the turn-off time of the device, there is no need of a dead zone and the output frequency f_o is the same as the resonant frequency f_r,

$$f_o = f_r = \frac{\omega_r}{2\pi} \tag{8.23}$$

where f_r is the resonant frequency of the series circuit in hertz. The minimum device switching time t_{sw} consists of delay time, rise time, fall time, and storage time, that is, $t_{sw} = t_d + t_r + t_f + t_s$. Thus, the maximum inverter frequency is given by

$$f_{s(\max)} = \frac{1}{2t_{sw}} \tag{8.24}$$

and f_o should be less than $f_{s(\max)}$.

If the switching device is a thyristor and t_q is its turn-off time, then maximum inverter frequency is given by

$$f_{s(\max)} = \frac{1}{2t_q} \tag{8.25}$$

The diode D_1 should be connected as close as possible to the thyristor and the connecting leads should be minimum to reduce any stray inductance in the loop formed by T_1 and D_1. Because the reverse voltage during the recovery time of thyristor T_1 is already low, typically 1 V, any inductance in the diode path would reduce the net reverse voltage across the terminals of T_1, and thyristor T_1 may not turn off. To overcome this problem, a *reverse conducting thyristor* (RCT) is normally used. An RCT is made by integration of an asymmetric thyristor and a fast-recovery diode into a single silicon chip, and RCTs are ideal for series resonant inverters.

The circuit diagram for the half-bridge version is shown in Figure 8.8a and the waveform for the load current and the conduction intervals of the power devices are shown in Figure 8.8b. The full-bridge configuration is shown in Figure 8.9a. The inverters can be operated in two different modes: nonoverlapping and overlapping. In a nonoverlapping mode, the firing of a transistor device is delayed until the last current oscillation through a diode has been completed, as in Figure 8.8b. In an overlapping mode, a device is fired, while the current in the diode of the other part is still conducting, as shown in Figure 8.9b. Although overlapping operation increases the output frequency, the output power is increased.

The maximum frequency of thyristor inverters are limited due to the turn-off or commutation requirements of thyristors, typically 12 to 20 μs, whereas transistors require only a microsecond or less. The transistor inverter can operate at the resonant frequency. A transistorized half-bridge inverter is shown in Figure 8.10 with a transformer-connected load. Transistor Q_2 can be turned on almost instantaneously after transistor Q_1 is turned off.

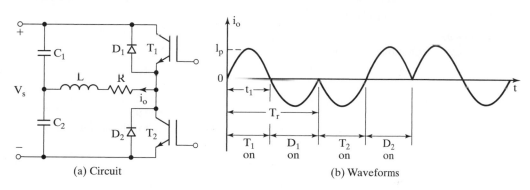

(a) Circuit   (b) Waveforms

FIGURE 8.8

Half-bridge series inverters with bidirectional switches.

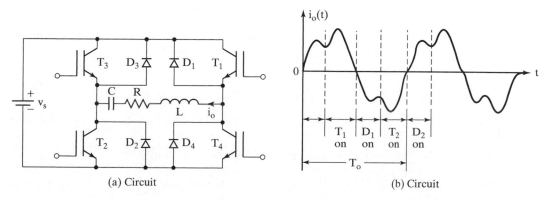

(a) Circuit   (b) Circuit

FIGURE 8.9

Full-bridge series inverters with bidirectional switches.

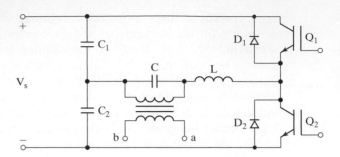

FIGURE 8.10

Half-bridge transistorized resonant inverter.

Example 8.3 Finding the Currents and Voltages of a Simple Resonant Inverter

The resonant inverter in Figure 8.7a has $C = 2\ \mu\text{F}$, $L = 20\ \mu\text{H}$, $R = 0$, and $V_s = 220$ V. The switching time of the transistor is $t_{sw} = 12\ \mu\text{s}$. The output frequency is $f_o = 20$ kHz. Determine (a) the peak supply current I_p, (b) the average device current I_A, (c) the rms device current I_R, (d) the peak-to-peak capacitor voltage V_{pp}, (e) the maximum permissible output frequency f_{max}, and (f) the average supply current I_s.

Solution
When device Q_1 is turned on, the current is described by

$$L \frac{di_0}{dt} + \frac{1}{C}\int i_0\, dt + v_c(t = 0) = V_s$$

with initial conditions $i_0(t = 0) = 0$, $v_c(t = 0) = V_c = 0$. Solving for the current gives

$$i_0(t) = V_s\sqrt{\frac{C}{L}}\sin \omega_r t \tag{8.26}$$

and the capacitor voltage is

$$v_c(t) = V_s(1 - \cos \omega_r t) \tag{8.27}$$

where

$$\omega_r = 1/\sqrt{LC}$$

$$\omega_r = \frac{10^6}{\sqrt{20 \times 2}} = 158{,}114 \text{ rad/s} \quad \text{and} \quad f_r = \frac{158{,}114}{2\pi} = 25{,}165 \text{ Hz}$$

$$T_r = \frac{1}{f_r} = \frac{1}{25{,}165} = 39.74\ \mu\text{s} \quad t_1 = \frac{T_r}{2} = \frac{39.74}{2} = 19.87\ \mu\text{s}$$

At $\omega_r t = \pi$,

$$v_c(\omega_r t = \pi) = V_{c1} = 2V_s = 2 \times 220 = 440 \text{ V}$$

$$v_c(\omega_r t = 0) = V_c = 0$$

a. $I_p = V_s\sqrt{C/L} = 220\sqrt{2/20} = 69.57$ A.

b. $I_A = f_o \int_0^\pi I_p \sin\theta\, d\theta = I_p f_o/(\pi f_r) = 69.57 \times 20{,}000/(\pi \times 25{,}165) = 17.6$ A

c. $I_R = I_p \sqrt{f_o t_1/2} = 69.57\sqrt{20{,}000 \times 19.87 \times 10^{-6}/2} = 31.01$ A.

d. The peak-to-peak capacitor voltage $V_{pp} = V_{c1} - V_c = 440$ V.

e. From Eq. (8.24), $f_{max} = 10^6/(2 \times 12) = 41.67$ kHz.

f. Because there is no power loss in the circuit, $I_s = 0$.

Example 8.4 Analysis of the Half-Bridge Resonant Inverter with Bidirectional Switches

The half-bridge resonant inverter in Figure 8.8a is operated at a frequency of $f_o = 3.5$ kHz. If $C_1 = C_2 = C = 3$ μF, $L_1 = L_2 = L = 50$ μH, $R = 2$ Ω, and $V_s = 220$ V, determine (a) the peak supply current I_P, (b) the average device current I_A, (c) the rms device current I_R, (d) the rms load current I_o, and (e) the average supply current I_s.

Solution

$V_s = 220$ V, $C_e = C_1 + C_2 = 6$ μF, $L = 50$ μH, $R = 2$ Ω, and $f_o = 3500$ Hz. The analysis of this inverter is similar to that of inverter in Figure 8.3. Instead of two current pulses, there are four pulses in a full cycle of the output voltage with one pulse through each of devices Q_1, D_1, Q_2, and D_2. Equation (8.20) is applicable. During the positive half-cycle, the current flows through Q_1; and during the negative half-cycle the current flows through D_1. In a nonoverlap control, there are two resonant cycles during the entire period of output frequency f_o. From Eq. (8.21),

$$\omega_r = 54{,}160 \text{ rad/s} \qquad f_r = \frac{54{,}160}{2\pi} = 8619.9 \text{ Hz}$$

$$T_r = \frac{1}{8619.9} = 116 \text{ μs} \qquad t_1 = \frac{116}{2} = 58 \text{ μs}$$

$$T_0 = \frac{1}{3500} = 285.72 \text{ μs}$$

The off-period of load current

$$t_d = T_0 - T_r = 285.72 - 116 = 169.72 \text{ μs}$$

Because t_d is greater than zero, the inverter would operate in the nonoverlap mode. From Eq. (8.14), $V_c = 100.4$ V and $V_{c1} = 220 + 100.4 = 320.4$ V.

a. From Eq. (8.7),

$$t_m = \frac{1}{54{,}160} \tan^{-1}\frac{54{,}160}{20{,}000} = 22.47 \text{ μs}$$

$$i_0(t) = \frac{V_s + V_c}{\omega_r L} e^{-\alpha t} \sin \omega_r t$$

and the peak load current becomes $I_p = i_0(t = t_m) = 70.82$ A.

b. A device conducts from a time of t_1. The average device current can be found from

$$I_A = f_o \int_0^{t_1} i_0(t) \, dt = 8.84 \text{ A}$$

c. The rms device current is

$$I_R = \left[f_o \int_0^{t_1} i_0^2(t)\, dt \right]^{1/2} = 22.05 \text{ A}$$

d. The rms load current $I_o = 2I_R = 2 \times 22.05 = 44.1$ A.

e. $P_o = 44.1^2 \times 2 = 3889$ W and the average supply current, $I_s = 3889/220 = 17.68$ A.

Note: With bidirectional switches, the current ratings of the devices are reduced. For the same output power, the average device current is half and the rms current is $1/\sqrt{2}$ of that for an inverter with unidirectional switches.

Example 8.5 Analysis of the Full-Bridge Resonant Inverter with Bidirectional Switches

The full-bridge resonant inverter in Figure 8.9a is operated at a frequency, $f_o = 3.5$ kHz. If $C = 6\ \mu$F, $L = 50\ \mu$H, $R = 2\ \Omega$, and $V_s = 220$ V, determine (a) the peak supply current I_p, (b) the average device current I_A, (c) the rms device current I_R, (d) the rms load current I_o, and (e) the average supply current I_s.

Solution

$V_s = 220$ V, $C = 6\ \mu$F, $L = 50\ \mu$H, $R = 2\ \Omega$, and $f_o = 3500$ Hz. From Eq. (8.21), $\omega_r = 54{,}160$ rad/s, $f_r = 54{,}160/(2\pi) = 8619.9$ Hz, $\alpha = 20{,}000$, $T_r = 1/8619.9 = 116\ \mu$s, $t_1 = 116/2 = 58\ \mu$s, and $T_0 = 1/3500 = 285.72\ \mu$s. The off-period of load current is $t_d = T_0 - T_r = 285.72 - 116 = 169.72\ \mu$s, and the inverter would operate in the nonoverlap mode.

Mode 1. This mode begins when Q_1 and Q_2 are fired. A resonant current flows through Q_1, Q_2, load, and supply. The equivalent circuit during mode 1 is shown in Figure 8.11a with an initial capacitor voltage indicated. The instantaneous current is described by

$$L\frac{di_0}{dt} + Ri_0 + \frac{1}{C}\int i_0\, dt + v_c(t = 0) = V_s$$

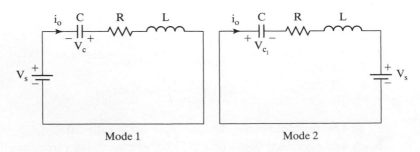

Mode 1   Mode 2

FIGURE 8.11

Mode equivalent circuits for a full-bridge resonant inverter.

with initial conditions $i_0(t = 0) = 0$, $v_{c1}(t = 0) = -V_c$ and the solution for the current gives

$$i_0(t) = \frac{V_s + V_c}{\omega_r L} e^{-\alpha t} \sin \omega_r t \tag{8.28}$$

$$v_c(t) = -(V_s + V_c)e^{-\alpha t}(\alpha \sin \omega_r t + \omega_r \cos \omega_r t) + V_s \tag{8.29}$$

Devices Q_1 and Q_2 are turned off at $t_1 = \pi/\omega_r$, when $i_1(t)$ becomes zero.

$$V_{c1} = v_c(t = t_1) = (V_s + V_c)e^{-\alpha\pi/\omega_r} + V_s \tag{8.30}$$

Mode 2. This mode begins when Q_3 and Q_4 are fired. A reverse resonant current flows through Q_3, Q_4, load, and supply. The equivalent circuit during mode 2 is shown in Figure 8.11b with an initial capacitor voltage indicated. The instantaneous load current is described by

$$L\frac{di_0}{dt} + Ri_0 + \frac{1}{C}\int i_0 \, dt + v_c(t = 0) = -V_s$$

with initial conditions $i_2(t = 0) = 0$ and $v_c(t = 0) = V_{c1}$, and the solution for the current gives

$$i_0(t) = -\frac{V_s + V_{c1}}{\omega_r L} e^{-\alpha t} \sin \omega_r t \tag{8.31}$$

$$v_c(t) = (V_s + V_{c1})e^{-\alpha t}(\alpha \sin \omega_r t + \omega_r \cos \omega_r t)/\omega_r - V_s \tag{8.32}$$

Devices Q_3 and Q_4 are turned off at $t_1 = \pi/\omega_r$, when $i_0(t)$ becomes zero.

$$V_c = -v_c(t = t_1) = (V_s + V_{c1})e^{-\alpha\pi/\omega_r} + V_s \tag{8.33}$$

Solving for V_c and V_{c1} from Eqs. (8.30) and (8.33) gives

$$V_c = V_{c1} = V_s \frac{e^z + 1}{e^z - 1} \tag{8.34}$$

where $z = \alpha\pi/\omega_r$. For $z = 20{,}000\pi/54{,}160 = 1.1601$, Eq. (8.34) gives $V_c = V_{c1} = 420.9$ V.

a. From Eq. (8.7),

$$t_m = \frac{1}{54{,}160} \tan^{-1} \frac{54{,}160}{20{,}000} = 22.47 \ \mu s$$

From Eq. (8.28), the peak load current $I_p = i_0(t = t_m) = 141.64$ A.

b. A device conducts from a time of t_1. The average device current can be found from Eq. (8.28):

$$I_A = f_o \int_0^{t_1} i_0(t) \, dt = 17.68 \text{ A}$$

c. The rms device current can be found from Eq. (8.28):

$$I_R = \left[f_o \int_0^{t_1} i_0^2(t) \, dt \right]^{1/2} = 44.1 \text{ A}$$

    d. The rms load current is $I_o = 2I_R = 2 \times 44.1 = 88.2$ A.

    e. $P_o = 88.2^2 \times 2 = 15{,}556$ W and the average supply current, $I_s = 15{,}556/220 = 70.71$ A.

Note: For the same circuit parameters, the output power is four times, and the device currents are twice that for a half-bridge inverter.

Key Points of Section 8.2

- For the same circuit parameters, the output power of a full-bridge inverter is four times and the device currents are twice that for a half-bridge inverter. For the same output power, the average device current of an inverter with bidirectional switches is half of that for an inverter with unidirectional switches. Thus, the half-bridge and full-bridge inverters with bidirectional switches are generally used.
- The basic inverter of Figure 8.1a describes the characteristics of a half-bridge inverter and Example 8.5 describes that of a full-bridge inverter.

8.3 FREQUENCY RESPONSE OF SERIES-RESONANT INVERTERS

It can be noticed from the waveforms of Figures 8.7b and 8.8b that varying the switching frequency $f_s(= f_o)$ can vary the output voltage. The frequency response of the voltage gain exhibits the gain limitations against the frequency variations [2]. There are three possible connections of the load resistance R in relation to the resonating components: (1) series, (2) parallel, and (3) series–parallel combination.

8.3.1 Frequency Response for Series Loaded

In Figures 8.4, 8.8, and 8.9a, the load resistance R forms a series circuit with the resonating components L and C. The equivalent circuit is shown in Figure 8.12a. The input voltage is a square wave whose peak fundamental component is $V_{i(\text{pk})} = 4V_s/\pi$, and its rms value is $V_i = 4V_s/\sqrt{2}\pi$. Using the voltage-divider rule in frequency domain, the voltage gain is given by

$$G(j\omega) = \frac{V_o}{V_i}(j\omega) = \frac{1}{1 + j\omega L/R - j/(\omega C R)}$$

Let $\omega_0 = 1/\sqrt{LC}$ be the resonant frequency, and $Q_s = \omega_0 L/R$ be the quality factor. Substituting L, C, and R in terms of Q_s and ω_0, we get

$$G(j\omega) = \frac{v_o}{v_i}(j\omega) = \frac{1}{1 + jQ_s(\omega/\omega_0 - \omega_0/\omega)} = \frac{1}{1 + jQ_s(u - 1/u)}$$

where $u = \omega/\omega_0$. The magnitude of $G(j\omega)$ can be found from

$$|G(j\omega)| = \frac{1}{[1 + Q_s^2(u - 1/u)^2]^{1/2}} \tag{8.35}$$

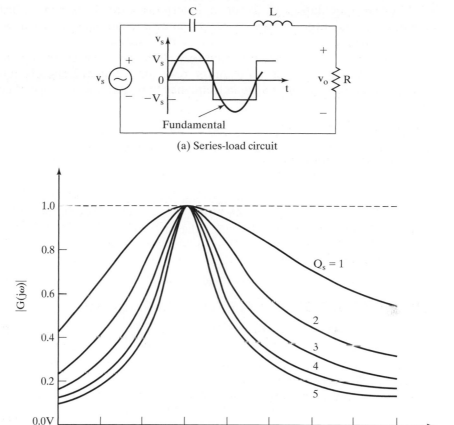

(a) Series-load circuit

(b) Frequency response

FIGURE 8.12

Frequency response for series loaded.

Figure 8.12b shows the magnitude plot of Eq. (8.35) for $Q_s = 1$ to 5. For a continuous output voltage, the switching frequency should be greater than the resonant frequency f_0. If the inverter operates near resonance and a short circuit occurs at the load, the current rises to a high value, especially at a high-load current. However, the output current can be controlled by raising the switching frequency. The current through the switching devices decreases as the load current decreases, thereby having lower on-state conduction losses and a high efficiency at a partial load. The series inverter is most suitable for high-voltage, low-current applications. The maximum output occurs at resonance, and the maximum gain for $u = 1$ is $|G(j\omega)|_{max} = 1$.

Under no-load conditions, $R = \infty$ and $Q_s = 0$. Thus, the curve would simply be a horizontal line. That is, for $Q_s = 1$, the characteristic has a poor "selectivity" and the output voltage changes significantly from no-load to full-load conditions, thereby yielding poor regulation. The resonant inverter is normally used in applications requiring only a fixed output voltage. However, some no-load regulations can be obtained by time ratio control at frequencies lower than the resonant frequency (e.g., in Figure 8.8b). This type of control has two disadvantages: (1) it limits how far the operating frequency can be varied up and down from the resonant frequency, and (2) due to a low Q-factor, it requires a large change in frequency to realize a wide range of output voltage control.

Example 8.6   Finding the Values of L and C for a Series-Loaded Resonant Inverter to Yield Specific Output Power

A series resonance inverter with series loaded delivers a load power of $P_L = 1$ kW at resonance. The load resistance is $R = 10 \, \Omega$. The resonant frequency is $f_0 = 20$ kHz. Determine (a) the dc input voltage V_s, (b) the quality factor Q_s if it is required to reduce the load power to 250 W by frequency control so that $u = 0.8$, (c) the inductor L, and (d) the capacitor C.

Solution

a. Because at resonance $u = 1$ and $|G(j\omega)|_{max} = 1$, the peak fundamental load voltage is $V_p = V_{i(pk)} = 4V_s/\pi$.

$$P_L = \frac{V_p^2}{2R} = \frac{4^2 V_s^2}{2R\pi^2} \quad \text{or} \quad 1000 = \frac{4^2 V_s^2}{2\pi^2 \times 10}$$

which gives $V_s = 110$ V.

b. To reduce the load power by $(1000/250 =)\ 4$, the voltage gain must be reduced by 2 at $u = 0.8$. That is, from Eq. (8.35), we get $1 + Q_s^2(u - 1/u)^2 = 2^2$, which gives $Q_s = 3.85$.

c. Q_s is defined by

$$Q_s = \frac{\omega_0 L}{R}$$

or

$$3.85 = \frac{2\pi \times 20 \text{ kHz} \times L}{10}$$

which gives $L = 306.37 \, \mu\text{H}$.

d. $f_0 = 1/2\pi\sqrt{LC}$ or $20 \text{ kHz} = 1/[2\pi\sqrt{(306.37 \, \mu\text{H} \times C)}]$, which gives $C = 0.2067 \, \mu\text{F}$.

8.3.2   Frequency Response for Parallel Loaded

With the load connected across the capacitor C directly (or through a transformer), as shown in Figure 8.7, the equivalent circuit is shown in Figure 8.13a. Using the voltage-divider rule in frequency domain, the voltage gain is given by

$$G(j\omega) = \frac{V_o}{V_i}(j\omega) = \frac{1}{1 - \omega^2 LC + j\omega L/R}$$

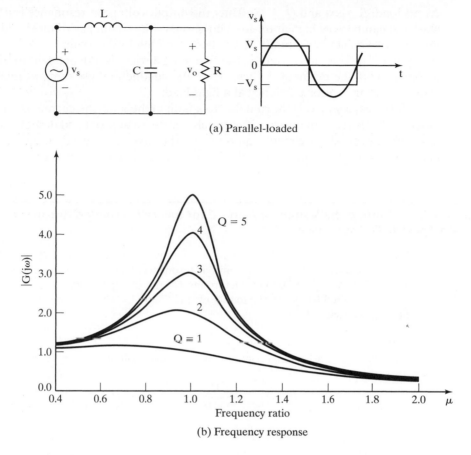

(a) Parallel-loaded

(b) Frequency response

FIGURE 8.13

Frequency response for series–parallel loaded.

Let $\omega_0 = 1/\sqrt{LC}$ be the resonant frequency, and $Q = 1/Q_s = R/\omega_o L$ be the quality factor. Substituting L, C, and R in terms of Q and ω_o, we get

$$G(j\omega) = \frac{V_o}{V_i}(j\omega) = \frac{1}{[1 - (\omega/\omega_o)^2] + j(\omega/\omega_o)/Q} = \frac{1}{(1 - u^2) + ju/Q}$$

where $u = \omega/\omega_o$. The magnitude of $G(j\omega)$ can be found from

$$|G(j\omega)| = \frac{1}{[(1 - u^2)^2 + (u/Q)^2]^{1/2}} \tag{8.36}$$

Figure 8.13b shows the magnitude plot of the voltage gain in Eq. (8.36) for $Q = 1$ to 5. The maximum gain occurs near resonance for $Q > 2$, and its value for $u = 1$ is

$$|G(j\omega)|_{\max} = Q \tag{8.37}$$

At no-load, $R = \infty$ and $Q = \infty$. Thus, the output voltage at resonance is a function of load and can be very high at no-load if the operating frequency is not raised. However, the output voltage is normally controlled at no-load by varying the frequency above resonance. The current carried by the switching devices is independent to the load, but increases with the dc input voltage. Thus, the conduction loss remains relatively constant, resulting in poor efficiency at a light load.

If the capacitor C is shorted due to a fault in the load, the current is limited by the inductor L. This type of inverter is naturally short-circuit proof and desirable for applications with severe short-circuit requirements. This inverter is mostly used in low-voltage, high-current applications, where the input voltage range is relatively narrow, typically up to $\pm 15\%$.

Example 8.7   Finding the Values of L and C for a Parallel-Loaded Resonant Inverter to Yield Specific Output Power

A series resonance inverter with parallel-loaded delivers a load power of $P_L = 1$ kW at a peak sinusoidal load voltage of $V_p = 330$ V and at resonance. The load resistance is $R = 10\ \Omega$. The resonant frequency is $f_0 = 20$ kHz. Determine (a) the dc input voltage V_s, (b) the frequency ratio u if it is required to reduce the load power to 250 W by frequency control, (c) the inductor L, and (d) the capacitor C.

Solution

a. The peak fundamental component of a square voltage is $V_p = 4V_s/\pi$.

$$P_L = \frac{V_p^2}{2R} = \frac{4^2 V_s^2}{2\pi^2 R} \quad \text{or} \quad 1000 = \frac{4^2 V_s^2}{2\pi^2 \times 10}$$

which gives $V_s = 110$ V. $V_{i(pk)} = 4V_s/\pi = 4 \times 110/\pi = 140.06$ V.

b. From Eq. (8.37), the quality factor is $Q = V_p/V_{i(pk)} = 330/140.06 = 2.356$. To reduce the load power by $(1000/250 =) 4$, the voltage gain must be reduced by 2. That is, from Eq. (8.36), we get

$$(1 - u^2)^2 + (u/2.356)^2 = 2^2$$

which gives $u = 1.693$.

c. Q is defined by

$$Q = \frac{R}{\omega_0 L} \quad \text{or} \quad 2.356 = \frac{R}{2\pi \times 20\ \text{kHz}\ L}$$

which gives $L = 33.78\ \mu\text{H}$.

d. $f_0 = 1/2\pi\sqrt{LC}$ or 20 kHz $= 1/2\pi\sqrt{(33.78\ \mu H \times C)}$, which gives $C = 1.875\ \mu\text{F}$.

8.3.3   Frequency Response for Series–Parallel Loaded

In Figure 8.10 the capacitor $C_1 = C_2 = C_s$ forms a series circuit and the capacitor C is in parallel with the load. This circuit is a compromise between the characteristics of a

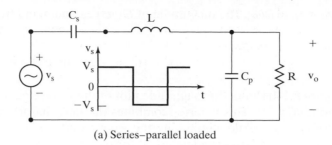

(a) Series–parallel loaded

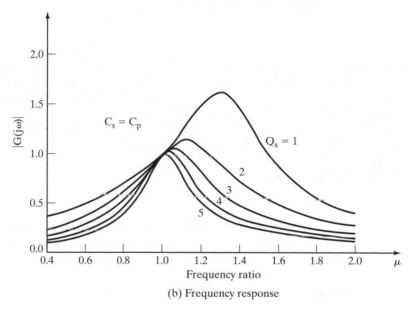

(b) Frequency response

FIGURE 8.14

Frequency response for series–parallel loaded.

series load and a parallel load. The equivalent circuit is shown in Figure 8.14a. Using the voltage-divider rule in frequency domain, the voltage gain is given by

$$G(j\omega) = \frac{V_o}{V_i}(j\omega) = \frac{1}{1 + C_p/C_s - \omega^2 LC_p + j\omega L/R - j/(\omega C_s R)}$$

Let $\omega_0 = 1/\sqrt{LC_s}$ be the resonant frequency, and $Q_s = \omega_0 L/R$ be the quality factor. Substituting L, C, and R in terms of Q_s and ω_0, we get

$$G(j\omega) = \frac{V_0}{V_i}(j\omega) = \frac{1}{1 + C_p/C_s - \omega^2 LC_p + jQ_s(\omega/\omega_0 - \omega_0/\omega)}$$

$$= \frac{1}{1 + (C_p/C_s)(1 - u^2) + jQ_s(u - 1/u)}$$

where $u = \omega/\omega_o$. The magnitude of $G(j\omega)$ can be found from

$$|G(j\omega)| = \frac{1}{\{[1 + (C_p/C_s)(1 - u^2)]^2 + Q_s^2(u - 1/u)^2\}^{1/2}} \qquad (8.38)$$

Figure 8.14b shows the magnitude plot of the voltage gain in Eq. (8.38) for $Q_s = 1$ to 5 and $C_p/C_s = 1$. This inverter combines the best characteristics of the series load and parallel load, while eliminating the weak points such as lack of regulation for series load and load current independent for parallel load.

As C_p gets smaller and smaller, the inverter exhibits the characteristics of series load. With a reasonable value of C_p, the inverter exhibits some of the characteristics of parallel load and can operate under no-load. As C_p becomes smaller, the upper frequency needed for a specified output voltage increases. Choosing $C_p = C_s$ is generally a good compromise between the efficiency at partial load and the regulation at no-load with a reasonable upper frequency. For making the current to decrease with the load to maintain a high efficiency at partial load, the full-load Q is chosen between 4 and 5. An inverter with series–parallel loaded can run over a wider input voltage and load ranges from no-load to full load, while maintaining excellent efficiency.

Key Points of Section 8.3

- The gain of a resonant inverter becomes maximum at $u = 1$. The resonant inverters are normally used in applications requiring a fixed output voltage.
- The series-loaded inverter is most suitable for high-voltage, low-current applications. The parallel-loaded inverter is mostly used in low-voltage, high-current applications. The series–parallel-loaded inverter can run over a wider input voltage and load ranges from no-load to full load.

8.4   PARALLEL RESONANT INVERTERS

A parallel resonant inverter is the dual of a series resonant inverter. It is supplied from a current source so that the circuit offers a high impedance to the switching current. A parallel resonant circuit is shown in Figure 8.15. Because the current is continuously controlled, this inverter gives a better short-circuit protection under fault conditions. Summing the currents through R, L, and C, gives

$$C\frac{dv}{dt} + \frac{v}{R} + \frac{1}{L}\int v\, dt = I_s$$

with initial condition $v(t = 0) = 0$ and $i_L(t = 0) = 0$. This equation is similar to Eq. (8.2) if i is replaced by v, R by $1/R$, L by C, C by L, and V_s by I_s. Using Eq. (8.5), the voltage v is given by

$$v = \frac{I_s}{\omega_r C} e^{-\alpha t} \sin \omega_r t \qquad (8.39)$$

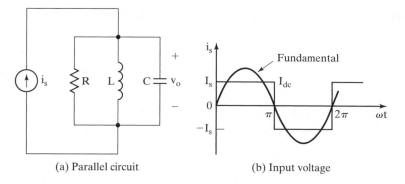

(a) Parallel circuit   (b) Input voltage

FIGURE 8.15

Parallel resonant circuit.

where $\alpha = 1/2RC$. The damped-resonant frequency ω_r is given by

$$\omega_r = \left(\frac{1}{LC} - \frac{1}{4R^2C^2}\right)^{1/2} \tag{8.40}$$

Using Eq. (8.7), the voltage v in Eq. (8.39) becomes maximum at t_m given by

$$t_m = \frac{1}{\omega_r}\tan^{-1}\frac{\omega_r}{\alpha} \tag{8.41}$$

which can be approximated to π/ω_r. The input impedance is given by

$$Z(j\omega) = \frac{V_o}{I_i}(j\omega) = R\,\frac{1}{1 + jR/\omega L + j\omega CR}$$

where I_i is the rms ac input current, and $I_i = 4I_s/\sqrt{2}\pi$. The quality factor Q_p is

$$Q_p = \omega_0 CR = \frac{R}{\omega_0 L} = R\sqrt{\frac{C}{L}} = 2\delta \tag{8.42}$$

when δ is the damping factor and $\delta = \alpha/\omega_0 = (R/2)\sqrt{C/L}$. Substituting L, C, and R in terms of Q_p and ω_0, we get

$$Z(j\omega) = \frac{V_o}{I_i}(j\omega) = \frac{1}{1 + jQ_p(\omega/\omega_0 - \omega_0/\omega)} = \frac{1}{1 + jQ_p(u - 1/u)}$$

where $u = \omega/\omega_o$. The magnitude of $Z(j\omega)$ can be found from

$$|Z(j\omega)| = \frac{1}{[1 + Q_p^2(u - 1/u)^2]^{1/2}} \tag{8.43}$$

which is identical to the voltage gain $|G(j\omega)|$ in Eq. (8.35). The magnitude gain plot is shown in Figure 8.12. A parallel resonant inverter is shown in Figure 8.16a. Inductor L_e acts as a current source and capacitor C is the resonating element. L_m is the mutual inductance of the transformer and acts as the resonating inductor. A constant current is

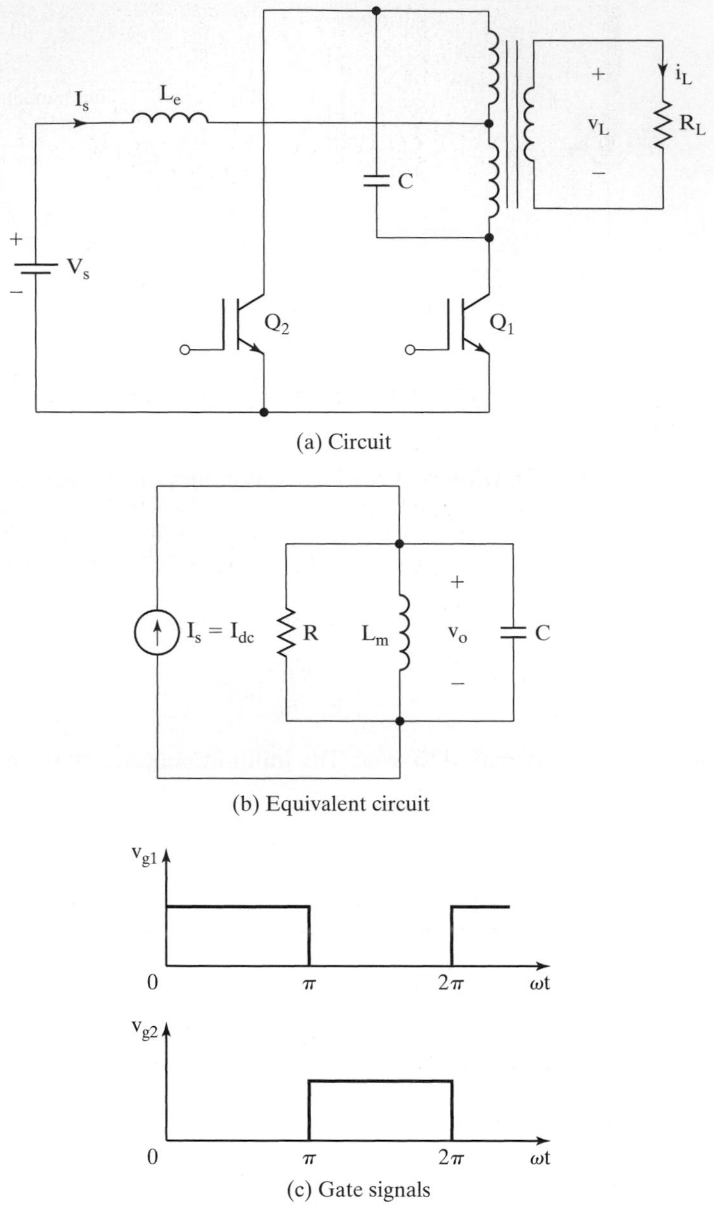

(a) Circuit

(b) Equivalent circuit

(c) Gate signals

FIGURE 8.16

Parallel resonant inverter.

switched alternatively into the resonant circuit by transistors Q_1 and Q_2. The gating signals are shown in Figure 8.16c. Referring the load resistance R_L into the primary side and neglecting the leakage inductances of the transformer, the equivalent circuit is shown in Figure 8.16b. A practical resonant inverter that supplies a fluorescent lamp is shown in Figure 8.17.

FIGURE 8.17

Practical resonant inverter. (Courtesy of Magnetek, Inc.)

Example 8.8 Finding the Values of L and C for a Parallel Resonant Inverter to Yield Specific Output Power

The parallel resonant inverter of Figure 8.16a delivers a load power of $P_L = 1$ kW at a peak sinusoidal load voltage of $V_p = 170$ V and at resonance. The load resistance is $R = 10\ \Omega$. The resonant frequency is $f_0 = 20$ kHz. Determine (a) the dc input current I_s, (b) the quality factor Q_p if it is required to reduce the load power to 250 W by frequency control so that $u = 1.25$, (c) the inductor L, and (d) the capacitor C.

Solution

a. Because at resonance $u = 1$ and $|Z(j\omega)|_{max} = 1$, the peak fundamental load current is $I_p = 4I_s/\pi$.

$$P_L = \frac{I_p^2 R}{2} = \frac{4^2 I_s^2 R}{2\pi^2} \quad \text{or} \quad 1000 = \frac{4^2 I_s^2 10}{2\pi^2}$$

which gives $I_s = 11.1$ A.

b. To reduce the load power by $(1000/250 =)\ 4$, the impedance must be reduced by 2 at $u = 1.25$. That is, from Eq. (8.43), we get $1 + Q_p^2(u - 1/u)^2 = 2^2$, which gives $Q_p = 3.85$.

c. Q_p is defined by $Q_p = \omega_0 CR$ or $3.85 = 2\pi \times 20$ kHz $\times C \times 10$, which gives $C = 3.06\ \mu$F.

d. $f_o = 1/2\pi\sqrt{LC}$ or 20 kHz $= 1/[2\pi\sqrt{(3.06\ \mu F \times L)}]$, which gives $L = 20.67\ \mu$H.

Key Points of Section 8.4

- A parallel resonant inverter is a dual of a series resonant inverter. A constant current is switched alternately into the resonant circuit and the load current become almost independent of the load impedance variations.

8.5 VOLTAGE CONTROL OF RESONANT INVERTERS

The quasi-resonant inverters (QRIs) [3] are normally used for output voltage control. QRIs can be considered as a hybrid of resonant and PWM converters. The underlying principle is to replace the power switch in PWM converters with the resonant switch. The switch current or voltage waveforms are forced to oscillate in a quasi-sinusoidal

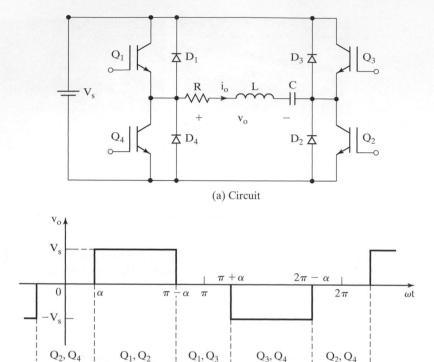

(a) Circuit

(b) Output voltage

FIGURE 8.18

Quasi-square voltage control for series resonant inverter.

manner. A large family of conventional converter circuits can be transformed into their resonant converter counterparts [4].

A bridge topology, as shown in Figure 8.18a, can be applied to achieve the output voltage control. The switching frequency f_s is kept constant at the resonant frequency f_o. By switching two devices simultaneously a *quasi-square wave*, as shown in Figure 8.18b, can be obtained. The rms fundamental input voltage is given by

$$V_i = \frac{4V_s}{\sqrt{2}\pi} \cos \alpha \qquad (8.44)$$

where α is the control angle. By varying α from 0 to $\pi/2$ at a constant frequency, the voltage V_i can be controlled from $4V_s/(\pi\sqrt{2})$ to 0.

The bridge topology in Figure 8.19a can control the output voltage. The switching frequency f_s is kept constant at the resonant frequency f_o. By switching two devices simultaneously a *quasi-square wave* as shown in Figure 8.19b can be obtained. The rms fundamental input current is given by

$$I_i = \frac{4I_s}{\sqrt{2}\pi} \cos \alpha \qquad (8.45)$$

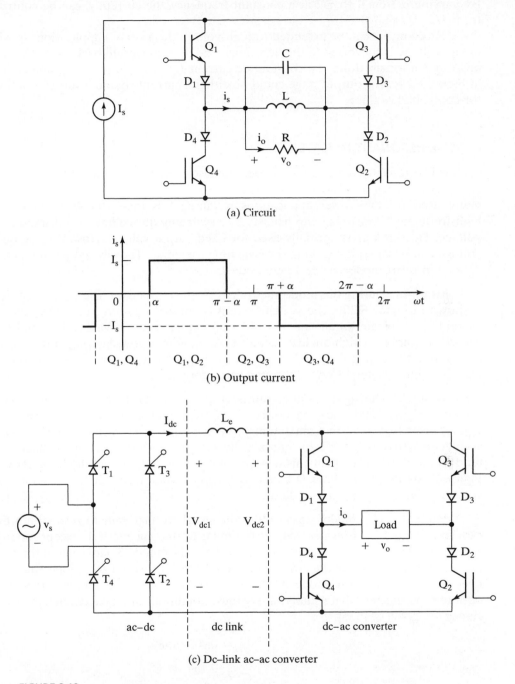

(a) Circuit

(b) Output current

(c) Dc–link ac–ac converter

FIGURE 8.19

Quasi-square current control for parallel resonant inverter.

By varying α from 0 to $\pi/2$ at a constant frequency, the current I_i can be controlled from $4I_s/(\sqrt{2}\pi)$ to 0.

This concept can be extended to high-voltage dc (HVDC) applications in which the ac voltage is converted to the dc voltage and then converted back to ac. The transmission is normally done at a constant dc current I_{dc}. A single-phase version is shown in Figure 8.19c. The output stage could be either a current-source inverter or a thyristor-controlled rectifier.

8.6   CLASS E RESONANT INVERTER

A class E resonant inverter uses only one transistor and has low-switching losses, yielding a high efficiency of more than 95%. The circuit is shown in Figure 8.20a. It is normally used for low-power applications requiring less than 100 W, particularly in high-frequency electronic lamp ballasts. The switching device has to withstand a high voltage. This inverter is normally used for fixed output voltage. However, the output voltage can be varied by varying the switching frequency. The circuit operation can be divided into two modes: mode 1 and mode 2.

Mode 1.   During this mode, the transistor Q_1 is turned on. The equivalent circuit is shown in Figure 8.20b. The switch current i_T consists of source current i_s and load current i_o. To obtain an almost sinusoidal output current, the values of L and C are chosen to have a high-quality factor, $Q \geq 7$, and a low-damping ratio, usually $\delta \leq 0.072$. The switch is turned off at zero voltage. When the switch is turned off, its current is immediately diverted through capacitor C_e.

Mode 2.   During this mode, transistor Q_1 is turned off. The equivalent circuit is shown in Figure 8.20b. The capacitor current i_c becomes the sum of i_s and i_o. The switch voltage rises from zero to a maximum value and falls to zero again. When the switch voltage falls to zero, $i_c = C_e \, dv_T/dt$ normally is negative. Thus, the switch voltage would tend to be negative. To limit this negative voltage, an antiparallel diode, as shown in Figure 8.20a by dashed lines, is connected. If the switch is an MOSFET, its negative voltage is limited by its built-in diode to a diode drop.

Mode 3.   This mode exists only if the switch voltage falls to zero with a finite negative slope. The equivalent circuit is similar to that for mode 1, except the initial conditions. The load current falls to zero at the end of mode 3. However, if the circuit parameters are such that the switch voltage falls to zero with a zero slope, there is no need for a diode and this mode would not exist. That is, $v_T = 0$, and $dv_T/dt = 0$. The optimum parameters that usually satisfy these conditions and give the maximum efficiency are given by [5, 6]:

$$L_e = 0.4001R/\omega_s$$

$$C_e = \frac{2.165}{R\omega_s}$$

$$\omega_s L - \frac{1}{\omega_s C} = 0.3533R$$

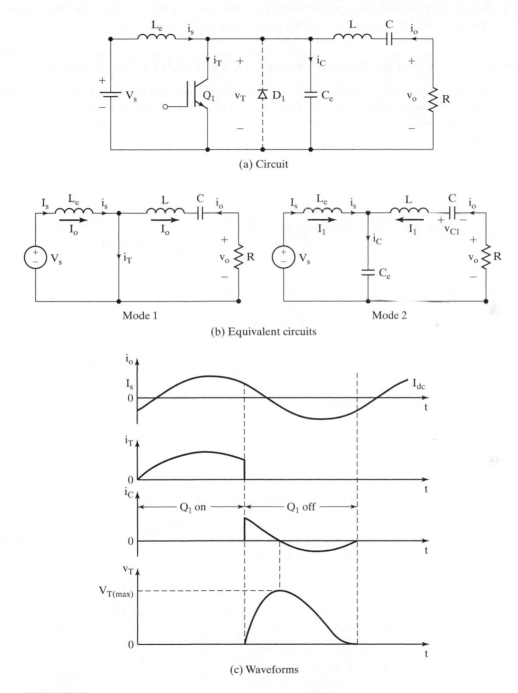

(a) Circuit

Mode 1   Mode 2

(b) Equivalent circuits

(c) Waveforms

FIGURE 8.20

Class E resonant inverter.

where ω_s is the switching frequency. The duty cycle is $k = t_{on}/T_s = 30.4\%$. The waveforms of the output current, switch current, and switch voltage are shown in Figure 8.20c.

Example 8.9   Finding the Optimum Values of C's and L's for a Class E Inverter

The class E inverter of Figure 8.20a operates at resonance and has $V_s = 12$ V and $R = 10\ \Omega$. The switching frequency is $f_s = 25$ kHz. (a) Determine the optimum values of L, C, C_e, and L_e. (b) Use PSpice to plot the output voltage v_o and the switch voltage v_T for $k = 0.304$. Assume that $Q = 7$.

Solution
$V_s = 12$ V, $R = 10\ \Omega$, and $\omega_s = 2\pi f_s = 2\pi \times 25$ kHz $= 157.1$ krad/s.

a.

$$L_e = \frac{0.4001R}{\omega_s} = 0.4001 \times \frac{10}{157.1\ \text{krad/s}} = 25.47\ \mu\text{H}$$

$$C_e = \frac{2.165}{R\omega_s} = \frac{2.165}{10 \times 157.1\ \text{krad/s}} = 1.38\ \mu\text{F}$$

$$L = \frac{QR}{\omega_s} = \frac{7 \times 10}{157.1\ \text{krad/s}} = 445.63\ \mu\text{H}$$

$\omega_s L - 1/\omega_s C = 0.3533\ R$ or $7 \times 10 - 1/\omega_s C = 0.3533 \times 10$, which gives $C = 0.0958\ \mu$F. The damping factor is

$$\delta = (R/2)\sqrt{C/L} = (10/2)\sqrt{0.0958/445.63} = 0.0733$$

which is very small, and the output current should essentially be sinusoidal. The resonant frequency is

$$f_0 = \frac{1}{2\pi\sqrt{LC}} = \frac{1}{2\pi\sqrt{(445.63\ \mu\text{H} \times 0.0958\ \mu\text{F})}} = 24.36\ \text{kHz}$$

b. $T_s = 1/f_s = 1/25$ kHz $= 40\ \mu$s, and $t_{on} = kT_s = 0.304 \times 40 = 12.24\ \mu$s. The circuit for PSpice simulation is shown in Figure 8.21a and the control voltage in Figure 8.21b. The list of the circuit file is as follows:

```
Example 8.9      Class  E Resonant Inverter
VS     1    0    DC      12V
VY     1    2    DC      0V    , Voltage source to measure input current
VG     8    0    PULSE (0V 20V    0     1NS      1NS      12.24US     40US)
RB     8    7    250            ; Transistor base-drive resistance
R      6    0    10
LE     2    3    25.47UH
CE     3    0    1.38UF
C      3    4    0.0958UF
L      5    6    445.63UH
VX     4    5    DC      0V    ; Voltage source to measure load current of L2
Q1     3    7    0    MODQ1                      ; BJT switch
```

```
.MODEL MODQ1 NPN (IS=6.734F BF=416.4 ISE=6.734F BR=.7371
+      CJE=3.638P  MJC=.3085 VJC=.75 CJE=4.493P MJE=.2593 VJE=.75
+      TR=239.5N   TF=301.2P)            ; Transistor model parameters
.TRAN    2US        300US    180US  1US UIC    ; Transient analysis
.PROBE                                   ; Graphics postprocessor
.OPTIONS ABSTOL = 1.00N RELTOL = 0.01 VNTOL = 0.1 ITL5=20000 ; convergence
.END
```

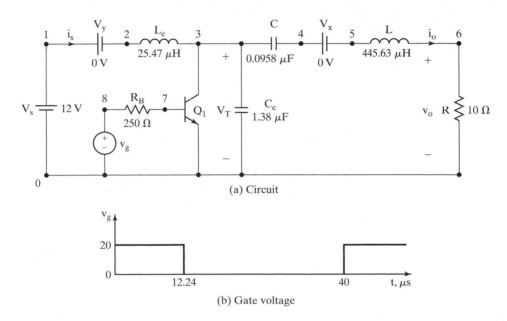

(a) Circuit

(b) Gate voltage

FIGURE 8.21

Class E resonant inverter for PSpice simulation.

The PSpice plots are shown in Figure 8.22, where $V(3)$ = switch voltage and $V(6)$ = output voltage. Using the PSpice cursor in Figure 8.22 gives $V_{o(pp)} = 29.18$ V, $V_{T(peak)} = 31.481$ V, and the output frequency $f_o = 1/(2 \times 19.656 \, \mu) = 25.44$ kHz (expected 24.36 kHz).

Key Points of Section 8.6

- A class E inverter that requires only one switching device is suitable for low-power applications requiring less than 100 W. It is normally used for fixed output voltage.

8.7   CLASS E RESONANT RECTIFIER

Because dc–dc converters generally consist of a dc–ac resonant inverter and dc–ac rectifier, a high-frequency diode rectifier suffers from disadvantages such as conduction and switching losses, parasitic oscillations, and high harmonic content of the input current. A

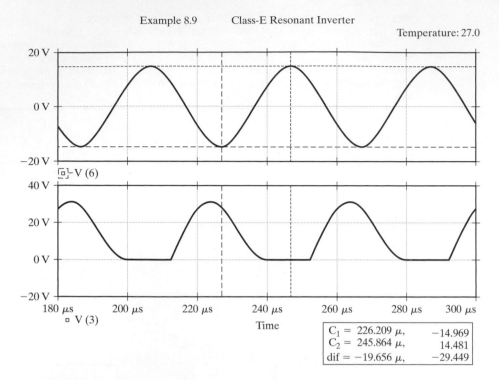

Example 8.9   Class-E Resonant Inverter

Temperature: 27.0

C$_1$ = 226.209 μ,	−14.969
C$_2$ = 245.864 μ,	14.481
dif = −19.656 μ,	−29.449

FIGURE 8.22

PSpice plots for Example 8.9.

class E resonant rectifier [7], as shown in Figure 8.23a, overcomes these limitations. It uses the principle of zero-voltage switching of the diode. That is, the diode turns off at zero voltage. The diode junction capacitance is included in the resonant capacitance C and therefore does not adversely affect the circuit operation. The circuit operation can be divided into two modes: mode 1 and mode 2. Let us assume that C_f is sufficiently large so that the dc output voltage V_o is constant. Let the input voltage $v_s = V_m \sin \omega t$.

Mode 1.   During this mode, the diode is off. The equivalent circuit is shown in Figure 8.23b. The values of L and C are such that $\omega L = 1/\omega C$ at the operating frequency f. The voltage appearing across L and C is $v_{(LC)} = V_s \sin \omega t - V_o$.

Mode 2.   During this mode, the diode is on. The equivalent circuit is shown in Figure 8.23b. The voltage appearing across L is $v_L = V_s \sin \omega t - V_o$. When the diode current i_D, which is the same as the inductor current i_L, reaches zero, the diode turns off. At turn-off, $i_D = i_L = 0$ and $v_D = v_C = 0$. That is, $i_c = C \, dv_c/dt = 0$, which gives $dv_c/dt = 0$. Therefore, the diode voltage is zero at turn-off, thereby reducing the switching losses. The inductor current can be expressed approximately by

$$i_L = I_m \sin(\omega t - \phi) - I_o \tag{8.46}$$

where $I_m = V_m/R$ and $I_o = V_o/R$. When the diode is on, the phase shift ϕ is 90°. When the diode is off, it is 0°, provided that $\omega L = 1/\omega C$. Therefore, ϕ has a value between 0° and 90°, and its value depends on the load resistance R. The peak-to-peak current is $2V_m/R$.

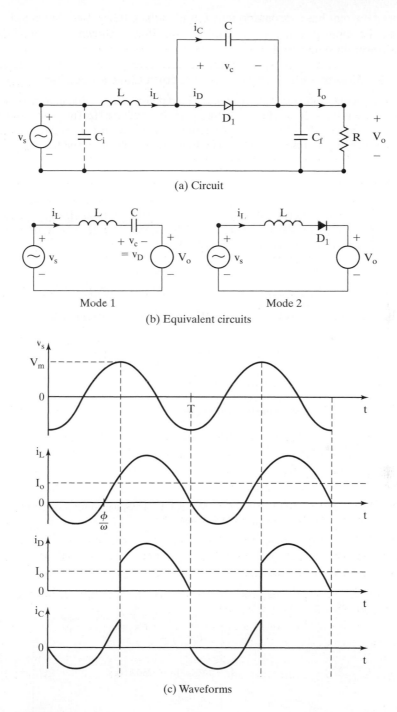

(a) Circuit

(b) Equivalent circuits

(c) Waveforms

FIGURE 8.23

Class E resonant rectifier.

The input current has a dc component I_o and a phase delay ϕ as shown in Figure 8.23c. To improve the input power factor, an input capacitor, as shown in Figure 8.23a by dashed lines, is normally connected.

Example 8.10 Finding the Values of L's and C's for a Class E Rectifier

The class E rectifier of Figure 8.23a supplies a load power of $P_L = 400$ mW at $V_o = 4$ V. The peak supply voltage is $V_m = 10$ V. The supply frequency is $f = 250$ kHz. The peak-to-peak ripple on the dc output voltage is $\Delta V_o = 40$ mV. (a) Determine the values of L, C, and C_f; and (b) the rms and dc currents of L and C. (c) Use PSpice to plot the output voltage v_o and the inductor current i_L.

Solution
$V_m = 10$ V, $V_o = 4$V, $\Delta V_o = 40$ mV, and $f = 250$ kHz.

a. Choose a suitable value of C. Let $C = 10$ nF. Let the resonant frequency be $f_o = f = 250$ kHz. 250 kHz $= f_o = 1/[2\pi\sqrt{(L \times 10 \text{ nF})}]$, which gives $L = 40.5$ μH. $P_L = V_o^2/R$ or 400 mW $= 4^2/R$, which gives $R = 40$ Ω. $I_o = V_o/R = 4/40 = 100$ mA. The value of capacitance C_f is given by

$$C_f = \frac{I_o}{2f\,\Delta V_o} = \frac{100 \text{ mA}}{2 \times 250 \text{ kHz} \times 40 \text{ mV}} = 5 \text{ μF}$$

b. $I_m = V_m/R = 10/40 = 250$ mA. The rms inductor current I_L is

$$I_{L(\text{rms})} = \sqrt{100^2 + \frac{250^2}{2}} = 203.1 \text{ mA}$$

$$I_{L(\text{dc})} = 100 \text{ mA}$$

The rms current of the capacitor C is

$$I_{C(\text{rms})} = \frac{250}{\sqrt{2}} = 176.78 \text{ mA}$$

$$I_{C(\text{dc})} = 0$$

c. $T = 1/f = 1/250$ kHz $= 4$ μs. The circuit for PSpice simulation is shown in Figure 8.24. The list of the circuit file is as follows:

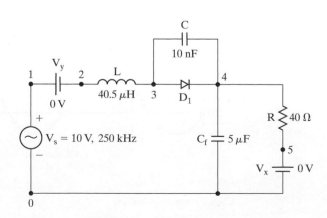

FIGURE 8.24

Class E resonant rectifier for PSpice simulation.

```
Example 8.10        Class E Resonant Rectifier
VS      1    0     SIN (0    10V     250KHZ)
VY      1    2     DC    0V    ; Voltage source to measure input current
R       4    5     40
L       2    3     40.5UH
C       3    4     10NF
CF      4    0     5UF
VX      5    0     DC    0V    ; Voltage source to measure current through R
D1      3    4     DMOD                      ; Rectifier diode
.MODEL DMOD D                                ; Diode default parameters
.TRAN 0.1US     1220US   1200US   0.1US  UIC  ; Transient analysis
.PROBE                                       ; Graphics postprocessor
.OPTIONS ABSTOL = 1.00N RETOL1 = 0.01 VNTOL = 0.1 ITL5=40000 ; convergence
.END
```

The PSpice plot is shown in Figure 8.25, where $I(L)$ = inductor current and $V(4)$ = output voltage. Using the PSpice cursor in Figure 8.25 gives $V_o = 3.98$ V, $\Delta V_o = 63.04$ mV, and $i_{L(\text{pp})} = 489.36$ mA.

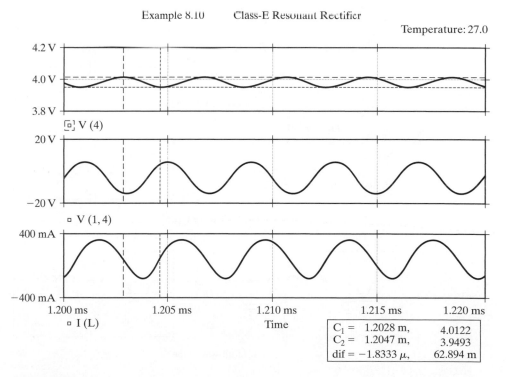

Example 8.10   Class-E Resonant Rectifier

FIGURE 8.25

PSpice plots for Example 11.10.

Key Points of Section 8.7

- A class E rectifier uses only one diode that is turned off at zero voltage. The diode conduction loss is reduced and the input current harmonic is low.

8.8 ZERO-CURRENT-SWITCHING RESONANT CONVERTERS

The switches of a zero-current-switching (ZCS) resonant converter turn on and off at zero current. The resonant circuit that consists of switch S_1, inductor L, and capacitor C is shown in Figure 8.26a. Inductor L is connected in series with a power switch S_1 to achieve ZCS. It is classified by Liu et al. [8] into two types: L type and M type. In both types, the inductor L limits the di/dt of the switch current, and L and C constitute a series-resonant circuit. When the switch current is zero, there is a current $i = C_j \, dv_T/dt$ flowing through the internal capacitance C_j due to a finite slope of the switch voltage at turn-off. This current flow causes power dissipation in the switch and limits the high-switching frequency.

The switch can be implemented either in a half-wave configuration as shown in Figure 8.26b, where diode D_1 allows unidirectional current flow, or in a full-wave configuration as shown in Figure 8.26c, where the switch current can flow bidirectionally.

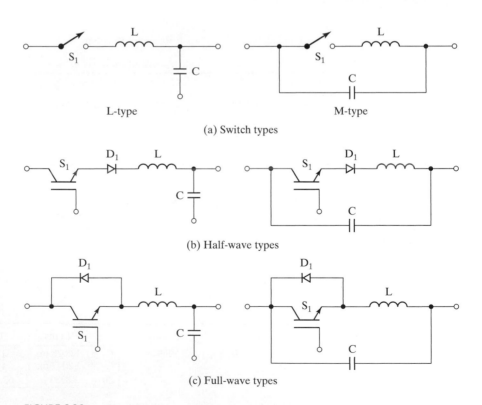

(a) Switch types

(b) Half-wave types

(c) Full-wave types

FIGURE 8.26

Switch configurations for ZCS resonant converters.

The practical devices do not turn off at zero current due to their recovery times. As a result, an amount of energy can be trapped in the inductor L of the M-type configuration, and voltage transients appear across the switch. This favors L-type configuration over the M-type one.

8.8.1   L-Type ZCS Resonant Converter

An L-type ZCS resonant converter is shown in Figure 8.27a. The circuit operation can be divided into five modes, whose equivalent circuits are shown in Figure 8.27b. We shall redefine the time origin, $t = 0$, at the beginning of each mode.

   Mode 1.   This mode is valid for $0 \le t \le t_1$. Switch S_1 is turned on and diode D_m conducts. The inductor current i_L, which rises linearly, is given by

$$i_L = \frac{V_s}{L} t \tag{8.47}$$

This mode ends at time $t = t_1$ when $i_L(t = t_1) = I_o$. That is, $t_1 = I_o L/V_s$.

   Mode 2.   This mode is valid for $0 \le t \le t_2$. Switch S_1 remains on, but diode D_m is off. The inductor current i_L is given by

$$i_L = I_m \sin \omega_0 t + I_o \tag{8.48}$$

where $I_m = V_s\sqrt{C/L}$, and $\omega_0 = 1/\sqrt{LC}$. The capacitor voltage v_c is given by

$$v_c = V_s(1 - \cos \omega_0 t)$$

The peak switch current, which occurs at $t = (\pi/2)\sqrt{LC}$, is

$$I_p = I_m + I_o$$

The peak capacitor voltage is

$$V_{c(\text{pk})} = 2V_s$$

This mode ends at $t = t_2$ when $i_L(t = t_2) = I_o$, and $v_c(t = t_2) = V_{c2} = 2V_s$. Therefore, $t_2 = \pi\sqrt{LC}$.

   Mode 3.   This mode is valid for $0 \le t \le t_3$. The inductor current that falls from I_o to zero is given by

$$i_L = I_o - I_m \sin \omega_0 t \tag{8.49}$$

The capacitor voltage is given by

$$v_c = 2V_s \cos \omega_0 t \tag{8.50}$$

This mode ends at $t = t_3$ when $i_L(t = t_3) = 0$ and $v_c(t = t_3) = V_{c3}$. Thus, $t_3 = \sqrt{LC} \sin^{-1}(1/x)$ where $x = I_m/I_o = (V_s/I_o)\sqrt{C/L}$.

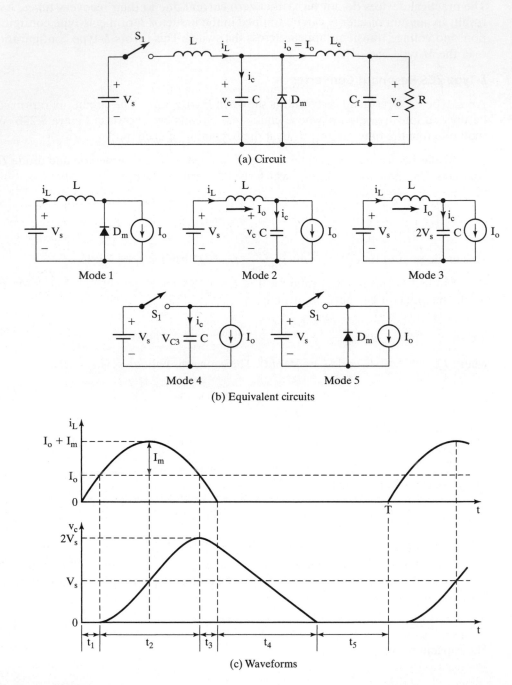

(a) Circuit

(b) Equivalent circuits

(c) Waveforms

FIGURE 8.27

L-type ZCS resonant converter.

Mode 4.   This mode is valid for $0 \leq t \leq t_4$. The capacitor supplies the load current I_o, and its voltage is given by

$$v_c = V_{c3} - \frac{I_o}{C} t \tag{8.51}$$

This mode ends at time $t = t_4$ when $v_c(t = t_4) = 0$. Thus, $t_4 = V_{c3}C/I_o$.

Mode 5.   This mode is valid for $0 \leq t \leq t_5$. When the capacitor voltage tends to be negative, the diode D_m conducts. The load current I_o flows through the diode D_m. This mode ends at time $t = t_5$ when the switch S_1 is turned on again, and the cycle is repeated. That is, $t_5 = T - (t_1 + t_2 + t_3 + t_4)$.

The waveforms for i_L and v_c are shown in Figure 8.27c. The peak switch voltage equals to the dc supply voltage V_s. Because the switch current is zero at turn-on and turn-off, the switching loss, which is the product of v and i, becomes very small. The peak resonant current I_m must be higher than the load current I_o, and this sets a limit on the minimum value of load resistance R. However, by placing an antiparallel diode across the switch, the output voltage can be made insensitive to load variations.

Example 8.11   Finding the Values of *L* and *C* for a Zero-Current-Switching Inverter

The ZCS resonant converter of Figure 8.27a delivers a maximum power of $P_L = 400$ mW at $V_o = 4$ V. The supply voltage is $V_s = 12$ V. The maximum operating frequency is $f_{max} = 50$ kHz. Determine the values of L and C. Assume that the intervals t_1 and t_3 are very small, and $x = 1.5$.

Solution
$V_s = 12$ V, $f = f_{max} = 50$ kHz, and $T = 1/50$ kHz $= 20$ μs. $P_L = V_o I_o$ or 400 mW $= 4I_o$, which gives $I_o = 100$ mA. The maximum frequency occurs when $t_5 = 0$. Because $t_1 = t_3 = t_5 = 0$, $t_2 + t_4 = T$. Substituting $t_4 = 2V_sC/I_m$ and using $x = (V_s/I_o)\sqrt{C/L}$ gives

$$\pi\sqrt{LC} + \frac{2V_sC}{I_o} = T \qquad \text{or} \qquad \frac{\pi V_s}{xI_o}C + \frac{2V_s}{I_o}C = T$$

which gives $C = 0.0407$ μF. Thus, $L = (V_s/xI_o)^2 C = 260.52$ μH.

8.8.2   *M*-Type ZCS Resonant Converter

An *M*-type ZCS resonant converter is shown in Figure 8.28a. The circuit operation can be divided into five modes, whose equivalent circuits are shown in Figure 8.28b. We shall redefine the time origin, $t = 0$, at the beginning of each mode. The mode equations are similar to those of an *L*-type converter, except the following.

Mode 2.   The capacitor voltage v_c is given by

$$v_c = V_s \cos \omega_0 t \tag{8.52}$$

The peak capacitor voltage is $V_{c(pk)} = V_s$. At the end of this mode at $t = t_2$, $v_c(t = t_2) = V_{c2} = -V_s$.

Mode 3.   The capacitor voltage is given by

$$v_c = -V_s \cos \omega_0 t \tag{8.53}$$

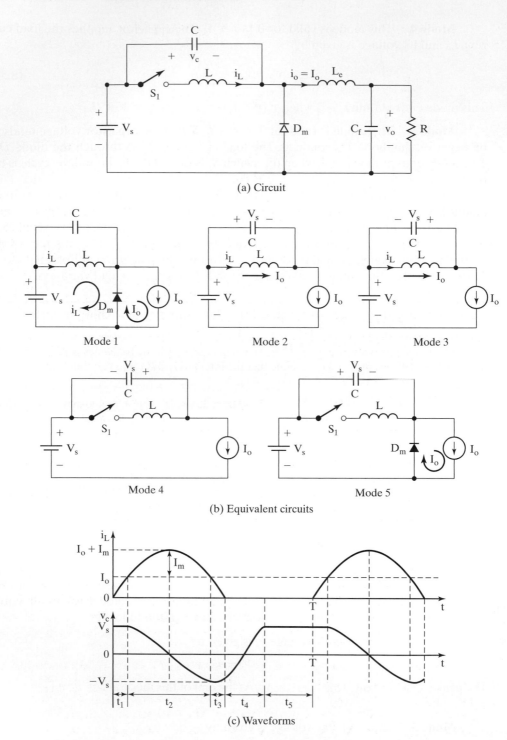

(a) Circuit

(b) Equivalent circuits

(c) Waveforms

FIGURE 8.28

M-type ZCS resonant converter.

At the end of this mode at $t = t_3$, $v_c(t = t_3) = V_{c3}$. It should be noted that V_{c3} can have a negative value.

Mode 4.   This mode ends at $t = t_4$ when $v_c(t = t_4) = V_s$. Thus, $t_4 = (V_s - V_{c3})C/I_o$. The waveforms for i_L and v_c are shown in Figure 8.28c.

Key Points of Section 8.8

- A zero-current (ZC) switch shapes the switch current waveform during its conduction time to create a ZC condition for the switch to turn-off.

8.9   ZERO-VOLTAGE-SWITCHING RESONANT CONVERTERS

The switches of a ZVS resonant converter turn on and off at zero voltage [9]. The resonant circuit is shown in Figure 8.29a. The capacitor C is connected in parallel with the switch S_1 to achieve ZVS. The internal switch capacitance C_j is added with the capacitor C, and it affects the resonant frequency only, thereby contributing no power dissipation in the switch. If the switch is implemented with a transistor Q_1 and an antiparallel diode D_1, as shown in Figure 8.29b, the voltage across C is clamped by D_1, and the switch is operated in a half-wave configuration. If the diode D_1 is connected in

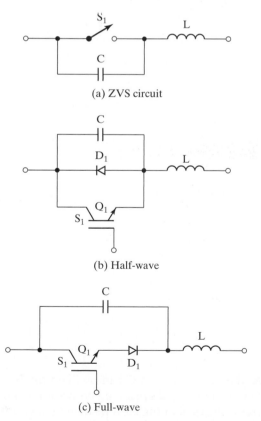

(a) ZVS circuit

(b) Half-wave

(c) Full-wave

FIGURE 8.29

Switch configurations for ZVS resonant converters.

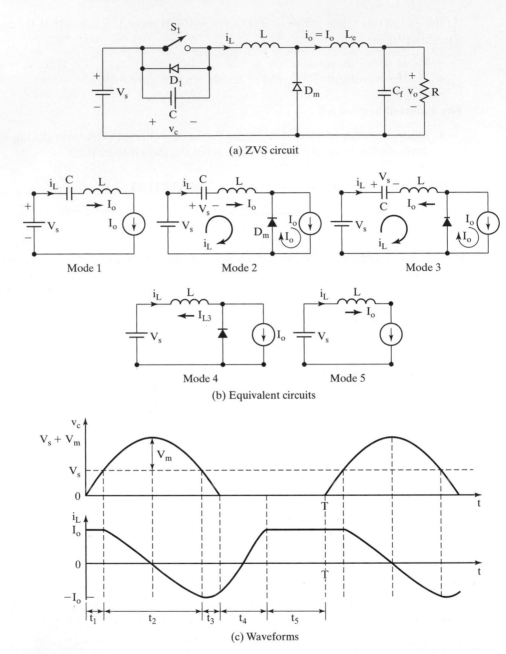

(a) ZVS circuit

(b) Equivalent circuits

(c) Waveforms

FIGURE 8.30

ZVS resonant converter.

series with Q_1, as shown in Figure 8.29c, the voltage across C can oscillate freely, and the switch is operated in a full-wave configuration. A ZVS resonant converter is shown in Figure 8.30a. A ZVS resonant converter is the dual of the ZCS resonant converter in Figure 8.28a. Equations for the M-type ZCS resonant converter can be applied if i_L is replaced by v_c and vice versa, L by C and vice versa, and V_s by I_o and vice versa. The circuit operation can be divided into five modes whose equivalent circuits are shown in Figure 8.30b. We shall redefine the time origin, $t = 0$, at the beginning of each mode.

Mode 1.   This mode is valid for $0 \le t \le t_1$. Both switch S_1 and diode D_m are off. Capacitor C charges at a constant rate of load current I_o. The capacitor voltage v_c, which rises, is given by

$$v_c = \frac{I_o}{C} t \tag{8.54}$$

This mode ends at time $t = t_1$ when $v_c(t = t_1) = V_s$. That is, $t_1 = V_sC/I_o$.

Mode 2.   This mode is valid for $0 \le t \le t_2$. The switch S_1 is still off, but diode D_m turns on. The capacitor voltage v_c is given by

$$v_c = V_m \sin \omega_0 t + V_s \tag{8.55}$$

where $V_m = I_o\sqrt{L/C}$. The peak switch voltage, which occurs at $t = (\pi/2)\sqrt{LC}$, is

$$V_{T(\text{pk})} = V_{C(\text{pk})} = I_o\sqrt{\frac{L}{C}} + V_s \tag{8.56}$$

The inductor current i_L is given by

$$i_L = I_o \cos \omega_0 t \tag{8.57}$$

This mode ends at $t = t_2$ when $v_c(t = t_2) = V_s$, and $i_L(t = t_2) = -I_o$. Therefore, $t_2 = \pi\sqrt{LC}$.

Mode 3.   This mode is valid for $0 \le t \le t_3$. The capacitor voltage that falls from V_s to zero is given by

$$v_c = V_s - V_m \sin \omega_0 t \tag{8.58}$$

The inductor current i_L is given by

$$i_L = -I_o \cos \omega_0 t \tag{8.59}$$

This mode ends at $t = t_3$ when $v_c(t = t_3) = 0$, and $i_L(t = t_3) = I_{L3}$. Thus,

$$t_3 = \sqrt{LC} \sin^{-1} x$$

where $x = V_s/V_m = (V_s/I_o)\sqrt{C/L}$.

Mode 4.   This mode is valid for $0 \le t \le t_4$. Switch S_1 is turned on, and diode D_m remains on. The inductor current, which rises linearly from I_{L3} to I_o, is given by

$$i_L = I_{L3} + \frac{V_s}{L} t \tag{8.60}$$

This mode ends at time $t = t_4$ when $i_L(t = t_4) = 0$. Thus, $t_4 = (I_o - I_{L3})(L/V_s)$. Note that I_{L3} is a negative value.

Mode 5.   This mode is valid for $0 \le t \le t_5$. Switch S_1 is on, but D_m is off. The load current I_o flows through the switch. This mode ends at time $t = t_5$, when switch S_1 is turned off again and the cycle is repeated. That is, $t_5 = T - (t_1 + t_2 + t_3 + t_4)$.

The waveforms for i_L and v_c are shown in Figure 8.30c. Equation (8.56) shows that the peak switch voltage $V_{T(\text{pk})}$ is dependent on the load current I_o. Therefore, a wide variation in the load current results in a wide variation of the switch voltage. For this reason, the ZVS converters are used only for constant-load applications. The switch must be turned on only at zero voltage. Otherwise, the energy stored in C can be dissipated in the switch. To avoid this situation, the antiparallel diode D_1 must conduct before turning on the switch.

Key Points of Section 8.9

- A ZVS shapes the switch voltage waveform during the off time to create a zero-voltage condition for the switch to turn on.

8.10   COMPARISONS BETWEEN ZCS AND ZVS RESONANT CONVERTERS

ZCS can eliminate the switching losses at turn-off and reduce the switching losses at turn-on. Because a relatively large capacitor is connected across the diode D_m, the inverter operation becomes insensitive to the diode's junction capacitance. When power MOSFETs are used for ZCS, the energy stored in the device's capacitance is dissipated during turn-on. This capacitive turn-on loss is proportional to the switching frequency. During turn-on, a high rate of change of voltage may appear in the gate drive circuit due to the coupling through the Miller capacitor, thus increasing switching loss and noise. Another limitation is that the switches are under high-current stress, resulting in higher conduction loss. It should, however, be noted that ZCS is particularly effective in reducing switching loss for power devices (such as IGBTs) with large tail current in the turn-off process.

By the nature of the resonant tank and ZCS, the peak switch current is much higher than that in a square wave. In addition, a high voltage becomes established across the switch in the off-state after the resonant oscillation. When the switch is turned on again, the energy stored in the output capacitor becomes discharged through the switch, causing a significant power loss at high frequencies and high voltages. This switching loss can be reduced by using ZVS.

ZVS eliminates the capacitive turn-on loss. It is suitable for high-frequency operation. Without any voltage clamping, the switches may be subjected to excessive voltage stress, which is proportional to the load.

For both ZCS and ZVS, the output voltage control can be achieved by varying the frequency. ZCS operates with a constant on-time control, whereas ZVS operates with a constant off-time control.

8.11   TWO-QUADRANT ZVS RESONANT CONVERTERS

The ZVS concept can be extended to a two-quadrant converter, as shown in Figure 8.31a, where the capacitors $C_+ = C_- = C/2$. The inductor L has such a value, so that it forms a resonant circuit. The resonant frequency is $f_o = 1/(2\pi\sqrt{LC})$, and it is much larger than the switching frequency f_s. Assuming the filter capacitance C_e to be large, the load is replaced by a dc voltage V_{dc}, as shown in Figure 8.31b. The circuit operations can be divided into six modes. The equivalent circuits for various modes are shown in Figure 8.31d.

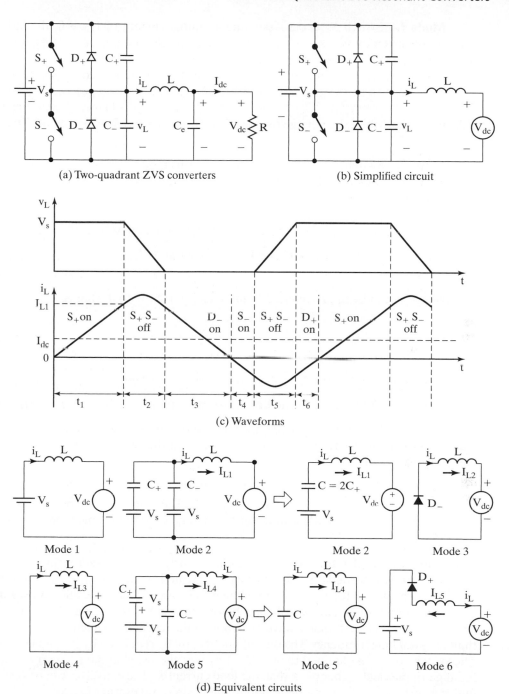

(a) Two-quadrant ZVS converters

(b) Simplified circuit

(c) Waveforms

Mode 1   Mode 2   Mode 2   Mode 3

Mode 4   Mode 5   Mode 5   Mode 6

(d) Equivalent circuits

FIGURE 8.31

Two-quadrant ZVS resonant converter.

Mode 1. Switch S_+ is on. Assuming an initial current of $I_{L0} = 0$, the inductor current i_L is given by

$$i_L = \frac{V_s}{L}t \tag{8.61}$$

This mode ends when the voltage on capacitor C_+ is zero and S_+ is turned off. The voltage on C_- is V_s.

Mode 2. Switches S_+ and S_- are both off. This mode begins with C_+ having zero voltage and C_- having V_s. The equivalent of this mode can be simplified to a resonant circuit of C and L with an initial inductor current I_{L1}. Current i_L can be approximately represented by

$$i_L = (V_s - V_{dc})\sqrt{\frac{L}{C}}\sin \omega_0 t + I_{L1} \tag{8.62}$$

The voltage v_o can be approximated to fall linearly from V_s to 0. That is,

$$v_o = V_s - \frac{V_s C}{I_{L1}}t \tag{8.63}$$

This mode ends when v_o becomes zero and diode D_- turns on.

Mode 3. Diode D_- is turned on. Current i_L falls linearly from $I_{L2}(= I_{L1})$ to 0.

Mode 4. Switch S_- is turned on when i_L and v_o become zero. Inductor current i_L continues to fall in the negative direction to I_{L4} until the switch voltage becomes zero, and S_- is turned off.

Mode 5. Switches S_+ and S_- are both off. This mode begins with C_- having zero voltage and C_+ having V_s, and is similar to mode 2. The voltage v_o can be approximated to rise linearly from 0 to V_s. This mode ends when v_o tends to become more than V_s and diode D_+ turns on.

Mode 6. Diode D_+ is turned on; i_L falls linearly from I_{L5} to zero. This mode ends when $i_L = 0$. S_+ is turned on, and the cycle is repeated.

The waveforms for i_L and v_o are shown in Figure 8.31c. For ZVS i_L must flow in either direction so that a diode conducts before its switch is turned on. The output voltage can be made almost square wave by choosing the resonant frequency f_o much larger than the switching frequency. The output voltage can be regulated by frequency control. The switch voltage is clamped to only V_s. However, the switches have to carry i_L, which has high ripples and higher peak than the load current I_o. The converter can be operated under a current-regulated mode to obtain the desired waveform of i_L.

The circuit in Figure 8.31a can be extended to a single-phase half-bridge inverter as shown in Figure 8.32. A three-phase version is shown in Figure 8.33a, where the load inductance L constitutes the resonant circuit. One arm of a three-phase circuit in which a separate resonant inductor [10] is used is shown in Figure 8.33b.

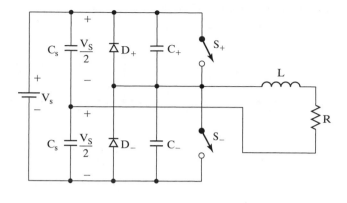

FIGURE 8.32

Single-phase ZVS resonant inverter.

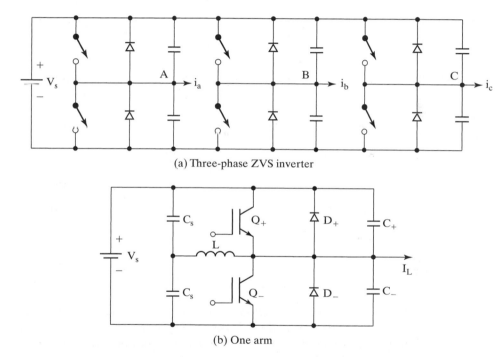

(a) Three-phase ZVS inverter

(b) One arm

FIGURE 8.33

Three-phase ZVS resonant inverter.

8.12   RESONANT DC-LINK INVERTERS

In resonant dc-link inverters, a resonant circuit is connected between the dc input voltage and the PWM inverter, so that the input voltage to the inverter oscillates between zero and a value slightly greater than twice the dc input voltage. The resonant link, which is similar to the class E inverter in Figure 8.20, is shown in Figure 8.34a, where I_o is the current drawn by the inverter. Assuming a lossless circuit and $R = 0$, the link voltage is

$$v_c = V_s(1 - \cos \omega_0 t) \tag{8.64}$$

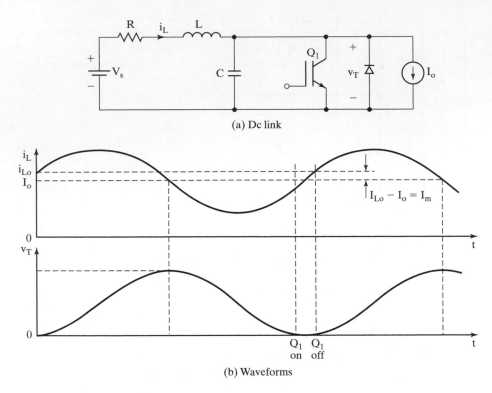

(a) Dc link

(b) Waveforms

FIGURE 8.34

Resonant dc link.

and the inductor current i_L is

$$i_L = V_s\sqrt{\frac{C}{L}}\sin \omega_0 t + I_o \tag{8.65}$$

Under lossless conditions the oscillation continues and there is no need to turn on switch S_1. However, in practice there is power loss in R, i_L is damped sinusoidal, and S_1 is turned on to bring the current to the initial level. The value of R is small and the circuit is underdamped. Under this condition, i_L and v_c can be shown [11] as

$$i_L \approx I_o + e^{\alpha t}\left[\frac{V_s}{\omega L}\sin \omega_0 t + (I_{Lo} - I_o)\cos \omega_0 t\right] \tag{8.66}$$

and the capacitor voltage v_c is

$$v_c \approx V_s + e^{-\alpha t}[\omega_o L(I_{Lo} - I_o)\sin \omega_0 t - V_s \cos \omega_0 t] \tag{8.67}$$

The waveforms for v_c and i_L are shown in Figure 8.34b. Switch S_1 is turned on when the capacitor voltage falls to zero and is turned off when the current i_L reaches the level of the initial current I_{Lo}. It can be noted that the capacitor voltage depends only on the difference $I_m(= I_{Lo} - I_o)$ instead of the load current I_o. Thus, the control circuit should monitor $(i_L - I_o)$ when the switch is conducting and turn off the switch when the desired value of I_m is reached.

A three-phase resonant dc-link inverter [12] is shown in Figure 8.35a. The six inverter devices are gated in such a way as to set up periodic oscillations on the dc-link *LC* circuit. The devices are turned on and off at zero-link voltages, thereby accomplishing lossless turn-on and turn-off of all devices. The waveforms for the link voltage and the inverter line-to-line voltage are shown in Figure 8.35b.

The dc-link resonant cycle is normally started with a fixed value of initial capacitor current. This causes the voltage across the resonant dc link to exceed $2V_s$, and all the inverter devices are subjected to this high-voltage stress. An active clamp [12], as shown in Figure 8.36a, can limit the link voltage shown in Figure 8.36b. The clamp factor k is related to the tank period T_k and resonant frequency $\omega_0 = 1/\sqrt{LC}$ by

$$T_k\omega_0 = 2\left[\cos^{-1}(1-k) + \frac{\sqrt{k(2-k)}}{k-1}\right] \qquad \text{for } 1 \le k \le 2 \qquad (8.68)$$

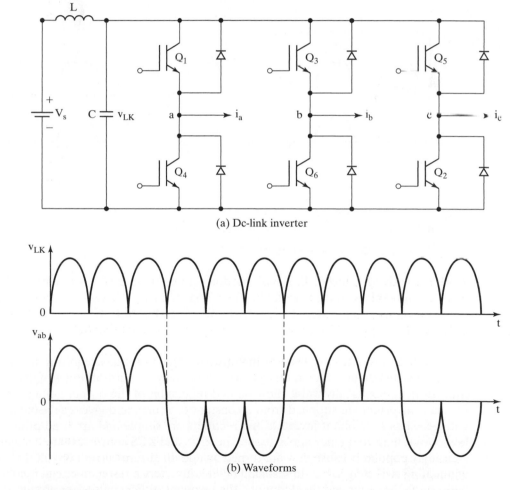

(a) Dc-link inverter

(b) Waveforms

FIGURE 8.35

Three-phase resonant dc-link inverter.

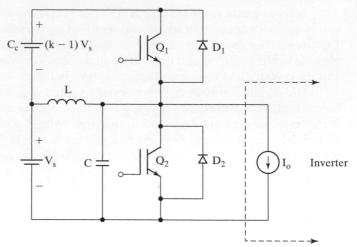

(a) Active clamp circuit

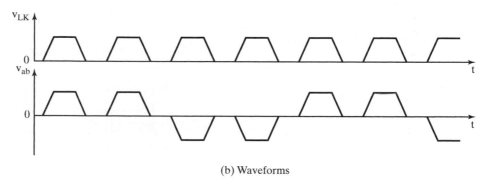

(b) Waveforms

FIGURE 8.36

Active clamped resonant dc-link inverter.

That is, for fixed value of k, T_k can be determined for a given resonant circuit. For $k = 1.5$, the tank period T_k should be $T_k = 7.65\sqrt{LC}$.

SUMMARY

The resonant inverters are used in high-frequency applications requiring fixed output voltage. The maximum resonant frequency is limited by the turn-off times of thyristors or transistors. Resonant inverters allow limited regulation of the output voltage. Parallel-resonant inverters are supplied from a constant dc source and give a sinusoidal output voltage. Class E resonant inverters and rectifiers are simple and are used primarily for low-power, high-frequency applications. The ZVS and ZCS converters are becoming increasingly popular because they are turned on and off at zero current or voltage, thereby eliminating switching losses. In resonant dc-link inverters, a resonant circuit is connected between the inverter and the dc supply. The resonant voltage pulses are produced at the input of the inverter, and the inverter devices are turned on and off at zero voltages.

REFERENCES

[1]   A. J. Forsyth, "Review of resonant techniques in power electronic systems," *IEE Power Engineering Journals*, 1996, pp. 110–120.

[2]   R. L. Steigerwald, "A compromise of half-bridge resonance converter topologies," *IEEE Transactions on Power Electronics*, Vol. PE3, No. 2, 1988, pp. 174–182.

[3]   K. Liu, R. Oruganti, and F. C. Y. Lee, "Quasi-resonant converters: Topologies and characteristics," *IEEE Transactions on Power Electronics*, Vol. PE2, No. 1, 1987, pp. 62–71.

[4]   R. S. Y. Hui and H. S. Chung, *Power Electronics Handbook*, edited by M. H. Rashid. San Diego, CA: Academic Press. 2001, Chapter 15—Resonant and Soft-Switching Converter.

[5]   N. O. Sokal and A. D. Sokal, "Class E: A new class of high-efficiency tuned single-ended switching power amplifiers," *IEEE Journal of Solid-State Circuits*, Vol. 10, No. 3, 1975, pp. 168–176.

[6]   R. E. Zuliski, "A high-efficiency self-regulated class-E power inverter/converter," *IEEE Transactions on Industrial Electronics*, Vol. IE-33, No. 3, 1986, pp. 340–342.

[7]   M. K. Kazimierczuk and I. Jozwik, "Class-E zero-voltage switching and zero-current switching rectifiers," *IEEE Transactions on Circuits and Systems*, Vol. CS-37, No. 3, 1990, pp. 436–444.

[8]   F. C. Lee, "High-Frequency Quasi-Resonant and Multi-Resonant Converter Technologies," *IEEE International Conference on Industrial Electronics*, 1988, pp. 509–521.

[9]   W. A. Tabisz and F. C. Lee, "DC Analysis and Design of Zero-Voltage Switched Multi-Resonant Converters," *IEEE Power Electronics Specialist Conference*, 1989, pp. 243–251.

[10]   C. P. Henze, H. C. Martin, and D. W. Parsley, "Zero-Voltage Switching in High Frequency Power Converters Using Pulse-Width Modulation," *IEEE Applied Power Electronics Conference*, 1988, pp. 33–40.

[11]   D. M. Devan, "The resonant DC link converter: A new concept in static power conversion," *IEEE Transactions on Industry Applications*, Vol. IA-25, No. 2, 1989, pp. 317–325.

[12]   D. M. Devan and G. Skibinski, "Zero-switching loss inverters for high power applications," *IEEE Transactions on Industry Applications*, Vol. IA-25, No. 4, 1989, pp. 634–643.

REVIEW QUESTIONS

8.1 What is the principle of series resonant inverters?
8.2 What is the dead zone of a resonant inverter?
8.3 What are the advantages and disadvantages of resonant inverters with bidirectional switches?
8.4 What are the advantages and disadvantages of resonant inverters with unidirectional switches?
8.5 What is the necessary condition for series resonant oscillation?
8.6 What is the purpose of coupled inductors in half-bridge resonant inverters?
8.7 What are the advantages of reverse-conducting thyristors in resonant inverters?
8.8 What is an overlap control of resonant inverters?
8.9 What is a nonoverlap control of inverters?
8.10 What are the effects of series loading in a series-resonant inverter?
8.11 What are the effects of parallel loading in a series-resonant inverter?
8.12 What are the effects of both series and parallel loading in a series-resonant inverter?
8.13 What are the methods for voltage control of series-resonant inverters?
8.14 What are the advantages of parallel-resonant inverters?
8.15 What is the class E resonant inverter?

8.16 What are the advantages and limitations of class E resonant inverters?

8.17 What is a class E resonant rectifier?

8.18 What are the advantages and limitations of class E resonant rectifiers?

8.19 What is the principle of zero-current-switching (ZCS) resonant converters?

8.20 What is the principle of zero-voltage-switching (ZVS) resonant converters?

8.21 What are the advantages and limitations of ZCS converters?

8.22 What are the advantages and limitations of ZVS converters?

PROBLEMS

8.1 The basic series resonant inverter in Figure 8.1a has $L_1 = L_2 = L = 25\ \mu H, C = 2\ \mu F$, and $R = 5\ \Omega$. The dc input voltage, $V_s = 220$ V and the output frequency, $f_o = 6.5$ kHz. The turn-off time of thyristors is $t_q = 15\ \mu s$. Determine **(a)** the available (or circuit) turn-off time t_{off}, **(b)** the maximum permissible frequency f_{max}, **(c)** the peak-to-peak capacitor voltage V_{pp}, and **(d)** the peak load current I_p. **(e)** Sketch the instantaneous load current $i_0(t)$; capacitor voltage $v_c(t)$; and dc supply current $I_s(t)$. Calculate **(f)** the rms load current I_o; **(g)** the output power P_o; **(h)** the average supply current I_s; and **(i)** the average, peak, and rms thyristor currents.

8.2 The half-bridge resonant inverter in Figure 8.3 uses nonoverlapping control. The inverter frequency is $f_0 = 8.5$ kHz. If $C_1 = C_2 = C = 2\ \mu F, L_1 = L_2 = L = 40\ \mu H, R = 2\ \Omega$, and $V_s = 220$ V. Determine **(a)** the peak supply current, **(b)** the average thyristor current I_A, and **(c)** the rms thyristor current I_R.

8.3 The resonant inverter in Figure 8.7a has $C = 2\ \mu F, L = 30\ \mu H, R = 0$, and $V_s = 220$ V. The turn-off time of thyristor, $t_q = 12\ \mu s$. The output frequency is, $f_o = 15$ kHz. Determine **(a)** the peak supply current I_{ps}, **(b)** the average thyristor current I_A, **(c)** the rms thyristor current I_R, **(d)** the peak-to-peak capacitor voltage V_c, **(e)** the maximum permissible output frequency f_{max}, and **(f)** the average supply current I_s.

8.4 The half-bridge resonant inverter in Figure 8.8a is operated at frequency, $f_0 = 3.5$ kHz in the nonoverlap mode. If $C_1 = C_2 = C = 2\ \mu F, L = 20\ \mu H, R = 1.5\ \Omega$, and $V_s = 220$ V, determine **(a)** the peak supply current I_{ps}, **(b)** the average thyristor current I_A, **(c)** the rms thyristor current I_R, **(d)** the rms load current I_o, and **(e)** the average supply current I_s.

8.5 Repeat Problem 8.4 with an overlapping control so that the firing of T_1 and T_2 are advanced by 50% of the resonant frequency.

8.6 The full-bridge resonant inverter in Figure 8.9a is operated at a frequency of $f_0 = 3.5$ kHz. If $C = 2\ \mu F, L = 20\ \mu H, R = 1.5\ \Omega$, and $V_s = 220$ V, determine **(a)** the peak supply current I_s, **(b)** the average thyristor current I_A, **(c)** the rms thyristor current I_R, **(d)** the rms load current I_o, and **(e)** the average supply current I_s.

8.7 A series resonant inverter with series-loaded delivers a load power of $P_L = 2$ kW at resonance. The load resistance is $R = 10\ \Omega$. The resonant frequency is $f_0 = 25$ kHz. Determine **(a)** the dc input voltage V_s, **(b)** the quality factor Q_s if it is required to reduce the load power to 500 W by frequency control so that $u = 0.8$, **(c)** the inductor L, and **(d)** the capacitor C.

8.8 A series resonant inverter with parallel-loaded delivers a load power of $P_L = 2$ kW at a peak sinusoidal load voltage of $V_p = 330$ V and at resonance. The load resistance is $R = 10\ \Omega$. The resonant frequency is $f_0 = 25$ kHz. Determine **(a)** the dc input voltage V_s, **(b)** the frequency ratio u if it is required to reduce the load power to 500 W by frequency control, **(c)** the inductor L, and **(d)** the capacitor C.

8.9 A parallel resonant inverter delivers a load power of $P_L = 2$ kW at a peak sinusoidal load voltage of $V_p = 170$ V and at resonance. The load resistance is $R = 10\ \Omega$. The resonant frequency is $f_0 = 25$ kHz. Determine **(a)** the dc input current I_s, **(b)** the quality factor Q_p

if it is required to reduce the load power to 500 W by frequency control so that $u = 1.25$, **(c)** the inductor L, and **(d)** the capacitor C.

8.10 The class E inverter of Figure 8.20 operates at resonance and has $V_s = 18$ V and $R = 10$ Ω. The switching frequency is $f_s = 50$ kHz. **(a)** Determine the optimum values of L, C, C_e, and L_e. **(b)** Use PSpice to plot the output voltage v_o and the switch voltage v_T for $k = 0.304$. Assume that $Q = 7$.

8.11 The class E rectifier of Figure 8.23a supplies a load power of $P_L = 1$ kW at $V_o = 5$ V. The peak supply voltage is $V_m = 12$ V. The supply frequency is $f = 350$ kHz. The peak-to-peak ripple on the output voltage is $\Delta V_o = 20$ mV. **(a)** Determine the values of L, C, and C_f, and **(b)** the rms and dc currents of L and C. **(c)** Use PSpice to plot the output voltage v_o and the inductor current i_L.

8.12 The ZCS resonant converter of Figure 8.27a delivers a maximum power of $P_L = 1$ kW at $V_o = 5$ V. The supply voltage is $V_s = 15$ V. The maximum operating frequency is $f_{max} = 40$ kHz. Determine the values of L and C. Assume that the intervals t_1 and t_3 are very small, and $x = I_m/I_o = 1.5$.

8.13 The ZVS resonant converter of Figure 8.30a supplies a load power of $P_L = 1$ kW at $V_o = 5$ V. The supply voltage is $V_s = 15$ V. The operating frequency is $f = 40$ kHz. The values of L and C are $L = 150$ μH and $C = 0.05$ μF. **(a)** Determine the peak switch voltage V_p and current I_p, and **(b)** the time durations of each mode.

8.14 For the active clamped of Figure 8.36, plot the f_o/f_k ratio for $1 < k \le 2$.

CHAPTER 9

Multilevel Inverters

The learning objectives of this chapter are as follows:

- To learn the switching technique for multilevel inverters and their types
- To study the operation and features of multilevel inverters
- To understand the advantages and disadvantages of multilevel inverters
- To learn about the control strategy to address capacitor voltage unbalancing
- To learn the potential applications of multilevel inverters

9.1 INTRODUCTION

The voltage source inverters produce an output voltage or a current with levels either 0 or $\pm V_{dc}$. They are known as the two-level inverter. To obtain a quality output voltage or a current waveform with a minimum amount of ripple content, they require high-switching frequency along with various pulse-width modulation (PWM) strategies. In high-power and high-voltage applications, these two-level inverters, however, have some limitations in operating at high frequency mainly due to switching losses and constraints of device ratings. Moreover, the semiconductor switching devices should be used in such a manner as to avoid problems associated with their series–parallel combinations that are necessary to obtain capability of handling high voltages and currents.

The multilevel inverters have drawn tremendous interest in the power industry. They present a new set of features that are well suited for use in reactive power compensation. It may be easier to produce a high-power, high-voltage inverter with the multilevel structure because of the way in which device voltage stresses are controlled in the structure. Increasing the number of voltage levels in the inverter without requiring higher ratings on individual devices can increase the power rating. The unique structure of multilevel voltage source inverters' allows them to reach high voltages with low harmonics without the use of transformers or series-connected synchronized-switching devices. As the number of voltage levels increases, the harmonic content of the output voltage waveform decreases significantly [1, 2].

9.2   MULTILEVEL CONCEPT

Let us consider a three-phase inverter system [4], as shown in Figure 9.1a, with a dc voltage V_{dc}. Series-connected capacitors constitute the energy tank for the inverter, providing some nodes to which the multilevel inverter can be connected. Each capacitor has the same voltage E_m, which is given by

$$E_m = \frac{V_{dc}}{m-1} \tag{9.1}$$

where m denotes the number of levels. The term *level* is referred to as the number of nodes to which the inverter can be accessible. An m-level inverter needs $(m-1)$ capacitors.

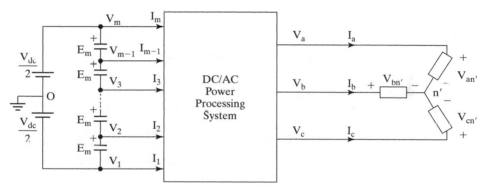

(a) Three-phase multilevel power processing system

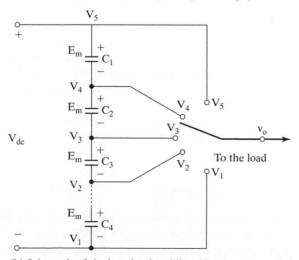

(b) Schematic of single pole of multilevel inverter by a switch

FIGURE 9.1

General topology of multilevel inverters.

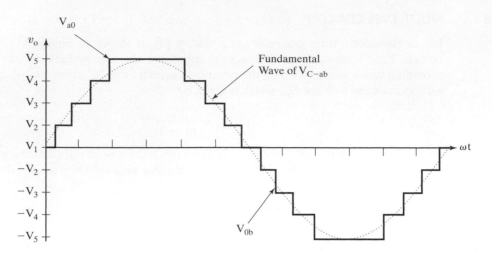

FIGURE 9.2

Typical output voltage of a five-level multilevel inverter.

Output phase voltages can be defined as voltages across output terminals of the inverter and the ground point denoted by o as shown in Figure 9.1a. Moreover, input node voltages and currents can be referred to input terminal voltages of the inverter with reference to ground point and the corresponding currents from each node of the capacitors to the inverter, respectively. For example, input node (dc) voltages are designated by V_1, V_2, etc. and the input node (dc) currents by I_1, I_2, etc., as shown in Figure 9.1a. V_a, V_b, and V_c, are the root-mean-square (rms) values of the line load voltages; I_a, I_b, and I_c are the rms values of the line load currents. Figure 9.1b shows the schematic of a pole in a multilevel inverter where v_a indicates an output phase voltage that can assume any voltage level depending on the selection of node (dc) voltage V_1, V_2, etc. Thus, a pole in a multilevel inverter can be regarded as a single-pole, multiple-throw switch. By connecting the switch to one node at a time, one can obtain the desired output. Figure 9.2 shows the typical output voltage of a five-level inverter.

The actual realization of the switch requires bidirectional switching devices for each node. The topological structure of multilevel inverter must (1) have less switching devices as far as possible, (2) be capable of withstanding very high input voltage for high-power applications, and (3) have lower switching frequency for each switching device.

9.3   TYPES OF MULTILEVEL INVERTERS

The general structure of the multilevel converter is to synthesize a near sinusoidal voltage from several levels of dc voltages, typically obtained from capacitor voltage sources. As the number of levels increases, the synthesized output waveform has more steps, which produce a staircase wave that approaches a desired waveform. Also, as more steps are added to the waveform, the harmonic distortion of the output wave decreases, approaching zero as the number of levels increases. As the number of levels increases, the voltage that can be spanned by summing multiple voltage levels also increases. The output voltage during the positive half-cycle can be found from

$$v_{ao} = \sum_{n=1}^{m} E_n SF_n \tag{9.2}$$

where SF_n is the switching or control function of nth node and it takes a value of 0 or 1. Generally, the capacitor terminal voltages $E_1, E_2, \ldots$ all have the same value E_m. Thus, the peak output voltage is $v_{ao(\text{peak})} = (m-1)E_m = V_{dc}$. To generate an output voltage with both positive and negative values, the circuit topology has another switch to produce the negative part v_{ob} so that $v_{ab} = v_{ao} + v_{ob} = v_{ao} - v_{bo}$.

The multilevel inverters can be classified into three types [5].

Diode-clamped multilevel inverter;

Flying-capacitors multilevel inverter;

Cascade multilevel inverter.

9.4   DIODE-CLAMPED MULTILEVEL INVERTER

A diode-clamped multilevel (m-level) inverter (DCMLI) typically consists of $(m-1)$ capacitors on the dc bus and produces m levels on the phase voltage. Figure 9.3a shows

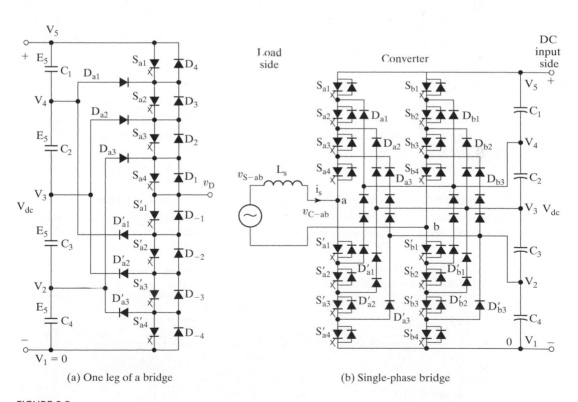

(a) One leg of a bridge   (b) Single-phase bridge

FIGURE 9.3

Diode-clamped five-level bridge multilevel inverter. [Ref. 4] one leg and Figure 9.3b shows a full-bridge five-level diode-clamped converter. The numbering order of the switches is $S_{a1}, S_{a2}, S_{a3}, S_{a4}, S'_{a1}, S'_{a2}, S'_{a3}$, and S'_{a4}. The dc bus consists of four capacitors, C_1, C_2, C_3, and C_4. For a dc bus voltage V_{dc}, the voltage across each capacitor is $V_{dc}/4$, and each device voltage stress is limited to one capacitor voltage level $V_{dc}/4$ through clamping diodes. An m-level inverter leg requires $(m-1)$ capacitors, $2(m-1)$ switching devices and $(m-1)(m-2)$ clamping diodes.

9.4.1   Principle of Operation

To produce a staircase-output voltage, let us consider only one leg of the five-level inverter, as shown in Figure 9.3a, as an example. A single-phase bridge with two legs is shown in Figure 9.3b. The *dc rail* 0 is the reference point of the output phase voltage. The steps to synthesize the five-level voltages are as follows:

1. For an output voltage level $v_{ao} = V_{dc}$, turn on all upper-half switches S_{a1} through S_{a4}.
2. For an output voltage level $v_{ao} = 3V_{dc}/4$, turn on three upper switches S_{a2} through S_{a4} and one lower switch S'_{a1}.
3. For an output voltage level $v_{ao} = V_{dc}/2$, turn on two upper switches S_{a3} through S_{a4} and two lower switches S'_{a1} and S'_{a2}.
4. For an output voltage level $v_{ao} = V_{dc}/4$, turn on one upper switch S_{a4} and three lower switches S'_{a1} through S'_{a3}.
5. For an output voltage level $v_{ao} = 0$, turn on all lower half switches S'_{a1} through S'_{a4}.

Table 9.1 shows the voltage levels and their corresponding switch states. State condition 1 means the switch is on, and state 0 means the switch is off. It should be noticed that each switch is turned on only once per cycle and there are four complementary switch pairs in each phase. These pairs for one leg of the inverter are (S_{a1}, S'_{a1}), (S_{a2}, S'_{a2}), (S_{a3}, S'_{a3}), and (S_{a4}, S'_{a4}). Thus, if one of the complementary switch pairs is turned on, the other of the same pair must be off. Four switches are always turned on at the same time.

Figure 9.4 shows the phase voltage waveform of the five-level inverter. The line voltage consists of the positive phase-leg voltage of terminal a and the negative phase-leg voltage of terminal b. Each phase-leg voltage tracks one-half of the sinusoidal wave. The

TABLE 9.1   Diode-Clamped Voltage Levels and Their Switch States

Output	Switch State							
V_{a0}	S_{a1}	S_{a2}	S_{a3}	S_{a4}	S'_{a1}	S'_{a2}	S'_{a3}	S'_{a4}
$V_5 = V_{dc}$	1	1	1	1	0	0	0	0
$V_4 = 3V_{dc}/4$	0	1	1	1	1	0	0	0
$V_3 = V_{dc}/2$	0	0	1	1	1	1	0	0
$V_2 = V_{dc}/4$	0	0	0	1	1	1	1	0
$V_1 = 0$	0	0	0	0	1	1	1	1

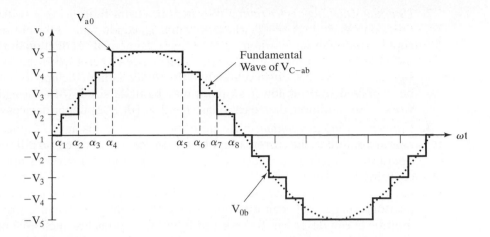

FIGURE 9.4

Phase and fundamental voltage waveforms of a five-level inverter.

resulting line voltage is a nine-level staircase wave. This implies that an m-level converter has an m-level output phase-leg voltage and a $(2m - 1)$-level output line voltage.

9.4.2   Features of Diode-Clamped Inverter

The main features are as follows:

1. *High-voltage rating for blocking diodes*: Although each switching device is only required to block a voltage level of $V_{dc}/(m - 1)$, the clamping diodes need to have different reverse voltage blocking ratings. For example, when all lower devices S'_{a1} through S'_{a4} are turned on, diode D'_{a1} needs to block three capacitor voltages, or $3V_{dc}/4$. Similarly, diodes D_{a2} and D'_{a2} need to block $2V_{dc}/4$, and D_{a3} needs to block $V_{dc}/4$. Even though each main switch is supposed to block the nominal blocking voltage, the blocking voltage of each clamping diode in the diode clamping inverter is dependent on its position in the structure. In an m-level leg, there can be two diodes, each seeing a blocking voltage of

$$V_D = \frac{m - 1 - k}{m - 1} V_{dc} \qquad (9.3)$$

where m is the number of levels;
$\quad$ k goes from 1 to $(m - 2)$;
$\quad$ V_{dc} is the total dc link voltage.

If the blocking voltage rating of each diode is the same as that of the switching device, the number of diodes required for each phase is $N_D = (m - 1) \times (m - 2)$. This number represents a quadratic increase in m. Thus, for $m = 5$, $N_D = (5 - 1) \times (5 - 2) = 12$. When m is sufficiently high, the number of diodes makes the system impractical to implement, which in effect limits the number of levels.

2. *Unequal switching device rating*: We can notice from Table 9.1 that switch S_{a1} conducts only during $v_{ao} = V_{dc}$, whereas switch S_{a4} conducts over the entire cycle except during the interval when $v_{ao} = 0$. Such an unequal conduction duty requires different current ratings for the switching devices. Therefore, if the inverter design uses the average duty cycle to find the device ratings, the upper switches may be oversized, and the lower switches may be undersized. If the design uses the worst-case condition, then each phase has $2 \times (m - 2)$ upper devices oversized.

3. *Capacitor voltage unbalance*: Because the voltage levels at the capacitor terminals are different, the currents supplied by the capacitors are also different. When operating at unity power factor, the discharging time for inverter operation (or charging time for rectifier operation) for each capacitor is different. Such a capacitor charging profile repeats every half-cycle, and the result is unbalanced capacitor voltages between different levels. This voltage unbalance problem in a multilevel converter can be resolved by using approaches such as replacing capacitors by a controlled constant dc voltage source, PWM voltage regulators, or batteries.

The major advantages of the diode-clamped inverter can be summarized as follows:

- When the number of levels is high enough, the harmonic content is low enough to avoid the need for filters.
- Inverter efficiency is high because all devices are switched at the fundamental frequency.
- The control method is simple.

The major disadvantages of the diode-clamped inverter can be summarized as follows:

- Excessive clamping diodes are required when the number of levels is high.
- It is difficult to control the real power flow of the individual converter in multi-converter systems.

9.4.3   Improved Diode-Clamped Inverter

The problem of multiple blocking voltages of the clamping diodes can be addressed by connecting an appropriate number of diodes in series, as shown in Figure 9.5. However, due to mismatches of the diode characteristics, the voltage sharing is not equal. An improved version of the diode-clamped inverter [6] is shown in Figure 9.6 for five levels. The numbering order of the switches is $S_1, S_2, S_3, S_4, S'_1, S'_2, S'_3$, and S'_4. There are a total of eight switches and 12 diodes of equal voltage rating, which are the same as the diode-clamping inverter with series-connected diodes. This pyramid architecture is extensible to any level, unless otherwise practically limited. A five-level inverter leg requires $(m - 1 =)$ 4 capacitors, $(2(m - 1) =)$ 8 switches and $((m - 1)(m - 2) =)$ 12 clamping diodes.

Principle of operation.   The modified diode-clamped inverter can be decomposed into two-level switching cells. For an m-level inverter, there are $(m - 1)$ switching cells. Thus, for $m = 5$, there are 4 cells: In cell 1, S_2, S_3, and S_4 are always on whereas S_1 and S'_1 are switched alternatively to produce an output voltage $V_{dc}/2$ and

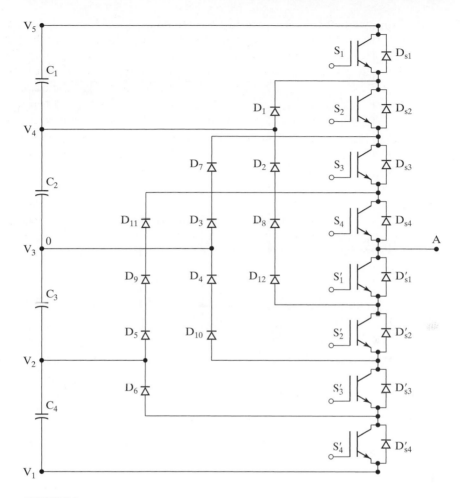

FIGURE 9.5

Diode-clamped multilevel inverter with diodes in series. [Ref. 6]

$V_{dc}/4$, respectively. Similarly, in cell 2, S_3, S_4, and S'_1 are always on whereas S_2 and S'_2 are switched alternatively to produce an output voltage $V_{dc}/4$ and 0, respectively. In cell 3, S_4, S'_1, and S'_2 are always on whereas S_3 and S'_3 are switched alternatively to produce an output voltage 0 and $-V_{dc}/2$, respectively. In final cell 4, S'_1, S'_2, and S'_3 are always on whereas S_4 and S'_4 are switched alternatively to produce an output voltage $-V_{dc}/4$ and $-V_{dc}/2$, respectively.

Each switching cell works actually as a normal two-level inverter, except that each forward or freewheeling path in the cell involves $(m - 1)$ devices instead of only one. Taking cell 2 as an example, the forward path of the up-arm involves D_1, S_2, S_2, and S_4, whereas the freewheeling path of the up-arm involves S'_1, D_{12}, D_8, and D_2, connecting the inverter output to $V_{dc}/4$ level for either positive or negative current flow. The forward path of the down-arm involves S'_1, S'_2, D_{10}, and D_4, whereas the freewheeling path of the down-arm involves D_3, D_7, S_3, and S_4, connecting the inverter output to

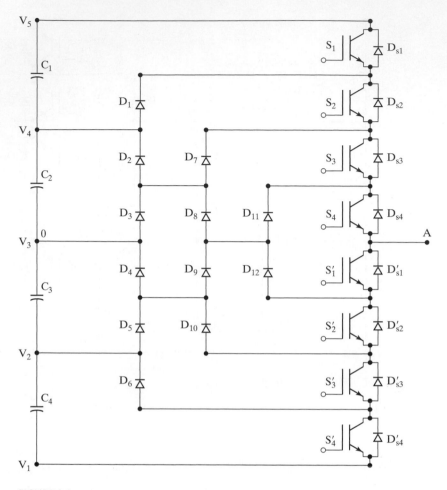

FIGURE 9.6

Modified diode-clamed inverter with distributed clamping diodes. [Ref. 6]

zero level for either positive or negative current flow. The following rules govern the switching of an m-level inverter:

1. At any moment, there must be $(m - 1)$ neighboring switches that are on.
2. For each two neighboring switches, the outer switch can only be turned on when the inner switch is on.
3. For each two neighboring switches, the inner switch can only be turned off when the outer switch is off.

9.5   FLYING-CAPACITORS MULTILEVEL INVERTER

Figure 9.7 shows a single-phase, full-bridge, five-level converter based on a flying-capacitors multilevel inverter (FCMLI) [5]. The numbering order of the switches is

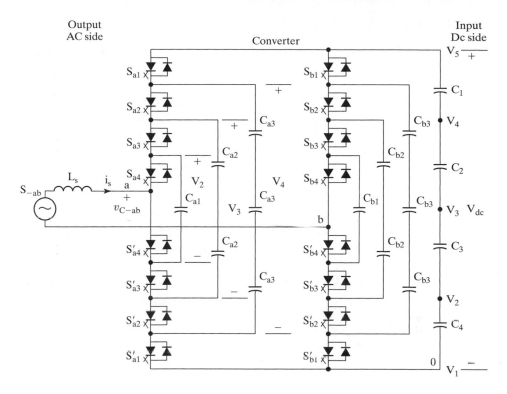

FIGURE 9.7

Circuit diagram of a five-level, flying-capacitors, single-phase inverter. [Ref. 5]

S_{a1}, S_{a2}, S_{a3}, S_{a4}, S'_{a4}, S'_{a3}, S'_{a2}, and S'_{a1}. Note that the order is numbered differently from that of the diode-clamped inverter in Figure 9.3. The numbering is immaterial as along as the switches are turned on and off in the right sequence to produce the desired output waveform. Each phase leg has an identical structure. Assuming that each capacitor has the same voltage rating, the series connection of the capacitors indicates the voltage level between the clamping points. Three inner-loop balancing capacitors (C_{a1}, C_{a2}, and C_{a3}) for phase-leg a are independent from those for phase-leg b. All phase legs share the same dc-link capacitors, C_1 through C_4.

The voltage level for the flying-capacitors converter is similar to that of the diode-clamped type of converter. That is, the phase voltage v_{ao} of an m-level converter has m levels (including the reference level), and the line voltage v_{ab} has $(2m - 1)$ levels. Assuming that each capacitor has the same voltage rating as the switching device, the dc bus needs $(m - 1)$ capacitors for an m-level converter. The number of capacitors required for each phase is $N_C = \sum_{j=1}^{m}(m - j)$. Thus, for $m = 5$, $N_C = 10$.

9.5.1 Principle of Operation

To produce a staircase-output voltage, let us consider the one leg of the five-level inverter shown in Figure 9.7 as an example. The dc rail 0 is the reference point of the output phase voltage. The steps to synthesize the five-level voltages are as follows:

1. For an output voltage level $v_{ao} = V_{dc}$, turn on all upper-half switches S_{a1} through S_{a4}.

2. For an output voltage level $v_{ao} = 3V_{dc}/4$, there are four combinations:
   a. $v_{ao} = V_{dc} - V_{dc}/4$ by turning on devices S_{a1}, S_{a2}, S_{a3}, and S'_{a4}.
   b. $v_{ao} = 3V_{dc}/4$ by turning on devices S_{a2}, S_{a3}, S_{a4}, and S'_{a1}.
   c. $v_{ao} = V_{dc} - 3V_{dc}/4 + V_{dc}/2$ by turning on devices S_{a1}, S_{a3}, S_{a4}, and S'_{a2}.
   d. $v_{ao} = V_{dc} - V_{dc}/2 + V_{dc}/4$ by turning on devices S_{a1}, S_{a2}, S_{a4}, and S'_{a3}.

3. For an output voltage level $v_{ao} = V_{dc}/2$, there are six combinations:
   a. $v_{ao} = V_{dc} - V_{dc}/2$ by turning on devices S_{a1}, S_{a2}, S'_{a3}, and S'_{a4}.
   b. $v_{ao} = V_{dc}/2$ by turning on devices S_{a3}, S_{a4}, S'_{a1}, and S'_{a2}.
   c. $v_{ao} = V_{dc} - 3V_{dc}/4 + V_{dc}/2 - V_{dc}/4$ by turning on devices S_{a1}, S_{a3}, S'_{a2}, and S'_{a4}.
   d. $v_{ao} = V_{dc} - 3V_{dc}/4 + V_{dc}/4$ by turning on devices S_{a1}, S_{a4}, S'_{a2}, and S'_{a3}.
   e. $v_{ao} = 3V_{dc}/4 - V_{dc}/2 + V_{dc}/4$ by turning on devices S_{a2}, S_{a4}, S'_{a1}, and S'_{a3}.
   f. $v_{ao} = 3V_{dc}/4 - V_{dc}/4$ by turning on devices S_{a2}, S_{a3}, S'_{a1}, and S'_{a4}.

4. For an output voltage level $v_{ao} = V_{dc}/4$, there are four combinations:
   a. $v_{ao} = V_{dc} - 3V_{dc}/4$ by turning on devices S_{a1}, S'_{a2}, S'_{a3}, and S'_{a4}.
   b. $v_{ao} = V_{dc}/4$ by turning on devices S_{a4}, S'_{a1}, S'_{a2}, and S'_{a3}.
   c. $v_{ao} = V_{dc}/2 - V_{dc}/4$ by turning on devices S_{a3}, S'_{a1}, S'_{a2}, and S'_{a4}.
   d. $v_{ao} = 3V_{dc}/4 - V_{dc}/2$ by turning on devices S_{a2}, S'_{a1}, S'_{a3}, and S'_{a4}.

5. For an output voltage level $v_{ao} = 0$, turn on all lower half switches S'_{a1} through S'_{a4}.

There are many possible switch combinations to generate the five-level output voltage. Table 9.2, however, lists a possible combination of the voltage levels and their corresponding switch states. Using such a switch combination requires each device to be switched only once per cycle. It can be noticed from Table 9.2 that the switching devices have unequal turn-on time. Like the diode-clamped inverter, the line voltage consists of the positive phase-leg voltage of terminal a and the negative phase-leg voltage of terminal b. The resulting line voltage is a nine-level staircase wave. This implies that an m-level converter has an m-level output phase-leg voltage and a $(2m - 1)$-level output line voltage.

TABLE 9.2   One Possible Switch Combination of the Flying-Capacitors Inverter

Output V_{a0}	Switch State							
	S_{a1}	S_{a2}	S_{a3}	S_{a4}	S'_{a4}	S'_{a3}	S'_{a2}	S'_{a1}
$V_5 = V_{dc}$	1	1	1	1	0	0	0	0
$V_4 = 3V_{dc}/4$	1	1	1	0	1	0	0	0
$V_3 = V_{dc}/2$	1	1	0	0	1	1	0	0
$V_2 = V_{dc}/4$	1	0	0	0	1	1	1	0
$V_1 = 0$	0	0	0	0	1	1	1	1

9.5.2   Features of Flying-Capacitors Inverter

The main features are as follows:

1. *Large number of capacitors*: The inverter requires a large number of storage capacitors. Assuming that the voltage rating of each capacitor is the same as that of a switching device, an m-level converter requires a total of $(m - 1) \times (m - 2)/2$ auxiliary capacitors per phase leg in addition to $(m - 1)$ main dc bus capacitors. On the contrary, an m-level diode-clamp inverter only requires $(m - 1)$ capacitors of the same voltage rating. Thus, for $m = 5$, $N_C = 4 \times 3/2 + 4 = 10$ compared with $N_C = 4$ for the diode-clamped type.

2. *Balancing capacitor voltages*: Unlike the diode-clamped inverter, the FCMLI has redundancy at its inner voltage levels. A voltage level is redundant if two or more valid switch combinations can synthesize it. The availability of voltage redundancies allows controlling the individual capacitor voltages. In producing the same output voltage, the inverter can involve different combinations of capacitors allowing preferential charging or discharging of individual capacitors. This flexibility makes it easier to manipulate the capacitor voltages and keep them at their proper values. It is possible to employ two or more switch combinations for middle voltage levels (i.e., $3V_{dc}/4$, $V_{dc}/2$, and $V_{dc}/4$) in one or several output cycle to balance the charging and discharging of the capacitors. Thus, by proper selection of switch combinations, the flying-capacitors multilevel converter may be used in real power conversions. However, when it involves real power conversions, the selection of a switch combination becomes very complicated, and the switching frequency needs to be higher than the fundamental frequency.

The major advantages of the flying-capacitors inverter can be summarized as follows:

- Large amounts of storage capacitors can provide capabilities during power outages.
- These inverters provide switch combination redundancy for balancing different voltage levels.
- Like the diode-clamp inverter with more levels, the harmonic content is low enough to avoid the need for filters.
- Both real and reactive power flow can be controlled.

The major disadvantages of the flying-capacitors inverter can be summarized as follows:

- An excessive number of storage capacitors is required when the number of levels is high. High-level inverters are more difficult to package with the bulky power capacitors and are more expensive too.
- The inverter control can be very complicated, and the switching frequency and switching losses are high for real power transmission.

9.6   CASCADED MULTILEVEL INVERTER

A cascaded multilevel inverter consists of a series of H-bridge (single-phase, full-bridge) inverter units. The general function of this multilevel inverter is to synthesize a

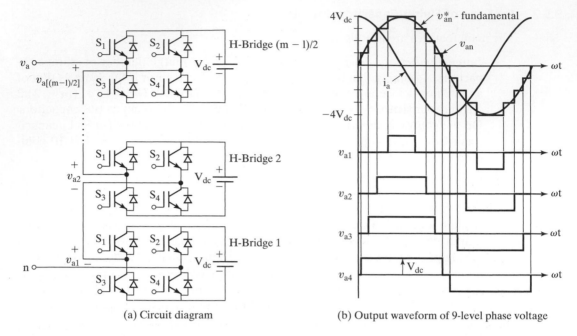

(a) Circuit diagram   (b) Output waveform of 9-level phase voltage

FIGURE 9.8

Single-phase multilevel cascaded H-bridge inverter. [Ref. 7]

desired voltage from several separate dc sources (SDCSs), which may be obtained from batteries, fuel cells, or solar cells. Figure 9.8a shows the basic structure of a single-phase cascaded inverter with SDCSs [7]. Each SDCS is connected to an H-bridge inverter. The ac terminal voltages of different level inverters are connected in series. Unlike the diode-clamp or flying-capacitors inverter, the cascaded inverter does not require any voltage-clamping diodes or voltage-balancing capacitors.

9.6.1 Principle of Operation

Figure 9.8b shows the synthesized phase voltage waveform of a five-level cascaded inverter with four SDCSs. The phase output voltage is synthesized by the sum of four inverter outputs, $v_{an} = v_{a1} + v_{a2} + v_{a3} + v_{a4}$. Each inverter level can generate three different voltage outputs, $+V_{dc}$, 0, and $-V_{dc}$, by connecting the dc source to the ac output side by different combinations of the four switches, S_1, S_2, S_3, and S_4. Using the top level as the example, turning on S_1 and S_4 yields $v_{a4} = +V_{dc}$. Turning on S_2 and S_3 yields $v_{a4} = -V_{dc}$. Turning off all switches yields $v_4 = 0$. Similarly, the ac output voltage at each level can be obtained in the same manner. If N_S is the number of dc sources, the output phase voltage level is $m = N_S + 1$. Thus, a five-level cascaded inverter needs four SDCSs and four full bridges. Controlling the conducting angles at different inverter levels can minimize the harmonic distortion of the output voltage.

The output voltage of the inverter is almost sinusoidal, and it has less than 5% total harmonic distribution (THD) with each of the H-bridges switching only at

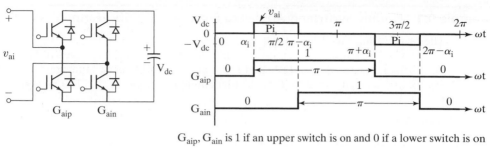

G_{aip}, G_{ain} is 1 if an upper switch is on and 0 if a lower switch is on

(a) One H-bridge   (b) Switching time

FIGURE 9.9

Generation of quasi-square waveform. [Ref. 7]

fundamental frequency. If the phase current i_a, as shown in Figure 9.8b, is sinusoidal and leads or lags the phase voltage v_{an} by 90°, the average charge to each dc capacitor is equal to zero over one cycle. Therefore, all SDCS capacitor voltages can be balanced.

Each H-bridge unit generates a quasi-square waveform by phase shifting its positive and negative phase-leg-switching timings. Figure 9.9 shows the switching timings to generate a quasi-square waveform of an H-bridge. It should be noted that each switching device always conducts for 180° (or half-cycle), regardless of the pulse width of the quasi-square wave. This switching method makes all of the switching device current stresses equal.

9.6.2   Features of Cascaded Inverter

The main features are as follows:

- For real power conversions from ac to dc and then dc to ac, the cascaded inverters need separate dc sources. The structure of separate dc sources is well suited for various renewable energy sources such as fuel cell, photovoltaic, and biomass.
- Connecting dc sources between two converters in a back-to-back fashion is not possible because a short circuit can be introduced when two back-to-back converters are not switching synchronously.

The major advantages of the cascaded inverter can be summarized as follows:

- Compared with the diode-clamped and flying-capacitors inverters, it requires the least number of components to achieve the same number of voltage levels.
- Optimized circuit layout and packaging are possible because each level has the same structure and there are no extra clamping diodes or voltage-balancing capacitors.
- Soft-switching techniques can be used to reduce switching losses and device stresses.

The major disadvantage of the cascaded inverter are as follows:

- It needs separate dc sources for real power conversions, thereby limiting its applications.

Example 9.1   Finding Switching Angles to Eliminate Specific Harmonics

The phase voltage waveform for a cascaded inverter is shown in Figure 9.10 for $m = 6$ (including 0 level). (a) Find the generalized Fourier series of the phase voltage. (b) Find the switching angles to eliminate 5th, 7th, 11th, and 13th harmonics if the peak fundamental phase voltage is 80% of its maximum value. (c) Find the fundamental component B_1, THD, and the distortion factor (DF).

Solution

a. For a cascaded inverter with m levels (including 0) per half-phase, the output voltage per leg is

$$v_{an} = v_{a1} + v_{a2} + v_{a3} + \cdots + v_{am-1} \tag{9.4}$$

Due to the quarter-wave symmetry along the x-axis, both Fourier coefficients A_0 and A_n are zero. We get B_n as

$$B_n = \frac{4V_{dc}}{\pi} \left[\int_{\alpha_1}^{\pi/2} \sin(n\omega t)d(\omega t) + \int_{\alpha_2}^{\pi/2} \sin(n\omega t)d(\omega t) + \cdots \right.$$

$$\left. + \int_{\alpha_{m-1}}^{\pi/2} \sin(n\omega t)d(\omega t) \right] \tag{9.5}$$

$$B_n = \frac{4V_{dc}}{n\pi} \left[\sum_{j=1}^{m-1} \cos(n\alpha_j) \right] \tag{9.6}$$

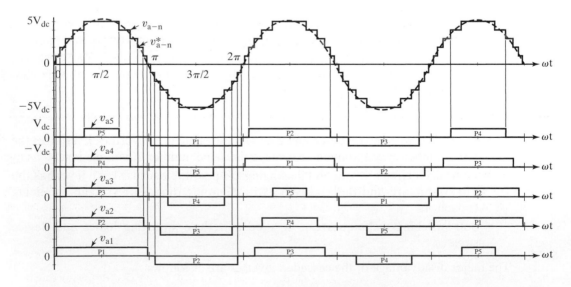

FIGURE 9.10

Switching pattern swapping of the cascade inverter for balancing battery charge. [Ref. 7]

which gives the instantaneous phase voltage v_{an} as

$$v_{an}(\omega t) = \frac{4V_{dc}}{n\pi} \left[\sum_{j=1}^{m-1} \cos(n\alpha_j) \right] \sin(n\omega t) \tag{9.7}$$

b. If the peak output phase voltage $V_{an(peak)}$ must equal to the carrier phase voltage, $V_{cr(peak)} = (m-1)V_{dc}$. Thus, the modulation index becomes

$$M = \frac{V_{cr(peak)}}{V_{an(peak)}} = \frac{V_{cr(peak)}}{(m-1)V_{dc}} \tag{9.8}$$

The conducting angles $\alpha_1, \alpha_2, \ldots, \alpha_{m-1}$ can be chosen such that the total harmonic distortion of the phase voltage is minimized. These angles are normally chosen so as to cancel some predominant lower frequency harmonics. Thus, to eliminate 5th, 7th, 11th, and 13th harmonics provided that the peak fundamental phase voltage is 80% of its maximum value, we must solve the following equations for modulation index $M = 0.8$.

$$\cos(5\alpha_1) + \cos(5\alpha_2) + \cos(5\alpha_3) + \cos(5\alpha_4) + \cos(5\alpha_5) = 0$$
$$\cos(7\alpha_1) + \cos(7\alpha_2) + \cos(7\alpha_3) + \cos(7\alpha_4) + \cos(7\alpha_5) = 0$$
$$\cos(11\alpha_1) + \cos(11\alpha_2) + \cos(11\alpha_3) + \cos(11\alpha_4) + \cos(11\alpha_5) = 0$$
$$\cos(13\alpha_1) + \cos(13\alpha_2) + \cos(13\alpha_3) + \cos(13\alpha_4) + \cos(13\alpha_5) = 0 \tag{9.9}$$
$$\cos(\alpha_1) + \cos(\alpha_2) + \cos(\alpha_3) + \cos(\alpha_4) + \cos(\alpha_5)$$
$$= (m-1)M$$
$$= 5 \times 0.8 = 4$$

This set of nonlinear transcendental equations can be solved by an iterative method such as the Newton–Raphson method. Using Mathcad, we get

$$\alpha_1 = 6.57°, \alpha_2 = 18.94°, \alpha_3 = 27.18°, \alpha_4 = 45.15°, \text{ and } \alpha_5 = 62.24°$$

Thus, if the inverter output is symmetrically switched during the positive half-cycle of the fundamental voltage to $+V_{dc}$ at 6.57°, $+2V_{dc}$ at 18.94°, $+3V_{dc}$ at 27.18°, $+4V_{dc}$ at 45.15°, and $+5V_{dc}$ at 62.24° and similarly in the negative half-cycle to $-V_{dc}$ at 186.57°, $-2V_{dc}$ at 198.94°, $-3V_{dc}$ at 207.18°, $-4V_{dc}$ at 225.15°, and $-5V_{dc}$ at 242.24°, the output voltage cannot contain the 5th, 7th, 11th, and 13th harmonics.

c. Using Mathcad, we get $B_1 = 5.093\%$, THD = 5.975%, and DF = 0.08%

Note: The duty cycle for each of the voltage levels is different. This means that the level-1 dc source discharges much sooner than the level-5 dc source. However, by using a switching pattern-swapping scheme among the various levels every half-cycle, as shown in Figure 9.10, all batteries can be equally used (discharged) or charged [7]. For example, if the first pulse sequence is $P_1, P_2, \ldots, P_5$, then the next sequence is P_2, P_3, P_4, P_5, P_1, and so on.

9.7 APPLICATIONS

There is considerable interest in applying voltage source inverters in high-power applications such as in utility systems for controlled sources of reactive power. In the steady-state operation, an inverter can produce a controlled reactive current and operates as a static volt–ampere reactive (VAR)-compensator (STATCON). Also, these inverters can reduce the physical size of the compensator and improve its performance during power system contingencies. The use of a high-voltage inverter makes possible direct connection to the high-voltage (e.g., 13-kV) distribution system, eliminating the distribution transformer and reducing system cost. In addition, the harmonic content of the inverter waveform can be reduced with appropriate control techniques and thus the efficiency of the system can be improved. The most common applications of multilevel converters include (1) reactive power compensation, (2) back-to-back intertie, and (3) variable speed drives.

9.7.1 Reactive Power Compensation

An inverter converts a dc voltage to an ac voltage; with a phase shift of 180°, the inverter can be operated as a dc–ac converter, that is, a controlled rectifier. With a purely capacitive load, the inverter operating as a dc–ac converter can draw reactive current from the ac supply. Figure 9.11 shows the circuit diagram of a multilevel converter directly connected to a power system for reactive power compensation. The load side is connected to the ac supply and the dc side is open, not connected to any dc voltage. For the control of the reactive power flow, the inverter gate control is phase shifted by 180°. The dc side capacitors act as the load.

When a multilevel converter draws pure reactive power, the phase voltage and current are 90° apart, and the capacitor charge and discharge can be balanced. Such a converter, when serving for reactive power compensation, is called a static-VAR generator (SVG). All three multilevel converters can be used in reactive power compensation without having the voltage unbalance problem.

The relationship of the source voltage vector $\mathbf{V_S}$ and the converter voltage vector $\mathbf{V_C}$ is simply $\mathbf{V_S} = \mathbf{V_C} + j\mathbf{I_C}X_S$, where $\mathbf{I_C}$ is the converter current vector, and X_S is

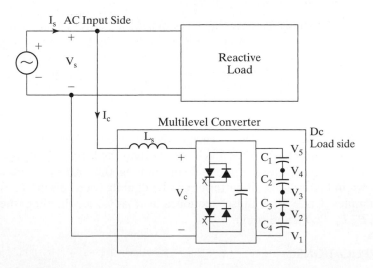

FIGURE 9.11

A multilevel converter connected to a power system for reactive power compensation. [Ref. 5]

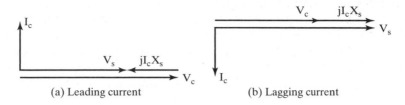

(a) Leading current   (b) Lagging current

FIGURE 9.12

Phasor diagrams of the source and the converter voltages for reactive power compensation.

the reactance of the inductor L_S. Figure 9.12a illustrates that the converter voltage is in phase with the source voltage with a leading reactive current, whereas Figure 9.12b illustrates a lagging reactive current. The polarity and the magnitude of the reactive current are controlled by the magnitude of the converter voltage $\mathbf{V_C}$, which is a function of the dc bus voltage and the voltage modulation index, as expressed by Eqs. (9.7) and (9.8).

9.7.2   Back-to-Back Intertie

Figure 9.13 shows two diode-clamped multilevel converters that are interconnected with a dc capacitor link. The left-hand side converter serves as the rectifier for utility interface, and the right-hand side converter serves as the inverter to supply the ac load. Each switch remains on once per fundamental cycle. The voltage across each capacitor remains well balanced, while maintaining the staircase voltage wave, because the unbalance capacitor voltages on both sides tend to compensate each other. Such a dc capacitor link is categorized as the back-to-back intertie.

The back-to-back intertie that connects two asynchronous systems can be regarded as (1) a frequency changer, (2) a phase shifter, or (3) a power flow controller. The power flow between two systems can be controlled bidirectionally. Figure 9.14 shows the phasor diagram for real power transmission from the source end to the load end. This diagram indicates that the source current can be leading or lagging the source voltage. The converter voltage is phase shifted from the source voltage with a power angle, δ. If the source voltage is constant, then the current or power flow can be

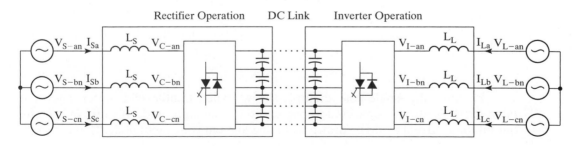

FIGURE 9.13

Back-to-back intertie system using two diode-clamped multilevel converters. [Ref. 5]

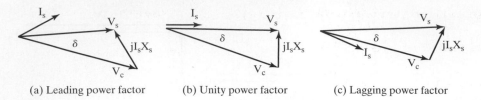

(a) Leading power factor   (b) Unity power factor   (c) Lagging power factor

FIGURE 9.14

Phasor diagram of the source voltage, converter voltage, and current showing real power conversions.

controlled by the converter voltage. For $\delta = 0$, the current is either 90° leading or lagging, meaning that only reactive power is generated.

9.7.3   Adjustable Speed Drives

The back-to-back intertie can be applied to a utility compatible adjustable speed drive (ASD) where the input is the constant frequency ac source from the utility supply and the output is the variable frequency ac load. For an ideal utility compatible system, it requires unity power factor, negligible harmonics, no electromagnetic interference (EMI), and high efficiency. The major differences, when using the same structure for ASDs and for back-to-back interties, are the control design and the size of the capacitor. Because the ASD needs to operate at different frequencies, the dc-link capacitor needs to be well sized to avoid a large voltage swing under dynamic conditions.

9.8   SWITCHING DEVICE CURRENTS

Let us take a three-level half-bridge inverter, as shown in Figure 9.15a, where V_o and I_0 indicate the rms load voltage and current, respectively. Assuming that the load inductance is sufficiently large and the capacitors maintain their voltages so that the output current is sinusoidal as given by

$$i_o = I_m \sin(\omega t - \phi) \tag{9.10}$$

where I_m is the peak value of the load current, and ϕ is the load impedance angle.

Figure 9.15b shows a typical current waveform of each switching device with a simple stepped control of output phase voltage. The most inner switches such as S_4 and S_1' carry more current than the most outer switches such as S_1 and S_4'.

Each input node current can be expressed as a function of the switching function SF_n as given by

$$i_n = SF_n i_o \qquad \text{for } n = 1, 2, \ldots, m \tag{9.11}$$

Because the single-pole multiple-throw switch multilevel inverter, shown in Figure 9.1b, is always connected to one and only one input node at every instant, the output load current could be drawn from one and only one input node. That is,

$$i_o = \sum_{n=1}^{m} i_n \tag{9.12}$$

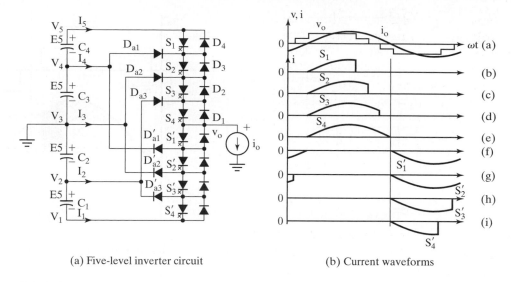

(a) Five-level inverter circuit   (b) Current waveforms

FIGURE 9.15

Half-bridge three-level diode-clamped inverter. [Ref. 4]

and the rms value of each current is expressed as

$$I_{o(\text{rms})}^2 = \sum_n^m I_{n(\text{rms})}^2 \tag{9.13}$$

where $I_{n(\text{rms})}$ is the rms current of the nth node given by

$$I_{n(\text{rms})} = \sqrt{\frac{1}{2\pi} \int_0^{2\pi} SF_n i_o^2 \, d(\omega t)} \qquad \text{for } n = 1, 2, \ldots, m \tag{9.14}$$

For balanced switching with respect to the ground level, we get

$$i_{1(\text{rms})}^2 = i_{5(\text{rms})}^2, \text{ and } i_{2(\text{rms})}^2 = i_{4(\text{rms})}^2 \tag{9.15}$$

It should be noted that by structure, the currents through the opposite switches such as $S_1', \ldots, S_4'$ would have the same rms current through $S_4, \ldots, S_1$ respectively.

9.9   DC-LINK CAPACITOR VOLTAGE BALANCING

The voltage balancing of capacitors acting as an energy tank is very important for the multilevel inverter to work satisfactorily. Figure 9.16a shows the schematic of a half-bridge inverter with five levels and Figure 9.16b illustrates the stepped-output voltage and the sinusoidal load current $i_o = I_m \sin(\omega t - \phi)$.

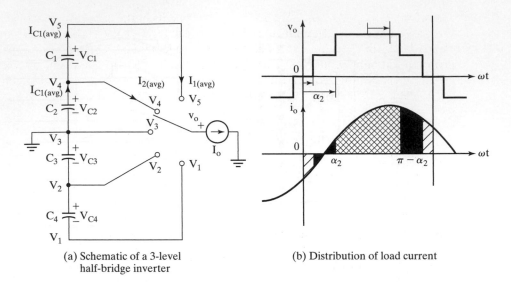

(a) Schematic of a 3-level
half-bridge inverter

(b) Distribution of load current

FIGURE 9.16

Charge distribution of capacitors. [Ref. 4]

The average value of the input node current i_1 is given by

$$I_{1(\text{avg})} = \frac{1}{2\pi} \int_{\alpha_2}^{\pi-\alpha_2} i_o d(\omega t) = \frac{1}{2\pi} \int_{\alpha_2}^{\pi-\alpha_2} I_m \sin(\omega t - \phi) d(\omega t)$$

$$= \frac{I_m}{\pi} \cos \phi \cos \alpha_2$$

(9.16)

Similarly, the average value of the input node current i_2 is given by

$$I_{2(\text{avg})} = \frac{1}{2\pi} \int_{\alpha_1}^{\alpha_2} i_o d(\omega t) = \frac{1}{2\pi} \int_{\alpha_1}^{\alpha_2} I_m \sin(\omega t - \phi) d(\omega t)$$

$$= \frac{I_m}{\pi} \cos \phi (\cos \alpha_1 - \cos \alpha_2)$$

(9.17)

By symmetry, $I_{3(\text{avg})} = 0$, $I_{4(\text{avg})} = -I_{2(\text{avg})}$, and $I_{5(\text{avg})} = -I_{1(\text{avg})}$. Thus, each capacitor voltage should be regulated so that each capacitor supplies the average current per cycle as follows;

$$I_{C1(\text{avg})} = I_{1(\text{avg})} = \frac{I_m}{\pi} \cos \phi \cos \alpha_2$$

(9.18)

$$I_{C2(\text{avg})} = I_{1(\text{avg})} + I_{2(\text{avg})} = \frac{I_m}{\pi} \cos \phi \cos \alpha_1$$

(9.19)

Therefore, $I_{C1(avg)} < I_{C2(avg)}$ for $\alpha_1 < \alpha_2$. This results in the capacitor charge unbalancing, and more change flows from the inner capacitor C_2 (or C_3) than that of the outer capacitor C_1 (or C_4). Thus, each capacitor voltage should be regulated to supply the appropriate amount of average current; otherwise, its voltage V_{C2} (or V_{C3}) goes to the ground level as times goes. Equations (9.18) and (9.19) can be extended to the nth capacitor of a multilevel converter as given by

$$I_{Cn(avg)} = \frac{I_m}{\pi} \cos \phi \cos \alpha_n \tag{9.20}$$

Equations (9.18) and (9.19) give

$$\frac{\cos \alpha_2}{\cos \alpha_1} = \frac{I_{C2(avg)}}{I_{C1(avg)}} \tag{9.21}$$

which can be generalized for the nth and $(n-1)$th capacitors

$$\frac{\cos \alpha_n}{\cos \alpha_{n-1}} = \frac{I_{Cn(avg)}}{I_{C(n-1)(avg)}} \tag{9.22}$$

which means that the capacitor charge unbalancing exists regardless of the load condition and it depends on the control strategy such as $\alpha_1, \alpha_2, \ldots, \alpha_n$. Applying control strategy that forces the energy transfer from the outer capacitors to the inner capacitors can solve this unbalancing problem [9–11].

9.10   FEATURES OF MULTILEVEL INVERTERS

A multilevel inverter can eliminate the need for the step-up transformer and reduce the harmonics produced by the inverter. Although the multilevel inverter structure was initially introduced as a means of reducing the output waveform harmonic content, it was found [1] that the dc bus voltage could be increased beyond the voltage rating of an individual power device by the use of a voltage clamping network consisting of diodes. A multilevel structure with more than three levels can significantly reduce the harmonic content [2, 3]. By using voltage-clamping techniques, the system KV rating can be extended beyond the limits of an individual device. The intriguing feature of the multilevel inverter structures is their ability to scale up the kilovolt–ampere (kVA)-rating and also to improve the harmonic performance greatly without having to resort to PWM techniques. The key features of a multilevel structure follow:

- The output voltage and power increase with number of levels. Adding a voltage level involves adding a main switching device to each phase.
- The harmonic content decreases as the number of levels increases and filtering requirements are reduced.
- With additional voltage levels, the voltage waveform has more free-switching angles, which can be preselected for harmonic elimination.
- In the absence of any PWM techniques, the switching losses can be avoided. Increasing output voltage and power does not require an increase in rating of individual device.

TABLE 9.3   Comparisons of Component Requirements per Leg of Three Multilevel Converters [Ref. 5]

Converter Type	Diode Clamp	Flying Capacitors	Cascaded Inverters
Main switching devices	$(m - 1) \times 2$	$(m - 1) \times 2$	$(m - 1) \times 2$
Main diodes	$(m - 1) \times 2$	$(m - 1) \times 2$	$(m - 1) \times 2$
Clamping diodes	$(m - 1) \times (m - 2)$	0	0
Dc bus capacitors	$(m - 1)$	$(m - 1)$	$(m - 1)/2$
Balancing capacitors	0	$(m - 1) \times (m - 2)/2$	0

- Static and dynamic voltage sharing among the switching devices is built into the structure through either clamping diodes or capacitors.
- The switching devices do not encounter any voltage-sharing problems. For this reason, multilevel inverters can easily be applied for high-power applications such as large motor drives and utility supplies.
- The fundamental output voltage of the inverter is set by the dc bus voltage V_{dc}, which can be controlled through a variable dc link.

9.11   COMPARISONS OF MULTILEVEL CONVERTERS

The multilevel converters [8] can replace the existing systems that use traditional multipulse converters without the need for transformers. For a three-phase system, the relationship between the number of levels m, and the number of pulses p, can be formulated by $p = 6 \times (m - 1)$. All three converters have the potential for applications in high-voltage, high-power systems such as an SVG without voltage unbalance problems because the SVG does not draw real power. The diode-clamped converter is most suitable for the back-to-back intertie system operating as a unified power flow controller. The other two types may also be suitable for the back-to-back intertie, but they would require more switching per cycle and more advanced control techniques to balance the voltage. The multilevel inverters can find potential applications in adjustable speed drives where the use of multilevel converters can not only solve harmonics and EMI problems but also avoid possible high-frequency switching dv/dt-induced motor failures.

Table 9.3 compares the component requirements per phase leg among the three multilevel converters. All devices are assumed to have the same voltage rating, but not necessarily the same current rating. The cascaded inverter uses a full bridge in each level as compared with the half-bridge version for the other two types. The cascaded inverter requires the least number of components and has the potential for utility interface applications because of its capabilities for applying modulation and soft-switching techniques.

SUMMARY

Multilevel converters can be applied to utility interface systems and motor drives. These converters offer a low output voltage THD, and a high efficiency and power factor. There are three types of multilevel converters: (1) diode clamped, (2) flying capacitors, and (3) cascaded. The main advantages of multilevel converters include the following:

- They are suitable for high-voltage and high-current applications.
- They have higher efficiency because the devices can be switched at a low frequency.
- Power factor is close to unity for multilevel inverters used as rectifiers to convert ac to dc.
- No EMI problem exists.
- No charge unbalance problem results when the converters are in either charge mode (rectification) or drive mode (inversion).

The multilevel converters require balancing the voltage across the series-connected dc-bus capacitors. Capacitors tend to overcharge or completely discharge, at which condition the multilevel converter reverts to a three-level converter unless an explicit control is devised to balance the capacitor charge. The voltage-balancing technique must be applied to the capacitor during the operations of the rectifier and the inverter. Thus, the real power flow into a capacitor must be the same as the real power flow out of the capacitor, and the net charge on the capacitor over one cycle remains the same.

REFERENCES

[1]   A. Nabae, I. Takahashi, and H. Akagi, "A new neutral-point clamped PWM inverter," *IEEE Transactions on Industry Applications*, Vol. IA-17, No. 5, September/October 1981, pp. 518–523.

[2]   P. M. Bhagwat and V. R. Stefanovic, "Generalized structure of a multilevel PWM inverter," *IEEE Transactions on Industry Applications*, Vol. 19, No. 6, November/December 1983, pp. 1057–1069.

[3]   M. Carpita and S. Teconi, "A novel multilevel structure for voltage source inverter," *Proc. European Power Electronics*, 1991, pp. 90–94.

[4]   N. S. Choi, L. G. Cho, and G. H. Cho, "A general circuit topology of multilevel inverter," *IEEE Power Electronics Specialist Conference*, 1991, pp. 96–103.

[5]   J.-S. Lai, and F. Z. Peng, "Multilevel converters—a new breed of power converters," *IEEE Transactions on Industry Applications*, Vol. 32, No. 3, May/June 1996, pp. 509–517.

[6]   X. Yuan and I. Barbi, "Fundamentals of a new diode clamping multilevel inverter," *IEEE Transactions on Power Electronics*, Vol. 15, No. 4, July 2000, pp. 711–718.

[7]   L. M. Tolbert, F. Z. Peng, and T. G. Habetler, "Multilevel converters for large electric drives," *IEEE Transactions on Industry Applications*, Vol. 35, No. 1, January/February 1999, pp. 36–44.

[8]   C. Hochgraf, R. I. Asseter, D. Divan, and T. A. Lipo, "Comparison of multilevel inverters for static-var compensation," IEEF-IAS Annual Meeting Record, 1994, pp. 921–928.

[9]   L. M. Tolbert and T. G. Habetler, "Novel multilevel inverter carrier-based PWM method," *IEEE Transactions on Industry Applications*, Vol. 35, No. 5, September/October 1999, pp. 1098–1107.

[10]   L. M. Tolbert, F. Z. Peng, and T. G. Habetler, "Multilevel PWM methods at low modulation indices," *IEEE Transactions on Power Electronics*, Vol. 15, No. 4, July 2000, pp. 719–725.

[11] J. H. Seo, C. H. Choi, and D. S. Hyun, "A new simplified space—vector PWM method for three-level inverters," *IEEE Transactions on Power Electronics*, Vol. 16, No. 4, July 2001, pp. 545–550.

REVIEW QUESTIONS

9.1 What is a multilevel converter?

9.2 What is the basic concept of multilevel converters?

9.3 What are the features of a multilevel converter?

9.4 What are the types of multilevel converters?

9.5 What is a diode-clamped multilevel inverter?

9.6 What are the advantages of a diode-clamped multilevel inverter?

9.7 What are the disadvantages of a diode-clamped multilevel inverter?

9.8 What are the advantages of a modified diode-clamped multilevel inverter?

9.9 What is a flying-capacitors multilevel inverter?

9.10 What are the advantages of a flying-capacitors multilevel inverter?

9.11 What are the disadvantages of a flying-capacitors multilevel inverter?

9.12 What is a cascaded multilevel inverter?

9.13 What are the advantages of a cascaded multilevel inverter?

9.14 What are the disadvantages of a cascaded multilevel inverter?

9.15 What is a back-to-back intertie system?

9.16 What does the capacitor voltage unbalancing mean?

9.17 What are the possible applications of multilevel inverters?

PROBLEMS

9.1 A single-phase diode-clamped inverter has $m = 5$. Find the generalized Fourier series and THD of the phase voltage.

9.2 A single-phase diode-clamped inverter has $m = 5$. Find the peak voltage and current ratings of diodes and switching devices if $V_{dc} = 5\,kV$ and $i_o = 50 \sin(\theta - \pi/3)$.

9.3 A single-phase diode-clamped inverter has $m = 5$. Find **(a)** instantaneous, average and rms currents of each node, and **(b)** average and rms capacitor current if $V_{dc} = 5\,kV$ and $i_o = 50 \sin(\theta - \pi/3)$.

9.4 A single-phase flying-capacitors multilevel inverter has $m = 5$. Find the generalized Fourier series and THD of the phase voltage.

9.5 A single-phase flying-capacitors multilevel inverter has $m = 5$. Find the number of capacitors, the peak voltage and current ratings of diodes and switching devices if $V_{dc} = 5\,kV$.

9.6 Compare the number of diodes and capacitors for diode clamp, flying capacitors and cascaded inverters if $m = 5$.

9.7 A single-phase cascaded multilevel inverter has $m = 5$. Find the peak voltage, and average and rms current ratings of H-bridge if $V_{dc} = 1\,kV$ and $i_o = 150 \sin(\theta - \pi/6)$.

9.8 A single-phase cascaded multilevel inverter has $m = 5$. Find the average current of each separate dc source (SDCS) if $V_{dc} = 1\,kV$ and $i_o = 150 \sin(\theta - \pi/6)$.

9.9 A single-phase cascaded multilevel inverter has $m = 5$. Find the generalized Fourier series and THD of the phase voltage. **(b)** Find the switching angles to eliminate 5th, 7th, 11th, and 13th harmonics.

9.10 A single-phase cascaded multilevel inverter has $m = 5$. **(a)** Find the generalized Fourier series and THD of the phase voltage. **(b)** Find the switching angles to eliminate 5th, 7th, and 11th harmonics if the peak fundamental phase voltage is 60% of its maximum value.

CHAPTER 10

Controlled Rectifiers

The learning objectives of this chapter are as follows:

- To understand the operation and characteristics of controlled rectifiers
- To learn the types of controlled rectifiers
- To understand the performance parameters of controlled rectifiers
- To learn the techniques for analysis and design of controlled rectifier circuits
- To learn the techniques for simulating controlled rectifiers by using SPICE
- To study effects of load inductance on the load current

10.1 INTRODUCTION

We have seen in Chapter 3 that diode rectifiers provide a fixed output voltage only. To obtain controlled output voltages, phase-control thyristors are used instead of diodes. The output voltage of thyristor rectifiers is varied by controlling the delay or firing angle of thyristors. A phase-control thyristor is turned on by applying a short pulse to its gate and turned off due to *natural* or *line commutation*; and in case of a highly inductive load, it is turned off by firing another thyristor of the rectifier during the negative half-cycle of input voltage.

These phase-controlled rectifiers are simple and less expensive; and the efficiency of these rectifiers is, in general, above 95%. Because these rectifiers convert from ac to dc, these controlled rectifiers are also called *ac–dc converters* and are used extensively in industrial applications, especially in variable-speed drives, ranging from fractional horsepower to megawatt power level.

The phase-control converters can be classified into two types, depending on the input supply: (1) single-phase converters, and (2) three-phase converters. Each type can be subdivided into (a) semiconverter, (b) full converter, and (c) dual converter. A *semiconverter* is a one-quadrant converter and it has one polarity of output voltage and current. A *full converter* is a two-quadrant converter and the polarity of its output voltage can be either positive or negative. However, the output current of full converter has one polarity only. A *dual converter* can operate in four quadrants; and both the output voltage and current can be either positive or negative. In some applications, converters are connected in series to operate at higher voltages and to improve the input power factor (PF).

The method of Fourier series similar to that of diode rectifiers can be applied to analyze the performances of phase-controlled converters with RL loads. However, to simplify the analysis, the load inductance can be assumed sufficiently high so that the load current is continuous and has negligible ripple.

10.2   PRINCIPLE OF PHASE-CONTROLLED CONVERTER OPERATION

Let us consider the circuit in Figure 10.1a with a resistive load. During the positive half-cycle of input voltage, the thyristor anode is positive with respect to its cathode and the thyristor is said to be *forward biased*. When thyristor T_1 is fired at $\omega t = \alpha$, thyristor T_1 conducts and the input voltage appears across the load. When the input voltage starts to be negative at $\omega t = \pi$, the thyristor anode is negative with respect to its cathode and thyristor T_1 is said to be *reverse biased*; and it is turned off. The time

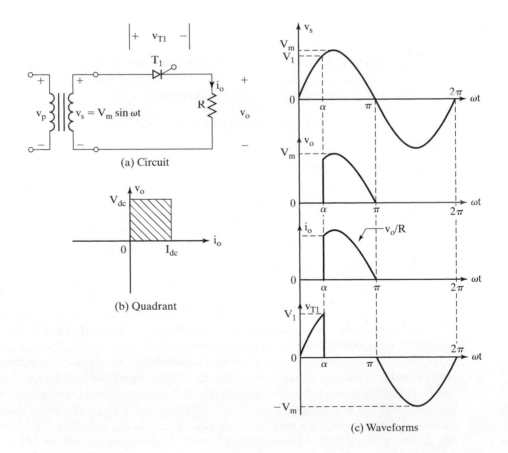

(a) Circuit

(b) Quadrant

(c) Waveforms

FIGURE 10.1

Single-phase thyristor converter with a resistive load.

after the input voltage starts to go positive until the thyristor is fired at $\omega t = \alpha$ is called the *delay* or *firing angle* α.

Figure 10.1b shows the region of converter operation, where the output voltage and current have one polarity. Figure 10.1c shows the waveforms for input voltage, output voltage, load current, and voltage across T_1. This converter is not normally used in industrial applications because its output has high ripple content and low ripple frequency. However, it explains the principle of the single-phase thyristor converter. If f_s is the frequency of input supply, the lowest frequency of output ripple voltage is f_s.

If V_m is the peak input voltage, the average output voltage V_{dc} can be found from

$$V_{dc} = \frac{1}{2\pi} \int_\alpha^\pi V_m \sin \omega t \, d(\omega t) = \frac{V_m}{2\pi} [-\cos \omega t]_\alpha^\pi$$

$$= \frac{V_m}{2\pi} (1 + \cos \alpha) \tag{10.1}$$

and V_{dc} can be varied from V_m/π to 0 by varying α from 0 to π. The average output voltage becomes maximum when $\alpha = 0$ and the maximum output voltage V_{dm} is

$$V_{dm} = \frac{V_m}{\pi} \tag{10.2}$$

Normalizing the output voltage with respect to V_{dm}, the normalized output voltage

$$V_n = \frac{V_{dc}}{V_{dm}} = 0.5(1 + \cos \alpha) \tag{10.3}$$

The root-mean-square (rms) output voltage is given by

$$V_{rms} = \left[\frac{1}{2\pi} \int_\alpha^\pi V_m^2 \sin^2 \omega t \, d(\omega t) \right]^{1/2} = \left[\frac{V_m^2}{4\pi} \int_\alpha^\pi (1 - \cos 2\omega t) \, d(\omega t) \right]^{1/2}$$

$$= \frac{V_m}{2} \left[\frac{1}{\pi} \left(\pi - \alpha + \frac{\sin 2\alpha}{2} \right) \right]^{1/2} \tag{10.4}$$

Gating sequence.   The gating sequence for the thyristor switch is as follows:

1. Generate a pulse-signal at the positive zero crossing of the supply voltage v_s.
2. Delay the pulse by the desired angle α and apply it between the gate and cathode terminals of T_1 through a gate-isolating circuit.

Note: Both the output voltage and input current are nonsinusoidal. The performance of a controlled rectifier can be measured by the same parameters as those of the diode rectifiers in Section 3.3 such as distortion factor (DF), total harmonic distortion (THD), PF, transformer utilization factor (TUF), and harmonic factor (HF).

Example 10.1 Finding the Performances of a Single-Phase Thyristor Converter

If the converter of Figure 10.1a has a purely resistive load of R and the delay angle is $\alpha = \pi/2$, determine (a) the rectification efficiency, (b) the form factor (FF), (c) the ripple factor (RF), (d) the TUF, and (e) the peak inverse voltage (PIV) of thyristor T_1.

Solution

The delay angle $\alpha = \pi/2$. From Eq. (10.1), $V_{dc} = 0.1592V_m$ and $I_{dc} = 0.1592V_m/R$. From Eq. (10.3), $V_n = 0.5$. From Eq. (10.4), $V_{rms} = 0.3536V_m$ and $I_{rms} = 0.3536V_m/R$. From Eq. (3.1), $P_{dc} = V_{dc}I_{dc} = (0.1592V_m)^2/R$ and from Eq. (3.2), $P_{ac} = V_{rms}I_{rms} = (0.3536V_m)^2/R$.

a. From Eq. (3.3) the rectification efficiency

$$\eta = \frac{P_{dc}}{P_{ac}} = \frac{(0.1592V_m)^2}{(0.3536V_m)^2} = 20.27\%$$

b. From Eq. (3.5), the FF

$$FF = \frac{V_{rms}}{V_{dc}} = \frac{0.3536V_m}{0.1592V_m} = 2.221 \quad \text{or} \quad 222.1\%$$

c. From Eq. (3.7), the RF $= \sqrt{(FF^2 - 1)} = (2.221^2 - 1)^{1/2} = 1.983$ or 198.3%.

d. The rms voltage of the transformer secondary, $V_s = V_m/\sqrt{2} = 0.707V_m$. The rms value of the transformer secondary current is the same as that of the load, $I_s = 0.3536V_m/R$. The volt-ampere rating (VA) of the transformer, $VA = V_sI_s = 0.707V_m \times 0.3536V_m/R$. From Eq. (3.8),

$$TUF = \frac{P_{dc}}{V_sI_s} = \frac{0.1592^2}{0.707 \times 0.3536} = 0.1014 \quad \text{and} \quad \frac{1}{TUF} = 9.86$$

The PF is approximately equal to TUF. Thus, PF $= 0.1014$.

e. The PIV $= V_m$.

Note: The performance of the converter is degraded at the higher range of delay angle α.

Key Points of Section 10.2

- Varying the delay angle α from 0 to π can vary the average output voltage from V_m/π to 0.
- The input transformer can carry dc current, thereby causing a magnetic saturation problem.

10.3   SINGLE-PHASE FULL CONVERTERS

The circuit arrangement of a single-phase full converter is shown in Figure 10.2a with a highly inductive load so that the load current is continuous and ripple free [10]. During the positive half-cycle, thyristors T_1 and T_2 are forward biased; and when these two thyristors are fired simultaneously at $\omega t = \alpha$, the load is connected to the input supply through T_1 and T_2. Due to the inductive load, thyristors T_1 and T_2 continue to conduct beyond $\omega t = \pi$, even though the input voltage is already negative. During the negative half-cycle of the input voltage, thyristors T_3 and T_4 are forward biased; and firing of thyristors T_3 and T_4 applies the supply voltage across thyristors T_1 and T_2 as reverse

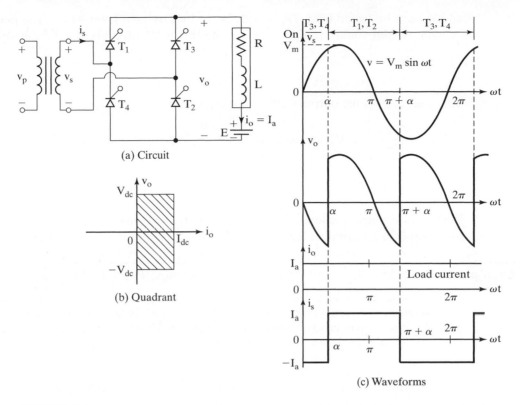

FIGURE 10.2

Single-phase full converter.

blocking voltage. T_1 and T_2 are turned off due to *line* or *natural commutation* and the load current is transferred from T_1 and T_2 to T_3 and T_4. Figure 10.2b shows the regions of converter operation and Figure 10.2c shows the waveforms for input voltage, output voltage, and input and output currents.

During the period from α to π, the input voltage v_s and input current i_s are positive; and the power flows from the supply to the load. The converter is said to be operated in *rectification* mode. During the period from π to $\pi + \alpha$, the input voltage v_s is negative and the input current i_s is positive; and reverse power flows from the load to the supply. The converter is said to be operated in *inversion mode*. This converter is extensively used in industrial applications up to 15 kW [1]. Depending on the value of α, the average output voltage could be either positive or negative and it provides two-quadrant operation.

The average output voltage can be found from

$$V_{dc} = \frac{2}{2\pi} \int_{\alpha}^{\pi+\alpha} V_m \sin \omega t \, d(\omega t) = \frac{2V_m}{2\pi} [-\cos \omega t]_{\alpha}^{\pi+\alpha}$$

$$= \frac{2V_m}{\pi} \cos \alpha$$

(10.5)

and V_{dc} can be varied from $2V_m/\pi$ to $-2V_m/\pi$ by varying α from 0 to π. The maximum average output voltage is $V_{dm} = 2V_m/\pi$ and the normalized average output voltage is

$$V_n = \frac{V_{dc}}{V_{dm}} = \cos\alpha \qquad (10.6)$$

The rms value of the output voltage is given by

$$V_{rms} = \left[\frac{2}{2\pi}\int_\alpha^{\pi+\alpha} V_m^2 \sin^2\omega t\, d(\omega t)\right]^{1/2} = \left[\frac{V_m^2}{2\pi}\int_\alpha^{\pi+\alpha}(1 - \cos 2\omega t)\, d(\omega t)\right]^{1/2}$$

$$= \frac{V_m}{\sqrt{2}} = V_s \qquad (10.7)$$

With a purely resistive load, thyristors T_1 and T_2 can conduct from α to π, and thyristors T_3 and T_4 can conduct from $\alpha + \pi$ to 2π.

Example 10.2   Finding the Input Power Factor of a Single-Phase Full Converter

The full converter in Figure 10.2a is connected to a 120-V, 60-Hz supply. The load current I_a is continuous and its ripple content is negligible. The turns ratio of the transformer is unity. (a) Express the input current in a Fourier series; determine the HF of the input current, DF, and input PF. (b) If the delay angle is $\alpha = \pi/3$, calculate V_{dc}, V_n, V_{rms}, HF, DF, and PF.

Solution

a.   The waveform for input current is shown in Figure 10.2c and the instantaneous input current can be expressed in a Fourier series as

$$i_s(t) = a_0 + \sum_{n=1,2,\ldots}^\infty (a_n \cos n\omega t + b_n \sin n\omega t)$$

where

$$a_0 = \frac{1}{2\pi}\int_\alpha^{2\pi+\alpha} i_s(t)\, d(\omega t) = \frac{1}{2\pi}\left[\int_\alpha^{\pi+\alpha} I_a\, d(\omega t) - \int_{\pi+\alpha}^{2\pi+\alpha} I_a\, d(\omega t)\right] = 0$$

$$a_n = \frac{1}{\pi}\int_\alpha^{2\pi+\alpha} i_s(t) \cos n\omega t\, d(\omega t)$$

$$= \frac{1}{\pi}\left[\int_\alpha^{\pi+\alpha} I_a \cos n\omega t\, d(\omega t) - \int_{\pi+\alpha}^{2\pi+\alpha} I_a \cos n\omega t\, d(\omega t)\right]$$

$$= -\frac{4I_a}{n\pi}\sin n\alpha \quad \text{for } n = 1, 3, 5,\ldots$$

$$= 0 \quad \text{for } n = 2, 4,\ldots$$

$$b_n = \frac{1}{\pi}\int_\alpha^{2\pi+\alpha} i(t) \sin n\omega t\, d(\omega t)$$

$$= \frac{1}{\pi}\left[\int_\alpha^{\pi+\alpha} I_a \sin n\omega t\, d(\omega t) - \int_{\pi+\alpha}^{2\pi+\alpha} I_a \sin n\omega t\, d(\omega t)\right]$$

$$= \frac{4I_a}{n\pi} \cos n\alpha \qquad \text{for } n = 1, 3, 5, \dots$$
$$= 0 \qquad \text{for } n = 2, 4, \dots$$

Because $a_0 = 0$, the input current can be written as

$$i_s(t) = \sum_{n=1,3,5\dots}^{\infty} \sqrt{2}\, I_n \sin(n\omega t + \phi_n)$$

where

$$\phi_n = \tan^{-1} \frac{a_n}{b_n} = -n\alpha \tag{10.8}$$

and ϕ_n is the displacement angle of the nth harmonic current. The rms value of the nth harmonic input current is

$$I_{sn} = \frac{1}{\sqrt{2}}(a_n^2 + b_n^2)^{1/2} = \frac{4I_a}{\sqrt{2}\, n\pi} = \frac{2\sqrt{2}\, I_a}{n\pi} \tag{10.9}$$

and the rms value of the fundamental current is

$$I_{s1} = \frac{2\sqrt{2}\, I_a}{\pi}$$

The rms value of the input current can be calculated from Eq. (10.9) as

$$I_s = \left(\sum_{n=1,3,5,\dots}^{\infty} I_{sn}^2 \right)^{1/2}$$

I_s can also be determined directly from

$$I_s = \left[\frac{2}{2\pi} \int_{\alpha}^{\pi+\alpha} I_a^2\, d(\omega t) \right]^{1/2} = I_a$$

From Eq. (3.10) the HF is found as

$$\text{HF} = \left[\left(\frac{I_s}{I_{s1}} \right)^2 - 1 \right]^{1/2} = 0.483 \quad \text{or} \quad 48.3\%$$

From Eqs. (3.9) and (10.8), the DF is

$$\text{DF} = \cos \phi_1 = \cos(-\alpha) \tag{10.10}$$

From Eq. (3.11) the PF is found as

$$\text{PF} = \frac{I_{s1}}{I_s} \cos(-\alpha) = \frac{2\sqrt{2}}{\pi} \cos \alpha \tag{10.11}$$

b. $\alpha = \pi/3$

$$V_{dc} = \frac{2V_m}{\pi} \cos \alpha = 54.02\ \text{V} \qquad \text{and} \qquad V_n = 0.5\ \text{pu}$$

$$V_{\text{rms}} = \frac{V_m}{\sqrt{2}} = V_s = 120 \text{ V}$$

$$I_{s1} = \left(2\sqrt{2}\,\frac{I_a}{\pi}\right) = 0.90032 I_a \quad \text{and} \quad I_s = I_a$$

$$\text{HF} = \left[\left(\frac{I_s}{I_{s1}}\right)^2 - 1\right]^{1/2} = 0.4834 \quad \text{or} \quad 48.34\%$$

$$\phi_1 = -\alpha \quad \text{and} \quad \text{DF} = \cos(-\alpha) = \cos\frac{-\pi}{3} = 0.5$$

$$\text{PF} = \frac{I_{s1}}{I_s}\cos(-\alpha) = 0.45 \text{ (lagging)}$$

Note: The fundamental component of input current is always 90.03% of I_a and the HF remains constant at 48.34%.

10.3.1 Single-Phase Full Converter with *RL* Load

The operation of the converter in Figure 10.2a can be divided into two identical modes: mode 1 when T_1 and T_2 conduct, and mode 2 when T_3 and T_4 conduct. The output currents during these modes are similar and we need to consider only one mode to find the output current i_L.

Mode 1 is valid for $\alpha \le \omega t \le (\alpha + \pi)$. If $v_s = \sqrt{2}\,V_s \sin \omega t$ is the input voltage, the load current i_L during mode 1 can be found from

$$L\frac{di_L}{dt} + Ri_L + E = \sqrt{2}\,V_s \sin \omega t \quad \text{for } i_L \ge 0$$

whose solution is of the form

$$i_L = \frac{\sqrt{2}\,V_s}{Z}\sin(\omega t - \theta) + A_1 e^{-(R/L)t} - \frac{E}{R} \quad \text{for } i_L \ge 0$$

where load impedance $Z = [R^2 + (\omega L)^2]^{1/2}$ and load angle $\theta = \tan^{-1}(\omega L/R)$.

Constant A_1, which can be determined from the initial condition: at $\omega t = \alpha$, $i_L = I_{Lo}$, is found as

$$A_1 = \left[I_{Lo} + \frac{E}{R} - \frac{\sqrt{2}\,V_s}{Z}\sin(\alpha - \theta)\right] e^{(R/L)(\alpha/\omega)}$$

Substitution of A_1 gives i_L as

$$i_L = \frac{\sqrt{2}\,V_s}{Z}\sin(\omega t - \theta) - \frac{E}{R}$$

$$+ \left[I_{Lo} + \frac{E}{R} - \frac{\sqrt{2}\,V_s}{Z}\sin(\alpha - \theta)\right] e^{(R/L)(\alpha/\omega - t)} \tag{10.12}$$

At the end of mode 1 in the steady-state condition $i_L(\omega t = \pi + \alpha) = I_{L1} = I_{Lo}$. Applying this condition to Eq. (10.12) and solving for I_{Lo}, we get

$$I_{Lo} = I_{L1} = \frac{\sqrt{2}\,V_s}{Z}\frac{-\sin(\alpha - \theta) - \sin(\alpha - \theta)e^{-(R/L)(\pi)/\omega}}{1 - e^{-(R/L)(\pi/\omega)}} - \frac{E}{R} \quad \text{for } I_{Lo} \geq 0$$

(10.13)

The critical value of α at which I_o becomes zero can be solved for known values of θ, R, L, E, and V_s by an iterative method. The rms current of a thyristor can be found from Eq. (10.12) as

$$I_R = \left[\frac{1}{2\pi}\int_{\alpha}^{\pi+\alpha} i_L^2\, d(\omega t)\right]^{1/2}$$

The rms output current can then be determined from

$$I_{\text{rms}} = (I_R^2 + I_R^2)^{1/2} = \sqrt{2}\,I_R$$

The average current of a thyristor can also be found from Eq. (10.12) as

$$I_A = \frac{1}{2\pi}\int_{\alpha}^{\pi+\alpha} i_L\, d(\omega t)$$

The average output current can be determined from

$$I_{\text{dc}} = I_A + I_A = 2I_A$$

Discontinuous load current. The critical value of α_c at which I_{Lo} becomes zero can be solved. Dividing Eq. (10.13) by $\sqrt{2}V_s/Z$, and substituting $R/Z = \cos\theta$ and $\omega L/R = \tan\theta$, we get

$$0 = \frac{V_s\sqrt{2}}{Z}\sin(\alpha - \theta)\left[\frac{1 + e^{-\left(\frac{R}{L}\right)\left(\frac{\pi}{\omega}\right)}}{1 - e^{-\left(\frac{R}{L}\right)\left(\frac{\pi}{\omega}\right)}}\right] + \frac{E}{R}$$

which can be solved for the critical value of α as

$$\alpha_c = \theta - \sin^{-1}\left[\frac{1 - e^{-\left(\frac{\pi}{\tan(\theta)}\right)}}{1 + e^{-\left(\frac{\pi}{\tan(\theta)}\right)}}\frac{x}{\cos(\theta)}\right]$$

(10.14)

where $x = E/\sqrt{2}V_s$ is the voltage ratio, and θ is the load impedance angle. For $\alpha \geq \alpha_c$, $I_{L0} = 0$. The load current that is described by Eq. (10.12) flows only during the period, $\alpha \leq \omega t \leq \beta$. At $\omega t = \beta$, the load current falls to zero again. The equations derived for the discontinuous case of diode rectifier in Section 3.5 are applicable to the controlled rectifier.

Gating sequence.   The gating sequence is as follows:

1. Generate a pulse signal at the positive zero crossing of the supply voltage v_s. Delay the pulse by the desired angle α and apply the same pulse between the gate and cathode terminals of T_1 and T_2 through gate-isolating circuits.
2. Generate another pulse of delay angle $\alpha + \pi$ and apply the same pulse between the gate and source terminals of T_3 and T_4 through gate-isolating circuits.

Example 10.3 Finding the Current Ratings of Single-Phase Full Converter with an *RL* load

The single-phase full converter of Figure 10.2a has a *RL* load having $L = 6.5$ mH, $R = 0.5\ \Omega$, and $E = 10$ V. The input voltage is $V_s = 120$ V at (rms) 60 Hz. Determine (a) the load current I_{Lo} at $\omega t = \alpha = 60°$, (b) the average thyristor current I_A, (c) the rms thyristor current I_R, (d) the rms output current I_{rms}, (e) the average output current I_{dc}, and (f) the critical delay angle α_c.

Solution

$\alpha = 60°$, $R = 0.5\ \Omega$, $L = 6.5$ mH, $f = 60$ Hz, $\omega = 2\pi \times 60 = 377$ rad/s, $V_s = 120$ V, and $\theta = \tan^{-1}(\omega L/R) = 78.47°$.

    a. The steady-state load current at $\omega t = \alpha$, $I_{Lo} = 49.34$ A.

    b. The numerical integration of i_L in Eq. (10.12) yields the average thyristor current as $I_A = 44.05$ A.

    c. By numerical integration of i_L^2 between the limits $\omega t = \alpha$ to $\pi + \alpha$, we get the rms thyristor current as $I_R = 63.71$ A.

    d. The rms output current $I_{rms} = \sqrt{2}\ I_R = \sqrt{2} \times 63.71 = 90.1$ A.

    e. The average output current $I_{dc} = 2I_A = 2 \times 44.04 = 88.1$ A.

From Eq. (10.14), by iteration we find the critical delay angle $\alpha_c = 73.23°$.

Key Points of Section 10.3

- Varying the delay angle α from 0 to π can vary the average output voltage from $2V_m/\pi$ to $-2V_m/\pi$, provided the load is highly inductive and its current is continuous.
- For a purely resistive load, the delay angle α can be varied from 0 to $\pi/2$ producing an output voltage ranging from $2V_m/\pi$ to 0.
- The full converter can operate in two quadrants for a highly inductive load and in one quadrant only for a purely resistive load.

10.4   SINGLE-PHASE DUAL CONVERTERS

We have seen in Section 10.3 that single-phase full converters with inductive loads allow only a two-quadrant operation. If two of these full converters are connected back to back, as shown in Figure 10.3a, both the output voltage and the load current flow can be reversed. The system provides a four-quadrant operation and is called a *dual converter*. Dual converters are normally used in high-power variable-speed drives. If α_1 and α_2 are the delay angles of converters 1 and 2, respectively, the corresponding

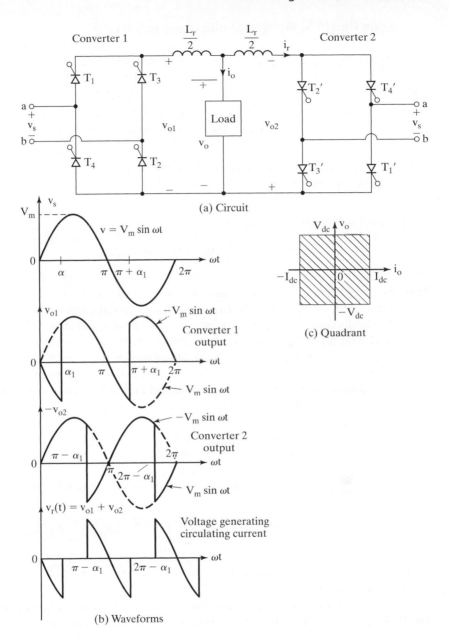

FIGURE 10.3

Single-phase dual converter.

average output voltages are V_{dc1} and V_{dc2}. The delay angles are controlled such that one converter operates as a rectifier and the other converter operates as an inverter; but both converters produce the same average output voltage. Figure 10.3b shows the output waveforms for two converters, where the two average output voltages are the same. Figure 10.3c shows the v–i characteristics of a dual converter.

From Eq. (10.5) the average output voltages are

$$V_{dc1} = \frac{2V_m}{\pi} \cos \alpha_1 \tag{10.15}$$

and

$$V_{dc2} = \frac{2V_m}{\pi} \cos \alpha_2 \tag{10.16}$$

Because one converter is rectifying and the other one is inverting,

$$V_{dc1} = -V_{dc2} \quad \text{or} \quad \cos \alpha_2 = -\cos \alpha_1 = \cos(\pi - \alpha_1)$$

Therefore,

$$\alpha_2 = \pi - \alpha_1 \tag{10.17}$$

Because the instantaneous output voltages of the two converters are out of phase, there can be an instantaneous voltage difference and this can result in circulating current between the two converters. This circulating current cannot flow through the load and is normally limited by a *circulating current reactor* L_r, as shown in Figure 10.3a.

If v_{o1} and v_{o2} are the instantaneous output voltages of converters 1 and 2, respectively, the circulating current can be found by integrating the instantaneous voltage difference starting from $\omega t = \pi - \alpha_1$. Because the two average output voltages during the interval $\omega t = \pi + \alpha_1$ to $2\pi - \alpha_1$ are equal and opposite, their contributions to the instantaneous circulating current i_r is zero.

$$
\begin{aligned}
i_r &= \frac{1}{\omega L_r} \int_{\pi - \alpha_1}^{\omega t} v_r \, d(\omega t) = \frac{1}{\omega L_r} \int_{\pi - \alpha_1}^{\omega t} (v_{o1} + v_{o2}) \, d(\omega t) \\
&= \frac{V_m}{\omega L_r} \left[\int_{2\pi - \alpha_1}^{\omega t} \sin \omega t \, d(\omega t) - \int_{2\pi - \alpha_1}^{\omega t} - \sin \omega t \, d(\omega t) \right] \\
&= \frac{2V_m}{\omega L_r} (\cos \alpha_1 - \cos \omega t) \quad i_r > 0 \quad \text{for} \quad 0 \le \alpha_1 < \frac{\pi}{2} \\
&\qquad\qquad\qquad\qquad\qquad\qquad i_r < 0 \quad \text{for} \quad \frac{\pi}{2} < \alpha_1 \le \pi
\end{aligned}
\tag{10.18}
$$

For $\alpha_1 = 0$, only the converter 1 operates; for $\alpha_1 = \pi$, only the converter 2 operates. For $0 \le \alpha_1 < \pi/2$, the converter 1 supplies a positive load current $+i_o$ and thus the circulating current can only be positive. For $\pi/2 < \alpha_1 \le \pi$, the converter 2 supplies a negative load current $-i_o$ and thus only a negative circulating current can flow. At $\alpha_1 = \pi/2$, the converter 1 supplies positive circulating during the first half-cycle, and the converter 2 supplies negative circulating during the second half-cycle.

The instantaneous circulating current depends on the delay angle. For $\alpha_1, = 0$, its magnitude becomes minimum when $\omega t = n\pi, n = 0, 2, 4, \ldots$, and maximum when $\omega t = n\pi, n = 1, 3, 5, \ldots$. If the peak load current is I_p, one of the converters that controls the power flow may carry a peak current of $(I_p + 4V_m/\omega L_r)$.

The dual converters can be operated with or without a circulating current. In case of operation without circulating current, only one converter operates at a time and carries the load current; and the other converter is completely blocked by inhibiting gate pulses. However, the operation with circulating current has the following advantages:

1. The circulating current maintains continuous conduction of both converters over the whole control range, independent of the load.
2. Because one converter always operates as a rectifier and the other converter operates as an inverter, the power flow in either direction at any time is possible.
3. Because both converters are in continuous conduction, the time response for changing from one quadrant operation to another is faster.

Gating sequence.   The gating sequence is as follows:

1. Gate the positive converter with a delay angle of $\alpha_1 = \alpha$.
2. Gate the negative converter with a delay angle of $\alpha_2 = \pi - \alpha$ through gate-isolating circuits.

Example 10.4 Finding the Peak Currents of a Single-Phase Dual Converter

The single-phase dual converter in Figure 10.3a is operated from a 120-V, 60-Hz supply and the load resistance is $R = 10\ \Omega$. The circulating inductance is $L_r = 40$ mH; delay angles are $\alpha_1 = 60°$ and $\alpha_2 = 120°$. Calculate the peak circulating current and the peak current of converter 1.

Solution
$\omega = 2\pi \times 60 = 377$ rad/s, $\alpha_1 = 60°$, $V_m = \sqrt{2} \times 120 = 169.7$ V, $f = 60$ Hz, and $L_r = 40$ mH. For $\omega t = 2\pi$ and $\alpha_1 = \pi/3$, Eq. (10.18) gives the peak circulating current

$$I_r(\text{max}) = \frac{2V_m}{\omega L_r}\,(1 - \cos\alpha_1) = \frac{169.7}{377 \times 0.04} = 11.25\ \text{A}$$

The peak load current is $I_p = 169.71/10 = 16.97$ A. The peak current of converter 1 is $(16.97 + 11.25) = 28.22$ A.

Key Points of Section 10.4

- The dual converter consists of two full converters: one converter producing positive output voltage and another converter producing negative output voltage. Varying the delay angle α from 0 to π can vary the average output voltage from $2V_m/\pi$ to $-2V_m/\pi$, provided the load is highly inductive and its current is continuous.
- For a highly inductive load, the dual converter can operate in four quadrants. The current can flow in and out of the load. A dc inductor is needed to reduce the circulating current.

10.5 PRINCIPLE OF THREE-PHASE HALF-WAVE CONVERTERS

Three-phase converters provide higher average output voltage, and in addition the frequency of the ripples on the output voltage is higher compared with that of single-phase converters. As a result, the filtering requirements for smoothing out the load current and load voltage are simpler. For these reasons, three-phase converters are used extensively in high-power variable-speed drives. Three single-phase half-wave converters in Figure 10.1a can be connected to form a three-phase half-wave converter, as shown in Figure 10.4a.

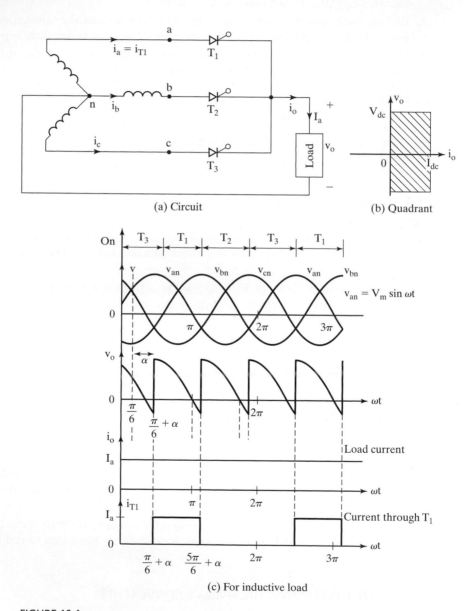

(a) Circuit   (b) Quadrant

(c) For inductive load

FIGURE 10.4

Three-phase half-wave converter.

When thyristor T_1 is fired at $\omega t = \pi/6 + \alpha$, the phase voltage v_{an} appears across the load until thyristor T_2 is fired at $\omega t = 5\pi/6 + \alpha$. When thyristor T_2 is fired, thyristor T_1 is reverse biased, because the line-to-line voltage, $v_{ab}(= v_{an} - v_{bn})$, is negative and T_1 is turned off. The phase voltage v_{bn} appears across the load until thyristor T_3 is fired at $\omega t = 3\pi/2 + \alpha$. When thyristor T_3 is fired, T_2 is turned off and v_{cn} appears across the load until T_1 is fired again at the beginning of next cycle. Figure 10.4b shows the $v-i$ characteristics of the load and this is a two-quadrant converter. Figure 10.4c shows the input voltages, output voltage, and the current through thyristor T_1 for a highly inductive load. For a resistive load and $\alpha > \pi/6$, the load current would be discontinuous and each thyristor is self-commutated when the polarity of its phase voltage is reversed. The frequency of output ripple voltage is $3f_s$. This converter is not normally used in practical systems, because the supply currents contain dc components. However, this converter explains the principle of the three-phase thyristor converter.

If the phase voltage is $v_{an} = V_m \sin \omega t$, the average output voltage for a continuous load current is

$$V_{dc} = \frac{3}{2\pi} \int_{\pi/6+\alpha}^{5\pi/6+\alpha} V_m \sin \omega t \, d(\omega t) = \frac{3\sqrt{3}\, V_m}{2\pi} \cos \alpha \qquad (10.19)$$

where V_m is the peak phase voltage. The maximum average output voltage that occurs at delay angle, $\alpha = 0$ is

$$V_{dm} = \frac{3\sqrt{3}\, V_m}{2\pi}$$

and the normalized average output voltage is

$$V_n = \frac{V_{dc}}{V_{dm}} = \cos \alpha \qquad (10.20)$$

The rms output voltage is found from

$$V_{rms} = \left[\frac{3}{2\pi} \left[\int_{\pi/6+\alpha}^{5\pi/6+\alpha} V_m^2 \sin^2 \omega t \, d(\omega t) \right] \right]^{1/2}$$

$$= \sqrt{3}\, V_m \left(\frac{1}{6} + \frac{\sqrt{3}}{8\pi} \cos 2\alpha \right)^{1/2} \qquad (10.21)$$

For a resistive load and $\alpha \geq \pi/6$:

$$V_{dc} = \frac{3}{2\pi} \int_{\pi/6+\alpha}^{\pi} V_m \sin \omega t \, d(\omega t) = \frac{3 V_m}{2\pi} \left[1 + \cos\left(\frac{\pi}{6} + \alpha \right) \right] \qquad (10.22)$$

$$V_n = \frac{V_{dc}}{V_{dm}} = \frac{1}{\sqrt{3}} \left[1 + \cos\left(\frac{\pi}{6} + \alpha \right) \right] \qquad (10.23)$$

$$V_{rms} = \left[\frac{3}{2\pi} \int_{\pi/6+\alpha}^{\pi} V_m^2 \sin^2 \omega t \, d(\omega t) \right]^{1/2}$$

$$= \sqrt{3}\, V_m \left[\frac{5}{24} - \frac{\alpha}{4\pi} + \frac{1}{8\pi} \sin\left(\frac{\pi}{3} + 2\alpha \right) \right]^{1/2} \qquad (10.24)$$

Gating sequence. The gating sequence is as follows:

1. Generate a pulse signal at the positive zero crossing of the phase voltage v_{an}. Delay the pulse by the desired angle $\alpha + \pi/6$ and apply it to the gate and cathode terminals of T_1 through a gate-isolating circuit.
2. Generate two more pulses of delay angles $\alpha + 5\pi/6$ and $\alpha + 9\pi/6$ for gating T_2 and T_3, respectively, through gate-isolating circuits.

Example 10.5 Finding the Performances of a Three-Phase Half-Wave Converter

A three-phase half-wave converter in Figure 10.4a is operated from a three-phase Y-connected 208-V, 60-Hz supply and the load resistance is $R = 10\ \Omega$. If it is required to obtain an average output voltage of 50% of the maximum possible output voltage, calculate (a) the delay angle α, (b) the rms and average output currents, (c) the average and rms thyristor currents, (d) the rectification efficiency, (e) the TUF, and (f) the input PF.

Solution
The phase voltage is $V_s = 208/\sqrt{3} = 120.1$ V, $V_m = \sqrt{2}V_s = 169.83$ V, $V_n = 0.5$, and $R = 10\ \Omega$. The maximum output voltage is

$$V_{dm} = \frac{3\sqrt{3}\,V_m}{2\pi} = 3\sqrt{3} \times \frac{169.83}{2\pi} = 140.45\text{ V}$$

The average output voltage, $V_{dc} = 0.5 \times 140.45 = 70.23$ V.

a. For a resistive load, the load current is continuous if $\alpha \le \pi/6$ and Eq. (10.20) gives $V_n \ge \cos(\pi/6) = 86.6\%$. With a resistive load and 50% output, the load current is discontinuous. From Eq. (10.23), $0.5 = (1/\sqrt{3})[1 + \cos(\pi/6 + \alpha)]$, which gives the delay angle as $\alpha = 67.7°$.

b. The average output current, $I_{dc} = V_{dc}/R = 70.23/10 = 7.02$ A. From Eq. (10.24), $V_{rms} = 94.74$ V and the rms load current, $I_{rms} = 94.74/10 = 9.47$ A.

c. The average current of a thyristor is $I_A = I_{dc}/3 = 7.02/3 = 2.34$ A and the rms current of a thyristor is $I_R = I_{rms}/\sqrt{3} = 9.47/\sqrt{3} = 5.47$ A.

d. From Eq. (3.3) the rectification efficiency is

$$\eta = V_{dc}I_{dc}/V_{rms}I_{rms} = 70.23 \times 7.02/(94.74 \times 9.47) = 54.95\%.$$

e. The rms input line current is the same as the thyristor rms current, and the input volt–ampere rating (VAR), $VI = 3V_sI_s = 3 \times 120.1 \times 5.47 = 1970.84$ W. From Eq. (3.8), TUF $= V_{dc}I_{dc}/VI = 70.23 \times 7.02/1970.84 = 0.25$ or 25%.

f. The output power $P_o = I_{rms}^2 R = 9.47^2 \times 10 = 896.81$ VA. The input PF $= P_o/VI = 896.81/1970.84 = 0.455$ (lagging).

Note: Due to the delay angle α, the fundamental component of input line current is also delayed with respect to the input phase voltage.

Key Points of Section 10.5

- The frequency of the output ripple is three times the supply frequency.
- For $\alpha > \pi/6$ with a resistive load, the current is discontinuous.
- This converter is not normally used in practical applications.

10.6   THREE-PHASE FULL CONVERTERS

Three-phase converters [2, 11] are extensively used in industrial applications up to the 120-kW level, where a two-quadrant operation is required. Figure 10.5a shows a full-converter circuit with a highly inductive load. This circuit is known as a three-phase bridge. The thyristors are fired at an interval of $\pi/3$. The frequency of output ripple voltage is $6f_s$ and the filtering requirement is less than that of half-wave converters. At $\omega t = \pi/6 + \alpha$, thyristor T_6 is already conducting and thyristor T_1 is turned on. During interval $(\pi/6 + \alpha) \leq \omega t \leq (\pi/2 + \alpha)$, thyristors T_1 and T_6 conduct and the line-to-line voltage $v_{ab} (= v_{an} - v_{bn})$ appears across the load. At $\omega t = \pi/2 + \alpha$, thyristor T_2 is fired and thyristor T_6 is reversed biased immediately. T_6 is turned off due to natural commutation. During interval $(\pi/2 + \alpha) \leq \omega t \leq (5\pi/6 + \alpha)$, thyristors T_1 and T_2 conduct and the line-to-line voltage v_{ac} appears across the load. If the thyristors are numbered, as shown in Figure 10.5a, the firing sequence is 12, 23, 34, 45, 56, and 61. Figure 10.5b shows the waveforms for input voltage, output voltage, input current, and currents through thyristors.

If the line-to-neutral voltages are defined as

$$v_{an} = V_m \sin \omega t$$

$$v_{bn} = V_m \sin\left(\omega t - \frac{2\pi}{3}\right)$$

$$v_{cn} = V_m \sin\left(\omega t + \frac{2\pi}{3}\right)$$

the corresponding line-to-line voltages are

$$v_{ab} = v_{an} - v_{bn} = \sqrt{3}\, V_m \sin\left(\omega t + \frac{\pi}{6}\right)$$

$$v_{bc} = v_{bn} - v_{cn} = \sqrt{3}\, V_m \sin\left(\omega t - \frac{\pi}{2}\right)$$

$$v_{ca} = v_{cn} - v_{an} = \sqrt{3}\, V_m \sin\left(\omega t + \frac{\pi}{2}\right)$$

The average output voltage is found from

$$V_{dc} = \frac{3}{\pi} \int_{\pi/6+\alpha}^{\pi/2+\alpha} v_{ab}\, d(\omega t) = \frac{3}{\pi} \int_{\pi/6+\alpha}^{\pi/2+\alpha} \sqrt{3}\, V_m \sin\left(\omega t + \frac{\pi}{6}\right) d(\omega t)$$

$$= \frac{3\sqrt{3}\, V_m}{\pi} \cos \alpha \tag{10.25}$$

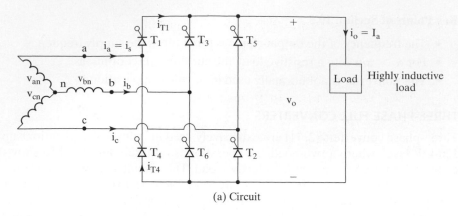

(a) Circuit

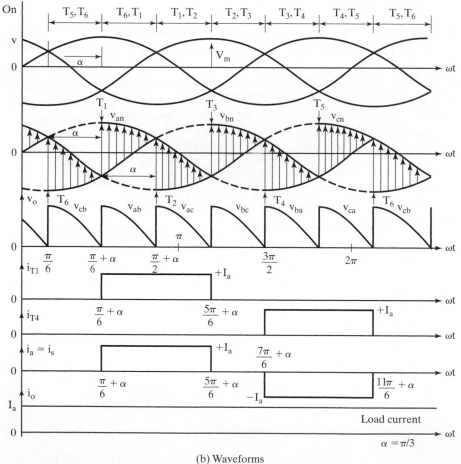

(b) Waveforms

FIGURE 10.5

Three-phase full converter.

The maximum average output voltage for delay angle, $\alpha = 0$, is

$$V_{dm} = \frac{3\sqrt{3}\,V_m}{\pi}$$

and the normalized average output voltage is

$$V_n = \frac{V_{dc}}{V_{dm}} = \cos\alpha \tag{10.26}$$

The rms value of the output voltage is found from

$$V_{rms} = \left[\frac{3}{\pi} \int_{\pi/6+\alpha}^{\pi/2+\alpha} 3V_m^2 \sin^2\!\left(\omega t + \frac{\pi}{6}\right) d(\omega t) \right]^{1/2}$$
$$= \sqrt{3}\,V_m \left(\frac{1}{2} + \frac{3\sqrt{3}}{4\pi}\cos 2\alpha\right)^{1/2} \tag{10.27}$$

Figure 10.5b shows the waveforms for $\alpha = \pi/3$. For $\alpha > \pi/3$, the instantaneous output voltage v_o has a negative part. Because the current through thyristors cannot be negative, the load current is always positive. Thus, with a resistive load, the instantaneous load voltage cannot be negative, and the full converter behaves as a semiconverter.

Gating sequence.   The gating sequence is as follows:

1. Generate a pulse signal at the positive zero crossing of the phase voltage v_{an}. Delay the pulse by the desired angle $\alpha + \pi/6$ and apply it to the gate and cathode terminals of T_1 through a gate-isolating circuit.
2. Generate five more pulses each delayed by $\pi/6$ from each other for gating $T_2, T_3, \ldots, T_6$, respectively, through gate isolating circuits.

Example 10.6   Finding the Performances of a Three-Phase Full-Wave Converter

Repeat Example 10.5 for the three-phase full converter in Figure 10.5a.

Solution

The phase voltage $V_s = 208/\sqrt{3} = 120.1$ V, $V_m = \sqrt{2}\,V_s = 169.83$, $V_n = 0.5$, and $R = 10\ \Omega$. The maximum output voltage $V_{dm} = 3\sqrt{3}\,V_m/\pi = 3\sqrt{3} \times 169.83/\pi = 280.9$ V. The average output voltage $V_{dc} = 0.5 \times 280.9 = 140.45$ V.

    a.   From Eq. (10.26), $0.5 = \cos\alpha$, and the delay angle $\alpha = 60°$.

    b.   The average output current $I_{dc} = V_{dc}/R = 140.45/10 = 14.05$ A. From Eq. (10.27),

$$V_{rms} = \sqrt{3} \times 169.83 \left[\frac{1}{2} + \frac{3\sqrt{3}}{4\pi}\cos(2 \times 60°)\right]^{1/2} = 159.29\ \text{V}$$

and the rms current $I_{rms} = 159.29/10 = 15.93$ A.

    c.   The average current of a thyristor $I_A = I_{dc}/3 = 14.05/3 = 4.68$ A, and the rms current of a thyristor $I_R = I_{rms}\sqrt{2/6} = 15.93\sqrt{2/6} = 9.2$ A.

d. From Eq. (3.15) the rectification efficiency is

$$\eta = \frac{V_{dc}I_{dc}}{V_{rms}I_{rms}} = \frac{140.45 \times 14.05}{159.29 \times 15.93} = 0.778 \quad \text{or} \quad 77.8\%$$

e. The rms input line current $I_s = I_{rms}\sqrt{4/6} = 13$ A and the input VAR rating $VI = 3V_sI_s = 3 \times 120.1 \times 13 = 4683.9$ VA. From Eq. (3.8), TUF $= V_{dc}I_{dc}/VI = 140.45 \times 14.05/4683.9 = 0.421$.

f. The output power $P_o = I_{rms}^2 R = 15.93^2 \times 10 = 2537.6$ W. The PF $= P_o/VI = 2537.6/4683.9 = 0.542$ (lagging).

Note: The PF is less than that of three-phase semiconverters, but higher than that of three-phase half-wave converters.

Example 10.7   Finding the Input Power Factor of a Three-Phase Full Converter

The load current of a three-phase full converter in Figure 10.5a is continuous with a negligible ripple content. (a) Express the input current in Fourier series, and determine the HF of input current, the DF, and the input PF. (b) If the delay angle $\alpha = \pi/3$, calculate V_n, HF, DF, and PF.

Solution

a. The waveform for input current is shown in Figure 10.5b and the instantaneous input current of a phase can be expressed in a Fourier series as

$$i_s(t) = a_0 + \sum_{n=1,2,\ldots}^{\infty} (a_n \cos n\omega t + b_n \sin n\omega t)$$

where

$$a_o = \frac{1}{2\pi} \int_0^{2\pi} i_s(t)\, d(\omega t) = 0$$

$$
\begin{aligned}
a_n &= \frac{1}{\pi} \int_0^{2\pi} i_s(t) \cos n\omega t\, d(\omega t) \\
&= \frac{1}{\pi}\left[\int_{\pi/6+\alpha}^{5\pi/6+\alpha} I_a \cos n\omega t\, d(\omega t) - \int_{7\pi/6+\alpha}^{11\pi/6+\alpha} I_a \cos n\omega t\, d(\omega t) \right] \\
&= -\frac{4I_a}{n\pi} \sin \frac{n\pi}{3} \sin n\alpha \quad \text{for } n = 1, 3, 5, \ldots \\
&= 0 \quad \text{for } n = 2, 4, 6, \ldots
\end{aligned}
$$

$$
\begin{aligned}
b_n &= \frac{1}{\pi} \int_0^{2\pi} i_s(t) \sin n\omega t\, d(\omega t) \\
&= \frac{1}{\pi}\left[\int_{\pi/6+\alpha}^{5\pi/6+\alpha} I_a \sin n\omega t\, d(\omega t) - \int_{7\pi/6+\alpha}^{11\pi/6+\alpha} I_a \sin n\omega t\, d(\omega t) \right] \\
&= \frac{4I_a}{n\pi} \sin \frac{n\pi}{3} \cos n\alpha \quad \text{for } n = 1, 3, 5, \ldots \\
&= 0 \quad \text{for } n = 2, 4, 6, \ldots
\end{aligned}
$$

Because $a_0 = 0$ and the triplen harmonic currents (for n = multiple of 3) will be absent in a balanced three-phase supply, the input current can be written as

$$i_s(t) = \sum_{n=1,3,5,\dots}^{\infty} \sqrt{2} I_{sn} \sin(n\omega t + \phi_n) \quad \text{for } n = 1, 5, 7, 11, 13, \dots$$

where

$$\phi_n = \tan^{-1} \frac{a_n}{b_n} = -n\alpha \tag{10.28}$$

The rms value of the nth harmonic input current is given by

$$I_{sn} = \frac{1}{\sqrt{2}} (a_n^2 + b_n^2)^{1/2} = \frac{2\sqrt{2} I_a}{n\pi} \sin \frac{n\pi}{3} \tag{10.29}$$

The rms value of the fundamental current is

$$I_{s1} = \frac{\sqrt{6}}{\pi} I_a = 0.7797 I_a$$

The rms input current

$$I_s = \left[\frac{2}{2\pi} \int_{\pi/6+\alpha}^{5\pi/6+\alpha} I_a^2 \, d(\omega t) \right]^{1/2} = I_a \sqrt{\frac{2}{3}} = 0.8165 I_a$$

$$\text{HF} = \left[\left(\frac{I_s}{I_{s1}} \right)^2 - 1 \right]^{1/2} = \left[\left(\frac{\pi}{3} \right)^2 - 1 \right]^{1/2} = 0.3108 \quad \text{or} \quad 31.08\%$$

$$\text{DF} = \cos \phi_1 = \cos(-\alpha)$$

$$\text{PF} = \frac{I_{s1}}{I_s} \cos(-\alpha) = \frac{3}{\pi} \cos \alpha = 0.9549 \text{ DF}$$

b. For $\alpha = \pi/3$, $V_n = \cos(\pi/3) = 0.5$ pu, $HF = 31.08\%$, DF $= \cos 60° = 0.5$, and PF $=$ 0.478 (lagging).

Note: If we compare the PF with that of Example 10.5, where the load is purely resistive, we can notice that the input PF depends on the PF of the load.

10.6.1 Three-Phase Full Converter with *RL* Load

From Figure 10.5b the output voltage is

$$v_o = v_{ab} = \sqrt{2} V_{ab} \sin \left(\omega t + \frac{\pi}{6} \right) \quad \text{for } \frac{\pi}{6} + \alpha \leq \omega t \leq \frac{\pi}{2} + \alpha$$

$$= \sqrt{2} V_{ab} \sin \omega t' \quad \text{for } \frac{\pi}{3} + \alpha \leq \omega t' \leq \frac{2\pi}{3} + \alpha$$

where $\omega t' = \omega t + \pi/6$, and V_{ab} is the line-to-line (rms) input voltage. Choosing v_{ab} as the time reference voltage, the load current i_L can be found from

$$L\frac{di_L}{dt} + Ri_L + E = \sqrt{2}\,V_{ab}\sin\omega t' \quad \text{for} \quad \frac{\pi}{3} + \alpha \leq \omega t' \leq \frac{2\pi}{3} + \alpha$$

whose solution from Eq. (10.12) is

$$i_L = \frac{\sqrt{2}\,V_{ab}}{Z}\sin(\omega t' - \theta) - \frac{E}{R}$$

$$+ \left[I_{L1} + \frac{E}{R} - \frac{\sqrt{2}\,V_{ab}}{Z}\sin\left(\frac{\pi}{3} + \alpha - \theta\right)\right]e^{(R/L)[(\pi/3+\alpha)/\omega - t']} \quad (10.30)$$

where $Z = [R^2 + (\omega L)^2]^{1/2}$ and $\theta = \tan^{-1}(\omega L/R)$. Under a steady-state condition, $i_L(\omega t' = 2\pi/3 + \alpha) = i_L(\omega t' = \pi/3 + \alpha) = I_{L1}$. Applying this condition to Eq. (10.30), we get the value of I_{L1} as

$$I_{L1} = \frac{\sqrt{2}\,V_{ab}}{Z}\,\frac{\sin(2\pi/3 + \alpha - \theta) - \sin(\pi/3 + \alpha - \theta)e^{-(R/L)(\pi/3\omega)}}{1 - e^{-(R/L)(\pi/3\omega)}}$$

$$- \frac{E}{R} \quad \text{for } I_{L1} \geq 0 \quad (10.31)$$

Discontinuous load current. By setting $I_{L1} = 0$ in Eq. (10.31), dividing by $\sqrt{2}V_s/Z$, and substituting $R/Z = \cos\theta$ and $\omega L/R = \tan\theta$, we get the critical value of voltage ratio $x = E/\sqrt{2}V_{ab}$ as

$$x = \left[\frac{\sin\left(\frac{2\pi}{3} + \alpha - \theta\right) - \sin\left(\frac{\pi}{3} + \alpha - \theta\right)e^{-\left(\frac{\pi}{3\tan(\theta)}\right)}}{1 - e^{-\left(\frac{\pi}{3\tan(\theta)}\right)}}\right]\cos(\theta) \quad (10.32)$$

which can be solved for the critical value of $\alpha = \alpha_c$ for known values of x and θ. For $\alpha \geq \alpha_c$, $I_{L1} = 0$. The load current that is described by Eq. (10.30) flows only during the period $\alpha \leq \omega t \leq \beta$. At $\omega t = \beta$, the load current falls to zero again. The equations derived for the discontinuous case of diode rectifier in Section 3.8 are applicable to the controlled rectifier.

Example 10.8 Finding the Current Ratings of Three-Phase Full-Converter with an *RL* Load

The three-phase full converter of Figure 10.5a has a load of $L = 1.5$ mH, $R = 2.5\ \Omega$, and $E = 10$ V. The line-to-line input voltage is $V_{ab} = 208$ V (rms), 60 Hz. The delay angle is $\alpha = \pi/3$. Determine (a) the steady-state load current I_{L1} at $\omega t' = \pi/3 + \alpha$ (or $\omega t = \pi/6 + \alpha$), (b) the average thyristor current I_A, (c) the rms thyristor current I_R, (d) the rms output current I_{rms}, and (e) the average output current I_{dc}.

Solution

$\alpha = \pi/3$, $R = 2.5\ \Omega$, $L = 1.5$ mH, $f = 60$ Hz, $\omega = 2\pi \times 60 = 377$ rad/s, $V_{ab} = 208$ V, $Z = [R^2 + (\omega L)^2]^{1/2} = 2.56\ \Omega$, and $\theta = \tan^{-1}(\omega L/R) = 12.74°$.

a. The steady-state load current at $\omega t' = \pi/3 + \alpha$, $I_{L1} = 20.49$ A.

b. The numerical integration of i_L in Eq. (10.30), between the limits $\omega t' = \pi/3 + \alpha$ to $2\pi/3 + \alpha$, gives the average thyristor current, $I_A = 17.42$ A.

c. By numerical integration of i_L^2, between the limits $\omega t' = \pi/3 + \alpha$ to $2\pi/3 + \alpha$, gives the rms thyristor current, $I_R = 31.32$ A.

d. The rms output current $I_{\text{rms}} = \sqrt{3} I_R = \sqrt{3} \times 31.32 = 54.25$ A.

e. The average output current $I_{\text{dc}} = 3 I_A = 3 \times 17.42 = 52.26$ A.

Key Points of Section 10.6

- The frequency of the output ripple is six times the supply frequency.
- The three-phase full converter is commonly used in practical applications.
- It can operate in two quadrants provided the load is highly inductive and maintains continuous current.

10.7   THREE-PHASE DUAL CONVERTERS

In many variable-speed drives, the four-quadrant operation is generally required and three-phase dual converters are extensively used in applications up to the 2000-kW level. Figure 10.6a shows three-phase dual converters where two three-phase converters are connected back to back. We have seen in Section 10.4 that due to the instantaneous voltage differences between the output voltages of converters, a circulating current flows through the converters. The circulating current is normally limited by circulating reactor L_r, as shown in Figure 10.6a. The two converters are controlled in such a way that if α_1 is the delay angle of converter 1, the delay angle of converter 2 is $\alpha_2 = \pi - \alpha_1$. Figure 10.6b shows the waveforms for input voltages, output voltages, and the voltage across inductor L_r. The operation of each converter is identical to that of a three-phase full converter. During the interval $(\pi/6 + \alpha_1) \le \omega t \le (\pi/2 + \alpha_1)$, the line-to-line voltage v_{ab} appears across the output of converter 1, and v_{bc} appears across converter 2.

If the line-to-neutral voltages are defined as

$$v_{an} = V_m \sin \omega t$$

$$v_{bn} = V_m \sin \left(\omega t - \frac{2\pi}{3} \right)$$

$$v_{cn} = V_m \sin \left(\omega t + \frac{2\pi}{3} \right)$$

the corresponding line-to-line voltages are

$$v_{ab} = v_{an} - v_{bn} = \sqrt{3}\, V_m \sin \left(\omega t + \frac{\pi}{6} \right)$$

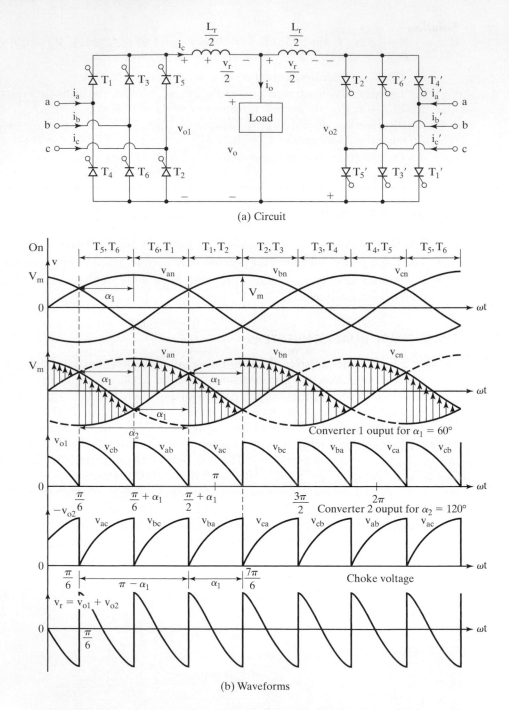

(a) Circuit

(b) Waveforms

FIGURE 10.6

Three-phase dual converter.

$$v_{bc} = v_{bn} - v_{cn} = \sqrt{3}\,V_m \sin\left(\omega t - \frac{\pi}{2}\right)$$

$$v_{ca} = v_{cn} - v_{an} = \sqrt{3}\,V_m \sin\left(\omega t + \frac{5\pi}{6}\right)$$

If v_{o1} and v_{o2} are the output voltages of converters 1 and 2, respectively, the instantaneous voltage across the inductor during interval $(\pi/6 + \alpha_1) \le \omega t \le (\pi/2 + \alpha_1)$ is

$$
\begin{aligned}
v_r = v_{o1} + v_{o2} &= v_{ab} - v_{bc} \\
&= \sqrt{3}\,V_m\left[\sin\left(\omega t + \frac{\pi}{6}\right) - \sin\left(\omega t - \frac{\pi}{2}\right)\right] \\
&= 3V_m \cos\left(\omega t - \frac{\pi}{6}\right)
\end{aligned}
\tag{10.33}
$$

The circulating current can be found from

$$
\begin{aligned}
i_r(t) &= \frac{1}{\omega L_r}\int_{\pi/6+\alpha_1}^{\omega t} v_r\,d(\omega t) = \frac{1}{\omega L_r}\int_{\pi/6+\alpha_1}^{\omega t} 3V_m \cos\left(\omega t - \frac{\pi}{6}\right) d(\omega t) \\
&= \frac{3V_m}{\omega L_r}\left[\sin\left(\omega t - \frac{\pi}{6}\right) - \sin\alpha_1\right]
\end{aligned}
\tag{10.34}
$$

The circulating current depends on delay angle α_1 and on inductance L_r. This current becomes maximum when $\omega t = 2\pi/3$ and $\alpha_1 = 0$. Even without any external load, the converters would be continuously running due to the circulating current as a result of ripple voltage across the inductor. This allows smooth reversal of load current during the changeover from one quadrant operation to another and provides fast dynamic responses, especially for electrical motor drives.

Gating sequence. The gating sequence is as follows:

1. Similar to the single-phase dual converter, gate the positive converter with a delay angle of $\alpha_1 = \alpha$.
2. Gate the negative converter with a delay angle of $\alpha_2 = \pi - \alpha$ through gate-isolating circuits.

Key Points of Section 10.7

- The three-phase dual converter is used for high-power applications up to 2000 kW.
- For a highly inductive load, the dual converter can operate in four quadrants. The current can flow in and out of the load.
- A dc inductor is needed to reduce the circulating current.

10.8   POWER FACTOR IMPROVEMENTS

The PF of phase-controlled converters depends on delay angle α, and is in general low, especially at the low output voltage range. These converters generate harmonics into the supply. Forced commutations can improve the input PF and reduce the harmonics levels. These forced-commutation techniques are becoming attractive to dc–ac conversion [3, 4]. With the advancement of power semiconductor devices (e.g., gate-turn-off thyristors [GTOs], insulated-gate bipolar transistors [IGBTs],), all of the forced-commutation techniques for inverters in Section 6.6 can be implemental for practical dc–ac converters [12–14]. In this section the basic techniques of forced commutation for dc–ac converters are discussed, and these can be classified as follows:

1. Extinction angle control
2. Symmetric angle control
3. Pulse-width modulation (PWM)
4. Single-phase sinusoidal PWM
5. Three-phase PWM control

10.8.1   Extinction Angle Control

Figure 10.7a shows a single-phase converter with two switches S_1 and S_2. A freewheeling diode D_m is connected across the load. The switching actions of S_1 and S_2 can be performed by GTO. The characteristics of GTOs (or IGBTs) are such that a GTO can be turned on by applying a short positive pulse to its gate as in the case of normal thyristors and can be turned off by applying a short negative pulse to its gate. An IGBT remains on as long as a gate voltage is applied to its gate terminal.

In an extinction angle control, switch S_1 is turned on at $\omega t = 0$ and is turned off by forced commutation at $\omega t = \pi - \beta$. Switch S_2 is turned on at $\omega t = \pi$ and is turned off at $\omega t = (2\pi - \beta)$. The output voltage is controlled by varying the extinction angle β. Figure 10.7b shows the waveforms for input voltage, output voltage, input current, and current through thyristor switches. The fundamental component of input current leads the input voltage, and the displacement factor (and PF) is leading. In some applications, this feature may be desirable to simulate a capacitive load and to compensate for line voltage drops.

The average output voltage is found from

$$V_{dc} = \frac{2}{2\pi} \int_0^{\pi-\beta} V_m \sin \omega t \, d(\omega t) = \frac{V_m}{\pi} (1 + \cos \beta) \tag{10.35}$$

and V_{dc} can be varied from $2V_m/\pi$ to 0 by varying β from 0 to π. The rms output voltage is given by

$$V_{rms} = \left[\frac{2}{2\pi} \int_0^{\pi-\beta} V_m^2 \sin^2 \omega t \, d(\omega t) \right]^{1/2}$$

$$= \frac{V_m}{\sqrt{2}} \left[\frac{1}{\pi} \left(\pi - \beta + \frac{\sin 2\beta}{2} \right) \right]^{1/2} \tag{10.36}$$

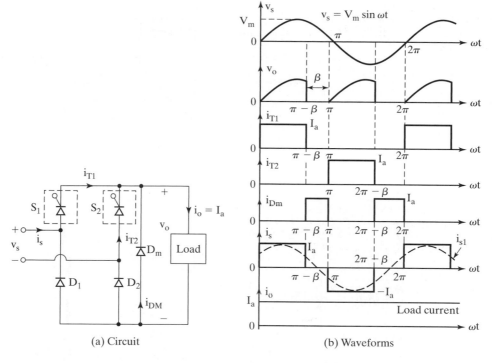

FIGURE 10.7

Single-phase forced-commutated semiconverter.

Figure 10.8a shows a single-phase full converter, where thyristors T_1, T_2, T_3, and T_4 are replaced by forced-commutated switches S_1, S_2, S_3, and S_4. Each switch conducts for 180°. Switches S_1 and S_2 are both on from $\omega t = 0$ to $\omega t = \pi - \beta$ and supply power to the load during the positive half-cycle of the input voltage. Similarly, switches S_3 and S_4 are both on from $\omega t = \pi$ to $\omega t = 2\pi - \beta$ and supply power to the load during the negative half-cycle of the input voltage. For an inductive load, the freewheeling path for the load current must be provided by switches $S_1 S_4$ or $S_3 S_2$. The firing sequence would be 12, 14, 34, and 32. Figure 10.8b shows the waveforms for input voltage, output voltage, input current, and current through switches. Each switch conducts for 180° and this converter is operated as a semiconverter. The freewheeling action is accomplished through two switches of the same arm. The average and rms output voltage are expressed by Eq. (10.35) and Eq. (10.36), respectively.

The performance of these converters with extinction angle control is similar to those with phase-angle control, except the PF is leading. With phase-angle control, the PF is lagging.

10.8.2 Symmetric Angle Control

The symmetric angle control allows one-quadrant operation and Figure 10.7a shows a single-phase semiconverter with forced-commutated switches S_1 and S_2. Switch S_1 is

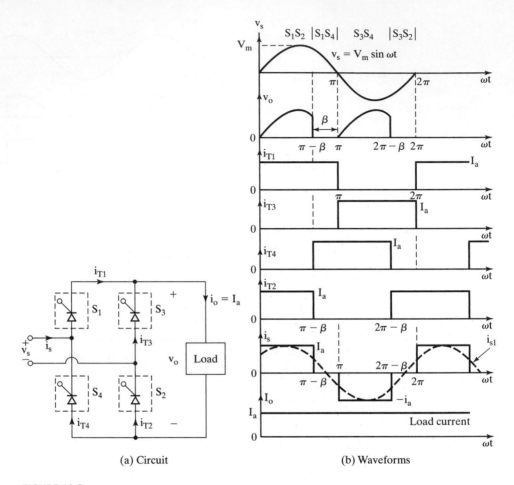

(a) Circuit

(b) Waveforms

FIGURE 10.8

Single-phase forced-commutated full converter.

turned on at $\omega t = (\pi - \beta)/2$ and is turned off at $\omega t = (\pi + \beta)/2$. Switch S_2 is turned on at $\omega t = (3\pi - \beta)/2$ and off at $\omega t = (3\pi + \beta)/2$. The output voltage is controlled by varying conduction angle β. The gate signals are generated by comparing half-sine waves with a dc signal, as shown in Figure 10.9b. Figure 10.9a shows the waveforms for input voltage, output voltage, input current, and current through switches. The fundamental component of input current is in phase with the input voltage and the DF is unity. Therefore, the PF is improved.

The average output voltage is found from

$$V_{dc} = \frac{2}{2\pi} \int_{(\pi-\beta)/2}^{(\pi+\beta)/2} V_m \sin \omega t \, d(\omega t) = \frac{2V_m}{\pi} \sin \frac{\beta}{2} \qquad (10.37)$$

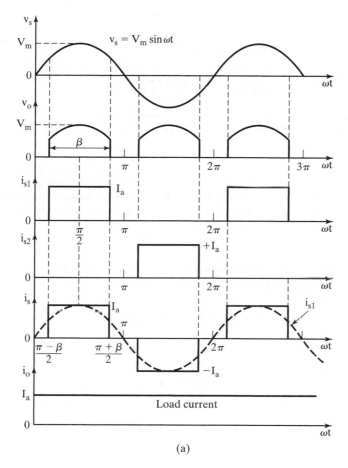

(a)

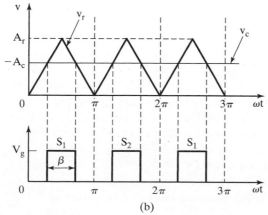

(b)

FIGURE 10.9

Symmetric angle control.

and V_{dc} can be varied from $2V_m/\pi$ to 0 by varying β from π to 0. The rms output voltage is given by

$$V_{rms} = \left[\frac{2}{2\pi} \int_{(\pi-\beta)/2}^{(\pi+\beta)/2} V_m^2 \sin^2 \omega t\, d(\omega t) \right]^{1/2}$$

$$= \frac{V_m}{\sqrt{2}} \left[\frac{1}{\pi} (\beta + \sin \beta) \right]^{1/2} \tag{10.38}$$

Example 10.9 Finding the Performances of a Single-Phase Full Converter with Symmetric Angle Control

The single-phase full converter in Figure 10.8a is operated with symmetric angle control. The load current with an average value of I_a, is continuous, where the ripple content is negligible. (a) Express the input current of converter in Fourier series, and determine the HF of input current, DF, and input PF. (b) If the conduction angle is $\beta = \pi/3$ and the peak input voltage is $V_m = 169.83$ V, calculate V_{dc}, V_{rms}, HF, DF, and PF.

Solution

a. The waveform for input current is shown in Figure 10.9a and the instantaneous input current can be expressed in Fourier series as

$$i_s(t) = a_0 + \sum_{n=1,2,\ldots}^{\infty} (A_n \cos n\omega t + B_n \sin n\omega t)$$

where

$$a_0 = \frac{1}{2\pi} \left[\int_{(\pi-\beta)/2}^{(\pi+\beta)/2} I_a\, d(\omega t) - \int_{(3\pi-\beta)/2}^{(3\pi+\beta)/2} I_a\, d(\omega t) \right] = 0$$

$$a_n = \frac{1}{\pi} \int_0^{2\pi} i_s(t) \cos n\omega t\, d(\omega t) = 0$$

$$b_n = \frac{1}{\pi} \int_0^{2\pi} i_s(t) \sin n\omega t\, d(\omega t) = \frac{4I_a}{n\pi} \sin \frac{n\beta}{2} \qquad \text{for } n = 1, 3, \ldots$$

$$= 0 \quad \text{for } n = 2, 4, \ldots$$

Because $a_0 = 0$, the input current can be written as

$$i_s(t) = \sum_{n=1,3,5,\ldots}^{\infty} \sqrt{2}\, I_n \sin(n\omega t + \phi_n) \tag{10.39}$$

where

$$\phi_n = \tan^{-1} \frac{a_n}{b_n} = 0 \tag{10.40}$$

The rms value of the nth harmonic input current is given as

$$I_{sn} = \frac{1}{\sqrt{2}} (a_n^2 + b_n^2)^{1/2} = \frac{2\sqrt{2}\, I_a}{n\pi} \sin \frac{n\beta}{2} \tag{10.41}$$

The rms value of the fundamental current is

$$I_{s1} = \frac{2\sqrt{2}\,I_a}{\pi} \sin \frac{\beta}{2} \qquad (10.42)$$

The rms input current is found as

$$I_s = I_a\sqrt{\frac{\beta}{\pi}} \qquad (10.43)$$

$$\text{HF} = \left[\left(\frac{I_s}{I_{s1}}\right)^2 - 1\right]^{1/2} = \left[\frac{\pi\beta}{4(1 - \cos \beta)} - 1\right]^{1/2} \qquad (10.44)$$

$$\text{DF} = \cos \phi_1 = 1 \qquad (10.45)$$

$$\text{PF} = \left(\frac{I_{s1}}{I_s}\right)\text{DF} = \frac{2\sqrt{2}}{\sqrt{\beta\pi}} \sin \frac{\beta}{2} \qquad (10.46)$$

b. $\beta = \pi/3$ and DF $= 1.0$. From Eq. (10.37),

$$V_{\text{dc}} = \left(2 \times \frac{169.83}{\pi}\right) \sin \frac{\pi}{6} = 54.06 \text{ V}$$

From Eq. (10.38),

$$V_{\text{rms}} = \frac{169.83}{\sqrt{2}}\left(\frac{\beta + \sin \beta}{\pi}\right)^{1/2} = 93.72 \text{ V}$$

$$I_{s1} = I_a\left(\frac{2\sqrt{2}}{\pi}\right) \sin \frac{\pi}{6} = 0.4502 I_a$$

$$I_s = I_a\sqrt{\frac{\beta}{\pi}} = 0.5774 I_a$$

$$\text{HF} = \left[\left(\frac{I_s}{I_{s1}}\right)^2 - 1\right]^{1/2} = 0.803 \quad \text{or} \quad 80.3\%$$

$$\text{PF} = \frac{I_{s1}}{I_s} = 0.7797 \text{ (lagging)}$$

Note: The PF is improved significantly. However, the HF is increased.

10.8.3 PWM Control

If the output voltage of single-phase converters is controlled by varying the delay angle, extinction angle, or symmetric angle, there is only one pulse per half-cycle in the input current of the converter, and as a result the lowest order harmonic is the third. It is difficult to filter out the lower order harmonic current. In PWM control, the converter switches are turned on and off several times during a half-cycle and the output voltage is controlled by varying the width of pulses [15–17]. The gate signals are generated by comparing a triangular wave with a dc signal, as shown in Figure 10.10b. Figure 10.10a shows the input voltage, output voltage, and input current. The lower order harmonics

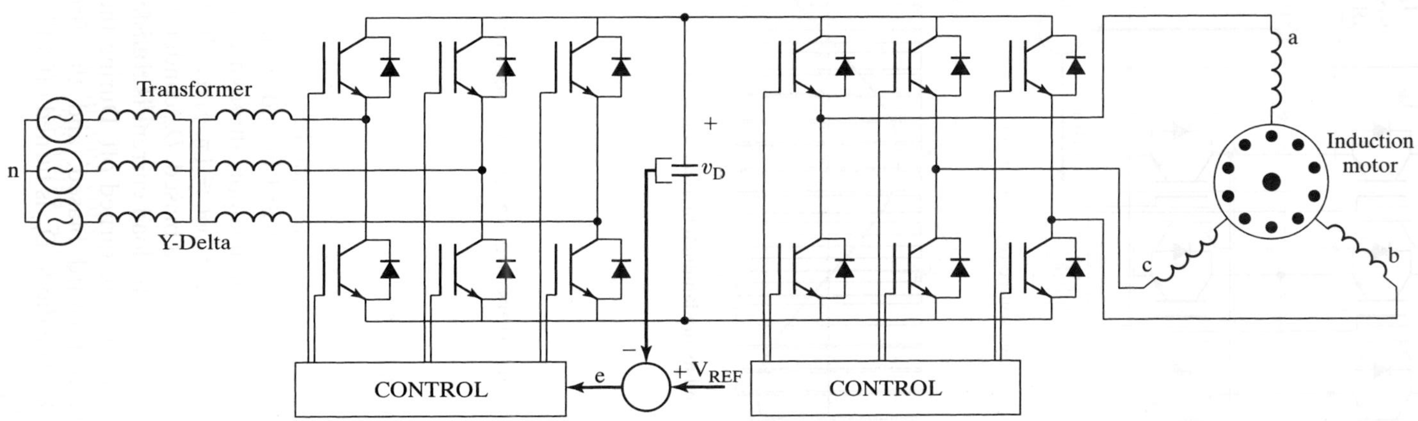

FIGURE 10.14

Two forced-commutated cascaded converters.

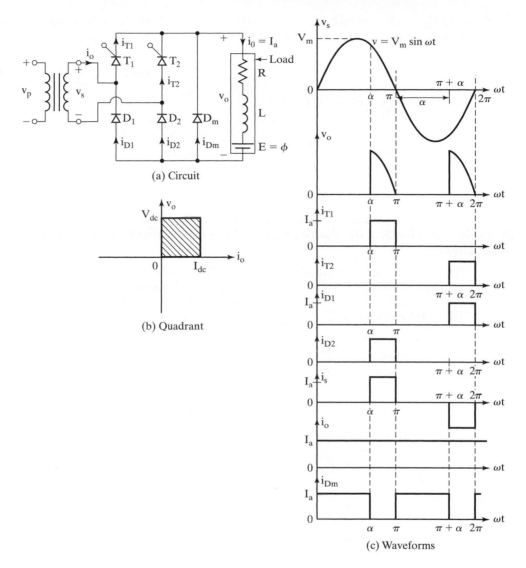

FIGURE 10.15

Single-phase semiconverter.

input voltage, output voltage, input current, and currents through T_1, T_2, D_1, and D_2. This converter has a better PF due to the freewheeling diode and is commonly used in applications up to 15 kW, where one-quadrant operation is acceptable.

The average output voltage can be found from

$$V_{dc} = \frac{2}{2\pi} \int_{\alpha}^{\pi} V_m \sin \omega t \, d(\omega t) = \frac{2V_m}{2\pi} [-\cos \omega t]_{\alpha}^{\pi}$$

$$= \frac{V_m}{\pi} (1 + \cos \alpha)$$

(10.52)

and V_{dc} can be varied from $2V_m/\pi$ to 0 by varying α from 0 to π. The maximum average output voltage is $V_{dm} = 2V_m/\pi$ and the normalized average output voltage is

$$V_n = \frac{V_{dc}}{V_{dm}} = 0.5(1 + \cos\alpha) \tag{10.53}$$

The rms output voltage is found from

$$V_{rms} = \left[\frac{2}{2\pi}\int_\alpha^\pi V_m^2 \sin^2\omega t\, d(\omega t)\right]^{1/2} = \left[\frac{V_m^2}{2\pi}\int_\alpha^\pi (1 - \cos 2\omega t)\, d(\omega t)\right]^{1/2}$$

$$= \frac{V_m}{\sqrt{2}}\left[\frac{1}{\pi}\left(\pi - \alpha + \frac{\sin 2\alpha}{2}\right)\right]^{1/2} \tag{10.54}$$

Example 10.10 Finding the Fourier Series of the Input Current and Input PF of 1-Phase Semiconverter

The semiconverter in Figure 10.15a is connected to a 120-V, 60-Hz supply. The load current I_a can be assumed to be continuous and its ripple content is negligible. The turns ratio of the transformer is unity. (a) Express the input current in a Fourier series; determine the input current HF, DF, and input PF. (b) If the delay angle is $\alpha = \pi/2$, calculate V_{dc}, V_n, V_{rms}, HF, DF, and PF.

Solution

a. The waveform for input current is shown in Figure 10.15c and the instantaneous input current can be expressed in a Fourier series as

$$i_s(t) = a_0 + \sum_{n=1,2,\ldots}^\infty (A_n \cos n\omega t + B_n \sin n\omega t) \tag{10.55}$$

where

$$a_0 = \frac{1}{2\pi}\int_\alpha^{2\pi} i_s(t)\, d(\omega t) = \frac{1}{2\pi}\left[\int_\alpha^\pi I_a\, d(\omega t) - \int_{\pi+\alpha}^{2\pi} I_a\, d(\omega t)\right] = 0$$

$$a_n = \frac{1}{\pi}\int_\alpha^{2\pi} i_s(t)\cos n\omega t\, d(\omega t)$$

$$= \frac{1}{\pi}\left[\int_\alpha^\pi I_a \cos n\omega t\, d(\omega t) - \int_{\pi+\alpha}^{2\pi} I_a \cos n\omega t\, d(\omega t)\right]$$

$$= -\frac{2I_a}{n\pi}\sin n\alpha \qquad \text{for } n = 1, 3, 5, \ldots$$

$$= 0 \qquad \text{for } n = 2, 4, 6, \ldots$$

$$b_n = \frac{1}{\pi}\int_\alpha^{2\pi} i_s(t)\sin n\omega t\, d(\omega t)$$

$$= \frac{1}{\pi}\left[\int_\alpha^\pi I_a \sin n\omega t\, d(\omega t) - \int_{\pi+\alpha}^{2\pi} I_a \sin n\omega t\, d(\omega t)\right]$$

$$= \frac{2I_a}{n\pi}(1 + \cos n\alpha) \qquad \text{for } n = 1, 3, 5, \ldots$$

$$= 0 \qquad \text{for } n = 2, 4, 6, \ldots$$

Because $a_0 = 0$, Eq. (10.55) can be written as

$$i_s(t) = \sum_{n=1,3,5,\ldots}^{\infty} \sqrt{2}\, I_{sn} \sin(n\omega t + \phi_n) \tag{10.56}$$

where

$$\phi_n = \tan^{-1} \frac{a_n}{b_n} = -\frac{n\alpha}{2} \tag{10.57}$$

The rms value of the nth harmonic component of the input current is derived as

$$I_{sn} = \frac{1}{\sqrt{2}} (a_n^2 + b_n^2)^{1/2} = \frac{2\sqrt{2}\, I_a}{n\pi} \cos \frac{n\alpha}{2} \tag{10.58}$$

From Eq. (10.58), the rms value of the fundamental current is

$$I_{s1} = \frac{2\sqrt{2}I_a}{\pi} \cos \frac{\alpha}{2}$$

The rms input current can be calculated from Eq. (10.58) as

$$I_s = \left(\sum_{n=1,2,\ldots}^{\infty} I_{sn} \right)^{1/2}$$

I_s can also be determined directly from

$$I_s = \left[\frac{2}{2\pi} \int_\alpha^\pi I_a^2\, d(\omega t) \right]^{1/2} = I_a \left(1 - \frac{\alpha}{\pi} \right)^{1/2}$$

From Eq. (3.10), HF $= [(I_s/I_{s1})^2 - 1]^{1/2}$ or

$$\text{HF} = \left[\frac{\pi(\pi - \alpha)}{4(1 + \cos \alpha)} - 1 \right]^{1/2} \tag{10.59}$$

From Eqs. (3.9) and (10.57),

$$\text{DF} = \cos \phi_1 = \cos \left(-\frac{\alpha}{2} \right) \tag{10.60}$$

From Eq. (3.11),

$$\text{PF} = \frac{I_{s1}}{I_s} \cos \frac{\alpha}{2} = \frac{\sqrt{2}(1 + \cos \alpha)}{[\pi(\pi - \alpha)]^{1/2}} \tag{10.61}$$

b. $\alpha = \pi/2$ and $V_m = \sqrt{2} \times 120 = 169.7$ V. From Eq. (10.52), $V_{dc} = (V_m/\pi)(1 + \cos \alpha) = 54.02$ V, from Eq. (10.53), $V_n = 0.5$ pu, and from Eq. (10.54),

$$V_{rms} = \frac{V_m}{\sqrt{2}} \left[\frac{1}{\pi} \left(\pi - \alpha + \frac{\sin 2\alpha}{2} \right) \right]^{1/2} = 84.57 \text{ V}$$

$$I_{s1} = \frac{2\sqrt{2}\, I_a}{\pi} \cos \frac{\pi}{4} = 0.6366 I_a$$

$$I_s = I_a \left(1 - \frac{\alpha}{\pi}\right)^{1/2} = 0.7071 I_a$$

$$\text{HF} = \left[\left(\frac{I_s}{I_{s1}}\right)^2 - 1\right]^{1/2} = 0.4835 \quad \text{or} \quad 48.35\%$$

$$\phi_1 = -\frac{\pi}{4} \quad \text{and} \quad \text{DF} = \cos\left(-\frac{\pi}{4}\right) = 0.7071$$

$$\text{PF} = \frac{I_{s1}}{I_s} \cos\frac{\alpha}{2} = 0.6366 \text{ (lagging)}$$

Note: The performance parameters of the converter depend on the delay angle α.

10.9.1 Single-Phase Semiconverter with *RL* Load

In practice, a load has a finite inductance. The load current depends on the values of load resistance R, load inductance L, and battery voltage E, as shown in Figure 10.15a. The converter operation can be divided into two modes: mode 1 and mode 2.

Mode 1. This mode is valid for $0 \le \omega t \le \alpha$, during which the freewheeling diode D_m conducts. The load current i_{L1} during mode 1 is described by

$$L \frac{di_{L1}}{dt} + R i_{L1} + E = 0 \qquad (10.62)$$

which, with initial condition $i_{L1}(\omega t = 0) = I_{Lo}$ in the steady state, gives

$$i_{L1} = I_{Lo} e^{-(R/L)t} - \frac{E}{R}(1 - e^{-(R/L)t}) \qquad \text{for } i_{L1} \ge 0 \qquad (10.63)$$

At the end of this mode at $\omega t = \alpha$, the load current becomes I_{L1}. That is,

$$I_{L1} = i_{L1}(\omega t = \alpha) = I_{Lo} e^{-(R/L)(\alpha/\omega)} - \frac{E}{R}[1 - e^{-(R/L)(\alpha/\omega)}] \qquad \text{for } I_{L1} \ge 0 \qquad (10.64)$$

Mode 2. This mode is valid for $\alpha \le \omega t \le \pi$ while thyristor T_1 conducts. If $v_s = \sqrt{2}\, V_s \sin \omega t$ is the input voltage, the load current i_{L2} during mode 2 can be found from

$$L \frac{di_{L2}}{dt} + R i_{L2} + E - \sqrt{2}\, V_s \sin \omega t \qquad \text{for } i_{L2} \ge 0 \qquad (10.65)$$

whose solution is of the form

$$i_{L2} = \frac{\sqrt{2}\, V_s}{Z} \sin(\omega t - \theta) + A_1 e^{-(R/L)t} - \frac{E}{R} \qquad \text{for } i_{L2} \ge 0$$

where load impedance $Z = [R^2 + (\omega L)^2]^{1/2}$ and load impedance angle $\theta = \tan^{-1}(\omega L/R)$.

Constant A_1, which can be determined from the initial condition: at $\omega t = \alpha$, $i_{L2} = I_{L1}$, is found as

$$A_1 = \left[I_{L1} + \frac{E}{R} - \frac{\sqrt{2}\,V_s}{Z} \sin(\alpha - \theta) \right] e^{(R/L)(\alpha/\omega)}$$

Substitution of A_1 yields

$$i_{L2} = \frac{\sqrt{2}\,V_s}{Z} \sin(\omega t - \theta) - \frac{E}{R} + \left[I_{L1} + \frac{E}{R} - \frac{\sqrt{2}\,V_s}{Z} \sin(\alpha - \theta) \right] e^{(R/L)(\alpha/\omega - t)}$$

$$\text{for } i_{L2} \geq 0 \qquad (10.66)$$

At the end of mode 2 in the steady-state condition: $I_{L2}(\omega t = \pi) = I_{L0}$. Applying this condition to Eq. (10.63) and solving for I_{L0}, we get

$$I_{L0} = \frac{\sqrt{2}\,V_s}{Z} \frac{\sin(\pi - \theta) - \sin(\alpha - \theta)e^{(R/L)(\alpha - \pi)/\omega}}{1 - e^{-(R/L)(\pi/\omega)}} - \frac{E}{R} \qquad (10.67)$$

$$\text{for } I_{L0} \geq 0 \text{ and } \theta \leq \alpha \leq \pi$$

The rms current of a thyristor can be found from Eq. (10.66) as

$$I_R = \left[\frac{1}{2\pi} \int_\alpha^\pi i_{L2}^2 \, d(\omega t) \right]^{1/2}$$

The average current of a thyristor can also be found from Eq. (10.66) as

$$I_A = \frac{1}{2\pi} \int_\alpha^\pi i_{L2} \, d(\omega t)$$

The rms output current can be found from Eqs. (10.63) and (10.66) as

$$I_{\text{rms}} = \left[\frac{2}{2\pi} \int_0^\alpha i_{L1}^2 \, d(\omega t) + \frac{2}{2\pi} \int_\alpha^\pi i_{L2}^2 \, d(\omega t) \right]^{1/2}$$

The average output current can be found from Eqs. (10.63) and (10.66) as

$$I_{\text{dc}} = \frac{1}{2\pi} \int_0^\alpha i_{L1} \, d(\omega t) + \frac{1}{2\pi} \int_\alpha^\pi i_{L2} \, d(\omega t)$$

Discontinuous load current. By setting $I_{L0} = 0$ in Eq. (10.67), dividing by $\sqrt{2}\,V_s/Z$, and substituting $R/Z = \cos\theta$ and $\omega L/R = \tan\theta$, we get the critical value of voltage ratio $x = E/\sqrt{2}\,V_{ab}$ as

$$x = \left[\frac{\sin(\pi - \theta) - \sin(\alpha - \theta)e^{\left(\frac{\alpha - \pi}{\tan(\theta)}\right)}}{1 - e^{-\left(\frac{\pi}{\tan(\theta)}\right)}} \right] \cos(\theta) \qquad (10.68)$$

which can be solved for the corresponding critical value of $\alpha = \alpha_c$ for known values of x and θ. For $\alpha \geq \alpha_c$, $I_{L1} = 0$. The load current that is described by Eq. (10.66) flows only during the period $\alpha \leq \omega t \leq \beta$. At $\omega t = \beta$, the load current falls to zero again.

Gating sequence.   The gating sequence is as follows:

1. Generate a pulse signal at the positive zero crossing of the supply voltage v_s.
2. Delay the pulse by the desired angles α and $\alpha + \pi$ for gating T_1 and T_2, respectively, through gate-isolating circuits.

Example 10.11 Finding the Current Ratings of Single-Phase Semiconverter with an *RL* Load

The single-phase semiconverter of Figure 10.15a has an *RL* load of $L = 6.5$ mH, $R = 2.5\ \Omega$, and $E = 10$ V. The input voltage is $V_s = 120$ V (rms) at 60 Hz. Determine (a) the load current I_{Lo} at $\omega t = 0$, and the load current I_{L1} at $\omega t = \alpha = 60°$, (b) the average thyristor current I_A, (c) the rms thyristor current I_R, (d) the rms output current I_{rms}, (e) the average output current I_{dc}, and (f) the critical value of delay angle α_c.

Solution
$R = 2.5\ \Omega$, $L = 6.5$ mH, $f = 60$ Hz, $\omega = 2\pi \times 60 = 377$ rad/s, $V_s = 120$ V, $\theta = \tan^{-1}$ $(\omega L/R) = 44.43°$, and $Z = 3.5\ \Omega$.

   a. The steady-state load current at $\omega t = 0$, $I_{Lo} = 29.77$ A. The steady-state load current at $\omega t = \alpha$, $I_{L1} = 7.6$ A.
   b. The numerical integration of i_{L2} in Eq. (10.66) yields the average thyristor current as $I_A = 11.42$ A.
   c. By numerical integration of i_{L2}^2 between the limits $\omega t = \alpha$ to π, we get the rms thyristor current as $I_R = 20.59$ A.
   d. The rms output current $I_{rms} = 33.79$ A.
   e. The average output current $I_{dc} = 32.74$ A.
   f. By iteration, Eq. (10.68) gives $\alpha_c = 158.2°$.

Key Points of Section 10.9

- A single-phase semiconverter uses a freewheeling diode across the load and it operates in the first quadrant.
- The freewheeling diode provides a path for the continuity of the load current and it has a better input PF than that of the full converter. A switching device conducts from α to π.

10.10   THREE-PHASE SEMICONVERTERS

Three-phase semiconverters are used in industrial applications up to the 120-kW level, where one-quadrant operation is required. The PF of this converter decreases as the delay angle increases, but it is better than that of three-phase half-wave converters. Figure 10.16a shows a three-phase semiconverter with a highly inductive load and the load current has a negligible ripple content.

Figure 10.16b shows the waveforms for input voltage, output voltage, input current, and current through thyristors and diodes. The frequency of output voltage is $3f_s$. The delay angle α can be varied from 0 to π. During the period $\pi/6 \leq \omega t < 7\pi/6$,

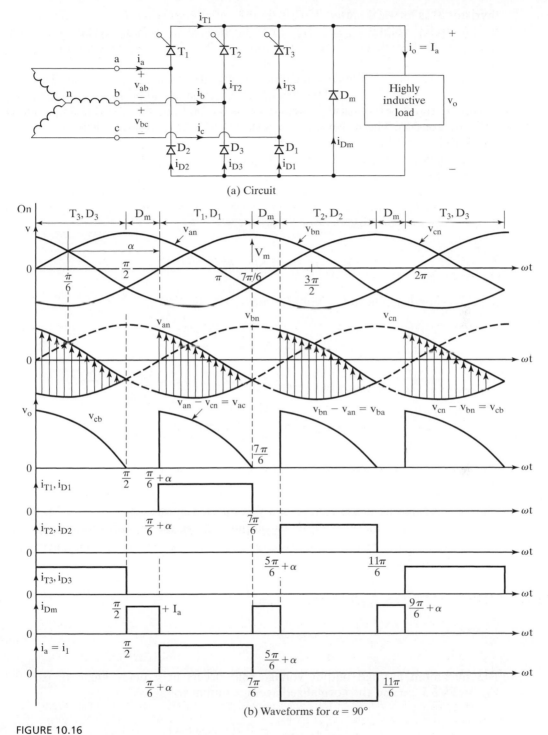

(a) Circuit

(b) Waveforms for α = 90°

FIGURE 10.16

Three-phase semiconverter.

thyristor T_1 is forward biased. If T_1 is fired at $\omega t = (\pi/6 + \alpha)$, T_1 and D_1 conduct and the line-to-line voltage v_{ac} appears across the load. At $\omega t = 7\pi/6$, v_{ac} starts to be negative and the freewheeling diode D_m conducts. The load current continues to flow through D_m; and T_1 and D_1 are turned off.

If there were no freewheeling diode, T_1 would continue to conduct until thyristor T_2 fired at $\omega t = 5\pi/6 + \alpha$ and the freewheeling action would be accomplished through T_1 and D_2. If $\alpha \leq \pi/3$, each thyristor conducts for $2\pi/3$ and the freewheeling diode D_m does not conduct. The waveforms for a three-phase semiconverter with $\alpha \leq \pi/3$ are shown in Figure 10.17.

If we define the three line-neutral voltages as follows:

$$v_{an} = V_m \sin \omega t$$

$$v_{bn} = V_m \sin \left(\omega t - \frac{2\pi}{3} \right)$$

$$v_{cn} = V_m \sin \left(\omega t + \frac{2\pi}{3} \right)$$

the corresponding line-to-line voltages are

$$v_{ac} = v_{an} - v_{cn} = \sqrt{3} V_m \sin \left(\omega t - \frac{\pi}{6} \right)$$

$$v_{ba} = v_{bn} - v_{an} = \sqrt{3} V_m \sin \left(\omega t - \frac{5\pi}{6} \right)$$

$$v_{cb} = v_{cn} - v_{bn} = \sqrt{3} V_m \sin \left(\omega t + \frac{\pi}{2} \right)$$

$$v_{ab} = v_{an} - v_{bn} = \sqrt{3} V_m \sin \left(\omega t + \frac{\pi}{6} \right)$$

where V_m is the peak phase voltage of a Y-connected source.

For $\alpha \geq \pi/3$, and discontinuous output voltage: the average output voltage is found from

$$V_{dc} = \frac{3}{2\pi} \int_{\pi/6+\alpha}^{7\pi/6} v_{ac}\, d(\omega t) = \frac{3}{2\pi} \int_{\pi/6+\alpha}^{7\pi/6} \sqrt{3}\, V_m \sin \left(\omega t - \frac{\pi}{6} \right) d(\omega t)$$

$$= \frac{3\sqrt{3}\, V_m}{2\pi} (1 + \cos \alpha) \tag{10.69}$$

The maximum average output voltage that occurs at a delay angle of $\alpha = 0$ is $V_{dm} = 3\sqrt{3}\, V_m/\pi$ and the normalized average output voltage is

$$V_n = \frac{V_{dc}}{V_{dm}} = 0.5(1 + \cos \alpha) \tag{10.70}$$

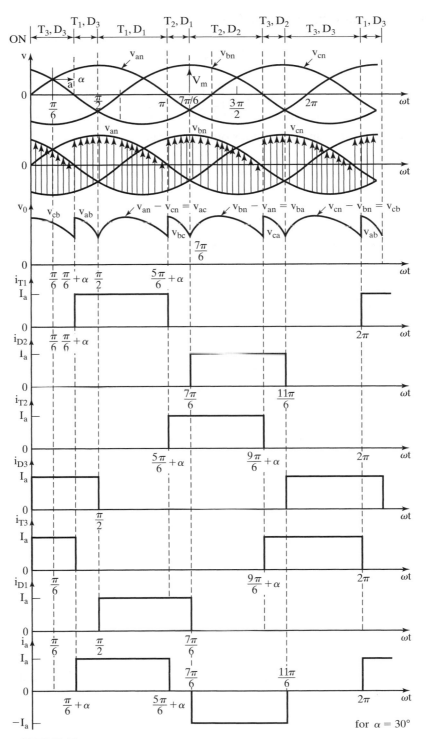

FIGURE 10.17

Three-phase semiconverter for $\alpha \leq \pi/3$.

and the normalized average output voltage is

$$V_n = \frac{V_{dc}}{V_{dm}} = 0.25(1 + \cos \alpha_1) \tag{10.78}$$

If both converters are operating: $\alpha_1 = 0$ and $0 \le \alpha_2 \le \pi$, then

$$V_{dc} = V_{dc1} + V_{dc2} = \frac{V_m}{\pi}(3 + \cos \alpha_2) \tag{10.79}$$

and the normalized average output voltage is

$$V_n = \frac{V_{dc}}{V_{dm}} = 0.25(3 + \cos \alpha_2) \tag{10.80}$$

Figure 10.19a shows two full converters that are connected in series and the turns ratio between the primary and secondary is $N_p/N_s = 2$. Due to the fact that there are no freewheeling diodes, one of the converters cannot be bypassed and both converters must operate at the same time.

In rectification mode, one converter is fully advanced ($\alpha_1 = 0$) and the delay angle of the other converter, α_2, is varied from 0 to π to control the dc output voltage. Figure 10.19b shows the input voltage, output voltage, input current to the converters, and input supply current. Comparing Figure 10.19b with Figure 10.15b, we can notice that the input current from the supply is similar to that of semiconverter. As a result the PF of this converter is improved, but the PF is less than that of series semiconverters.

In the inversion mode, one converter is fully retarded, $\alpha_2 = \pi$, and the delay angle of the other converter, α_1, is varied from 0 to π to control the average output voltage. Figure 10.19d shows the v–i characteristics of series-full converters.

From Eq. (10.5), the average output voltages of two full converters are

$$V_{dc1} = \frac{2V_m}{\pi} \cos \alpha_1$$

$$V_{dc2} = \frac{2V_m}{\pi} \cos \alpha_2$$

The resultant average output voltage is

$$V_{dc} = V_{dc1} + V_{dc2} = \frac{2V_m}{\pi}(\cos \alpha_1 + \cos \alpha_2) \tag{10.81}$$

The maximum average output voltage for $\alpha_1 = \alpha_2 = 0$ is $V_{dm} = 4V_m/\pi$. In the rectification mode, $\alpha_1 = 0$ and $0 \le \alpha_2 \le \pi$; then

$$V_{dc} = V_{dc1} + V_{dc2} = \frac{2V_m}{\pi}(1 + \cos \alpha_2) \tag{10.82}$$

and the normalized dc output voltage is

$$V_n = \frac{V_{dc}}{V_{dm}} = 0.5(1 + \cos \alpha_2) \tag{10.83}$$

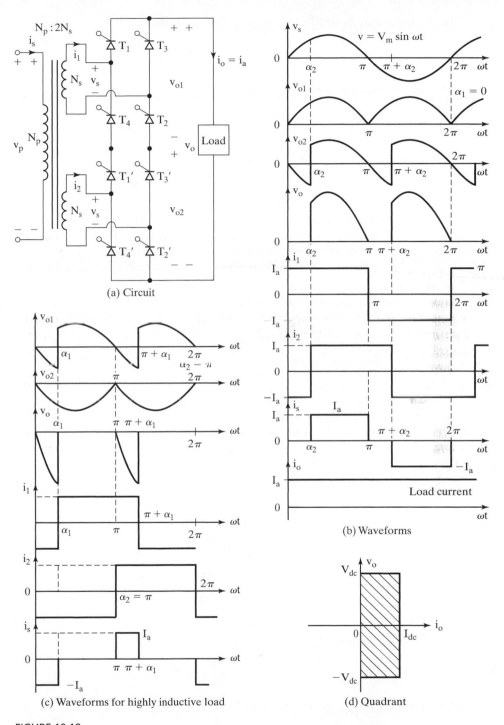

(a) Circuit

(b) Waveforms

(c) Waveforms for highly inductive load

(d) Quadrant

FIGURE 10.19

Single-phase full converters.

In the inversion mode, $0 \le \alpha_1 \le \pi$ and $\alpha_2 = \pi$; then

$$V_{dc} = V_{dc1} + V_{dc2} = \frac{2V_m}{\pi}(\cos \alpha_1 - 1) \tag{10.84}$$

and the normalized average output voltage is

$$V_n = \frac{V_{dc}}{V_{dm}} = 0.5(\cos \alpha_1 - 1) \tag{10.85}$$

Gating sequence.   The gating sequence is as follows:

1. Generate a pulse signal at the positive zero crossing of the phase voltage v_s.
2. Delay the pulse by the desired angles $\alpha_1 = 0$ and $\alpha_2 = \alpha$ for gating converter 1 and converter 2, respectively, through gate-isolating circuits.

Example 10.13 Finding the Input Power Factor of a Series Single-Phase Full Converter

The load current (with an average value of I_a) of series-full converters in Figure 10.19a is continuous and the ripple content is negligible. The turns ratio of the transformer is $N_p/N_s = 2$. The converters operate in rectification mode such that $\alpha_1 = 0$ and α_2 varies from 0 to π. (a) Express the input supply current in Fourier series, determine the input current HF, DF, and input PF. (b) If the delay angle is $\alpha_2 = \pi/2$ and the peak input voltage is $V_m = 162$ V, calculate V_{dc}, V_n, V_{rms}, HF, DF, and PF.

Solution

a.   The waveform for input current is shown in Figure 10.19b and the instantaneous input supply current can be expressed in a Fourier series as

$$i_s(t) = \sum_{n=1,2,\ldots}^{\infty} \sqrt{2}\, I_n \sin(n\omega t + \phi_n) \tag{10.86}$$

where $\phi_n = -n\alpha_2/2$. Equation (10.58) gives the rms value of the nth harmonic input current

$$I_{sn} = \frac{4I_a}{\sqrt{2}\, n\pi}\cos\frac{n\alpha_2}{2} = \frac{2\sqrt{2}I_a}{n\pi}\cos\frac{n\alpha_2}{2} \tag{10.87}$$

The rms value of fundamental current is

$$I_{s1} = \frac{2\sqrt{2}I_a}{\pi}\cos\frac{\alpha_2}{2} \tag{10.88}$$

The rms input current is found as

$$I_s = I_a\left(1 - \frac{\alpha_2}{\pi}\right)^{1/2} \tag{10.89}$$

From Eq. (3.10),

$$\mathrm{HF} = \left[\frac{\pi(\pi - \alpha_2)}{4(1 + \cos \alpha_2)} - 1\right]^{1/2} \tag{10.90}$$

From Eqs. (3.9),

$$DF = \cos \phi_1 = \cos \left(-\frac{\alpha_2}{2} \right) \tag{10.91}$$

From Eq. (3.11),

$$PF = \frac{I_{s1}}{I_s} \cos \frac{\alpha_2}{2} = \frac{\sqrt{2}\,(1 + \cos \alpha_2)}{[\pi(\pi - \alpha_2)]^{1/2}} \tag{10.92}$$

b. $\alpha_1 = 0$ and $\alpha_2 = \pi/2$. From Eq. (10.81),

$$V_{dc} = \left(2 \times \frac{162}{\pi} \right) \left(1 + \cos \frac{\pi}{2} \right) = 103.13 \text{ V}$$

From Eq. (10.83), $V_n = 0.5$ pu and

$$V_{rms}^2 = \frac{2}{2\pi} \int_{\alpha 2}^{\pi} (2V_m)^2 \sin^2 \omega t \, d(\omega t)$$

$$V_{rms} = \sqrt{2}\, V_m \left[\frac{1}{\pi} \left(\pi - \alpha_2 + \frac{\sin 2\alpha_2}{2} \right) \right]^{1/2} = V_m = 162 \text{ V}$$

$$I_{s1} = I_a \frac{2\sqrt{2}}{\pi} \cos \frac{\pi}{4} = 0.6366 I_a \qquad \text{and} \qquad I_s = 0.7071 I_a$$

$$HF = \left[\left(\frac{I_s}{I_{s1}} \right)^2 - 1 \right]^{1/2} = 0.4835 \quad \text{or} \quad 48.35\%$$

$$\phi_1 = -\frac{\pi}{4} \qquad \text{and} \qquad DF = \cos \left(-\frac{\pi}{4} \right) = 0.7071$$

$$PF = \frac{I_{s1}}{I_s} \cos(-\phi_1) = 0.6366 \text{ (lagging)}$$

Note: The performance of series-full converters is similar to that of single-phase semiconverters.

Key Point of Section 10.11

- Semiconverters and full converters can be connected in series to share the voltage and also to improve the input PF.

10.12   TWELVE-PULSE CONVERTERS

A three-phase bridge gives a six-pulse output voltage. For high-power applications such as high-voltage dc transmission and dc motor drives, a 12-pulse output is generally required to reduce the output ripples and to increase the ripple frequencies. Two 6-pulse bridges can be combined either in series or in parallel to produce an effective 12-pulse output. Two configurations are shown in Figure 10.20. A 30° phase shift between secondary windings can be accomplished by connecting one secondary in Y and the other in delta (Δ).

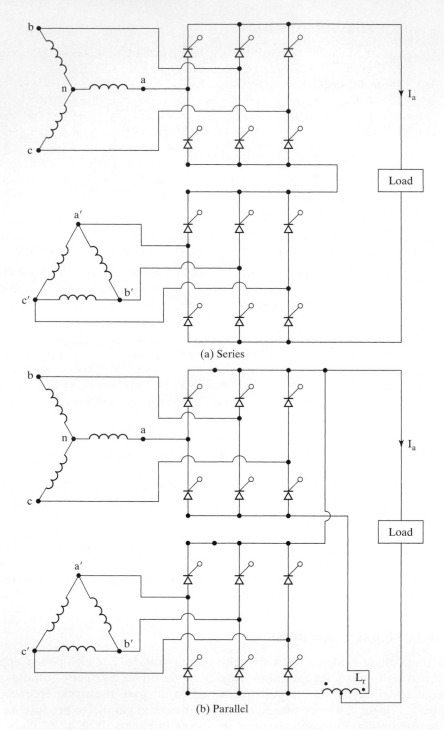

(a) Series

(b) Parallel

FIGURE 10.20

Configurations for 12-pulse output.

10.13 DESIGN OF CONVERTER CIRCUITS

The design of converter circuits requires determining the ratings of switching devices (e.g., thyristors) and diodes. The switches and diodes are specified by the average current, rms current, peak current, and peak inverse voltage. In the case of controlled rectifiers, the current ratings of devices depend on the delay (or control) angle. The ratings of power devices must be designed under the worst-case condition and this occurs when the converter delivers the maximum average output voltage V_{dm}.

The output of converters contains harmonics that depend on the control (or delay) angle and the worst-case condition generally prevails under the minimum output voltage. Input and output filters must be designed under the minimum output voltage condition. The steps involved in designing the converters and filters are similar to those of rectifier circuit design in Section 3-9.

Example 10.14 Finding the Thyristor Ratings of a Three-Phase Full Converter

A three-phase full converter is operated from a three-phase 230-V, 60-Hz supply. The load is highly inductive and the average load current is $I_a = 150$ A with negligible ripple content. If the delay angle is $\alpha = \pi/3$, determine the ratings of thyristors.

Solution

The waveforms for thyristor currents are shown in Figure 10.5b. $V_s = 230/\sqrt{3} = 132.79$ V, $V_m = 187.79$ V, and $\alpha = \pi/3$. From Eq. (10.27), $V_{dc} = 3(\sqrt{3}/\pi) \times 187.79 \times \cos(\pi/3) = 155.3$ V. The output power $P_{dc} = 155.3 \times 150 = 23,295$ W. The average current through a thyristor $I_A = 150/3 = 50$ A. The rms current through a thyristor $I_R = 150\sqrt{2/6} = 86.6$ A. The peak current through a thyristor $I_{PT} = 150$ A. The peak inverse voltage is the peak amplitude of line-to-line voltage PIV $= \sqrt{3} V_m = \sqrt{3} \times 187.79 = 325.27$ V.

Example 10.15 Finding the Value of an Output C Filter for a Single-Phase Full Converter

A single-phase full converter, as shown in Figure 10.21, uses delay-angle control and is supplied from a 120-V, 60-Hz supply. (a) Use the method of Fourier series to obtain expressions for output voltage $v_o(t)$ and load current $i_o(t)$ as a function of delay angle α. (b) If $\alpha = \pi/3$, $E = 10$ V, $L = 20$ mH, and $R = 10\ \Omega$, determine the rms value of lowest order harmonic current in the load. (c) If in (b), a filter capacitor is connected across the load, determine the capacitor value to reduce the lowest order harmonic current to 10% of the value without the capacitor. (d) Use PSpice to plot the output voltage and the load current and to compute the THD of the load current and the input PF with the output filter capacitor in (c).

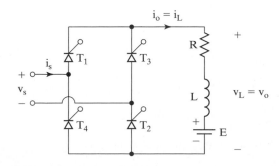

FIGURE 10.21

Single-phase full converter with RL load.

Solution

a. The waveform for output voltage is shown in Figure 10.2c. The frequency of output voltage is twice that of the main supply. The instantaneous output voltage can be expressed in a Fourier series as

$$v_o(t) = V_{dc} + \sum_{n=2,4,\ldots}^{\infty} (a_n \cos n\omega t + b_n \sin n\omega t) \tag{10.93}$$

where

$$V_{dc} = \frac{1}{2\pi} \int_{\alpha}^{2\pi+\alpha} V_m \sin \omega t \, d(\omega t) = \frac{2V_m}{\pi} \cos \alpha$$

$$a_n = \frac{2}{\pi} \int_{\alpha}^{\pi+\alpha} V_m \sin \omega t \cos n\omega t \, d(\omega t) = \frac{2V_m}{\pi} \left[\frac{\cos(n+1)\alpha}{n+1} - \frac{\cos(n-1)\alpha}{n-1} \right]$$

$$b_n = \frac{2}{\pi} \int_{\alpha}^{\pi+\alpha} V_m \sin \omega t \sin n\omega t \, d(\omega t) = \frac{2V_m}{\pi} \left[\frac{\sin(n+1)\alpha}{n+1} - \frac{\sin(n-1)\alpha}{n-1} \right]$$

The load impedance

$$Z = R + j(n\omega L) = [R^2 + (n\omega L)^2]^{1/2} \underline{/\theta_n}$$

and $\theta_n = \tan^{-1}(n\omega L/R)$. Dividing $v_o(t)$ of Eq. (10.93) by load impedance Z and simplifying the sine and cosine terms give the instantaneous load current as

$$i_o(t) = I_{dc} + \sum_{n=2,4,\ldots}^{\infty} \sqrt{2} I_n \sin(n\omega t + \phi_n - \theta_n) \tag{10.94}$$

where $I_{dc} = (V_{dc} - E)/R$, $\phi_n = \tan^{-1}(A_n/B_n)$, and

$$I_n = \frac{1}{\sqrt{2}} \frac{(a_n^2 + b_n^2)^{1/2}}{\sqrt{R^2 + (n\omega L)^2}}$$

b. If $\alpha = \pi/3$, $E = 10$ V, $L = 20$ mH, $R = 10\ \Omega$, $\omega = 2\pi \times 60 = 377$ rad/s, $V_m = \sqrt{2} \times 120 = 169.71$ V, and $V_{dc} = 54.02$ V.

$$I_{dc} = \frac{54.02 - 10}{10} = 4.40 \text{ A}$$

$a_2 = -0.833$, $b_2 = -0.866$, $\phi_2 = -223.9°$, $\theta_2 = 56.45°$

$a_4 = 0.433$, $b_4 = -0.173$, $\phi_4 = -111.79°$, $\theta_4 = 71.65°$

$a_6 = -0.029$, $b_6 = 0.297$, $\phi_6 = -5.5°$, $\theta_6 = 77.53°$

$$i_L(t) = 4.4 + \frac{2V_m}{\pi[R^2 + (n\omega L)^2]^{1/2}} [1.2 \sin(2\omega t + 223.9° - 56.45°) + 0.47 \sin(4\omega t$$

$$+ 111.79° - 71.65°) + 0.3 \sin(6\omega t - 5.5° - 77.53°) + \cdots]$$

$$= 4.4 + \frac{2 \times 169.71}{\pi[10^2 + (7.54n)^2]^{1/2}} [1.2 \sin(2\omega t + 167.45°)$$

$$\tag{10.95}$$

$$+ 0.47 \sin(4\omega t + 40.14°) + 0.3 \sin(6\omega t - 80.03°) + \cdots]$$

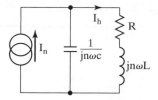

FIGURE 10.22

Equivalent circuit for harmonics.

The second harmonic is the lowest one and its rms value is

$$I_2 = \frac{2 \times 169.71}{\pi[10^2 + (7.54 \times 2)^2]^{1/2}} \left(\frac{1.2}{\sqrt{2}}\right) = 5.07 \text{ A}$$

c. Figure 10.22 shows the equivalent circuit for the harmonics. Using the current-divider rule, the harmonic current through the load is given by

$$\frac{I_h}{I_n} = \frac{1/(n\omega C)}{\{R^2 + [n\omega L - 1/(n\omega C)]^2\}^{1/2}}$$

For $n = 2$ and $\omega = 377$,

$$\frac{I_h}{I_n} = \frac{1/(2 \times 377C)}{\{10^2 + [2 \times 7.54 - 1/(2 \times 377C)]^2\}^{1/2}} = 0.1$$

and this gives $C = -670 \ \mu\text{F}$ or $793 \ \mu\text{F}$. Thus, $C = 793 \ \mu\text{F}$.

d. The peak supply voltage $V_m = 169.7$ V. For $\alpha_1 = 60°$, time delay $t_1 = (60/360) \times (1000/60 \text{ Hz}) \times 1000 = 2777.78 \ \mu\text{s}$ and time delay $t_2 = (240/360) \times (1000/60 \text{ Hz}) \times 1000 = 11{,}111.1 \ \mu\text{s}$. The single-phase full-converter circuit for PSpice simulation is shown in Figure 10.23a. The gate voltages V_{g1}, V_{g2}, V_{g3}, and V_{g4} for thyristors are shown in Figure 10.23b. The subcircuit definition for the thyristor model silicon-controlled rectifier (SCR) is described in Section 7.11.

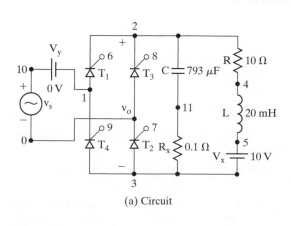

(a) Circuit

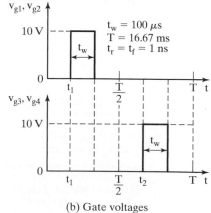

(b) Gate voltages

FIGURE 10.23

Single-phase full converter for PSpice simulation.

The list of the circuit file is as follows:

```
Example 10.15 Single-Phase Full Converter
VS    10    0   SIN (0    169. 7V    60HZ)
Vg1    6    2   PULSE (0V   10V    2777.8US    1NS    1NS    100US    16666.7US)
Vg2    7    0   PULSE (0V   10V    2777.8US    1NS    1NS    100US    16666.7US)
Vg3    8    2   PULSE (0V   10V    11111.1US   1NS    1NS    100US    16666.7US)
Vg4    9    1   PULSE (0V   10V    11111.1US   1NS    1NS    100US    16666.7US)
R      2    4   10
L      4    5   20MH
C      2   11   793UF
RX    11    3   0.1                 ; Added to help convergence
VX     5    3   DC      10V         ; Load battery voltage
VY    10    1   DC      0V          ; Voltage source to measure supply current
*    Subcircuit calls for thyristor model
XT1    1    6   2     SCR           ;    Thyristor T1
XT3    0    8   2     SCR           ;    Thyristor T3
XT2    3    7   0     SCR           ;    Thyristor T2
XT4    3    9   1     SCR           ;    Thyristor T4
*    Subcircuit  SCR which is missing must be inserted
.TRAN     10US     35MS     16.67MS          ; Transient analysis
.PROBE                                       ; Graphics postprocessor
.options   abstol  = 1.00u reltol = 1.0 m vntol  = 0.1 ITL5=10000
.FOUR     120HZ     I(VX)                    ; Fourier analysis
.END
```

The PSpice plots of the output voltage V (2, 3) and the load current I (VX) are shown in Figure 10.24.

The Fourier components of the load current are:

FOURIER COMPONENTS OF TRANSIENT RESPONSE I (VX)
DC COMPONENT = 1.147163E+01

HARMONIC NO	FREQUENCY (HZ)	FOURIER COMPONENT	NORMALIZED COMPONENT	PHASE (DEG)	NORMALIZED PHASE (DEG)
1	1.200E+02	2.136E+00	1.000E+00	−1.132E+02	0.000E+00
2	2.400E+02	4.917E−01	2.302E−01	1.738E+02	2.871E+02
3	3.600E+02	1.823E−01	8.533E−02	1.199E+02	2.332E+02
4	4.800E+02	9.933E−02	4.650E−02	7.794E+01	1.912E+02
5	6.000E+02	7.140E−02	3.342E−02	2.501E+01	1.382E+02
6	7.200E+02	4.339E−02	2.031E−02	−3.260E+01	8.063E+01
7	8.400E+02	2.642E−02	1.237E−02	−7.200E+01	4.123E+01
8	9.600E+02	2.248E−02	1.052E−02	−1.126E+02	6.192E+01
9	1.080E+03	2.012E−02	9.420E−03	−1.594E+02	−4.617E+01

TOTAL HARMONIC DISTORTION = 2.535750E+01 PERCENT

To find the input PF, we need to find the Fourier components of the input current, which are the same as the current through source VY.

FOURIER COMPONENTS OF TRANSIENT RESPONSE I (VY)
DC COMPONENT = 1.013355E−02

HARMONIC NO	FREQUENCY (HZ)	FOURIER COMPONENT	NORMALIZED COMPONENT	PHASE (DEG)	NORMALIZED PHASE (DEG)
1	6.000E+01	2.202E+01	1.000E+00	5.801E+01	0.000E+00
2	1.200E+02	2.073E−02	9.415E−04	4.033E+01	−1.768E+01

3	1.800E+02	1.958E+01	8.890E−01	−3.935E+00	−6.194E+01
4	2.400E+02	2.167E−02	9.841E−04	−1.159E+01	−6.960E+01
5	3.000E+02	1.613E+01	7.323E−01	−5.968E+01	−1.177E+02
6	3.600E+02	2.218E−02	1.007E−03	−6.575E+01	−1.238E+02
7	4.200E+02	1.375E+01	6.243E−01	−1.077E+02	−1.657E+02
8	4.800E+02	2.178E−02	9.891E−04	−1.202E+02	−1.783E+02
9	5.400E+02	1.317E+01	5.983E−01	−1.542E+02	−2.122E+02

TOTAL HARMONIC DISTORTION = 1.440281E+02 PERCENT

THD $= 144\% = 1.44$

Displacement angle $\phi_1 = 58.01°$

$$\text{DF} = \cos \phi_1 = \cos(-58.01) = 0.53 \text{ (lagging)}$$

$$\text{PF} = \frac{I_{s1}}{I_s} \cos \phi_1 = \frac{1}{[1 + (\%\text{THD}/100)^2]^{1/2}} \cos \phi_1 \qquad (10.96)$$

$$= \frac{1}{(1 + 1.44^2)^{1/2}} \times 0.53 = 0.302 \text{ (lagging)}$$

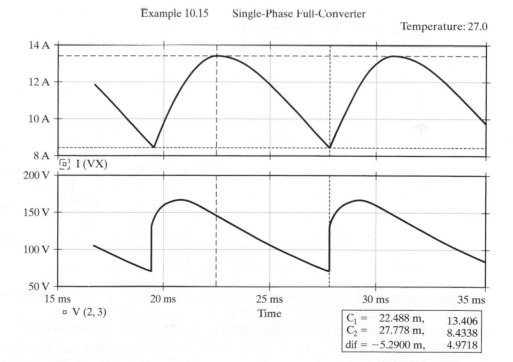

Example 10.15   Single-Phase Full-Converter

Temperature: 27.0

$C_1 =$	22.488 m,	13.406
$C_2 =$	27.778 m,	8.4338
dif $=$	−5.2900 m,	4.9718

FIGURE 10.24

SPICE plots for Example 10.15.

Notes:

1. The preceding analyses are valid only if the delay angle α is greater than α_0, which is given by

$$\alpha_0 = \sin^{-1} \frac{E}{V_m} = \sin^{-1} \frac{10}{169.71} = 3.38°$$

2. Due to the filter capacitor C, a high peak charging current flows from the source, and the THD of the input current has a high value of 144%.

3. Without the capacitor C, the load current becomes discontinuous, the peak second harmonic load current is $i_{2(\text{peak})} = 5.845$ A, I_{dc} is 6.257 A, the THD of the load current is 14.75%, and the THD of the input current is 15.66%.

Key Points of Section 10–13

- The design of a converter circuit requires (a) calculating the voltage and current ratings of the power devices, (b) finding the Fourier series of the output voltage and the input current and the (c) calculating the values of the input and output filters under worst-case conditions.

10.14   EFFECTS OF LOAD AND SOURCE INDUCTANCES

We can notice from Eq. (10.95) that the load current harmonics depend on load inductances. In Example 10.6 the input PF is calculated for a purely resistive load and in Example 10.7 for a highly inductive load. We can also notice that the input PF depends on the load PF.

In the derivations of output voltages and the performance criteria of converters, we have assumed that the source has no inductances and resistances. Normally, the values of line resistances are small and can be neglected. The amount of voltage drop due to source inductances is equal to that of rectifiers and does not change due to the phase control. Equation (3.79) can be applied to calculate the voltage drop due to the line commutating reactance L_c. If all the line inductances are equal, Eq. (3.80) gives the voltage drop as $V_{6x} = 6fL_cI_{\text{dc}}$ for a three-phase full converter.

The voltage drop is not dependent on delay angle α_1 under normal operation. However, the commutation (or overlap) angle μ varies with the delay angle. As the delay angle is increased, the overlap angle becomes smaller. This is illustrated in Figure 10.25. The volt-time integral as shown by crosshatched areas is equal to $I_{\text{dc}}L_c$ and is independent of voltages. As the commutating phase voltage increases, the time required to commutate gets smaller, but the "volt-seconds" remain the same.

If V_x is the average voltage drop per commutation due to overlap and V_y is the average voltage reduction due to phase-angle control, the average output voltage for a delay angle of α is

$$V_{\text{dc}}(\alpha) = V_{\text{dc}}(\alpha = 0) - V_y = V_{dm} - V_y \tag{10.97}$$

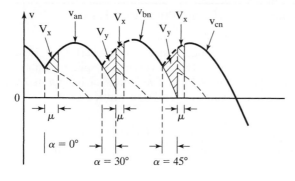

FIGURE 10.25

Relationship between delay angle and overlap angle.

and

$$V_y = V_{dm} - V_{dc}(\alpha) \tag{10.98}$$

where V_{dm} = maximum possible average output voltage. The average output voltage with overlap angle μ and two commutations is

$$V_{dc}(\alpha + \mu) = V_{dc}(\alpha = 0) - 2V_x - V_y = V_{dm} - 2V_x - V_y \tag{10.99}$$

Substituting V_y from Eq. (10.98) into Eq. (10.99), we can write the voltage drop due to overlap as

$$2V_x = 2f_s I_{dc} L_c = V_{dc}(\alpha) - V_{dc}(\alpha + \mu) \tag{10.100}$$

The overlap angle μ can be determined from Eq. (10.100) for known values of load current I_{dc}, commutating inductance L_c, and delay angle α. It should be noted that Eq. (10.100) is applicable only to a single-phase full converter.

Example 10.16 Finding the Overlap Angle for a Three-Phase Full Converter

A three-phase full converter is supplied from a three-phase 230-V, 60-Hz supply. The load current is continuous and has negligible ripple. If the average load current $I_{dc} = 150$ A and the commutating inductance $L_c = 0.1$ mH, determine the overlap angle when (a) $\alpha = 10°$, (b) $\alpha = 30°$, and (c) $\alpha = 60°$.

Solution
$V_m = \sqrt{2} \times 230/\sqrt{3} = 187.79$ V and $V_{dm} = 3\sqrt{3} \, V_m/\pi = 310.61$ V. From Eq. (10.25), $V_{dc}(\alpha) = 310.6 \cos \alpha$ and

$$V_{dc}(\alpha + \mu) = 310.61 \cos(\alpha + \mu)$$

For a three-phase converter, Eq. (10.100) can be modified to

$$6V_x = 6f_s I_{dc} L_c = V_{dc}(\alpha) - V_{dc}(\alpha + \mu)$$

$$6 \times 60 \times 150 \times 0.1 \times 10^{-3} = 310.61[\cos \alpha - \cos(\alpha + \mu)] \tag{10.101}$$

    a. For $\alpha = 10°$, $\mu = 4.66°$;

    b. For $\alpha = 30°$, $\mu = 1.94°$;

    c. For $\alpha = 60°$, $\mu = 1.14°$.

Example 10.17 Finding the Minimum Value of Gate Pulse Width for a Single-Phase Full Converter

The holding current of thyristors in the single-phase full converter of Figure 10.2a is $I_H = 500$ mA and the delay time is $t_d = 1.5$ μs. The converter is supplied from a 120-V, 60-Hz supply and has a load of $L = 10$ mH and $R = 10$ Ω. The converter is operated with a delay angle of $\alpha = 30°$. Determine the minimum value of gate pulse width t_G.

Solution

$I_H = 500$ mA $= 0.5$ A, $t_d = 1.5$ μs, $\alpha = 30° = \pi/6$, $L = 10$ mH, and $R = 10$ Ω. The instantaneous value of the input voltage is $v_s(t) = V_m \sin \omega t$, where $V_m = \sqrt{2} \times 120 = 169.7$ V.

At $\omega t = \alpha$,

$$V_1 = v_s(\omega t = \alpha) = 169.7 \times \sin \frac{\pi}{6} = 84.85 \text{ V}$$

The rate of rise of anode current di/dt at the instant of triggering is approximately

$$\frac{di}{dt} = \frac{V_1}{L} = \frac{84.85}{10 \times 10^{-3}} = 8485 \text{ A/s}$$

If di/dt is assumed constant for a short time after the gate triggering, the time t_1 required for the anode current to rise to the level of holding current is calculated from $t_1 \times (di/dt) = I_H$ or $t_1 \times 8485 = 0.5$ and this gives $t_1 = 0.5/8485 = 58.93$ μs. Therefore, the minimum width of the gate pulse is

$$t_G = t_1 + t_d = 58.93 + 1.5 = 60.43 \text{ μs}$$

Key Points of Section 10–14

- The load current harmonics and the input PF depend of the load PF.
- A practical supply will have a source reactance. As a result, the transfer of current from one device to another one will not be instantaneous. There will be an overlap known as commutation or overlap angle, which will lower the effective output voltage of the converter.

SUMMARY

In this chapter we have seen that average output voltage (and output power) of ac–dc converters can be controlled by varying the conduction time of power devices. Depending on the types of supply, the converters could be single phase or three phase. For each type of supply, they can be half-wave, semi-, or full converters. The semi- and full converters are used extensively in practical applications. Although semiconverters provide better input PF than that of full converters, these converters are only suitable for a one-quadrant operation. Full converters and dual converters allow two-quadrant and four-quadrant operations, respectively. Three-phase converters are normally used in high-power applications and the frequency of output ripples is higher.

The input PF, which is dependent on the load, can be improved and the voltage rating can be increased by series connection of converters. With forced commutations, the PF can be further improved and certain lower order harmonics can be reduced or eliminated.

The load current could be continuous or discontinuous depending on the load-time constant and delay angle. For the analysis of converters, the method of Fourier series is used. However, other techniques (e.g., transfer function approach or spectrum multiplication of switching function) can be used for the analysis of power-switching circuits. The delay-angle control does not affect the voltage drop due to commutating inductances, and this drop is the same as that of normal diode rectifiers.

REFERENCES

[1]   J. Rodríguez and A. Weinstein, *Power Electronics Handbook*, edited by M. H. Rashid. San Diego, CA: Academic Press. 2001, Chapter 11—Single-Phase Controlled Rectifiers.

[2]   J. Dixon, *Power Electronics Handbook,* edited by M. H. Rashid. San Diego, CA: Academic Press. 2001, Chapter 12—Three-Phase Controlled Rectifiers.

[3]   P. D. Ziogas, L. Morán, G. Joos, and D. Vincenti, "A refined PWM scheme for voltage and current source converters," *IEEE-IAS Annual Meeting*, 1990, pp. 997–983.

[4]   R. Wu, S. B. Dewan, and G.R. Slemon, "Analysis of an AC-to-DC voltage source converter using PWM with phase and amplitude control," *IEEE Transactions on Industry Applications,* Vol. 27, No. 2, March/April 1991, pp. 355–364.

[5]   B.-H. Kwon and B.-D. Min, "A fully software-controlled PWM rectifier with current link," *IEEE Transactions on Industrial Electronics,* Vol. 40, No. 3, June 1993, pp. 355–363.

[6]   C.-T. Pan and J.-J. Shieh, "A new space-vector control-strategies for three-phase step-up/down ac/dc converter," *IEEE Transactions on Industrial Electronics,* Vol. 47, No. 1, February 2000, pp. 25–35.

[7]   P. N. Enjeti and A. Rahman, "A new single-phase to three-phase converter with active input current shaping for low cost AC motor drives," *IEEE Transactions on Industry Applications,* Vol. 29, No. 4, July/August 1993, pp. 806–813.

[8]   C.-T. Pan and J.-J. Shieh, "A single-stage three-phase boost-buck AC/DC converter based on generalized zero-space vectors," *IEEE Transactions on Power Electronics,* Vol. 14, No. 5, September 1999, pp. 949–958.

[9]   H.-Taek and T. A. Lipo, "VSI-PWM rectifier/inverter system with reduced switch count," *IEEE Transactions on Industry Applications,* Vol. 32, No. 6, November/December 1996, pp. 1331–1337.

[10] J. Rodríguez and A. Weinstein, *Power Electronics Handbook,* edited by M. H. Rashid. San Diego, CA: Academic Press. 2001, Chapter 11—Single-Phase Controlled Rectifiers.

[11] J. Dixon, *Power Electronics Handbook,* edited by M. H. Rashid. San Diego, CA: Academic Press. 2001, Chapter 12—Three-Phase Controlled Rectifiers.

[12] P. D. Ziogas, "Optimum voltage and harmonic control PWM techniques for 3-phase static UPS systems," *IEEE Transactions on Industry Applications,* Vol. IA-I6, No. 4, 1980, pp. 542–546.

[13] P. D. Ziogas, L. Morán, G. Joos, and D. Vincenti, "A refined PWM scheme for voltage and current source converters," *IEEE-IAS Annual Meeting,* 1990, pp. 997–983.

[14] M. A. Boost and P. Ziogas, "State-of-the-Art PWM techniques, a critical evaluation," *IEEE Transactions on Industry Applications,* Vol. 24, No. 2, March/April 1988, pp. 271–280.

[15] X. Ruan, L. Zhou, and Y. Yan, "Soft-switching PWM three-level converters," *IEEE Transactions on Power Electronics,* Vol. 16, No. 5, September 2001, pp. 612–622.

[16] R. Wu, S. B. Dewan, and G. R. Slemon, "A PWM AC-to-DC converter with fixed switching frequency," *IEEE Transactions on Industry Applications,* Vol. 26, No. 5, September–October 1990, pp. 880–885.

[17] J. W. Dixon, and B.-T. Ooi, "Indirect current control of a unity power factor sinusoidal current boost type three-phase rectifier," *IEEE Transactions on Industrial Electronics,* Vol. 35, No. 4, November 1988, pp. 508–515.

[18] R. Wu, S. B. Dewan, and G. R. Slemon, "Analysis of an AC-to-DC voltage source converter using PWM with phase and amplitude control," *IEEE Transactions on Industry Applications,* Vol. 27, No. 2, March/April 1991, pp. 355–364.

[19] R. Itoh and K. Ishizaka, "Three-phase flyback AC–DC convertor with sinusoidal supply currents," *IEE Proceedings Electric Power Applications, Part B,* Vol. 138, No. 3, May 1991, pp. 143–151.

[20] C. T. Pan and T. C. Chen, "Step-up/down three-phase AC to DC convertor with sinusoidal input current and unity power factor," *IEE Proceedings Electric Power Applications,* Vol. 141, No. 2, March 1994, pp. 77–84.

[21] C.-T. Pan and J.-J. Shieh, "A new space-vector control strategies for three-phase step-up/down ac/dc converter," *IEEE Transactions on Industrial Electronics,* Vol. 47, No. 1, February 2000, pp. 25–35.

[22] J. T. Boys, and A. W. Green, "Current-forced single-phase reversible rectifier," *IEE Proceedings Electric Power Applications,* Part B, Vol. 136, No. 5, September 1989, pp. 205–211.

[23] P. N. Enjeti and A. Rahman, "A new single-phase to three-phase converter with active input current shaping for low cost AC motor drives," *IEEE Transactions on Industry Applications,* Vol. 29, No. 4, July/August 1993, pp. 806–813.

[24] G. A. Covic, G. L. Peters, and J. T. Boys, "An improved single phase to three phase converter for low cost AC motor drives," *International Conference on Power Electronics and Drive Systems,* 1995, Vol. 1, pp. 549–554.

[25] C.-T. Pan and J.-J. Shieh, "A single-stage three-phase boost-buck AC/DC converter based on generalized zero-space vectors," *IEEE Transactions on Power Electronics,* Vol. 14, No. 5, September 1999, pp. 949–958.

[26] H.-Taek and T. A. Lipo, "VSI-PWM rectifier/inverter system with reduced switch count," *IEEE Transactions on Industry Applications,* Vol. 32, No. 6, November/December 1996, pp. 1331–1337.

REVIEW QUESTIONS

10.1 What is a natural or line commutation?

10.2 What is a controlled rectifier?

10.3 What is a converter?

10.4 What is a delay-angle control of converters?

10.5 What is a semiconverter? Draw two semiconverter circuits.

10.6 What is a full converter? Draw two full-converter circuits.

10.7 What is a dual converter? Draw two dual-converter circuits.

10.8 What is the principle of phase control?

10.9 What are the effects of removing the free-wheeling diode in single-phase semiconverters?

10.10 Why is the power factor of semiconverters better than that of full converters?

10.11 What is the cause of circulating current in dual converters?

10.12 Why is a circulating current inductor required in dual converters?

10.13 What are the advantages and disadvantages of series converters?

10.14 How is the delay angle of one converter related to the delay angle of the other converter in a dual-converter system?

10.15 What is the inversion mode of converters?

10.16 What is the rectification mode of converters?

10.17 What is the frequency of the lowest order harmonic in three-phase semiconverters?

10.18 What is the frequency of the lowest order harmonic in three-phase full converters?

10.19 What is the frequency of the lowest order harmonic in a single-phase semiconverter?

10.20 How are gate-turn-off thyristors turned on and off?

10.21 How is a phase-control thyristor turned on and off?

10.22 What is a forced commutation? What are the advantages of forced commutation for ac–dc converters?

10.23 What is extinction-angle control of converters?

10.24 What is symmetric-angle control of converters?

10.25 What is pulse-width-modulation control of converters?

10.26 What is sinusoidal pulse-width-modulation control of a converter?

10.27 What is the modulation index?

10.28 How is the output voltage of a phase-control converter varied?

10.29 How is the output voltage of a sinusoidal PWM control converter varied?

10.30 Does the commutation angle depend on the delay angle of converters?

10.31 Does the voltage drop due to commutating inductances depend on the delay angle of converters?

10.32 Does the input power factor of converters depend on the load power factor?

10.33 Do the output ripple voltages of converters depend on the delay angle?

PROBLEMS

10.1 A single-phase half-wave converter in Figure 10.1a is operated from a 120-V, 60-Hz supply. If the load resistive load is $R = 10\ \Omega$ and the delay angle is $\alpha = \pi/3$, determine (a) the efficiency, (b) the form factor, (c) the ripple factor, (d) the transformer utilization factor, and (e) the peak inverse voltage (PIV) of thyristor T_1.

10.2 A single-phase half-wave converter in Figure 10.1a is operated from a 120-V, 60-Hz supply and the load resistive load is $R = 10\ \Omega$. If the average output voltage is 25% of the maximum possible average output voltage, calculate (a) the delay angle, (b) the rms and average output currents, (c) the average and rms thyristor currents, and (d) the input power factor.

10.3 A single-phase half-converter in Figure 10.1a is supplied from a 120-V, 60-Hz supply and a freewheeling diode is connected across the load. The load consists of series-connected resistance $R = 10\ \Omega$; inductance $L = 5$ mH; and battery voltage $E = 20$ V. (a) Express the instantaneous output voltage in a Fourier series, and (b) determine the rms value of the lowest order output harmonic current.

10.4 A single-phase semiconverter in Figure 10.15 is operated from a 120-V, 60-Hz supply. The load current with an average value of I_a is continuous with negligible ripple content. The turns ratio of the transformer is unity. If the delay angle is $\alpha = \pi/3$, calculate (a) the harmonic factor of input current, (b) the displacement factor, and (c) the input power factor.

10.5 Repeat Problem 10.2 for the single-phase semiconverter in Figure 10.15a.

10.6 The single-phase semiconverter in Figure 10.15a is operated from a 120-V, 60-Hz supply. The load consists of series-connected resistance $R = 10\ \Omega$, inductance $L = 5$ mH, and battery voltage $E = 20$ V. **(a)** Express the output voltage in a Fourier series, and **(b)** determine the rms value of the lowest order output harmonic current.

10.7 Repeat Problem 10.4 for the single-phase full converter in Figure 10.2a.

10.8 Repeat Problem 10.2 for the single-phase full converter in Figure 10.2a.

10.9 Repeat Problem 10.6 for the single-phase full converter in Figure 10.2a.

10.10 The dual converter in Figure 10.3a is operated from a 120-V, 60-Hz supply and delivers ripple-free average current of $I_{dc} = 20$ A. The circulating inductance is $L_r = 5$ mH, and the delay angles are $\alpha_1 - 30°$ and $\alpha_2 = 150°$. Calculate the peak circulating current and the peak current of converter 1.

10.11 A single-phase series semiconverter in Figure 10.18a is operated from a 120-V, 60-Hz supply and the load resistance is $R = 10$ Ω. If the average output voltage is 75% of the maximum possible average output voltage, calculate **(a)** the delay angles of converters, **(b)** the rms and average output currents, **(c)** the average and rms thyristor currents, and **(d)** the input power factor.

10.12 A single-phase series semiconverter in Figure 10.18a is operated from a 120-V, 60-Hz supply. The load current with an average value of I_a is continuous and the ripple content is negligible. The turns ratio of the transformer is $N_p/N_s = 2$. If delay angles are $\alpha_1 = 0$ and $\alpha_2 = \pi/3$, calculate **(a)** the harmonic factor of input current, **(b)** the displacement factor, and **(c)** the input power factor.

10.13 Repeat Problem 10.11 for the single-phase series full converter in Figure 10.19a.

10.14 Repeat Problem 10.12 for the single-phase series full converter in Figure 10.19a.

10.15 The three-phase half-wave converter in Figure 10.4a is operated from a three-phase Y-connected 220-V, 60-Hz supply and a freewheeling diode is connected across the load. The load current with an average value of I_a is continuous and the ripple content is negligible. If the delay angle $\alpha = \pi/3$, calculate **(a)** the harmonic factor of input current, **(b)** the displacement factor, and **(c)** the input power factor.

10.16 The three-phase half-wave converter in Figure 10.4a is operated from a three-phase Y-connected 220-V, 60-Hz supply and the load resistance is $R = 10$ Ω. If the average output voltage is 25% of the maximum possible average output voltage, calculate **(a)** the delay angle, **(b)** the rms and average output currents, **(c)** the average and rms thyristor currents, **(d)** the rectification efficiency, **(e)** the transformer utilization factor, and **(f)** the input power factor.

10.17 The three-phase half-wave converter in Figure 10.4a is operated from a three-phase Y-connected 220-V, 60-Hz supply and a freewheeling diode is connected across the load. The load consists of series-connected resistance $R = 10$ Ω, inductance $L = 5$ mH, and battery voltage $E = 20$ V. **(a)** Express the instantaneous output voltage in a Fourier series, and **(b)** determine the rms value of the lowest order harmonic on the output current.

10.18 The three-phase semiconverter in Figure 10.16a is operated from a three-phase Y-connected 220-V, 60-Hz supply. The load current with an average value of I_a is continuous with negligible ripple content. The turns ratio of the transformer is unity. If the delay angle is $\alpha = 2\pi/3$, calculate **(a)** the harmonic factor of input current, **(b)** the displacement factor, and **(c)** the input power factor.

10.19 Repeat Problem 10.16 for the three-phase semiconverter in Figure 10.16a.

10.20 Repeat Problem 10.19 if the average output voltage is 90% of the maximum possible output voltage.

10.21 Repeat Problem 10.17 for the three-phase semiconverter in Figure 10.16a.

10.22 Repeat Problem 10.18 for the three-phase full converter in Figure 10.5a.

10.23 Repeat Problem 10.16 for the three-phase full converter in Figure 10.5a.

10.24 Repeat Problem 10.17 for the three-phase full converter in Figure 10.5a.

10.25 The three-phase dual converter in Figure 10.6a is operated from a three-phase Y-connected 220-V, 60-Hz supply and the load resistance $R = 10\ \Omega$. The circulating inductance $L_r = 5$ mH, and the delay angles are $\alpha_1 = 60°$ and $\alpha_2 = 120°$. Calculate the peak circulating current and the peak current of converters.

10.26 The single-phase semiconverter of Figure 10.15a has an RL load of $L = 1.5$ mH, $R = 1.5\ \Omega$, and $E = 0$ V. The input voltage is $V_s = 120$ V (rms) at 60 Hz. **(a)** Determine (1) the load current I_o at $\omega t = 0$, and the load current I_1 at $\omega t = \alpha = 30°$, (2) the average thyristor current I_A, (3) the rms thyristor current I_R, (4) the rms output current I_{rms}, and (5) the average output current I_{dc}. **(b)** Use SPICE to check your results.

10.27 The single-phase full converter of Figure 10.2a has an RL load having $L = 4.5$ mH, $R = 1.5\ \Omega$, and $E = 10$ V. The input voltage is $V_s = 120$ V at (rms) 60 Hz. **(a)** Determine (1) the load current I_o at $\omega t = \alpha = 30°$, (2) the average thyristor current I_A, (3) the rms thyristor current I_R, (4) the rms output current I_{rms}, and (5) the average output current I_{dc}. **(b)** Use SPICE to check your results.

10.28 The three-phase full converter of Figure 10.5a has a load of $L = 1.5$ mH, $R = 1.5\ \Omega$, and $E = 0$ V. The line-to-line input voltage is $V_{ab} = 208$ V (rms), 60 Hz. The delay angle is $\alpha = \pi/6$. **(a)** Determine (1) the steady-state load current I_1 at $\omega t' = \pi/3 + \alpha$ (or $\omega t = \pi/6 + \alpha$), (2) the average thyristor current I_A, (3) the rms thyristor current I_R, (4) the rms output current I_{rms}, and (5) the average output current I_{dc}. **(b)** Use SPICE to check your results.

10.29 The single-phase semiconverter in Figure 10.7a is operated from a 120-V, 60-Hz supply and uses an extinction angle control. The load current with an average value of I_a is continuous and has negligible ripple content. If the extinction angle is $\beta = \pi/3$, calculate **(a)** the outputs V_{dc} and V_{rms}, **(b)** the harmonic factor of input current, **(c)** the displacement factor, and **(d)** the input power factor.

10.30 Repeat Problem 10.29 for the single-phase full converter in Figure 10.8a.

10.31 Repeat Problem 10.18 if symmetric-angle control is used.

10.32 Repeat Problem 10.18 if extinction-angle control is used.

10.33 The single-phase semiconverter in Figure 10.7a is operated with a sinusoidal PWM control and is supplied from a 120-V, 60-Hz supply. The load current with an average value of I_a is continuous with negligible ripple content. There are five pulses per half-cycle and the pulses are $\alpha_1 = 7.93°, \delta_1 = 5.82°; \alpha_2 = 30°, \delta_2 = 16.25°; \alpha_3 = 52.07°, \delta_3 = 127.93°; \alpha_4 = 133.75°, \delta_4 = 16.25°;$ and $\alpha_5 = 166.25°, \delta_5 = 5.82°$. Calculate **(a)** the V_{dc} and V_{rms}, **(b)** the harmonic factor of input current, **(c)** the displacement factor, and **(d)** the input power factor.

10.34 Repeat Problem 10.33 for five pulses per half-cycle with equal pulse width, $M = 0.8$.

10.35 A three-phase semiconverter is operated from a three-phase Y-connected 220-V, 60-Hz supply. The load current is continuous and has negligible ripple. The average load current is $I_{dc} = 150$ A and commutating inductance per phase is $L_c = 0.5$ mH. Determine the overlap angle if **(a)** $\alpha = \pi/6$, and **(b)** $\alpha = \pi/3$.

10.36 The holding current of thyristors in the three-phase full converter in Figure 10.5a is $I_H = 200$ mA and the delay time is 2.5 μs. The converter is supplied from a three-phase Y-connected 208-V, 60-Hz supply and has a load of $L = 8$ mH and $R = 2\ \Omega$; it is operated with a delay angle of $\alpha = 60°$. Determine the minimum width of gate pulse width t_G.

10.37 Repeat Problem 10.36 if $L = 0$.

C H A P T E R   1 1

Ac Voltage Controllers

The learning objectives of this chapter are as follows:

- To understand the operation and characteristics of ac voltage controllers
- To understand the operation of matrix converters
- To learn the types of ac voltage controllers
- To understand the performance parameters of ac voltage controllers
- To learn the techniques for analysis and design of ac voltage controllers
- To learn the techniques for simulating controlled rectifiers by using SPICE
- To study effects of load inductance on the load current

11.1   INTRODUCTION

If a thyristor switch is connected between ac supply and load, the power flow can be controlled by varying the rms value of ac voltage applied to the load; and this type of power circuit is known as an *ac voltage controller*. The most common applications of ac voltage controllers are: industrial heating, on-load transformer connection changing, light controls, speed control of polyphase induction motors, and ac magnet controls. For power transfer, two types of control are normally used:

1. On–off control
2. Phase-angle control

In on–off control, thyristor switches connect the load to the ac source for a few cycles of input voltage and then disconnect it for another few cycles. In phase control, thyristor switches connect the load to the ac source for a portion of each cycle of input voltage.

The ac voltage controllers can be classified into two types: (1) single-phase controllers and (2) three-phase controllers, with each type subdivided into (a) unidirectional or half-wave control and (b) bidirectional or full-wave control. There are various configurations of three-phase controllers depending on the connections of thyristor switches.

Because the input voltage is ac, thyristors are line commutated; and phase-control thyristors, which are relatively inexpensive and slower than fast-switching thyristors, are normally used. For applications up to 400 Hz, if TRIACs are available to meet the voltage and current ratings of a particular application, TRIACs are more commonly used.

Due to line or natural commutation, there is no need of extra commutation circuitry and the circuits for ac voltage controllers are very simple. Due to the nature of output waveforms, the analysis for the derivations of explicit expressions for the performance parameters of circuits is not simple, especially for phase-angle-controlled converters with *RL* loads. For the sake of simplicity, resistive loads are considered in this chapter to compare the performances of various configurations. However, the practical loads are of the *RL* type and should be considered in the design and analysis of ac voltage controllers.

11.2 PRINCIPLE OF ON–OFF CONTROL

The principle of on–off control can be explained with a single-phase full-wave controller, as shown in Figure 11.1a. The thyristor switch connects the ac supply to load for a time t_n; the switch is turned off by a gate pulse inhibiting for time t_0. The on-time t_n usually consists of an integral number of cycles. The thyristors are turned on at the

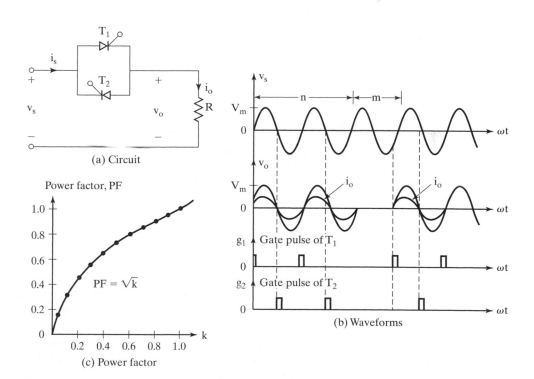

FIGURE 11.1

On–off control.

zero-voltage crossings of ac input voltage. The gate pulses for thyristors T_1 and T_2 and the waveforms for input and output voltages are shown in Figure 11.1b.

This type of control is applied in applications that have a high mechanical inertia and high thermal time constant (e.g., industrial heating and speed control of motors). Due to zero-voltage and zero-current switching of thyristors, the harmonics generated by switching actions are reduced.

For a sinusoidal input voltage, $v_s = V_m \sin \omega t = \sqrt{2}\, V_s \sin \omega t$. If the input voltage is connected to load for n cycles and is disconnected for m cycles, the rms output (or load) voltage can be found from

$$V_o = \left[\frac{n}{2\pi(n + m)} \int_0^{2\pi} 2V_s^2 \sin^2 \omega t \; d(\omega t) \right]^{1/2}$$

$$= V_s \sqrt{\frac{n}{m + n}} = V_s \sqrt{k} \tag{11.1}$$

where $k = n/(m + n)$ and k is called the *duty cycle*. V_s is the rms phase voltage. The circuit configurations for on–off control are similar to those of phase control and the performance analysis is also similar. For these reasons, the phase-control techniques are only discussed and analyzed in this chapter.

Example 11.1   Finding the Performances of an Ac Voltage Controller with On–Off Control

An ac voltage controller in Figure 11.1a has a resistive load of $R = 10\ \Omega$ and the root-mean-square (rms) input voltage is $V_s = 120$ V, 60 Hz. The thyristors switch is on for $n = 25$ cycles and is off for $m = 75$ cycles. Determine (a) the rms output voltage V_o, (b) the input power factor (PF), and (c) the average and rms current of thyristors.

Solution
$R = 10\ \Omega$, $V_s = 120$ V, $V_m = \sqrt{2} \times 120 = 169.7$ V, and $k = n/(n + m) = 25/100 = 0.25$.

    a. From Eq. (11.1), the rms value of output voltage is

$$V_o = V_s \sqrt{k} = V_s \sqrt{\frac{n}{m + n}} = 120 \sqrt{\frac{25}{100}} = 60 \text{ V}$$

and the rms load current is $I_o = V_o/R = 60/10 = 6.0$ A.

    b. The load power is $P_o = I_o^2 R = 6^2 \times 10 = 360$ W. Because the input current is the same as the load current, the volt–ampere (VA) input is

$$\text{VA} = V_s I_s = V_s I_o = 120 \times 6 = 720 \text{ W}$$

The input PF is

$$\text{PF} = \frac{P_o}{\text{VA}} = \sqrt{\frac{n}{m + n}} = \sqrt{k}$$

$$= \sqrt{0.25} = \frac{360}{720} = 0.5 \text{ (lagging)}$$

$$\tag{11.2}$$

c. The peak thyristor current is $I_m = V_m/R = 169.7/10 = 16.97$ A. The average current of thyristors is

$$I_A = \frac{n}{2\pi(m + n)} \int_0^\pi I_m \sin \omega t \, d(\omega t) = \frac{I_m n}{\pi(m + n)} = \frac{kI_m}{\pi}$$

(11.3)

$$= \frac{16.97}{\pi} \times 0.25 = 1.33 \text{ A}$$

The rms current of thyristors is

$$I_R = \left[\frac{n}{2\pi(m + n)} \int_0^\pi I_m^2 \sin^2 \omega t \, d(\omega t)\right]^{1/2} = \frac{I_m}{2}\sqrt{\frac{n}{m + n}} = \frac{I_m \sqrt{k}}{2}$$

(11.4)

$$= \frac{16.97}{2} \times \sqrt{0.25} = 4.24 \text{ A}$$

Notes:

1. The PF and output voltage vary with the square root of the duty cycle. The PF is poor at the low value of the duty cycle k and is shown in Figure 11.1c.
2. If T is the period of the input voltage, $(m + n)T$ is the period of on–off control. $(m + n)T$ should be less than the mechanical or thermal time constant of the load, and is usually less than 1 s, but not in hours or days. The sum of m and n is generally around 100.
3. If Eq. (11.2) is used to determine the PF with m and n in days, it can give erroneous results. For example, if $m = 3$ days and $n = 3$ days, Eq. (11.2) gives PF $= [3/(3 + 3)]^{1/2} = 0.707$, which is not physically possible. Because if the controller is on for 3 days and off for 3 days, the PF becomes independent on the load impedance angle θ.

Key Points of Section 11.2

- The on-off control switches the ac supply to the load for an integral number (m) of supply-frequency cycles and then disconnect the load for a certain number (n) of supply cycles. This type of control is applied in applications that have a high mechanical inertia and high thermal constant.

11.3 PRINCIPLE OF PHASE CONTROL

The principle of phase control can be explained with reference to Figure 11.2a. The power flow to the load is controlled by delaying the firing angle of thyristor T_1. Figure 11.2b illustrates the gate pulses of thyristor T_1 and the waveforms for the input and output voltages. Due to the presence of diode D_1, the control range is limited and the effective rms output voltage can only be varied between 70.7 and 100%. The output voltage and input current are asymmetric and contain a dc component. If there is an

11.4   SINGLE-PHASE BIDIRECTIONAL CONTROLLERS WITH RESISTIVE LOADS

The problem of dc input current can be prevented by using bidirectional (or full-wave) control, and a single-phase full-wave controller with a resistive load is shown in Figure 11.3a. During the positive half-cycle of input voltage, the power flow is controlled by varying the delay angle of thyristor T_1; and thyristor T_2 controls the power flow during the negative half-cycle of input voltage. The firing pulses of T_1 and T_2 are kept 180° apart. The waveforms for the input voltage, output voltage, and gating signals for T_1 and T_2 are shown in Figure 11.3b.

If $v_s = \sqrt{2}\, V_s \sin \omega t$ is the input voltage, and the delay angles of thyristors T_1 and T_2 are equal ($\alpha_2 = \pi + \alpha_1$), the rms output voltage can be found from

$$
\begin{aligned}
V_o &= \left\{ \frac{2}{2\pi} \int_\alpha^\pi 2 V_s^2 \sin^2 \omega t \; d(\omega t) \right\}^{1/2} \\
&= \left\{ \frac{4 V_s^2}{4\pi} \int_\alpha^\pi (1 - \cos 2\omega t)\, d(\omega t) \right\}^{1/2} \\
&= V_s \left[\frac{1}{\pi} \left(\pi - \alpha + \frac{\sin 2\alpha}{2} \right) \right]^{1/2}
\end{aligned}
\tag{11.8}
$$

By varying α from 0 to π, V_o can be varied from V_s to 0.

In Figure 11.3a, the gating circuits for thyristors T_1 and T_2 must be isolated. It is possible to have a common cathode for T_1 and T_2 by adding two diodes, as shown in Figure 11.4. Thyristor T_1 and diode D_1 conduct together during the positive half-cycle;

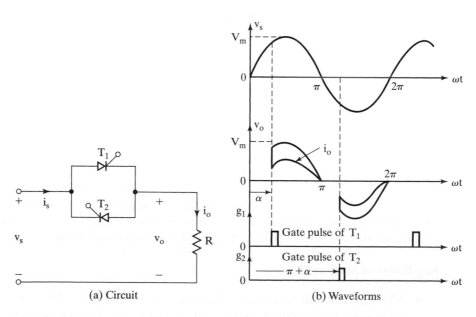

(a) Circuit   (b) Waveforms

FIGURE 11.3

Single-phase full-wave controller.

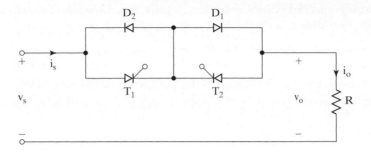

FIGURE 11.4

Single-phase full-wave controller with common cathode.

and thyristor T_2 and diode D_2 conduct during the negative half-cycle. Because this circuit can have a common terminal for gating signals of T_1 and T_2, only one isolation circuit is required, but at the expense of two power diodes. Due to two power devices conducting at the same time, the conduction losses of devices would increase and efficiency would be reduced.

A single-phase full-wave controller can also be implemented with one thyristor and four diodes, as shown in Figure 11.5a. The four diodes act as a bridge rectifier. The voltage across thyristor T_1 and its current are always unidirectional. With a resistive load, the thyristor current would fall to zero due to natural commutation in every half-cycle, as shown in Figure 11.5b. However, if there is a large inductance in the circuit, thyristor T_1 may not be turned off in every half-cycle of input voltage, and this may result in a loss of control. It would require detecting the zero crossing of the load current to guarantee turn-off of the conducting thyristor before firing the next one. Three power devices conduct at the same time and the efficiency is also reduced. The bridge

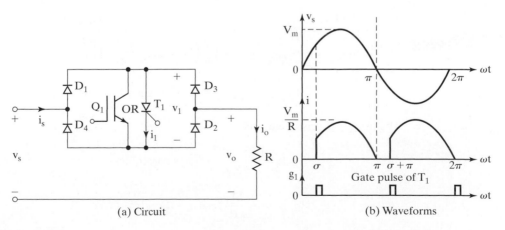

(a) Circuit   (b) Waveforms

FIGURE 11.5

Single-phase full-wave controller with one thyristor.

rectifier and thyristor (or transistor) act as a *bidirectional switch*, which is commercially available as a single device with a relatively low on-state conduction loss.

Gating sequence.   The gating sequence is as follows:

1. Generate a pulse signal at the positive zero crossing of the supply voltage v_s.
2. Delay the pulse by the desired angle α for gating T_1 through a gate-isolating circuit.
3. Generate another pulse of delay angle $\alpha + \pi$ for gating T_2.

Example 11.3 Finding the Performance Parameters of a Single-Phase Full-Wave Controller

A single-phase full-wave ac voltage controller in Figure 11.3a has a resistive load of $R = 10\ \Omega$ and the input voltage is $V_s = 120$ V (rms), 60 Hz. The delay angles of thyristors T_1 and T_2 are equal: $\alpha_1 = \alpha_2 = \alpha = \pi/2$. Determine (a) the rms output voltage V_o, (b) the input PF, (c) the average current of thyristors I_A, and (d) the rms current of thyristors I_R.

Solution
$R = 10\ \Omega$, $V_s = 120$ V, $\alpha = \pi/2$, and $V_m = \sqrt{2} \times 120 = 169.7$ V.

a. From Eq. (11.8), the rms output voltage

$$V_o = \frac{120}{\sqrt{2}} = 84.85 \text{ V}$$

b. The rms value of load current is $I_o = V_o/R = 84.85/10 = 8.485$ A and the load power is $P_o = I_o^2 R = 8.485^2 \times 10 = 719.95$ W. Because the input current is the same as the load current, the input VA rating is

$$\text{VA} = V_s I_s = V_s I_o = 120 \times 8.485 = 1018.2 \text{ W}$$

The input PF is

$$\text{PF} = \frac{P_o}{\text{VA}} = \frac{V_o}{V_s} = \left[\frac{1}{\pi} \left(\pi - \alpha + \frac{\sin 2\alpha}{2} \right) \right]^{1/2} \tag{11.9}$$

$$= \frac{1}{\sqrt{2}} = \frac{719.95}{1018.2} = 0.707 \text{ (lagging)}$$

c. The average thyristor current

$$I_A = \frac{1}{2\pi R} \int_{\alpha}^{\pi} \sqrt{2}\, V_s \sin \omega t \; d(\omega t)$$

$$= \frac{\sqrt{2}\, V_s}{2\pi R} (\cos \alpha + 1) \tag{11.10}$$

$$= \sqrt{2} \times \frac{120}{2\pi \times 10} = 2.7 \text{ A}$$

d. The rms value of the thyristor current

$$I_R = \left[\frac{1}{2\pi R^2} \int_\alpha^\pi 2V_s^2 \sin^2 \omega t \, d(\omega t) \right]^{1/2}$$

$$= \left[\frac{2V_s^2}{4\pi R^2} \int_\alpha^\pi (1 - \cos 2\omega t) \, d(\omega t) \right]^{1/2}$$

$$= \frac{V_s}{\sqrt{2}\,R} \left[\frac{1}{\pi} \left(\pi - \alpha + \frac{\sin 2\alpha}{2} \right) \right]^{1/2}$$ (11.11)

$$= \frac{120}{2 \times 10} = 6\text{A}$$

Key Points of Section 11.4

- Varying the delay angle α from 0 to π can vary the rms output voltage from V_s to 0.
- The output of this controller contains no dc component.

11.5 SINGLE-PHASE CONTROLLERS WITH INDUCTIVE LOADS

Section 11.4 deals with the single-phase controllers with resistive loads. In practice, most loads are inductive to a certain extent. A full-wave controller with an RL load is shown in Figure 11.6a. Let us assume that thyristor T_1 is fired during the positive half-cycle and carries the load current. Due to inductance in the circuit, the current of thyristor T_1 would not fall to zero at $\omega t = \pi$, when the input voltage starts to be negative. Thyristor T_1 continues to conduct until its current i_1 falls to zero at $\omega t = \beta$. The conduction angle of thyristor T_1 is $\delta = \beta - \alpha$ and depends on the delay angle α and the PF angle of load θ. The waveforms for the thyristor current, gating pulses, and input voltage are shown in Figure 11.6b.

If $v_s = \sqrt{2}\,V_s \sin \omega t$ is the instantaneous input voltage and the delay angle of thyristor T_1 is α, thyristor current i_1 can be found from

$$L \frac{di_1}{dt} + Ri_1 = \sqrt{2}\,V_s \sin \omega t$$ (11.12)

The solution of Eq. (11.12) is of the form

$$i_1 = \frac{\sqrt{2}\,V_s}{Z} \sin(\omega t - \theta) + A_1 e^{-(R/L)t}$$ (11.13)

where load impedance $Z = [R^2 + (\omega L)^2]^{1/2}$ and load angle $\theta = \tan^{-1}(\omega L/R)$.

The constant A_1 can be determined from the initial condition: at $\omega t = \alpha$, $i_1 = 0$. From Eq. (11.13) A_1 is found as

$$A_1 = -\frac{\sqrt{2}\,V_s}{Z} \sin(\alpha - \theta) e^{(R/L)(\alpha/\omega)}$$ (11.14)

Substitution of A_1 from Eq. (11.14) in Eq. (11.13) yields

$$i_1 = \frac{\sqrt{2}\,V_s}{Z} [\sin(\omega t - \theta) - \sin(\alpha - \theta) e^{(R/L)(\alpha/\omega - t)}]$$ (11.15)

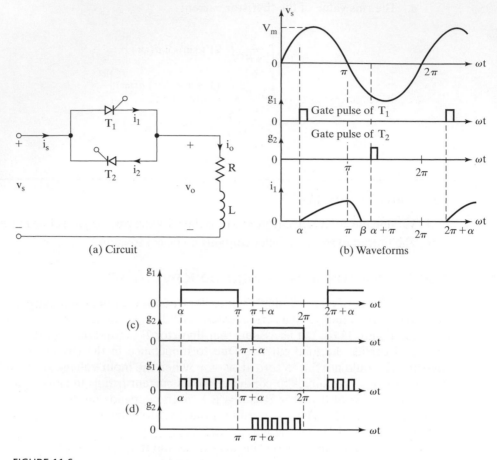

(a) Circuit

(b) Waveforms

(c)

(d)

FIGURE 11.6

Single-phase full-wave controller with RL load.

The angle β, when current i_1 falls to zero and thyristor T_1 is turned off, can be found from the condition $i_1(\omega t = \beta) = 0$ in Eq. (11.15) and is given by the relation

$$\sin(\beta - \theta) = \sin(\alpha - \theta)e^{(R/L)(\alpha-\beta)/\omega} \qquad (11.16)$$

The angle β, which is also known as an *extinction angle*, can be determined from this transcendental equation and it requires an iterative method of solution. Once β is known, the conduction angle δ of thyristor T_1 can be found from

$$\delta = \beta - \alpha \qquad (11.17)$$

The rms output voltage

$$V_o = \left[\frac{2}{2\pi} \int_{\alpha}^{\beta} 2V_s^2 \sin^2 \omega t \, d(\omega t) \right]^{1/2}$$

$$= \left[\frac{4V_s^2}{4\pi} \int_{\alpha}^{\beta} (1 - \cos 2\omega t) \, d(\omega t) \right]^{1/2}$$

$$= V_s \left[\frac{1}{\pi} \left(\beta - \alpha + \frac{\sin 2\alpha}{2} - \frac{\sin 2\beta}{2} \right) \right]^{1/2} \tag{11.18}$$

The rms thyristor current can be found from Eq. (11.15) as

$$I_R = \left[\frac{1}{2\pi} \int_{\alpha}^{\beta} i_1^2 \, d(\omega t) \right]^{1/2}$$

$$= \frac{V_s}{Z} \left[\frac{1}{\pi} \int_{\alpha}^{\beta} \{ \sin(\omega t - \theta) - \sin(\alpha - \theta) e^{(R/L)(\alpha/\omega - t)} \}^2 \, d(\omega t) \right]^{1/2} \tag{11.19}$$

and the rms output current can then be determined by combining the rms current of each thyristor as

$$I_o = (I_R^2 + I_R^2)^{1/2} = \sqrt{2} \, I_R \tag{11.20}$$

The average value of thyristor current can also be found from Eq. (11.15) as

$$I_A = \frac{1}{2\pi} \int_{\alpha}^{\beta} i_1 \, d(\omega t)$$

$$= \frac{\sqrt{2} \, V_s}{2\pi Z} \int_{\alpha}^{\beta} [\sin(\omega t - \theta) - \sin(\alpha - \theta) e^{(R/L)(\alpha/\omega - t)}] \, d(\omega t) \tag{11.21}$$

The gating signals of thyristors could be short pulses for a controller with resistive loads. However, such short pulses are not suitable for inductive loads. This can be explained with reference to Figure 11.6b. When thyristor T_2 is fired at $\omega t = \pi + \alpha$, thyristor T_1 is still conducting due to load inductance. By the time the current of thyristor T_1 falls to zero and T_1 is turned off at $\omega t = \beta = \alpha + \delta$, the gate pulse of thyristor T_2 has already ceased and consequently, T_2 cannot be turned on. As a result, only thyristor T_1 operates, causing asymmetric waveforms of output voltage and current. This difficulty can be resolved by using continuous gate signals with a duration of $(\pi - \alpha)$, as shown in Figure 11.6c. As soon as the current of T_1 falls to zero, thyristor T_2 (with gate pulses as shown in Figure 11.6c) would be turned on. However, a continuous gate pulse increases the switching loss of thyristors and requires a larger isolating transformer for the gating circuit. In practice, a train of pulses with short durations as shown in Figure 11.6d are normally used to overcome these problems.

The waveforms for the output voltage v_o, output current i_o, and the voltage across T_1, v_{T1} are shown in Figure 11.7 for an RL load. There may be a short hold-off angle γ after the zero crossing of negative going current.

Equation (11.15) indicates that the load voltage (and current) can be sinusoidal if the delay angle α is less than the load angle θ. If α is greater than θ, the load current would be discontinuous and nonsinusoidal.

11.6 THREE-PHASE FULL-WAVE CONTROLLERS

The unidirectional controllers, which contain dc input current and higher harmonic content due to the asymmetric nature of the output voltage waveform, are not normally used in ac motor drives; a three-phase bidirectional control is commonly used. The circuit diagram of a three-phase full-wave (or bidirectional) controller is shown in Figure 11.8 with a Y-connected resistive load. The firing sequence of thyristors is $T_1, T_2, T_3, T_4, T_5, T_6$.

If we define the instantaneous input phase voltages as

$$v_{AN} = \sqrt{2}\, V_s \sin \omega t$$

$$v_{BN} = \sqrt{2}\, V_s \sin\left(\omega t - \frac{2\pi}{3} \right)$$

$$v_{CN} = \sqrt{2}\, V_s \sin\left(\omega t - \frac{4\pi}{3} \right)$$

the instantaneous input line voltages are

$$v_{AB} = \sqrt{6}\, V_s \sin\left(\omega t + \frac{\pi}{6} \right)$$

$$v_{BC} = \sqrt{6}\, V_s \sin\left(\omega t - \frac{\pi}{2} \right)$$

$$v_{CA} = \sqrt{6}\, V_s \sin\left(\omega t - \frac{7\pi}{6} \right)$$

The waveforms for the input voltages, conduction angles of thyristors, and output phase voltages are shown in Figure 11.9 for $\alpha = 60°$ and $\alpha = 120°$. For $0 \leq \alpha < 60°$,

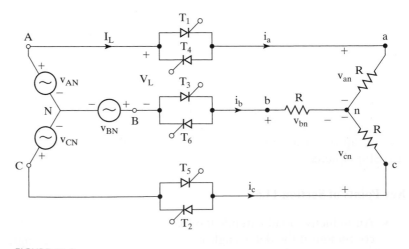

FIGURE 11.8

Three-phase bidirectional controller.

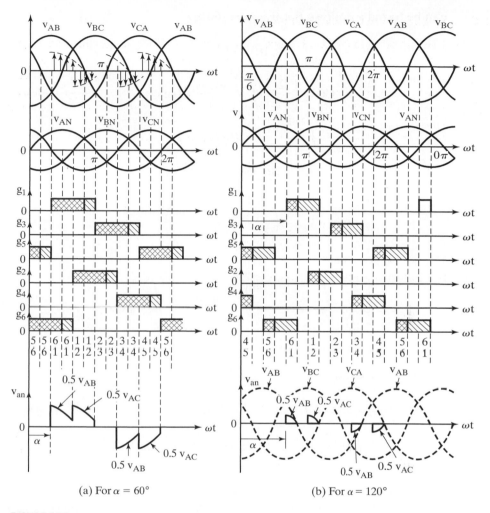

FIGURE 11.9

Waveforms for three-phase bidirectional controller.

immediately before the firing of T_1, two thyristors conduct. Once T_1 is fired, three thyristors conduct. A thyristor turns off when its current attempts to reverse. The conditions alternate between two and three conducting thyristors.

For $60° \leq \alpha < 90°$, only two thyristors conduct at any time. For $90° \leq \alpha < 150°$, although two thyristors conduct at any time, there are periods when no thyristors are on. For $\alpha \geq 150°$, there is no period for two conducting thyristors and the output voltage becomes zero at $\alpha = 150°$. The range of delay angle is

$$0 \leq \alpha \leq 150° \tag{11.25}$$

Similar to half-wave controllers, the expression for the rms output phase voltage depends on the range of delay angles. The rms output voltage for a Y-connected load can be found as follows. For $0 \leq \alpha < 60°$:

$$V_o = \left[\frac{1}{2\pi} \int_0^{2\pi} v_{an}^2 \, d(\omega t) \right]^{1/2}$$

$$= \sqrt{6} \, V_s \left\{ \frac{2}{2\pi} \left[\int_\alpha^{\pi/3} \frac{\sin^2 \omega t}{3} \, d(\omega t) + \int_{\pi/4}^{\pi/2+\alpha} \frac{\sin^2 \omega t}{4} \, d(\omega t) \right. \right.$$

$$+ \int_{\pi/3+\alpha}^{2\pi/3} \frac{\sin^2 \omega t}{3} \, d(\omega t) + \int_{\pi/2}^{\pi/2+\alpha} \frac{\sin^2 \omega t}{4} \, d(\omega t) \tag{11.26}$$

$$\left. \left. + \int_{2\pi/3+\alpha}^{\pi} \frac{\sin^2 \omega t}{3} \, d(\omega t) \right] \right\}^{1/2}$$

$$= \sqrt{6} \, V_s \left[\frac{1}{\pi} \left(\frac{\pi}{6} - \frac{\alpha}{4} + \frac{\sin 2\alpha}{8} \right) \right]^{1/2}$$

For $60° \leq \alpha < 90°$:

$$V_o = \sqrt{6} \, V_s \left[\frac{2}{2\pi} \left\{ \int_{\pi/2-\pi/3+\alpha}^{5\pi/6-\pi/3+\alpha} \frac{\sin^2 \omega t}{4} \, d(\omega t) + \int_{\pi/2-\pi/3+\alpha}^{5\pi/6-\pi/3+\alpha} \frac{\sin^2 \omega t}{4} \, d(\omega t) \right\} \right]^{1/2}$$

$$= \sqrt{6} \, V_s \left[\frac{1}{\pi} \left(\frac{\pi}{12} + \frac{3 \sin 2\alpha}{16} + \frac{\sqrt{3} \cos 2\alpha}{16} \right) \right]^{1/2} \tag{11.27}$$

For $90° \leq \alpha \leq 150°$:

$$V_0 = \sqrt{6} \, V_s \left\{ \frac{2}{2\pi} \left[\int_{\pi/2-\pi/3+\alpha}^{\pi} \frac{\sin^2 \omega t}{4} \, d(\omega t) + \int_{\pi/2-\pi/3+\alpha}^{\pi} \frac{\sin^2 \omega t}{4} \, d(\omega t) \right] \right\}^{1/2}$$

$$= \sqrt{6} \, V_s \left[\frac{1}{\pi} \left(\frac{5\pi}{24} - \frac{\alpha}{4} + \frac{\sin 2\alpha}{16} + \frac{\sqrt{3} \cos 2\alpha}{16} \right) \right]^{1/2} \tag{11.28}$$

The power devices of a three-phase bidirectional controller can be connected together, as shown in Figure 11.10. This arrangement is also known as *tie control* and

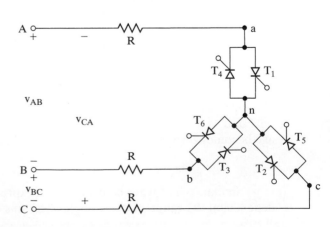

FIGURE 11.10

Arrangement for three-phase bidirectional tie control.

allows assembly of all thyristors as one unit. However, this arrangement is not possible for motor control because the terminals of the motor windings are not normally accessible.

Gating Sequence.   The gating sequence is as follows:

1. Generate a pulse signal at the positive zero crossing of the supply phase voltage v_{an}.
2. Delay the pulse by angles α, $\alpha + 2\pi/3$, and $\alpha + 4\pi/3$ for gating T_1, T_3, and T_5 through gate-isolating circuits.
3. Similarly, generate pulses with delay angles $\pi + \alpha$, $5\pi/3 + \alpha$, and $7\pi/3 + \alpha$ for gating T_2, T_4, and T_6.

Example 11.5   Finding the Performance Parameters of a Three-Phase Full-Wave Controller

The three-phase full-wave controller in Figure 11.8 supplies a Y-connected resistive load of $R = 10\ \Omega$ and the line-to-line input voltage is 208 V (rms), 60 Hz. The delay angle is $\alpha = \pi/3$. Determine (a) the rms output phase voltage V_o, (b) the input PF, and (c) the expression for the instantaneous output voltage of phase a.

Solution
$V_L = 208$ V, $V_s = V_L/\sqrt{3} = 208/\sqrt{3} = 120$ V, $\alpha = \pi/3$, and $R = 10\ \Omega$.

a.   From Eq. (11.26) the rms output phase voltage is $V_o = 100.9$ V.

b.   The rms phase current of the load is $I_a = 100.9/10 = 10.09$ A and the output power is

$$P_o = 3I_a^2 R = 3 \times 10.09^2 \times 10 = 3054.24\ \text{W}$$

Because the load is connected in Y, the phase current is equal to the line current, $I_L = I_a = 10.09$ A. The input VA

$$\text{VA} = 3 V_s I_L = 3 \times 120 \times 10.09 = 3632.4\ \text{VA}$$

The PF is

$$\text{PF} = \frac{P_o}{\text{VA}} = \frac{3054.24}{3632.4} = 0.84\ \text{(lagging)}$$

c.   If the input phase voltage is taken as the reference and is $v_{AN} = 120\sqrt{2}\sin \omega t = 169.7 \sin \omega t$, the instantaneous input line voltages are

$$v_{AB} = 208\sqrt{2}\sin\left(\omega t + \frac{\pi}{6}\right) = 294.2 \sin\left(\omega t + \frac{\pi}{6}\right)$$

$$v_{BC} = 294.2 \sin\left(\omega t - \frac{\pi}{2}\right)$$

$$v_{CA} = 294.2 \sin\left(\omega t - \frac{7\pi}{6}\right)$$

The instantaneous output phase voltage v_{an}, which depends on the number of conducting devices, can be determined from Figure 11.9a as follows:

For $0 \leq \omega t < \pi/3$: $v_{an} = 0$

For $\pi/3 \leq \omega t < 2\pi/3$: $v_{an} = v_{AB}/2 = 147.1 \sin(\omega t + \pi/6)$

For $2\pi/3 \le \omega t < \pi$: $v_{an} = v_{AC}/2 = -v_{CA}/2 = 147.1 \sin(\omega t - 7\pi/6 - \pi)$

For $\pi \le \omega t < 4\pi/3$: $v_{an} = 0$

For $4\pi/3 \le \omega t < 5\pi/3$: $v_{an} = v_{AB}/2 = 147.1 \sin(\omega t + \pi/6)$

For $5\pi/3 \le \omega t < 2\pi$: $v_{an} = v_{AC}/2 = 147.1 \sin(\omega t - 7\pi/6 - \pi)$

Note: The PF, which depends on the delay angle α, is in general poor compared with that of the half-wave controller.

Key Points of Section 11.6

- Varying the delay angle α from 0 to $5\pi/6$ can vary the rms output phase voltage from V_s to 0.
- The arrangement of the tie control is not suitable for motor control.

11.7 THREE-PHASE BIDIRECTIONAL DELTA-CONNECTED CONTROLLERS

If the terminals of a three-phase system are accessible, the control elements (or power devices) and load may be connected in delta, as shown in Figure 11.11. Because the phase current in a normal three-phase system is only $1/\sqrt{3}$ of the line current, the current ratings of thyristors would be less than that if thyristors (or control elements) were placed in the line.

Let us assume that the instantaneous line-to-line voltages are

$$v_{AB} = v_{ab} = \sqrt{2}\, V_s \sin \omega t$$

$$v_{BC} = v_{bc} = \sqrt{2}\, V_s \sin\left(\omega t - \frac{2\pi}{3}\right)$$

$$v_{CA} = v_{ca} = \sqrt{2}\, V_s \sin\left(\omega t - \frac{4\pi}{3}\right)$$

The input line voltages, phase and line currents, and thyristor gating signals are shown in Figure 11.12 for $\alpha = 120°$ and a resistive load.

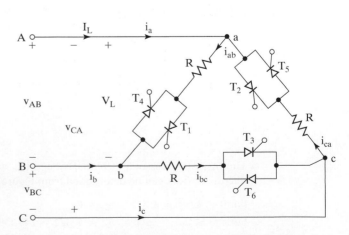

FIGURE 11.11

Delta-connected three-phase controller.

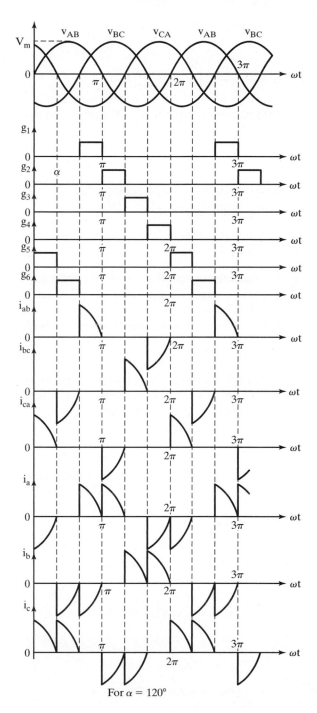

For $\alpha = 120°$

FIGURE 11.12

Waveforms for delta-connected controller.

For resistive loads, the rms output phase voltage can be determined from

$$V_o = \left[\frac{1}{2\pi} \int_\alpha^{2\pi} v_{ab}^2 \, d(\omega t) \right]^{1/2} = \left[\frac{2}{2\pi} \int_\alpha^{\pi} 2 \, V_s^2 \sin \omega t \, d(\omega t) \right]^{1/2}$$

$$= V_s \left[\frac{1}{\pi} \left(\pi - \alpha + \frac{\sin 2\alpha}{2} \right) \right]^{1/2} \tag{11.29}$$

The maximum output voltage would be obtained when $\alpha = 0$, and the control range of delay angle is

$$0 \le \alpha \le \pi \tag{11.30}$$

The line currents, which can be determined from the phase currents, are

$$i_a = i_{ab} - i_{ca}$$
$$i_b = i_{bc} - i_{ab}$$
$$i_c = i_{ca} - i_{bc} \tag{11.31}$$

We can notice from Figure 11.12 that the line currents depend on the delay angle and may be discontinuous. The rms value of line and phase currents for the load circuits can be determined by numerical solution or Fourier analysis. If I_n is the rms value of the nth harmonic component of a phase current, the rms value of phase current can be found from

$$I_{ab} = (I_1^2 + I_3^2 + I_5^2 + I_7^2 + I_9^2 + I_{11}^2 + \cdots + I_n^2)^{1/2} \tag{11.32}$$

Due to the delta connection, the triplen harmonic components (i.e., those of order $n = 3m$, where m is an odd integer) of the phase currents would flow around the delta and would not appear in the line. This is due to the fact that the zero-sequence harmonics are in phase in all three phases of load. The rms line current becomes

$$I_a = \sqrt{3} \, (I_1^2 + I_5^2 + I_7^2 + I_{11}^2 + \cdots + I_n^2)^{1/2} \tag{11.33}$$

As a result, the rms value of line current would not follow the normal relationship of a three-phase system such that

$$I_a < \sqrt{3} \, I_{ab} \tag{11.34}$$

An alternative form of delta-connected controllers that requires only three thyristors and simplifies the control circuitry is shown in Figure 11.13. This arrangement is also known as a *neutral-point controller*.

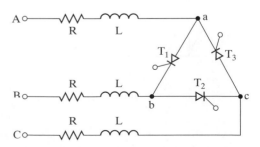

FIGURE 11.13

Three-phase three-thyristor controller.

Example 11.6 Finding the Performance Parameters of a Three-Phase Delta-Connected Controller

The three-phase bidirectional delta-connected controller in Figure 11.11 has a resistive load of $R = 10\ \Omega$. The line-to-line voltage is $V_s = 208$ V (rms), 60 Hz, and the delay angle is $\alpha = 2\pi/3$. Determine (a) the rms output phase voltage V_o; (b) the expressions for instantaneous currents i_a, i_{ab}, and i_{ca}; (c) the rms output phase current I_{ab} and rms line current I_a; (d) the input PF; and (e) the rms current of a thyristor I_R.

Solution

$V_L = V_s = 208$ V, $\alpha = 2\pi/3$, $R = 10\ \Omega$, and peak value of phase current, $I_m = \sqrt{2} \times 208/10 = 29.4$ A.

a. From Eq. (11.29), $V_o = 92$ V.

b. Assuming i_{ab} as the reference phasor and $i_{ab} = I_m \sin \omega t$, the instantaneous currents are:

For $0 \le \omega t < \pi/3$:

$$I_{ab} = 0$$
$$i_{ca} = I_m \sin(\omega t - 4\pi/3)$$
$$i_a = i_{ab} - i_{ca} = -I_m \sin(\omega t - 4\pi/3)$$

For $\pi/3 < \omega t < 2\pi/3$:

$$i_{ab} = i_{ca} = i_a = 0$$

For $2\pi/3 < \omega t < \pi$:

$$i_{ab} = I_m \sin \omega t$$
$$i_{ca} = 0$$
$$i_a = i_{ab} - i_{ca} = I_m \sin \omega t$$

For $\pi < \omega t < 4\pi/3$:

$$i_{ab} = 0$$
$$i_{ca} = I_m \sin(\omega t - 4\pi/3)$$
$$i_a = i_{ab} - i_{ca} = -I_m \sin(\omega t - 4\pi/3)$$

For $4\pi/3 < \omega t < 5\pi/3$:

$$i_{ab} = i_{ca} = i_a = 0$$

For $5\pi/3 < \omega t < 2\pi$:

$$i_{ab} = I_m \sin \omega t$$
$$i_{ca} = 0$$
$$i_a = i_{ab} - i_{ca} = I_m \sin \omega t$$

c. The rms values of i_{ab} and i_a are determined by numerical integration using a Mathcad program. Students are encouraged to verify the results.

$$I_{ab} = 9.2 \text{ A} \qquad I_L = I_a = 13.01 \text{ A} \qquad \frac{I_a}{I_{ab}} = \frac{13.01}{9.2} = 1.1414 \neq \sqrt{3}$$

d. The output power

$$P_o = 3I_{ab}^2 R = 3 \times 9.2^2 \times 10 = 2537$$

The VA is found by

$$VA = 3V_s I_{ab} = 3 \times 208 \times 9.2 = 5739$$

The PF is

$$PF = \frac{P_o}{VA} = \frac{2537}{5739} = 0.442 \text{ (lagging)}$$

e. The thyristor current can be determined from the phase current

$$I_R = \frac{I_{ab}}{\sqrt{2}} = \frac{9.2}{\sqrt{2}} = 6.5 \text{ A}$$

Note: For the ac voltage controller of Figure 11.13, the line current I_a is not related to the phase current I_{ab} by a factor of $\sqrt{3}$. This is due to the discontinuity of the load current in the presence of the ac voltage controller.

Key Point of Section 11.7

- Although, the delta-connected controller has lower current ratings than those of a full-wave controller, it is not used for motor control.

11.8   SINGLE-PHASE TRANSFORMER CONNECTION CHANGERS

Thyristors can be used as static switches for on-load changing of transformer connections. The static connection changers have the advantage of very fast switching action. The changeover can be controlled to cope with load conditions and is smooth. The circuit diagram of a single-phase transformer changer is shown in Figure 11.14. Although a transformer may have multiple secondary windings, only two secondary windings are shown, for the sake of simplicity.

The turns ratio of the input transformer are such that if the primary instantaneous voltage is

$$v_p = \sqrt{2}\,V_s \sin \omega t = \sqrt{2}\,V_p \sin \omega t$$

the secondary instantaneous voltages are

$$v_1 = \sqrt{2}\,V_1 \sin \omega t$$

and

$$v_2 = \sqrt{2}\,V_2 \sin \omega t$$

A connection changer is most commonly used for resistive heating loads. When only thyristors T_3 and T_4 are alternately fired with a delay angle of $\alpha = 0$, the load voltage is held at a reduced level of $V_o = V_1$. If full output voltage is required, only thyristors T_1 and T_2 are alternately fired with a delay angle of $\alpha = 0$ and the full voltage is $v_o = V_1 + V_2$.

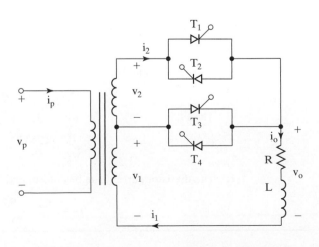

FIGURE 11.14

Single-phase transformer connection changer.

The gating pulses of thyristors can be controlled to vary the load voltage. The rms value of load voltage V_o can be varied within three possible ranges:

$$0 < V_o < V_1$$
$$0 < V_o < (V_1 + V_2)$$

and

$$V_1 < V_o < (V_1 + V_2)$$

Control range 1: $0 \le V_o \le V_1$.   To vary the load voltage within this range, thyristors T_1 and T_2 are turned off. Thyristors T_3 and T_4 can operate as a single-phase voltage controller. The instantaneous load voltage v_0 and load current i_0 are shown in Figure 11.15c for a resistive load. The rms load voltage that can be determined from

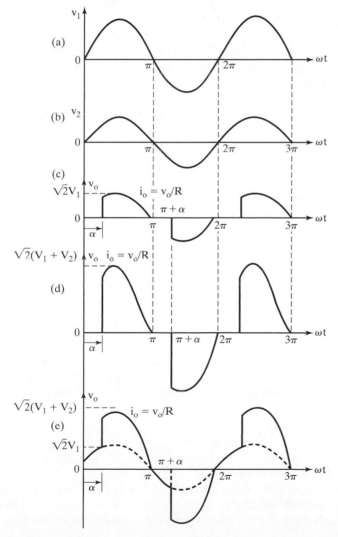

FIGURE 11.15

Waveforms for transformer connection changer.

Eq. (11.8) load is

$$V_o = V_1 \left[\frac{1}{\pi} \left(\pi - \alpha + \frac{\sin 2\alpha}{2} \right) \right]^{1/2} \tag{11.35}$$

and the range of delay angle is $0 \le \alpha \le \pi$.

Control range 2: $0 \le V_o \le (V_1 + V_2)$. Thyristors T_3 and T_4 are turned off. Thyristors T_1 and T_2 operate as a single-phase voltage controller. Figure 11.15d shows the load voltage v_0 and load current i_0 for a resistive load. The rms load voltage can be found from

$$V_o = (V_1 + V_2) \left[\frac{1}{\pi} \left(\pi - \alpha + \frac{\sin 2\alpha}{2} \right) \right]^{1/2} \tag{11.36}$$

and the range of delay angle is $0 \le \alpha \le \pi$.

Control range 3: $V_1 < V_o < (V_1 + V_2)$. Thyristor T_3 is turned on at $\omega t = 0$ and the secondary voltage v_1 appears across the load. If thyristor T_1 is turned on at $\omega t = \alpha$, thyristor T_3 is reverse biased due to secondary voltage v_2, and T_3 is turned off. The voltage appearing across the load is $(v_1 + v_2)$. At $\omega t = \pi$, T_1 is self-commutated and T_4 is turned on. The secondary voltage v_1 appears across the load until T_2 is fired at $\omega t = \pi + \alpha$. When T_2 is turned on at $\omega t = \pi + \alpha$, T_4 is turned off due to reverse voltage v_2 and the load voltage is $(v_1 + v_2)$. At $\omega t = 2\pi$, T_2 is self-commutated, T_3 is turned on again, and the cycle is repeated. The instantaneous load voltage v_0 and load current i_0 are shown in Figure 11.15e for a resistive load.

A connection changer with this type of control is also known as a *synchronous connection changer*. It uses two-step control. A part of secondary voltage v_2 is superimposed on a sinusoidal voltage v_1. As a result, the harmonic contents are less than those that would be obtained by a normal phase delay, as discussed earlier for control range 2. The rms load voltage can be found from

$$
\begin{aligned}
V_o &= \left[\frac{1}{2\pi} \int_0^{2\pi} v_0^2 \, d(\omega t) \right]^{1/2} \\
&= \left\{ \frac{2}{2\pi} \left[\int_0^{\alpha} 2V_1^2 \sin^2 \omega t \, d(\omega t) + \int_\alpha^{\pi} 2(V_1 + V_2)^2 \sin^2 \omega t \, d(\omega t) \right] \right\}^{1/2} \\
&= \left[\frac{V_1^2}{\pi} \left(\alpha - \frac{\sin 2\alpha}{2} \right) + \frac{(V_1 + V_2)^2)}{\pi} \left(\pi - \alpha + \frac{\sin 2\alpha}{2} \right) \right]^{1/2}
\end{aligned} \tag{11.37}
$$

With RL loads, the gating circuit of a synchronous connection changer requires a careful design. Let us assume that thyristors T_1 and T_2 are turned off, while thyristors T_3 and T_4 are turned on during the alternate half-cycle at the zero crossing of the load current. The load current would then be

$$i_o = \frac{\sqrt{2} \, V_1}{Z} \sin(\omega t - \theta)$$

where $Z = [R^2 + (\omega L)^2]^{1/2}$ and $\theta = \tan^{-1}(\omega L/R)$.

The instantaneous load current i_0 is shown in Figure 11.16a. If T_1 is then turned on at $\omega t = \alpha$, where $\alpha < \theta$, the second winding of the transformer is short circuited because thyristor T_3 is still conducting and carrying current due to the inductive load.

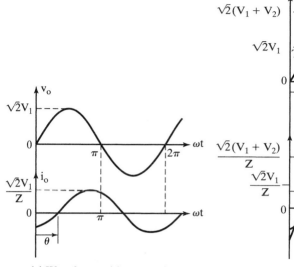

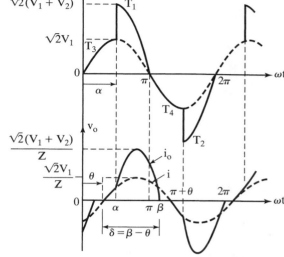

(a) Waveforms without tap changer

(b) Synchronous changer

FIGURE 11.16

Voltage and current waveforms for *RL* load.

Therefore, the control circuit should be designed so that T_1 is not turned on until T_3 turns off and $i_0 \geq 0$. Similarly, T_2 should not be turned on until T_4 turns off and $i_0 \leq 0$. The waveforms of load voltage v_0 and load current i_0 are shown in Figure 11.16b for $\alpha > \theta$.

Gating sequence. The gating sequence is as follows:

1. For output voltage $0 \leq V_o \leq V_1$, gate T_3 and T_4 at delay angles α and $\pi + \alpha$, respectively, while turning off gate signals for T_1 and T_2.

2. For output voltage $0 \leq V_o \leq (V_1 + V_2)$, gate T_1 and T_2 at delay angles α and $\pi + \alpha$, respectively, while turning off gate signals for T_3 and T_4.

Example 11.7 Finding the Performance Parameters of a Single-Phase Connection Changer

The circuit in Figure 11.14 is controlled as a synchronous connection changer. The primary voltage is 240 V (rms), 60 Hz. The secondary voltages are $V_1 = 120$ V and $V_2 = 120$ V. If the load resistance is $R = 10\ \Omega$ and the rms load voltage is 180 V, determine (a) the delay angle of thyristors T_1 and T_2, (b) the rms current of thyristors T_1 and T_2, (c) the rms current of thyristors T_3 and T_4, and (d) the input PF.

Solution

$V_o = 180$ V, $V_p = 240$ V, $V_1 = 120$ V, $V_2 = 120$ V, and $R = 10\ \Omega$.

a. The required value of delay angle α for $V_o = 180$ V can be found from Eq. (11.37) in two ways: (1) plot V_o against α and find the required value of α, or (2) use an iterative method of solution. A Mathcad program is used to solve Eq. (11.37) for α by iteration and this gives $\alpha = 98°$.

Eq. (11.8), the input PF,

$$PF = \frac{P_o}{V_s I_s} = \frac{V_o \cos\theta}{V_s} = \cos\theta \left[\frac{1}{\pi}\left(\pi - \alpha + \frac{\sin 2\alpha}{2}\right)\right]^{1/2}$$

$$= \frac{279.35}{897.6} = 0.311 \text{ (lagging)}$$

(11.42)

Note: Equation (11.42) does not include the harmonic content on the output voltage and gives the approximate value of PF. The actual value is less than that given by Eq. (11.42). Equations (11.41) and (11.42) are also valid for resistive loads.

11.9.2 Three-Phase Cycloconverters

The circuit diagram of a three-phase/single-phase cycloconverter is shown in Figure 11.19a. The two ac–dc converters are three-phase controlled rectifiers. The synthesis of output waveform for an output frequency of 12 Hz is shown in Figure 11.19b. The positive converter operates for half the period of output frequency and the negative converter operates for the other half period. The analysis of this cycloconverter is similar to that of single-phase/single-phase cycloconverters.

The control of ac motors requires a three-phase voltage at a variable frequency. The cycloconverter in Figure 11.19a can be extended to provide three-phase output by having 6 three-phase converters, as shown in Figure 11.20a. Each phase consists of 6 thyristors, as shown in Figure 11.20b, and a total of 18 thyristors are required. If six full-wave three-phase converters are used, 36 thyristors would be required.

Gating sequence.   The gating sequence [1] is as follows:

1. During the first half period of the output frequency $T_o/2$, operate converter P as a normal three-phase controlled rectifier (in Section 11.6) with a delay angle of $\alpha_p = \alpha$.
2. During the second half period $T_o/2$, operate converter N as a normal controlled three-phase rectifier with a delay angle of $\alpha_N = \pi - \alpha$.

11.9.3 Reduction of Output Harmonics

We can notice from Figures 11.17c and 11.19b that the output voltage is not purely sinusoidal, and as a result the output voltage contains harmonics. Equation (11.42) shows that the input PF depends on the delay angle of thyristors and is poor, especially at the low output voltage range.

The output voltage of cycloconverters is basically made up of segments of input voltages and the average value of a segment depends on the delay angle for that segment. If the delay angles of segments were varied in such a way that the average values of segments correspond as closely as possible to the variations of desired sinusoidal output voltage, the harmonics on the output voltage can be minimized [2, 3]. Equation (11.6) indicates that the average output voltage of a segment is a cosine function of delay angle. The delay angles for segments can be generated by comparing a cosine signal at source frequency ($v_c = \sqrt{2}\, V_s \cos\omega_s t$) with an ideal sinusoidal reference

Key

•

•

11.9   CY(

The
out
out
fror
able
Cha
dia
at c
teri

put

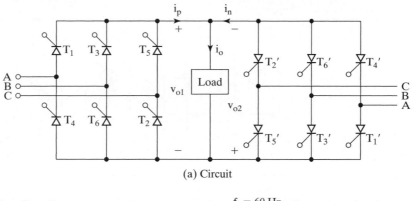

(a) Circuit

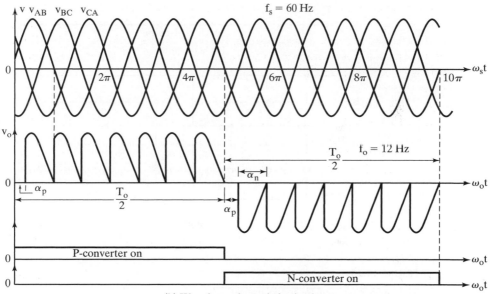

(b) Waveforms for resistive load

FIGURE 11.19

Three-phase/single-phase cycloconverter.

voltage at the output frequency ($v_r = \sqrt{2}\, V_r \sin \omega_0 t$). Figure 11.21 shows the generation of gating signals for the thyristors of the cycloconverter in Figure 11.19a.

The maximum average voltage of a segment (which occurs for $\alpha_p = 0$) should be equal to the peak value of output voltage; for example, from Eq. (10.5),

$$V_p = \frac{2\sqrt{2}\, V_s}{\pi} = \sqrt{2}\, V_o \tag{11.43}$$

which gives the rms value of output voltage as

$$V_o = \frac{2V_s}{\pi} = \frac{2V_p}{\pi} \tag{11.44}$$

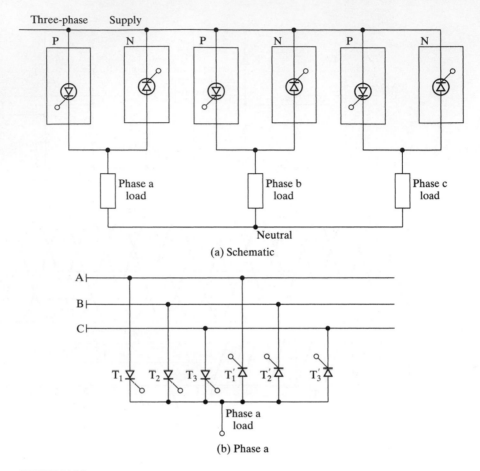

Three-phase   Supply

Phase a
load

Phase b
load

Phase c
load

Neutral

(a) Schematic

A

B

C

T_1 T_2 T_3 T_1' T_2' T_3'

Phase a
load

(b) Phase a

FIGURE 11.20

Three-phase/three-phase cycloconverter.

Example 11.9   Finding the Performance Parameters of a Single-Phase Cycloconverter with a Cosine Reference Signal

Repeat Example 11.8 if the delay angles of the cycloconverter are generated by comparing a cosine signal at the source frequency with a sinusoidal signal at the output frequency as shown in Figure 11.21.

Solution

$V_s = 120$ V, $f_s = 60$ Hz, $f_o = 20$ Hz, $R = 5 \, \Omega$, $L = 40$ mH, $\alpha_p = 2\pi/3$, $\omega_0 = 2\pi \times 20 = 125.66$ rad/s, and $X_L = \omega_0 L = 5.027 \, \Omega$.

a.   From Eq. (11.44), the rms value of the output voltage

$$V_o = \frac{2V_s}{\pi} = 0.6366V_s = 0.6366 \times 120 = 76.39 \text{ V}$$

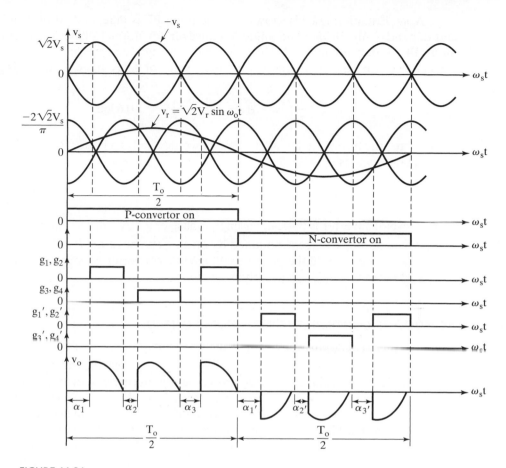

FIGURE 11.21

Generation of thyristor gating signals.

b. $Z = [R^2 + (\omega_0 L)^2]^{1/2} = 7.09\ \Omega$ and $\theta = \tan^{-1}(\omega_0 L/R) = 45.2°$. The rms load current is $I_o = V_o/Z = 76.39/7.09 = 10.77$ A. The rms current through each converter, $I_p = I_N = I_L/\sqrt{2} = 7.62$ A, and the rms current through each thyristor is $I_R = I_p/\sqrt{2} = 5.39$ A.

c. The rms input current $I_s = I_o = 10.77$ A, the VA rating is VA $= V_s I_s = 1292.4$ VA, and the output power is

$$P_o = V_o I_o \cos\theta = 0.6366 V_s I_o \cos\theta = 579.73\ \text{W}.$$

The input PF is

$$\begin{aligned} \text{PF} &= 0.6366 \cos\theta \\ &= \frac{579.73}{1292.4} = 0.449\ \text{(lagging)} \end{aligned} \tag{11.45}$$

Note: Equation (11.45) shows that the input PF is independent of delay angle α and depends only on the load angle θ. However, for normal phase-angle control, the input PF is dependent on both delay angle α and load angle θ. If we compare Eq. (11.42) with Eq. (11.45), there is a critical value of delay angle α_c, which is given by

$$\left[\frac{1}{\pi}\left(\pi - \alpha_c + \frac{\sin 2\alpha_c}{2}\right)\right]^{1/2} = 0.6366 \qquad (11.46)$$

For $\alpha < \alpha_c$, the normal delay-angle control would exhibit a better PF and the solution of Eq. (11.46) yields $\alpha_c = 98.59°$.

Key Points of Section 11.9

- A cycloconverter is basically a single-phase or a three-phase dual converter.
- An ac output voltage is obtained by turning on converter P only during the first $T_o/2$ period to yield the positive voltage and converter N only during the second $T_o/2$ period to yield the negative voltage.

11.10   AC VOLTAGE CONTROLLERS WITH PWM CONTROL

It has been shown in Section 11.8 that the input PF of controlled rectifiers can be improved by the pulse-width-modulation (PWM) type of control. The naturally commutated thyristor controllers introduce lower order harmonics in both the load and supply side and have low-input PF. The performance of ac voltage controllers can be improved by PWM control [4]. The circuit configuration of a single-phase ac voltage controller for PWM control is shown in Figure 11.22a. The gating signals of the switches are shown in Figure 11.22b. Switches S_1 and S_2 are turned on and off several times during the positive and negative half-cycles of the input voltage, respectively. S_1' and S_2' provide the freewheeling paths for the load current, whereas S_1 and S_2, respectively,

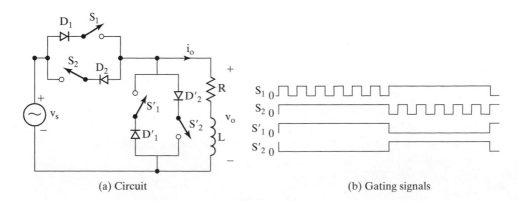

(a) Circuit   (b) Gating signals

FIGURE 11.22

Ac voltage controller for PWM control.

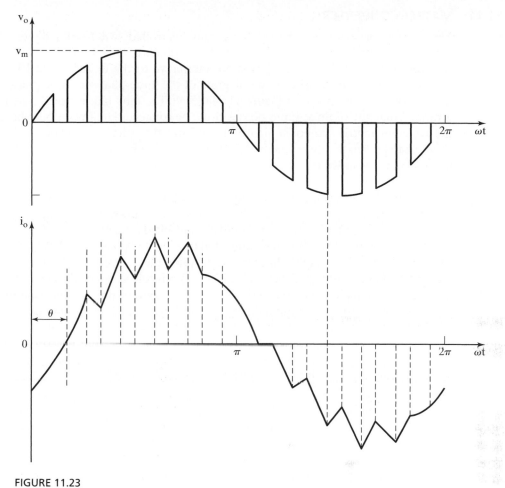

FIGURE 11.23

Output voltage and load current of ac voltage controller.

are in the off-state. The diodes prevent reverse voltages from appearing across the switches.

The output voltage is shown in Figure 11.23a. For a resistive load, the load current resembles the output voltage. With an RL load, the load current rises in the positive or negative direction when switch S_1 or S_2 is turned on, respectively. Similarly, the load current falls when either S_1' or S_2' is turned on. The load current is shown in Figure 11.23b with an RL load.

Key Point of Section 11.10

- Using fast switching devices, PWM techniques can be applied to the ac voltage controllers for producing variable output voltage with a better input PF.

11.11   MATRIX CONVERTER

The matrix converter uses bidirectional fully controlled switches for direct conversion from ac to ac. It is a single-stage converter that requires only nine switches for three-phase to three-phase conversion [5–7]. It is an alternative to the double-sided PWM voltage source rectifier inverter. The circuit diagram of the three-phase to three-phase (3ϕ–3ϕ) matrix converter is shown in Figure 11.24a [8, 9]. The nine bidirectional switches are so arranged that any of three input phases could be connected to any output phase through the switching matrix symbol in Figure 11.24b. Thus, the voltage at any input terminal may be made to appear at any output terminal or terminals whereas the current in any phase of the load may be drawn from any phase or phases of the input supply. An ac input LC filter is normally used to eliminate harmonic currents in the input side and the load is suffi-ciently inductive to maintain continuity of the output currents [10]. The term *matrix* is due to the fact that it uses exactly one switch for each of the possible connections between the input and the output. The switches should be controlled in such a way that, at any time, one and only one of the three switches connected to an output phase must be closed to prevent short circuiting of the supply lines or interrupting the load current flow in an inductive load. With these constraints, there are $512 (=2^9)$ possible states of the converter, but only 27 switch combinations are allowed to produce the output line voltages and input phase currents. With a given set of input three-phase voltages, any desired set of three-phase out-put voltages can be synthesized by adopting a suitable switching strategy [11, 12].

The matrix converter can connect any input phase (A, B, and C) to any output phase (a, b, and c) at any instant. When connected, the voltages v_{an}, v_{bn}, v_{cn} at the out-put terminals are related to the input voltages v_{AN}, v_{BN}, v_{CN} as

$$\begin{bmatrix} V_{an} \\ V_{bn} \\ V_{cn} \end{bmatrix} = \begin{bmatrix} S_{Aa} & S_{Ba} & S_{Ca} \\ S_{Ab} & S_{Bb} & S_{cb} \\ S_{Ac} & S_{Bc} & S_{Cc} \end{bmatrix} \begin{bmatrix} V_{AN} \\ V_{BN} \\ V_{CN} \end{bmatrix} \tag{11.47}$$

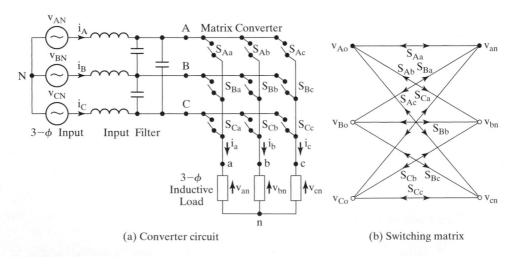

(a) Converter circuit   (b) Switching matrix

FIGURE 11.24

(a) Matrix (3ϕ–3ϕ) converter circuit with input filter (b) Switching matrix symbol for converter.

where S_{Aa} through S_{Cc} are the switching variables of the corresponding switches. For a balanced linear Y-connected load at the output terminals, the input phase currents are related to the output phase currents by

$$
\begin{bmatrix} i_A \\ i_B \\ i_C \end{bmatrix} = \begin{bmatrix} S_{Aa} & S_{Ab} & S_{Ac} \\ S_{Ba} & S_{Bb} & S_{Bc} \\ S_{Ca} & S_{Cb} & S_{Cc} \end{bmatrix}^T \begin{bmatrix} i_a \\ i_b \\ i_c \end{bmatrix}
\tag{11.48}
$$

where the matrix of the switching variables in Eq. (11.48) is a transpose of the respective matrix in Eq. (11.47). The matrix converter should be controlled using a specific and appropriately timed sequence of the values of the switching variables, which result in balanced output voltages having the desired frequency and amplitude, whereas the input currents are balanced and in phase with respect to the input voltages. However, the maximum peak-to-peak output voltage cannot be greater than the minimum voltage difference between two phases of the input. Regardless of the switching strategy, there is a physical limit on the achievable output voltage and the maximum voltage transfer ratio is 0.866. The control methods for matrix converters must have the ability for independent control of the output voltages and input currents. Three types of methods [12] are commonly used: (1) Venturini method that is based on a mathematical approach of transfer function analysis [5], (2) PWM, and (3) space vector modulation [3].

The matrix converter has the advantages of (1) inherent bidirectional power flow, (2) sinusoidal input–output waveforms with moderate switching frequency, (3) possibility of compact design due to absence of dc-link reactive components and (4) controllable input PF independent of the output load current. However, the practical applications of the matrix converters are very limited. The main reasons are (1) non-availability of the bilateral fully controlled monolithic switches capable of high-frequency operation (2) complex control law implementation (3) an intrinsic limitation of the output–input voltage ratio and (4) commutation and protection of the switches. With space vector PWM control using overmodulation, the voltage transfer ratio may be increased to 1.05 at the expense of more harmonics and large filter capacitors [13].

Key Point of Section 11.11

- The matrix converter is a single-stage converter. It uses bi-directional fully controlled switches for direct conversion from ac to ac. It is an alternative to the double-sided PWM voltage source rectifier-inverter.

11.12   DESIGN OF AC VOLTAGE-CONTROLLER CIRCUITS

The ratings of power devices must be designed for the worst-case condition, which occurs when the converter delivers the maximum rms value of output voltage V_o. The input and output filters must also be designed for worst-case conditions. The output of a power controller contains harmonics, and the delay angle for the worst-case condition of a particular circuit arrangement should be determined. The steps involved in designing the power circuits and filters are similar to those of rectifier circuit design in Section 3.10.

Example 11.10 Finding the Device Ratings of the Single-Phase Full-Wave Controller

A single-phase full-wave ac voltage controller in Figure 11.3a controls power flow from a 230-V, 60-Hz ac source into a resistive load. The maximum desired output power is 10 kW. Calculate (a) the maximum rms current rating of thyristors I_{RM}, (b) the maximum average current rating of thyristors I_{AM}, (c) the peak current of thyristors I_p, and (d) the peak value of thyristor voltage V_p.

Solution

$P_o = 10{,}000$ W, $V_s = 230$ V, and $V_m = \sqrt{2} \times 230 = 325.3$ V. The maximum power can be delivered when the delay angle is $\alpha = 0$. From Eq. (11.8), the rms value of output voltage $V_o = V_s = 230$ V, $P_o = V_o^2/R = 230^2/R = 10{,}000$, and load resistance is $R = 5.29 \ \Omega$.

   a. The maximum rms value of load current $I_{oM} = V_o/R = 230/5.29 = 43.48$ A, and the maximum rms value of thyristor current $I_{RM} = I_{oM}/\sqrt{2} = 30.75$ A.

   b. From Eq. (11.10), the maximum average current of thyristors,

$$I_{AM} = \frac{\sqrt{2} \times 230}{\pi \times 5.29} = 19.57 \text{ A}$$

   c. The peak thyristor current $I_p = V_m/R = 325.3/5.29 = 61.5$ A.

   d. The peak thyristor voltage $V_p = V_m = 325.3$ V.

Example 11.11 Finding the Harmonic Voltages and Currents of a Single-Phase Full-Wave Controller

A single-phase full-wave controller in Figure 11.6a controls power to an RL load and the source voltage is 120 V (rms), 60 Hz. (a) Use the method of Fourier series to obtain expressions for output voltage $v_o(t)$ and load current $i_o(t)$ as a function of delay angle α. (b) Determine the delay angle for the maximum amount of lowest order harmonic current in the load. (c) If $R = 5 \ \Omega$, $L = 10$ mH, and $\alpha = \pi/2$, determine the rms value of third harmonic current. (d) If a capacitor is connected across the load (Figure 11.25a), calculate the value of capacitance to reduce the third harmonic current to 10% of the value without the capacitor.

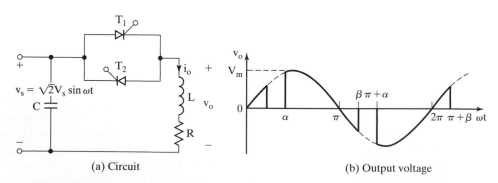

(a) Circuit   (b) Output voltage

FIGURE 11.25

Single-phase full converter with RL load.

Solution

a. The waveform for the input voltage is shown in Figure 11.6b. The instantaneous output voltage as shown in Figure 11.25b can be expressed in Fourier series as

$$v_o(t) = V_{dc} + \sum_{n=1,2,\ldots}^{\infty} a_n \cos n\omega t + \sum_{n=1,2,\ldots}^{\infty} b_n \sin n\omega t \tag{11.49}$$

where

$$V_{dc} = \frac{1}{2\pi} \int_0^{2\pi} V_m \sin \omega t \, d(\omega t) = 0$$

$$a_n = \frac{1}{\pi} \left[\int_\alpha^\beta \sqrt{2} V_s \sin \omega t \cos n\omega t \, d(\omega t) + \int_{\pi+\alpha}^{\pi+\beta} \sqrt{2} V_s \sin \omega t \cos n\omega t \, d(\omega t) \right]$$

$$= \frac{\sqrt{2} V_s}{2\pi} \left[\frac{\begin{array}{c} \cos(1-n)\alpha - \cos(1-n)\beta + \cos(1-n)(\pi+\alpha) \\ -\cos(1-n)(\pi+\beta) \end{array}}{1-n} \right.$$

$$\left. + \frac{\begin{array}{c} \cos(1+n)\alpha - \cos(1+n)\beta + \cos(1+n)(\pi+\alpha) \\ -\cos(1+n)(\pi+\beta) \end{array}}{1+n} \right]$$

$$\text{for } n = 3, 5, \ldots \tag{11.50}$$

$$= 0 \quad \text{for } n = 2, 4, \ldots$$

$$b_n = \frac{1}{\pi} \left[\int_\alpha^\beta \sqrt{2} V_s \sin \omega t \sin n\omega t \, d(\omega t) + \int_{\pi+\alpha}^{\pi+\beta} \sqrt{2} V_s \sin \omega t \sin n\omega t \, d(\omega t) \right]$$

$$= \frac{\sqrt{2} V_s}{2\pi} \left[\frac{\begin{array}{c} \sin(1-n)\beta - \sin(1-n)\alpha + \sin(1-n)(\pi+\beta) \\ -\sin(1-n)(\pi+\alpha) \end{array}}{1-n} \right.$$

$$\left. - \frac{\begin{array}{c} \sin(1+n)\beta - \sin(1+n)\alpha + \sin(1+n)(\pi+\beta) \\ -\sin(1+n)(\pi+\alpha) \end{array}}{1+n} \right]$$

$$\text{for } n = 3, 5, \ldots \tag{11.51}$$

$$= 0 \quad \text{for } n = 2, 4, \ldots$$

$$a_1 = \frac{1}{\pi} \left[\int_\alpha^\beta \sqrt{2} V_s \sin \omega t \cos \omega t \, d(\omega t) + \int_{\pi+\alpha}^{\pi+\beta} \sqrt{2} V_s \sin \omega t \cos \omega t \, d(\omega t) \right]$$

$$= \frac{\sqrt{2} V_s}{2\pi} [\sin^2 \beta - \sin^2 \alpha + \sin^2(\pi+\beta) - \sin^2(\pi+\alpha)] \quad \text{for } n = 1 \tag{11.52}$$

$$b_1 = \frac{1}{\pi} \left[\int_\alpha^\beta \sqrt{2} V_s \sin^2 \omega t \, d(\omega t) + \int_{\pi+\alpha}^{\pi+\beta} \sqrt{2} V_s \sin^2 \omega t \, d(\omega t) \right]$$

$$= \frac{\sqrt{2} V_s}{2\pi} \left[2(\beta - \alpha) - \frac{\sin 2\beta - \sin 2\alpha + \sin 2(\pi+\beta) - \sin 2(\pi+\alpha)}{2} \right]$$

$$\text{for } n = 1 \tag{11.53}$$

The load impedance

$$Z = R + j(n\omega L) = [R^2 + (n\omega L)^2]^{1/2}\underline{/\theta_n}$$

and $\theta_n = \tan^{-1}(n\omega L/R)$. Dividing $v_o(t)$ in Eq. (11.49) by load impedance Z and simplifying the sine and cosine terms give the load current as

$$i_o(t) = \sum_{n=1,3,5,\ldots}^{\infty} \sqrt{2}\, I_n \sin(n\omega t - \theta_n + \phi_n) \tag{11.54}$$

where $\phi_n = \tan^{-1}(a_n/b_n)$ and

$$I_n = \frac{1}{\sqrt{2}} \frac{(a_n^2 + b_n^2)^{1/2}}{[R^2 + (n\omega L)^2]^{1/2}}$$

b. The third harmonic is the lowest order harmonic. The calculation of third harmonic for various values of delay angle shows that it becomes maximum for $\alpha = \pi/2$. The harmonic distortion increases and the quality of the input current decreases with an increase in firing angles. The variations of low-order harmonics with the firing angle are shown in Figure 11.26. Only odd harmonics exist in the input current because of half-wave symmetry.

c. For $\alpha = \pi/2$, $L = 6.5$ mH, $R = 2.5\ \Omega$, $\omega = 2\pi \times 60 = 377$ rad/s, and $V_s = 120$ V. From Example 11.4, we get the extinction angle as $\beta = 220.43°$. For known values of α, β, R, L, and V_s, a_n, and b_n of the Fourier series in Eq. (11.49) and the load current i_o of Eq. (11.54) can be calculated. The load current is given by

$$i_o(t) = 28.93 \sin(\omega t - 44.2° - 18°) + 7.96 \sin(3\omega t - 71.2° + 68.7°)$$
$$+ 2.68 \sin(5\omega t - 78.5° - 68.6°) + 0.42 \sin(7\omega t - 81.7° + 122.7°)$$
$$+ 0.59 \sin(9\omega t - 83.5° - 126.3°) + \cdots$$

The rms value of third harmonic current is

$$I_3 = \frac{7.96}{\sqrt{2}} = 5.63\ \text{A}$$

d. Figure 11.27 shows the equivalent circuit for harmonic current. Using the current-divider rule, the harmonic current through load is given by

$$\frac{I_h}{I_n} = \frac{X_c}{[R^2 + (n\omega L - X_c)^2]^{1/2}}$$

where $X_c = 1/(n\omega C)$. For $n = 3$ and $\omega = 377$,

$$\frac{I_h}{I_n} = \frac{X_c}{[2.5^2 + (3 \times 0.377 \times 6.5 - X_c)^2]^{1/2}} = 0.1$$

which yields $X_c = -0.858$ or 0.7097. Because X_c cannot be negative, $X_c = 0.7097 = 1/(3 \times 377\ C)$ or $C = 1245.94\ \mu$F.

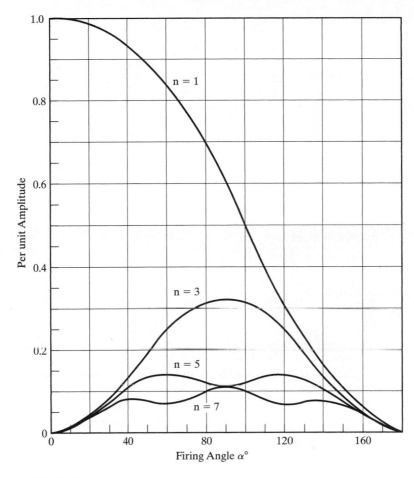

FIGURE 11.26

Harmonic content as a function of the firing angle for a single-phase voltage controller with *RL* load.

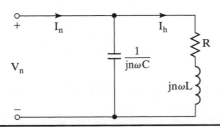

FIGURE 11.27

Equivalent circuit for harmonic current.

Example 11.12 PSpice Simulation of the Single-Phase Full-Wave Controller

The single-phase ac voltage controller in Figure 11.6a has a load of $R = 2.5 \; \Omega$ and $L = 6.5$ mH. The supply voltage is 120 V (rms), 60 Hz. The delay angle is $\alpha = \pi/2$. Use PSpice to plot the output voltage and the load current and to compute the total harmonic distortion (THD) of the output voltage and output current and the input PF.

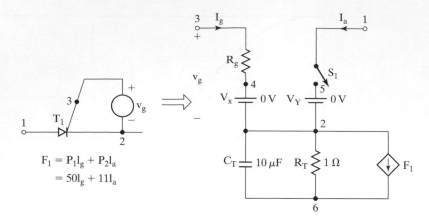

FIGURE 11.28

Ac thyristor SPICE model.

Solution

The load current of ac voltage controllers is the ac type, and the current of a thyristor is always reduced to zero. There is no need for the diode D_T in Figure 7.34b, and the thyristor model can be simplified to Figure 11.28. This model can be used as a subcircuit.

The subcircuit definition for the thyristor model silicon-controlled rectifier (SCR) can be described as follows:

```
*    Subcircuit for ac thyristor model
.SUBCKT        SCR      1            3            2
*             model    anode        +control     cathode
*             name                  voltage
S1      1     5     6     2     SMOD            ; Switch
RG      3     4     50
VX      4     2     DC        0V
VY      5     2     DC        0V
RT      2     6     1
CT      6     2     10UF
F1      2     6     POLY(2)     VX     VY     0     50     11
.MODEL      SMOD      VSWITCH (RON=0.01   ROFF=10E+5   VON=0.1V   VOFF=0V)
.ENDS  SCR                                  ; Ends subcircuit definition
```

The peak supply voltage $V_m = 169.7$ V. For $\alpha_1 = \alpha_2 = 90°$, time delay $t_1 = (90/360) \times (1000/60 \text{ Hz}) \times 1000 = 4166.7$ μs. A series snubber with $C_s = 0.1$ μF and $R_s = 750$ Ω is connected across the thyristor to cope with the transient voltage due to the inductive load. The single-phase ac voltage controller for PSpice simulation is shown in Figure 11.29a. The gate voltages V_{g1} and V_{g2} for thyristors are shown in Figure 11.29b.

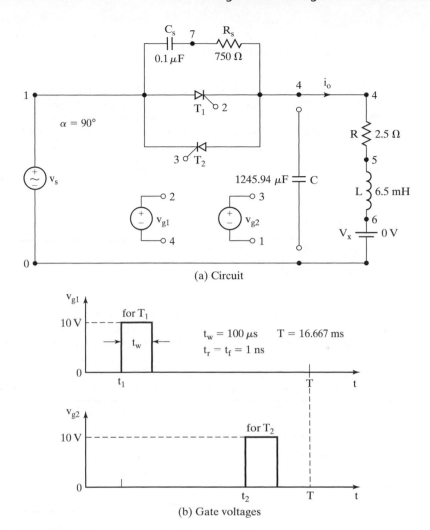

(a) Circuit

(b) Gate voltages

FIGURE 11.29

Single-phase ac voltage controller for PSpice simulation.

The list of the circuit file is as follows:

```
Example 11.12   Single-Phase AC Voltage Controller
VS    1    0    SIN   (0 169.7V    60HZ)
Vg1   2    4    PULSE (0V   10V  4166.7US 1NS  1NS  100US 16666. 7US)
Vg2   3    1    PULSE (0V   10V 12500.0US 1NS  1NS  100US 16666. 7US)
R     4    5    2.5
L     5    6    6.5MH
VX    6    0    DC .   0V    ; Voltage source to measure the load current
* C   4    0       1245.94UF  ; Output filter capacitance    ; Load filter
```

```
CS    1    7    0.1UF
RS    7    4    750
*  Subcircuit call for thyristor model
XT1   1    2    4    SCR                   ; Thyristor T1
XT2   4    3    1    SCR                   ; Thyristor T2
*  Subcircuit SCR which is missing must  be inserted
.TRAN       10US   33.33MS                 ; Transient analysis
.PROBE                                     ; Graphics postprocessor
.options abstol = 1.00n reltol = 1.0m vntol = 1.0m ITL5=10000
.FOUR  60HZ   V(4)                         ; Fourier analysis
.END
```

The PSpice plots of instantaneous output voltage V (4) and load current I (VX) are shown in Figure 11.30.

The Fourier components of the output voltage are as follows:

FOURIER COMPONENTS OF TRANSIENT RESPONSE V (4)

DC COMPONENT = 1.784608E−03

HARMONIC NO	FREQUENCY (HZ)	FOURIER COMPONENT	NORMALIZED COMPONENT	PHASE (DEG)	NORMALIZED PHASE (DEG)
1	6.000E+01	1.006E+02	1.000E+00	−1.828E+01	0.000E+00
2	1.200E+02	2.764E−03	2.748E−05	6.196E+01	8.024E+01
3	1.800E+02	6.174E+01	6.139E−01	6.960E+01	8.787E+01

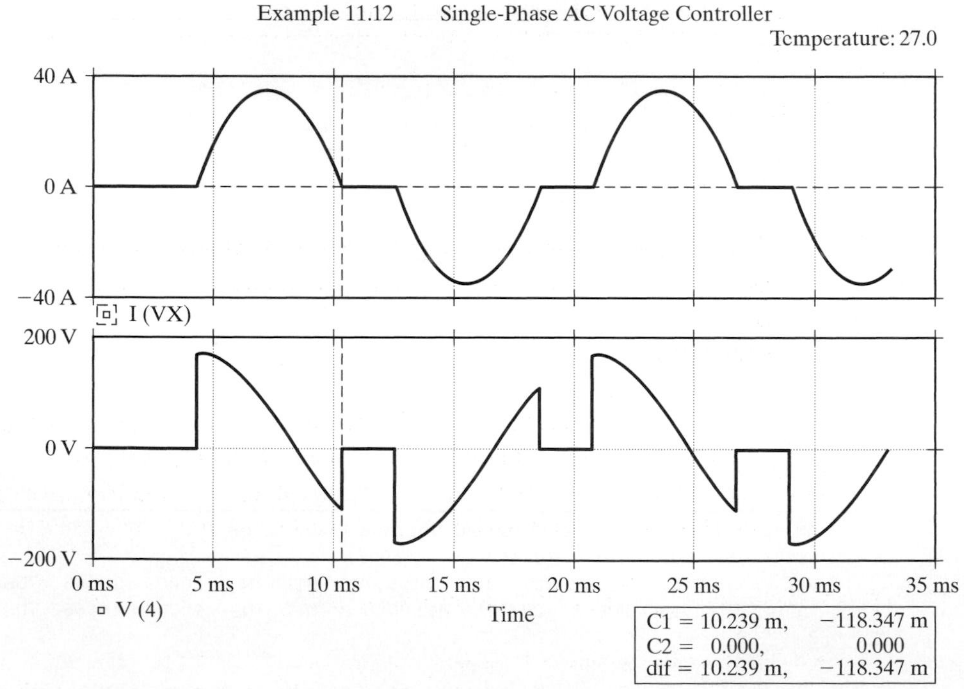
Example 11.12 Single-Phase AC Voltage Controller
Temperature: 27.0
40 A
0 A
−40 A
I (VX)
200 V
0 V
−200 V
0 ms
5 ms
10 ms
15 ms
20 ms
25 ms
30 ms
35 ms
V (4)
Time
C1 = 10.239 m, −118.347 m
C2 = 0.000, 0.000
dif = 10.239 m, −118.347 m

FIGURE 11.30

Plots for Example 11.12.

4	2.400E+02	1.038E−03	1.033E−05	6.731E+01	8.559E+01
5	3.000E+02	3.311E+01	3.293E−01	−6.771E+01	−4.943E+01
6	3.600E+02	1.969E−03	1.958E−05	1.261E+02	1.444E+02
7	4.200E+02	6.954E+00	6.915E−02	1.185E+02	1.367E+02
8	4.800E+02	3.451E−03	3.431E−05	1.017E+02	1.199E+02
9	5.400E+02	1.384E+01	1.376E−01	−1.251E+02	−1.068E+02

TOTAL HARMONIC DISTORTION =   7.134427E+01 PERCENT

The Fourier components of the output current, which is the same as the input current, are as follows:

```
FOURIER COMPONENTS OF TRANSIENT RESPONSE I (VX)
DC COMPONENT   =   -2.557837E-03
```

HARMONIC NO	FREQUENCY (HZ)	FOURIER COMPONENT	NORMALIZED COMPONENT	PHASE (DEG)	NORMALIZED PHASE (DEG)
1	6.000E+01	2.869E+01	1.000E+00	−6.253E+01	0.000E+00
2	1.200E+02	4.416E−03	1.539E−04	−1.257E+02	−6.319E+01
3	1.800E+02	7.844E+00	2.735E−01	−2.918E+00	5.961E+01
4	2.400E+02	3.641E−03	1.269E−04	−1.620E+02	−9.948E+01
5	3.000E+02	2.682E+00	9.350E−02	−1.462E+02	−8.370E+01
6	3.600E+02	2.198E−03	7.662E−05	1.653E+02	2.278E+02
7	4.200E+02	4.310E−01	1.503E−02	4.124E+01	1.038E+02
8	4.800E+02	1.019E−03	3.551E−05	1.480E+02	2.105E+02
9	5.400E+02	6.055E−01	2.111E−02	1.533E+02	2.158E+02

TOTAL HARMONIC DISTORTION =   2.901609E+01 PERCENT

Input current THD = 29.01% = 0.2901

Displacement angle $\phi_1 = -62.53°$

DF = $\cos \phi_1 = \cos(-62.53) = 0.461$ (logging)

From Eq. (10.96), the input PF

$$PF = \frac{1}{(1 + THD^2)^{1/2}} \cos \phi_1 = \frac{1}{(1 + 0.2901^2)^{1/2}} \times 0.461 = 0.443 \text{ (lagging)}$$

Key Points of Section 11.12

- The design of an ac voltage controller requires determining the device ratings and the ratings of filter components at the input and output sides.
- Filters are needed to smooth out the output voltage and the input current to reduce the amount of harmonic injection to the input supply by ac filters.

11.13   EFFECTS OF SOURCE AND LOAD INDUCTANCES

In the derivations of output voltages, we have assumed that the source has no inductance. The effect of any source inductance would be to delay the turn-off of thyristors. Thyristors would not turn off at the zero crossing of input voltage, as shown in Figure 11.31b, and gate pulses of short duration may not be suitable. The harmonic contents on the output voltage would also increase.

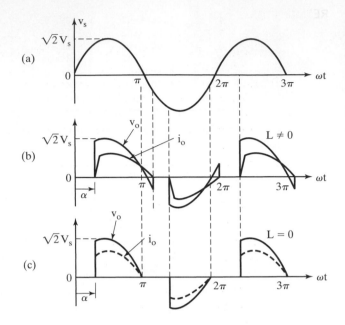

FIGURE 11.31

Effects of load inductance on load current and voltage.

We have seen in Section 11.5 that the load inductance plays a significant part on the performance of power controllers. Although the output voltage is a pulsed waveform, the load inductance tries to maintain a continuous current flow, as shown in Figures 11.6b and 11.31b. We can also notice from Eqs. (11.42) and (11.45) that the input PF of a power converter depends on the load PF. Due to the switching characteristics of thyristors, any inductance in the circuit makes the analysis more complex.

SUMMARY

The ac voltage controller can use on–off control or phase-angle control. The on–off control is more suitable for systems having a high time constant. Due to the dc component on the output of unidirectional controllers, bidirectional controllers are normally used in industrial applications. Due to the switching characteristics of thyristors, an inductive load makes the solutions of equations describing the performance of controllers more complex and an iterative method of solution is more convenient. The input PF of controllers, which varies with delay angle, is generally poor, especially at the low-output range. The ac voltage controllers can be used as transformer static connection changers.

The voltage controllers provide an output voltage at a fixed frequency. Two phase-controlled rectifiers connected as dual converters can be operated as direct-frequency changers known as *cycloconverters*. With the development of fast-switching power devices, the forced commutation of cycloconverters is possible; however, it requires synthesizing the switching functions for power devices.

REFERENCES

[1] A. K. Chattopadhyay, *Power Electronics Handbook*, edited by M. H. Rashid. San Diego, CA: Academic Press. 2001, Chapter 16—AC-AC Converters.

[2] A. Ishiguru, T. Furuhashi, and S. Okuma, "A novel control method of forced-commutated cycloconverters using instantaneous values of input line voltages," *IEEE Transactions on Industrial Electronics*, Vol. 38, No. 3, June 1991, pp. 166–172.

[3] L. Huber, D. Borojevic, and N. Burany, "Analysis, design and implementation of the space-vector modulator for forced-commutated cycloconverters," *IEE Proc B,* Vol. 139, No. 2, March 1992, pp. 103–113.

[4] K. E. Ad'doweesh, "An exact analysis of an ideal static ac chopper," *International Journal of Electronics*, Vol. 75, No. 5, 1993, pp. 999–1013.

[5] M. Venturini, "A new sine-wave in sine-wave out conversion technique eliminates reactive elements," *Proceedings Powercon 7*, 1980, pp. E3.1–3.13.

[6] A. Alesina and M. Venturini, "Analysis and design of optimum amplitude nine-switch direct ac–ac converters," *IEEE Transactions on Power Electronics*, Vol. 4, No. 1, January 1989, pp. 101–112.

[7] P. D. Ziogas, S. I. Khan, and M. Rashid, "Some improved forced commutated cycloconverter structures," *IEEE Transactions on Industry Applications*, Vol. 21, July/August 1985, pp. 1242–1253.

[8] P. D. Ziogas, S. I. Khan, and M. Rashid, "Analysis and design of forced-commutated cycloconverter structures and improved transfer characteristics," *IEEE Transactions on Industrial Electronics*, Vol. 3, No. 3, August 1986, pp. 271–280.

[9] D. G. Holmes and T. A. Lipo, "Implementation of a controlled rectifier using ac–ac matrix converter theory," *IEEE Transactions on Power Electronics*, Vol. 7, No. 1, January 1992, pp. 240–250.

[10] L. Huber and D. Borojevic, "Space vector modulated three-phase to three-phase matrix converter with input power factor correction," *IEEE Transactions on Industry Applications*, Vol. 31, November/December 1995, pp. 1234–1246.

[11] L. Zhang, C. Watthanasarn, and W. Shepherd, "Analysis and comparison of control strategies for ac–ac matrix converters," *IEE Proceedings of Electric Power Applications*, Vol. 145, No. 4, July 1998, pp. 284–294.

[12] P. Wheeler and D. Grant, "Opimised input filter design and low-loss switching techniques for a practical matrix converter," *IEE Proceedings of Electric Power Applications*, Vol. 144, No 1, January 1997, pp. 53–59.

[13] J. Mahlein, O. Simon, and M. Braun, "A matrix-converter with space-vector control enabling overmodulation," *Conference Proc. EPE'99*, Laussanne, September 1999, pp. 1–11.

REVIEW QUESTIONS

11.1 What are the advantages and disadvantages of on–off control?
11.2 What are the advantages and disadvantages of phase-angle control?
11.3 What are the effects of load inductance on the performance of ac voltage controllers?
11.4 What is the extinction angle?
11.5 What are the advantages and disadvantages of unidirectional controllers?
11.6 What are the advantages and disadvantages of bidirectional controllers?
11.7 What is a tie control arrangement?
11.8 What is a matrix converter?
11.9 What are the steps involved in determining the output voltage waveforms of three-phase bidirectional controllers?

11.10 What are the advantages and disadvantages of delta-connected controllers?

11.11 What is the control range of the delay angle for single-phase unidirectional controllers?

11.12 What is the control range of the delay angle for single-phase bidirectional controllers?

11.13 What are the advantages and disadvantages of a matrix converter?

11.14 What is the control range of the delay angle for three-phase bidirectional controllers?

11.15 What are the advantages and disadvantages of transformer connection changers?

11.16 What are the methods for output voltage control of transformer connection changers?

11.17 What is a synchronous connection changer?

11.18 What is a cycloconverter?

11.19 What are the advantages and disadvantages of cycloconverters?

11.20 What are the advantages and disadvantages of ac voltage controllers?

11.21 What is the principle of operation of cycloconverters?

11.22 What are the effects of load inductance on the performance of cycloconverters?

11.23 What are the three possible arrangements for a single-phase full-wave ac voltage controller?

11.24 What are the advantages of sinusoidal harmonic reduction techniques for cycloconverters?

11.25 What are the gate signal requirements of thyristors for voltage controllers with RL loads?

11.26 What are the effects of source and load inductances?

11.27 What are the conditions for the worst-case design of power devices for ac voltage controllers?

11.28 What are the conditions for the worst-case design of load filters for ac voltage controllers?

PROBLEMS

11.1 The ac voltage controller in Figure 11.1a is used for heating a resistive load of $R = 5\,\Omega$ and the input voltage is $V_s = 120$ V (rms), 60 Hz. The thyristor switch is on for $n = 125$ cycles and is off for $m = 75$ cycles. Determine **(a)** the rms output voltage V_o, **(b)** the input power factor (PF), and **(c)** the average and rms thyristor currents.

11.2 The ac voltage controller in Figure 11.1a uses on–off control for heating a resistive load of $R = 4\,\Omega$ and the input voltage is $V_s = 208$ V (rms), 60 Hz. If the desired output power is $P_o = 3$ kW, determine **(a)** the duty cycle k, and **(b)** the input PF.

11.3 The single-phase half-wave ac voltage controller in Figure 11.2a has a resistive load of $R = 5\,\Omega$ and the input voltage is $V_s = 120$ V (rms), 60 Hz. The delay angle of thyristor T_1 is $\alpha = \pi/3$. Determine **(a)** the rms output voltage V_o, **(b)** the input PF, and **(c)** the average input current.

11.4 The single-phase half-wave ac voltage controller in Figure 11.2a has a resistive load of $R = 5\,\Omega$ and the input voltage is $V_s = 208$ V (rms), 60 Hz. If the desired output power is $P_o = 2$ kW, calculate **(a)** the delay angle α, and **(b)** the input PF.

11.5 The single-phase full-wave ac voltage controller in Figure 11.3a has a resistive load of $R = 5\,\Omega$ and the input voltage is $V_s = 120$ V (rms), 60 Hz. The delay angles of thyristors T_1 and T_2 are equal: $\alpha_1 = \alpha_2 = \alpha = 2\pi/3$. Determine **(a)** the rms output voltage V_o, **(b)** the input PF, **(c)** the average current of thyristors I_A, and **(d)** the rms current of thyristors I_R.

11.6 The single-phase full-wave ac voltage controller in Figure 11.3a has a resistive load of $R = 1.5\,\Omega$ and the input voltage is $V_s = 120$ V (rms), 60 Hz. If the desired output power is $P_o = 7.5$ kW, determine **(a)** the delay angles of thyristors T_1 and T_2, **(b)** the rms output voltage V_o, **(c)** the input PF, **(d)** the average current of thyristors I_A, and **(e)** the rms current of thyristors I_R.

11.7 The load of an ac voltage controller is resistive, with $R = 1.5\,\Omega$. The input voltage is $V_s = 120$ V (rms), 60 Hz. Plot the PF against the delay angle for single-phase half-wave and full-wave controllers.

11.8 The single-phase full-wave controller in Figure 11.6a supplies an RL load. The input voltage is $V_s = 120$ V (rms) at 60 Hz. The load is such that $L = 5$ mH and $R = 5\,\Omega$. The delay angles of thyristor T_1 and thyristor T_2 are equal, where $\alpha = \pi/3$. Determine **(a)** the conduction angle of thyristor T_1, δ; **(b)** the rms output voltage V_o; **(c)** the rms thyristor current I_R; **(d)** the rms output current I_o; **(e)** the average current of a thyristor I_A; and **(f)** the input PF.

11.9 The single-phase full-wave controller in Figure 11.6a supplies an RL load. The input voltage is $V_s = 120$ V at 60 Hz. Plot the PF, against the delay angle, α, for **(a)** $L = 5$ mH and $R = 5 \ \Omega$, and **(b)** $R = 5 \ \Omega$ and $L = 0$.

11.10 The three-phase unidirectional controller in Figure P11.10 supplies a Y-connected resistive load with $R = 5 \ \Omega$ and the line-to-line input voltage is 208 V (rms), 60 Hz. The delay angle is $\alpha = \pi/6$. Determine **(a)** the rms output phase voltage V_o, **(b)** the input power, and **(c)** the expressions for the instantaneous output voltage of phase a.

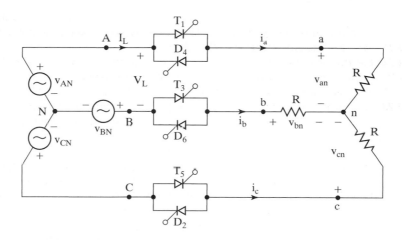

FIGURE 11.P10

Three-phase unidirectional controller.

11.11 The three-phase unidirectional controller in Figure P11.10 supplies a Y-connected resistive load with $R = 2.5 \ \Omega$ and the line-to-line input voltage is 208 V (rms), 60 Hz. If the desired output power is $P_o = 12$ kW, calculate **(a)** the delay angle α, **(b)** the rms output phase voltage V_o, and **(c)** the input PF.

11.12 The three-phase unidirectional controller in Figure P11.10 supplies a Y-connected resistive load with $R = 5 \ \Omega$ and the line-to-line input voltage is 208 V (rms), 60 Hz. The delay angle is $\alpha = 2\pi/3$. Determine **(a)** the rms output phase voltage V_o, **(b)** the input PF, and **(c)** the expressions for the instantaneous output voltage of phase a.

11.13 Repeat Problem 11.10 for the three-phase bidirectional controller in Figure 11.8.

11.14 Repeat Problem 11.11 for the three-phase bidirectional controller in Figure 11.8.

11.15 Repeat Problem 11.12 for the three-phase bidirectional controller in Figure 11.8.

11.16 The three-phase bidirectional controller in Figure 11.8 supplies a Y-connected load of $R = 5 \ \Omega$ and $L = 10$ mH. The line-to-line input voltage is 208 V, 60 Hz. The delay angle is $\alpha = \pi/2$. Plot the line current for the first cycle after the controller is switched on.

11.17 A three-phase ac voltage controller supplies a Y-connected resistive load of $R = 5 \ \Omega$ and the line-to-line input voltage is $V_s = 208$ V at 60 Hz. Plot the PF against the delay angle α for **(a)** the half-wave controller in Figure P11.10, and **(b)** the full-wave controller in Figure 11.8.

11.18 A three-phase bidirectional delta-connected controller in Figure 11.11 has a resistive load of $R = 5\ \Omega$. If the line-to-line voltage is $V_s = 208$ V, 60 Hz and the delay angle $\alpha = \pi/3$, determine **(a)** the rms output phase voltage V_o, **(b)** the expressions for instantaneous currents i_a, i_{ab}, and i_{ca}; **(c)** the rms output phase current I_{ab} and rms output line current I_a; **(d)** the input PF; and **(e)** the rms current of thyristors I_R.

11.19 The circuit in Figure 11.14 is controlled as a synchronous connection changer. The primary voltage is 208 V, 60 Hz. The secondary voltages are $V_1 = 120$ V and $V_2 = 88$ V. If the load resistance is $R = 5\ \Omega$ and the rms load voltage is 180 V, determine **(a)** the delay angles of thyristors T_1 and T_2, **(b)** the rms current of thyristors T_1 and T_2, **(c)** the rms current of thyristors T_3 and T_4, and **(d)** the input PF.

11.20 The input voltage to the single-phase/single-phase cycloconverter in Figure 11.17a is 120 V, 60 Hz. The load resistance is 2.5 Ω and load inductance is $L = 40$ mH. The frequency of output voltage is 20 Hz. If the delay angle of thyristors is $\alpha_p = 2\pi/4$, determine **(a)** the rms output voltage, **(b)** the rms current of each thyristor, and **(c)** the input PF.

11.21 Repeat Problem 11.20 if $L = 0$.

11.22 For Problem 11.20, plot the power factor against the delay angle α.

11.23 Repeat Problem 11.20 for the three-phase/single-phase cycloconverter in Figure 11.19a, $L = 0$.

11.24 Repeat Problem 11.20 if the delay angles are generated by comparing a cosine signal at source frequency with a sinusoidal reference signal at output frequency as shown in Figure 11.21.

11.25 For Problem 11.24, plot the input power factor against the delay angle.

11.26 The single-phase full-wave ac voltage controller in Figure 11.5a controls the power from a 208-V, 60-Hz ac source into a resistive load. The maximum desired output power is 10 kW. Calculate **(a)** the maximum rms current rating of thyristor, **(b)** the maximum average current rating of thyristor, and **(c)** the peak thyristor voltage.

11.27 The three-phase full-wave ac voltage controller in Figure P11.10 is used to control the power from a 2300-V, 60-Hz ac source into a delta-connected resistive load. The maximum desired output power is 100 kW. Calculate **(a)** the maximum rms current rating of thyristors I_{RM}, **(b)** the maximum average current rating of thyristors I_{AM}, and **(c)** the peak value of thyristor voltage V_p.

11.28 The single-phase full-wave controller in Figure 11.6a controls power to an RL load and the source voltage is 208 V, 60 Hz. The load is $R = 5\ \Omega$ and $L = 6.5$ mH. **(a)** Determine the rms value of third harmonic current. **(b)** If a capacitor is connected across the load, calculate the value of capacitance to reduce the third harmonic current in the load to 5% of the load current, $\alpha = \pi/3$. **(c)** Use PSpice to plot the output voltage and the load current, and to compute the total harmonic distortion (THD) of the output voltage and output current, and the input PF with and without the output filter capacitor in (b).

CHAPTER 12

Static Switches

The learning objectives of this chapter are as follows:

- To understand the operation and characteristics of static switches and microelectronic relays (MERs)
- To learn the different types of static switches and MERs

12.1 INTRODUCTION

Thyristors that can be turned on and off within a few microseconds may be operated as fast-acting switches to replace mechanical and electromechanical circuit breakers. For low-power dc applications, power transistors can also be used as switches. The static switches have many advantages (e.g., very high switching speeds, no moving parts, and no contact bounce on closing).

In addition to applications as static switches, the thyristor (or transistor) circuits can be designed to provide time-delay, latching, over- and undercurrent, and voltage detections. The transducers, for detecting mechanical, electrical, position, proximity, and so on, can generate the gating or control signals for the thyristors (or transistors).

The static switches can be classified into two types: (1) ac switches, and (2) dc switches. The ac switches can be subdivided into (a) single phase, and (b) three phase. For ac switches, the thyristors are line or natural commutated and the switching speed is limited by the frequency of the ac supply and the turn-off time of thyristors. The dc switches are forced commutated and the switching speed depends on the turn-on time and the turn-off time of the devices.

12.2 SINGLE-PHASE AC SWITCHES

The circuit diagram of a single-phase full-wave switch is shown in Figure 12.1a, where the two thyristors are connected in inverse parallel. Thyristor T_1 is fired at $\omega t = 0$ and thyristor T_2 is fired at $\omega t = \pi$. The output voltage is the same as the input voltage. The thyristors act like switches and are line commutated. The waveforms for the input voltage, output voltage, and output current are shown in Figure 12.1b.

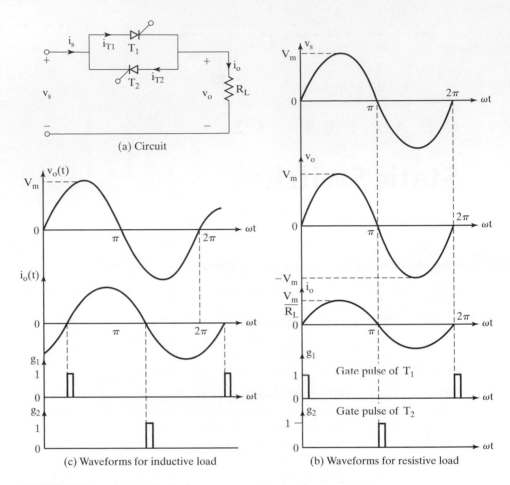

FIGURE 12.1

Single-phase thyristor ac switch.

With an inductive load, thyristor T_1 should be fired when the current passes through the zero crossing after the positive half-cycle of input voltage and thyristor T_2 should be fired when the current passes through the zero crossing after the negative half-cycle of input voltage. The triggering pulses for T_1 and T_2 are shown in Figure 12.1c. A TRIAC may be used instead of two thyristors, as shown in Figure 12.2.

If the instantaneous line current is $i_s(t) = I_m \sin \omega t$, the rms line current is

$$I_s = \left[\frac{2}{2\pi} \int_0^\pi I_m^2 \sin^2 \omega t \; d(\omega t) \right]^{1/2} = \frac{I_m}{\sqrt{2}} \qquad (12.1)$$

Because each thyristor carries current for only one-half cycle, the average current through each thyristor is

$$I_A = \frac{1}{2\pi} \int_0^\pi I_m \sin \omega t \; d(\omega t) = \frac{I_m}{\pi} \qquad (12.2)$$

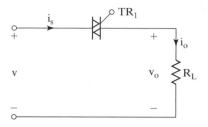

FIGURE 12.2

Single-phase TRIAC ac switch.

and the root-mean-square (rms) current of each thyristor is

$$I_R = \left[\frac{1}{2\pi} \int_0^\pi I_m^2 \sin^2 \omega t \, d(\omega t) \right]^{1/2} = \frac{I_m}{2} \tag{12.3}$$

The circuit in Figure 12.1a can be modified, as shown in Figure 12.3a, where the two thyristors have a common cathode and the gating signals have a common terminal. Thyristor T_1 and diode D_1 conduct during the positive half-cycle and thyristor T_2 and diode D_2 conduct during the negative half-cycle.

A diode bridge rectifier and a thyristor T_1, as shown in Figure 12.4a, can perform the same function as that of Figure 12.1a. The current through the load is ac and that through thyristor T_1 is dc. A transistor can replace thyristor T_1. The unit comprising the transistor (or thyristor or gate-turn-off thyristor [GTO]) and bridge rectifier is known as a *bidirectional switch*.

Key Points of Section 12.2

- The switches are turned at the zero crossing of the input voltage.
- The circuit operation is similar to the single-phase ac voltage controller with a delay angle $\alpha = 0$.

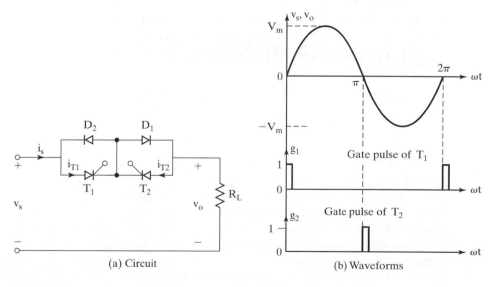

(a) Circuit

(b) Waveforms

FIGURE 12.3

Single-phase bridge diode and thyristor ac switch.

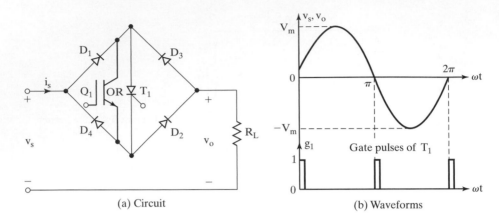

(a) Circuit   (b) Waveforms

FIGURE 12.4

Single-phase bridge rectifier and thyristor ac switch.

12.3   THREE-PHASE AC SWITCHES

The concept of single-phase ac switching can be extended to three-phase applications. Three single-phase switches in Figure 12.1a can be connected to form a three-phase switch, as shown in Figure 12.5a. The gating signals for thyristors and the current through T_1 are shown in Figure 12.5b. The load can be connected in either Y or delta.

To reduce the number of thyristors and costs, a diode and a thyristor can also be used to form a three-phase switch, as shown in Figure 12.6. In case of two thristors connected in back to back, there is the possibility of stopping the current flow in every half-cycle (e.g., 8.33 ms for a 60-Hz supply). However, with a diode and a thyristor, the current flow can only be stopped in every cycle of input voltage and the reaction time becomes slow (e.g., 16.67 ms for a 60-Hz supply).

Key Point of Section 12.3

- The circuit operation is similar to the three-phase ac voltage controller with a delay angle $\alpha = 0$.

12.4   THREE-PHASE REVERSING SWITCHES

The reversal of three-phase power supplied to a load can be achieved by adding two more single-phase switches to the three-phase switch in Figure 12.5a. This is shown in Figure 12.7. Under a normal operation, thyristors T_7 through T_{10} are turned off by gate pulse inhibiting (or suppression) and thyristors T_1 through T_6 are turned on. Line A feeds terminal a, line B feeds terminal b, and line C feeds terminal c. Under the phase-reversing operation, thyristors T_2, T_3, T_5, and T_6 are turned off by gate pulse inhibiting and thyristors T_7 through T_{10} are operative. Line B feeds terminal c and line C feeds terminal b, resulting in a phase reversal of the voltage applied to the load. To obtain phase reversal, all the devices must be thyristors. A combination of thyristors and

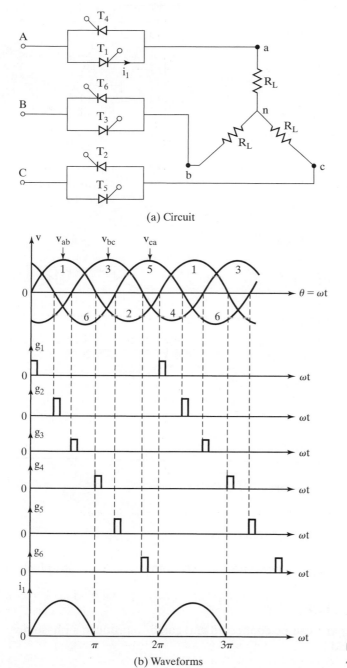

(a) Circuit

(b) Waveforms

FIGURE 12.5

Three-phase thyristor ac switch.

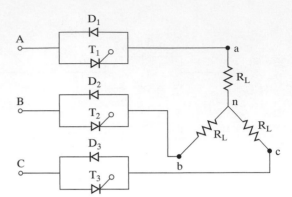

FIGURE 12.6

Three-phase diode and thyristor ac switch.

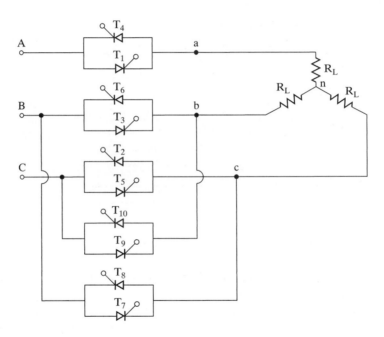

FIGURE 12.7

Three-phase reversing thyristor ac switch.

diodes, as shown in Figure 12.6, cannot be used; otherwise, phase-to-phase short circuits would occur.

Key Points of Section 12.4

- Five-thyristor pairs can be connected and gated to produce a phase reversal to a
  three-phase load.

12.5   AC SWITCHES FOR BUS TRANSFER

The static switches can be used for bus transfer from one source to another. In a practical supply system, it is sometimes required to switch the load from the normal source

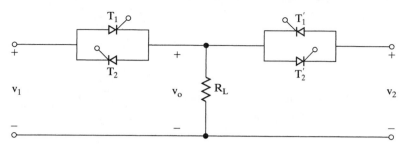

FIGURE 12.8

Single-phase bus transfer.

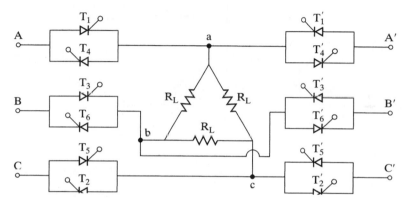

FIGURE 12.9

Three-phase bus transfer.

to an alternative source in case of (1) unavailability of normal source, and (2) under-voltage or overvoltage condition of normal source. Figure 12.8 shows a single-phase bus transfer switch. When thyristors T_1 and T_2 are operative, the load is connected to the normal source; and for transfer to an alternative source, thyristors T_1' and T_2' are operative, while T_1 and T_2 are turned off by gate signal inhibiting. The extension of single-phase bus transfer to three-phase transfer is shown in Figure 12.9.

Key Points of Section 12.5

- Two or more thyristor pairs can be connected in back-to-back for bus transfer from one supply source to another. This type of transfer is used in the case of switching the load to an alternate supply source(s).

12.6 DC SWITCHES

In the case of dc switches, the input voltage is dc and power transistors or fast-switching thyristors or GTOs can be used. A single-pole transistor switch is shown in Figure 12.10, with a resistive load; and in case of an inductive load, a diode (as shown by dashed lines)

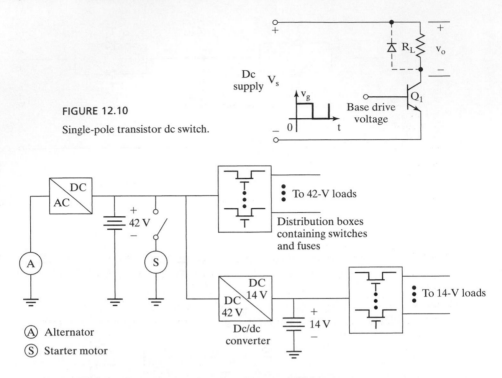

FIGURE 12.10

Single-pole transistor dc switch.

FIGURE 12.11

Automotive powering systems [Ref. 6] with 42 V.

must be connected across the load to protect the transistor from the transient voltage during switch-off. The single-pole switches can also be applied to bus transfer from one source to another.

The power demand for automotives is continuously increasing [3]. Reasons for this include the emergence of new features such as heated windshields and preheating of catalytic converters as well as financial reasons such as reducing gasoline consumption and cutting emissions. Power electronics find potential applications for power conversion as well as static switches in the automotive industry. Figure 12.11 shows the block diagram of a 42-V automotive powering system with outlets for both 42 and 14 V. The dc–dc converter steps down from 42 to 14 V and the metal oxide semiconductor field-effect transistor (MOSFET) switches connect the loads to either 42- or 14-V supply. Due to their inherent advantages as a voltage-controlled device with a negligible gate current, power MOSFETs are increasingly used in applications such as static switches. Figure 12.12 shows an *N*-channel MOSFET switch. The introduction of the additional 42-V electrical systems in the automotive industry opens opportunities for the use of power semiconductors. The electromechanical systems are replacing the mechanical ones, for example, in mechanical water pumps and power steering.

Smart power switches. As microelectronics take on more and more applications, especially in automation and automotive technology, higher performance is

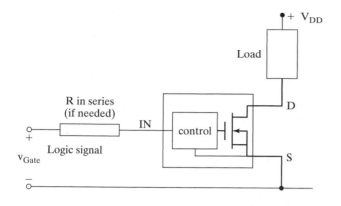

FIGURE 12.12

N-channel power MOSFET switch.

demanded from power switches. In many cases, these requirements can no longer be satisfied with conventional electromechanical solutions. These include an intelligent driver integrated circuit (IC) for power transistors as well as a range of smart power switches. As a rule, high-side switches are required for loads connected to ground. Figure 12.13 shows the block diagram of a smart power switch including all internal features [4].

Thyristor switches. Thyristor switches are most suited for high-voltage and high-current applications beyond the range of transistor ratings. Once a thyristor is turned on, it must be turned off by forced commutation. GTOs, however, can be turned off by gate control.

If forced commutated thyristors are used, the commutation circuit is an integral part of the switch and a dc switch for high-power applications is shown in Figure 12.14. If thyristor T_3 is fired, the capacitor C is charged through the supply V_s, L, and T_3. The charging current i and the capacitor voltage v_c can be expressed as

$$i(t) = V_s \sqrt{\frac{C}{L}} \sin \omega t \qquad (12.4)$$

and

$$v_c(t) = V_s(1 - \cos \omega t) \qquad (12.5)$$

where $\omega = 1/\sqrt{LC}$. After time $t = t_0 = \pi\sqrt{LC}$, the charging current becomes zero and the capacitor is charged to $2V_s$. If thyristor T_1 is conducting and supplies power to the load, thyristor T_2 is fired to switch T_1 off. T_3 is self-commutated. Firing T_2 causes a resonant pulse of current of the capacitor C through capacitor C, inductor L, and thyristor T_2. As the resonant current increases, the current through thyristor T_1 reduces. When the resonant current rises to the load current I_L, the current of thyristor T_1 falls to zero and thyristor T_1 is turned off. The capacitor discharges its remaining charge through the load resistance R_L. T_2 is self-commutated. A freewheeling diode, D_m, across the load is necessary for an inductive load. The capacitor must be discharged completely in every switching action; and a negative voltage on the capacitor can be

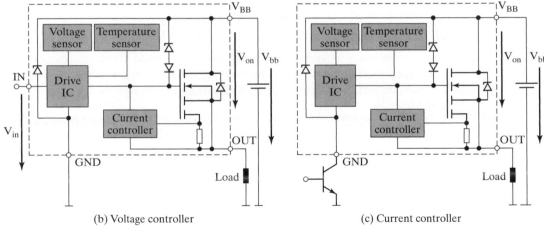

FIGURE 12.13

Block diagram of a single-channel switch [Ref. 4]. (Courtesy of Siemens Semiconductor Group, Germany.)

prevented by connecting a resistor and a diode, as shown in Figure 12.14 by dashed lines. It is not very easy to turn-off dc circuits, and dc static switches require additional circuits for turning off.

Dc switches can be applied for control of power flow in very high-voltage and high-current applications (e.g., fusion reactor) [1] and these can also be used as a fast-acting current breaker [2]. Instead of transistors, GTOs can be used. A GTO is turned on by the application of a short positive pulse to its gate similar to normal thyristors; however, a GTO can be turned off by applying a short negative pulse to its gate and it does not require any commutation circuitry. A single-pole GTO switch is shown in Figure 12.15.

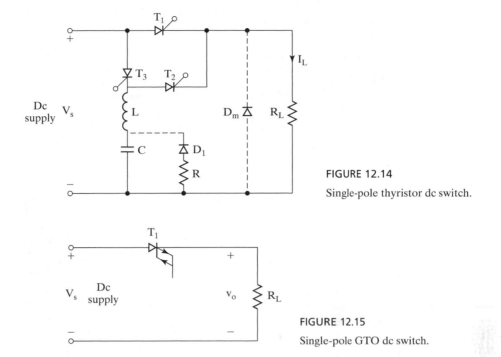

FIGURE 12.14

Single-pole thyristor dc switch.

FIGURE 12.15

Single-pole GTO dc switch.

Key Points of Section 12.6

- The power semiconductor devices such as BJTs, IGBTs, MOSFETs, GTOs, and thyristors can be operated at dc switches and can also replace electromechanical switches.

12.7 SOLID-STATE RELAYS

Static switches can be used as solid-state relays (SSRs), which are used for the control of ac and dc power. SSRs are used for many applications in industrial control (e.g., control of motor loads, transformers, resistance heating) to replace electromechanical relays. For ac applications, thyristors or TRIACs can be used; and for dc applications, transistors are used. The SSRs are normally isolated electrically between the control circuit and the load circuit by reed relay, transformer, or optocoupler.

Figure 12.16 shows two basic circuits for dc SSRs, one with a reed relay isolation and the other with an optocoupler. Although the single-phase circuit in Figure 12.1a can be operated as an SSR, the circuit in Figure 12.2 with a TRIAC is normally used for ac power, because of the requirement of only one gating circuit for a TRIAC. Figure 12.17 shows SSRs with a reed relay, a transformer isolation, and an optocoupler. If the application requirements demand thyristors for high-power levels, the circuit in Figure 12.1a can also be used to operate as an SSR even though the complexity of the gating circuit would increase.

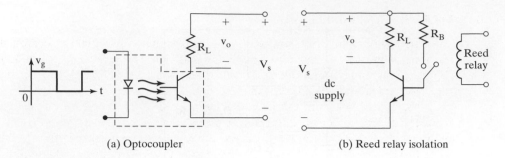

(a) Optocoupler　　　　　　　　　(b) Reed relay isolation

FIGURE 12.16

Dc solid-state relays.

Key Points of Section 12.7

- Solid-state relays (SSRs) are used for many applications in industrial control to replace electromechanical relays. They are normally isolated electrically between the control circuit and the load.

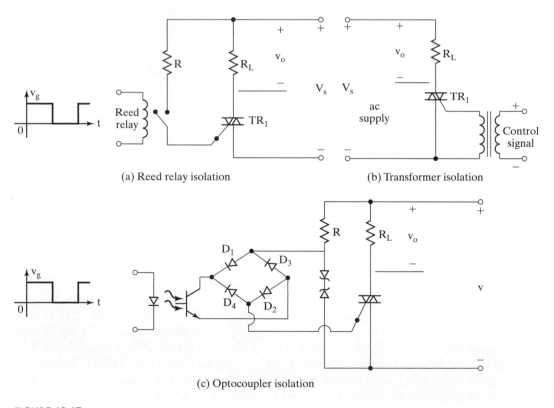

(a) Reed relay isolation　　　　　　　　　(b) Transformer isolation

(c) Optocoupler isolation

FIGURE 12.17

Ac solid-state relays.

12.8   MICROELECTRONIC RELAYS

The MERs are normally open, solid-state replacements for electromechanical relays for general-purpose switching of analog signals [5]. There are two types: (1) photovoltaic relays, and (2) photovoltaic isolators. They use power MOSFETs or insulated-gate bipolar transistors (IGBTs) on input side.

An economically affordable and sufficiently compact photovoltaic generator (PVG) produces only a weak output. A practical PVG can generate several volts into an open circuit load, but only microamperes of output currents. The output of the PVG is used to drive a MOSFET. A modern power MOSFET requires only several volts of signal for full conduction, but it requires essentially zero steady-state current. Only transient energy to charge the gate capacitance is required to turn on and then hold the MOSFET in conduction. A charging current of only a few microamperes can turn on a typical MOSFET in a small fraction of a millisecond, a fast response relative to electromechanical switching times. Therefore, the photo voltage generator is used in conjunction with a metal oxide semiconductor (MOS) gate. Thyristors and bipolar transistors both require too much drive current and are not normally used in MERs.

12.8.1   Photovoltaic Relay

Figure 12.18a shows the schematic of a photovoltaic relay (PVR). The light-emitting diode (LED) that provides the electrooptical isolation energizes a PVG consisting of a

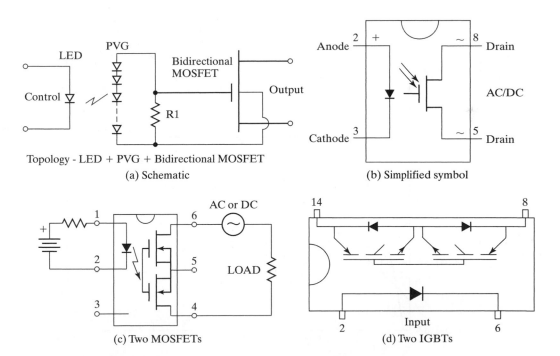

FIGURE 12.18

Photovoltaic relay [5].

series connection of silicon *pn*-junctions. The signal from the PVG, in turn, activates a bidirectional MOSFET device. The symbol is shown in Figure 12.18b.

An LED can control more than one device, as shown in Figure 12.18c, with two MOSFETs and in Figure 12.18d with IGBTs. The operating parameters of photo-voltaic relays are ideal for switching low-level signal loads in instrumentation and data acquisition to medium power loads in industrial controls and process automation (i.e., from microvolts to 400 V (ac peak or dc) and microamperes up to 4.5 A of load current at a contact resistance as low as 40 mΩ. The PVRs overcome the limitations of both conventional electromechanical and reed relays by offering the solid-state advantages of long life, fast-operating speed, low pick up power, bounce-free operation, noninductive input, insensitivity to position and magnetic fields, extreme shock and vibration resistance, low thermal offset voltages, and miniature package for applications (such as process control, multiplexing, automatic test equipment, and data acquisition).

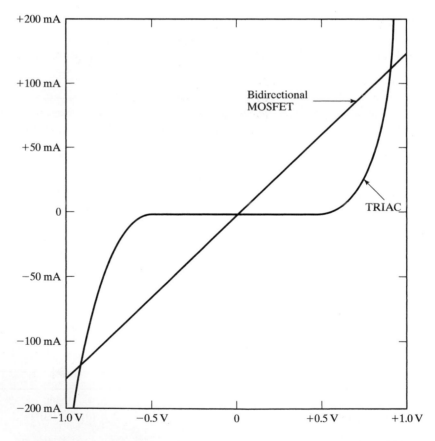

FIGURE 12.19

Output characteristics of bidirectional MOSFET and TRIAC [Ref. 5].

MOSFET versus thyristor or bipolar junction transistor (BJT) output. The MOS technology provides a much better analog of an ideal electromechanical switch than does thyristor or bipolar transistor technology that is used dominantly as the output contacts for SSRs. Compared to thyristors, the MOSFET displays a linear on-state resistance instead of a 0.6-V threshold in forward conduction, as shown in Figure 12.19. An inverse series connection of two MOSFETs, as shown in Figures 12.18c and 12.18d, can switch dc or ac at frequencies well into the radio frequency (RF) range. Compared to bipolar transistors, MOSFETs display lower on-state voltages, much lower off-state leakage currents, and most importantly have essentially infinite static forward current gain (i.e., MOSFETs are voltage-controlled devices).

MOSFET devices have limitations in high-power ac control applications where operational characteristics of thyristors are suitable. Thus, PVRs replace thyristor output SSRs in applications requiring fast switching of signals from microvolts to several hundred volts of either dc polarity or ac through the radio frequency range. These applications have commonly been served by reed-capsule relays. PVRs can provide a functional equivalent of the reed relay and have the advantages of solid-state implementation.

12.8.2 Photovoltaic Isolators

A photovoltaic isolator (PI), as shown in Figure 12.20, uses a series connection of photodiodes as a photoreceptor to form an isolated PVG. This type of isolator actually transforms energy across the isolation barrier and creates an isolated voltage source. PIs utilize monolithic integrated circuit PVGs as their outputs. These outputs are controlled by radiation from an LED, which is optically isolated from the PVG. They are ideally suited for applications requiring high-current or high-voltage switching with optical isolation between the low-level driving circuitry and high-energy or high-voltage load circuits. These devices can be used for directly driving the gates of discrete power MOSFETs or IGBTs, giving designers the flexibility of creating their own, custom-made solid-state relays capable of controlling loads well over 1000 V and 100 A.

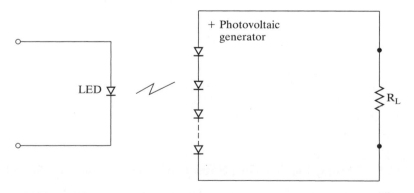

FIGURE 12.20

Photoisolator.

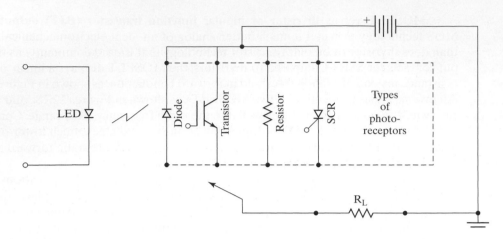

FIGURE 12.21

Photoisolated device [5].

PIs versus photocouplers. Photoisolated bipolar transistor (in Figure 12.17b) and thyristor-type SSRs use photoconductive-type isolators, as shown in Figure 12.21. These photo-isolators (also known as the photocouplers) receive optical radiation from a dielectrically isolated LED. This radiation modulates the conductivity of the photoreceptor that can be a resistor (cadmium sulfide cell), a diode, a transistor, a Darlington transistor, or a thyristor. The increased conductivity allows a current to flow through the load resistance R_L.

Key Points of Section 12.8

- MERs are normally open, solid-state replacement for electromechanical relays for general purpose switching of small analog signals. There are two types: (a) photovoltaic relays, and (b) photovoltaic isolators. They are consisted of MOSFETs or IGBTs.

12.9 DESIGN OF STATIC SWITCHES

The solid-state switches are available commercially with limited voltage and current ratings ranging up to 440 V and from 1 to 50 A. If it is necessary to design SSRs to meet specific requirements, the design is simple and requires determining the voltage and current ratings of power semiconductor devices. The design procedures can be illustrated by examples.

Example 12.1 Finding the Thyristor Ratings of a Single-Phase Ac Switch

A single-phase ac switch with configuration in Figure 12.1a is used between a 120-V, 60-Hz supply and an inductive load. The load power is 5 kW at a power factor (PF) of 0.88 lagging. Determine (a) the voltage and current ratings of thyristors, and (b) the firing angles of thyristors.

Solution

$P_o = 5000$ W, PF $= 0.88$, and $V_s = 120$ V.

    a. The peak load current $I_m = \sqrt{2} \times 5000/(120 \times 0.88) = 66.96$ A. From Eq. (12.2) the average current $I_A = 66.96/\pi = 21.31$ A and from Eq. (12.3) the rms current is $I_R = 66.96/2 = 33.48$ A. The peak inverse voltage (PIV) $= \sqrt{2} \times 120 = 169.7$ V.

    b. $\cos \theta = 0.88$ or $\theta = 28.36°$. Thus, the firing angle of T_1 is $\alpha_1 = 28.36°$ and for thyristor T_2, $\alpha_2 = 180° + 28.36° = 208.36°$.

Example 12.2   Finding the Thyristor Ratings of a Three-Phase Ac Switch

A three-phase ac switch with configuration in Figure 12.5a is used between a three-phase 440-V, 60-Hz supply and a three-phase Y-connected load. The load power is 20 kW at a PF of 0.707 lagging. Determine the voltage and current ratings of thyristors.

Solution

$P_o = 20,000$ W, PF $= 0.707$, $V_L = 440$ V, and $V_s = 440/\sqrt{3} = 254.03$ V. The line current is calculated from the power as

$$I_s = \frac{20,000}{\sqrt{3} \times 440 \times 0.707} = 37.119 \text{ A}$$

The peak current of a thyristor $I_m = \sqrt{2} \times 37.119 = 52.494$ A. The average current of a thyristor $I_A = 52.494/\pi = 16.71$ A. The rms current of a thyristor $I_R = 52.494/2 = 26.247$ A. The PIV of a thyristor $= \sqrt{2} \times 440 = 622.3$ V.

Key Points of Section 12.9

- The design of a static switch requires calculating the voltage and current ratings of the power devices.

SUMMARY

Solid-state ac and dc switches have a number of advantages over conventional electromechanical switches and relays. With the developments of power semiconductor devices and integrated circuits, static switches are used in a wide range of applications in industrial control. Static switches can be interfaced with digital or computer control systems. MERs that use a PVG are used for applications for switching low-level signal loads in instrumentation and data acquisition to medium power loads in industrial controls and process automation.

REFERENCES

[1]   W. F. Praeg, "Detailed design of a 13-kA, 13-kV DC solid state turn-off switch," *IEEE Industry Applications Conference Record*, 1985, pp. 1221–1226.

[2]   P. F. Dawson, L. E. Lansing, and S. B. Dewan, "A fast dc current breaker," *IEEE Transactions on Industry Applications*, Vol. IA21, No. 5, 1985, pp. 1176–1181.

[3]   A. Graf, D. Vogel, J. Gantioler, and F. Klotz, *Intelligent Power Semiconductors for Future Automotive Electrical Systems*, 17th Meeting "Elektronik im Kraftfahrzeug," Munich, June 3–4, 1997, pp. 1–14. www.infineon.com

[4]   PROFET: Functional Description and Applications Notes, Siemens Semiconductor Group, Munich, Germany, March 4, 1997. www.infineon.com

[5]   *Micro Electronic Relay: Designer's Manual*, International Rectifier, El Segundo, CA, 2000. www.irf.com

[6]   J. G. Kassakian, J. H. Miller and N. Traub, *Automotive electronics power up. The IEEE Spectrum*, May 2000, pp. 34–39.

REVIEW QUESTIONS

12.1   What is a static switch?

12.2   What are the differences between ac and dc switches?

12.3   What are the advantages of static switches over mechanical or electromechanical switches?

12.4   What are the advantages and disadvantages of inverse-parallel thyristor ac switches?

12.5   What are the advantages and disadvantages of TRIAC ac switches?

12.6   What are the advantages and disadvantages of diode and thyristor ac switches?

12.7   What are the advantages and disadvantages of bridge rectifier and thyristor ac switches?

12.8   What are the effects of load inductance on the gating requirements of ac switches?

12.9   What is the principle of operation of SSRs?

12.10   What are the methods of isolating the control circuit from the load circuit of SSRs?

12.11   What are the factors involved in the design of dc switches?

12.12   What are the factors involved in the design of ac switches?

12.13   What type of commutation is required for dc switches?

12.14   What type of commutation is required for ac switches?

12.15   What is a microelectronic relay?

12.16   What are the advantages and limitations of a microelectronic relay?

12.17   What is a photovoltaic relay?

12.18   What is a photovoltaic isolator?

PROBLEMS

12.1   A single-phase ac switch with configuration in Figure 12.1a is used between a 120-V, 60-Hz supply and an inductive load. The load power is 15 kW at a power factor of 0.90 lagging. Determine the voltage and current ratings of thyristors.

12.2   Determine the firing angles of thyristors T_1 and T_2 in Problem 12.1.

12.3   A single-phase ac switch with configuration in Figure 12.3a is used between a 120-V, 60-Hz supply and an inductive load. The load power is 15 kW at a power factor of 0.90 lagging. Determine the voltage and current ratings of diodes and thyristors.

12.4   A single-phase ac switch with configuration in Figure 12.4a is used between a 120-V, 60-Hz supply and an inductive load. The load power is 15 kW at a power factor of 0.90 lagging. Determine the voltage and current ratings of the thyristor and diodes in the bridge rectifier.

12.5   Determine the firing angles of thyristor T_1 in Problem 12.4.

12.6   A three-phase ac switch with configuration in Figure 12.5a is used between a three-phase 440-V, 60-Hz supply and a three-phase Y-connected load. The load power is 20 kW at a power factor of 0.86 lagging. Determine the voltage and current ratings of thyristors.

12.7   Determine the firing angles of thyristors in Problem 12.6.

12.8 Repeat Problem 12.6 for a delta-connected load.

12.9 A three-phase ac switch with configuration in Figure 12.6 has a three-phase 440-V, 60-Hz supply and a three-phase Y-connected load. The load power is 20 kW at a power factor of 0.86 lagging. Determine the voltage and current ratings of diodes and thyristors.

12.10 The thyristor dc switch in Figure 12.14 has a load resistance $R_L = 5\ \Omega$, dc supply voltage $V_s = 220$ V, inductance $L = 40\ \mu$H, and capacitance $C = 40\ \mu$F. Determine **(a)** the peak current through thyristor T_3, and **(b)** the time required to reduce the current of thyristor T_1 from the steady-state value to 1.0 A.

12.11 For Problem 12.10, determine the time required for the capacitor to discharge from $2V_s$ to zero after the firing of thyristor T_2.

12.12 The thyristor dc switch in Figure 12.14 has a load resistance $R_L = 2\ \Omega$, supply voltage $V_s = 220$ V, inductance $L = 40\ \mu$H, and capacitance $C = 80\ \mu$F. If the switch is operated at a frequency of 60 Hz, determine **(a)** the peak, rms, and average currents of thyristors T_1, T_2, and T_3; and **(b)** the rms current rating of capacitor C.

12.13 For Problem 12.12, determine the time required for the capacitor to discharge from $2V_s$ to 1.0 A after the firing of thyristor T_2.

C H A P T E R   1 3

Flexible Ac Transmission Systems

The learning objectives of this chapter are as follows:

- To study the types of compensation techniques for transmission lines and explain their operation and characteristics
- To learn techniques for implementing the compensation by switching power electronics for controlling power flow
- To know the advantages and disadvantages of a particular compensator for a particular application and to estimate its component values

13.1   INTRODUCTION

The operation of an ac power transmission line is generally constrained by limitations of one or more network parameters (such as line impedance) and operating variables (such as voltages and currents). As a result, the power line is unable to direct power flow among generating stations. Therefore, other parallel transmission lines that have an adequate capability of carrying additional amounts of power may not be able to supply the power demand. *Flexible ac transmission systems* (FACTS) is a new emerging technology and its principal role is to enhance controllability and power transfer capability in ac systems. FACTS technology uses switching power electronics to control power flow in the range of a few tens to a few hundreds of megawatts.

FACTS devices that have an integrated control function are known as the *FACT controllers*. These may consist of thyristor devices with only gate turn-on and no gate turn-off, or with power devices with gate turn-off capability. FACTS controllers are capable of controlling the interrelated line parameters and other operating variables that govern the operation of transmission systems including series impedance, shunt impedance, current, voltage, phase angle, and damping of oscillations at various frequencies below the rated frequency. By providing added flexibility, FACTS controllers can enable a transmission line to carry power closer to its thermal rating.

FACTS technology opens up new opportunities for controlling power and enhancing the usable capacity of present, new, and upgraded lines. The possibility that current through a line can be controlled at a reasonable cost creates a large potential of increasing the capacity of existing lines with larger conductors and with the use of one of the FACTS controllers to enable corresponding power to flow through such lines under normal and contingency conditions.

The philosophy of FACTS is to use power electronics for controlling power flow in a transmission network, thereby allowing the transmission line to be loaded to its full capability. Power electronic controlled devices, such as static volt–ampere reactive (VAR) compensators, have been used in transmission networks for many years. However, Dr. N. Hingorani [1] introduced the concept of FACTS as a total network control philosophy.

13.2 PRINCIPLE OF POWER TRANSMISSION

To model the operation, a transmission line can be represented by a series reactance and with the sending and receiving end voltages. This is shown in Figure 13.1a for one phase of a three-phase system. Therefore, all quantities such as voltages and currents are defined per phase. V_s and V_r are the per-phase sending-end voltage and receiving-end voltage, respectively. They represent Thevenin equivalents with respect to the midpoint. The equivalent impedance $(jX/2)$ of each Thevenin equivalent represents the

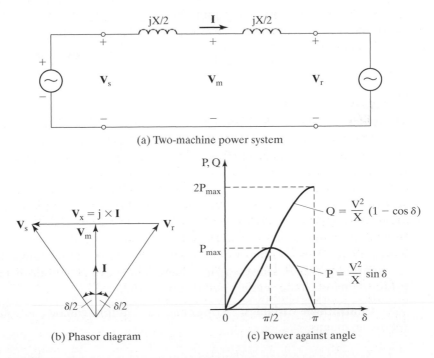

(a) Two-machine power system

(b) Phasor diagram

(c) Power against angle

FIGURE 13.1

Power flow in a transmission line [Ref. 3].

"short-circuit impedance" located on the right or left side of that midpoint. As shown in the phasor diagram in Figure 13.1b, δ is the phase angle between them.

For the sake of simplicity, let us assume that the magnitudes of the terminal voltages remain constant, equal to V. That is, $V_s = V_r = V_m = V$. The two terminal voltages can be expressed in phasor notations in rectangular coordinates as follows:

$$\mathbf{V}_s = Ve^{j\delta/2} = V\left(\cos\frac{\delta}{2} + j\sin\frac{\delta}{2}\right) \tag{13.1}$$

$$\mathbf{V}_r = Ve^{-j\delta/2} = V\left(\cos\frac{\delta}{2} - j\sin\frac{\delta}{2}\right) \tag{13.2}$$

where δ is the angle between $\mathbf{V}_s$ and $\mathbf{V}_r$. Thus, the midpoint phasor voltage $\mathbf{V}_m$ is the average value of $\mathbf{V}_s$ and $\mathbf{V}_r$ as given by

$$\mathbf{V}_m = \frac{\mathbf{V}_s + \mathbf{V}_r}{2} = V_m e^{j0} = V\cos\frac{\delta}{2}\angle 0° \tag{13.3}$$

The line current phasor is given by

$$\mathbf{I} = \frac{\mathbf{V}_s - \mathbf{V}_r}{X} = \frac{2V}{X}\sin\frac{\delta}{2}\angle 90° \tag{13.4}$$

where the magnitude of $|\mathbf{I}|$ is $I = 2V/X \sin\delta/2$. For a lossless line, the power is the same at both ends and at the midpoint. Thus, we get the active (real) power P as given by

$$P = |\mathbf{V}_m||\mathbf{I}| = \left(V\cos\frac{\delta}{2}\right)\times\left(\frac{2V}{X}\sin\frac{\delta}{2}\right) = \frac{V^2}{X}\sin\delta \tag{13.5}$$

The reactive power at the receiving-end Q_r is equal and opposite of the reactive power Q_s supplied by the sources. Thus, the reactive power Q for the line is given by

$$Q = Q_s = -Q_r = V|\mathbf{I}|\sin\frac{\delta}{2} = V\times\left(\frac{2V}{X}\sin\frac{\delta}{2}\right)\times\sin\frac{\delta}{2} = \frac{V^2}{X}\left(1-\cos\delta\right) \tag{13.6}$$

Active power P in Eq. (13.5) becomes the maximum $P_{max} = V^2/X$ at $\delta = 90°$, and reactive power Q in Eq. (13.6) becomes the maximum $Q_{max} = 2V^2/X$ at $\delta = 180°$. The plots of the active power P and the reactive power Q against the angle δ are shown in Figure 13.1c. For a constant value of line reactance X, varying angle δ can control the transmitted power P. However, any changes in active power also change the reactive power demand on the sending and receiving ends.

Controllable variables.   Power and current flow can be controlled by one of the following means:

1. Applying a voltage in the midpoint can also increase or decrease the magnitude of power.

2. Applying a voltage in series with the line, and in phase quadrature with the current flow, can increase or decrease the magnitude of current flow. Because the current flow lags the voltage by 90°, there is injection of reactive power in series.

3. If a voltage with a variable magnitude and a phase is applied in series, then varying the amplitude and phase angle can control both the active and reactive current flows. This requires injection of both active power and reactive power in series.

4. Increasing and decreasing the value of the reactance X cause a decrease and an increase of the power height of the curves, respectively, as shown in Figure 13.1c. For a given power flow, varying X correspondingly varies the angle δ between the terminal voltages.

5. Power flow can also be controlled by regulating the magnitude of sending and receiving end voltages V_s and V_r. This type of control has much more influence over the reactive power flow than the active power flow.

Therefore, we can conclude that the power flow in a transmission line can be controlled by (1) applying a shunt voltage V_m at the midpoint, (2) varying the reactance X, and (3) applying a voltage with a variable magnitude in series with line.

Key Point of Section 13.2

- Varying the line impedance X, the angle δ, and the voltage difference can control the power flow in a transmission line.

13.3 PRINCIPLE OF SHUNT COMPENSATION

The ultimate objective of applying shunt compensation in a transmission system is to supply reactive power to increase the transmittable power and to make it more compatible with the prevailing load demand. Thus, the shunt compensator should be able to minimize the line overvoltage under light load conditions, and maintain voltage levels under heavy load conditions. An ideal shunt compensator is connected at the midpoint of the transmission line, as shown in Figure 13.2a. The compensator voltage that is in phase with the midpoint voltage $\mathbf{V}_m$ has amplitude of V identical to that of the sending- and receiving-end voltages. That is, $V_m = V_s = V_r = V$. The midpoint compensator in effect segments the transmission line into two independent parts: (1) the first segment, with an impedance of $jX/2$, carries power from the sending-end to the midpoint, and (2) the second segment, also with an impedance of $jX/2$, carries power from the midpoint to the receiving end.

An ideal compensator is lossless. That is, the active power is the same at the sending end, midpoint, and receiving end. Using the phasor diagram, as shown in Figure 13.2b, we get the magnitudes of the voltage component from Eq. (13.3) and the current component from Eq. (13.4) as

$$V_{sm} = V_{mr} = V \cos \frac{\delta}{4} \tag{13.7a}$$

$$I_{sm} = I_{mr} = I = \frac{4V}{X} \sin \frac{\delta}{4} \tag{13.7b}$$

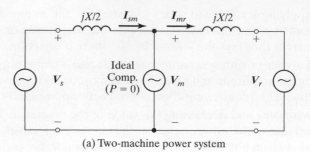

(a) Two-machine power system

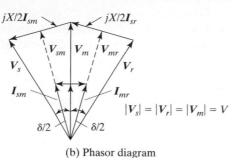

(b) Phasor diagram

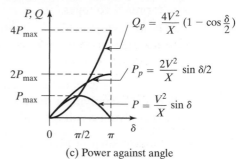

(c) Power against angle

FIGURE 13.2

Ideal shunt compensated transmission line [Ref. 2].

Using Eqs. (13.7a) and (13.7b), the transmitted active P_p for shunt compensation is given by

$$P_p = V_{sm}I_{sm} = V_{mr}I_{mr} = V_mI_{sm}\cos\frac{\delta}{4} = VI\cos\frac{\delta}{4}$$

which, after substituting for I from Eq. (13.7b), becomes

$$P_p = \frac{4V^2}{X}\sin\frac{\delta}{4}\times\cos\frac{\delta}{4} = \frac{2V^2}{X}\sin\frac{\delta}{2} \tag{13.8}$$

The reactive power Q_s at the sending end, which is equal and opposite to that at the receiving end Q_r, is given by

$$Q_s = -Q_r = VI\sin\frac{\delta}{4} = \frac{4V^2}{X}\sin^2\left(\frac{\delta}{4}\right) = \frac{2V^2}{X}\left(1-\cos\frac{\delta}{2}\right) \tag{13.9}$$

The reactive power Q_p supplied by the shunt compensation is given by

$$Q_p = 2VI\sin\frac{\delta}{4} = \frac{8V^2}{X}\sin^2\left(\frac{\delta}{4}\right)$$

which can rewritten as

$$Q_p = \frac{4V^2}{X}\left(1-\cos\frac{\delta}{2}\right) \tag{13.10}$$

Thus, P_p becomes the maximum $P_{p(\text{max})} = 2V^2/X$ at $\delta = 180°$ and Q_p becomes the maximum $Q_{p(\text{max})} = 4V^2/X$ at $\delta = 180°$. The plots of the reactive power P_p and the reactive power Q_p against the angle δ are shown in Figure 13.2c. The maximum transmitted power $P_{p(\text{max})}$ is increased significantly to twice the uncompensated value P_{max} in Eq. (13.5) for $\delta = 90°$, but at the expense of increasing reactive power demand $Q_{p(\text{max})}$ on the shunt compensator and also on the end terminals.

It should be noted that the midpoint of the transmission line is the best location for the shunt compensator. This is because the voltage sag (or drop) along the uncompensated transmission line is the largest at the midpoint. Also, the compensation at the midpoint breaks the transmission line into two equal segments for each of which the maximum transmittable power is the same. For unequal segments, the transmittable power of the longer segment would clearly determine the overall transmission limit.

Key Point of Section 13.3

- Applying a voltage at the midpoint and in quadrature with the line current can increase the transmittable power, but at the expense of increasing the reactive power demand.

13.4   SHUNT COMPENSATORS

In the shunt compensation, a current is injected into the system at the point of connection. This can be implemented by varying a shunt impedance, a voltage source, or a current source. As long as the injected current is in phase quadrature with the line voltage, the shunt compensator only supplies or consumes variable reactive power [2, 3]. Power converters using thyristors, gate-turn-off thyristors (GTOs), Mos-controlled thyristors (MCTs), or Gate-insulated bi-polar transistors (IGBTs) can be used to control the injected current or the compensating voltage.

13.4.1   Thyristor-controlled Reactor

A thyristor-controlled reactor (TCR) consists of a fixed (usually air-cored) reactor of inductance L and a bidirectional thyristor switch SW, as shown in Figure 13.3a. The current through the reactor can be controlled from zero (when the switch is open) to maximum (when the switch is closed) by varying the delay angle α of the thyristor firing. This is shown in Figure 13.3b where σ is the conduction angle of the thyristor switch such that $\sigma = \pi - 2\alpha$. When $\alpha = 0$, the switch is permanently closed and it has no effect of the inductor current. If the gating of the switch is delayed by angle α with respect to crest (or peak) V_m of the supply voltage, $v(t) = V_m \cos \omega t = \sqrt{2}\, V \cos \omega t$, the instantaneous inductor current can be expressed as a function of α as follows:

$$i_L(t) = \frac{1}{L} \int_\alpha^{\omega t} v(t)\, dt = \frac{V_m}{\omega L}(\sin \omega t - \sin \alpha) \qquad (13.11)$$

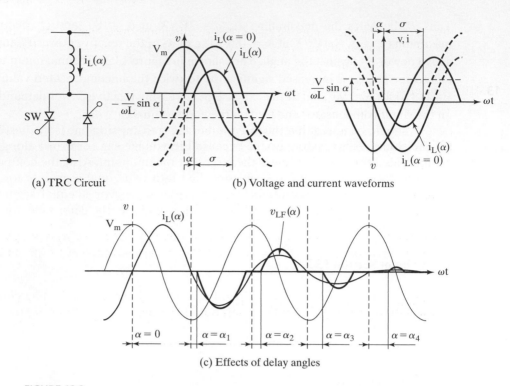

(a) TRC Circuit   (b) Voltage and current waveforms

(c) Effects of delay angles

FIGURE 13.3

Thyristor-controlled reactor (TCR) Ref [2].

which is valid for $\alpha \le \omega t \le \pi - \alpha$. For the subsequent negative half-cycle interval, the sign of the terms in Eq. (13.11) becomes opposite. The term $(V_m/\omega L)\sin \alpha$ in Eq. (13.11) is simply an α-dependent constant by which the sinusoidal current obtained at $\alpha = 0$ is offset, shifted down for positive, and up for negative half-current during half-cycles. The current $i_L(t)$ is at maximum when $\alpha = 0$ and it is zero when $\alpha = \pi/2$. The waveforms of $i_L(t)$ for various values of $\alpha(\alpha_1, \alpha_2, \alpha_3, \alpha_4)$ are shown in Figure 13.3c. Using Eq. (13.11), the fundamental root-mean-square (rms) current of the reactor current can be found as

$$I_{LF}(\alpha) = \frac{V}{\omega L}\left(1 - \frac{2}{\pi}\alpha - \frac{1}{\pi}\sin 2\alpha\right) \tag{13.12}$$

which gives the admittance as a function of α as

$$Y_L(\alpha) = \frac{I_{LF}}{V} = \frac{1}{\omega L}\left(1 - \frac{2}{\pi}\alpha - \frac{1}{\pi}\sin 2\alpha\right) \tag{13.13}$$

Thus, the compensator can vary the impedance, $Z_L(\alpha) = 1/Y_L(\alpha)$, and hence the compensating current. Due to the phase-angle control, harmonic currents of low order also appear. Passive filters may be necessary to eliminate these harmonics. Transformers with Y–delta connections are normally used at the sending end to avoid harmonic injection to the ac supply line.

13.4.2 Thyristor-Switched Capacitor

Thyristor-switched capacitor (TSC) consists of a fixed capacitance C, a bidirectional thyristor switch SW, and a relatively small surge-limiting reactor L. This is shown in Figure 13.4a. The switch is operated to turn either on or off the capacitor. Using KVL in Laplace's domain of s, we get

$$V(s) = \left(Ls + \frac{1}{Cs} \right) I(s) + \frac{V_{co}}{s}$$ (13.14)

where V_{co} is the initial capacitor voltage. Assuming sinusoidal voltage of $v = V_m \sin(\omega t + \alpha)$, Eq. (13.14) can be solved for the instantaneous current $i(t)$ as given by

$$i(t) = V_m \frac{n^2}{n^2 - 1} \omega C \cos(\omega t + \alpha) - n\omega C \left(V_{co} - \frac{n^2 V_m}{n^2 - 1} \sin \alpha \right)$$

$$\times \sin \omega_n t - V_m \omega C \cos \alpha \cos \omega_n t$$ (13.15)

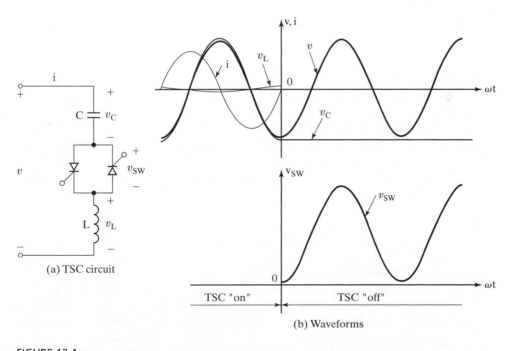

(a) TSC circuit

(b) Waveforms

FIGURE 13.4

Thyristor-switched capacitor (TSC) [Ref. 2].

where ω_n is the natural frequency of the LC circuit as given by

$$\omega_n = \frac{1}{\sqrt{LC}} = n\omega \tag{13.16}$$

$$n = \frac{1}{\sqrt{\omega^2 LC}} = \sqrt{\frac{X_C}{X_L}} \tag{13.17}$$

To obtain transient-free switching, the last two terms on the right-hand side of Eq. (13.15) must equal to zero; that is, the following conditions must be satisfied:

Condition 1

$$\cos \alpha = 0, \text{ or } \sin \alpha = 1 \tag{13.18a}$$

Condition 2

$$V_{co} = \pm V_m \frac{n^2}{n^2 - 1} \tag{13.18b}$$

The first condition implies that the capacitor is gated at the supply voltage peak. The second condition means that the capacitor must be charged to a voltage higher than the supply voltage prior to gating. Thus, for a transient-free operation, the steady-state current (when the TSC is closed) is given by

$$i(t) = V_m \frac{n^2}{n^2 - 1} \omega C \cos(\omega t + 90°) = -V_m \frac{n^2}{n^2 - 1} \omega C \sin \omega t \tag{13.19}$$

The TSC can be disconnected at zero current by prior removal of the thyristor-gating signal. At the current zero crossing, the capacitor voltage, however, reaches its peak value of $V_{co} = \pm V_m n^2/(n^2 - 1)$. The disconnected capacitor stays charged to this voltage and, consequently, the voltage across the nonconducting TSC varies between zero and the peak-to-peak value of the applied ac voltage, as shown in Figure 13.4b.

If the voltage across the disconnected capacitor remained unchanged, the TSC could be switched on again, without any transient, at the appropriate peak of the applied ac voltage; this is shown in Figure 13.5a for a positively charged capacitor, and in Figure 13.5b for a negatively charged capacitor. In practice the capacitor voltage slowly discharges between gating (or switching) periods and the system voltage and impedance can change abruptly, making any control strategy problematic. Thus, the capacitor should be reconnected at some residual voltage between zero and $\pm V_m n^2/(n^2 - 1)$. This can be accomplished with the minimum of possible transient disturbance if the TSC is turned on at those instants at which the capacitor residual voltage and the applied ac voltage are equal. Thus, the TSC should be switched on when the voltage across it becomes zero, that is, zero-voltage switching (ZVS). Otherwise, there are switching transients. These transients are caused by the nonzero dv/dt at the instant of

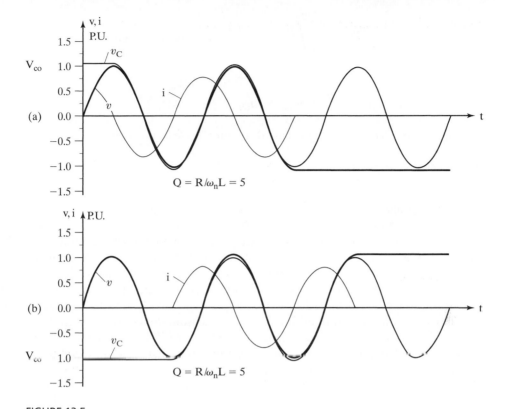

FIGURE 13.5

Transient-free switching of thyristor-switched capacitor [Ref. 2].

switching, which, without the series reactor, would result in an instantaneous current of $i = C\,dv/dt$ through the capacitor.

The *rules* for transient-free switching are:

1. If the residual capacitor voltage V_{co} is lower than the peak ac voltage V_m (i.e., $V_{co} < V_m$), then the TSC should be switched on when the instantaneous ac voltage $v(i)$ becomes equal to the capacitor voltage $v(t) = V_{co}$.

2. If the residual capacitor voltage V_{co} is equal or higher than the peak ac voltage (i.e., $V_{co} \geq V_m$), then the TSC should be switched on when the instantaneous ac voltage is at its peak $v(t) = V_m$ so that the voltage across the TSC is minimum (i.e., $V_{co} - V_m$).

If the switch is turned on for m_{on} cycles and off for m_{off} cycles of the input voltage, the rms capacitor current can be found from

$$I_c = \left[\frac{m_{on}}{2\pi(m_{on} + m_{off})} \int_0^{2\pi} i^2(t)\,d(\omega t) \right]^{1/2}$$

$$= \left[\frac{m_{\text{on}}}{2\pi(m_{\text{on}} + m_{\text{off}})} \int_0^{2\pi} \left(-V_m \frac{n^2}{n^2 - 1} \omega C \sin \omega t \right)^2 d(\omega t) \right]^{1/2}$$

$$= \frac{n^2 V_m}{(n^2 - 1)\sqrt{2}} \omega C \sqrt{\frac{m_{\text{on}}}{m_{\text{on}} + m_{\text{off}}}} = \frac{n^2 V_m}{(n^2 - 1)\sqrt{2}} \omega C \sqrt{k} \tag{13.20}$$

where $k = m_{\text{on}}/(m_{\text{on}} + m_{\text{off}})$ is called the *duty cycle* of the switch.

13.4.3 Static VAR Compensator

The use of either TCR or TSC would allow only capacitive or inductive compensation. However, in most of the applications, it is desirable to have the possibilities of both capacitive and inductive compensation. A static VAR compensator (SVC) consists of TCRs in parallel with one or more TSCs [4, 7]. The general arrangement of an SVC is shown in Figure 13.6. The reactive elements of the compensator are connected to the transmission line through a transformer to prevent the elements having to withstand full system voltage. A control system determines the exact gating instants of reactors according to a predetermined strategy. The strategy usually aims to maintain the transmission line voltage at a fixed level. For this reason, the control system has a system voltage input taken through a potential transformer (PT); additionally, other input parameters (or variables) to the control system may exist. The control system ensures

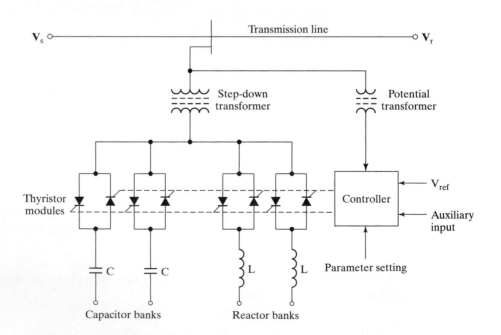

FIGURE 13.6

General arrangement of static VAR compensator [Ref. 4].

that the compensator voltage remains more or less constant by adjusting the conduction angle [5, 6].

13.4.4 Advanced Static VAR Compensator

An advanced static VAR compensator is essentially a voltage-sourced converter, as shown in Figure 13.7. A current-source inverter can also be substituted [11]. It is simply known as the *static compensator, STATCOM.* If the line voltage $\mathbf{V}$ is in phase with the converter output voltage $\mathbf{V}_o$ and has the same magnitude so that $V \angle 0° = V_o \angle 0°$, there can be no current flowing into or out of the compensator and no exchange of reactive power with the line. If the converter voltage is now increased, the voltage difference between $\mathbf{V}$ and $\mathbf{V}_o$ appears across the leakage reactance of the step-down transformer. As a result, a leading current with respect to $\mathbf{V}$ is drawn and the compensator behaves as a capacitor, generating VARs. Conversely, if $V > V_o$, then the compensator draws a lagging current, behaving as an inductor, and absorbs VARs. This compensator operates essentially like a synchronous compensator where the excitation may be greater or less than the terminal voltage. This operation allows continuous control of reactive power, but at a far higher speed, especially with a forced-commutated converter using GTOs, MCTs, or IGBTs.

The main features of a STATCOM are: (1) wide operating range providing full capacitive reactance even at a low voltage, (2) lower rating than its conventional counterpart SVC to achieve the same stability, and (3) increased transient rating and superior capability to handle dynamic system disturbances. If a dc storage device such as a superconducting coil arrangement replaces the capacitor, it would be possible to exchange both active and reactive power with the system. Under conditions of low

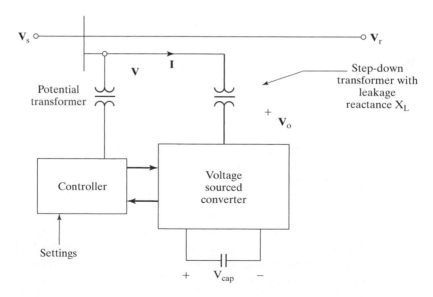

FIGURE 13.7

General arrangement of advanced shunt static-VAR compensator (STATCOM) [Ref. 4].

demand, the superconducting coil can supply power, which can be released into the system under contingency conditions.

Key Point of Section 13.4

- Shunt compensators generally consist of thyristors, GTOs, MCTs, or IGBTs.
- There are four types: (1) TCRs, (2) TSCs, (3) SVCs, and (4) advanced SVCs (STATCOM).

Example 13.1 Finding the Inductive Reactance and the Delay Angle of TCR

The particulars of a transmission line with a TCR, as shown in Figure 13.3a, are $V = 220$ V, $f = 60$ Hz, $X = 1.2\ \Omega$, and $P_p = 56$ kW. The maximum current of the TCR is $I_{L(mx)} = 100$ A. Find (a) the phase-angle δ, (b) the line current I, (c) the reactive power Q_p of the shunt compensator, (d) the current through the TCR, (e) the inductance reactance X_L, and (f) the delay angle of the TCR if the I_L is 60% of the maximum current.

Solution

$V = 220$, $f = 60$ Hz, $X = 1.2\ \Omega$, $\omega = 2\pi f = 377$ rad/s, $P_p = 56$ kW, $I_{L(mx)} = 100$ A, $k = 0.6$

a. Using Eq. (13.8), $\delta = 2\sin^{-1}\left(\dfrac{XP_p}{2V^2}\right) = 2\sin^{-1}\left(\dfrac{1.2 \times 56 \times 10^3}{2 \times 220^2}\right) = 87.93°$.

b. Using Eq. (13.7b), $I = \dfrac{4V}{X}\sin\dfrac{\delta}{4} = \dfrac{4 \times 220}{1.2} \times \sin\dfrac{87.93}{4} = 274.5$ A.

c. Using Eq. (13.10), $Q_p = \dfrac{4V^2}{X}\left(1 - \cos\dfrac{\delta}{2}\right) = \dfrac{4 \times 220^2}{1.2} \times \left(1 - \cos\dfrac{87.93}{2}\right)$
$= 45.21 \times 10^3$ A.

d. The current through TCR, $I_Q = \dfrac{Q_p}{V} = \dfrac{45.21 \times 10^3}{220} = 205.504$ A.

e. The inductance reactance $X_L = \dfrac{V}{I_{L(max)}} = \dfrac{220}{100} = 2.2\ \Omega$.

f. $I_L = kI_{L(max)} = 0.6 \times 100 = 60$ A.

Using (13.12), $60 = 220/2.2 \times \left(1 - \dfrac{2}{\pi}\alpha - \dfrac{1}{\pi}\sin 2\alpha\right)$, which by using Mathcad gives the delay angle $\alpha = 18.64°$.

13.5 PRINCIPLE OF SERIES COMPENSATION

A voltage in series with the transmission line can be introduced to control the current flow and thereby the power transmissions from the sending end to the receiving end. An ideal series compensator, represented by the voltage source $\mathbf{V}_c$, is connected in the middle of a transmission line, as shown in Figure 13.8. The current flowing through the transmission line is given by:

$$\mathbf{I} = \frac{\mathbf{V}_s - \mathbf{V}_r - \mathbf{V}_c}{jX} \tag{13.21}$$

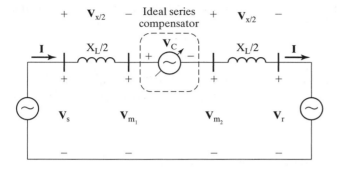

FIGURE 13.8

Ideal series compensation of a transmission line.

If the series applied voltage V_c is in quadrature with respect to the line current, the series compensator cannot supply or absorb active power. That is, the power at the source V_c terminals can be only reactive. This means that capacitive or inductive equivalent impedance may replace the voltage source V_c. The transmission line equivalent impedance is given by:

$$X_{eq} = X - X_{comp} = X(1 - r) \tag{13.22}$$

where

$$r = \frac{X_{comp}}{X} \tag{13.23}$$

and r is the degree of series compensation, $0 \le r \le 1$. X_{comp} is the series equivalent compensation reactance, which is positive if it is capacitive and negative if it is inductive. Using Eq. (13.4), the magnitude of the current through the line is given by

$$I = \frac{2V}{(1 - r)X} \sin \frac{\delta}{2} \tag{13.24}$$

Using Eq. (13.5), the active power flowing through the transmission line is given by

$$P_c = V_c I = \frac{V^2}{(1 - r)X} \sin \delta \tag{13.25}$$

Using Eq. (13.6), the reactive power Q_c at the source V_c terminals is given by

$$Q_c = I^2 X_{comp} = \frac{2V^2}{X} \times \frac{r}{(1 - r)^2} (1 - \cos \delta) \tag{13.26}$$

If the source V_c is compensating only capacitive reactive power, the line current is leading the voltage V_c by 90°. For inductive compensation, the line current is lagging the voltage V_c by 90°. The inductive compensation may be used when it is necessary to decrease the power flowing in the line. In both capacitive and inductive compensations,

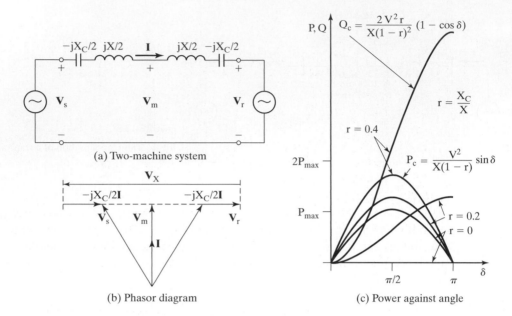

FIGURE 13.9

Series capacitor compensation [Ref. 2].

no active power is absorbed or generated by the source $\mathbf{V}_c$. The capacitive compensation is, however, more commonly used.

Series capacitive impedance can decrease the overall effective series transmission impedance from the sending end to the receiving end, and thereby increase the transmittable power. A series capacitor compensated line with two identical segments is shown in Figure 13.9a. Assuming that the magnitudes of the terminal voltages remain constant, equal to V. For $V_s = V_r = V$, the corresponding voltage and current phasors are shown in Figure 13.9b. Assuming the same end voltages, the magnitude of the total voltage across the series line inductance $V_x = 2V_{x/2}$ is increased by the magnitude of the opposite voltage across the series capacitor $-V_c$. This results in an increase in the line current.

Equation (13.25) shows that the transmitted power can be increased considerably by varying the degree of series compensation r. The plots of the active power P_C and the reactive power Q_C against angle δ are shown in Figure 13.9c. The transmitted power P_C increases rapidly with the degree of series compensation r. Also, the reactive power Q_C supplied by the series capacitor increases sharply with s and varies with angle δ in a manner similar to that of the line reactive power P_C.

According to Eq. (13.5), a large series reactive impedance of a long transmission line can limit the power transmission. In such cases, the impedance of the series compensating capacitor can cancel a portion of the actual line reactance and thereby the effective transmission impedance is reduced as if the line was physically shortened.

Key Points of Section 13.5

- A series voltage that is in quadrature with respect to the line current is applied, thereby increasing the current and the transmittable power.
- The series compensator does not supply or absorb active power.

13.6   SERIES COMPENSATORS

A series compensator, in principle, injects a voltage in series with the line. Even, variable impedance multiplied by the current flow through it represents an applied series voltage in the line. As long as the voltage is in phase quadrature with the line current, the series compensator supplies or consumes variable reactive power only. Therefore, the series compensator could be a variable impedance (such as a capacitor or a reactor) or a power electronics-based variable source of main frequency, and subsynchronous and harmonic frequencies (or a combination) to meet the desired control strategy.

13.6.1   Thyristor-Switched Series Capacitor

A thyristor-switched series capacitor (TSSC) consists of a number of capacitors in series, each shunted by a switch composed of two antiparallel thyristors. The circuit arrangement is shown in Figure 13.10a. A capacitor is inserted by turning off, and it is bypassed by turning on the corresponding thyristor switch. Thus, if all switches are off, the equivalent capacitance of the string becomes $C_{eq} = C/m$ and if all switches are turned on simultaneously, $C_{eq} = 0$. The amount of effective capacitance and hence the

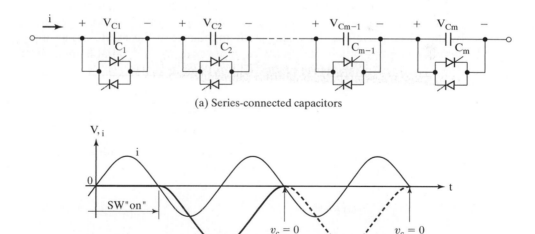

(a) Series-connected capacitors

(b) Switching at zero current and capacitor offset voltage

FIGURE 13.10

Thyristor-switched series capacitor [Ref. 2].

degree of series compensation are controlled in a steplike manner by increasing or decreasing the number of series capacitors inserted.

A thyristor is commutated "naturally", that is, it turns off when the current crosses zero. Thus, a capacitor can be inserted into the line only at the zero crossings of the line current, that is, zero current switching (ZCS). Because the insertion can take place only at the zero line current, the capacitor can be charged from zero to maximum during the full half-cycle of the line current and it can discharge from this maximum to zero by the successive line current of opposite polarity during the next full half-cycle. This results in a dc-offset voltage, which is equal to the amplitude of the ac capacitor voltage, as shown in Figure 13.10b.

To minimize the initial surge current through the switch and the resulting transient due to $v_C = C\, dv/dt$ condition, the thyristors should be turned on for bypass only when the capacitor voltage is zero. The dc offset and the requirement of $v_C = 0$ may cause a delay of up to one full cycle, which would set the theoretical limit for the attainable response time of the TSSC. Due to the *di/dt* limitation of thyristors, it would necessitate, in practice, the use of a current-limiting reactor in series with the thyristor switch. A reactor in series with the switch can result in a new power circuit known as a *thyristor-controlled series capacitor* (TCSC) (see Section 13.6.2) that can significantly improve the operating and performance characteristics of the TSSC.

13.6.2 Thyristor-Controlled Series Capacitor

The TCSC consists of the series-compensating capacitor shunted by a thyristor-controlled reactor (TCR), as shown in Figure 13.11. This arrangement is similar in structure to the TSSC. If the impedance of the reactor X_L is sufficiently smaller than that of the capacitor X_C, it can be operated in an on–off manner like the TSSC. Varying the delay angle α can vary the inductive impedance of the TCR. Thus, the TCSC can provide a continuously variable capacitor by means of partially canceling the effective compensating capacitance by the TCR. Therefore, the steady-state impedance of the TCSC is that of a parallel LC circuit, consisting of a fixed capacitive impedance X_C and variable inductive impedance X_L. The effective impedance of the TCSC is given by

$$X_T(\alpha) = \frac{X_C X_L(\alpha)}{X_L(\alpha) - X_C} \tag{13.27a}$$

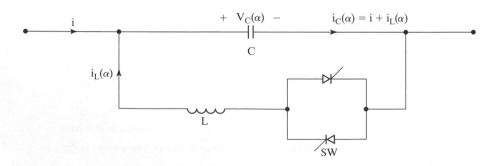

FIGURE 13.11

Thyristor-controlled series capacitor (TCSC).

where $X_L(\alpha)$ that can be found from Eq. (13.13) is given by

$$X_L(\alpha) = X_L \frac{\pi}{\pi - 2\alpha - \sin 2\alpha} \qquad \text{for } X_L \le X_L(\alpha) \le \infty \qquad (13.27\text{b})$$

where $X_L = \omega L$ and α is the delay angle measured from the crest of the capacitor voltage or the zero crossing of the line current.

The TCSC behaves as a tunable parallel LC-circuit to the line current. As the impedance of the controlled reactor $X_L(\alpha)$ is varied from its maximum (infinity) toward its minimum (ωL), the TCSC increases its minimum capacitive impedance, $X_{T(\min)} = X_C = 1/\omega C$, until parallel resonance occurs at $X_C = X_L(\alpha)$, and $X_{T(\max)}$ theoretically becomes infinite. Decreasing $X_L(\alpha)$ further, the impedance $X_T(\alpha)$ becomes inductive, reaching its minimum value of $X_C X_L/(X_L - X_C)$ at $\alpha = 0$; that is, the capacitor is in effect bypassed by the TCR. In general, the impedance of reactor X_C is smaller than that of capacitor X_C. Angle α has two limiting values (1) one for inductive $\alpha_{L(\lim)}$ and (2) one for capacitive $\alpha_{C(\lim)}$. The TCSC has two operating ranges around its internal circuit resonance: (1) one is the $\alpha_{C(\lim)} \le \alpha \le \pi/2$ range, where $X_T(\alpha)$ is capacitive, and (2) the other is the $0 \le \alpha \le \alpha_{L(\lim)}$ range, where $X_T(\alpha)$ is inductive.

13.6.3 Forced-Commutation-Controlled Series Capacitor

The forced-commutation-controlled series capacitor (FCSC) consists of a fixed capacitor in parallel with a forced-commutation type of device such as a GTO, an MCT, or an IGBT. A GTO circuit arrangement is shown in Figure 13.12a. It is similar to the TSC,

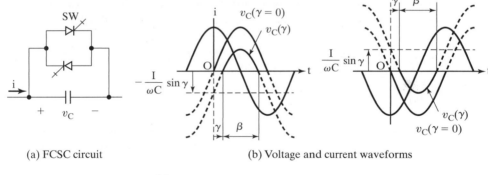

(a) FCSC circuit   (b) Voltage and current waveforms

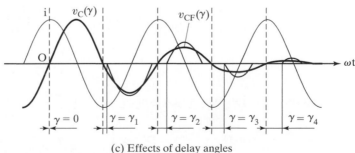

(c) Effects of delay angles

FIGURE 13.12

Forced-commutation-controlled series capacitor (FCSC) Ref [2].

except the bidirectional thyristor switch is replaced by a bidirectional forced commutated device. When the GTO switch SW is closed, the voltage across the capacitor v_C is zero; when the switch is open, v_C becomes a maximum. The switch can control the ac voltage v_C across the capacitor at a given line current i. Therefore, by closing and opening of the switch in each half-cycle in synchronism with the ac system frequency can control the capacitor voltage.

The GTO is turned on whenever the capacitor voltage crosses zero and is turned off at a delay angle $\gamma(0 \leq \gamma \leq \pi/2)$ measured with respect to the peak of the line current or the zero crossing of the line voltage. Figure 13.12b shows the line current i, and the capacitor voltage v_C at a delay angle γ for a positive and a negative half-cycle. The switch SW is on from 0 to γ and off from $\pi - \gamma$ and π. For $\gamma = 0$, the switch is permanently open and it has no effect on the resultant capacitor voltage v_C.

If the opening of the switch is delayed by an angle γ with respect to the line current $i = I_m \cos \omega t = \sqrt{2}I \cos \omega t$, the capacitor voltage can be expressed as the function of γ as follows:

$$v_C(t) = \frac{1}{C} \int_{\gamma}^{\omega t} i(t)dt = \frac{I_m}{\omega C}(\sin \omega t - \sin \gamma) \tag{13.28}$$

which is valid for $\gamma \leq \omega t \leq \pi - \gamma$. For the subsequent negative half-cycle interval, the sign of the terms in Eq. (13.28) becomes opposite. The term $(I_m/\omega L) \sin \gamma$ in Eq. (13.28) is simply a γ-dependent constant by which the sinusoidal voltage obtained at $\gamma = 0$ is offset, shifted down for positive, and up for negative half-cycles.

Turning on the GTO at the instant of voltage zero crossing controls the nonconducting (blocking) interval (or angle) λ. That is, the turn-off delay angle γ defines the prevailing blocking angle $\beta = \pi - 2\gamma$. Thus, as the turn-off delay angle γ increases, the correspondingly increasing offset results in the reduction of the blocking angle β of the switch, and the consequent reduction of the capacitor voltage. At the maximum delay of $\gamma = \pi/2$, the offset also reaches its maximum of $I_m/\omega C$, at which both the blocking angle λ and the capacitor voltage $v_C(t)$ become zero. The voltage $v_C(t)$ is at maximum when $\gamma = 0$, and it becomes zero when $\gamma = \pi/2$. Therefore, the magnitude of the capacitor voltage can be varied continuously from maximum $I_m/\omega C$ to zero by varying the turn-off delay from $\gamma = 0$ to $\gamma = \pi/2$. The waveforms of $v_C(t)$ for various values of $\gamma(\gamma_1, \gamma_2, \gamma_3, \gamma_4)$ are shown in Figure 13.12c.

Equation (13.28) is identical to Eq. (13.11), and hence the FCSC is the dual of the TCR. Similar to Eq. (13.12), the fundamental capacitor voltage can be found from

$$V_{CF}(\gamma) = \frac{I}{\omega C}\left(1 - \frac{2}{\pi}\gamma - \frac{1}{\pi}\sin 2\gamma\right) \tag{13.29}$$

which gives the impedance as a function of γ as

$$X_C(\gamma) = \frac{V_{CF}(\gamma)}{I} = \frac{1}{\omega C}\left(1 - \frac{2}{\pi}\gamma - \frac{1}{\pi}\sin 2\gamma\right) \tag{13.30}$$

where $I = I_m/\sqrt{2}$ is the rms line current. Thus, the FCSC behaves as a variable capacitive impedance, whereas the TCR behaves as a variable inductive impedance.

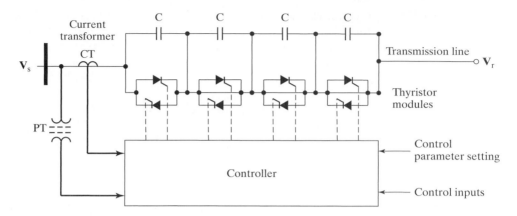

FIGURE 13.13

General arrangement of series static VAR compensator [Ref. 4].

13.6.4 Series Static VAR Compensator

The use of TSC, TCSC, or FCSC would allow capacitive series compensation. A series static VAR compensator (SSVC) consists of one of the series compensators. The general arrangement of an SSVC is shown in Figure 13.13 with a TCSC. The control system receives a system voltage input taken from a PT and a system current input taken from a current transformer (CT). There may be other additional input parameters to the control system. The control strategy of the series compensator is typically based on achieving an objective line power flow in addition to the capability of damping power oscillations.

13.6.5 Advanced SSVC

This series compensator is the dual circuit of the shunt version in Figure 13.7. Figure 13.14 shows the general arrangement of an advanced series compensator. It uses the voltage-source inverter (VSI) with a capacitor in its dc converter to replace the switched capacitors of the conventional series compensators. The converter output is arranged to appear in series with the transmission line via the use of the series transformer. The converter output voltage $\mathbf{V}_c$, which can be set to any relative phase, and any magnitude within its operating limits, is adjusted to appear to lead the line current by $90°$, thus behaving as a capacitor. If the angle between $\mathbf{V}_c$ and the line current was not $90°$, then this would imply that the series compensator exchanges active power with the transmission line, which is clearly impossible because the compensator in Figure 13.14 has no active power source.

This type of series compensation can provide a continuous degree of series compensation by varying the magnitude of $\mathbf{V}_c$. Also, it can reverse the phase of $\mathbf{V}_c$, thereby increasing the overall line reactance; this may be desirable to limit fault current, or to dampen power oscillations. In general, the controllable series compensator can be used to increase transient stability, to dampen subsynchronous resonance where other fixed capacitors are used, and to increase line power capability.

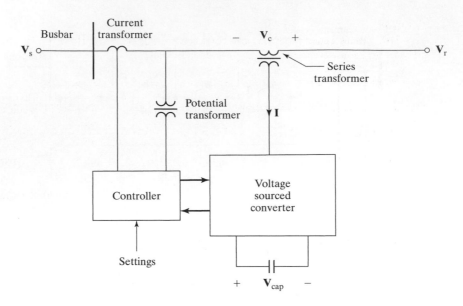

FIGURE 13.14

General arrangement of advanced series static VAR compensator [Ref. 4].

Any variation in the line current does not lead to a variation in $\mathbf{V}_c$. Thus, the converter exhibits nearly zero impedance at the fundamental power system frequency. The voltage applied into the line by the converter is not derived from a real capacitance reactance, and is not able to resonate. Therefore, this compensator can be used to produce subsynchronous resonance, that is, resonance between the series capacitor and the line inductance.

Key Points of Section 13.6

- Series compensators generally consist of thyristors, GTOs, MCTs, or IGBTs.
- There are four types: (1) TSSCs, (2) TCSCs, (3) FCSCs, (4) SSVCs, and (5) advanced series static VAR compensator (SSTATCOM).

Example 13.2   Finding the Series Compensating Reactance and the Delay Angle of TCSC

The particulars of a transmission line are $V = 220$ V, $f = 60$ Hz, $X = 12\ \Omega$, and $P_p = 56$ kW. The particulars of the TCSC are $\delta = 80°$, $C = 20\ \mu$F, and $L = 0.4$ mH. Find (a) the degree of compensation r, (b) the compensating capacitance reactance X_{comp}, (c) the line current I, (d) the reactive power Q_c, (e) the delay angle α of the TCSC if the effective capacitive reactance is $X_T = -50\ \Omega$, and (f) plot $X_L(\alpha)$ and $X_T(\alpha)$ against the delay angle α.

Solution

$V = 220$, $f = 60$ Hz, $X = 12\ \Omega$, $\omega = 2\pi f = 377$ rad/s, $P_s = 56$ kW, $C = 20\ \mu$F, and $L = 0.4$ mH, $X_C = -1/\omega C = -132.63\ \Omega$, $X_L = \omega L = 0.151\ \Omega$.

a. Using (13.25), $r = 1 - \dfrac{V^2}{X P_c} \sin \delta = 1 - \dfrac{220^2}{12 \times 56 \times 10^3} = 0.914$.

b. The compensating capacitive reactance $X_{comp} = r \times X = 0.914 \times 12 = 10.7\ \Omega$.

c. Using Eq. (13.24), $I = \dfrac{2V}{(1-r)X} \sin \dfrac{\delta}{2} = \dfrac{2 \times 220}{(1 - 0.914) \times 1.2} \times \sin \dfrac{80}{2} = 317.23$ A.

d. Using Eq. (13.26),

$$Q_c = \frac{2V^2}{X} \times \frac{r}{(1-r)^2}(1 - \cos \delta) = \frac{2 \times 220^2 \times 0.914}{12 \times (1 - 0.914)^2} \times (1 - \cos 80°) = 1.104 \times 10^6$$

e. Using Eq. (13.27), $X_L(\alpha) = X_L \dfrac{\pi}{\pi - 2\alpha - \sin 2\alpha}$

$$X_T(\alpha) = -50 = \frac{X_C X_L(\alpha)}{X_L(\alpha) - X_C}$$

which, by Mathcad, gives the delay angle $\alpha = 77.707°$.

f. The plot $X_L(\alpha)$ and $X_T(\alpha)$ against the delay angle α is shown in Figure 13.15.

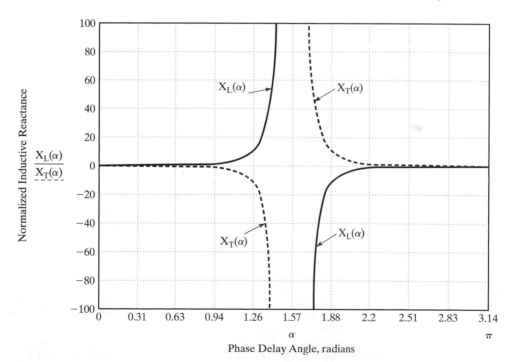

FIGURE 13.15

Impedance of the TCR and the effective impedance versus delay angle.

13.7   PRINCIPLE OF PHASE-ANGLE COMPENSATION

Phase-angle compensation is a special case of the series compensator in Figure 13.8. The power flow is controlled by phase angle. The phase compensator is inserted between the sending-end generator and the transmission line. This compensator is an ac voltage source with controllable amplitude and phase angle. An ideal phase compensator is shown Figure 13.16a. The compensator controls the phase difference between the two ac systems and thereby can control the power exchanged between the two ac systems. The effective sending-end voltage is the sum of sending-end voltage $\mathbf{V}_s$ and the compensator voltage $\mathbf{V}_\sigma$, as shown in the phasor diagram in Figure 13.16b. The angle σ between $\mathbf{V}_s$ and $\mathbf{V}_\sigma$ can be varied in such a way that the angle σ change does not result in a magnitude change. That is,

$$\mathbf{V}_{seff} = \mathbf{V}_s + \mathbf{V}_\sigma \tag{13.31a}$$

$$|\mathbf{V}_{seff}| = |\mathbf{V}_s| = V_{seff} = V_s = V \tag{13.31b}$$

By controlling the angle σ independently, it is possible to keep the transmitted power at the desired level, independent of the transmission angle δ. Thus, for example, the

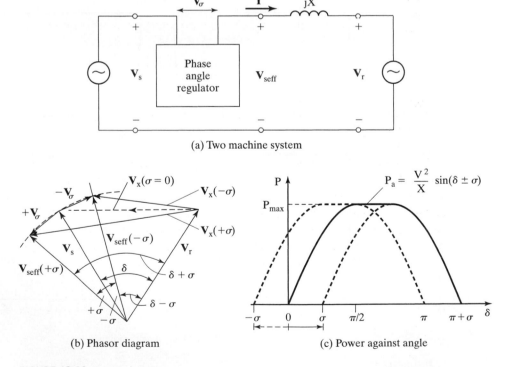

(a) Two machine system

(b) Phasor diagram

(c) Power against angle

FIGURE 13.16

Phase-angle compensation [Ref. 2].

power can be kept at its peak value after angle δ exceeds the peak power angle $\pi/2$ by controlling the amplitude of compensation voltage so that the effective phase angle $(\delta - \sigma)$ between the sending-end and receiving-end voltages stays at $\pi/2$. From the phasor diagram, the transmitted power with phase compensation is given by

$$P_a = \frac{V^2}{X} \sin(\delta - \sigma) \tag{13.32}$$

The transmitted reactive power with phase compensation is given by

$$Q_a = \frac{2V^2}{X} [1 - \cos(\delta - \sigma)] \tag{13.33}$$

Unlike other shunt and series compensators, the angle compensator needs to be able to handle both active and reactive power. This assumes that the magnitudes of the terminal voltages remain constant, equal to V. That is, $V_{seff} = V_s = V_r = V$. We can find the magnitudes of V_σ and I from the phasor diagram in Figure 13.16b as given by

$$V_\sigma = 2V \sin \frac{\sigma}{2} \tag{13.34}$$

$$I = \frac{2V}{X} \sin \frac{\delta}{2} \tag{13.35}$$

The apparent (volt–ampere [VA]) power through the phase compensator is given by

$$VA_a = V_a I = \frac{4V^2}{X} \sin\left(\frac{\delta}{2}\right) \sin\left(\frac{\sigma}{2}\right) \tag{13.36}$$

The plot of the active power P_a against angle δ for $\pm\sigma$ is shown in Figure 13.16c. The flat-topped curve indicates the range of the action of the phase compensation. This type of compensation does not increase the transmittable power of the uncompensated line. The active power P_a and reactive power Q_a remain the same as those for the uncompensated system with equivalent transmission angle δ. It is, however, theoretically possible to keep the power at its maximum value at any angle δ in the range $\pi/2 < \delta < \pi/2 + \sigma$ by, in effect, shifting the P_a versus δ curve to the right. The P_a versus δ curve can also be shifted to the left by inserting the voltage of the angle compensation with an opposite polarity. Therefore, the power transfer can be increased and the maximum power can be reached at a generator angle less than $\pi/2$, that is, at $\delta = \pi/2 - \sigma$. The effect of connecting the phase compensator in reverse is shown by the broken curve.

If angle σ of phasor $\mathbf{V}_\sigma$ relative to phasor V_s is maintained fixed at $\pm90°$, the phase compensator becomes a *quadrature booster* (QB) that has the following relationships:

$$\mathbf{V}_{seff} = \mathbf{V}_s + \mathbf{V}_\sigma \tag{13.37a}$$

$$|\mathbf{V}_{seff}| = V_{seff} = \sqrt{V_s^2 + V_\sigma^2} \tag{13.37b}$$

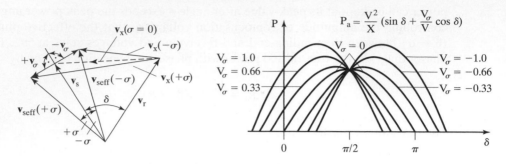

FIGURE 13.17

Phasor diagram and transmitted power of a quadrature booster [Ref. 2].

The phasor diagram of the QB-type angle compensator is shown in Figure 13.17a and its transmitted power P_b with the booster compensator is given by

$$P_b = \frac{V^2}{X}\left(\sin\delta + \frac{V_\sigma}{V}\cos\delta\right) \tag{13.38}$$

The transmitted power P_b versus angle δ as a parametric function of the applied quadrature voltage V_σ is shown in Figure 13.17b. The maximum transmittable power increases with the applied voltage V_σ, because in contrast to the phase-angle compensator, the QB increases the magnitude of the effective sending-end voltage.

Key Points of Section 13.7

- The phase compensator is inserted between the sending-end generator and the transmission line.
- This compensator is an ac voltage source with controllable amplitude and phase angle.

13.8   PHASE-ANGLE COMPENSATOR

When a thyristor is used for phase-angle compensation, it is called the phase shifter. Figure 13.18a shows the general arrangement of a phase shifter. The shunt-connected *excitation transformer* may have either identical or nonidentical separate windings per phase. The thyristor switches are connected, forming an on-load tap changer. Thyristors are connected in antiparallel, forming bidirectional naturally commutated switches. Thyristor tap-changing unit controls the voltage to the series transformer secondary.

Using phase control may control the magnitude of the series voltage $\mathbf{V}_q$. To avoid excessive harmonic generation, various taps are used. The tap changer can connect the excitation winding either completely or not at all; this allows the series voltage $\mathbf{V}_q$ to assume 1 of 27 different voltage values depending on the state of the 12-thyristor switches in the tap changer [4]. It should be noted that the transforming arrangement between the excitation and series transformers ensures that $\mathbf{V}_q$ *is always at 90° to* $\mathbf{V}$, the excitation transformer primary voltage as shown in Figure 13.18b. Thus, it is called the

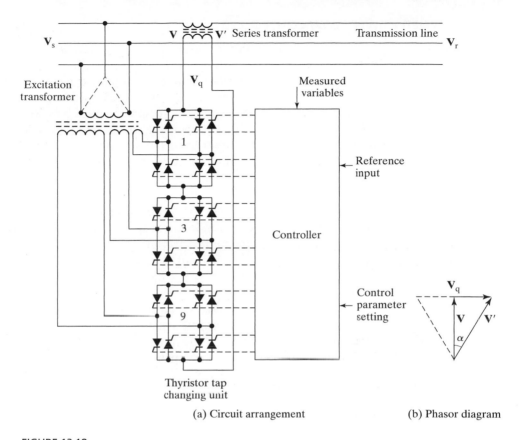

FIGURE 13.18

General arrangement of a thyristor phase shifter [Ref. 4].

quadrature booster. One important characteristic of the phase shifter is that the active power can only flow from the shunt to the series transformers. Thus, reverse power flow is not possible.

The phase shifter controls the magnitude of V_q and thus the phase shift α to the sending-end voltage [8]. This control can be achieved either by sensing the generator angle, or from using power measurements. The controller can also be set to dampen power oscillations. Phase shifters like series capacitor compensators allow control of power through the network and power sharing between parallel circuits. Series capacitors are more suitable for long-distance lines because unlike phase shifters, they effectively reduce line reactance and hence reduce the reactive power and voltage control problems associated with long-distance transmission. Phase shifters are more suitable for power flow control in compact high power density networks.

Key Points of Section 13.8

- A thyristor connection changer unit is used as a phase shifter.
- This unit controls the series voltage through a series transformer secondary.

13.9 UNIFIED POWER FLOW CONTROLLER

A *unified power flow controller* (UPFC) consists of an advanced shunt and series compensator with a common DC link, as shown in Figure 13.19a. The energy-storing capacity of the dc capacitor is generally small. Therefore, the active power drawn (generated) by the shunt converter should be equal to the active power generated (drawn) by the series converter. Otherwise, the dc-link voltage may increase or decrease with respect to the rated voltage, depending on the net power absorbed and generated by both converters. On the other hand, the reactive power in the shunt or series converter can be chosen independently, giving a greater flexibility to the power flow control [9].

Power control is achieved by adding the series voltage $\mathbf{V}_{inj}$ to $\mathbf{V}_s$, thus giving the line voltage $\mathbf{V}_L$, as shown in Figure 13.19b. With two converters, the UPFC can supply active power in addition to reactive power. Because any active power requirement can be supplied through the shunt-connected converter, the applied voltage $\mathbf{V}_{inj}$ can assume any phase with respect to the line current. Because there is no restriction on $\mathbf{V}_{inj}$, the locus of $\mathbf{V}_{inj}$ becomes a circle centered at $\mathbf{V}_s$ with a maximum radius that equals the maximum magnitude of $\mathbf{V}_{inj} = |\mathbf{V}_{inj}|$.

UPFC is a more complete compensator and can function in any of the compensating modes, hence, its name. It should be noted that the UPFC shown in Figure 13.19a is valid for power flowing from $\mathbf{V}_s$ to $\mathbf{V}_L$. If the power flow is reverted, it may be necessary to change the connection of the shunt compensator. In a more general UPFC with bidirectional power flow, it would be necessary to have two shunt converters: one at the sending end and one at the receiving end.

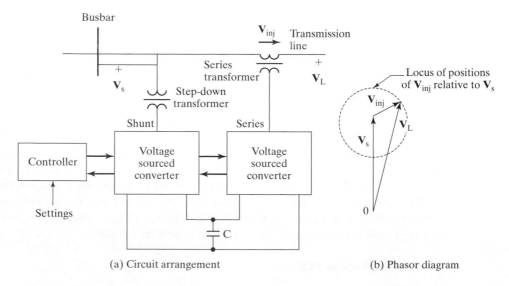

(a) Circuit arrangement   (b) Phasor diagram

FIGURE 13.19

Unified power flow controller [Ref. 4].

Key Points of Section 13.9

- UPFC is a complete compensator.
- This compensator can function in any of the compensating modes.

13.10 COMPARISONS OF COMPENSATORS

The shunt controller is like a current source that draws or injects current into the line. The shunt controller is, therefore, a good way to control the voltage at and around the point of connection. It can inject reactive leading or lagging current only, or a combination of active and reactive current for a more effective voltage control and damping of voltage oscillations. A shunt controller is independent of the other line and much more effective in maintaining a desired voltage at a substation node.

The series controller impacts the driving voltage and hence the current and power flow directly. Therefore, if the purpose is to control the current or power flow and damp oscillations, the series controller for the same million volt–ampere (MVA) size is several times more powerful than the shunt controller. The MVA size of a series controller is small compared with that of the shunt controller. However, the shunt controller does not provide control of the power flow in the lines.

A series compensation, as shown in Figure 13.8, is just a particular case of the phase-angle compensator depicted in Figure 13.16, with the difference that the latter may supply active power, whereas the series compensator supplies or absorbs only reactive power.

Normally, the phase-angle controller is connected close to the transmission line at the sending or receiving end whereas the series compensator is connected in the middle of the line. If the objective is to control the active power flow through the transmission line, the location of the compensator is just a question of convenience. The basic difference is that the phase-angle compensator may need a power source whereas the series compensator does not.

Figure 13.20 shows the active power transfer characteristics for ac systems without compensation, with shunt and series compensation, and phase shifter compensation [10]. Depending on the compensation level, the series compensator is the best choice to increase power transfer capability. The phase shifter is important for connecting two systems with excessive or uncontrollable phase difference. The shunt compensator is the best option to increase the stability margin. In fact, for a given operating point, if a transient fault occurs, all three compensations presented considerably increase the stability margin. However, this is especially true for the shunt compensation.

UPFC combines the characteristics of three compensators to produce a more complete compensator. It, however, requires two voltage sources: one in series and the other in shunt connection. These two sources may operate separately as a series or a shunt reactive compensator and they may also compensate active power. Current-source converters based on thyristors with no gate turn-off capability only consume, but cannot supply reactive power, whereas the voltage-sourced converters with gate turn-off devices can supply reactive power. The most dominant converters needed in FACTS controllers are the voltage-source converters. Such converters are based on devices with gate turn-off capability.

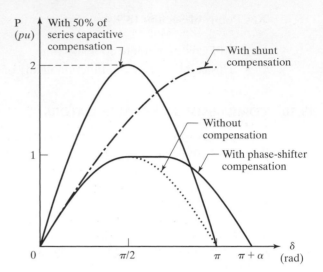

FIGURE 13.20

Power transfer characteristics with compensations and no compensation [Ref. 10].

Key Points of Section 13.10

- Each compensator performs separate functions and is suited for a specific application.
- However, UPFC combines the characteristics of three compensators to produce a more complete compensator.

SUMMARY

The amount of transfer power from the sending end to the receiving end is limited by the operating parameters of the transmission line such as line impedance, phase angle between the sending and receiving voltages, and the magnitude of the voltages. The transferable power can be increased by one of four compensation methods: shunt, series, phase-angle, and series-shunt compensations. These methods are generally implemented by switching power electronics together with appropriate control strategy. The compensators are known as FACTS controllers.

REFERENCES

[1]   N. G. Hingorani, "Power electronics in electric utilities: Role of power electronics in future power systems," *Proceedings of the IEEE,* Vol. 76, No. 4, April 1988.

[2]   N. G. Hingorani and L. Gyugyi, *Understanding FACTS: Concepts and Technology of Flexible AC Transmission Systems.* Piscataway, NJ: IEEE Press. 2000.

[3]   Y. H. Song and A. T. Johns, *Flexible AC Transmission Systems*. London, United Kingdom: IEE Press. 1999.

[4]   P. Moore and P. Ashmole, "Flexible ac transmission systems: Part 4–advanced FACTS controllers," *Power Engineering Journal,* April 1998, pp. 95–100.

[5]   E. H. Watanabe, R. M. Stephan, and M. Aredes, "New concepts of instantaneous active and reactive power for three phase system and generic loads," *IEEE Transactions on Power Delivery,* Vol. 8, No. 2, April 1993.

[6]   S. Mori, K. Matsuno, M. Takeda, and M. Seto, "Development of a large static VAR generator using self-commutated inverters for improving power system stability," *IEEE Transactions on Power Delivery,* Vol. 8, No. 1, February 1993.

[7]   C. Schauder, M. Gernhardt, E. Stacey, T. Lemak, L. Gyugyi, T. W. Cease, and A. Edris, "Development of a ±100 Mvar Static Condenser for Voltage Control of Transmission System,"*IEEE Transactions on Power Delivery,* Vol. 10, No. 3, July 1995.

[8]   B. T. Ooi, S. Z. Dai, and F. D. Galiana, "A solid-state PWM phase shifter," *IEEE Transactions on Power Delivery,* Vol. 8, No. 2, April 1993.

[9]   L. Gyugyi, "Unified power-flow control concept for flexible AC transmission systems," *IEE Proceedings-C,* Vol. 139, No. 4, July 1992.

[10]   E. H. Watanabe and P. G. Barbosa, "Principles of Operation of Facts Devices," *Workshop on FACTS*—Cigré Brazil CE 38/14, Rio de Janeiro, Brazil, November 6–9, 1995, pp. 1–12.

[11]   B. M. Han and S. I. Moon, "Static reactive-power compensator using soft-switching current-source inverter," *IEEE Transactions on Power Electronics,* No. 6, Vol. 48, December 2001, pp. 1158–1165.

REVIEW QUESTIONS

13.1 What are the parameters for controlling power in a transmission line?
13.2 What is the basic principle of shunt compensation?
13.3 What is a thyristor-controlled reactor (TCR)?
13.4 What is a thyristor-switched capacitor (TSC)?
13.5 What are the rules for transient-free switching of thyristor-switched capacitor?
13.6 What is a static VAR compensator (SVC)?
13.7 What is a STATCOM?
13.8 What is the basic principle of series compensation?
13.9 What is a thyristor-switched series capacitor (TSSC)?
13.10 What is a thyristor-controlled series capacitor (TCSC)?
13.11 What is a forced-commutation-controlled series capacitor (FCSC)?
13.12 What is a series static VAR compensator (SSVC)?
13.13 What is a series STATCOM?
13.14 What is the basic principle of phase-angle compensation?
13.15 What is a phase shifter?
13.16 What is a quadrature booster (QB)?
13.17 What is a unified power flow controller (UPFC)?

PROBLEMS

13.1 The particulars of the uncompensated transmission line in Figure 13.1a are $V = 220$ V, $f = 60$ Hz, $X = 1.8\ \Omega$, and $\delta = 70°$. Find **(a)** the line current I, **(b)** the active power P, and **(c)** the reactive power Q.

13.2 The particulars of the shunt compensated transmission line in Figure 13.2a are $V = 220$ V, $f = 60$ Hz, $X = 1.8\ \Omega$, and $\delta = 70°$. Find **(a)** the line current I, **(b)** the active power P_p, and **(c)** the reactive power Q_p.

13.3 The particulars of a shunt compensator with a TCR, as shown in Figure 13.3a, are $V = 480$ V, $f = 60$ Hz, $X = 1.8\ \Omega$, and $P_p = 96$ kW. The maximum current of the TCR is $I_{L(mx)} = 150$ A. Find **(a)** the phase angle δ, **(b)** the line current I, **(c)** the reactive power Q_p, **(d)** the current through the TCR, **(e)** the inductance reactance X_L, and **(f)** the delay angle of the TCR if the I_L is 60% of the maximum current.

13.4 The particulars of a shunt compensator with a TSC, as shown in Figure 4.4a, are $V = 480$ V, $f = 60$ Hz, $X = 1.8\ \Omega$, $\delta = 70°$, $C = 20\ \mu$F, and $L = 200\ \mu$H. The thyristor switch is operated with $m_{on} = 2$ and $m_{off} = 1$. Find **(a)** the capacitor voltage V_{co} at switching, **(b)** the peak-to-peak capacitor voltage $V_{c(pp)}$, **(c)** the rms capacitor current I_c, and **(d)** the peak switch current $I_{sw(pk)}$.

13.5 The particulars of the series compensated transmission line in Figure 13.9a are $V = 220$ V, $f = 60$ Hz, $X = 18\ \Omega$, and $\delta = 70°$. The degree of compensation is $r = 70\%$. Find **(a)** the line current I, **(b)** the active power P_p, and **(c)** the reactive power Q_p.

13.6 The particulars of a series compensator with a TCSC, as shown in Figure 4.11a, are $V = 480$ V, $f = 60$ Hz, $X = 16\ \Omega$, and $P_p = 96$ kW. The particulars of the TCSC are $\delta = 80°$, $C = 25\ \mu$F, and $L = 0.4$ mH. Find **(a)** the degree of compensation r, **(b)** the compensating capacitance reactance X_{comp} **(c)** the line current I, **(d)** the reactive power Q_c, **(e)** the delay angle α of the TCSC if the effective capacitive reactance is $X_T = -40\ \Omega$, and **(f)** plot $X_L(\alpha)$ and $X_T(\alpha)$ against the delay angle α.

13.7 The particulars of a series compensator with an FCSC, as shown in Figure 13.12a, are $V = 480$ V, $I = 150$ A, $f = 60$ Hz, $X = 18\ \Omega$, and $P_c = 96$ kW. The maximum voltage across the capacitor FCSC is $V_{C(mx)} = 50$ V. Find **(b)** the phase angle δ, **(a)** the degree of compensation r, **(c)** the capacitance C, and **(d)** the capacitive reactance X_C and the delay angle of the FCSC.

CHAPTER 14

Power Supplies

The learning objectives of this chapter are as follows:

- To understand the operation and analysis of power supplies
- To learn the types and circuit topologies of power supplies
- To learn the parameters of magnetic circuits
- To learn the techniques for designing transformers and inductors

14.1 INTRODUCTION

Power supplies, which are used extensively in industrial applications, are often required to meet all or most of the following specifications:

1. Isolation between the source and the load.
2. High-power density for reduction of size and weight.
3. Controlled direction of power flow.
4. High conversion efficiency.
5. Input and output waveforms with a low total harmonic distortion for small filters.
6. Controlled power factor (PF) if the source is an ac voltage.

The single-stage dc–dc or dc–ac or ac–dc or ac–ac converters discussed in Chapters 5, 6, 10, and 11 respectively, do not meet most of these specifications [13], and multistage conversions are normally required. There are various possible conversion topologies, depending on the permissible complexity and the design requirements. Only the basic topologies are discussed in this chapter. Depending on the type of output voltages, the power supplies can be categorized into two types:

1. Dc power supplies
2. Ac power supplies

14.2   DC POWER SUPPLIES

The ac–dc converters in Chapter 10 can provide the isolation between the input and output through an input transformer, but the harmonic contents are high. The switched-mode regulators in Section 5.8 do not provide the necessary isolation and the output power is low. The common practice is to use two-stage conversions, dc–ac and ac–dc. In the case of ac input, it is three-stage conversions, ac–dc, dc–ac, and ac–dc. The isolation is provided by an interstage transformer. The dc–ac conversion can be accomplished by pulse-width modulation (PWM) or resonant inverter. Based on the type of conversion techniques and the direction of power control, the dc power supplies can be subdivided into three types:

1. Switched-mode power supplies
2. Resonant power supplies
3. Bidirectional power supplies

14.2.1   Switched-Mode Dc Power Supplies

The switching mode supplies have high efficiency and can supply a high-load current at a low voltage. There are four common configurations for the switched-mode or PWM operation of the inverter (or dc–ac converter) stage: flyback forward, push–pull, half-bridge, and full-bridge [1, 2]. The output of the inverter, which is varied by a PWM technique, is converted to a dc voltage by a diode rectifier. Because the inverter can operate at a very high frequency, the ripples on the dc output voltage can easily be filtered out with small filters. To select an appropriate topology for an application, it is necessary to understand the merits and drawbacks of each topology and the requirements of the application. Basically, most topologies can work for different applications [3, 9].

14.2.2   Flyback Converter

Figure 14.1a shows the circuit of a flyback converter. There are two modes of operation: (1) mode 1 when switch Q_1 is turned on, and (2) mode 2 when Q_1 is turned off. Figure 14.1b shows the steady-state waveforms under a *discontinuous-mode* operation.

Mode 1.   This mode begins when switch Q_1 is turned on and it is valid for $0 < t \leq kT$, where k is the duty-cycle ratio, and T is the switching period. The voltage across the primary winding of the transformer is V_s. The primary current i_p starts to build up and stores energy in the primary winding. Due to the opposite polarity arrangement between the input and output windings of the transformer, diode D_1 is reverse biased. There is no energy transferred from the input to load R_L. The output filter capacitor C maintains the output voltage and supplies the load current i_L. The primary current i_p that increases linearly is given by

$$i_p = \frac{V_s t}{L_p} \tag{14.1}$$

where L_p is the primary magnetizing inductance. At the end of this mode at $t = kT$, the peak primary current reaches a value equal to $I_{p(pk)}$ as given by

$$I_{p(pk)} = i_p(t = kT) = \frac{V_s kT}{L_p} \tag{14.2}$$

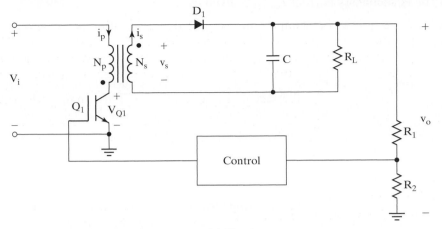

(a) Circuit

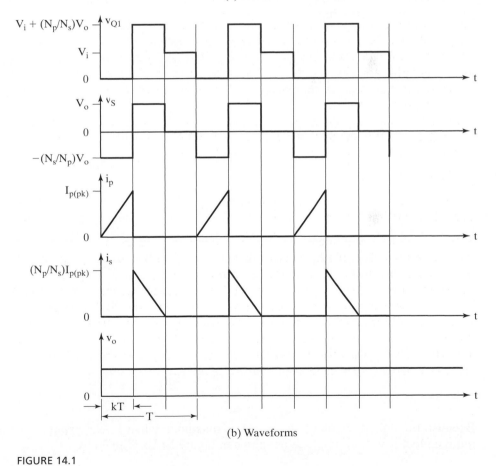

(b) Waveforms

FIGURE 14.1

Flyback converter.

The peak secondary current $I_{se(pk)}$ is given by

$$I_{se(pk)} = \left(\frac{N_p}{N_s}\right) I_{p(pk)} \qquad (14.3)$$

Mode 2.   This mode begins when switch Q_1 is turned off. The polarity of the windings reverses due to the fact that i_p cannot change instantaneously. This causes diode D_1 to turn on and charges the output capacitor C and also delivers current to R_L. The secondary current that decreases linearly is given by

$$i_{se} = I_{se(pk)} - \frac{V_o}{L_s} t \qquad (14.4)$$

where L_s is the secondary magnetizing inductance. Under the discontinuous-mode operation, i_{se} decreases linearly to zero before the start of the next cycle.

Because energy is transferred from the source to the output during the time interval 0 to kT only, the input power is given by

$$P_i = \frac{\frac{1}{2}L_p I_{p(pk)}^2}{T} = \frac{(kV_s)^2}{2fL_p} \qquad (14.5)$$

For an efficiency of η, the output power P_o can be found from

$$P_o = \eta P_i = \frac{\eta(V_s k)^2}{2fL_p} \qquad (14.6)$$

which can be equated to $P_o = V_o^2/R_L$ so that we can find the output voltage V_o as

$$V_o = V_s k \sqrt{\frac{\eta R_L}{2fL_p}} \qquad (14.7)$$

Thus, V_o can be maintained constant by keeping the product $V_s kT$ constant. Because the maximum duty-cycle k_{max} occurs at minimum supply voltage $V_{s(min)}$, the allowable k_{max} for the discontinuous mode can be found from Eq. (14.7) as

$$k_{max} = \frac{V_o}{V_{s(min)}} \sqrt{\frac{2fL_p}{\eta R_L}} \qquad (14.8)$$

Therefore, V_o at k_{max} is then given by

$$V_o = V_{s(min)} k_{max} \sqrt{\frac{\eta R_L}{2fL_p}} \qquad (14.9)$$

Because the collector voltage V_{Q1} of Q_1 is maximum when V_s is maximum, the maximum collector voltage $V_{Q1(max)}$, as shown in Figure 14.1b, is given by

$$V_{Q1(max)} = V_{s(max)} + \left(\frac{N_p}{N_s}\right) V_o \qquad (14.10)$$

The peak primary current $I_{p(pk)}$, which is the same as the maximum collector current $I_{C(max)}$ of the power switch Q_1, is given by

$$I_{C(max)} = I_{p(pk)} = \frac{2P_i}{kV_s} = \frac{2P_o}{\eta V_s k} \tag{14.11}$$

The flyback converter is used mostly in applications below 100 W. It is widely used for high-output voltage at relatively low power. Its essential features are simplicity and low cost. The switching device must be capable of sustaining a voltage $V_{Q1(max)}$ in Eq. (14.10). If the voltage is too high, the double-ended flyback converter, as shown in Figure 14.2, may be used. The two devices are switched on or off simultaneously. Diodes D_1 and D_2 are used to limit the maximum switch voltage to V_s.

Continuous versus discontinuous mode of operation. In a continuous mode of operation, switch Q_1 is turned on before the secondary current falls to zero. The continuous mode can provide higher power capability for the same value of peak current $I_{p(pk)}$. It means that, for the same output power, the peak currents in the discontinuous mode are much higher than those in continuous mode. As a result, a more expensive power transistor with a higher current rating is needed. Moreover, the higher secondary peak currents in the discontinuous mode can have a larger transient spike at the instant of turn-off. However, despite all these problems, the discontinuous mode is still more preferred than the continuous mode. There are two main reasons. First, the

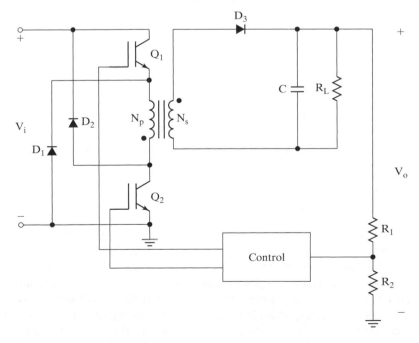

FIGURE 14.2

Double-ended flyback converter.

inherently smaller magnetizing inductance in the discontinuous mode has a quicker response and a lower transient output voltage spike to sudden change in load current or input voltage. Second, the continuous mode has a right-half-plane zero in its transfer function, thereby making the feedback control circuit more difficult to design [10, 11].

Example 14.1   Finding the Performance Parameters of a Flyback Converter

The average (or dc) output voltage of the flyback circuit in Figure 14.1a is $V_o = 24$ V at a resistive load of $R = 0.8\ \Omega$. The duty-cycle ratio is $k = 50\%$ and the switching frequency is $f = 1$ kHz. The on-state voltage drops of transistors and diodes are $V_t = 1.2$ V and $V_d = 0.7$ V, respectively. The turns ratio of the transformer is $a = N_s/N_p = 0.25$. Determine (a) the average input current I_s, (b) the efficiency η, (c) the average transistor current I_A, (d) the peak transistor current I_p, (e) the rms transistor current I_R, (f) the open-circuit transistor voltage V_{oc}, and (g) the primary magnetizing inductor L_p. Neglect the losses in the transformer, and the ripple current of the load.

Solution
$a = N_s/N_p = 0.25$ and $I_o = V_o/R = 24/0.8 = 30$ A.

a. The output power $P_o = V_o I_o = 24 \times 30 = 720$ W. The secondary voltage $V_2 = V_o + V_d = 24 + 0.7 = 24.7$ V. The primary voltage $V_1 = V_2/a = 24.7/0.25 = 98.8$ V. The input voltage $V_s = V_1 + V_t = 98.8 + 1.2 = 100$ and the input power is

$$P_i = V_s I_s = 1.2 I_A + V_d I_o + P_o$$

Substituting $I_A = I_s$ gives

$$I_s(100 - 1.2) = 0.7 \times 30 + 720$$
$$I_s = \frac{741}{98.8} = 7.5 \text{ A}$$

b. $P_i = V_s I_s = 100 \times 7.5 = 750$ W. The efficiency $\eta = 7.5/750 = 96.0\%$.

c. $I_A = I_s = 7.5$ A.

d. $I_p = 2I_A/k = 2 \times 7.5/0.5 = 30$ A.

e. $I_R = \sqrt{k/3}\, I_p = \sqrt{0.5/3} \times 30 = 12.25$ A, for 50% duty cycle.

f. $V_{oc} = V_s + V_2/a = 100 + 24.7/0.25 = 198.8$ V.

g. Using Eq. (14.2) for I_p gives $L_p = V_s k/f I_p = 100 \times 0.5/(1 \times 10^{-3} \times 30) = 1.67$ mH.

14.2.3 Forward Converter

The forward converter is similar to the flyback. The transformer core is reset by reset winding, as shown in Figure 14.3a, where the energy stored in the transformer core is returned to the supply and the efficiency is increased. The dot on the secondary winding of the transformer is so arranged that the output diode D_2 is forward biased when the voltage across the primary is positive, that is, when the transistor is on. Thus, energy is not stored in the primary inductance as it was for the flyback. The transformer acts strictly as an ideal transformer. Unlike the flyback, the forward converter is operated in the continuous mode. In the discontinuous mode, the forward converter is more difficult to control because of a double pole existing at the output filter. There are two modes of

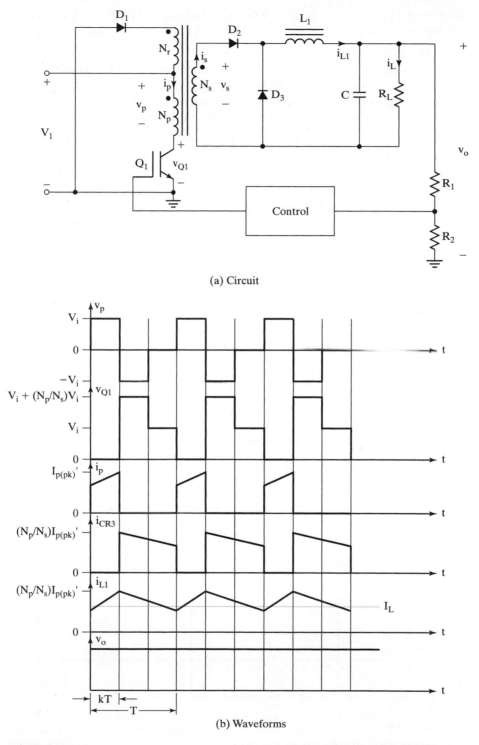

(a) Circuit

(b) Waveforms

FIGURE 14.3

Forward converter.

operation: (1) mode 1 when switch Q_1 is turned on, and (2) mode 2 when Q_1 is turned off. Figure 14.3b shows the steady-state waveforms in the continuous mode operation.

Mode 1.   This mode begins when switch Q_1 turns on. The voltage across the primary winding is V_s. The primary current i_p starts to build up and transfers energy from the primary winding to the secondary and onto the L_1C filter and the load R_L through the rectifier diode D_2, which is forward biased.

The secondary current I_{se} is reflected into the primary as I_p, as shown in Figure 14.4, given by

$$i_p = \frac{N_s}{N_p} i_{se} \tag{14.12}$$

The primary magnetizing current i_{mag} that rises linearly is given by

$$I_{mag} = \frac{V_s}{L_p} t \tag{14.13}$$

Thus, the total primary current i'_p is given by

$$i'_p = i_p + i_{mag} = \frac{N_s}{N_p} i_{se} + \frac{V_s}{L_p} t \tag{14.14}$$

At the end of mode 1 at $t = kT$, the total primary current reaches a peak value $I'_{p(pk)}$ as given by

$$I'_{p(pk)} = I_{p(pk)} + \frac{V_s kT}{L_p} \tag{14.15}$$

where $I_{p(pk)}$ is reflected peak current of the output inductor L_1 from the secondary as given by

$$I_{p(pk)} = \left(\frac{N_p}{N_s}\right) I_{L1(pk)} \tag{14.16}$$

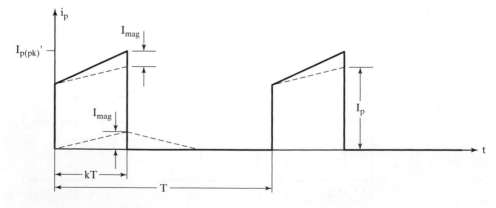

FIGURE 14.4

Current components in the primary winding.

The voltage developed across the secondary winding is

$$V_{se} = \frac{N_s}{N_p} V_s \tag{14.17}$$

Because the voltage across the output inductor L_1 is $V_{se} - V_o$, its current i_{L1} increases linearly at a rate of

$$\frac{di_{L1}}{dt} = \frac{V_s - V_o}{L_1}$$

which gives the peak output inductor current $I_{L1(pk)}$ at $t = kT$ as

$$I_{L1(pk)} = I_{L1}(0) + \frac{(V_s - V_o)kT}{L_1} \tag{14.18}$$

Mode 2. This mode begins when Q_1 turns off. The polarity of the transformer voltage reverses. This causes D_2 to turn off and D_1 and D_3 to turn on. While D_3 is conducting, energy is delivered to R_L through the inductor L_1. Diode D_1 and the tertiary winding provide a path for the magnetizing current returning to the input. The current i_{L1} through inductor L_1, which is equal to the current i_{D3} through diode D_3, decreases linearly as given by

$$i_{L1} = i_{D3} = I_{L1(pk)} - \frac{V_o}{L_1} t \quad \text{for } 0 < t \le (1 - k)T \tag{14.19}$$

which gives $I_{L1}(0) = i_{L1}(t = (1 - k)T) = I_{L1(pk)} - V_0(1 - k)T/L_1$ in the continuous mode operation. The output voltage V_o, which is the time integral of the secondary winding voltage, is given

$$V_o = \frac{1}{T} \int_0^{kT} \frac{N_s}{N_p} V_s \, dt = \frac{N_s}{N_p} V_s k \tag{14.20}$$

The maximum collector current $I_{C(max)}$ during turn-on is equal to $I'_{p(pk)}$ as given by

$$I_{C(max)} = I'_{p(pk)} = \left(\frac{N_p}{N_s} \right) I_{L1(pk)} + \frac{V_s kT}{L_p} \tag{14.21}$$

The maximum collector voltage $V_{Q1(max)}$ at turn-off, which is equal to the maximum input voltage $V_{i(max)}$ plus the maximum voltage $V_{r(max)}$ across the tertiary, is given by

$$V_{Q1(max)} = V_{s(max)} + V_{r(max)} = V_{s(max)} \left(1 + \frac{N_p}{N_r} \right) \tag{14.22}$$

Equating the time integral of the input voltage when Q_1 is on to the clamping voltage V_r when Q_1 is off gives

$$V_s kT = V_r (1 - k)T \tag{14.23}$$

which, after replacing V_r/V_s by N_r/N_p, give the maximum duty-cycle k_{max} as

$$k_{max} = \frac{1}{1 + N_r/N_p} \tag{14.24}$$

Thus, k_{max} depends on the turns ratio between the resetting winding and the primary one. The duty-cycle k must be kept below the maximum duty-cycle k_{max} to avoid saturating the transformer. The transformer magnetizing current must be reset to zero at the end of each cycle. Otherwise, the transformer can be driven into saturation, which can cause damage to the switching device. A tertiary winding, as shown in Figure 14.3a, is added to the transformer so that the magnetizing current can return to the input source V_s when the transistor turns off.

The forward converter is widely used with output power below 200 W, though it can be easily constructed with a much higher output power. The limitations are due to the inability of the power transistor to handle the voltage and current stresses. Figure 14.5 shows a double-ended forward converter. The circuit uses two transistors that are switched on and off simultaneously. The diodes are used to restrict the maximum collector voltage to V_s. Therefore, the transistors with low-voltage rating can be used.

Flyback versus forward converter.   Unlike the flyback, the forward converter requires a minimum load at the output. Otherwise, excess output voltage can be produced. To avoid this situation, a large load resistance is permanently connected across

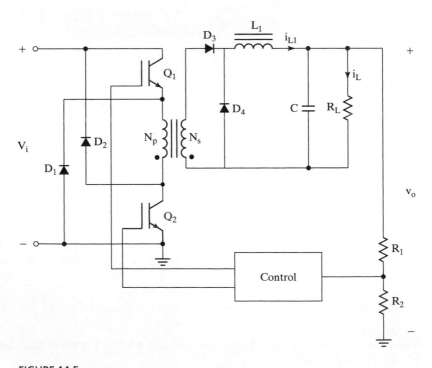

FIGURE 14.5

Double-ended forward converter.

the output terminals. Because the forward converter does not store energy in the transformer, for the same output power level, the size of the transformer can be made smaller than that for the flyback. The output current is reasonably constant due to the action of the output inductor and the freewheeling diode D_3. As a result, the output filter capacitor can be made smaller and its ripple current rating can be much lower than that required for the flyback.

Example 14.2   Finding the Performance Parameters of a Forward Converter

The average (or dc) output voltage of the push–pull circuit in Figure 14.3a is $V_o = 24$ V at a resistive load of $R = 0.8$ Ω. The on-state voltage drops of transistors and diodes are $V_t = 1.2$ V and $V_d = 0.7$ V, respectively. The duty cycle is $k = 40\%$ and the switching frequency is $f = 1$ kHz. The dc supply voltage $V_s = 12$ V. The turns ratio of the transformer is $a = N_s/N_p = 0.25$. Determine (a) the average input current I_s, (b) the efficiency η, (c) the average transistor current I_A, (d) the peak transistor current I_p, (e) the rms transistor current I_R, (f) the open-circuit transistor voltage V_{oc}, (g) the primary magnetizing inductor L_p for maintaining the peak-to-peak ripple current to 5% of the average input dc current, and (h) the output inductor L_1 for maintaining the peak-to-peak ripple current to 4% of its average value. Neglect the losses in the transformer and the ripple content of the output voltage is 3%.

Solution

$a = N_s/N_p = 0.25$ and $I_o = V_o/R = 24/0.8 = 30$ A.

a. The output power $P_o = V_o I_o = 24 \times 30 = 720$ W. The secondary voltage $V_2 = V_o + V_d = 24 + 0.7 = 24.7$ V. The primary voltage is $V_1 = V_s - V_t = 12 - 1.2 = 10.8$ V. The turns ratio is $a = V_2/V_1 = 24.7/10.8 = 2.287$. The input power is $P_i = V_s I_s = V_t k I_s + V_d (1 - k) I_s + V_d I_o + P_o$ which gives

$$I_s = \frac{V_d I_o + P_o}{V_s - V_t k - V_d (1 - k)} = \frac{0.7 \times 30 + 720}{12 - 1.2 \times 0.4 - 0.7 \times 0.6} = 66.76 \text{ A}$$

b. $P_i = V_s I_s = 12 \times 66.756 = 801$ W. The efficiency $\eta = 720/801 = 89.9\%$.

c. $I_A = k I_s = 0.4 \times 66.76 = 26.7$ A.

d. $\Delta I_p = 0.05 \times I_s = 0.05 \times 66.76 = 3.353$ A.

e. $I_R = \sqrt{k} [I_p^2 + {}^{\Delta I_p}/_3 + \Delta I_p I_p]^{1/2} = \sqrt{0.4} \times [66.76^2 + 3.35/3 + 3.35 \times 66.76]^{1/2} = 44.3$ A.

f. $V_{oc} = V_s + V_2/a = 22.8$ V.

g. $\Delta I_{L1} = 0.04 \times I_o = 0.04 \times 30 = 1.2$ A and $\Delta V_o = 0.03 \times V_o = 0.03 \times 24 = 0.72$ V. Using Eq. (14.18), $L_1 = \dfrac{\Delta V_o k}{f \Delta I_{L1}} = \dfrac{0.72 \times 0.4}{1 \times 10^3 \times 1.2} = 0.24$ mH

h. Using (14.15), $\Delta I_p = a \times \Delta I_{L1} + (V_s - V_t) k T / L_p$, which gives

$$L_p = \frac{(V_s - V_t) k}{f (\Delta I_p - a \times \Delta I_{L1})} = \frac{(12 - 1.2) \times 0.4}{1 \times 10^3 \times (3.353 - 2.287 \times 1.2)} = 7.28 \text{ mH}$$

14.2.4 Push–Pull Converter

The push–pull configuration is shown in Figure 14.6. When Q_1 is turned on, V_s appears across one-half of the primary. When Q_2 is turned on, V_s is applied across the other half of the transformer. The voltage of a primary winding swings from $-V_s$ to V_s. The

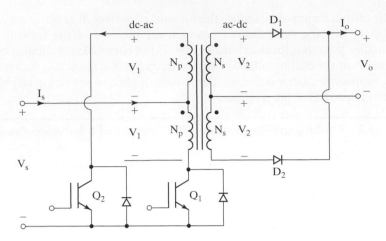

FIGURE 14.6

Push–pull converter configuration.

average current through the transformer should ideally be zero. The average output voltage is

$$V_o = V_2 = \frac{N_s}{N_p}V_1 = aV_1 = aV_s \tag{14.25}$$

Transistors Q_1 and Q_2 operate with a 50% duty cycle. The open-circuit voltage is $V_{oc} = 2V_s$, the average current of a transistor is $I_A = I_s/2$, and the peak transistor current is $I_p = I_s$. Because the open-circuit transistor voltage is twice the supply voltage, this configuration is suitable for low-voltage applications.

The push–pull converter is often driven by a constant current source I_s such that primary current is a square wave that produces a secondary voltage.

Example 14.3   Finding the Performance Parameters of a Push–Pull Converter

The average (or dc) output voltage of the push–pull circuit in Figure 14.6 is $V_o = 24$ V at a resistive load of $R = 0.8\ \Omega$. The on-state voltage drops of transistors and diodes are $V_t = 1.2$ V and $V_d = 0.7$ V, respectively. The turns ratio of the transformer is $a = N_s/N_p = 0.25$. Determine (a) the average input current I_s, (b) the efficiency η, (c) the average transistor current I_A, (d) the peak transistor current I_p, (e) the rms transistor current I_R, and (f) the open-circuit transistor voltage V_{oc}. Neglect the losses in the transformer, and the ripple current of the load and input supply is negligible. Assume duty cycle, $k = 0.5$.

Solution
$a = N_s/N_p = 0.25$ and $I_o = V_o/R = 24/0.8 = 30$ A.

    a. The output power $P_o = V_oI_o = 24 \times 30 = 720$ W. The secondary voltage $V_2 = V_o + V_d = 24 + 0.7 = 24.7$ V. The primary voltage $V_1 = V_2/a = 24.7/0.25 = 98.8$ V. The input voltage $V_s = V_1 + V_t = 98.8 + 1.2 = 100$ and the input power is

$$P_i = V_sI_s = 1.2I_A + 1.2I_A + V_dI_o + P_o$$

Substituting $I_A = I_s/2$ gives

$$I_s(100 - 1.2) = 0.7 \times 30 + 720$$

$$I_s = \frac{741}{98.8} = 7.5 \text{ A}$$

b. $P_i = V_s I_s = 100 \times 7.5 = 750$ W. The efficiency $\eta = 720/750 = 96.0\%$.

c. $I_A = I_s/2 = 7.5/2 = 3.75$ A.

d. $I_p = I_s = 7.5$ A.

e. $I_R = \sqrt{k} I_p = \sqrt{0.5} \times 7.5 = 5.30$ A, for 50% duty cycle.

f. $V_{oc} = 2V_s = 2 \times 100 = 200$ V.

14.2.5 Half-Bridge Converter

Figure 14.7a shows the basic configuration of a half-bridge converter. This converter can be viewed as two back-to-back forward converters that are fed by the same input voltage, each delivering power to the load at each alternate half-cycle. The capacitors C_1 and C_2 are placed across the input terminals such that the voltage across the primary winding always is half of the input voltage $V_s/2$.

There are four modes of operation: (1) mode 1 when switch Q_1 is turned on and switch Q_2 is off, (2) mode 2 when both Q_1 and Q_2 are off, (3) mode 3 when switch Q_1 is off and switch Q_2 is turned on, and (4) mode 4 when both Q_1 and Q_2 are off again. Switches Q_1 and Q_2 are switched on and off accordingly to produce a square-wave ac at the primary side of the transformer. This square wave is either stepped down or up by the isolation transformer and then rectified by diodes D_1 and D_2. The rectified voltage is, subsequently, filtered to produce the output voltage V_o. Figure 14.7b shows the steady-state waveforms in the continuous mode operation.

Mode 1. During this mode Q_1 is on and Q_2 is off, D_1 conducts and D_2 is reverse biased. The primary voltage V_p is $V_s/2$. The primary current i_p starts to build up and stores energy in the primary winding. This energy is forward transferred to the secondary and onto the L_1C filter and the load R_L through the rectifier diode D_1.

The voltage across the secondary winding is given by

$$V_{se} = \frac{N_{s1}}{N_p}\left(\frac{V_s}{2}\right) \tag{14.26}$$

The voltage across the output inductor is then given by

$$v_{L1} = \frac{N_{s1}}{N_p}\left(\frac{V_s}{2}\right) - V_o \tag{14.27}$$

The inductor current i_{L1} increases linearly at a rate of

$$\frac{di_{L1}}{dt} = \frac{v_{L1}}{L_1} = \frac{1}{L_1}\left[\frac{N_{s1}}{N_p}\left(\frac{V_s}{2}\right) - V_o\right]$$

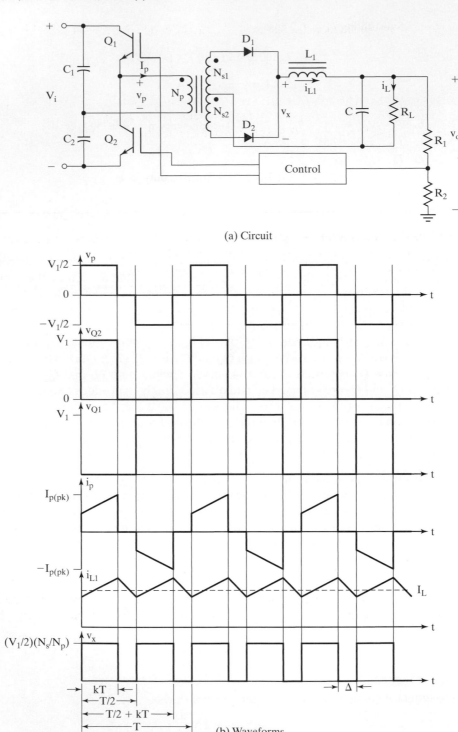

(a) Circuit

(b) Waveforms

FIGURE 14.7

Half-bridge converter.

which gives the peak inductor current $I_{L1(pk)}$ at the end of this mode at $t = kT$ as given by

$$I_{L1(pk)} = I_{L1}(0) + \frac{1}{L_1}\left[\frac{N_{s1}}{N_p}\left(\frac{V_s}{2}\right) - V_o\right]kT \tag{14.28}$$

Mode 2. This mode is valid for $kT < t \le T/2$. During this mode both Q_1 and Q_2 are off, and D_1 and D_2 are forced to conduct the magnetizing current that resulted during mode 1. Redefining the time origin at the beginning of this mode, the rate of fall of i_{L1} is given by

$$\frac{di_{L1}}{dt} = -\frac{V_o}{L_1} \qquad \text{for } 0 < t \le (0.5 - k)T \tag{14.29}$$

which gives $I_{L1}(0) = i_{l,1}[t = (0.5 - k)T] = I_{L1(pk)} - V_o(0.5 - k)T/L_1$.

Modes 3 and 4. During mode 3, Q_2 is on and Q_1 off, D_1 is reverse biased, and D_2 conducts. The primary voltage V_p is now $-V_s/2$. The circuit operates in the same manner as in mode 1 followed by mode 4 that is similar to mode 2.

The output voltage V_o can be found from the time integral of the inductor voltage v_{L1} over the switching period T. That is,

$$V_o - 2 \times \frac{1}{T}\left[\int_0^{kT}\left(\frac{N_{s1}}{N_p}\left(\frac{V_s}{2}\right) - V_o\right)dt + \int_{T/2}^{T/2+kT} - V_o\,dt\right]$$

which gives V_o as

$$V_o = \frac{N_{s1}}{N_p}V_s k \tag{14.30}$$

The output power P_o is given by

$$P_o = V_o I_L = \eta P_i = \eta \frac{V_s I_{p(\text{avg})}k}{2}$$

which gives

$$I_{p(\text{avg})} = \frac{2P_o}{\eta V_s k} \tag{14.31}$$

where $I_{p(\text{avg})}$ is the average primary current. Assuming that the secondary load current reflected to the primary side is much greater than the magnetizing current, then the maximum collector currents for Q_1 and Q_2 are given by

$$I_{C(\text{max})} = I_{p(\text{avg})} = \frac{2P_o}{\eta V_s k_{\text{max}}} \tag{14.32}$$

The maximum collector voltages for Q_1 and Q_2 during turn-off are given by

$$V_{C(\text{max})} = V_{s(\text{max})} \tag{14.33}$$

The maximum duty-cycle k can never be greater than 50%. The half-bridge converter is widely used for medium-power applications. Because of its core-balancing feature, the half-bridge converter becomes the predominant choice for output power ranging from 200 to 400 W.

Forward versus half-bridge converter.   In a half-bridge converter, the voltage stress imposed on the power transistors is subject to only the input voltage and is only half of that in a forward converter. Thus, the output power of a half-bridge is double to that of a forward converter for the same semiconductor devices and magnetic core. Because the half-bridge is more complex, for application below 200 W, the flyback or forward converter is considered to be a better choice and more cost-effective. Above 400 W, the primary and switch currents of the half-bridge become very high. Thus, it becomes unsuitable for high-power applications.

Note: The emitter of Q_1 is not at ground level, but is at a high-ac level. Thus, the gate driver circuit must be isolated from the ground through transformers or other coupling devices.

14.2.6 Full-Bridge Converter

Figure 14.8a shows the basic configuration of a full-bridge converter with four power switches [4]. There are four modes of operation: (1) mode 1 when switches Q_1 and Q_4 are on, while Q_2 and Q_3 are off; (2) mode 2 when all switches are off; (3) mode 3 when switches Q_1 and Q_4 are off, while Q_2 and Q_3 are on; and (4) mode 4 when all switches are off. Switches are switched on and off accordingly to produce a square-wave ac at the primary side of the transformer. The output voltage is stepped up (or down), rectified, and then filtered to produce a dc output voltage. The capacitor C_1 is used to balance the volt-second integrals during the two half-cycles and prevent the transformer from becoming driven into saturation. Figure 14.8b shows the steady-state waveforms in the continuous mode operation.

Mode 1.   During this mode both Q_1 and Q_4 are turned on. The voltage across the secondary winding is

$$V_{se} = \frac{N_s}{N_p} V_s \qquad (14.34)$$

The voltage across the output inductor L_1 is then given by

$$v_{L1} = \frac{N_s}{N_p} V_s - V_o \qquad (14.35)$$

The inductor current i_{L1} increases linearly at a rate of

$$\frac{di_{L1}}{dt} = \frac{v_{L1}}{L_1} = \frac{1}{L_1}\left[\frac{N_s}{N_p} V_s - V_o\right] \qquad (14.36)$$

which gives the peak inductor current $I_{L1(pk)}$ at the end of this mode at $t = kT$ as given by

$$I_{L1(pk)} = I_{L1}(0) + \frac{1}{L_1}\left[\frac{N_s}{N_p} V_s - V_o\right] kT \qquad (14.37)$$

Mode 2.   This mode is valid for $kT < t \leq T/2$. During this mode all switching devices are off, while D_1 and D_2 are forced to conduct the magnetizing current at the

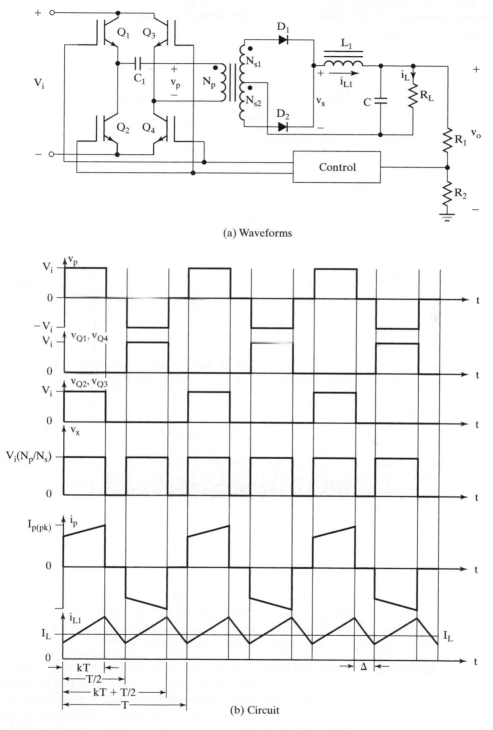

(a) Waveforms

(b) Circuit

FIGURE 14.8

Full-bridge converter.

end of mode 1. Redefining the time origin at the beginning of this mode, the rate of fall of i_{L1} is given by

$$\frac{di_{L1}}{dt} = -\frac{V_o}{L_1} \qquad \text{for } 0 < t \le (0.5 - k)T \tag{14.38}$$

which gives $I_{L1}(0) = i_{L1}[t = (0.5 - k)T] = I_{L1(pk)} - V_0(0.5 - k)T/L_1$.

Modes 3 and 4. During mode 3, Q_2 and Q_3 are on, while Q_1 and Q_4 are off. D_1 is reverse biased and D_2 conducts. The voltage across the primary V_p is V_s. The circuit operates in the same manner as in mode 1 followed by mode 4 that is similar to mode 2.

The output voltage V_o can be found from the time integral of the inductor voltage v_{L1} over the switching period T. That is,

$$V_o = 2 \times \frac{1}{T} \left[\int_0^{kT} \left(\frac{N_s}{N_p} V_s - V_o \right) dt + \int_{T/2}^{T/2 + kT} -V_o \, dt \right]$$

which gives V_o as

$$V_o = \frac{N_s}{N_p} 2V_s k \tag{14.39}$$

The output power P_o is given by

$$P_o = \eta P_i = \eta V_s I_{p(\text{avg})} k$$

which gives

$$I_{p(\text{avg})} = \frac{P_o}{\eta V_s k} \tag{14.40}$$

where $I_{p(\text{avg})}$ has the average primary current. Neglecting the magnetizing current, the maximum collector currents for $Q_1, Q_2, Q_3,$ and Q_4 are given by

$$I_{C(\text{max})} = I_{p(\text{avg})} = \frac{P_o}{\eta V_s k_{\text{max}}} \tag{14.41}$$

The maximum collector voltage for $Q_1, Q_2, Q_3,$ and Q_4 during turn off is given by

$$V_{C(\text{max})} = V_{s(\text{max})} \tag{14.42}$$

The full-bridge regulator is used for high-power applications ranging from several hundred to thousand kilowatts. It has the most efficient use of magnetic core and semiconductor switches. The full-bridge is complex and therefore expensive to build, and is only justified for high-power applications, typically over 500 W.

Half-bridge versus full-bridge converter. The full-bridge uses four power switches instead of two, as in the half-bridge. Therefore, it requires two more gate drivers and secondary windings in the pulse transformer for the gate control circuit. Comparing Eq. (14.41) with Eq. (14.31), for the same output power, the maximum collector current of a full bridge is only half that of the half bridge. Thus, the output power of a full bridge is twice that of a half bridge with the same input voltage and current.

Note: The emitter terminals of Q_1 and Q_3 are not at ground level, but are at high-ac levels. Thus, the gate driver circuit must be isolated from the ground through transformers or other coupling devices.

14.2.7 Resonant Dc Power Supplies

If the variation of the dc output voltage is not wide, resonant pulse inverters can be used. The inverter frequency, which could be the same as the resonant frequency, is very high, and the inverter output voltage is almost sinusoidal [12]. Due to resonant oscillation, the transformer core is always reset and there are no dc saturation problems. The half-bridge and full-bridge configurations of the resonant inverters are shown in Figure 14.9. The sizes of the transformer and output filter are reduced due to high inverter frequency.

Example 14.4   Finding the Performance Parameters of a Half-Bridge Resonant Inverter

The average output voltage of the half-bridge resonant circuit in Figure 14.9a is $V_o = 24$ V at a resistive load of $R_L - 0.8\ \Omega$. The inverter operates at the resonant frequency. The circuit parameters are $C_1 = C_2 = C = 1\ \mu F$, $L = 20\ \mu H$, and $R = 0$. The dc input voltage is $V_s = 100$ V. The on-state voltage drops of transistors and diodes are negligible. The turns ratio of the transformer is $a = N_s/N_p = 0.25$. Determine (a) the average input current I_s, (b) the average transistor current I_A, (c) the peak transistor current I_p, (d) the rms transistor current I_s, and (e) the open-circuit transistor voltage V_{oc}. Neglect the losses in the transformer, and the effect of the load on the resonant frequency is negligible.

Solution

$C_e = C_1 + C_2 = 2C$. The resonant frequency $\omega_r = 10^6/\sqrt{2 \times 20} = 158,113.8$ rad/s or $f_r = 25,164.6$ Hz, a = $N_s/N_p = 0.25$, and $I_o = V_o/R = 24/0.8 = 30$ A.

a. The output power $P_o = V_o I_o = 24 \times 30 = 720$ W. From Eq. (3.21), the rms secondary voltage $V_2 = \pi V_o/(2\sqrt{2}) = 1.1107 V_o = 26.66$ V. The average input current $I_s = 720/100 = 7.2$ A.

b. The average transistor current $I_A = I_s = 7.2$ A.

c. For a sinusoidal pulse of current through the transistor, $I_A = I_p/\pi$ and the peak transistor current $I_p = 7.2\pi = 22.62$ A.

d. With a sinusoidal pulse of current with 180° conduction, the rms transistor current $I_R = I_p/2 = 11.31$ A.

e. $V_{oc} = V_s = 100$ V.

14.2.8 Bidirectional Power Supplies

In some applications, such as battery charging and discharging, it is desirable to have bidirectional power flow capability. A bidirectional power supply is shown in Figure 14.10. The direction of the power flow depends on the values of V_o, V_s, and turns ratio $(a = N_s/N_p)$. For power flow from the source to the load, the inverter operates in the inversion mode if

$$V_o < aV_s \qquad (14.43)$$

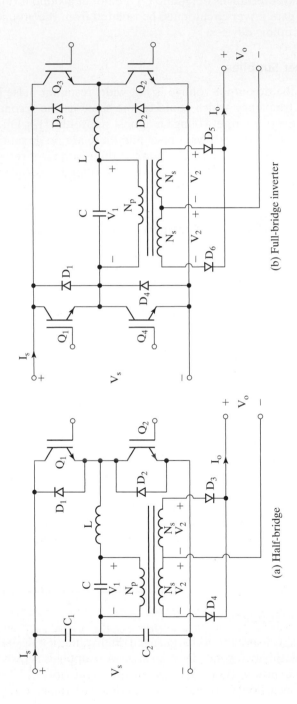

(a) Half-bridge

(b) Full-bridge inverter

FIGURE 14.9

Configurations for resonant dc power supplies.

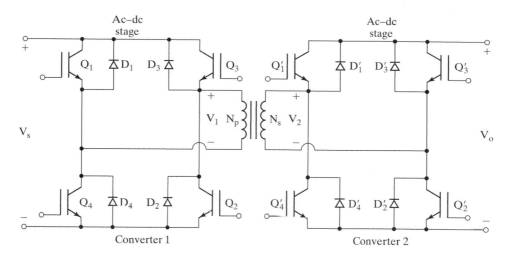

FIGURE 14.10

Bidirectional dc power supply.

For power flow from the output to the input, the inverter operates as a rectifier if

$$V_o > aV_s \tag{14.44}$$

The bidirectional converters allow the inductive current to flow in either direction and the current flow becomes continuous.

Key Points of Section 14.2

- Although most of the converters can be used to meet the dc output requirements, the ratings of the switching device and the ratings and sizes of the transformer limit their applications to a specific output power. The choice of the converter depends on the output power requirement.
- The push–pull converter is often driven by a constant current source such that primary current is a square wave that produces a secondary voltage. The size of the transformer and output inductor is reduced in resonant power supplies.

14.3   AC POWER SUPPLIES

The ac power supplies are commonly used as standby sources for critical loads and in applications where normal ac supplies are not available. The standby power supplies are also known as uninterruptible power supply (UPS) systems. The two configurations commonly used in UPSs are shown in Figure 14.11. The load in the configuration of Figure 14.11a is normally supplied from the ac main supply and the rectifier maintains the full charge of the battery. If the supply fails, the load is switched to the output of the inverter, which then takes over the main supply. This configuration requires breaking the circuit momentarily and the transfer by a solid-state switch usually takes 4 to 5 ms.

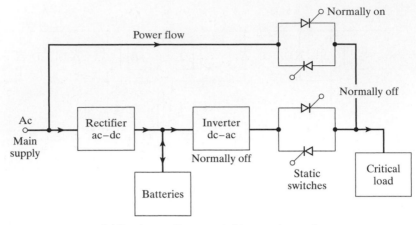

(a) Load normally connected to ac main supply

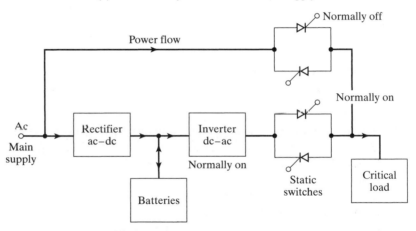

(b) Load normally connected to inverter

FIGURE 14.11

UPS configurations.

The switchover by a mechanical contactor may take 30 to 50 ms. The inverter runs only during the time when the supply failure occurs.

The inverter in the configuration of Figure 14.11b operates continuously and its output is connected to the load. There is no need for breaking the supply in the event of supply failure. The rectifier supplies the inverter and maintains the charge on the standby battery. The inverter can be used to condition the supply to the load, to protect the load from the transients in the main supply, and to maintain the load frequency at the desired value. In case of inverter failure, the load is switched to the main supply.

The standby battery is normally either a nickel–cadmium or a lead–acid type. A nickel–cadmium battery is preferable to a lead–acid battery, because the electrolyte of a nickel–cadmium battery is noncorrosive and does not emit explosive gas. It has a longer life due to its ability to withstand overheating or discharging. However, its cost

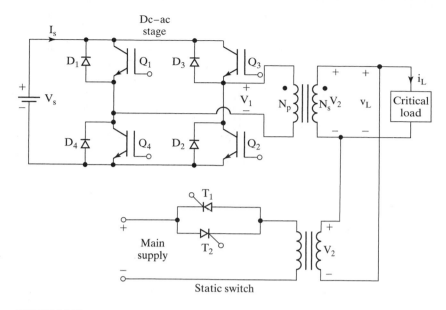

FIGURE 14.12

Arrangement of UPS systems.

is at least three times that of a lead–acid battery. An alternative arrangement of a UPS system is shown in Figure 14.12, which consists of a battery, an inverter, and a static switch. In case of power failure, the battery supplies the inverter. When the main supply is on, the inverter operates as a rectifier and charges the battery. In this arrangement the inverter has to operate at the fundamental output frequency. Consequently, the high-frequency capability of the inverter is not utilized in reducing the size of the transformer. Similar to dc power supplies, ac power supplies can be categorized into three types:

1. Switched-mode ac power supplies
2. Resonant ac power supplies
3. Bidirectional ac power supplies

14.3.1   Switched-Mode Ac Power Supplies

The size of the transformer in Figure 14.12 can be reduced by the addition of a high-frequency dc link, as shown in Figure 14.13. There are two inverters. The input-side inverter operates with a PWM control at a very high frequency to reduce the size of the transformer and the dc filter at the input of the output-side inverter. The output-side inverter operates at the output frequency.

14.3.2   Resonant Ac Power Supplies

The input-stage inverter in Figure 14.13 can be replaced by a resonant inverter, as shown in Figure 14.14. The output-side inverter operates with a PWM control at the output frequency.

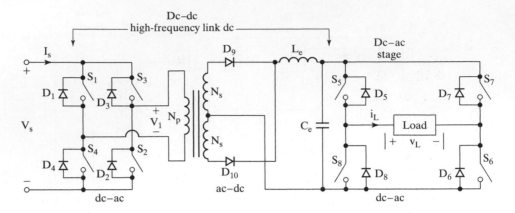

FIGURE 14.13

Switched-mode ac power supplies.

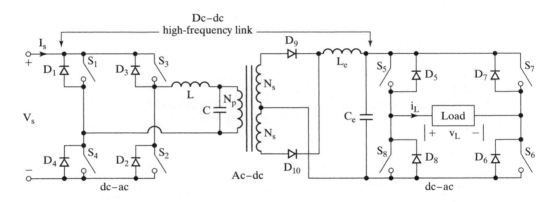

FIGURE 14.14

Resonant ac power supply.

14.3.3 Bidirectional Ac Power Supplies

The diode rectifier and the output inverter can be combined by a cycloconverter with bidirectional switches, as shown in Figure 14.15. The cycloconverter converts the high-frequency ac to a low-frequency ac. The power flow can be controlled in either direction.

Example 14.5 Finding the Performance Parameters of an Ac Power Supply with a PWM Control

The load resistance of the ac power supply in Figure 14.13 is $R = 2.5\ \Omega$. The dc input voltage is $V_s = 100$ V. The input inverter operates at a frequency of 20 kHz with one pulse per half-cycle. The on-state voltage drops of transistor switches and diodes are negligible. The turns ratio of the transformer is $a = N_s/N_p = 0.5$. The output inverter operates with a uniform PWM of four pulses

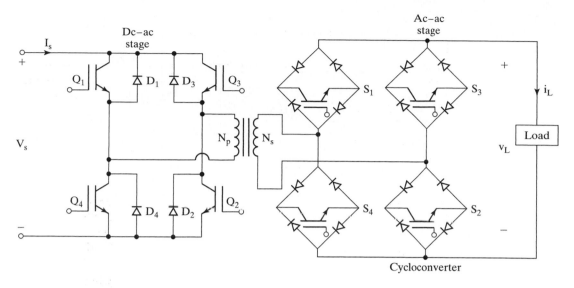

FIGURE 14.15

Bidirectional ac power supplies.

per half-cycle. The width of each pulse is $\delta = 18°$. Determine the rms load current. The ripple voltage on the output of the rectifier is negligible. Neglect the losses in the transformer, and the effect of the load on the resonant frequency is negligible.

Solution

The rms output voltage of the input inverter is $V_1 = V_s = 100$ V. The rms transformer secondary voltage is $V_2 = aV_1 = 0.5 \times 100 = 50$ V. The dc voltage of the rectifier is $V_o = V_2 = 50$ V. With the pulse width of $\delta = 18°$, Eq. (6.31) gives the rms load voltage, $V_L = V_o\sqrt{(p\delta/\pi)} = 50\sqrt{4 \times 18/180} = 31.6$ V. The rms load current is $I_L = V_L/R = 31.6/2.5 = 12.64$ A.

Key Points of Section 14.3

- The ac power supplies are commonly used as standby sources for critical loads and in applications where normal ac supplies are not available.
- They use switched mode, resonant, or bidirectional types of conversion.

14.4 MULTISTAGE CONVERSIONS

If the input is an ac source, an input-stage rectifier is required as shown in Figure 14.16 and there are four conversions: ac–dc–ac–dc–ac. The rectifier and inverter pair can be replaced by a converter with bidirectional ac switches as shown in Figure 14.17. The switching functions of this converter can be synthesized to combine the functions of the rectifier and inverter. This converter, which converts ac–ac directly, is called the forced commutated cycloconverter. The ac–dc–ac–dc–ac conversions in Figure 14.16 can be performed by two forced-commutated cycloconverters as shown in Figure 14.17.

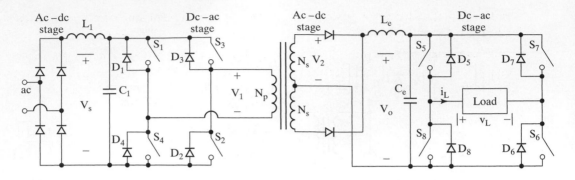

FIGURE 14.16

Multistage conversions.

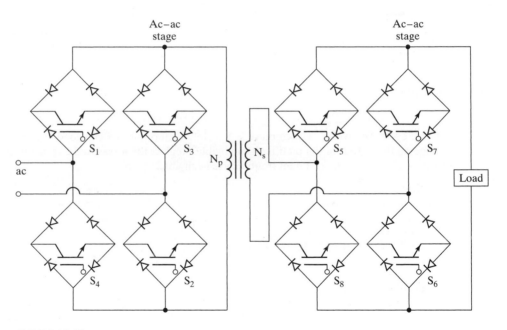

FIGURE 14.17

Cycloconverters with bilateral switches.

14.5 CONTROL CIRCUITS

Varying the duty-cycle k can control the output voltage of a converter. There are commercially available PWM integrated circuit (IC) controllers that have all the features to build a PWM switching power supply using a minimum number of components. A PWM controller consists of four main functional components: an adjustable clock for setting the switching frequency, an output voltage error amplifier, a sawtooth generator for providing a sawtooth signal that is synchronized to the clock, and a comparator that compares the output error signal with the sawtooth signal. The output of the comparator is the signal that drives the power switch. Either voltage-mode control or current-mode control is normally applied.

Voltage-mode control.   Figure 14.18a shows a simple PWM controlled forward converter operating at fixed frequency [1]. The duration of the on-time is determined by the time between the reset of the sawtooth generator and the intersection of the error voltage with the positive-going ramp signal.

The error voltage v_E is given by

$$v_e = \left(1 + \frac{Z_2}{Z_1}\right) V_{\text{REF}} - \frac{Z_2}{Z_1} v_A \qquad (14.45)$$

which can be separated into two components: $v_E = V_E + \Delta v_e$ due to the feedback voltage $v_A = V_A + \Delta v_a$. The dc operating point is given by

$$V_E = \left(1 + \frac{Z_2}{Z_1}\right) V_{\text{REF}} - \frac{Z_2}{Z_1} V_A \qquad (14.46)$$

The small-signal term can be separated from the dc operating point as

$$\Delta v_e = -\frac{Z_2}{Z_1} \Delta v_a \qquad (14.47)$$

The duty cycle k as shown in Figure 14.18b is related to the error voltage by

$$k = \frac{v_e}{V_{cr}} \qquad (14.48)$$

where V_{cr} is the peak voltage of the sawtooth carrier signal. Hence, the small-signal duty cycle is related to the small-signal error voltage by

$$\Delta k = \frac{\Delta v_e}{V_{cr}} \qquad (14.49)$$

When the output is lower than the nominal dc output value, a high-error voltage is produced. This means that Δv_e is positive. Hence, Δk is positive. The duty cycle is increased to cause a subsequent increase in output voltage in the *voltage-mode control*. The feedback dynamics is determined by the error amplifier circuit consisting of Z_1 and Z_2.

Current-mode control.   The *current-mode control* uses the current as the feedback signal to achieve output voltage control [5]. It consists of an inner loop that samples the primary current value and turns the switches off as soon as the current reaches a certain value set by the outer voltage loop. This way the current control achieves faster response than the voltage mode. The primary current waveform acts as the sawtooth wave. The voltage analog of the current may be provided by a small resistance, or by a current transformer. Figure 14.19a shows a current-mode controlled flyback converter, where the switch current is used as the carrier signal. Turn-on is synchronized with the clock pulse, and turn-off is determined by the instant at which the input current equals the error voltage.

Because of its inherent peak current limiting capability, current-mode control can enhance reliability of power switches. The dynamic performance is improved because

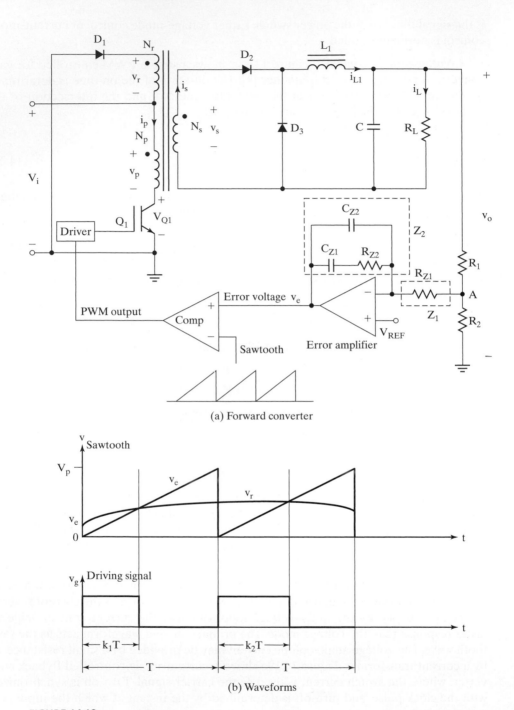

(a) Forward converter

(b) Waveforms

FIGURE 14.18

Voltage-mode control of forward converter.

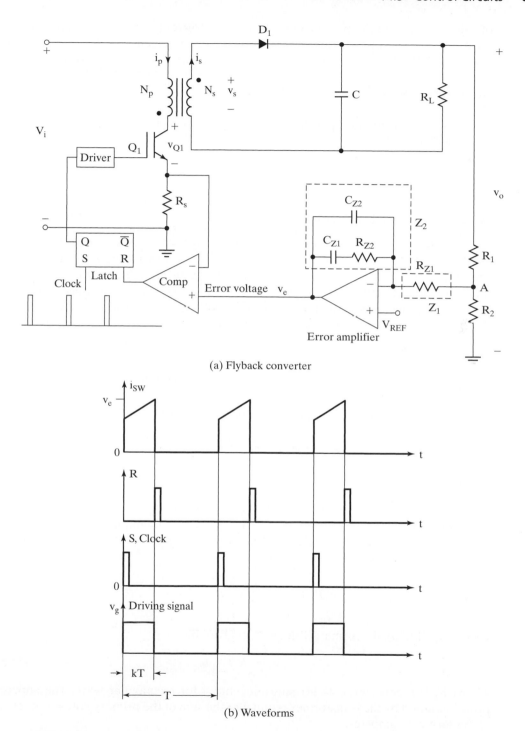

(a) Flyback converter

(b) Waveforms

FIGURE 14.19

A current-mode controlled flyback regulator.

of the use of the additional current information. Current-mode control effectively reduces the system to first order by forcing the inductor current to be related to the output voltage, thus achieving faster response. Figure 14.18b shows the waveforms.

Key Points of Section 14.5

- The converters are operated with feedback loop in a voltage-mode or current-mode control.
- The PWM technique is used to vary the duty cycle to maintain the output voltage at a desired value.

14.6   MAGNETIC DESIGN CONSIDERATIONS

Transformers are commonly used to step up or down voltages, and inductors are used as storage during energy transfer. An inductor often carries a dc current while trying to supply a current constant. A high dc current may saturate the magnetic core, making the inductor ineffective. The magnetic flux is the key element for voltage transformation and offering inductance. For a sinusoidal voltage $e = E_m \sin(\omega t) = \sqrt{2}E \sin(\omega t)$, the flux also varies sinusoidally, $\phi = \phi_m \sin(\omega t)$. The instantaneous primary voltage, according to Faraday's law, is given by:

$$e = N\frac{d\phi}{dt} = -N\phi_m\omega \cos(\omega t) = N\phi_m\omega \sin(\omega t - 90°)$$

which gives $E_m = N\phi_m\omega$ and its rms value becomes (see Appendix B: Magnetic Circuits)

$$E = \frac{E_m}{\sqrt{2}} = \frac{2\pi f N\phi_m}{\sqrt{2}} = 4.44 f N\phi_m \tag{14.50}$$

14.6.1   Transformer Design

The apparent transformer power P_t, which is the sum of the input P_i and output power P_o, depends on the converter circuit [6], as shown in Figure 14.20. For a transformer efficiency η, P_t is related to P_o as given by

$$P_t = P_i + P_o = \frac{P_o}{\eta} + P_o = \left(1 + \frac{1}{\eta}\right) P_o \tag{14.51}$$

Using Eq. (14.50), the primary voltage V_1 is given by

$$V_1 = K_t f N_1 \phi_m \tag{14.52}$$

where K_t is a constant, 4.44 for sinusoids and 4 for rectangular wave. The apparent power handled by the transformer is equal to the sum of the primary volt–amperes and secondary volt–amperes.

$$P_t = V_1 I_1 + V_2 I_2$$

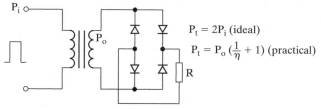

$P_t = 2P_i$ (ideal)

$P_t = P_o (\frac{1}{\eta} + 1)$ (practical)

(a) Half bridge with bridge rectifier

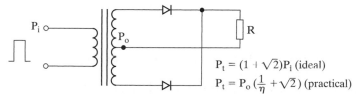

$P_t = (1 + \sqrt{2})P_i$ (ideal)

$P_t = P_o (\frac{1}{\eta} + \sqrt{2})$ (practical)

(b) Half bridge with center-tap rectifier

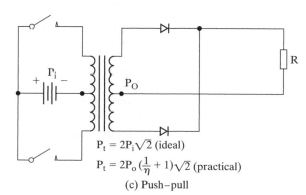

$P_t = 2P_i\sqrt{2}$ (ideal)

$P_t = 2P_o (\frac{1}{\eta} + 1)\sqrt{2}$ (practical)

(c) Push–pull

FIGURE 14.20

Transformer apparent power for various converter circuits.

Thus, for $N_1 = N_2 = N$ and $I_1 = I_2 = I$, the primary or secondary volt–amperes are given by

$$P_t = VI = K_t f N \phi_m I$$
$$= K_t f B_m A_c N I \tag{14.53}$$

where A_c is the cross-sectional area of the flux path and B_m is the peak flux density.

The number of ampere–turns (NI) is related to the current density J as given by

$$NI = K_u W_a J \tag{14.54}$$

where W_a is the window area and K_u is the fill factor between 0.4 and 0.6.

Substituting NI from Eq. (14.54) into Eq. (14.53) gives the area product as

$$A_p = W_a A_c = \frac{P_t}{K_t f B_c K_u J} \tag{14.55}$$

The current density J is related to A_p by [1]

$$J = K_j A_p^x \tag{14.56}$$

TABLE 14.1 Core Configuration Constants

Core Type	K_j @ 25° C	K_j @ 50° C	x, Exponent	Core Losses
Pot core	433	632	−0.17	$P_{cu} = P_{fe}$
Powder core	403	590	−0.12	$P_{cu} \gg P_{fe}$
Lamination core	366	534	−0.12	$P_{cu} = P_{fe}$
C-core	323	468	−0.14	$P_{cu} = P_{fe}$
Single-coil	395	569	−0.14	$P_{cu} \gg P_{fe}$
Tape-wound core	250	365	−0.13	$P_{cu} = P_{fe}$

where K_j and x are constants that depend on the magnetic core, as given in Table 14.1. P_{cu} is the copper loss and P_{fe} is the core loss.

Substituting J from Eq. (14.56) into Eq. (14.55), we can find A_p as given by

$$A_p = \left[\frac{P_t \times 10^4}{K_t f B_m K_u K_j} \right]^{\frac{1}{1-x}} \quad (cm^4) \qquad (14.57)$$

where B_m is in flux density/cm². Equation (14.57) relates the core area to the transformer power requirement. That is, the amount of copper wire, and the amount of iron ferrite or other core material determine the transformer power capability P_t. From the calculated value of A_p, the core type can be selected and the core characteristics and dimensions can be found from the manufacturer's data. Figure 14.21 shows the core area A_c for various core types.

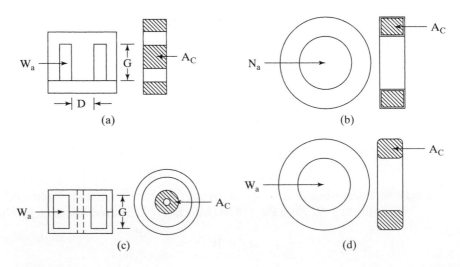

FIGURE 14.21

Core area for various core types.

Example 14.6   Design of a Transformer

An ac–dc converter steps down the voltage through a transformer and supplies the load through a bridge rectifier, as shown in Figure 14.20a. Design a 60-Hz power transformer of the specifications: primary voltage $V_1 = 120$ V, 60-Hz (square-wave), secondary voltage output $V_o = 40$ V, and secondary output current $I_o = 6.5$ A. Assume transformer efficiency $\eta = 95\%$, and window factor $K_u = 0.4$. Use E-core.

Solution

$K_t = 4.44$ for square wave, and $P_o = 40 \times 6.5 = 260$ W. Using Eq. (14.51),

$$P_t = \left(1 + \frac{1}{0.95}\right) 260 = 533.7 \text{ W.}$$

From Table 14.1 for E-core, $K_j = 366$ and $x = -0.14$. Let us $B_m = 1.4$. From Eq. (14.57),

$$A_p = \left[\frac{533.7 \times 10^4}{0.4 \times 60 \times 1.4 \times 0.4 \times 366}\right]^{\frac{1}{1+0.14}} = 206.1 \text{ cm}^4$$

Choose the E-core type, core2-138EI with $A_p = 223.39$ cm^4, core weight $W_t = 3.901$ kg, core area $A_c = 24.4$ cm^2, and mean length of a turn $l_{mt} = 27.7$ cm.

Using Eq. (14.51), the primary number of turns is $N_p = \dfrac{V_1 \times 10^4}{K_t f B_m A_c} = 132$.

Using Eq. (14.52), the primary number of turns is

$$N_p = \frac{V_1 \times 10^4}{K_t f B_m A_c}$$

$$= \frac{120 \times 10^4}{4.44 \times 60 \times 1.4 \times 24.4} = 132 \tag{14.58}$$

The secondary number of turns is

$$N_s = \frac{N_p}{V_1} V_o$$

$$= \frac{132 \times 40}{120} = 44 \tag{14.59}$$

From Eq. (14.56), $J = K_j A_p^x = 366 \times 206.7^{-0.14} = 173.6$ A/cm^2. The primary current is $I_1 = (P_t - P_o)/V_1 = (533.7 - 260)/120 = 2.28$ A. The primary bare-wire cross-sectional area is $A_{wp} = I_1/J = 0.28$ A$/173.6 = 0.013$ cm^2.

From the wire-size Table B.2 (Appendix B), we find the primary AWG number 16 with $\sigma_p = 131.8$ $\mu\Omega$/cm. The primary winding resistance is $R_p = l_{mt} N_p \sigma_p \times 10^{-6} = 27.7 \times /132 \times 131.8 = 0.48$ Ω. The primary copper loss is $P_p = I_p^2 R_p = 2.28^2 \times 0.48 = 2.5$ W. The secondary bare-wire cross-sectional area is $A_{ws} = I_o/J = 6.5/173.6 = 0.037$ cm^2.

From the wire-size Table B.1 (Appendix B), we find the secondary AWG number 16 with $\sigma_s = 41.37 \ \mu\Omega/\text{cm}$. The secondary winding resistance is $R_s = l_{mt}N_s\sigma_s \times 10^{-6} = 27.7 \times /44 \times 41.37 = 0.05 \ \Omega$. The secondary copper loss is $P_s = I_o^2 R_s = 6.5^2 \times 0.05 = 2.1$ W.

Using Figure B.1 (Appendix B), the transformer core loss is $P_{fe} = 0.557 \times 10^{-3} \times f^{1.68} \times B_m^{1.86} = 0.557 \times 10^{-3} \times 60^{1.68} \times 1.4^{1.86} = 3.95$ W. The transformer efficiency is $\eta = P_o/(P_o + P_p + P_s + P_{fe}) = 260/(260 + 2.5 + 2.1 + 3.9) = 96\%$.

14.6.2 Dc Inductor

The dc inductor is the most essential component in a power converter and it is used in all power converters and in input filters as well as output filters. Using Eq. (B.11) (Appendix B), the inductance L is related to the number of turns as given by

$$L = \frac{N^2}{\mathfrak{R}} = \frac{\mu_o\mu_r A_c}{l_c} \times N^2 \tag{14.60}$$

which relates the inductance L to the square of the number of turns for distributed air-gap cores. The core manufacturer often quotes the inductance value for a specific number of turns [7]. With a finite air-gap length l_g, Eq. (14.60) becomes

$$L = \frac{N^2}{\mathfrak{R}_c + \mathfrak{R}_g} = \frac{\mu_o A_c}{l_g + \dfrac{l_c}{\mu_r}} \times N^2 \tag{14.61}$$

Using Eq. (B.10) (Appendix B), we get

$$N = \frac{LI}{\phi} = \frac{LI}{B_c A_c}$$

which, after multiplying both sides by I, gives

$$NI = \frac{LI^2}{B_c A_c} \times 10^4 \tag{14.62}$$

Substituting NI from Eq. (14.53) into Eq. (14.62) gives the area product

$$A_p = W_a A_c = \frac{LI^2 \times 10^4}{B_c K_u J} \tag{14.63}$$

Substituting J from Eq. (14.56) in Eq. (14.63), we can find A_p as given by

$$A_p = \left[\frac{LI^2 \times 10^4}{B_c K_u K_j}\right]^{\frac{1}{1-x}} \quad (\text{cm}^4) \tag{14.64}$$

where B_c is in flux density/cm^2. Equation (14.64) relates directly to the energy storage capability of the inductor ($W_L = 1/2 \, LI^2$). That is, the amount of copper wire, and the amount of iron ferrite or other core material) determine the inductor's energy storage capability W_L. From the calculated value of A_p, the core type can be selected and the core characteristics and dimensions can be found from the manufacturer's data.

Example 14.7   Designing a Dc Inductor

Design a dc inductor of $L = 450\ \mu H$. The dc current is $I_L = 7.2$ A with a ripple of $\Delta I = 1$ A. Assume window factor $K_u = 0.4$. Use power core with a graded air gap.

Solution

The peak inductor current $I_m = I_L + \Delta I = 7.2 + 1 = 8.2$ A. The inductor energy, $W_t = \frac{1}{2} \times LI_m^2 = \frac{1}{2} \times 450 \times 10^{-6} \times 8.2^2 = 15$ mJ. From Table 14.1 for power core, $K_j = 403$ and $x = -0.12$. Let us choose $B_m = 0.3$. From Eq. (14.64),

$$A_p = \left[\frac{2 \times 0.015 \times 10^4}{0.4 \times 403 \times 0.3} \right]^{\frac{1}{1+0.12}} = 8.03 \text{ cm}^4$$

Choose the power-core type, 55090-A2 (Magnetics, Inc.) with $A_p = 8.06$ cm^4, core weight $W_t = 0.131$ kg, core area $A_c = 1.32$ cm^2, magnetic path length $l_c = 11.62$ cm, and mean length of a turn $l_{mt} = 6.66$ cm.

From Eq. (14.56), $J = K_j A_p^x = 403 \times 8.03^{-0.12} = 313.8$ A/cm^2.

Substituting NI from Eq. (14.54) into $B_m = \mu_o \mu_r H = \mu_o \mu_r NI/l_c$ and simplifying, we get

$$\mu_r = \frac{B_m l_c \times 10^5}{\mu_o W_a J K_u} \tag{14.65}$$

$$= \frac{0.3 \times 11.62 \times \times 10^5}{4\pi \times 6.11 \times 313.8 \times 0.4} = 36.2$$

Find the material with $\mu_r \geq 36.2$. Let us select type MPP-330T (Magnetics, Inc.) that gives $L_c = 86$ mH with $N_c = 1000$ turns. Thus, the number of required turns is

$$N = N_c \sqrt{\frac{L}{L_c}} \tag{14.66}$$

$$= 1000 \sqrt{\frac{450 \times 10^{-6}}{86 \times 10^{-3}}} = 72$$

The bare-wire cross-sectional area is $A_w = I_m/J = 8.2/313.8 = 0.026$ cm^2.

From the wire-size Table B.1 (Appendix B), we find the primary AWG number 14 with $A_w = 0.0208$ cm^2 and $\sigma = 82.8\ \mu\Omega$/cm. The winding resistance is $R = l_{mt} N\sigma \times 10^{-6} = 6.66 \times 72 \times 82.8 = 0.04\ \Omega$. The copper loss is $P_{cu} = I_L^2 R = 7.2^2 \times 0.04 = 2.1$ W.

Note: Using Eq. (14.61), the air-gap length l_g with a discrete air gap is given by

$$l_g = \frac{\mu_o A_c N^2}{L} - \frac{l_c}{\mu_r} \tag{14.67}$$

$$= \left[\frac{4 \times \pi \times 10^{-7} \times 1.32 \times 72^2}{450 \times 10^{-6}} - \frac{11.62}{36.2} \right] \times 10^{-2} = 0.19 \text{ cm}$$

14.6.3   Magnetic Saturation

If there is any dc imbalance, the transformer or inductor core may saturate, resulting in a high-magnetizing current. An ideal core should exhibit a very high relative permeability

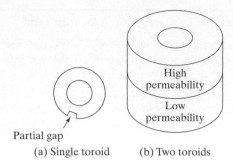

FIGURE 14.22

Cores with two permeability regions.

Partial gap

(a) Single toroid   (b) Two toroids

in the normal operating region and under dc imbalance conditions [8], it should not go into hard saturation. This problem of saturation can be minimized by having two permeability regions in the core, low and high permeability. An air gap may be inserted, as shown in a toroid in Figure 14.22, where the inner part has high permeability and the outer part has relatively low permeability. Under normal operation the flux flows through the inner part. In the case of saturation, the flux has to flow through the outer region, which has a lower permeability due to the air gap, and the core does not go into hard saturation. Two toroids with high and low permeability can be combined, as shown in Figure 14.22b.

Key Points of Section 14.6

- The design of the magnetic components is greatly influenced by the presence of dc imbalances in transformers and inductors.
- Any dc component may cause the magnetic saturation problem, thereby requiring a larger core.

SUMMARY

Industrial power supplies are of two types: dc supplies and ac supplies. In a single-stage conversion, the isolating transformer has to operate at the output frequency. To reduce the size of the transformer and meet the industrial specifications, multi-stage conversions are normally required. There are various power supply topologies, depending on the output power requirement and acceptable complexity. Converters with bidirectional switches, which allow the control of power flow in either direction, require synthesizing the switching functions to obtain the desired output waveforms.

REFERENCES

[1]   Y. M. Lai, *Power Electronics Handbook*, edited by M. H. Rashid. San Diego, CA: Academic Press. 2001, Chapter 20—Power Supplies.

[2]   J. Hancock, *Application Note AN-CoolMOS-08: SMPS Topologies Overview*, Infineon Technologies AG, Munich, Germany. June 2000. www.infineon.com.

[3] R. L. Steigerwald, R. W. De Doncker, and M. H. Kheraluwala, "A comparison of high-power dc–dc soft-switched converter topologies," *IEEE Transactions on Industry Applications*, Vol. 32, No. 5, September/October 1996, pp. 1139–1145.

[4] J. Goo, J. A. Sabate, G. Hua, and F. C. Lee, "Zero-voltage and zero-current-switching full-bridge PWM converter for high-power applications," *IEEE Transactions on Power Electronics*, Vol. 11, No. 4, July 1996, pp. 622–627.

[5] C. M. Liaw, S. J. Chiang, C. Y. Lai, K. H. Pan, G. C. Leu, and G. S. Hsu, "Modeling and controller design of a current-mode controlled converter," *IEEE Transactions on Industrial Electronics,* Vol. 41, No. 2, April 1994, pp. 231–240.

[6] K. K. Sum, *Switch Mode Power Conversion: Basic Theory and Design*. New York: Marcel Dekker. 1984.

[7] K. Billings, *Switch Mode Power Supply Handbook*. New York: McGraw-Hill. 1989.

[8] W. A. Roshen, R. L. Steigerwald, R. J. Charles, W. G. Earls, G. S. Clayton, and C. F. Saj, "High-efficiency, high-density MHz magnetic components for low profile converter," *IEEE Transactions on Industry Applications*, Vol. 31, No. 4, July/August 1995, pp. 869–877.

[9] H. Zöllinger and R. Kling, *Application Note AN-SMPS-1683X-1: Off-Line Switch Mode Power Supplies*, Infineon Technologies AG, Munich, Germany, June 2000. www.infineon.com.

[10] A. I. Pressman, *Switching Power Supply Design*, 2nd edition. New York: McGraw-Hill. 1999.

[11] M. Brown, *Practical Switching Power Supply Design*, 2nd edition. New York: McGraw-Hill. 1999.

[12] G. Chryssis, *High Frequency Switching Power Supplies*. New York: McGraw-Hill. 1984.

[13] "Standard Publication/No. PE 1-1983: Uninterruptible Power Systems," National Electrical Manufacturer's Association (NEMA), 1983.

REVIEW QUESTIONS

14.1 What are the normal specifications of power supplies?
14.2 What are the types of power supplies in general?
14.3 Name three types of dc power supplies.
14.4 Name three types of ac power supplies.
14.5 What are the advantages and disadvantages of single-stage conversion?
14.6 What are the advantages and disadvantages of switched-mode power supplies?
14.7 What are the advantages and disadvantages of resonant power supplies?
14.8 What are the advantages and disadvantages of bidirectional power supplies?
14.9 What are the advantages and disadvantages of flyback converters?
14.10 What are the advantages and disadvantages of push–pull converters?
14.11 What are the advantages and disadvantages of half-bridge converters?
14.12 What are the various configurations of resonant dc power supplies?
14.13 What are the advantages and disadvantages of high-frequency link power supplies?
14.14 What is the general arrangement of UPS systems?
14.15 What are the problems of the transformer core?
14.16 What are two commonly used control methods for power supplies?
14.17 Why is the design of a dc inductor different from that of an ac inductor?

PROBLEMS

14.1 The dc output voltage of the push–pull circuit in Figure 14.6 is $V_o = 24$ V at a resistive load of $R = 0.4$ Ω. The on-state voltage drops of transistors and diodes are $V_t = 1.2$ V and

$V_d = 0.7$ V, respectively. The turns ratio of the transformer is $a = N_s/N_p = 0.5$. Determine **(a)** the average input current I_s, **(b)** the efficiency η, **(c)** the average transistor current I_A, **(d)** the peak transistor current I_p, **(e)** the rms transistor current I_R, and **(f)** the open-circuit transistor voltage V_{oc}. Neglect the losses in the transformer, and the ripple current of the load and input supply is negligible.

14.2 Repeat Problem 14.1 for the circuit in Figure P14.2, for $k = 0.5$.

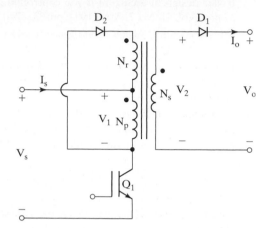

FIGURE P14.2

Flyback converter with reset winding.

14.3 Repeat Problem 14.1 for the circuit in Figure P14.3.

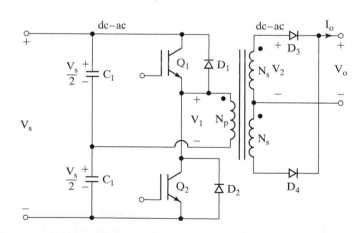

FIGURE P14.3

Half-bridge converter.

14.4 Repeat Problem 14.1 for the circuit in Figure P14.4.

14.5 The dc output voltage of the half-bridge circuit in Figure 14.9 is $V_o = 24$ V at a load resistance of $R = 0.8\ \Omega$. The inverter operates at the resonant frequency. The circuit parameters are $C_1 = C_2 = C = 2\ \mu\text{F}$, $L = 5\ \mu\text{H}$, and $R = 0$. The dc input voltage, $V_s = 50$ V. The on-state voltage drops of transistors and diodes are negligible. The turns ratio of the transformer, $a = N_s/N_p = 0.5$. Determine **(a)** the average input current I_s, **(b)** the average

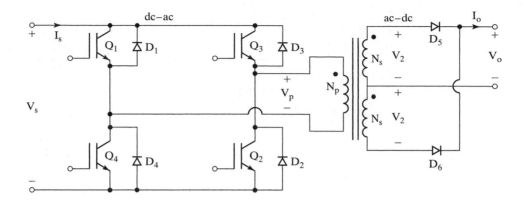

FIGURE P14.4

Full-bridge converter.

transistor current I_A, **(c)** the peak transistor current I_p, **(d)** the rms transistor current I_R, and **(e)** the open-circuit transistor voltage V_{oc}. Neglect the losses in the transformer, and the effect of the load on the resonant frequency is negligible.

14.6 Repeat Problem 14.2 for the full-bridge circuit in Figure 14.9b.

14.7 The load resistance of the ac power supplied in Figure 14.12 is $R = 1.5 \, \Omega$. The dc input voltage is $V_s = 24$ V. The input inverter operates at a frequency of 400 Hz with a uniform PWM of eight pulses per half-cycle and the width of each pulse is $\delta = 20°$. The on-state voltage drops of transistor switches and diodes are negligible. The turns ratio of the transformer, $a = N_s/N_p = 4$. Determine the rms load current. Neglect the losses in the transformer, and the effect of the load on the resonant frequency is negligible.

14.8 The load resistance of the ac power supplies in Figure 14.13 is $R = 1.5 \, \Omega$. The dc input voltage is $V_s = 24$ V. The input inverter operates at a frequency of 20 kHz with a uniform PWM of four pulses per half-cycle and the width of each pulse is $\delta_i = 40°$. The on-state voltage drops of transistor switches and diodes are negligible. The turns ratio of the transformer, $a = N_s/N_p = 0.5$. The output inverter operates with a uniform PWM of eight pulses per half-cycle and the width of each pulse is $\delta_0 = 20°$. Determine the rms load current. The ripple voltage on the output of the rectifier is negligible. Neglect the losses in the transformer, and the effect of the load on the resonant frequency is negligible.

14.9 An ac–dc converter steps down the voltage through a transformer and supplies the load through a bridge rectifier, as shown in Figure 14.20b. Design a 60-Hz power transformer of the specifications: primary voltage $V_1 = 120$ V, 60 Hz (square wave), secondary voltage output $V_o = 40$ V, and secondary output current $I_o = 6.5$ A. Assume transformer efficiency $\eta = 95\%$, and window factor $K_u = 0.4$. Use E-core.

14.10 Design a 110-W flyback transformer. The switching frequency is 30 kHz, period $T = 33 \, \mu s$, and duty-cycle $k = 50\%$. The primary voltage $V_1 = 100$ V (square wave), secondary voltage output $V_o = 6.2$ V, and auxiliary voltage $V_t = 12$ V. Assume transformer efficiency $\eta = 95\%$, and window factor $K_u = 0.4$. Use an E-core.

14.11 Design a dc inductor of $L = 650 \, \mu H$. The dc current is $I_L = 7.2$ A with a ripple current of $\pm \Delta I = 1$ A. Assume window factor $K_u = 0.4$. Use power core with a graded air gap.

14.12 Design a dc inductor of $L = 90 \, \mu H$. The dc current is $I_L = 8.5$ A with a ripple current of $\pm \Delta I = 1.5$ A. Assume window factor $K_u = 0.4$. Use power core with a graded air gap.

CHAPTER 15

Dc Drives

The learning objectives of this chapter are as follows:

- To learn the basic characteristics of dc motors and their control parameters
- To understand the types and operating modes of dc drives
- To learn about the control requirements of four-quadrant drives
- To understand parameters of the transfer function of converter-fed dc motors for closed control of motor speed and torque

15.1 INTRODUCTION

Direct current (dc) motors have variable characteristics and are used extensively in variable-speed drives. Dc motors can provide a high starting torque and it is also possible to obtain speed control over a wide range. The methods of speed control are normally simpler and less expensive than those of ac drives. Dc motors play a significant role in modern industrial drives. Both series and separately excited dc motors are normally used in variable-speed drives, but series motors are traditionally employed for traction applications. Due to commutators, dc motors are not suitable for very high speed applications and require more maintenance than do ac motors. With the recent advancements in power conversions, control techniques, and micro-computers, the ac motor drives are becoming increasingly competitive with dc motor drives. Although the future trend is toward ac drives, dc drives are currently used in many industries. It might be a few decades before the dc drives are completely replaced by ac drives.

Controlled rectifiers provide a variable dc output voltage from a fixed ac voltage, whereas a dc–dc converter can provide a variable dc voltage from a fixed dc voltage. Due to their ability to supply a continuously variable dc voltage, controlled rectifiers and dc–dc converters made a revolution in modern industrial control equipment and variable-speed drives, with power levels ranging from fractional horsepower to several megawatts. Controlled rectifiers are generally used for the speed control of dc motors, as shown in Figure 15.1a. The alternative form would be a diode rectifier followed by

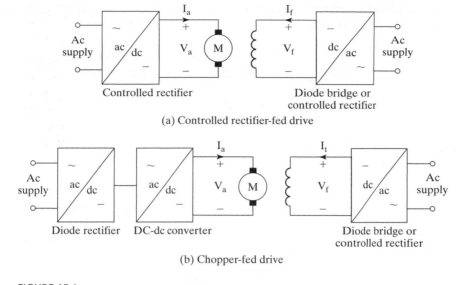

(a) Controlled rectifier-fed drive

(b) Chopper-fed drive

FIGURE 15.1

Controller rectifier- and dc-dc converter-fed drives.

dc–dc converter, as shown in Figure 15.1b. Dc drives can be classified, in general, into three types:

1. Single-phase drives
2. Three-phase drives
3. Dc–dc converter drives

15.2 BASIC CHARACTERISTICS OF DC MOTORS

The equivalent circuit for a separately excited dc motor is shown in Figure 15.2 [1]. When a separately excited motor is excited by a field current of i_f and an armature current of i_a flows in the armature circuit, the motor develops a back electromotive force (emf) and a torque to balance the load torque at a particular speed. The field current i_f

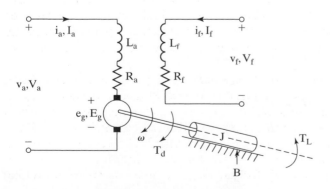

FIGURE 15.2

Equivalent circuit of separately excited dc motors.

of a separately excited motor is independent of the armature current i_a and any change in the armature current has no effect on the field current. The field current is normally much less than the armature current.

The equations describing the characteristics of a separately excited motor can be determined from Figure 15.2. The instantaneous field current i_f is described as

$$v_f = R_f i_f + L_f \frac{di_f}{dt}$$

The instantaneous armature current can be found from

$$v_a = R_a i_a + L_a \frac{di_a}{dt} + e_g$$

The motor back emf, which is also known as *speed voltage*, is expressed as

$$e_g = K_v \omega i_f$$

The torque developed by the motor is

$$T_d = K_t i_f i_a$$

The developed torque must be equal to the load torque:

$$T_d = J \frac{d\omega}{dt} + B\omega + T_L$$

where ω = motor angular speed, or rotor angular frequency, rad/s;
B = viscous friction constant, N · m/rad/s;
K_v = voltage constant, V/A-rad/s;
K_t = torque constant, which equals voltage constant, k_v;
L_a = armature circuit inductance, H;
L_f = field circuit inductance, H;
R_a = armature circuit resistance, Ω;
R_f = field circuit resistance, Ω;
T_L = load torque, N · m.

Under steady-state conditions, the time derivatives in these equations are zero and the steady-state average quantities are

$$V_f = R_f I_f \tag{15.1}$$

$$E_g = K_v \omega I_f \tag{15.2}$$

$$V_a = R_a I_a + E_g$$

$$= R_a I_a + K_v \omega I_f \tag{15.3}$$

$$T_d = K_t I_f I_a \tag{15.4}$$

$$= B\omega + T_L \tag{15.5}$$

The developed power is

$$P_d = T_d \omega \tag{15.6}$$

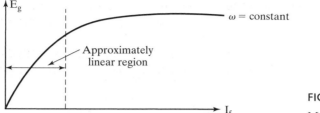

FIGURE 15.3

Magnetization characteristic.

The relationship between the field current I_f and the back emf E_g is nonlinear due to magnetic saturation. The relationship, which is shown in Figure 15.3, is known as *magnetization characteristic* of the motor. From Eq. (15.3), the speed of a separately excited motor can be found from

$$\omega = \frac{V_a - R_a I_a}{K_v I_f} = \frac{V_a - R_a I_a}{K_v V_f / R_f} \tag{15.7}$$

We can notice from Eq. (15.7) that the motor speed can be varied by (1) controlling the armature voltage V_a, known as *voltage control*; (2) controlling the field current I_f, known as *field control*; or (3) torque demand, which corresponds to an armature current I_a, for a fixed field current I_f. The speed, which corresponds to the rated armature voltage, rated field current and rated armature current, is known as the *rated* (or *base*) *speed*.

In practice, for a speed less than the base speed, the armature current and field currents are maintained constant to meet the torque demand, and the armature voltage V_a is varied to control the speed. For speed higher than the base speed, the armature

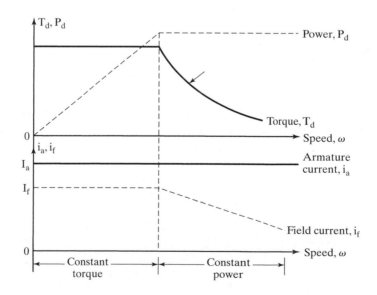

FIGURE 15.4

Characteristics of separately excited motors.

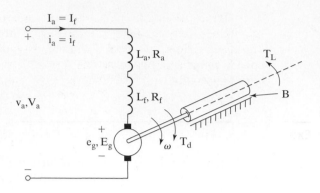

FIGURE 15.5

Equivalent circuit of dc series motors.

voltage is maintained at the rated value and the field current is varied to control the speed. However, the power developed by the motor (= torque × speed) remains constant. Figure 15.4 shows the characteristics of torque, power, armature current, and field current against the speed.

The field of a dc motor may be connected in series with the armature circuit, as shown in Figure 15.5, and this type of motor is called a *series motor*. The field circuit is designed to carry the armature current. The steady-state average quantities are

$$E_g = K_v \omega I_a \tag{15.8}$$

$$V_a = (R_a + R_f)I_a + E_g \tag{15.9}$$

$$= (R_a + R_f)I_a + K_v \omega I_f \tag{15.10}$$

$$T_d = K_t I_a I_f$$

$$= B\omega + T_L \tag{15.11}$$

The speed of a series motor can be determined from Eq. (15.10):

$$\omega = \frac{V_a - (R_a + R_f)I_a}{K_v I_f} \tag{15.12}$$

The speed can be varied by controlling the (1) armature voltage V_a; or (2) armature current, which is a measure of the torque demand. Equation (15.11) indicates that a series motor can provide a high torque, especially at starting; and for this reason, series motors are commonly used in traction applications.

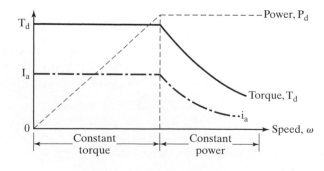

FIGURE 15.6

Characteristics of dc series motors.

For a speed up to the base speed, the armature voltage is varied and the torque is maintained constant. Once the rated armature voltage is applied, the speed–torque relationship follows the natural characteristic of the motor and the power ($=$ torque $\times$ speed) remains constant. As the torque demand is reduced, the speed increases. At a very light load, the speed could be very high and it is not advisable to run a dc series motor without a load. Figure 15.6 shows the characteristics of dc series motors.

Example 15.1   Finding the Voltage and Current of a Separately Excited Motor

A 15-hp, 220-V, 2000-rpm separately excited dc motor controls a load requiring a torque of $T_L = 45$ N·m at a speed of 1200 rpm. The field circuit resistance is $R_f = 147\ \Omega$, the armature circuit resistance is $R_a = 0.25\ \Omega$, and the voltage constant of the motor is $K_v = 0.7032$ V/A rad/s. The field voltage is $V_f = 220$ V. The viscous friction and no-load losses are negligible. The armature current may be assumed continuous and ripple free. Determine (a) the back emf E_g, (b) the required armature voltage V_a, and (c) the rated armature current of the motor.

Solution

$R_f = 147\ \Omega$, $R_a = 0.25\ \Omega$, $K_v = K_t = 0.7032$ V/A rad/s, $V_f = 220$ V, $T_d = T_L = 45$ N·m, $\omega = 1200\ \pi/30 = 125.66$ rad/s, and $I_f = 220/147 = 1.497$ A.

   a. From Eq. (15.4), $I_a = 45/(0.7032 \times 1.497) = 42.75$ A. From Eq. (15.2), $E_g = 0.7032 \times 125.66 \times 1.497 = 132.28$ V.

   b. From Eq. (15.3), $V_a = 0.25 \times 42.75 + 132.28 = 142.97$ V.

   c. Because 1 hp is equal to 746 W, $I_{\text{rated}} = 15 \times 746/220 = 50.87$ A.

Key Points of Section 15.2

- The speed of a dc motor can be varied by controlling (1) the armature voltage, (2) the field current, or (3) the armature current that is a measure of the torque demand.
- For a speed less than the rated speed (also known as base speed), the armature voltage is varied to control the speed, while the armature and field currents are maintained constant. For a speed higher than the rated speed, the field current is varied to control the speed, while the armature voltage is maintained at the rated value.

15.3   OPERATING MODES

In variable-speed applications, a dc motor may be operating in one or more modes: motoring, regenerative braking, dynamic braking, plugging, and four quadrants [2, 3].

    Motoring.   The arrangements for motoring are shown in Figure 15.7a. Back emf E_g is less than supply voltage V_a. Both armature and field currents are positive. The motor develops torque to meet the load demand.

    Regenerative braking.   The arrangements for regenerative braking are shown in Figure 15.7b. The motor acts as a generator and develops an induced voltage E_g. E_g must be greater than supply voltage V_a. The armature current is negative, but the field current is positive. The kinetic energy of the motor is returned to the supply. A series

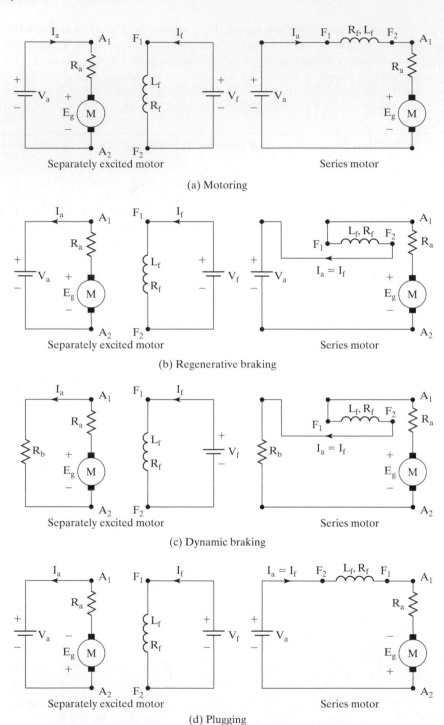

FIGURE 15.7

Operating modes.

motor is usually connected as a self-excited generator. For self-excitation, it is necessary that the field current aids the residual flux. This is normally accomplished by reversing the armature terminals or the field terminals.

Dynamic braking.   The arrangements shown in Figure 15.7c are similar to those of regenerative braking, except the supply voltage V_a is replaced by a braking resistance R_b. The kinetic energy of the motor is dissipated in R_b.

Plugging.   Plugging is a type of braking. The connections for plugging are shown in Figure 15.7d. The armature terminals are reversed while running. The supply voltage V_a and the induced voltage E_g act in the same direction. The armature current is reversed, thereby producing a braking torque. The field current is positive. For a series motor, either the armature terminals or field terminals should be reversed, but not both.

Four quadrants.   Figure 15.8 shows the polarities of the supply voltage V_a, back emf E_g, and armature current I_a for a separately excited motor. In forward motoring (quadrant I), V_a, E_g, and I_a are all positive. The torque and speed are also positive in this quadrant.

During forward braking (quadrant II), the motor runs in the forward direction and the induced emf E_g continues to be positive. For the torque to be negative and the direction of energy flow to reverse, the armature current must be negative. The supply voltage V_a should be kept less than E_g.

In reverse motoring (quadrant III), V_a, E_g, and I_a are all negative. The torque and speed are also negative in this quadrant. To keep the torque negative and the energy flow from the source to the motor, the back emf E_g must satisfy the condition

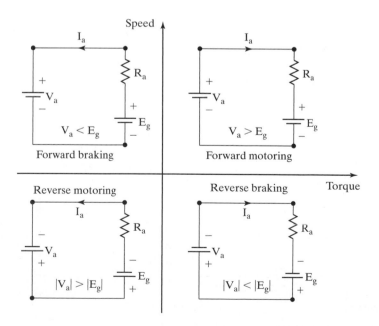

FIGURE 15.8

Conditions for four quadrants.

$|V_a| > |E_g|$. The polarity of E_g can be reversed by changing the direction of field current or by reversing the armature terminals.

During reverse braking (quadrant IV), the motor runs in the reverse direction. V_a and E_g continue to be negative. For the torque to be positive and the energy to flow from the motor to the source, the armature current must be positive. The induced emf E_g must satisfy the condition $|V_a| < |E_g|$.

Key Points of Section 15.3

- A motor drive should be capable of four quadrant operations: forward motoring, forward braking, reverse motoring, or reverse braking.
- For operations in the reverse direction, the field excitation must be reversed to reverse the polarity of the back emf.

15.4 SINGLE-PHASE DRIVES

If the armature circuit of a dc motor is connected to the output of a single-phase controlled rectifier, the armature voltage can be varied by varying the delay angle of the converter α_a. The forced-commutated ac–dc converters can also be used to improve the power factor (PF) and to reduce the harmonics. The basic circuit agreement for a single-phase converter-fed separately excited motor is shown in Figure 15.9. At a low delay angle, the armature current may be discontinuous, and this would increase the losses in the motor. A smoothing inductor, L_m, is normally connected in series with the armature circuit to reduce the ripple current to an acceptable magnitude. A converter is also applied in the field circuit to control the field current by varying the delay angle α_f. For operating the motor in a particular mode, it is often necessary to use contactors for reversing the armature circuit, as shown in Figure 15.10a or the field circuit, as shown in Figure 15.10b. To avoid inductive voltage surges, the field or the armature reversing is performed at a zero armature current. The delay (or firing) angle is normally adjusted to give a zero current; and additionally, a dead time of typically 2 to 10 ms is provided to ensure that the armature current becomes zero. Because of a relatively

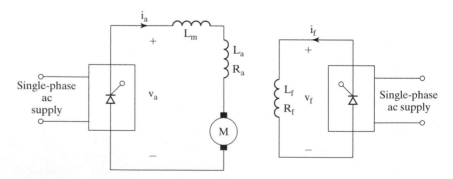

FIGURE 15.9

Basic circuit arrangement of a single-phase dc drive.

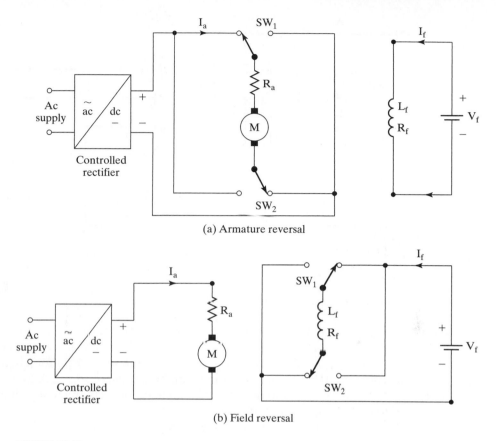

(a) Armature reversal

(b) Field reversal

FIGURE 15.10

Field and armature reversals using contactors.

large time constant of the field winding, the field reversal takes a longer time. A semi- or full converter can be used to vary the field voltage, but a full converter is preferable. Due to the ability to reverse the voltage, a full converter can reduce the field current much faster than a semiconverter. Depending on the type of single-phase converters, single-phase drives [4, 5] may be subdivided into:

1. Single-phase half-wave-converter drives.
2. Single-phase semiconverter drives.
3. Single-phase full-converter drives.
4. Single-phase dual-converter drives.

15.4.1 Single-Phase Half-Wave-Converter Drives

A single-phase half-wave converter feeds a dc motor, as shown in Figure 15.11a. The armature current is normally discontinuous unless a very large inductor is connected in

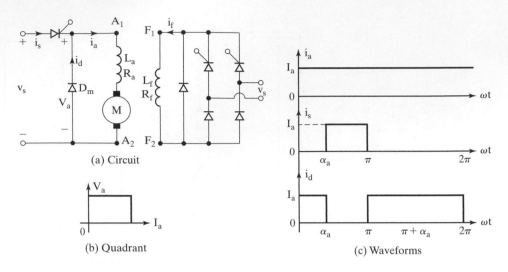

(a) Circuit

(b) Quadrant

(c) Waveforms

FIGURE 15.11

Single-phase half-wave-converter drive.

the armature circuit. A freewheeling diode is always required for a dc motor load and it is a one-quadrant drive, as shown in Figure 15.11b. The applications of this drive are limited to the $\frac{1}{2}$ kW power level. Figure 15.11c shows the waveforms for a highly inductive load. The converter in the field circuit can be a semiconverter. A half-wave converter in the field circuit would increase the magnetic losses of the motor due to a high ripple content on the field excitation current.

With a single-phase half-wave converter in the armature circuit, Eq. (10.1) gives the average armature voltage as

$$V_a = \frac{V_m}{2\pi} (1 + \cos \alpha_a) \qquad \text{for } 0 \le \alpha_a \le \pi \tag{15.13}$$

where V_m is the peak voltage of the ac supply. With a semiconverter in the field circuit, Eq. (10.52) gives the average field voltage as

$$V_f = \frac{V_m}{\pi} (1 + \cos \alpha_f) \qquad \text{for } 0 \le \alpha_f \le \pi \tag{15.14}$$

15.4.2 Single-Phase Semiconverter Drives

A single-phase semiconverter feeds the armature circuit, as shown in Figure 15.12a. It is a one-quadrant drive, as shown in Figure 15.12b, and is limited to applications up to 15 kW. The converter in the field circuit can be a semiconverter. The current waveforms for a highly inductive load are shown in Figure 15.12c.

With a single-phase semiconverter in the armature circuit, Eq. (10.52) gives the average armature voltage as

$$V_a = \frac{V_m}{\pi} (1 + \cos \alpha_a) \qquad \text{for } 0 \le \alpha_a \le \pi \tag{15.15}$$

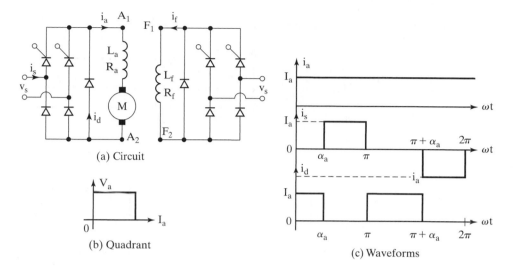

FIGURE 15.12

Single-phase semiconverter drive.

With a semiconverter in the field circuit, Eq. (10.52) gives the average field voltage as

$$V_f = \frac{V_m}{\pi}(1 + \cos \alpha_f) \qquad \text{for } 0 \le \alpha_f \le \pi \tag{15.16}$$

15.4.3 Single-Phase Full-Converter Drives

The armature voltage is varied by a single-phase full-wave converter, as shown in Figure 15.13a. It is a two-quadrant drive, as shown in Figure 15.13b, and is limited to applications up to 15 kW. The armature converter gives $+V_a$ or $-V_a$, and allows operation in the first and fourth quadrants. During regeneration for reversing the direction of power flow, the back emf of the motor can be reversed by reversing the field excitation. The converter in the field circuit could be a semi-, a full, or even a dual converter. The reversal of the armature or field allows operation in the second and third quadrants. The current waveforms for a highly inductive load are shown in Figure 15.13c for powering action. A 9.5-kW, 40-A single-phase full-converter drive is shown in Figure 15.14, where the power stack is on the back of the panel and the control signals are implemented by analog electronics.

With a single-phase full-wave converter in the armature circuit, Eq. (10.5) gives the average armature voltage as

$$V_a = \frac{2V_m}{\pi} \cos \alpha_a \qquad \text{for } 0 \le \alpha_a \le \pi \tag{15.17}$$

With a single-phase full-converter in the field circuit, Eq. (10.5) gives the field voltage as

$$V_f = \frac{2V_m}{\pi} \cos \alpha_f \qquad \text{for } 0 \le \alpha_f \le \pi \tag{15.18}$$

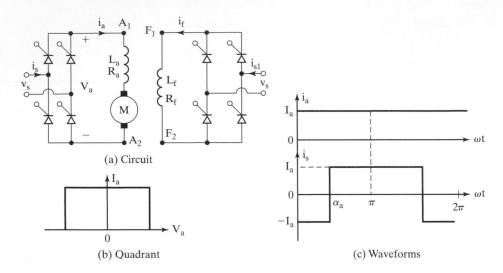

(a) Circuit

(b) Quadrant

(c) Waveforms

FIGURE 15.13

Single-phase full-converter drive.

FIGURE 15.14

A 9.5-kW analog-based single-phase full-wave drive. (Reproduced by permission of Brush Electrical Machines Ltd., England.)

15.4.4 Single-Phase Dual-Converter Drives

Two single-phase full-wave converters are connected, as shown in Figure 15.15. Either converter 1 operates to supply a positive armature voltage, V_a, or converter 2 operates to supply a negative armature voltage, $-V_a$. Converter 1 provides operation in the first and fourth quadrants, and converter 2, in the second and third quadrants. It is a four-quadrant drive and permits four modes of operation: forward powering, forward braking (regeneration), reverse powering, and reverse braking (regeneration). It is limited to applications up to 15 kW. The field converter could be a full-wave, a semi-, or a dual converter.

If converter 1 operates with a delay angle of α_{a1}, Eq. (10.15) gives the armature voltage as

$$V_a = \frac{2V_m}{\pi} \cos \alpha_{a1} \qquad \text{for } 0 \le \alpha_{a1} \le \pi \qquad (15.19)$$

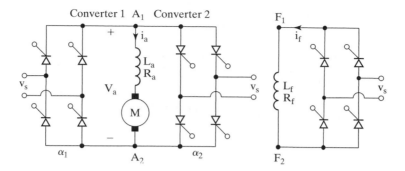

FIGURE 15.15

Single-phase dual-converter drive.

If converter 2 operates with a delay angle of α_{a2}, Eq. (10.16) gives the armature voltage as

$$V_a = \frac{2V_m}{\pi} \cos \alpha_{a2} \qquad \text{for } 0 \le \alpha_{a2} \le \pi \tag{15.20}$$

where $\alpha_{a2} = \pi - \alpha_{a1}$. With a full converter in the field circuit, Eq. (10.5) gives the field voltage as

$$V_f = \frac{2V_m}{\pi} \cos \alpha_f \qquad \text{for } 0 \le \alpha_f \le \pi \tag{15.21}$$

Example 15.2   Finding the Performance Parameters of a Single-Phase Semiconverter Drive

The speed of a separately excited motor is controlled by a single-phase semiconverter in Figure 15.12a. The field current, which is also controlled by a semiconverter, is set to the maximum possible value. The ac supply voltage to the armature and field converters is one phase, 208 V, 60 Hz. The armature resistance is $R_a = 0.25\ \Omega$, the field resistance is $R_f = 147\ \Omega$, and the motor voltage constant is $K_v = 0.7032$ V/A rad/s. The load torque is $T_L = 45$ N · m at 1000 rpm. The viscous friction and no-load losses are negligible. The inductances of the armature and field circuits are sufficient enough to make the armature and field currents continuous and ripple free. Determine (a) the field current I_f; (b) the delay angle of the converter in the armature circuit α_a; and (c) the input power factor (PF) of the armature circuit converter.

Solution
$V_s = 208$ V, $V_m = \sqrt{2} \times 208 = 294.16$ V, $R_a = 0.25\ \Omega$, $R_f = 147\ \Omega$, $T_d = T_L = 45$ N · m, $K_v = 0.7032$ V/A rad/s, and $\omega = 1000\ \pi/30 = 104.72$ rad/s.

    a. From Eq. (15.16), the maximum field voltage (and current) is obtained for a delay angle of $\alpha_f = 0$ and

$$V_f = \frac{2V_m}{\pi} = \frac{2 \times 294.16}{\pi} = 187.27 \text{ V}$$

The field current is

$$I_f = \frac{V_f}{R_f} = \frac{187.27}{147} = 1.274 \text{ A}$$

b. From Eq. (15.4),

$$I_a = \frac{T_d}{K_v I_f} = \frac{45}{0.7032 \times 1.274} = 50.23 \text{ A}$$

From Eq. (15.2),

$$E_g = K_v \omega I_f = 0.7032 \times 104.72 \times 1.274 = 93.82 \text{ V}$$

From Eq. (15.3), the armature voltage is

$$V_a = 93.82 + I_a R_a = 93.82 + 50.23 \times 0.25 = 93.82 + 12.56 = 106.38 \text{ V}$$

From Eq. (15.15), $V_a = 106.38 = (294.16/\pi) \times (1 + \cos \alpha_a)$ and this gives the delay angle as $\alpha_a = 82.2°$.

c. If the armature current is constant and ripple free, the output power is $P_o = V_a I_a = 106.38 \times 50.23 = 5343.5$ W. If the losses in the armature converter are neglected, the power from the supply is $P_a = P_o = 5343.5$ W. The rms input current of the armature converter, as shown in Figure 15.12, is

$$I_{sa} = \left(\frac{2}{2\pi} \int_{\alpha_a}^{\pi} I_a^2 \, d\theta \right)^{1/2} = I_a \left(\frac{\pi - \alpha_a}{\pi} \right)^{1/2}$$

$$= 50.23 \left(\frac{180 - 82.2}{180} \right)^{1/2} = 37.03 \text{ A}$$

and the input volt–ampere (VA) rating is $VI = V_s I_{sa} = 208 \times 37.03 = 7702.24$. Assuming negligible harmonics, the input PF is approximately

$$PF = \frac{P_o}{VI} = \frac{5343.5}{7702.24} = 0.694 \text{ (lagging)}$$

From Eq. (10.61),

$$PF = \frac{\sqrt{2}(1 + \cos 82.2°)}{[\pi(\pi - 82.2°)]^{1/2}} = 0.694 \text{ (lagging)}$$

Example 15.3 Finding the Performance Parameters of a Single-Phase Full Converter Drive

The speed of a separately excited dc motor is controlled by a single-phase full-wave converter in Figure 15.13a. The field circuit is also controlled by a full converter and the field current is set to the maximum possible value. The ac supply voltage to the armature and field converters is one phase, 440 V, 60 Hz. The armature resistance is $R_a = 0.25 \ \Omega$, the field circuit resistance is $R_f = 175 \ \Omega$, and the motor voltage constant is $K_v = 1.4$ V/A rad/s. The armature current corresponding to the load demand is $I_a = 45$ A. The viscous friction and no-load losses are negligible. The inductances of the armature and field circuits are sufficient to make the armature and field currents continuous and ripple free. If the delay angle of the armature converter is $\alpha_a = 60°$ and the armature current is $I_a = 45$ A, determine (a) the torque developed by the motor T_d, (b) the speed ω, and (c) the input PF of the drive.

Solution

$V_s = 440$ V, $V_m = \sqrt{2} \times 440 = 622.25$ V, $R_a = 0.25\ \Omega$, $R_f = 175\ \Omega$, $\alpha_a = 60°$, and $K_v = 1.4$ V/A rad/s.

a. From Eq. (15.18), the maximum field voltage (and current) would be obtained for a delay angle of $\alpha_f = 0$ and

$$V_f = \frac{2V_m}{\pi} = \frac{2 \times 622.25}{\pi} = 396.14 \text{ V}$$

The field current is

$$I_f = \frac{V_f}{R_f} = \frac{396.14}{175} = 2.26 \text{ A}$$

From Eq. (15.4), the developed torque is

$$T_d = T_L = K_v I_f I_a = 1.4 \times 2.26 \times 45 = 142.4 \text{ N} \cdot \text{m}$$

From Eq. (15.17), the armature voltage is

$$V_a = \frac{2V_m}{\pi} \cos 60° = \frac{2 \times 622.25}{\pi} \cos 60° = 198.07 \text{ V}$$

The back emf is

$$E_g = V_a - I_a R_a = 198.07 - 45 \times 0.25 = 186.82 \text{ V}$$

b. From Eq. (15.2), the speed is

$$\omega = \frac{E_g}{K_v I_f} = \frac{186.82}{1.4 \times 2.26} = 59.05 \text{ rad/s or 564 rpm}$$

c. Assuming lossless converters, the total input power from the supply is

$$P_i = V_a I_a + V_f I_f = 198.07 \times 45 + 396.14 \times 2.26 = 9808.4 \text{ W}$$

The input current of the armature converter for a highly inductive load is shown in Figure 15.13b and its rms value is $I_{sa} = I_a = 45$ A. The rms value of the input current of field converter is $I_{sf} = I_f = 2.26$ A. The effective rms supply current can be found from

$$I_s = (I_{sa}^2 + I_{sf}^2)^{1/2}$$
$$= (45^2 + 2.26^2)^{1/2} = 45.06 \text{ A}$$

and the input VA rating, $VI = V_s I_s = 440 \times 45.06 = 19{,}826.4$. Neglecting the ripples, the input power factor is approximately

$$PF = \frac{P_i}{VI} = \frac{9808.4}{19{,}826.4} = 0.495 \text{ (lagging)}$$

From Eq. (10.11),

$$PF = \left(\frac{2\sqrt{2}}{\pi}\right) \cos \alpha_a = \left(\frac{2\sqrt{2}}{\pi}\right) \cos 60° = 0.45 \text{ (lagging)}$$

Example 15.4 Finding the Delay Angle and Feedback Power in Regenerative Braking

If the polarity of the motor back emf in Example 15.3 is reversed by reversing the polarity of the field current, determine (a) the delay angle of the armature circuit converter, α_a, to maintain the armature current constant at the same value of $I_a = 45$ A; and (b) the power fed back to the supply due to regenerative braking of the motor.

Solution

a. From part (b) of Example 15.3, the back emf at the time of polarity reversal is $E_g = 186.82$ V and after polarity reversal $E_g = -186.82$ V. From Eq. (15.3),

$$V_a = E_g + I_a R_a = -186.82 + 45 \times 0.25 = -175.57 \text{ V}$$

From Eq. (15.17),

$$V_a = \frac{2V_m}{\pi} \cos \alpha_a = \frac{2 \times 622.25}{\pi} \cos \alpha_a = -175.57 \text{ V}$$

and this yields the delay angle of the armature converter as $\alpha_a = 116.31°$.

b. The power fed back to the supply is $P_a = V_a I_a = 175.57 \times 45 = 7900.7$ W.

Note: The speed and back emf of the motor decrease with time. If the armature current is to be maintained constant at $I_a = 45$ A during regeneration, the delay angle of the armature converter has to be reduced. This would require a closed-loop control to maintain the armature current constant and to adjust the delay angle continuously.

Key Points of Section 15.4

- A single-phase drive uses a single-phase converter. The type of single-phase converters classifies the single-phase drive.
- A semiconverter drive operates in one quadrant; a full-converter drive in two quadrants; and a dual converter, in four quadrants. The field excitation is normally supplied from a full converter.

15.5 THREE-PHASE DRIVES

The armature circuit is connected to the output of a three-phase controlled rectifier or a forced-commutated three-phase ac–dc converter. Three-phase drives are used for high-power applications up to megawatt power levels. The ripple frequency of the armature voltage is higher than that of single-phase drives and it requires less inductance in the armature circuit to reduce the armature ripple current. The armature current is mostly continuous, and therefore the motor performance is better compared with that of single-phase drives. Similar to the single-phase drives, three-phase drives [2, 6] may also be subdivided into:

1. Three-phase half-wave-converter drives.
2. Three-phase semiconverter drives.
3. Three-phase full-converter drives.
4. Three-phase dual-converter drives.

15.5.1 Three-Phase Half-Wave-Converter Drives

A three-phase half-wave converter-fed dc motor drive operates in one quadrant and could be used in applications up to a 40-kW power level. The field converter could be a single-phase or three-phase semiconverter. This drive is not normally used in industrial applications because the ac supply contains dc components.

With a three-phase half-wave converter in the armature circuit, Eq. (10.19) gives the armature voltage as

$$V_a = \frac{3\sqrt{3}V_m}{2\pi} \cos \alpha_a \qquad \text{for } 0 \le \alpha_a \le \pi \qquad (15.22)$$

where V_m is the peak phase voltage of a Y-connected three-phase ac supply. With a three-phase semiconverter in the field circuit, Eq. (10.69) gives the field voltage as

$$V_f = \frac{3\sqrt{3}V_m}{2\pi}(1 + \cos \alpha_f) \qquad \text{for } 0 \le \alpha_f \le \pi \qquad (15.23)$$

15.5.2 Three-Phase Semiconverter Drives

A three-phase semiconverter-fed drive is a one-quadrant drive without field reversal, and is limited to applications up to 115 kW. The field converter should also be a single-phase or a three-phase semiconverter.

With a three-phase semiconverter in the armature circuit, Eq. (10.69) gives the armature voltage as

$$V_a = \frac{3\sqrt{3}V_m}{2\pi}(1 + \cos \alpha_a) \qquad \text{for } 0 \le \alpha_a \le \pi \qquad (15.24)$$

With a three-phase semiconverter in the field circuit, Eq. (10.69) gives the field voltage as

$$V_f = \frac{3\sqrt{3}V_m}{2\pi}(1 + \cos \alpha_f) \qquad \text{for } 0 \le \alpha_f \le \pi \qquad (15.25)$$

15.5.3 Three-Phase Full-Converter Drives

A three-phase full-wave-converter drive is a two-quadrant drive without any field reversal, and is limited to applications up to 1500 kW. During regeneration for reversing the direction of power flow, the back emf of the motor is reversed by reversing the field excitation. The converter in the field circuit should be a single- or three-phase full converter. A 68-kW, 170-A microprocessor-based three-phase full-converter dc drive is shown in Figure 15.16, where the power semiconductor stacks are at the back of the panel.

FIGURE 15.16

A 68-kW microprocessor-based three-phase full converter. (Reproduced by permission of Brush Electrical Machines Ltd., England.)

With a three-phase full-wave converter in the armature circuit, Eq. (10.25) gives the armature voltage as

$$V_a = \frac{3\sqrt{3}V_m}{\pi} \cos\alpha_a \qquad \text{for } 0 \le \alpha_a \le \pi \qquad (15.26)$$

With a three-phase full converter in the field circuit, Eq. (10.25) gives the field voltage as

$$V_f = \frac{3\sqrt{3}V_m}{\pi} \cos\alpha_f \qquad \text{for } 0 \le \alpha_f \le \pi \qquad (15.27)$$

15.5.4 Three-Phase Dual-Converter Drives

Two three-phase full-wave converters are connected in an arrangement similar to Figure 15.15a. Either converter 1 operates to supply a positive armature voltage, V_a, or converter 2 operates to supply a negative armature voltage, $-V_a$. It is a four-quadrant drive and is limited to applications up to 1500 kW. Similar to single-phase drives, the field converter can be a full-wave converter or a semiconverter.

A 12-pulse ac–dc converter for a 360-kW motor for driving a cement kiln is shown in Figure 15.17, where the control electronics are mounted on the cubicle door and the pulse drive boards are mounted on the front of the thyristor stacks. The cooling fans are mounted at the top of each stack. If converter 1 operates with a delay angle of

FIGURE 15.17

A 360-kW, 12-pulse ac–dc converter for dc drives. (Reproduced by permission of Brush Electrical Machines Ltd., England.)

α_{a1}, Eq. (10.25) gives the average armature voltage as

$$V_a = \frac{3\sqrt{3}V_m}{\pi}\cos\alpha_{a1} \qquad \text{for } 0 \le \alpha_{a1} \le \pi \qquad (15.28)$$

If converter 2 operates with a delay angle of α_{a2}, Eq. (10.25) gives the average armature voltage as

$$V_a = \frac{3\sqrt{3}V_m}{\pi}\cos\alpha_{a2} \qquad \text{for } 0 \le \alpha_{a2} \le \pi \qquad (15.29)$$

With a three-phase full converter in the field circuit, Eq. (10.25) gives the average field voltage as

$$V_f = \frac{3\sqrt{3}V_m}{\pi}\cos\alpha_f \qquad \text{for } 0 \le \alpha_f \le \pi \qquad (15.30)$$

Example 15.5   Finding the Performance Parameters of a Three-Phase Full-Converter Drive

The speed of a 20-hp, 300-V, 1800-rpm separately excited dc motor is controlled by a three-phase full-converter drive. The field current is also controlled by a three-phase full converter and is set to the maximum possible value. The ac input is a three-phase, Y-connected, 208-V, 60-Hz supply. The armature resistance is $R_a = 0.25\ \Omega$, the field resistance is $R_f = 245\ \Omega$, and the motor voltage constant is $K_v = 1.2$ V/A rad/s. The armature and field currents can be assumed to be continuous and ripple free. The viscous friction is negligible. Determine (a) the delay angle of the armature converter α_a, if the motor supplies the rated power at the rated speed; (b) the no-load speed if the delay angles are the same as in (a) and the armature current at no load is 10% of the rated value; and (c) the speed regulation.

Solution

$R_a = 0.25\ \Omega$, $R_f = 245\ \Omega$, $K_v = 1.2$ V/A rad/s, $V_L = 208$ V, and $\omega = 1800\ \pi/30 = 188.5$ rad/s. The phase voltage is $V_p = V_L/\sqrt{3} = 208/\sqrt{3} = 120$ V and $V_m = 120 \times \sqrt{2} = 169.7$ V. Because 1 hp is equal to 746 W, the rated armature current is $I_{rated} = 20 \times 746/300 = 49.73$ A; and for maximum possible field current, $\alpha_f = 0$. From Eq. (15.27),

$$V_f = 3\sqrt{3} \times \frac{169.7}{\pi} = 280.7 \text{ V}$$

$$I_f = \frac{V_f}{R_f} = \frac{280.7}{245} = 1.146 \text{ A}$$

a. $I_a = I_{rated} = 49.73$ A and

$$E_g = K_v I_f \omega = 1.2 \times 1.146 \times 188.5 = 259.2 \text{ V}$$
$$V_a = 259.2 + I_a R_a = 259.2 + 49.73 \times 0.25 = 271.63 \text{ V}$$

From Eq. (15.26),

$$V_a = 271.63 = \frac{3\sqrt{3}\,V_m}{\pi} \cos\alpha_a = \frac{3\sqrt{3} \times 169.7}{\pi} \cos\alpha_a$$

and this gives the delay angle as $\alpha_a = 14.59°$.

b. $I_a = 10\%$ of $49.73 = 4.973$ A and

$$E_{go} = V_a - R_a I_a = 271.63 - 0.25 \times 4.973 = 270.39 \text{ V}$$

From Eq. (15.4), the no-load speed is

$$\omega_0 = \frac{E_{go}}{K_v I_f} = \frac{270.39}{1.2 \times 1.146} = 196.62 \text{ rad/s} \quad \text{or} \quad 196.62 \times \frac{30}{\pi} = 1877.58 \text{ rpm}$$

c. The speed regulation is defined as

$$\frac{\text{no-load speed} - \text{full-load speed}}{\text{full-load speed}} = \frac{1877.58 - 1800}{1800} = 0.043 \quad \text{or} \quad 4.3\%$$

Example 15.6 Finding the Performance of a Three-Phase Full-Converter Drive with Field Control

The speed of a 20-hp, 300-V, 900-rpm separately excited dc motor is controlled by a three-phase full converter. The field circuit is also controlled by a three-phase full converter. The ac input to the armature and field converters is three-phase, Y-connected, 208 V, 60 Hz. The armature resistance is $R_a = 0.25 \ \Omega$, the field circuit resistance is $R_f = 145 \ \Omega$, and the motor voltage constant is $K_v = 1.2$ V/A rad/s. The viscous friction and no-load losses can be considered negligible. The armature and field currents are continuous and ripple free. (a) If the field converter is operated at the maximum field current and the developed torque is $T_d = 116$ N·m at 900 rpm, determine the delay angle of the armature converter α_a. (b) If the field circuit converter is set for the maximum field current, the developed torque is $T_d = 116$ N·m, and the delay angle of the armature converter is $\alpha_a = 0$, determine the speed of the motor. (c) For the same load demand as in (b), determine the delay angle of the field converter if the speed has to be increased to 1800 rpm.

Solution

$R_a = 0.25 \ \Omega$, $R_f = 145 \ \Omega$, $K_v = 1.2$ V/A rad/s, and $V_L = 208$ V. The phase voltage is $V_p = 208/\sqrt{3} = 120$ V and $V_m = \sqrt{2} \times 120 = 169.7$ V.

a. $T_d = 116$ N·m and $\omega = 900 \ \pi/30 = 94.25$ rad/s. For maximum field current, $\alpha_f = 0$. From Eq. (15.27),

$$V_f = \frac{3 \times \sqrt{3} \times 169.7}{\pi} = 280.7 \text{ V}$$

$$I_f = \frac{280.7}{145} = 1.936 \text{ A}$$

From Eq. (15.4),

$$I_a = \frac{T_d}{K_v I_f} = \frac{116}{1.2 \times 1.936} = 49.93 \text{ A}$$

$$E_g = K_v I_f \omega = 1.2 \times 1.936 \times 94.25 = 218.96 \text{ V}$$

$$V_a = E_g + I_a R_a = 218.96 + 49.93 \times 0.25 = 231.44 \text{ V}$$

From Eq. (15.26),

$$V_a = 231.44 = \frac{3 \times \sqrt{3} \times 169.7}{\pi} \cos \alpha_a$$

which gives the delay angle as $\alpha_a = 34.46°$.

b. $\alpha_a = 0$ and

$$V_a = \frac{3 \times \sqrt{3} \times 169.7}{\pi} = 280.7 \text{ V}$$

$$E_g = 280.7 - 49.93 \times 0.25 = 268.22 \text{ V}$$

and the speed

$$\omega = \frac{E_g}{K_v I_f} = \frac{268.22}{1.2 \times 1.936} = 115.45 \text{ rad/s} \quad \text{or} \quad 1102.5 \text{ rpm}$$

c. $\omega = 1800 \ \pi/30 = 188.5$ rad/s

$$E_g = 268.22 \text{ V} = 1.2 \times 188.5 \times I_f \quad \text{or} \quad I_f = 1.186 \text{ A}$$

$$V_f = 1.186 \times 145 = 171.97 \text{ V}$$

From Eq. (15.27),

$$V_f = 171.97 = \frac{3 \times \sqrt{3} \times 169.7}{\pi} \cos \alpha_f$$

which gives the delay angle as $\alpha_f = 52.2°$.

Key Points of Section 15.5

- A three-phase drive uses a three-phase converter. The type of three-phase converters classifies the three-phase drive. In general, a semiconverter drive operates in one quadrant; a full-converter drive, in two quadrants; and a dual-converter, in four quadrants. The field excitation is normally supplied from a full converter.

15.6   DC–DC CONVERTER DRIVES

Dc–dc converter (or simply chopper) drives are widely used in traction applications all over the world. A dc–dc converter is connected between a fixed-voltage dc source and a dc motor to vary the armature voltage. In addition to armature voltage control, a dc–dc converter can provide regenerative braking of the motors and can return energy back to the supply. This energy-saving feature is particulary attractive to transportation systems with frequent stops such as mass rapid transit (MRT). Dc–dc converter drives are also used in battery electric vehicles (BEVs). A dc motor can be operated in one of the four quadrants by controlling the armature or field voltages (or currents). It is often required to reverse the armature or field terminals to operate the motor in the desired quadrant.

If the supply is nonreceptive during the regenerative braking, the line voltage would increase and regenerative braking may not be possible. In this case, an alternative form of braking is necessary, such as rheostatic braking. The possible control modes of a dc–dc converter drive are:

1. Power (or acceleration) control.
2. Regenerative brake control.
3. Rheostatic brake control.
4. Combined regenerative and rheostatic brake control.

15.6.1   Principle of Power Control

The dc–dc converter is used to control the armature voltage of a dc motor. The circuit arrangement of a converter-fed dc separately excited motor is shown in Figure 15.18a. The dc–dc converter switch could be a transistor or forced-commutated thyristor dc–dc converter, as discussed in Section 5.3. This is a one-quadrant drive, as shown in Figure 18.18b. The waveforms for the armature voltage, load current, and input current are shown in Figure 15.18c, assuming a highly inductive load.

The average armature voltage is

$$V_a = kV_s \tag{15.31}$$

where k is the duty cycle of the dc–dc converter. The power supplied to the motor is

$$P_o = V_a I_a = kV_s I_a \tag{15.32}$$

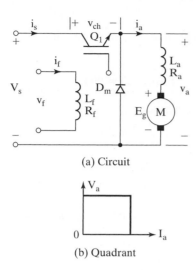

(a) Circuit

(b) Quadrant

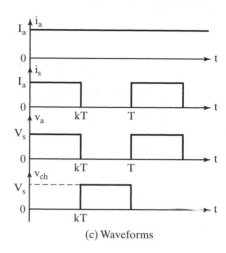

(c) Waveforms

FIGURE 15.18

Converter-fed dc drive in power control.

where I_a is the average armature current of the motor and it is ripple free. Assuming a lossless dc–dc converter, the input power is $P_i - P_0 = kV_sI_s$. The average value of the input current is

$$I_s = kI_a \tag{15.33}$$

The equivalent input resistance of the dc–dc converter drive seen by the source is

$$R_{eq} = \frac{V_s}{I_s} = \frac{V_s}{I_a}\frac{1}{k} \tag{15.34}$$

By varying the duty-cycle k, the power flow to the motor (and speed) can be controlled. For a finite armature circuit inductance, Eq. (5.21) can be applied to find the maximum peak-to-peak ripple current as

$$\Delta I_{max} = \frac{V_s}{R_m} \tanh \frac{R_m}{4fL_m} \tag{15.35}$$

where R_m and L_m are the total armature circuit resistance and inductance, respectively. For a separately excited motor, $R_m = R_a +$ any series resistance, and $L_m = L_a +$ any series inductance. For a series motor, $R_m = R_a + R_f +$ any series resistance, and $L_m = L_a + L_f +$ any series inductance.

Example 15.7   Finding the Performance Parameters of a Dc–dc Converter Drive

A dc separately excited motor is powered by a dc–dc converter, (as shown in Figure 15.18a), from a 600-V dc source. The armature resistance is $R_a = 0.05 \ \Omega$. The back emf constant of the motor is $K_v = 1.527$ V/A rad/s. The average armature current is $I_a = 250$ A. The field current is $I_f = 2.5$ A. The armature current is continuous and has negligible ripple. If the duty cycle of the dc–dc converter is 60%, determine (a) the input power from the source, (b) the equivalent input resistance of the dc–dc converter drive, (c) the motor speed, and (d) the developed torque.

Solution

$V_s = 600$ V, $I_a = 250$ A, and $k = 0.6$. The total armature circuit resistance is $R_m = R_a = 0.05 \ \Omega$.

a. From Eq. (15.32),

$$P_i = kV_sI_a = 0.6 \times 600 \times 250 = 90 \ \text{kW}$$

b. From Eq. (15.34) $R_{eq} = 600/(250 \times 0.6) = 4 \ \Omega$.

c. From Eq. (15.31), $V_a = 0.6 \times 600 = 360$ V. The back emf is

$$E_g = V_a - R_mI_m = 360 - 0.05 \times 250 = 347.5 \ \text{V}$$

From Eq. (15.2), the motor speed is

$$\omega = \frac{347.5}{1.527 \times 2.5} = 91.03 \ \text{rad/s} \quad \text{or} \quad 91.03 \times \frac{30}{\pi} = 869.3 \ \text{rpm}$$

d. From Eq. (15.4),

$$T_d = 1.527 \times 250 \times 2.5 = 954.38 \ \text{N} \cdot \text{m}$$

15.6.2 Principle of Regenerative Brake Control

In regenerative braking, the motor acts as a generator and the kinetic energy of the motor and load is returned back to the supply. The principle of energy transfer from one dc source to another of higher voltage is discussed in Section 5.4, and this can be applied in regenerative braking of dc motors.

The application of dc–dc converters in regenerative braking can be explained with Figure 15.19a. It requires rearranging the switch from powering mode to regenerative braking. Let us assume that the armature of a separately excited motor is rotating due to the inertia of the motor (and load); and in case of a transportation system, the kinetic energy of the vehicle or train would rotate the armature shaft. Then if the transistor is switched on, the armature current rises due to the short-circuiting of the motor terminals. If the dc–dc converter is turned off, diode D_m would be turned on and the energy stored in the armature circuit inductances would be transferred to the supply, provided that the supply is receptive. It is a one-quadrant drive and operates in the second quadrant, as shown in Figure 15.19b. Figure 15.19c shows the voltage and current waveforms assuming that the armature current is continuous and ripple free.

The average voltage across the dc–dc converter is

$$V_{ch} = (1 - k)V_s \tag{15.36}$$

If I_a is the average armature current, the regenerated power can be found from

$$P_g = I_aV_s(1 - k) \tag{15.37}$$

The voltage generated by the motor acting as a generator is

$$E_g = K_vI_f\omega$$

$$= V_{ch} + R_mI_a = (1 - k)V_s + R_mI_a \tag{15.38}$$

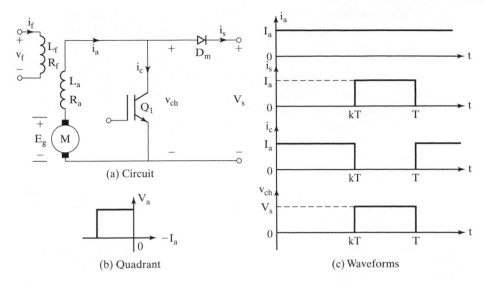

FIGURE 15.19

Regenerative braking of dc separately excited motors.

where K_v is machine constant and ω is the machine speed in rads per second. Therefore, the equivalent load resistance of the motor acting as a generator is

$$R_{eq} = \frac{E_g}{I_a} = \frac{V_s}{I_a}(1 - k) + R_m \tag{15.39}$$

By varying the duty-cycle k, the equivalent load resistance seen by the motor can be varied from R_m to $(V_s/I_a + R_m)$ and the regenerative power can be controlled.

From Eq. (5.30), the conditions for permissible potentials and polarity of the two voltages are

$$0 \leq (E_g - R_m I_a) \leq V_s \tag{15.40}$$

which gives the minimum braking speed of the motor as

$$E_g = K_v \omega_{min} I_f = R_m I_a$$

or

$$\omega_{min} = \frac{R_m I_a}{K_v I_f} \tag{15.41}$$

and $\omega \geq \omega_{min}$. The maximum braking speed of a series motor can be found from Eq. (15.40):

$$K_v \omega_{max} I_f - R_m I_a = V_s$$

or

$$\omega_{max} = \frac{V_s}{K_v I_f} + \frac{R_m I_a}{K_v I_f} \tag{15.42}$$

and $\omega \leq \omega_{max}$.

The regenerative braking would be effective only if the motor speed is between these two speed limits (e.g., $\omega_{min} < \omega < \omega_{max}$). At any speed less than ω_{min}, an alternative braking arrangement would be required.

Although dc series motors are traditionally used for traction applications due to their high-starting torque, a series-excited generator is unstable when working into a fixed voltage supply. Thus, for running on the traction supply, a separate excitation control is required and such an arrangement of series motor is, normally, sensitive to supply voltage fluctuations and a fast dynamic response is required to provide an adequate brake control. The application of a dc–dc converter allows the regenerative braking of dc series motors due to its fast dynamic response.

A separately excited dc motor is stable in regenerative braking. The armature and field can be controlled independently to provide the required torque during starting. A dc–dc converter-fed series and separately excited dc motors are both suitable for traction applications.

Example 15.8   Finding the Performance of a Dc–dc Converter Fed Drive in Regenerative Braking

A dc–dc converter is used in regenerative braking of a dc series motor similar to the arrangement shown in Figure 15.19a. The dc supply voltage is 600 V. The armature resistance is $R_a = 0.02\ \Omega$ and the field resistance is $R_f = 0.03\ \Omega$. The back emf constant is $K_v = 15.27$ mV/A rad/s. The average armature current is maintained constant at $I_a = 250$ A. The armature current is continuous and has negligible ripple. If the duty cycle of the dc–dc converter is 60%, determine (a) the average voltage across the dc–dc converter V_{ch}; (b) the power regenerated to the dc supply P_g; (c) the equivalent load resistance of the motor acting as a generator, R_{eq}; (d) the minimum permissible braking speed ω_{min}; (e) the maximum permissible braking speed ω_{max}; and (f) the motor speed.

Solution
$V_s = 600$ V, $I_a = 250$ A, $K_v = 0.01527$ V/A rad/s, $k = 0.6$. For a series motor $R_m = R_a + R_f = 0.02 + 0.03 = 0.05\ \Omega$.

a. From Eq. (15.36), $V_{ch} = (1 - 0.6) \times 600 = 240$ V.

b. From Eq. (15.37), $P_g = 250 \times 600 \times (1 - 0.6) = 60$ kW.

c. From Eq. (15.39), $R_{eq} = (600/250)(1 - 0.6) + 0.05 = 1.01\ \Omega$.

d. From Eq. (15.41), the minimum permissible braking speed,

$$\omega_{min} = \frac{0.05}{0.01527} = 3.274 \text{ rad/s} \quad \text{or} \quad 3.274 \times \frac{30}{\pi} = 31.26 \text{ rpm}$$

e. From Eq. (15.42), the maximum permissible braking speed,

$$\omega_{max} = \frac{600}{0.01527 \times 250} + \frac{0.05}{0.01527} = 160.445 \text{ rad/s} \quad \text{or} \quad 1532.14 \text{ rpm}$$

f. From Eq. (15.38), $E_g = 240 + 0.05 \times 250 = 252.5$ V and the motor speed,

$$\omega = \frac{252.5}{0.01527 \times 250} = 66.14 \text{ rad/s or } 631.6 \text{ rpm}$$

Note: The motor speed would decrease with time. To maintain the armature current at the same level, the effective load resistance of the series generator should be adjusted by varying the duty cycle of the dc–dc converter.

15.6.3 Principle of Rheostatic Brake Control

In a rheostatic braking, the energy is dissipated in a rheostat and it may not be a desirable feature. In MRT systems, the energy may be used in heating the trains. The rheostatic braking is also known as *dynamic braking*. An arrangement for the rheostatic braking of a dc separately excited motor is shown in Figure 15.20a. This is a one-quadrant drive and operates in the second quadrant, as shown in Figure 15.20b. Figure 15.20c shows the waveforms for the current and voltage, assuming that the armature current is continuous and ripple free.

The average current of the braking resistor,

$$I_b = I_a(1 - k) \tag{15.43}$$

and the average voltage across the braking resistor,

$$V_b - R_b I_a(1 - k) \tag{15.44}$$

The equivalent load resistance of the generator,

$$R_{eq} = \frac{V_b}{I_a} = R_b(1 - k) + R_m \tag{15.45}$$

The power dissipated in the resistor R_b is

$$P_b = I_a^2 R_b(1 - k) \tag{15.46}$$

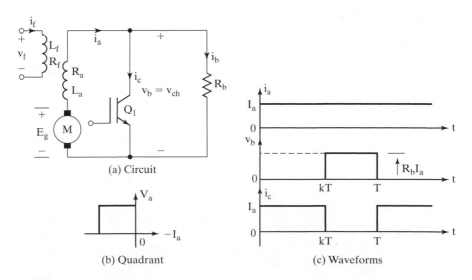

(a) Circuit

(b) Quadrant

(c) Waveforms

FIGURE 15.20

Rheostatic braking of dc separately excited motors.

By controlling the duty-cycle k, the effective load resistance can be varied from R_m to $R_m + R_b$; and the braking power can be controlled. The braking resistance R_b determines the maximum voltage rating of the dc–dc converter.

Example 15.9 Finding the Performance of a Dc–dc Converter-Fed Drive in Rheostatic Braking

A dc–dc converter is used in rheostatic braking of a dc separately excited motor, as shown in Figure 15.20a. The armature resistance is $R_a = 0.05\ \Omega$. The braking resistor is $R_b = 5\ \Omega$. The back emf constant is $K_v = 1.527$ V/A rad/s. The average armature current is maintained constant at $I_a = 150$ A. The armature current is continuous and has negligible ripple. The field current is $I_f = 1.5$ A. If the duty cycle of the dc–dc converter is 40%, determine (a) the average voltage across the dc–dc converter V_{ch}; (b) the power dissipated in the braking resistor P_b; (c) the equivalent load resistance of the motor acting as a generator, R_{eq}; (d) the motor speed; and (e) the peak dc–dc converter voltage V_p.

Solution
$I_a = 150$ A, $K_v = 1.527$ V/A rad/s, $k = 0.4$, and $R_m = R_a = 0.05\ \Omega$.

a. From Eq. (15.44), $V_{ch} = V_b = 5 \times 150 \times (1 - 0.4) = 450$ V.

b. From Eq. (15.46), $P_b = 150 \times 150 \times 5 \times (1 - 0.4) = 67.5$ kW.

c. From Eq. (15.45), $R_{eq} = 5 \times (1 - 0.4) + 0.05 = 3.05\ \Omega$.

d. The generated emf $E_g = 450 + 0.05 \times 150 = 457.5$ V and the braking speed,

$$\omega = \frac{E_g}{K_v I_f} = \frac{457.5}{1.527 \times 1.5} = 199.74\ \text{rad/s} \qquad \text{or} \qquad 1907.4\ \text{rpm}$$

e. The peak dc–dc converter voltage $V_p = I_a R_b = 150 \times 5 = 750$ V.

15.6.4 Principle of Combined Regenerative and Rheostatic Brake Control

Regenerative braking is energy-efficient braking. On the other hand, the energy is dissipated as heat in rheostatic braking. If the supply is partly receptive, which is normally the case in practical traction systems, a combined regenerative and rheostatic brake control would be the most energy efficient. Figure 15.21 shows an arrangement in which rheostatic braking is combined with regenerative braking.

During regenerative brakings, the line voltage is sensed continuously. If it exceeds a certain preset value, normally 20% above the line voltage, the regenerative braking is

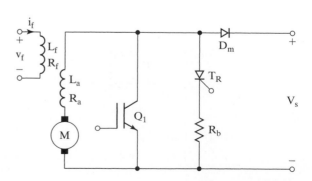

FIGURE 15.21

Combined regenerative and rheostatic braking.

removed and a rheostatic braking is applied. It allows an almost instantaneous transfer from regenerative to rheostatic braking if the line becomes nonreceptive, even momentarily. In every cycle, the logic circuit determines the receptivity of the supply. If it is nonreceptive, thyristor T_R is turned on to divert the motor current to the resistor R_b. Thyristor T_R is self-commutated when transistor Q_1 is turned on in the next cycle.

15.6.5   Two- and Four-Quadrant Dc–dc Converter Drives

During power control, a dc–dc converter-fed drive operates in the first quadrant, where the armature voltage and armature current are positive, as shown in Figure 15.18b. In a regenerative braking, the dc–dc converter drive operates in the second quadrant, where the armature voltage is positive and the armature current is negative, as shown in Figure 15.19b. Two-quadrant operation, as shown in Figure 15.22a, is required to allow power and regenerative braking control. The circuit arrangement of a transistorized two-quadrant drive is shown in Figure 15.22b.

Power control.   Transistor Q_1 and diode D_2 operate. When Q_1 is turned on, the supply voltage V_s is connected to the motor terminals. When Q_1 is turned off, the armature current that flows through the freewheeling diode D_2 decays.

Regenerative control.   Transistor Q_2 and diode D_1 operate. When Q_2 is turned on, the motor acts as a generator and the armature current rises. When Q_2 is turned off, the motor, acting as a generator, returns energy to the supply through the regenerative diode D_1. In industrial applications, four-quadrant operation, as shown in Figure 15.23a, is required. A transistorized four-quadrant drive is shown in Figure 15.23b.

Forward power control.   Transistors Q_1 and Q_2 operate. Transistors Q_3 and Q_4 are off. When Q_1 and Q_2 are turned on together, the supply voltage appears across the motor terminals and the armature current rises. When Q_1 is turned off and Q_2 is still turned on, the armature current decays through Q_2 and D_4. Alternatively, both Q_1 and Q_2 can be turned off, while the armature current is forced to decay through D_3 and D_4.

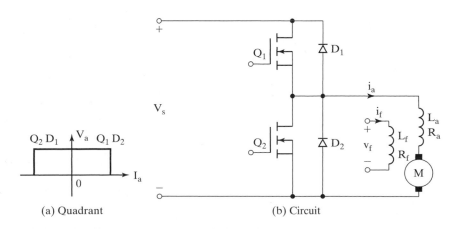

(a) Quadrant   (b) Circuit

FIGURE 15.22

Two-quadrant transistorized dc–dc converter drive.

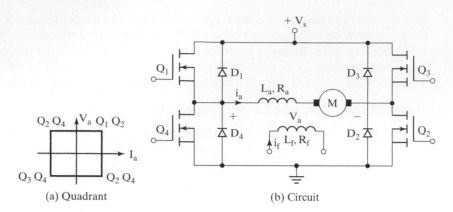

FIGURE 15.23

Four-quadrant transistorized dc–dc converter drive.

Forward regeneration. Transistors Q_1, Q_2, and Q_3 are turned off. When transistor Q_4 is turned on, the armature current, which rises, flows through Q_4 and D_2. When Q_4 is turned off, the motor, acting as a generator, returns energy to the supply through D_1 and D_2.

Reverse power control. Transistors Q_3 and Q_4 operate. Transistors Q_1 and Q_2 are off. When Q_3 and Q_4 are turned on together, the armature current rises and flows in the reverse direction. When Q_3 is turned off and Q_4 is turned on, the armature current falls through Q_4 and D_2. Alternatively, both Q_3 and Q_4 can be turned off, while forcing the armature current to decay through D_1 and D_2.

Reverse regeneration. Transistors Q_1, Q_3, and Q_4 are off. When Q_2 is turned on, the armature current rises through Q_2 and D_4. When Q_2 is turned off, the armature current falls and the motor returns energy to the supply through D_3 and D_4.

15.6.6 Multiphase Dc–dc Converters

If two or more dc–dc converters are operated in parallel and are phase shifted from each other by π/u, as shown in Figure 15.24a, the amplitude of the load current ripples decreases and the ripple frequency increases [7, 8]. As a result, the dc–dc converter-generated harmonic currents in the supply are reduced. The sizes of the input filter are also reduced. The multiphase operation allows the reduction of smoothing chokes, which are normally connected in the armature circuit of dc motors. Separate inductors in each phase are used for current sharing. Figure 15.24b shows the waveforms for the currents with u dc–dc converters.

For u dc–dc converters in multiphase operation, it can be proved that Eq. (5.21) is satisfied when $k = 1/2u$ and the maximum peak-to-peak load ripple current becomes

$$\Delta I_{\max} = \frac{V_s}{R_m} \tanh \frac{R_m}{4uf L_m} \tag{15.47}$$

where L_m and R_m are the total armature inductance and resistance, respectively. For $4uf L_m \gg R_m$, the maximum peak-to-peak load ripple current can be approximated to

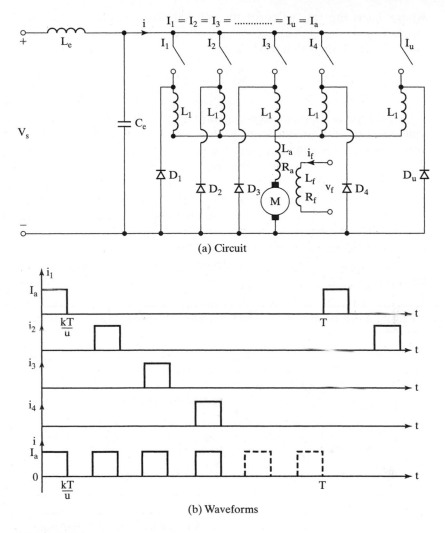

(a) Circuit

(b) Waveforms

FIGURE 15.24

Multiphase dc–dc converters.

$$\Delta I_{max} = \frac{V_s}{4ufL_m} \tag{15.48}$$

If an LC-input filter is used, Eq. (5.125) can be applied to find the rms nth harmonic component of dc–dc converter-generated harmonics in the supply

$$I_{ns} = \frac{1}{1 + (2n\pi uf)^2 L_e C_e} I_{nh}$$

$$= \frac{1}{1 + (nuf/f_0)^2} I_{nh} \tag{15.49}$$

where I_{nh} is the rms value of the nth harmonic component of the dc–dc converter current and $f_0 = [1/(2\pi L_e C_e)]$ is the resonant frequency of the input filter. If $(nuf/f_0) \gg 1$, the nth harmonic current in the supply becomes

$$I_{ns} = I_{nh} \left(\frac{f_0}{nuf} \right)^2 \tag{15.50}$$

Multiphase operations are advantageous for large motor drives, especially if the load current requirement is large. However, considering the additional complexity involved in increasing the number of dc–dc converters, there is not much reduction in the dc–dc converter-generated harmonics in the supply line if more than two dc–dc converters are used [7]. In practice, both the magnitude and frequency of the line current harmonics are important factors to determine the level of interferences into the signaling circuits. In many rapid transit systems, the power and signaling lines are very close; in three-line systems, they even share a common line. The signaling circuits are sensitive to particular frequencies and the reduction in the magnitude of harmonics by using a multiphase operation of dc–dc converters could generate frequencies within the sensitivity band—which might cause more problems than it solves.

Example 15.10   Finding the Peak Load Current Ripple of Two Multiphase Dc–dc Converters

Two dc–dc converters control a dc separately excited motor and they are phase shifted in operation by $\pi/2$. The supply voltage to the dc–dc converter drive is $V_s = 220$ V, the total armature circuit resistance is $R_m = 4\ \Omega$, the total armature circuit inductance is $L_m = 15$ mH, and the frequency of each dc–dc converter is $f = 350$ Hz. Calculate the maximum peak-to-peak load ripple current.

Solution

The effective chopping frequency is $f_e = 2 \times 350 = 700$ Hz, $R_m = 4\ \Omega$, $L_m = 15$ mH, $u = 2$, and $V_s = 220$ V. $4uf L_m = 4 \times 2 \times 350 \times 15 \times 10^{-3} = 42$. Because $42 \gg 4$, approximate Eq. (15.48) can be used, and this gives the maximum peak-to-peak load ripple current, $\Delta I_{max} = 220/42 = 5.24$ A.

Example 15.11   Finding the Line Harmonic Current of Two Multiphase Dc–dc Converters with an Input Filter

A dc separately excited motor is controlled by two multiphase dc–dc converters. The average armature current is $I_a = 100$ A. A simple LC-input filter with $L_e = 0.3$ mH and $C_e = 4500\ \mu$F is used. Each dc–dc converter is operated at a frequency of $f = 350$ Hz. Determine the rms fundamental component of the dc–dc converter-generated harmonic current in the supply.

Solution

$I_a = 100$ A, $u = 2$, $L_e = 0.3$ mH, $C_e = 4500\ \mu$F, and $f_0 = 1/(2\pi\sqrt{L_e C_e}) = 136.98$ Hz. The effective dc–dc converter frequency is $f_e = 2 \times 350 = 700$ Hz. From the results of Example 9.13, the rms value of the fundamental component of the dc–dc converter current is $I_{1h} = 45.02$ A. From Eq. (15.49), the fundamental component of the dc–dc converter-generated harmonic current is

$$I_{1s} = \frac{45.02}{1 + (2 \times 350/136.98)^2} = 1.66 \text{ A}$$

Key Points of Section 15.6

- It is possible to arrange the dc–dc converter in powering the motor, or in regenerative or rheostatic braking. A dc–dc converter-fed drive can operate in four quadrants.
- Multiphase dc–dc converters are often used to supply load current that cannot be handled by one dc–dc converter. Multiphase dc–dc converters have the advantage of increasing the effective converter frequency, thereby reducing the component values and size of the input filter.

15.7   CLOSED-LOOP CONTROL OF DC DRIVES

The speed of dc motors changes with the load torque. To maintain a constant speed, the armature (and or field) voltage should be varied continuously by varying the delay angle of ac–dc converters or duty cycle of dc–dc converters. In practical drive systems it is required to operate the drive at a constant torque or constant power; in addition, controlled acceleration and deceleration are required. Most industrial drives operate as closed-loop feedback systems. A closed-loop control system has the advantages of improved accuracy, fast dynamic response, and reduced effects of load disturbances and system nonlinearities [9].

The block diagram of a closed-loop converter-fed separately excited dc drive is shown in Figure 15.25. If the speed of the motor decreases due to the application of additional load torque, the speed error V_e increases. The speed controller responses with an increased control signal V_c, change the delay angle or duty cycle of the converter, and increase the armature voltage of the motor. An increased armature voltage develops more torque to restore the motor speed to the original value. The drive normally passes through a transient period until the developed torque is equal to the load torque.

15.7.1   Open-Loop Transfer Function

The steady-state characteristics of dc drives, which are discussed in the preceding sections, are of major importance in the selection of dc drives and are not sufficient when the drive is in closed-loop control. Knowledge of the dynamic behavior, which is normally expressed in the form of a transfer function, is also important.

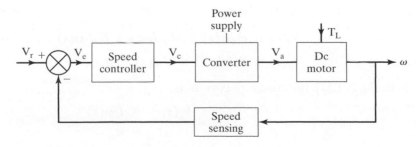

FIGURE 15.25

Block diagram of a closed-loop converter-fed dc motor drive.

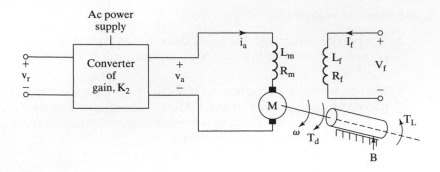

FIGURE 15.26

Converter-fed separately excited dc motor drive.

The circuit arrangement of a converter-fed separately excited dc motor drive with open-loop control is shown in Figure 15.26. The motor speed is adjusted by setting reference (or control) voltage v_r. Assuming a linear power converter of gain K_2, the armature voltage of the motor is

$$v_a = K_2 v_r \tag{15.51}$$

Assuming that the motor field current I_f and the back emf constant K_v remain constant during any transient disturbances, the system equations are

$$e_g = K_v I_f \omega \tag{15.52}$$

$$v_a = R_m i_a + L_m \frac{di_a}{dt} + e_g = R_m i_a + L_m \frac{di_a}{dt} + K_v I_f \omega \tag{15.53}$$

$$T_d = K_t I_f i_a \tag{15.54}$$

$$T_d = K_t I_f i_a = J \frac{d\omega}{dt} + B\omega + T_L \tag{15.55}$$

The transient behavior may be analyzed by changing the system equations into the Laplace transforms with zero initial conditions. Transforming Eqs. (15.51), (15.53), and (15.55) yields

$$V_a(s) = K_2 V_r(s) \tag{15.56}$$

$$V_a(s) = R_m I_a(s) + s L_m I_a(s) + K_v I_f \omega(s) \tag{15.57}$$

$$T_d(s) = K_t I_f I_a(s) = s J \omega(s) + B \omega(s) + T_L(s) \tag{15.58}$$

From Eq. (15.57), the armature current is

$$I_a(s) = \frac{V_a(s) - K_v I_f \omega(s)}{s L_m + R_m} \tag{15.59}$$

$$= \frac{V_a(s) - K_v I_f \omega(s)}{R_m(s \tau_a + 1)} \tag{15.60}$$

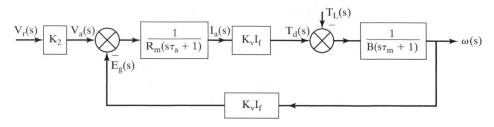

FIGURE 15.27

Open-loop block diagram of separately excited dc motor drive.

where $\tau_a = L_m/R_m$ is known as the *time constant* of motor armature circuit. From Eq. (15.58), the motor speed is

$$\omega(s) = \frac{T_d(s) - T_L(s)}{sJ + B} \tag{15.61}$$

$$= \frac{T_d(s) - T_L(s)}{B(s\tau_m + 1)} \tag{15.62}$$

where $\tau_m = J/B$ is known as the *mechanical time constant* of the motor. Equations (15.56), (15.60), and (15.62) can be used to draw the open-loop block diagram as shown in Figure 15.27. Two possible disturbances are control voltage V_r and load torque T_L. The steady-state responses can be determined by combining the individual response due to V_r and T_L.

The response due to a step change in the reference voltage is obtained by setting T_L to zero. From Figure 15.27, we can obtain the speed response due to reference voltage as

$$\frac{\omega(s)}{V_r(s)} = \frac{K_2 K_v I_f/(R_m B)}{s^2(\tau_a\tau_m) + s(\tau_a + \tau_m) + 1 + (K_v I_f)^2/R_m B} \tag{15.63}$$

The response due to a change in load torque T_L can be obtained by setting V_r to zero. The block diagram for a step change in load torque disturbance is shown in Figure 15.28.

$$\frac{\omega(s)}{T_L(s)} = -\frac{(1/B)(s\tau_a + 1)}{s^2(\tau_a\tau_m) + s(\tau_a + \tau_m) + 1 + (K_v I_f)^2/R_m B} \tag{15.64}$$

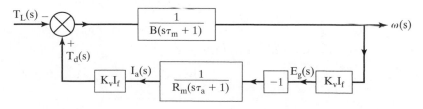

FIGURE 15.28

Open-loop block diagram for torque disturbance input.

Using the final value theorem, the steady-state relationship of a change in speed $\Delta\omega$, due to a step change in control voltage ΔV_r, and a step change in load torque ΔT_L, can be found from Eqs. (15.63) and (15.64), respectively, by substituting $s = 0$.

$$\Delta\omega = \frac{K_2 K_v I_f}{R_m B + (K_v I_f)^2} \Delta V_r \tag{15.65}$$

$$\Delta\omega = -\frac{R_m}{R_m B + (K_v I_f)^2} \Delta T_L \tag{15.66}$$

The dc series motors are used extensively in traction applications where the steady-state speed is determined by the friction and gradient forces. By adjusting the armature voltage, the motor may be operated at a constant torque (or current) up to the base speed, which corresponds to the maximum armature voltage. A dc–dc converter-controlled dc series motor drive is shown in Figure 15.29.

The armature voltage is related to the control (or reference) voltage by a linear gain of dc–dc converter K_2. Assuming that the back emf constant K_v does not change with the armature current and remains constant, the system equations are

$$v_a = K_2 v_r \tag{15.67}$$

$$e_g = K_v i_a \omega \tag{15.68}$$

$$v_a = R_m i_a + L_m \frac{di_a}{dt} + e_g \tag{15.69}$$

$$T_d = K_t i_a^2 \tag{15.70}$$

$$T_d = J\frac{d\omega}{dt} + B\omega + T_L \tag{15.71}$$

Equation (15.70) contains a product of variable-type nonlinearities, and as a result the application of transfer function techniques would no longer be valid. However, these equations can be linearized by considering a small perturbation at the operating point.

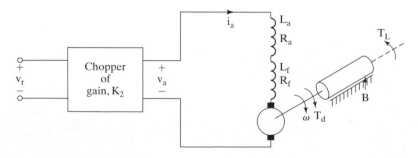

FIGURE 15.29

DC–dc converter-fed dc series motor drive.

Let us define the system parameters around the operating point as

$$e_g = E_{g0} + \Delta e_g \quad i_a = I_{a0} + \Delta i_a \quad v_a = V_{a0} + \Delta v_a \quad T_d = T_{d0} + \Delta T_d$$
$$\omega = \omega_0 + \Delta \omega \quad v_r = V_{r0} + \Delta v_r \quad T_L = T_{L0} + \Delta T_L$$

Recognizing that $\Delta i_a\, \Delta \omega$ and $(\Delta i_a)^2$ are very small, tending to zero, Eqs. (15.67) to (15.71) can be linearized to

$$\Delta v_a = K_2\, \Delta v_r$$

$$\Delta e_g = K_v(I_{a0}\, \Delta \omega + \omega_0\, \Delta i_a)$$

$$\Delta v_a = R_m\, \Delta i_a + L_m\, \frac{d(\Delta i_a)}{dt} + \Delta e_g$$

$$\Delta T_d = 2K_v I_{a0}\, \Delta i_a$$

$$\Delta T_d = J\, \frac{d(\Delta \omega)}{dt} + B\, \Delta \omega + \Delta T_L$$

Transforming these equations in the Laplace domain gives us

$$\Delta V_a(s) = K_2\, \Delta V_r(s) \tag{15.72}$$

$$\Delta E_g(s) = K_v[I_{a0}\, \Delta \omega(s) + \omega_0\, \Delta I_a(s)] \tag{15.73}$$

$$\Delta V_a(s) = R_m\, \Delta I_a(s) + sL_m\, \Delta I_a(s) + \Delta E_g(s) \tag{15.74}$$

$$\Delta T_d(s) = 2K_v I_{a0}\, \Delta I_a(s) \tag{15.75}$$

$$\Delta T_d(s) = sJ\, \Delta \omega(s) + B\, \Delta \omega(s) + \Delta T_L(s) \tag{15.76}$$

These five equations are sufficient to establish the block diagram of a dc series motor drive, as shown in Figure 15.30. It is evident from Figure 15.30 that any change in either reference voltage or load torque can result in a change of speed. The block diagram for a change in reference voltage is shown in Figure 15.31a and that for a change in load torque is shown in Figure 15.31b.

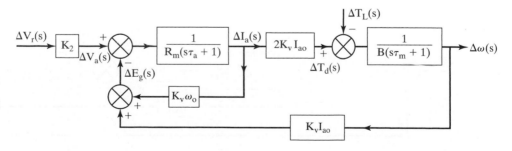

FIGURE 15.30

Open-loop block diagram of dc–dc converter-fed dc series drive.

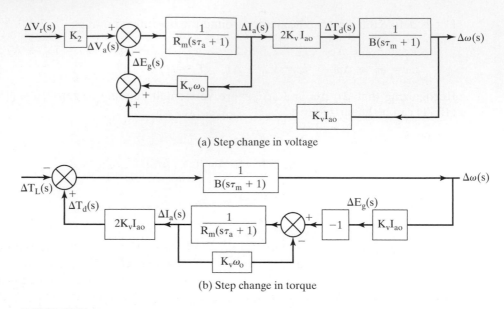

(a) Step change in voltage

(b) Step change in torque

FIGURE 15.31

Block diagram for reference voltage and load torque distributions.

15.7.2 Closed-Loop Transfer Function

Once the models for motors are known, feedback paths can be added to obtain the desired output response. To change the open-loop arrangement in Figure 15.26 into a closed-loop system, a speed sensor is connected to the output shaft. The output of the sensor, which is proportional to the speed, is amplified by a factor of K_1 and is compared with the reference voltage V_r to form the error voltage V_e. The complete block diagram is shown in Figure 15.32.

The closed-loop step response due to a change in reference voltage can be found from Figure 15.28 with $T_L = 0$. The transfer function becomes

$$\frac{\omega(s)}{V_r(s)} = \frac{K_2 K_v I_f / (R_m B)}{s^2(\tau_a \tau_m) + s(\tau_a + \tau_m) + 1 + [(K_v I_f)^2 + K_1 K_2 K_v I_f]/R_m B} \quad (15.77)$$

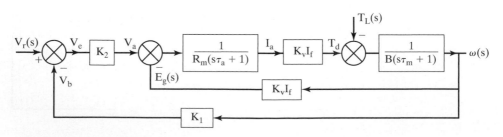

FIGURE 15.32

Block diagram for closed-loop control of separately excited dc motor.

The response due to a change in the load torque T_L can also be obtained from Figure 15.32 by setting V_r to zero. The transfer function becomes

$$\frac{\omega(s)}{T_L(s)} = -\frac{(1/B)(s\tau_a + 1)}{s^2(\tau_a\tau_m) + s(\tau_a + \tau_m) + 1 + [(K_vI_f)^2 + K_1K_2K_vI_f]/R_mB} \quad (15.78)$$

Using the final value theorem, the steady-state change in speed $\Delta\omega$, due to a step change in control voltage ΔV_r and a step change in load torque ΔT_L, can be found from Eqs. (15.77) and (15.78), respectively, by substituting $s = 0$.

$$\Delta\omega = \frac{K_2K_vI_f}{R_mB + (K_vI_f)^2 + K_1K_2K_vI_f}\Delta V_r \quad (15.79)$$

$$\Delta\omega = -\frac{R_m}{R_mB + (K_vI_f)^2 + K_1K_2K_vI_f}\Delta T_L \quad (15.80)$$

Figure 15.32 uses a speed feedback only. In practice, the motor is required to operate at a desired speed, but it has to meet the load torque, which depends on the armature current. While the motor is operating at a particular speed, if a load is applied suddenly, the speed falls and the motor takes time to come up to the desired speed. A speed feedback with an inner current loop, as shown in Figure 15.33, provides faster response to any disturbances in speed command, load torque, and supply voltage.

The current loop is used to cope with a sudden torque demand under transient condition. The output of the speed controller e_c is applied to the current limiter, which sets the current reference $I_{a(ref)}$ for the current loop. The armature current I_a is sensed by a current sensor, filtered normally by an active filter to remove ripple, and compared with the current reference $I_{a(ref)}$. The error current is processed through a current controller whose output v_c adjusts the firing angle of the converter and brings the motor speed to the desired value.

Any positive speed error caused by an increase in either speed command or load torque demand can produce a high reference current $I_{a(ref)}$. The motor accelerates to correct the speed error, and finally settles at a new $I_{a(ref)}$, which makes the motor torque equal to the load torque, resulting in a speed error close to zero. For any large positive speed error, the current limiter saturates and limits the reference current $I_{a(ref)}$ to a maximum value $I_{a(max)}$. The speed error is then corrected at the maximum permissible armature current $I_{a(max)}$ until the speed error becomes small and the current limiter comes out of saturation. Normally, the speed error is corrected with I_a less than the permissible value $I_{a(max)}$.

The speed control from zero to base speed is normally done at the maximum field by armature voltage control, and control above the base speed should be done by field weakening at the rated armature voltage. In the field control loop, the back emf $E_g(= V_a - R_aI_a)$ is compared with a reference voltage $E_{g(ref)}$, which is generally between 0.85 to 0.95 of the rated armature voltage. For speeds below the base speed, the field error e_f is large and the field controller saturates, thereby applying the maximum field voltage and current.

When the speed is close to the base speed, V_a is almost near to the rated value and the field controller comes out of saturation. For a speed command above the base speed, the speed error causes a higher value of V_a. The motor accelerates, the back emf

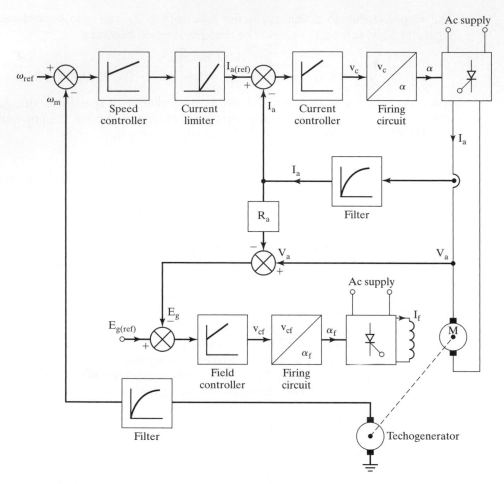

FIGURE 15.33

Closed-loop speed control with inner current loop and field weakening.

E_g increases, and the field error e_f decreases. The field current then decreases and the motor speed continues to increase until the motor speed reaches the desired speed. Thus, the speed control above the base speed is obtained by the field weakening while the armature terminal voltage is maintained at near the rated value. In the field weakening mode, the drive responds very slowly due to the large field time constant. A full converter is normally used in the field, because it has the ability to reverse the voltage, thereby reducing the field current much faster than a semiconverter.

A dc–dc converter-fed dc drives can operate from a rectified dc supply or a battery. These can also operate in one, two, or four quadrants, offering a few choices to meet application requirements [9]. Servo drive systems normally use the full four-quadrant converter, as shown in Figure 15.34, which allows bidirectional speed control with regenerative breaking capabilities. For forward driving, the transistors T_1 and T_4 and diode D_2 are used as a buck converter that supplies a variable voltage, v_a, to the

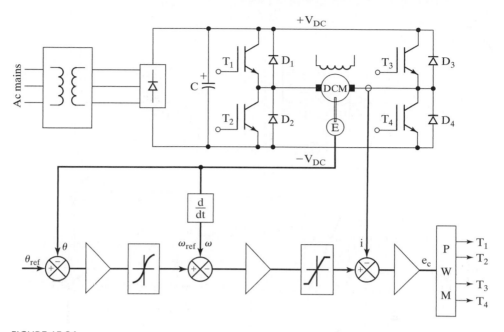

FIGURE 15.34

Dc–dc converter-fed speed and position control system with an IGBT bridge.

armature given by

$$v_a = \delta V_{DC} \tag{15.81}$$

where V_{DC} is the dc supply voltage to the converter and δ is the duty cycle of the transistor T_1.

During regenerative braking in the forward direction, transistor T_2 and diode D_4 are used as a boost converter, which regulates the breaking current through the motor by automatically adjusting the duty cycle of T_2. The energy of the overhauling motor now returns to the dc supply through diode D_1, aided by the motor back emf and the dc supply. The braking converter, composed of T_2 and D_1, may be used to maintain the regenerative braking current at the maximum allowable level right down to zero speed. Figure 15.35 shows a typical acceleration–deceleration profile of a dc–dc converter-fed drive under the closed-loop control.

Example 15.12 Finding the Speed and Torque Response of a Converter-Fed Drive

A 50-kW, 240-V, 1700-rpm separately excited dc motor is controlled by a converter, as shown in the block diagram Figure 15.32. The field current is maintained constant at $I_f = 1.4$ A and the machine back emf constant is $K_v = 0.91$ V/A rad/s. The armature resistance is $R_m = 0.1\ \Omega$ and the viscous friction constant is $B = 0.3$ N · m/rad/s. The amplification of the speed sensor is $K_1 = 95$ mV/rad/s and the gain of the power controller is $K_2 = 100$. (a) Determine the rated torque of the motor. (b) Determine the reference voltage V_r to drive the motor at the rated speed. (c) If the reference voltage is kept unchanged, determine the speed at which the motor

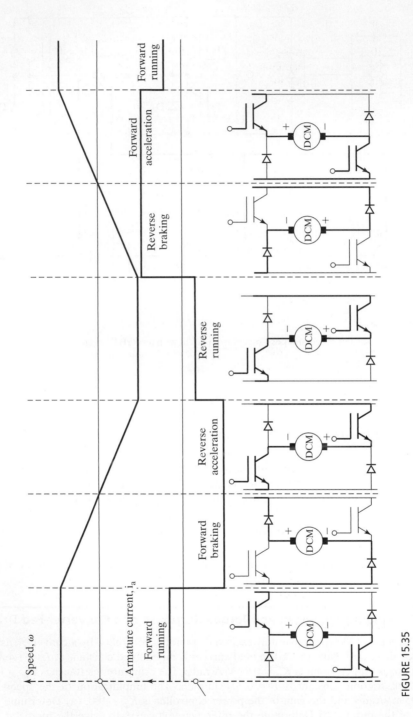

FIGURE 15.35

Typical profile of a dc–dc converter-fed four-quadrant drive.

develops the rated torque. (d) If the load torque is increased by 10% of the rated value, determine the motor speed. (e) If the reference voltage is reduced by 10%, determine the motor speed. (f) If the load torque is increased by 10% of the rated value and the reference voltage is reduced by 10%, determine the motor speed. (g) If there was no feedback in an open-loop control, determine the speed regulation for a reference voltage of $V_r = 2.31$ V. (h) Determine the speed regulation with a closed-loop control.

Solution

$I_f = 1.4$ A, $K_v = 0.91$ V/A rad/s, $K_1 = 95$ mV/rad/s, $K_2 = 100$, $R_m = 0.1$ Ω, $B = 0.3$ N·m/rad/s, and $\omega_{rated} = 1700\ \pi/30 = 178.02$ rad/s.

a.   The rated torque is $T_L = 50,000/178.02 = 280.87$ N·m.

b.   Because $V_a = K_2 V_r$, for open-loop control Eq. (15.65) gives

$$\frac{\omega}{V_a} = \frac{\omega}{K_2 V_r} = \frac{K_v I_f}{R_m B + (K_v I_f)^2} = \frac{0.91 \times 1.4}{0.1 \times 0.3 + (0.91 \times 1.4)^2} = 0.7707$$

At rated speed,

$$V_a = \frac{\omega}{0.7707} = \frac{178.02}{0.7707} = 230.98 \text{ V}$$

and feedback voltage,

$$V_b = K_1 \omega = 95 \times 10^{-3} \times 178.02 = 16.912 \text{ V}$$

With closed-loop control, $(V_r - V_b)K_2 = V_a$ or $(V_r - 16.912) \times 100 = 230.98$, which gives the reference voltage, $V_r = 19.222$ V.

c.   For $V_r = 19.222$ V and $\Delta T_L = 280.87$ N·m, Eq. (15.80) gives

$$\Delta \omega = -\frac{0.1 \times 280.86}{0.1 \times 0.3 + (0.91 \times 1.4)^2 + 95 \times 10^{-3} \times 100 \times 0.91 \times 1.4}$$

$$= -2.04 \text{ rad/s}$$

The speed at rated torque,

$$\omega = 178.02 - 2.04 = 175.98 \text{ rad/s}\quad \text{or}\quad 1680.5 \text{ rpm}$$

d. $\Delta T_L = 1.1 \times 280.87 = 308.96$ N·m and Eq. (15.80) gives

$$\Delta \omega = -\frac{0.1 \times 308.96}{0.1 \times 0.3 + (0.91 \times 1.4)^2 + 95 \times 10^{-3} \times 100 \times 0.91 \times 1.4}$$

$$= -2.246 \text{ rad/s}$$

The motor speed

$$\omega = 178.02 - 2.246 = 175.774 \text{ rad/s}\quad \text{or}\quad 1678.5 \text{ rpm}$$

e. $\Delta V_r = -0.1 \times 19.222 = -1.9222$ V and Eq. (15.79) gives the change in speed,

$$\Delta \omega = -\frac{100 \times 0.91 \times 1.4 \times 1.9222}{0.1 \times 0.3 + (0.91 \times 1.4)^2 + 95 \times 10^{-3} \times 100 \times 0.91 \times 1.4}$$

$$= -17.8 \text{ rad/s}$$

The motor speed is

$$\omega = 178.02 - 17.8 = 160.22 \text{ rad/s} \quad \text{or} \quad 1530 \text{ rpm}$$

f. The motor speed can be obtained by using superposition:

$$\omega = 178.02 - 2.246 - 17.8 = 158 \text{ rad/s} \quad \text{or} \quad 1508.5 \text{ rpm}$$

g. $\Delta V_r = 2.31$ V and Eq. (15.65) gives

$$\Delta\omega = \frac{100 \times 0.91 \times 1.4 \times 2.31}{0.1 \times 0.3 + (0.91 \times 1.4)^2} = 178.02 \text{ rad/s} \quad \text{or} \quad 1700 \text{ rpm}$$

and the no-load speed is $\omega = 178.02$ rad/s or 1700 rpm. For full load, $\Delta T_L = 280.87$ N $\cdot$ m, Eq. (15.66) gives

$$\Delta\omega = -\frac{0.1 \times 280.87}{0.1 \times 0.3 + (0.91 \times 1.4)^2} = -16.99 \text{ rad/s}$$

and the full-load speed

$$\omega = 178.02 - 16.99 = 161.03 \text{ rad/s} \quad \text{or} \quad 1537.7 \text{ rpm}$$

The speed regulation with open-loop control is

$$\frac{1700 - 1537.7}{1537.7} = 10.55\%$$

h. Using the speed from (c), the speed regulation with closed-loop control is

$$\frac{1700 - 1680.5}{1680.5} = 1.16\%$$

Note: By closed-loop control, the speed regulation is reduced by a factor of approximately 10, from 10.55% to 1.16%.

15.7.3 Phase-Locked-Loop Control

For precise speed control of servo systems, closed-loop control is normally used. The speed, which is sensed by analog sensing devices (e.g., tachometer), is compared with the reference speed to generate the error signal and to vary the armature voltage of the motor. These analog devices for speed sensing and comparing signals are not ideal and the speed regulation is more than 0.2%. The speed regulator can be improved if digital phase-locked-loop (PLL) control is used [10]. The block diagram of a converter-fed dc motor drive with PLL control is shown in Figure 15.36a and the transfer function block diagram is shown in Figure 15.36b.

In a PLL control system, the motor speed is converted to a digital pulse train by using a speed encoder. The output of the encoder acts as the speed feedback signal of frequency f_0. The phase detector compares the reference pulse train (or frequency) f_r with the feedback frequency f_0 and provides a pulse-width-modulated (PWM) output voltage V_e that is proportional to the difference in phases and frequencies of the reference and

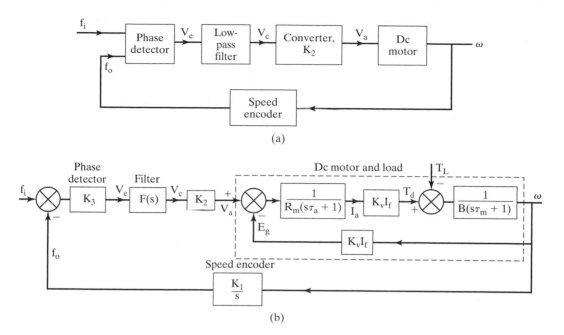

FIGURE 15.36

Phase-locked-loop control system.

feedback pulse trains. The phase detector (or comparator) is available in integrated circuits. A low-pass loop filter converts the pulse train V_e to a continuous dc level V_c, which varies the output of the power converter and in turn the motor speed.

When the motor runs at the same speed as the reference pulse train, the two frequencies would be synchronized (or locked) together with a phase difference. The output of the phase detector would be a constant voltage proportional to the phase difference and the steady-state motor speed would be maintained at a fixed value irrespective of the load on the motor. Any disturbances contributing to the speed change would result in a phase difference and the output of the phase detector would respond immediately to vary the speed of the motor in such a direction and magnitude as to retain the locking of the reference and feedback frequencies. The response of the phase detector is very fast. As long as the two frequencies are locked, the speed regulation should ideally be zero. However, in practice the speed regulation is limited to 0.002%, and this represents a significant improvement over the analog speed control system.

15.7.4 Microcomputer Control of Dc Drives

The analog control scheme for a converter-fed dc motor drive can be implemented by hardwired electronics. An analog control scheme has several disadvantages: nonlinearity of speed sensor, temperature dependency, drift, and offset. Once a control circuit is built to meet certain performance criteria, it may require major changes in the hardwired logic circuits to meet other performance requirements.

A microcomputer control reduces the size and costs of hardwired electronics, improving reliability and control performance. This control scheme is implemented in the software and is flexible to change the control strategy to meet different performance characteristics or to add extra control features. A micro-computer control system can also perform various desirable functions: on and off of the main power supply, start and stop of the drive, speed control, current control, monitoring the control variables, initiating protection and trip circuit, diagnostics for built-in fault finding, and communication with a supervisory central computer. Figure 15.37 shows a schematic diagram for a microcomputer control of a converter-fed four-quadrant dc drive.

The speed signal is fed into the microcomputer using an analog-to-digital (A/D) converter. To limit the armature current of the motor, an inner current-control loop is used. The armature current signal can be fed into the microcomputer through an A/D converter or by sampling the armature current. The line synchronizing circuit is required to synchronize the generation of the firing pulses with the supply line frequency. Although the microcomputer can perform the functions of gate pulse generator and logic circuit, these are shown outside the microcomputer. The pulse amplifier provides the necessary isolation and produces gate pulses of required magnitude and duration. A microprocessor-controlled drive has become a norm. Analog control has become almost obsolete.

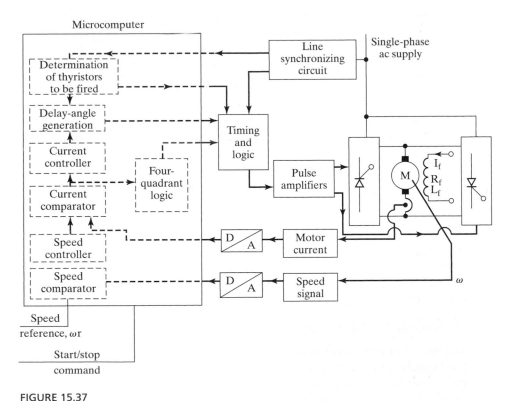

FIGURE 15.37

Schematic diagram of computer-controlled four-quadrant dc drive.

Key Points of Section 15.7

- The transfer function describes the speed response for any changes in torque or reference signal.
- A motor drive is required to operate at a desired speed at a certain load torque.
- An additional current loop is applied to the feedback path to provide faster response to any disturbances in speed command, load torque, and supply voltage.

SUMMARY

In dc drives, the armature and field voltages of dc motors are varied by either ac–dc converters or dc–dc converters. The ac–dc converter-fed drives are normally used in variable-speed applications, whereas dc–dc converter-fed drives are more suited for traction applications. Dc series motors are mostly used in traction applications, due to their capability of high-starting torque.

Dc drives can be classified broadly into three types depending on the input supply: (1) single-phase drives, (2) three-phase drives, and (3) dc–dc converter drives. Again each drive could be subdivided into three types depending on the modes of operation: (a) one-quadrant drives, (b) two-quadrant drives, and (c) four-quadrant drives. The energy-saving feature of dc–dc converter-fed drives is very attractive for use in transportation systems requiring frequent stops.

Closed-loop control, which has many advantages, is normally used in industrial drives. The speed regulation of dc drives can be significantly improved by using PLL control. The analog control schemes, which are hardwired electronics, are limited in flexibility and have certain disadvantages, whereas microcomputer control drives, which are implemented in software, are more flexible and can perform many desirable functions.

REFERENCES

[1] J. F. Lindsay and M. H. Rashid, *Electromechanics and Electrical Machinery*. Englewood Cliffs, NJ: Prentice-Hall. 1986.

[2] G. K. Dubey, *Power Semiconductor Controlled Drives*. Englewood Cliffs, NJ: Prentice-Hall. 1989.

[3] M. A. El-Sharkawi, *Fundamentals of Electric Drives*. Boston, MA: International Thompson Publishing. 2000.

[4] P. C. Sen, *Thyristor DC Drives*. New York: John Wiley & Sons. 1981.

[5] V. Subrahmanyam, *Electric Drives; Concepts and Applications*. New York: McGraw-Hill. 1994.

[6] W. Leonard, *Control of Electric Drives*. Germany: Springer-Verlag. 1985.

[7] E. Reimers, "Design analysis of multiphase dc–dc converter motor drive," *IEEE Transactions on Industry Applications*, Vol. IA8, No. 2, 1972, pp. 136–144.

[8] M. H. Rashid, "Design of LC input filter for multiphase dc–dc converters," *Proceedings IEE*, Vol. B130, No. 1, 1983, pp. 310–344.

[9] M. F. Rahman, D. Patterson, A. Cheok, and R. Betts, *Power Electronics Handbook*, edited by M. H. Rashid. San Diego, CA: Academic Press. 2001, Chapter 27—Motor Drives.

[10] D. F. Geiger, *Phaselock Loops for DC Motor Speed Control*. New York: John Wiley & Sons. 1981.

REVIEW QUESTIONS

15.1 What are the three types of dc drives based on the input supply?

15.2 What is the magnetization characteristic of dc motors?

15.3 What is the purpose of a converter in dc drives?

15.4 What is a base speed of dc motors?

15.5 What are the parameters to be varied for speed control of separately excited dc motors?

15.6 Which are the parameters to be varied for speed control of dc series motors?

15.7 Why are the dc series motors mostly used in traction applications?

15.8 What is a speed regulation of dc drives?

15.9 What is the principle of single-phase full-converter-fed dc motor drives?

15.10 What is the principle of three-phase semiconverter-fed dc motor drives?

15.11 What are the advantages and disadvantages of single-phase full-converter-fed dc motor drives?

15.12 What are the advantages and disadvantages of single-phase semiconverter-fed dc motor drives?

15.13 What are the advantages and disadvantages of three-phase full-converter-fed dc motor drives?

15.14 What are the advantages and disadvantages of three-phase semiconverter-fed dc motor drives?

15.15 What are the advantages and disadvantages of three-phase dual-converter-fed dc motor drives?

15.16 Why is it preferable to use a full converter for field control of separately excited motors?

15.17 What is a one-quadrant dc drive?

15.18 What is a two-quadrant dc drive?

15.19 What is a four-quadrant dc drive?

15.20 What is the principle of regenerative braking of dc–dc converter-fed dc motor drives?

15.21 What is the principle of rheostatic braking of dc–dc converter-fed dc motor drives?

15.22 What are the advantages of dc–dc converter-fed dc drives?

15.23 What are the advantages of multiphase dc–dc converters?

15.24 What is the principle of closed-loop control of dc drives?

15.25 What are the advantages of closed-loop control of dc drives?

15.26 What is the principle of phase-locked-loop control of dc drives?

15.27 What are the advantages of phase-locked-loop control of dc drives?

15.28 What is the principle of microcomputer control of dc drives?

15.29 What are the advantages of microcomputer control of dc drives?

15.30 What is a mechanical time constant of dc motors?

15.31 What is an electrical time constant of dc motors?

PROBLEMS

15.1 A separately excited dc motor is supplied from a dc source of 600 V to control the speed of a mechanical load and the field current is maintained constant. The armature resistance and losses are negligible. **(a)** If the load torque is $T_L = 550$ N · m at 1500 rpm, determine the armature current I_a. **(b)** If the armature current remains the same as that in (a) and the field current is reduced such that the motor runs at a speed of 2800, determine the load torque.

15.2 Repeat Problem 15.1 if the armature resistance is $R_a = 0.12\ \Omega$. The viscous friction and the no-load losses are negligible.

15.3 A 30-hp, 440-V, 2000-rpm separately excited motor dc controls a load requiring a torque of $T_L = 85$ N · m at 1200 rpm. The field circuit resistance is $R_f = 294\ \Omega$, the armature circuit resistance is $R_a = 0.12\ \Omega$, and the motor voltage constant is $K_v = 0.7032$ V/A rad/s. The field voltage is $V_f = 440$ V. The viscous friction and the no-load losses are negligible. The armature current may be assumed continuous and ripple free. Determine **(a)** the back emf E_g, **(b)** the required armature voltage V_a, **(c)** the rated armature current of the motor, and **(d)** the speed regulation at full load.

15.4 A 120-hp, 600-V, 1200-rpm dc series motor controls a load requiring a torque of $T_L = 185$ N $\cdot$ m at 1100 rpm. The field circuit resistance is $R_f = 0.06\ \Omega$, the armature circuit resistance is $R_a = 0.04\ \Omega$, and the voltage constant is $K_v = 32$ mV/A rad/s. The viscous friction and the no-load losses are negligible. The armature current is continuous and ripple free. Determine **(a)** the back emf E_g, **(b)** the required armature voltage V_a, **(c)** the rated armature current, and **(d)** the speed regulation at full speed.

15.5 The speed of a separately excited motor is controlled by a single-phase semiconverter in Figure 15.12a. The field current is also controlled by a semiconverter and the field current is set to the maximum possible value. The ac supply voltage to the armature and field converter is one phase, 208 V, 60 Hz. The armature resistance is $R_a = 0.12\ \Omega$, the field resistance is $R_f = 220\ \Omega$, and the motor voltage constant is $K_v = 1.055$ V/A rad/s. The load torque is $T_L = 75$ N $\cdot$ m at a speed of 700 rpm. The viscous friction and no-load losses are negligible. The armature and field currents are continuous and ripple free. Determine **(a)** the field current I_f; **(b)** the delay angle of the converter in the armature circuit α_a; and **(c)** the input power factor (PF) of the armature circuit.

15.6 The speed of a separately excited dc motor is controlled by a single-phase full-wave converter in Figure 15.13a. The field circuit is also controlled by a full converter and the field current is set to the maximum possible value. The ac supply voltage to the armature and field converters is one phase, 208 V, 60 Hz. The armature resistance is $R_a = 0.50\ \Omega$, the field circuit resistance is $R_f = 345\ \Omega$, and the motor voltage constant is $K_v = 0.71$ V/A rad/s. The viscous friction and no-load losses are negligible. The armature and field currents are continuous and ripple free. If the delay angle of the armature converter is $\alpha_a = 45°$ and the armature current of the motor is $I_a = 55$ A, determine **(a)** the torque developed by the motor T_d, **(b)** the speed ω, and **(c)** the input PF of the drive.

15.7 If the polarity of the motor back emf in Problem 15.6 is reversed by reversing the polarity of the field current, determine **(a)** the delay angle of the armature circuit converter α_a to maintain the armature current constant at the same value of $I_a = 55$ A; and **(b)** the power fed back to the supply during regenerative braking of the motor.

15.8 The speed of a 20-hp, 300-V, 1800-rpm separately excited dc motor is controlled by a three-phase full-converter drive. The field current is also controlled by a three-phase full-converter and is set to the maximum possible value. The ac input is three-phase, Y-connected, 208 V, 60 Hz. The armature resistance is $R_a = 0.35\ \Omega$, the field resistance is $R_f = 250\ \Omega$, and the motor voltage constant is $K_v = 1.15$ V/A rad/s. The armature and field currents are continuous and ripple free. The viscous friction and no-load losses are negligible. Determine **(a)** the delay angle of the armature converter α_a, if the motor supplies the rated power at the rated speed; **(b)** the no-load speed if the delay angles are the same as in (a) and the armature current at no-load is 10% of the rated value; and **(c)** the speed regulation.

15.9 Repeat Problem 15.8 if both armature and field circuits are controlled by three-phase semiconverters.

15.10 The speed of a 20-hp, 300-V, 900-rpm separately excited dc motor is controlled by a three-phase full converter. The field circuit is also controlled by a three-phase full converter. The ac input to armature and field converters is three-phase, Y-connected, 208 V, 60 Hz. The armature resistance $R_a = 0.15\ \Omega$, the field circuit resistance $R_f = 145\ \Omega$, and the motor voltage constant $K_v = 1.15$ V/A rad/s. The viscous friction and no-load losses are negligible. The armature and field currents are continuous and ripple free. **(a)** If the field converter is operated at the maximum field current and the developed torque is $T_d = 106$ N · m at 750 rpm, determine the delay angle of the armature converter α_a. **(b)** If the field circuit converter is set to the maximum field current, the developed torque is $T_d = 108$ N · m, and the delay angle of the armature converter is $\alpha_a = 0$, determine the speed. **(c)** For the same load demand as in (b), determine the delay angle of the field circuit converter if the speed has to be increased to 1800 rpm.

15.11 Repeat Problem 15.10 if both the armature and field circuits are controlled by three-phase semiconverters.

15.12 A dc–dc converter controls the speed of a dc series motor. The armature resistance $R_a = 0.04$ Ω, field circuit resistance $R_f = 0.06$ Ω, and back emf constant $K_v = 35$ mV/rad/s. The dc input voltage of the dc–dc converter $V_s = 600$ V. If it is required to maintain a constant developed torque of $T_d = 547$ N · m, plot the motor speed against the duty-cycle k of the dc–dc converter.

15.13 A dc–dc converter controls the speed of a separately excited motor. The armature resistance is $R_a = 0.05$ Ω. The back emf constant is $K_v = 1.527$ V/A rad/s. The rated field current is $I_f = 2.5$ A. The dc input voltage to the dc–dc converter is $V_s = 600$ V. If it is required to maintain a constant developed torque of $T_d = 547$ N · m, plot the motor speed against the duty-cycle k of the dc–dc converter.

15.14 A dc series motor is powered by a dc–dc converter, as shown in Figure 15.18a, from a 600-V dc source. The armature resistance is $R_a = 0.03$ Ω and the field resistance is $R_f = 0.05$ Ω. The back emf constant of the motor is $K_v = 15.27$ mV/A rad/s. The average armature current $I_a = 450$ A. The armature current is continuous and has negligible ripple. If the duty cycle of the dc–dc converter is 75%, determine **(a)** the input power from the source, **(b)** the equivalent input resistance of the dc–dc converter drive, **(c)** the motor speed, and **(d)** the developed torque of the motor.

15.15 The drive in Figure 15.19a is operated in regenerative braking of a dc series motor. The dc supply voltage is 600 V. The armature resistance is $R_a = 0.03$ Ω and the field resistance is $R_f = 0.05$ Ω. The back emf constant of the motor is $K_v = 12$ mV/A rad/s. The average armature current is maintained constant at $I_a = 350$ A. The armature current is continuous and has negligible ripple. If the duty cycle of the dc–dc converter is 50%, determine **(a)** the average voltage across the dc–dc converter V_{ch}; **(b)** the power regenerated to the dc supply P_g; **(c)** the equivalent load resistance of the motor acting as a generator, R_{eq}; **(d)** the minimum permissible braking speed ω_{min}; **(e)** the maximum permissible braking speed ω_{max}; and **(f)** the motor speed.

15.16 A dc–dc converter is used in rheostatic braking of a dc series motor, as shown in Figure 15.20. The armature resistance $R_a = 0.03$ Ω and the field resistance $R_f = 0.05$ Ω. The braking resistor $R_b = 5$ Ω. The back emf constant $K_v = 14$ mV/A rad/s. The average armature current is maintained constant at $I_a = 250$ A. The armature current is continuous and has negligible ripple. If the duty cycle of the dc–dc converter is 60%, determine **(a)** the average voltage across the dc–dc converter V_{ch}; **(b)** the power dissipated in the resistor P_b; **(c)** the equivalent load resistance of the motor acting as a generator, R_{eq}; **(d)** the motor speed; and **(e)** the peak dc–dc converter voltage V_p.

15.17 Two dc–dc converters control a dc motor, as shown in Figure 15.24a, and they are phase shifted in operation by π/m, where m is the number of multiphase dc–dc converters. The supply voltage $V_s = 440$ V, total armature circuit resistance $R_m = 8$ Ω, armature circuit inductance $L_m = 12$ mH, and the frequency of each dc–dc converter $f = 250$ Hz. Calculate the maximum value of peak-to-peak load ripple current.

15.18 For Problem 15.17 plot the maximum value of peak-to-peak load ripple current against the number of multiphase dc–dc converters.

15.19 A dc motor is controlled by two multiphase dc–dc converters. The average armature current is $I_a = 250$ A. A simple LC-input filter with $L_e = 0.35$ mH and $C_e = 5600$ μF is used. Each dc–dc converter is operated at a frequency $f = 250$ Hz. Determine the rms fundamental component of the dc–dc converter-generated harmonic current in the supply.

15.20 For Problem 15.19, plot the rms fundamental component of the dc–dc converter-generated harmonic current in the supply against the number of multiphase dc–dc converters.

15.21 A 40-hp, 230-V, 3500-rpm separately excited dc motor is controlled by a linear converter of gain $K_2 = 200$. The moment of inertia of the motor load is $J = 0.156$ N · m/rad/s, viscous friction constant is negligible, total armature resistance is $R_m = 0.045$ Ω, and total armature inductance is $L_m = 730$ mH. The back emf constant is $K_v = 0.502$ V/A rad/s and the field current is maintained constant at $I_f = 1.25$ A. **(a)** Obtain the open-loop transfer function $\omega(s)/V_r(s)$ and $\omega(s)/T_L(s)$ for the motor. **(b)** Calculate the motor steady-state speed if the reference voltage is $V_r = 1$ V and the load torque is 60% of the rated value.

15.22 Repeat Problem 15.21 with a closed-loop control if the amplification of speed sensor is $K_1 = 3$ mV/rad/s.

15.23 The motor in Problem 15.21 is controlled by a linear converter of gain K_2 with a closed-loop control. If the amplification of speed sensor is $K_1 = 3$ mV/rad/s, determine the gain of the converter K_2 to limit the speed regulation at full load to 1%.

15.24 A 60-hp, 230-V, 1750-rpm separately excited dc motor is controlled by a converter, as shown in the block diagram in Figure 15.32. The field current is maintained constant at $I_f = 1.25$ A and the machine back emf constant is $K_v = 0.81$ V/A rad/s. The armature resistance is $R_a = 0.02$ Ω and the viscous friction constant is $B = 0.3$ N · m/rad/s. The amplification of the speed sensor is $K_1 = 96$ mV/rad/s and the gain of the power controller is $K_2 = 150$. **(a)** Determine the rated torque of the motor. **(b)** Determine the reference voltage V_r to drive the motor at the rated speed. **(c)** If the reference voltage is kept unchanged, determine the speed at which the motor develops the rated torque.

15.25 Repeat Problem 15.24. **(a)** If the load torque is increased by 20% of the rated value, determine the motor speed. **(b)** If the reference voltage is reduced by 10%, determine the motor speed. **(c)** If the load torque is reduced by 15% of the rated value and the reference voltage is reduced by 20%, determine the motor speed. **(d)** If there was no feedback, as in an open-loop control, determine the speed regulation for a reference voltage $V_r = 1.24$ V. **(e)** Determine the speed regulation with a closed-loop control.

15.26 A 40-hp, 230-V, 3500-rpm series excited dc motor is controlled by a linear converter of gain $K_2 = 200$. The moment of inertia of the motor load $J = 0.156$ N · m/rad/s, viscous friction constant is negligible, total armature resistance is $R_m = 0.045$ Ω, and total armature inductance is $L_m = 730$ mH. The back emf constant is $K_v = 340$ mV/A rad/s. The field resistance is $R_f = 0.035$ Ω and field inductance is $L_f = 450$ mH. **(a)** Obtain the open-loop transfer function $\omega(s)/V_r(s)$ and $\omega(s)/T_L(s)$ for the motor. **(b)** Calculate the motor steady-state speed if the reference voltage, $V_r = 1$ V, and the load torque is 60% of the rated value.

15.27 Repeat Problem 15.26 with closed-loop control if the amplification of speed sensor $K_1 = 3$ mV/rad/s.

CHAPTER 16

Ac Drives

The learning objectives of this chapter are as follows:

- To study the control characteristics of induction motors and the methods for speed control
- To study the principle of vector or field-oriented control for induction motors
- To understand the types of synchronous motors
- To study the control characteristics of synchronous motors and the methods for speed control
- To understand the methods for speed control of stepper motors

16.1 INTRODUCTION

Ac motors exhibit highly coupled, nonlinear, and multivariable structures as opposed to much simpler decoupled structures of separately excited dc motors. The control of ac drives generally requires complex control algorithms that can be performed by microprocessors or microcomputers along with fast-switching power converters.

The ac motors have a number of advantages; they are lightweight (20 to 40% lighter than equivalent dc motors), are inexpensive, and have low maintenance compared with dc motors. They require control of frequency, voltage, and current for variable-speed applications. The power converters, inverters, and ac voltage controllers can control the frequency, voltage, or current to meet the drive requirements. These power controllers, which are relatively complex and more expensive, require advanced feedback control techniques such as model reference, adaptive control, sliding mode control, and field-oriented control. However, the advantages of ac drives outweigh the disadvantages. There are two types of ac drives:

1. Induction motor drives
2. Synchronous motor drives

Ac drives are replacing dc drives and are used in many industrial and domestic applications [1, 2].

16.2 INDUCTION MOTOR DRIVES

Three-phase induction motors are commonly used in adjustable-speed drives [1] and they have three-phase stator and rotor windings. The stator windings are supplied with balanced three-phase ac voltages, which produce induced voltages in the rotor windings due to transformer action. It is possible to arrange the distribution of stator windings so that there is an effect of multiple poles, producing several cycles of magnetomotive force (mmf) (or field) around the air gap. This field establishes a spatially distributed sinusoidal flux density in the air gap. The speed of rotation of the field is called the *synchronous speed*, which is defined by

$$\omega_s = \frac{2\omega}{p} \tag{16.1}$$

where p is the number of poles and ω is the supply frequency in rads per second.

If a stator phase voltage, $v_s = \sqrt{2}\,V_s \sin \omega t$, produces a flux linkage (in the rotor) given by

$$\phi(t) = \phi_m \cos(\omega_m t + \delta - \omega_s t) \tag{16.2}$$

the induced voltage per phase in the rotor winding is

$$
\begin{aligned}
e_r &= N_r \frac{d\phi}{dt} = N_r \frac{d}{dt}\left[\phi_m \cos(\omega_m t + \delta - \omega_s t)\right] \\
&= -N_r \phi_m (\omega_s - \omega_m) \sin[(\omega_s - \omega_m)t - \delta] \\
&= -s E_m \sin(s\omega_s t - \delta) \\
&= -s\sqrt{2}\,E_r \sin(s\omega_s t - \delta)
\end{aligned}
\tag{16.3}
$$

where N_r = number of turns on each rotor phase;
$\quad\omega_m$ = angular rotor speed or frequency, Hz;
$\quad\ \delta$ = relative position of the rotor;
$\quad E_r$ = rms value of the induced voltage in the rotor per phase, V;
$\quad E_m$ = peak induced voltage in the rotor per phase, V.

and s is the slip, defined as

$$s = \frac{\omega_s - \omega_m}{\omega_s} \tag{16.4}$$

which gives the motor speed as $\omega_m = \omega_s(1 - s)$. The equivalent circuit for one phase of the rotor is shown in Figure 16.1a,

where R_r is the resistance per phase of the rotor windings;
$\quad X_r$ is the leakage reactance per phase of the rotor at the supply frequency;
$\quad E_r$ represents the induced rms phase voltage when the speed is zero
$\quad$ (or $s = 1$).

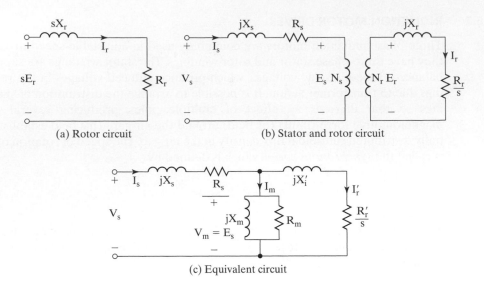

(a) Rotor circuit   (b) Stator and rotor circuit

(c) Equivalent circuit

FIGURE 16.1

Circuit model of induction motors.

The rotor current is given by

$$I_r = \frac{sE_r}{R_r + jsX_r} \tag{16.5}$$

$$= \frac{E_r}{R/s + jX_r} \tag{16.5a}$$

where R_r and X_r are referred to the rotor winding.

The per phase circuit model of induction motors is shown in Figure 16.1b, where R_s and X_s are the per phase resistance and leakage reactance of the stator winding. The complete circuit model with all parameters referred to the stator is shown in Figure 16.1c, where R_m represents the resistance for excitation (or core) loss and X_m is the magnetizing reactance. R_r' and X_r' are the rotor resistance and reactance referred to the stator. I_r' is the rotor current referred to the stator. There will be stator core loss, when the supply is connected and the rotor core loss depends on the slip. The friction and windage loss $P_{\text{no load}}$ exists when the machine rotates. The core loss P_c may be included as a part of rotational loss $P_{\text{no load}}$.

16.2.1 Performance Characteristics

The rotor current I_r and stator current I_s can be found from the circuit model in Figure 16.1c where R_r and X_r are referred to the stator windings. Once the values of I_r and I_s are known, the performance parameters of a three-phase motor can be determined as follows:

Stator copper loss

$$P_{su} = 3I_s^2 R_s \tag{16.6}$$

Rotor copper loss

$$P_{ru} = 3(I'_r)^2 R'_r \qquad (16.7)$$

Core loss

$$P_c = \frac{3V_m^2}{R_m} \approx \frac{3V_s^2}{R_m} \qquad (16.8)$$

Gap power (power passing from the stator to the rotor through the air gap)

$$P_g = 3(I'_r)^2 \frac{R'_r}{s} \qquad (16.9)$$

Developed power

$$P_d = P_g - P_{ru} = 3(I'_r)^2 \frac{R'_r}{s}(1 - s) \qquad (16.10)$$

$$= P_g(1 - s) \qquad (16.11)$$

Developed torque

$$T_d = \frac{P_d}{\omega_m} \qquad (16.12)$$

$$= \frac{P_g(1 - s)}{\omega_s(1 - s)} = \frac{P_g}{\omega_s} \qquad (16.12a)$$

Input power

$$P_i = 3 V_s I_s \cos \theta_m \qquad (16.13)$$

$$= P_c + P_{su} + P_g \qquad (16.13a)$$

where θ_m is the angle between I_s and V_s. Output power

$$P_o = P_d - P_{\text{no load}}$$

Efficiency

$$\eta = \frac{P_o}{P_i} = \frac{P_d - P_{\text{no load}}}{P_c + P_{su} + P_g} \qquad (16.14)$$

If $P_g \gg (P_c + P_{su})$ and $P_d \gg P_{\text{no load}}$, the efficiency becomes approximately

$$\eta \approx \frac{P_d}{P_g} = \frac{P_g(1 - s)}{P_g} = 1 - s \qquad (16.14a)$$

The value of X_m is normally large and R_m, which is much larger, can be removed from the circuit model to simplify the calculations. If $X_m^2 \gg (R_s^2 + X_s^2)$, then $V_s \approx V_m$, and the magnetizing reactance X_m may be moved to the stator winding to simplify further; this is shown in Figure 16.2.

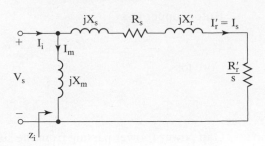

FIGURE 16.2

Approximate perphase equivalent circuit.

The input impedance of the motor becomes

$$\mathbf{Z}_i = \frac{-X_m(X_s + X_r') + jX_m(R_s + R_r'/s)}{R_s + R_r'/s + j(X_m + X_s + X_r')} \tag{16.15}$$

and the power factor (PF) angle of the motor

$$\theta_m = \pi - \tan^{-1} \frac{R_s + R_r'/s}{X_s + X_r'} + \tan^{-1} \frac{X_m + X_s + X_r'}{R_s + R_r'/s} \tag{16.16}$$

From Figure 16.2, the rms rotor current

$$I_r' = \frac{V_s}{[(R_s + R_r'/s)^2 + (X_s + X_r')^2]^{1/2}} \tag{16.17}$$

Substituting I_r from Eq. (16.17) in Eq. (16.9) and then P_g in Eq. (16.12a) yields

$$T_d = \frac{3 R_r' V_s^2}{s\omega_s[(R_s + R_r'/s)^2 + (X_s + X_r')^2]} \tag{16.18}$$

If the motor is supplied from a fixed voltage at a constant frequency, the developed torque is a function of the slip and the torque–speed characteristics can be determined from Eq. (16.18). A typical plot of developed torque as a function of slip or speed is shown in Figure 16.3. The operation in the reverse motoring and regenerative braking is obtained by the reversal of the phase sequence of the motor terminals. The reverse speed–torque characteristics are shown by dashed lines. There are three regions of operation: (1) motoring or powering, $0 \le s \le 1$; (2) regeneration, $s < 0$; and (3) plugging, $1 \le s \le 2$. In motoring, the motor rotates in the same direction as the field; and as the slip increases, the torque also increases while the air-gap flux remains constant. Once the torque reaches its maximum value, T_m at $s = s_m$, the torque decreases, with the increase in slip due to reduction of the air-gap flux.

In regeneration, the speed ω_m is greater than the synchronous speed ω_s with ω_m and ω_s in the same direction, and the slip is negative. Therefore, R_r/s is negative. This means that power is fed back from the shaft into the rotor circuit and the motor operates as a generator. The motor returns power to the supply system. The torque–speed characteristic is similar to that of motoring, but having negative value of torque.

In reverse plugging, the speed is opposite to the direction of the field and the slip is greater than unity. This may happen if the sequence of the supply source is reversed while forward motoring, so that the direction of the field is also reversed. The developed

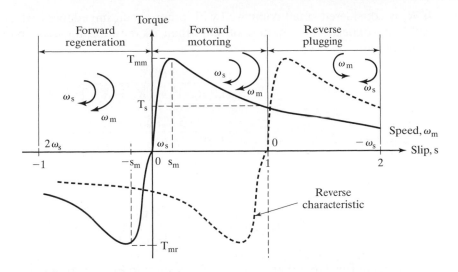

FIGURE 16.3

Torque–speed characteristics.

torque, which is in the same direction as the field, opposes the motion and acts as braking torque. Because $s > 1$, the motor currents are high, but the developed torque is low. The energy due to a plugging brake must be dissipated within the motor and this may cause excessive heating of the motor. This type of braking is not normally recommended.

At starting, the machine speed is $\omega_m - 0$ and $s = 1$. The starting torque can be found from Eq. (16.18) by setting $s = 1$ as

$$T_s = \frac{3 R_r' V_s^2}{\omega_s[(R_s + R_r')^2 + (X_s + X_r')^2]} \tag{16.19}$$

The slip for maximum torque s_m can be determined by setting $dT_d/ds = 0$ and Eq. (16.18) yields

$$s_m = \pm \frac{R_r'}{[R_s^2 + (X_s + X_r')^2]^{1/2}} \tag{16.20}$$

Substituting $s = s_m$ in Eq. (16.18) gives the maximum developed torque during motoring, which is also called *pull-out torque*, or *breakdown torque*,

$$T_{mm} = \frac{3 V_s^2}{2\omega_s[R_s + \sqrt{R_s^2 + (X_s + X_r')^2}]} \tag{16.21}$$

and the maximum regenerative torque can be found from Eq. (16.18) by letting

$$s = -s_m$$

$$T_{mr} = \frac{3 V_s^2}{2\omega_s[-R_s + \sqrt{R_s^2 + (X_s + X_r')^2}]} \tag{16.22}$$

If R_s is considered small compared with other circuit impedances, which is usually a valid approximation for motors of more than 1-kW rating, the corresponding expressions become

$$T_d = \frac{3 R_r' V_s^2}{s\omega_s[(R_r'/s)^2 + (X_s + X_r')^2]} \tag{16.23}$$

$$T_s = \frac{3 R_r' V_s^2}{\omega_s[(R_r')^2 + (X_s + X_r')^2]} \tag{16.24}$$

$$s_m = \pm \frac{R_r'}{X_s + X_r'} \tag{16.25}$$

$$T_{mm} = -T_{mr} = \frac{3 V_s^2}{2\omega_s(X_s + X_r')} \tag{16.26}$$

Normalizing Eqs. (16.23) and (16.24) with respect to Eq. (16.26) gives

$$\frac{T_d}{T_{mm}} = \frac{2R_r'(X_s + X_r')}{s[(R_r'/s)^2 + (X_s + X_r')^2]} = \frac{2ss_m}{s_m^2 + s^2} \tag{16.27}$$

and

$$\frac{T_s}{T_{mm}} = \frac{2R_r'(X_s + X_r')}{(R_r')^2 + (X_s + X_r')^2} = \frac{2s_m}{s_m^2 + 1} \tag{16.28}$$

If $s < 1$, $s^2 \ll s_m^2$ and Eq. (16.27) can be approximated to

$$\frac{T_d}{T_{mm}} = \frac{2s}{s_m} = \frac{2(\omega_s - \omega_m)}{s_m\omega_s} \tag{16.29}$$

which gives the speed as a function of torque,

$$\omega_m = \omega_s \left(1 - \frac{s_m}{2T_{mm}} T_d \right) \tag{16.30}$$

It can be noticed from Eqs. (16.29) and (16.30) that if the motor operates with small slip, the developed torque is proportional to slip and the speed decreases with torque. The rotor current, which is zero at the synchronous speed, increases due to the decrease in R_r/s as the speed is decreased. The developed torque also increases until it becomes maximum at $s = s_m$. For $s < s_m$, the motor operates stably on the portion of the speed–torque characteristic. If the rotor resistance is low, s_m is low. That is, the change of motor speed from no-load to rated torque is only a small percentage. The motor operates essentially at a constant speed. When the load torque exceeds the breakdown torque, the motor stops and the overload protection must immediately disconnect the source to prevent damage due to overheating. It should be noted that for $s > s_m$, the torque decreases despite an increase in the rotor current and the operation is unstable for most motors. The speed and torque of induction motors can be varied by one of the following means [3–8]:

1. Stator voltage control
2. Rotor voltage control
3. Frequency control
4. Stator voltage and frequency control
5. Stator current control
6. Voltage, current, and frequency control

To meet the torque–speed duty cycle of a drive, the voltage, current, and frequency control are normally used.

Example 16.1   Finding the Performance Parameters of a Three-Phase Induction Motor

A three-phase, 460-V, 60-Hz, four-pole Y-connected induction motor has the following equivalent-circuit parameters: $R_s = 0.42\ \Omega$, $R'_r = 0.23\ \Omega$, $X_s = X'_r = 0.82\ \Omega$, and $X_m = 22\ \Omega$. The no-load loss, which is $P_{no\ load} = 60$ W, may be assumed constant. The rotor speed is 1750 rpm. Use the approximate equivalent circuit in Figure 16.2 to determine (a) the synchronous speed ω_s; (b) the slip s; (c) the input current I_i; (d) the input power P_i; (e) the input PF of the supply, PF_s; (f) the gap power P_g; (g) the rotor copper loss P_{ru}; (h) the stator copper loss P_{su}; (i) the developed torque T_d; (j) the efficiency; (k) the starting current I_{rs} and starting torque T_s; (l) the slip for maximum torque s_m; (m) the maximum developed torque in motoring, T_{mm}; (n) the maximum regenerative developed torque T_{mr}; and (o) T_{mm} and T_{mr} if R_s is neglected.

Solution

$f = 60$ Hz, $p = 4$, $R_s = 0.42\ \Omega$, $R'_r = 0.23\ \Omega$, $X_s = X'_r = 0.82\ \Omega$, $X_m = 22\ \Omega$, and $N = 1750$ rpm. The phase voltage is $V_s = 460/\sqrt{3} = 265.58$ V, $\omega = 2\pi \times 60 = 377$ rad/s, and $\omega_m = 1750\ \pi/30 = 183.26$ rad/s.

   a. From Eq. (16.1), $\omega_s = 2\omega/p = 2 \times 377/4 = 188.5$ rad/s.

   b. From Eq. (16.4), $s = (188.5 - 183.26)/188.5 = 0.028$.

   c. From Eq. (16.15),

$$\mathbf{Z}_i = \frac{-22 \times (0.82 + 0.82) + j22 \times (0.42 + 0.23/0.028)}{0.42 + 0.23/0.028 + j(22 + 0.82 + 0.82)} = 7.732\ \underline{/30.88^\circ}$$

$$\mathbf{I}_i = \frac{V_s}{Z_i} = \frac{265.58}{7.732}\ \underline{/-30.8^\circ} = 34.35\ \underline{/-30.88^\circ}\ \text{A}$$

   d. The PF of the motor is

$$PF_m = \cos(-30.88^\circ) = 0.858\ \text{(lagging)}$$

From Eq. (16.13),

$$P_i = 3 \times 265.58 \times 34.35 \times 0.858 = 23{,}482\ \text{W}$$

   e. The PF of the input supply is $PF_s = PF_m = 0.858$ (lagging) which is the same as the motor PF, PF_m, because the supply is sinusoidal.

f. From Eq. (16.17), the rms rotor current is

$$I'_r = \frac{265.58}{[(0.42 + 0.23/0.028)^2 + (0.82 + 0.82)^2]^{1/2}} = 30.1 \text{ A}$$

From Eq. (16.9),

$$P_g = \frac{3 \times 30.1^2 \times 0.23}{0.028} = 22{,}327 \text{ W}$$

g. From Eq. (16.7), $P_{ru} = 3 \times 30.1^2 \times 0.23 = 625$ W.

h. The stator copper loss is $P_{su} = 3 \times 30.1^2 \times 0.42 = 1142$ W.

i. From Eq. (16.12a), $T_d = 22{,}327/188.5 = 118.4$ N·m.

j. $P_0 = P_g - P_{ru} - P_{\text{no load}} = 22{,}327 - 625 - 60 = 21{,}642$ W.

k. For $s = 1$, Eq. (16.17) gives the starting rms rotor current

$$I_{rs} = \frac{265.58}{[(0.42 + 0.23)^2 + (0.82 + 0.82)^2]^{1/2}} = 150.5 \text{ A}$$

From Eq. (16.19),

$$T_s = \frac{3 \times 0.23 \times 150.5^2}{188.5} = 82.9 \text{ N·m}$$

l. From Eq. (16.20), the slip for maximum torque (or power)

$$s_m = \pm\frac{0.23}{[0.42^2 + (0.82 + 0.82)^2]^{1/2}} = \pm 0.1359$$

m. From Eq. (16.21), the maximum developed torque

$$T_{mm} = \frac{3 \times 265.58^2}{2 \times 188.5 \times [0.42 + \sqrt{0.42^2 + (0.82 + 0.82)^2}]}$$

$$= 265.64 \text{ N·m}$$

n. From Eq. (16.22), the maximum regenerative torque is

$$T_{mr} = -\frac{3 \times 265.58^2}{2 \times 188.5 \times [-0.42 + \sqrt{0.42^2 + (0.82 + 0.82)^2}]}$$

$$= -440.94 \text{ N·m}$$

o. From Eq. (16.25),

$$s_m = \pm\frac{0.23}{0.82 + 0.82} = \pm 0.1402$$

From Eq. (16.26),

$$T_{mm} = -T_{mr} = \frac{3 \times 265.58^2}{2 \times 188.5 \times (0.82 + 0.82)} = 342.2 \text{ N·m}$$

Note: R_s spreads the difference between T_{mm} and T_{mr}. For $R_s = 0$, $T_{mm} = -T_{mr} = 342.2 \text{ N} \cdot \text{m}$, as compared with $T_{mm} = 265.64 \text{ N} \cdot \text{m}$ and $T_{mr} = -440.94 \text{ N} \cdot \text{m}$.

16.2.2   Stator Voltage Control

Equation (16.18) indicates that the torque is proportional to the square of the stator supply voltage and a reduction in stator voltage can produce a reduction in speed. If the terminal voltage is reduced to bV_s, Eq. (16.18) gives the developed torque

$$T_d = \frac{3R'_r(bV_s)^2}{s\omega_s[(R_s + R'_r/s)^2 + (X_s + X'_r)^2]}$$

where $b \le 1$.

Figure 16.4 shows the typical torque–speed characteristics for various values of b. The points of intersection with the load line define the stable operating points. In any magnetic circuit, the induced voltage is proportional to flux and frequency, and the rms air-gap flux can be expressed as

$$V_a = bV_s = K_m\omega\phi$$

or

$$\phi = \frac{V_a}{K_m\omega} = \frac{bV_s}{K_m\omega} \tag{16.31}$$

where K_m is a constant and depends on the number of turns of the stator winding. As the stator voltage is reduced, the air-gap flux and the torque are also reduced. At a lower voltage, the current can be peaking at a slip of $s_a = \frac{1}{3}$. The range of speed control depends on the slip for maximum torque s_m. For a low-slip motor, the speed range is very narrow. This type of voltage control is not suitable for a constant-torque load and is normally applied to applications requiring low-starting torque and a narrow range of speed at a relatively low slip.

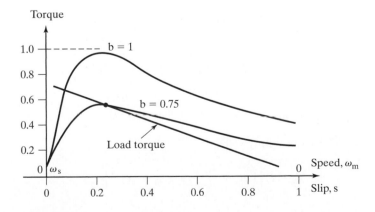

FIGURE 16.4

Torque–speed characteristics with variable stator voltage.

The stator voltage can be varied by three-phase (1) ac voltage controllers, (2) voltage-fed variable dc-link inverters, or (3) pulse-width modulation (PWM) inverters. However, due to limited speed range requirements, the ac voltage controllers are normally used to provide the voltage control. The ac voltage controllers are very simple. However, the harmonic contents are high and the input PF of the controllers is low. They are used mainly in low-power applications, such as fans, blowers, and centrifugal pumps, where the starting torque is low. They are also used for starting high-power induction motors to limit the in-rush current.

Example 16.2   Finding the Performance Parameters of a Three-Phase Induction Motor with Stator Voltage Control

A three-phase, 460-V, 60-Hz, four-pole Y-connected induction motor has the following parameters: $R_s = 1.01\ \Omega$, $R_r' = 0.69\ \Omega$, $X_s = 1.3\ \Omega$, $X_r' = 1.94\ \Omega$, and $X_m = 43.5\ \Omega$. The no-load loss, $P_{\text{no load}}$, is negligible. The load torque, which is proportional to the speed squared, is 41 N·m at 1740 rpm. If the motor speed is 1550 rpm, determine (a) the load torque T_L, (b) the rotor current I_r, (c) the stator supply voltage V_a, (d) the motor input current I_i, (e) the motor input power P_i, (f) the slip for maximum current s_a, (g) the maximum rotor current $I_{r(\max)}$, (h) the speed at maximum rotor current ω_a, and (i) the torque at the maximum current T_a.

Solution

$p = 4$, $f = 60\ \text{Hz}$, $V_s = 460/\sqrt{3} = 265.58\ \text{V}$, $R_s = 1.01\ \Omega$, $R_r' = 0.69\ \Omega$, $X_s = 1.3\ \Omega$, $X_r' = 1.94\ \Omega$, $X_m = 43.5\ \Omega$, $\omega = 2\pi \times 60 = 377\ \text{rad/s}$, and $\omega_s = 377 \times 2/4 = 188.5\ \text{rad/s}$. Because torque is proportional to speed squared,

$$T_L = K_m \omega_m^2 \tag{16.32}$$

At $\omega_m = 1740\ \pi/30 = 182.2\ \text{rad/s}$, $T_L = 41\ \text{N·m}$, and Eq. (16.32) yields $K_m = 41/182.2^2 = 1.235 \times 10^{-3}$ and $\omega_m = 1550\ \pi/30 = 162.3\ \text{rad/s}$. From Eq. (16.4), $s = (188.5 - 162.3)/188.500 = 0.139$.

a. From Eq. (16.32), $T_L = 1.235 \times 10^{-3} \times 162.3^2 = 32.5\ \text{N·m}$.

b. From Eqs. (16.10) and (16.12),

$$P_d = 3(I_r')^2\, \frac{R_r'}{s}\,(1 - s) = T_L \omega_m + P_{\text{no load}} \tag{16.33}$$

For negligible no-load loss,

$$I_r = \left[\frac{sT_L\omega_m}{3R_r'(1-s)}\right]^{1/2} \tag{16.34}$$

$$= \left[\frac{0.139 \times 32.5 \times 162.3}{3 \times 0.69(1 - 0.139)}\right]^{1/2} = 20.28\ \text{A}$$

c. The stator supply voltage

$$V_a = I_r'\left[\left(R_s + \frac{R_r'}{s}\right)^2 + (X_s + X_r')^2\right]^{1/2} \tag{16.35}$$

$$= 20.28 \times \left[\left(1.01 + \frac{0.69}{0.139}\right)^2 + (1.3 + 1.94)^2\right]^{1/2} = 137.82$$

d. From Eq. (16.15),

$$\mathbf{Z}_i = \frac{-43.5 \times (1.3 + 1.94) + j43.5 \times (1.01 + 0.69/0.139)}{1.01 + 0.69/0.139 + j(43.5 + 1.3 + 1.94)} = 6.27 \underline{/35.82°}$$

$$\mathbf{I}_i = \frac{V_a}{\mathbf{Z}_i} = \frac{137.82}{6.27} \underline{/-144.26°} = 22 \underline{/-35.82°} \text{ A}$$

e. $PF_m = \cos(-35.82°) = 0.812$ (lagging). From Eq. (16.13),

$$P_i = 3 \times 137.82 \times 22.0 \times 0.812 = 7386 \text{ W}$$

f. Substituting $\omega_m = \omega_s(1 - s)$ and $T_L = K_m\omega_m^2$ in Eq. (16.34) yields

$$I'_r = \left[\frac{sT_L\omega_m}{3R'_r(1-s)}\right]^{1/2} = (1-s)\omega_s\left(\frac{sK_m\omega_s}{3R'_r}\right)^{1/2} \tag{16.36}$$

The slip at which I_r becomes maximum can be obtained by setting $dI_r/ds = 0$, and this yields

$$s_a = \tfrac{1}{3} \tag{16.37}$$

g. Substituting $s_a = \tfrac{1}{3}$ in Eq. (16.36) gives the maximum rotor current

$$I'_{r(max)} = \omega_s\left(\frac{4K_m\omega_s}{81R'_r}\right)^{1/2}$$

$$= 188.5 \times \left(\frac{4 \times 1.235 \times 10^{-3} \times 188.5}{81 \times 0.69}\right)^{1/2} = 24.3 \text{ A} \tag{16.38}$$

h. The speed at the maximum current

$$\omega_a = \omega_s(1 - s_a) = (2/3)\omega_s = 0.6667\omega_s$$

$$= 188.5 \times 2/3 = 125.27 \text{ rad/s} \quad \text{or} \quad 1200 \text{ rpm} \tag{16.39}$$

i. From Eqs. (16.9), (16.12a), and (16.36),

$$T_a = 9I'^2_{r(max)}\frac{R_r}{\omega_s}$$

$$= 9 \times 24.3^2 \times \frac{0.69}{188.5} = 19.45 \text{ N} \cdot \text{m} \tag{16.40}$$

16.2.3 Rotor Voltage Control

In a wound-rotor motor, an external three-phase resistor may be connected to its slip rings, as shown in Figure 16.5a. The developed torque may be varied by varying the resistance R_x. If R_x is referred to the stator winding and added to R_r, Eq. (16.18) may be applied to determine the developed torque. The typical torque–speed characteristics for variations in rotor resistance are shown in Figure 16.5b. This method increases the

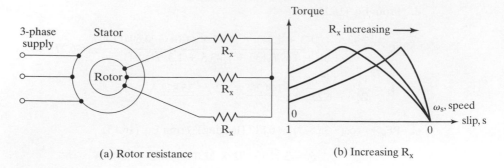

(a) Rotor resistance   (b) Increasing R_x

FIGURE 16.5

Speed control by motor resistance.

starting torque while limiting the starting current. However, this is an inefficient method and there would be imbalances in voltages and currents if the resistances in the rotor circuit are not equal. A wound-rotor induction motor is designed to have a low-rotor resistance so that the running efficiency is high and the full-load slip is low. The increase in the rotor resistance does not affect the value of maximum torque but increases the slip at maximum torque. The wound-rotor motors are widely used in applications requiring frequent starting and braking with large motor torques (e.g., crane hoists). Because of the availability of rotor windings for changing the rotor resistance, the wound rotor offers greater flexibility for control. However, it increases the cost and needs maintenance due to slip rings and brushes. The wound-rotor motor is less widely used as compared with the squirrel-case motor.

The three-phase resistor may be replaced by a three-phase diode rectifier and a dc converter, as shown in Figure 16.6a, where the gate-turn-off thyristor (GTO) or an insulated-gate bipolar transistor (IGBT) operates as a dc converter switch. The inductor L_d acts as a current source I_d and the dc converter varies the effective resistance, which can be found from Eq. (15.45):

$$R_e = R(1 - k) \tag{16.41}$$

where k is the duty cycle of the dc converter. The speed can be controlled by varying the duty cycle. The portion of the air-gap power, which is not converted into mechanical power, is called *slip power*. The slip power is dissipated in the resistance R.

The slip power in the rotor circuit may be returned to the supply by replacing the dc converter and resistance R, with a three-phase full converter, as shown in Figure 16.6b. The converter is operated in the inversion mode with delay range of $\pi/2 \leq \alpha \leq \pi$, thereby returning energy to the source. The variation of the delay angle permits PF and speed control. This type of drive is known as a *static Kramer* drive. Again, by replacing the bridge rectifiers by three three-phase dual converters (or cycloconverters), as shown in Figure 16.6c, the slip PF in either direction is possible and this arrangement is called a *static Scherbius* drive. The static Kramer and Scherbius drives are used in large power pump and blower applications where limited range of speed control is required. Because the motor is connected directly to the source, the PF of these drives is generally high.

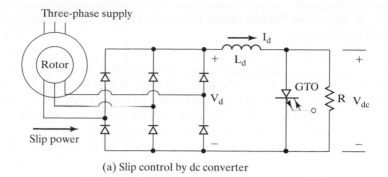

(a) Slip control by dc converter

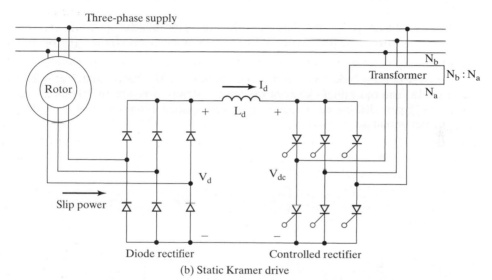

(b) Static Kramer drive

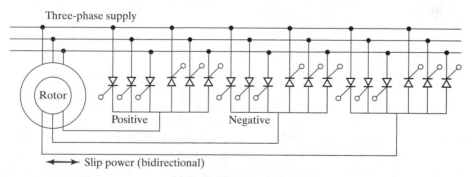

(c) Static Scherbius drive

FIGURE 16.6

Slip power control.

Example 16.3 Finding the Performance Parameters of a Three-Phase Induction Motor with Rotor Voltage Control

A three-phase, 460-V, 60-Hz, six-pole Y-connected wound-rotor induction motor whose speed is controlled by slip power as shown in Figure 16.6a, has the following parameters: $R_s = 0.041 \ \Omega$, $R'_r = 0.044 \ \Omega$, $X_s = 0.29 \ \Omega$, $X'_r = 0.44 \ \Omega$, and $X_m = 6.1 \ \Omega$. The turns ratio of the rotor to stator windings is $n_m = N_r/N_s = 0.9$. The inductance L_d is very large and its current I_d has negligible ripple. The values of R_s, R_r, X_s, and X_r for the equivalent circuit in Figure 16.2 can be considered negligible compared with the effective impedance of L_d. The no-load loss of the motor is negligible. The losses in the rectifier, inductor L_d, and the GTO dc converter are also negligible.

The load torque, which is proportional to speed squared, is 750 N · m at 1175 rpm. (a) If the motor has to operate with a minimum speed of 800 rpm, determine the resistance R. With this value of R, if the desired speed is 1050 rpm, calculate (b) the inductor current I_d, (c) the duty cycle of the dc converter k, (d) the dc voltage V_d, (e) the efficiency, and (f) the input PF_s of the drive.

Solution

$V_a = V_s = 460/\sqrt{3} = 265.58$ V, $\quad p = 6, \omega = 2\pi \times 60 = 377$ rad/s, $\quad$ and $\quad \omega_s = 2 \times 377/6 = 125.66$ rad/s. The equivalent circuit of the drive is shown in Figure 16.7a, which is reduced to Figure 16.7b provided the motor parameters are neglected. From Eq. (16.41), the dc voltage at the rectifier output is

$$V_d = I_d R_e = I_d R(1 - k) \tag{16.42}$$

and

$$E_r = sV_s \frac{N_r}{N_s} = sV_s n_m \tag{16.43}$$

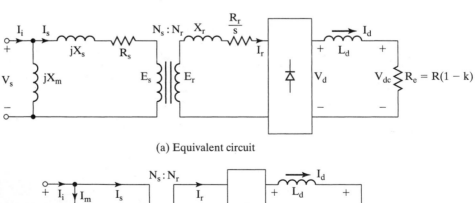

(a) Equivalent circuit

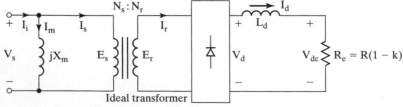

Ideal transformer

(b) Approximate equivalent circuit

FIGURE 16.7

Equivalent circuits for Example 16.3.

For a three-phase rectifier, Eq. (3.40) relates E_r and V_d as

$$V_d = 1.654 \times \sqrt{2} \, E_r = 2.3394 E_r$$

Using Eq. (16.43),

$$V_d = 2.3394 s V_s n_m \tag{16.44}$$

If P_r is the slip power, Eq. (16.9) gives the gap power

$$P_g = \frac{P_r}{s}$$

and Eq. (16.10) gives the developed power as

$$P_d = 3(P_g - P_r) = 3\left(\frac{P_r}{s} - P_r\right) = \frac{3 P_r(1 - s)}{s} \tag{16.45}$$

Because the total slip power is $3 P_r = V_d I_d$ and $P_d = T_L \omega_m$, Eq. (16.45) becomes

$$P_d = \frac{(1 - s) V_d I_d}{s} = T_L \omega_m = T_L \omega_s (1 - s) \tag{16.46}$$

Substituting V_d from Eq. (16.44) in Eq. (16.46) and solving for I_d gives

$$I_d = \frac{T_L \omega_s}{2.3394 V_s n_m} \tag{16.47}$$

which indicates that the inductor current is independent of the speed. Equating Eq. (16.42) to Eq. (16.44) gives

$$2.3394 s V_s n_m = I_d R(1 - k)$$

which gives

$$s = \frac{I_d R(1 - k)}{2.3394 \, V_s n_m} \tag{16.48}$$

The speed can be found from Eq. (16.48) as

$$\omega_m = \omega_s(1 - s) = \omega_s\left[1 - \frac{I_d R(1 - k)}{2.3394 V_s n_m}\right] \tag{16.49}$$

$$= \omega_s\left[1 - \frac{T_L \omega_s R(1 - k)}{(2.3394 V_s n_m)^2}\right] \tag{16.50}$$

which shows that for a fixed duty cycle, the speed decreases with load torque. By varying k from 0 to 1, the speed can be varied from a minimum value to ω_s.

   a. $\omega_m = 800 \, \pi/30 = 83.77$ rad/s. From Eq. (16.32) the torque at 900 rpm is

$$T_L = 750 \times \left(\frac{800}{1175}\right)^2 = 347.67 \text{ N} \cdot \text{m}$$

From Eq. (16.47), the corresponding inductor current is

$$I_d = \frac{347.67 \times 125.66}{2.3394 \times 265.58 \times 0.9} = 78.13 \text{ A}$$

The speed is minimum when the duty-cycle k is zero and Eq. (16.49) gives the minimum speed,

$$83.77 = 125.66 \left(1 - \frac{78.13R}{2.3394 \times 265.58 \times 0.9} \right)$$

and this yields $R = 2.3856 \ \Omega$.

b.   At 1050 rpm

$$T_L = 750 \times \left(\frac{1050}{1175} \right)^2 = 598.91 \text{ N} \cdot \text{m}$$

$$I_d = \frac{598.91 \times 125.66}{2.3394 \times 265.58 \times 0.9} = 134.6 \text{ A}$$

c. $\omega_m = 1050 \ \pi/30 = 109.96$ rad/s and Eq. (16.49) gives

$$109.96 = 125.66 \left[1 - \frac{134.6 \times 2.3856(1 - k)}{2.3394 \times 265.58 \times 0.9} \right]$$

which gives $k = 0.782$.

d.   Using Eq. (16.4), the slip is

$$s = \frac{125.66 - 109.96}{125.66} = 0.125$$

From Eq. (16.44),

$$V_d = 2.3394 \times 0.125 \times 265.58 \times 0.9 = 69.9 \text{ V}$$

e.   The power loss,

$$P_1 = V_d I_d = 69.9 \times 134.6 = 9409 \text{ W}$$

The output power,

$$P_o = T_L \omega_m = 598.91 \times 109.96 = 65,856 \text{ W}$$

The rms rotor current referred to the stator is

$$I_r' = \sqrt{\frac{2}{3}} \ I_d n_m = \sqrt{\frac{2}{3}} \times 134.6 \times 0.9 = 98.9 \text{ A}$$

The rotor copper loss is $P_{ru} = 3 \times 0.044 \times 98.9^2 = 1291$ W, and the stator copper loss is $P_{su} = 3 \times 0.041 \times 98.9^2 = 1203$ W. The input power is

$$P_i = 65,856 + 9409 + 1291 + 1203 = 77,759 \text{ W}$$

The efficiency is 65,856/77,759 = 85%.

f.   From Eq. (10.29) for $n = 1$, the fundamental component of the rotor current referred to the stator is

$$I'_{r1} = 0.7797 I_d \frac{N_r}{N_s} = 0.7797 I_d n_m$$

$$= 0.7797 \times 134.6 \times 0.9 = 94.45 \text{ A}$$

and the rms current through the magnetizing branch is

$$I_m = \frac{V_a}{X_m} = \frac{265.58}{6.1} = 43.54 \text{ A}$$

The rms fundamental component of the input current is

$$I_{i1} = \left[(0.7797 I_d n_m)^2 + \left(\frac{V_a}{X_m} \right)^2 \right]^{1/2}$$

$$= (94.45^2 + 43.54^2)^{1/2} = 104 \text{ A}$$

(16.51)

The PF angle is given approximately by

$$\theta_m = -\tan^{-1} \frac{V_a / X_m}{0.7797 I_d n_m}$$

$$= -\tan^{-1} \frac{43.54}{94.45} = \angle -24.74°$$

(16.52)

The input PF is $PF_s = \cos(-24.74°) = 0.908$ (lagging).

Example 16.4   Finding the Performance Parameters of a Static Kramer Drive

The induction motor in Example 16.3 is controlled by a static Kramer drive, as shown in Figure 16.6b. The turns ratio of the converter ac voltage to supply voltage is $n_c = N_a/N_b = 0.40$. The load torque is 750 N · m at 1175 rpm. If the motor is required to operate at a speed of 1050 rpm, calculate (a) the inductor current I_d; (b) the dc voltage V_d; (c) the delay angle of the converter α; (d) the efficiency; and (e) the input PF of the drive, PF_s. The losses in the diode rectifier, converter, transformer, and inductor L_d are negligible.

Solution
$V_a = V_s = 460/\sqrt{3} = 265.58 \text{ V}$, $p = 6$, $\omega = 2\pi \times 60 = 377 \text{ rad/s}$, $\omega_s = 2 \times 377/6 = 125.66$ rad/s, and $\omega_m = 1050 \, \pi/30 = 109.96$ rad/s. Then

$$s = \frac{125.66 - 109.96}{125.66} = 0.125$$

$$T_L = 750 \times \left(\frac{1050}{1175} \right)^2 = 598.91 \text{ N} \cdot \text{m}$$

a.   The equivalent circuit of the drive is shown in Figure 16.8, where the motor parameters are neglected. From Eq. (16.47), the inductor current is

$$I_d = \frac{598.91 \times 125.66}{2.3394 \times 265.58 \times 0.9} = 134.6 \text{ A}$$

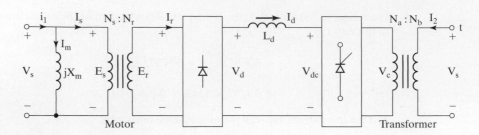

FIGURE 16.8

Equivalent circuit for static Kramer drive.

b. From Eq. (16.44),

$$V_d = 2.3394 \times 0.125 \times 265.58 \times 0.9 = 69.9 \text{ V}$$

c. Because the ac input voltage to the converter is $V_c = n_c V_s$, Eq. (10.25) gives the average voltage at the dc side of the converter as

$$V_{dc} = -\frac{3\sqrt{3}\sqrt{2}\, n_c V_s}{\pi} \cos\alpha = -2.3394 n_c V_s \cos\alpha \qquad (16.53)$$

Because $V_d = V_{dc}$, Eqs. (16.44) and (16.53) give

$$2.3394 s V_s n_m = -2.3394 n_c V_s \cos\alpha$$

which gives

$$s = \frac{-n_c \cos\alpha}{n_m} \qquad (16.54)$$

The speed, which is independent of torque, becomes

$$\omega_m = \omega_s(1 - s) = \omega_s\left(1 + \frac{n_c \cos\alpha}{n_m}\right)$$

$$109.96 = 125.66 \times \left(1 + \frac{0.4 \cos\alpha}{0.9}\right) \qquad (16.55)$$

which gives the delay angle, $\alpha = 106.3°$.

d. The power fed back

$$P_1 = V_d I_d = 69.9 \times 134.6 = 9409 \text{ W}$$

The output power

$$P_o = T_L \omega_m = 598.91 \times 109.96 = 65,856 \text{ W}$$

The rms rotor current referred to the stator is

$$I'_r = \sqrt{\frac{2}{3}}\, I_d n_m = \sqrt{\frac{2}{3}} \times 134.6 \times 0.9 = 98.9 \text{ A}$$

$$P_{ru} = 3 \times 0.044 \times 98.9^2 = 1291 \text{ W}$$

$$P_{su} = 3 \times 0.041 \times 98.9^2 = 1203 \text{ W}$$

$$P_i = 65{,}856 + 1291 + 1203 = 68{,}350 \text{ W}$$

The efficiency is $65{,}856/68{,}350 = 96\%$.

e. From (f) in Example 16.3, $I'_{r1} = 0.7797 I_d n_m = 94.45$ A, $I_m = 265.58/6.1 = 43.54$ A, and $\mathbf{I}_{i1} = 104 \underline{/-24.74°}$. From Example 10.7, the rms current fed back to the supply is

$$\mathbf{I}_{i2} = \sqrt{\frac{2}{3}} I_d n_c \underline{/-\alpha} = \sqrt{\frac{2}{3}} \times 134.6 \times 0.4 \underline{/-\alpha} = 41.98 \underline{/-106.3°}$$

The effective input current of the drive is

$$\mathbf{I}_i = \mathbf{I}_{i1} + \mathbf{I}_{i2} = 104 \underline{/-24.74°} + 41.98 \underline{/-106.3°} - 117.7 \underline{/-45.4°} \text{ A}$$

The input PF is $PF_s = \cos(-45.4°) = 0.702$ (lagging).

Note: The efficiency of this drive is higher than that of the rotor resistor control by a dc converter. The PF is dependent on the turns ratio of the transformer (e.g., if $n_c = 0.9$, $\alpha = 97.1°$, and $PF_s = 0.5$; if $n_c = 0.2$, $\alpha - 124.2°$, and $PF_s = 0.8$).

16.2.4 Frequency Control

The torque and speed of induction motors can be controlled by changing the supply frequency. We can notice from Eq. (16.31) that at the rated voltage and rated frequency, the flux is the rated value. If the voltage is maintained fixed at its rated value while the frequency is reduced below its rated value, the flux increases. This would cause saturation of the air-gap flux, and the motor parameters would not be valid in determining the torque–speed characteristics. At low frequency, the reactances decrease and the motor current may be too high. This type of frequency control is not normally used.

If the frequency is increased above its rated value, the flux and torque would decrease. If the synchronous speed corresponding to the rated frequency is called the *base speed* ω_b, the synchronous speed at any other frequency becomes

$$\omega_s = \beta \omega_b$$

and

$$s = \frac{\beta \omega_b - \omega_m}{\beta \omega_b} = 1 - \frac{\omega_m}{\beta \omega_b} \tag{16.56}$$

The torque expression in Eq. (16.18) becomes

$$T_d = \frac{3 R'_r V_a^2}{s \beta \omega_b [(R_s + R'_r/s)^2 + (\beta X_s + \beta X'_r)^2]} \tag{16.57}$$

The typical torque–speed characteristics are shown in Figure 16.9 for various values of β. The three-phase inverter in Figure 6.5a can vary the frequency at a fixed voltage. If

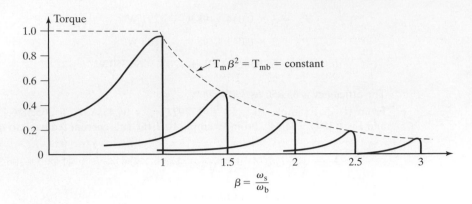

FIGURE 16.9

Torque characteristics with frequency control.

R_s is negligible, Eq. (16.26) gives the maximum torque at the base speed as

$$T_{mb} = \frac{3V_a^2}{2\omega_b(X_s + X_r')}$$ (16.58)

The maximum torque at any other frequency is

$$T_m = \frac{3}{2\omega_b(X_s + X_r')}\left(\frac{V_a}{\beta}\right)^2$$ (16.59)

and from Eq. (16.25), the corresponding slip is

$$s_m = \frac{R_r'}{\beta(X_s + X_r')}$$ (16.60)

Normalizing Eq. (16.59) with respect to Eq. (16.58) yields

$$\frac{T_m}{T_{mb}} = \frac{1}{\beta^2}$$ (16.61)

and

$$T_m\beta^2 = T_{mb}$$ (16.62)

Thus, from Eqs. (16.61) and (16.62), it can be concluded that the maximum torque is inversely proportional to frequency squared, and $T_m\beta^2$ remains constant, similar to the behavior of dc series motors. In this type of control, the motor is said to be operated in a *field-weakening mode*. For $\beta > 1$, the motor is operated at a constant terminal voltage and the flux is reduced, thereby limiting the torque capability of the motor. For $1 < \beta < 1.5$, the relation between T_m and β can be considered approximately linear. For $\beta < 1$, the motor is normally operated at a constant flux by reducing the terminal voltage V_a along with the frequency so that the flux remains constant.

Example 16.5 Finding the Performance Parameters of a Three-Phase Induction Motor with Frequency Control

A three-phase, 11.2-kW, 1750-rpm, 460-V, 60-Hz, four-pole Y-connected induction motor has the following parameters: $R_s = 0$, $R'_r = 0.38 \ \Omega$, $X_s = 1.14 \ \Omega$, $X'_r = 1.71 \ \Omega$, and $X_m = 33.2 \ \Omega$. The motor is controlled by varying the supply frequency. If the breakdown torque requirement is 35 N·m, calculate (a) the supply frequency, and (b) the speed ω_m at the maximum torque.

Solution

$V_a = V_s = 460/\sqrt{3} = 258 \times 58$ V, $\omega_b = 2\pi \times 60 = 377$ rad/s, $p = 4$, $P_0 = 11{,}200$ W, $T_{mb} \times 1750 \ \pi/30 = 11{,}200$, $T_{mb} = 61.11$ N·m, and $T_m = 35$ N·m.

a. From Eq. (16.62),

$$\beta = \sqrt{\frac{T_{mb}}{T_m}} = \sqrt{\frac{61.11}{35}} = 1.321$$

$$\omega_s = \beta \omega_b = 1.321 \times 377 = 498.01 \ \text{rad/s}$$

From Eq. (16.1), the supply frequency is

$$\omega = \frac{4 \times 498.01}{2} = 996 \ \text{rad/s} \quad \text{or} \quad 158.51 \ \text{Hz}$$

b. From Eq. (16.60), the slip for maximum torque is

$$s_m = \frac{R'_r/\beta}{X_s + X'_r} = \frac{0.38/1.321}{1.14 + 1.71} = 0.101$$

$$\omega_m = 498.01 \times (1 - 0.101) = 447.711 \ \text{rad/s} \quad \text{or} \quad 4275 \ \text{rpm}$$

16.2.5 Voltage and Frequency Control

If the ratio of voltage to frequency is kept constant, the flux in Eq. (16.31) remains constant. Equation (16.59) indicates that the maximum torque, which is independent of frequency, can be maintained approximately constant. However, at a high frequency, the air-gap flux is reduced due to the drop in the stator impedance and the voltage has to be increased to maintain the torque level. This type of control is usually known as *volts/hertz* control.

If $\omega_s = \beta \omega_b$, and the voltage-to-frequency ratio is constant so that

$$\frac{V_a}{\omega_s} = d \tag{16.63}$$

The ratio d, which is determined from the rated terminal voltage V_s and the base speed ω_b, is given by

$$d = \frac{V_s}{\omega_b} \tag{16.64}$$

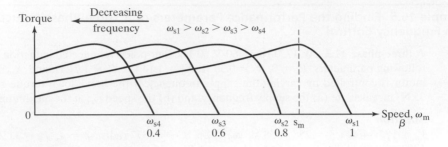

FIGURE 16.10

Torque–speed characteristics with volts/hertz control.

Substituting V_a from Eq. (16.63) into Eq. (16.57) yields the torque T_d, and the slip for maximum torque is

$$s_m = \frac{R'_r}{[R_s^2 + \beta^2(X_s + X'_r)^2]^{1/2}} \qquad (16.65)$$

The typical torque–speed characteristics are shown in Figure 16.10. As the frequency is reduced, β decreases and the slip for maximum torque increases. For a given torque demand, the speed can be controlled according to Eq. (16.64) by changing the frequency. Therefore, by varying both the voltage and frequency, the torque and speed can be controlled. The torque is normally maintained constant while the speed is varied. The voltage at variable frequency can be obtained from three-phase inverters or cycloconverters. The cycloconverters are used in very large power applications (e.g., locomotives and cement mills), where the frequency requirement is one-half or one-third of the line frequency.

Three possible circuit arrangements for obtaining variable voltage and frequency are shown in Figure 16.11. In Figure 16.11a, the dc voltage remains constant and the PWM techniques are applied to vary both the voltage and frequency within the inverter. Due to diode rectifier, regeneration is not possible and the inverter would generate harmonics into the ac supply. In Figure 16.11b, the dc-dc converter varies the dc voltage to the inverter and the inverter controls the frequency. Due to the dc converter, the harmonic injection into the ac supply is reduced. In Figure 16.11c, the dc voltage is varied by the dual converter and frequency is controlled within the inverter. This arrangement permits regeneration; however, the input PF of the converter is low, especially at a high delay angle.

Example 16.6   Finding the Performance Parameters of a Three-Phase Induction Motor with Voltage and Frequency Control

A three-phase, 11.2-kW, 1750-rpm, 460-V, 60-Hz, four-pole, Y-connected induction motor has the following parameters: $R_s = 0.66\ \Omega$, $R'_r = 0.38\ \Omega$, $X_s = 1.14\ \Omega$, $X'_r = 1.71\ \Omega$, and $X_m = 33.2\ \Omega$. The motor is controlled by varying both the voltage and frequency. The volts/hertz ratio, which corresponds to the rated voltage and rated frequency, is maintained constant. (a) Calculate the

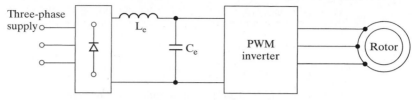

(a) Fixed dc and PWM inverter drive

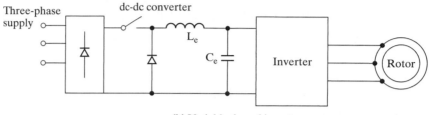

(b) Variable dc and inverter

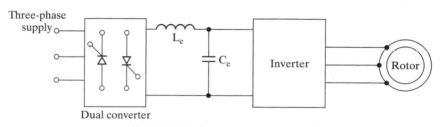

(c) Variable dc from dual converter and inverter

FIGURE 16.11

Voltage-source induction motor drives.

maximum torque T_m and the corresponding speed ω_m for 60 and 30 Hz. (b) Repeat (a) if R_s is negligible.

Solution
$p = 4$, $V_a = V_s = 460/\sqrt{3} = 265.58$ V, $\omega = 2\pi \times 60 = 377$ rad/s, and from Eq. (16.1), $\omega_b = 2 \times 377/4 = 188.5$ rad/s. From Eq. (16.63), $d = 265.58/188.5 = 1.409$.

a. At 60 Hz, $\omega_b = \omega_s = 188.5$ rad/s, $\beta = 1$, and $V_a = d\omega_s = 1.409 \times 188.5 = 265.58$ V. From Eq. (16.65),

$$s_m = \frac{0.38}{[0.66^2 + (1.14 + 1.71)^2]^{1/2}} = 0.1299$$

$$\omega_m = 188.5 \times (1 - 0.1299) = 164.01 \text{ rad/s} \quad \text{or} \quad 1566 \text{ rpm}$$

From Eq. (16.21), the maximum torque is

$$T_m = \frac{3 \times 265.58^2}{2 \times 188.5 \times [0.66 + \sqrt{0.66^2 + (1.14 + 1.71)^2}]} = 156.55 \text{ N} \cdot \text{m}$$

At 30 Hz, $\omega_s = 2 \times 2 \times \pi\, 30/4 = 94.25$ rad/s, $\beta = 30/60 = 0.5$, and $V_a = d\omega_s = 1.409 \times 94.25 = 132.79$ V. From Eq. (16.65), the slip for maximum torque is

$$s_m = \frac{0.38}{[0.66^2 + 0.5^2 \times (1.14 + 1.71)^2]^{1/2}} = 0.242$$

$$\omega_m = 94.25 \times (1 - 0.242) = 71.44 \text{ rad/s} \quad \text{or} \quad 682 \text{ rpm}$$

$$T_m = \frac{3 \times 132.79^2}{2 \times 94.25 \times [0.66 + \sqrt{0.66^2 + 0.5^2 \times (1.14 + 1.71)^2}]} = 125.82 \text{ N} \cdot \text{m}$$

b. At 60 Hz, $\omega_b = \omega_s = 188.5$ rad/s and $V_a = 265.58$ V. From Eq. (16.60),

$$s_m = \frac{0.38}{1.14 + 1.71} = 0.1333$$

$$\omega_m = 188.5 \times (1 - 0.1333) = 163.36 \text{ rad/s} \quad \text{or} \quad 1560 \text{ rpm}$$

From Eq. (16.59), the maximum torque is $T_m = 196.94$ N $\cdot$ m.
At 30 Hz, $\omega_s = 94.25$ rad/s, $\beta = 0.5$, and $V_a = 132.79$ V. From Eq. (16.60),

$$s_m = \frac{0.38/0.5}{1.14 + 1.71} = 0.2666$$

$$\omega_m = 94.25 \times (1 - 0.2666) = 69.11 \text{ rad/s} \quad \text{or} \quad 660 \text{ rpm}$$

From Eq. (16.59), the maximum torque is $T_m = 196.94$ N $\cdot$ m.

Note: Neglecting R_s may introduce a significant error in the torque estimation, especially at a low frequency.

16.2.6 Current Control

The torque of induction motors can be controlled by varying the rotor current. The input current, which is readily accessible, is varied instead of the rotor current. For a fixed input current, the rotor current depends on the relative values of the magnetizing and rotor circuit impedances. From Figure 16.2, the rotor current can be found as

$$\bar{I}_r' = \frac{jX_m I_i}{R_s + R_r'/s + j(X_m + X_s + X_r')} = I_r' \underline{/\theta_1} \tag{16.66}$$

From Eqs. (16.9) and (16.12a), the developed torque is

$$T_d = \frac{3R_r'(X_m I_i)^2}{s\omega_s[(R_s + R_r'/s)^2 + (X_m + X_s + X_r')^2]} \tag{16.67}$$

and the starting torque at $s = 1$ is

$$T_s = \frac{3R_r'(X_m I_i)^2}{\omega_s[(R_s + R_r')^2 + (X_m + X_s + X_r')^2]} \tag{16.68}$$

The slip for maximum torque is

$$s_m = \pm \frac{R'_r}{[R_s^2 + (X_m + X_s + X'_r)^2]^{1/2}} \tag{16.69}$$

In a real situation, as shown in Figure 16.1b and 16.1c, the stator current through R_s and X_s is constant at I_i. Generally, X_m is much greater than X_s and R_s, which can be neglected for most applications. Neglecting the values of R_s and X_s, Eq. (16.69) becomes

$$s_m = \pm \frac{R'_r}{X_m + X'_r} \tag{16.70}$$

and at $s = s_m$, Eq. (16.67) gives the maximum torque,

$$T_m = \frac{3 X_m^2}{2\omega_s(X_m + X'_r)} I_i^2 = \frac{3 L_m^2}{2(L_m + L'_r)} I_i^2 \tag{16.71}$$

It can be noticed from Eq. (16.71) that the maximum torque depends on the square of the current and is approximately independent of the frequency. The typical torque–speed characteristics are shown in Figure 16.12 for increasing values of stator current. Because X_m is large as compared with X_s and X'_r, the starting torque is low. As the speed increases (or slip decreases), the stator voltage rises and the torque increases. The starting current is low due to the low values of flux (as I_m is low and X_m is large) and rotor current compared with their rated values. The torque increases with the speed due to the increase in flux. A further increase in speed toward the positive slope of the characteristics increases the terminal voltage beyond the rated value. The flux and the magnetizing current are also increased, thereby saturating the flux. The torque can be controlled by the stator current and slip. To keep the air-gap flux constant and to avoid saturation due to high voltage, the motor is normally operated on the negative slope of the equivalent torque–speed characteristics with voltage control. The negative slope is in the unstable region and the motor must be operated in closed-loop control. At a low slip, the terminal voltage could be excessive and the flux would saturate. Due to saturation, the torque peaking, as shown in Figure 16.12, is less than that as shown.

The constant current can be supplied by three-phase current source inverters. The current-fed inverter has the advantages of fault current control and the current is less sensitive to the motor parameter variations. However, they generate harmonics

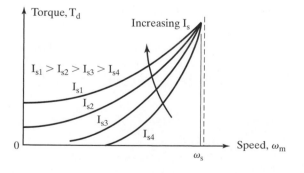

FIGURE 16.12

Torque–speed characteristics by current control.

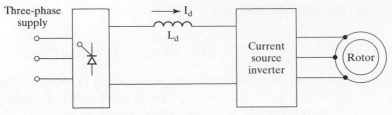

(a) Controlled rectifier-fed current source

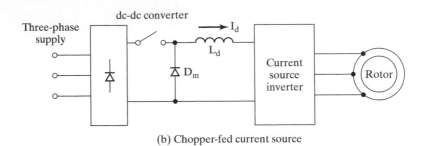

(b) Chopper-fed current source

FIGURE 16.13

Current-source inductor motor drive.

and torque pulsation. Two possible configurations of current-fed inverter drives are shown in Figure 16.13. In Figure 16.13a the inductor acts as a current source and the controlled rectifier controls the current source. The input PF of this arrangement is very low. In Figure 16.13b the dc-dc converter controls the current source and the input PF is higher.

Example 16.7 Finding the Performance Parameters of a Three-Phase Induction Motor with Current Control

A three-phase, 11.2-kW, 1750-rpm, 460-V, 60-Hz, four-pole, Y-connected induction motor has the following parameters: $R_s = 0.66\ \Omega$, $R'_r = 0.38\ \Omega$, $X_s = 1.14\ \Omega$, $X'_r = 1.71\ \Omega$, and $X_m = 33.2\ \Omega$. The no-load loss is negligible. The motor is controlled by a current-source inverter and the input current is maintained constant at 20 A. If the frequency is 40 Hz and the developed torque is 55 N·m, determine (a) the slip for maximum torque s_m and maximum torque T_m, (b) the slip s, (c) the rotor speed ω_m, (d) the terminal voltage per phase V_a, and (e) the PF_m.

Solution

$V_{a(\text{rated})} = 460/\sqrt{3} = 265.58$ V, $I_i = 20$ A, $T_L = T_d = 55$ N·m, and $p = 4$. At 40 Hz, $\omega = 2\pi \times 40 = 251.33$ rad/s, $\omega_s = 2 \times 251.33/4 = 125.66$ rad/s, $R_s = 0.66\ \Omega$, $R'_r = 0.38\ \Omega$, $X_s = 1.14 \times 40/60 = 0.76\ \Omega$, $X'_r = 1.71 \times 40/60 = 1.14\ \Omega$, and $X_m = 33.2 \times 40/60 = 22.13\ \Omega$.

a. From Eq. (16.69),

$$s_m = \frac{0.38}{[0.66^2 + (22.13 + 0.78 + 1.14)^2]^{1/2}} = 0.0158$$

From Eq. (16.67), $T_m = 94.68$ N·m.

b. From Eq. (16.67),

$$T_d = 55 = \frac{3(R_r/s)(22.13 \times 20)^2}{125.66 \times [(0.66 + R_r/s)^2 + (22.13 + 0.76 + 1.14)^2]}$$

which gives $(R_r/s)^2 - 83.74(R_r'/s) + 578.04 = 0$, and solving for R_r/s yields

$$\frac{R_r'}{s} = 76.144 \quad \text{or} \quad 7.581$$

and $s = 0.00499$ or 0.0501. Because the motor is normally operated with a large slip in the negative slope of the torque–speed characteristic,

$$s = 0.0501$$

c. $\omega_m = 125.656 \times (1 - 0.0501) = 119.36$ rad/s or 1140 rpm.

d. From Figure 16.2, the input impedance can be derived as

$$\overline{Z}_i = R_i + jX_i = (R_i^2 + X_i^2)^{1/2} \, \underline{/\theta_m} = Z_i \, \underline{/\theta_m}$$

where

$$R_i = \frac{X_m^2(R_s + R_r/s)}{(R_s + R_r/s)^2 + (X_m + X_s + X_r)^2}$$

(16.72)

$$- 6.26 \, \Omega$$

$$X_i = \frac{X_m[(R_s + R_r/s)^2 + (X_s + X_r)(X_m + X_s + X_r)]}{(R_s + R_r/s)^2 + (X_m + X_s + X_r)^2}$$

(16.73)

$$= 3.899 \, \Omega$$

and

$$\theta_m = \tan^{-1}\frac{X_i}{R_i}$$

$$= 31.9°$$

(16.74)

$$Z_i = (6.26^2 + 3.899^2)^{1/2} = 7.38 \, \Omega$$

$$V_a = Z_i I_i = 7.38 \times 20 = 147.6 \text{ V}$$

e. $PF_m = \cos(31.9°) = 0.849$ (lagging).

Note: If the maximum torque is calculated from Eq. (16.71), $T_m = 100.49$ and V_a (at $s = s_m$) is 313 V. For a supply frequency of 90 Hz, recalculations of the values give ω_s = 282.74 rad/s, X_s = 1.71 Ω, X_r' = 2.565 Ω, X_m = 49.8 Ω, s_m = 0.00726, T_m = 96.1 N·m, s = 0.0225, V_a = 316 V, and V_a (at s = s_m) = 699.6 V. It is evident that at a high frequency and a low slip, the terminal voltage would exceed the rated value and saturate the air-gap flux.

16.2.7 Voltage, Current, and Frequency Control

The torque–speed characteristics of induction motors depend on the type of control. It may be necessary to vary the voltage, frequency, and current to meet the torque–speed requirements, as shown in Figure 16.14, where there are three regions. In the first region, the speed can be varied by voltage (or current) control at constant torque. In the second region, the motor is operated at constant current and the slip is varied. In the third region, the speed is controlled by frequency at a reduced stator current.

The torque and power variations for a given stator current and frequencies below the rated frequency are shown by dots in Figure 16.15. For β < 1, the motor operates at a constant flux. For β > 1, the motor is operated by frequency control, but at a constant voltage. Therefore, the flux decreases in the inverse ratio of per-unit frequency, and the motor operates in the field weakening mode.

When motoring, a decrease in speed command decreases the supply frequency. This shifts the operation to regenerative braking. The drive decelerates under the influence of braking torque and load torque. For speed below rate value ω_b, the voltage and frequency are reduced with speed to maintain the desired V to f ratio or constant flux and to keep the operation on the speed–torque curves with a negative slope by limiting the slip speed. For speed above ω_b, the frequency alone is reduced with the speed to maintain the operation on the portion of the speed–torque curves with a negative slope. When close to the desired speed, the operation shifts to motoring operation and the drive settles at the desired speed.

When motoring, an increase in the speed command increases the supply frequency. The motor torque exceeds the load torque and the motor accelerates. The operation is maintained on the portion of the speed–torque curves with a negative slope by limiting the slip speed. Finally, the drive settles at the desired speed.

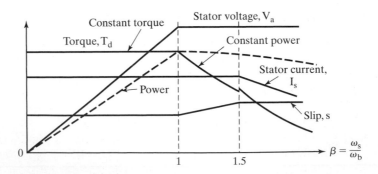

FIGURE 16.14

Control variables versus frequency.

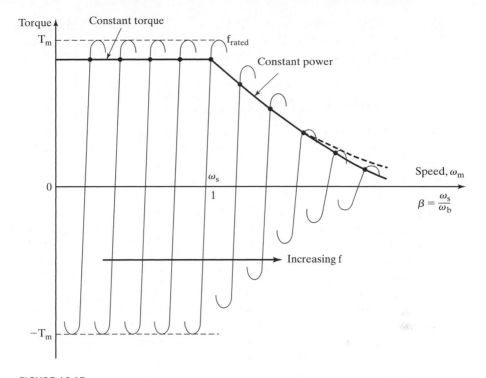

FIGURE 16.15

Torque–speed characteristics for variable frequency control.

Key Points of Section 16.2

- The speed and torque of induction motors can be varied by (1) stator voltage control; (2) rotor voltage control; (3) frequency control; (4) stator voltage and frequency control; (5) stator current control; or (6) voltage, current, and frequency control.

- To meet the torque–speed duty cycle of a drive, the voltage, current, and frequency are normally controlled such that the flux or the V to f ratio remains constant.

16.3   CLOSED-LOOP CONTROL OF INDUCTION MOTORS

A closed-loop control is normally required to satisfy the steady-state and transient performance specifications of ac drives [9, 10]. The control strategy can be implemented by (1) *scalar control*, where the control variables are dc quantities and only their magnitudes are controlled; (2) *vector control*, where both the magnitude and phase of the control variables are controlled; or (3) *adaptive control*, where the parameters of the controller are continuously varied to adapt to the variations of the output variables.

   The dynamic model of induction motors differs significantly from that of Figure 16.1c and is more complex than dc motors. The design of feedback-loop parameters requires complete analysis and simulation of the entire drive. The control and modeling of ac drives are beyond the scope of this book [2, 5, 17, 18]; and only some of the basic scalar feedback techniques are discussed in this section.

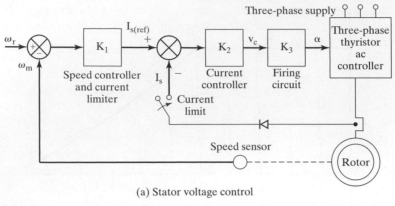

(a) Stator voltage control

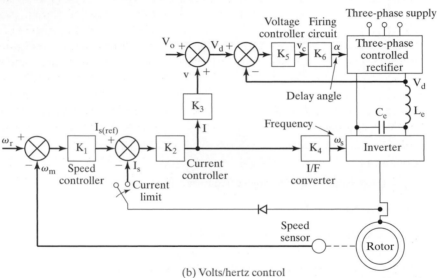

(b) Volts/hertz control

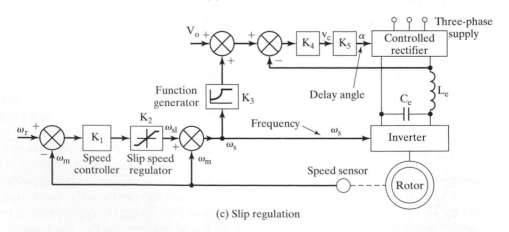

(c) Slip regulation

FIGURE 16.16

Closed-loop control of induction motors.

A control system is generally characterized by the hierarchy of the control loops, where the outer loop controls the inner loops. The inner loops are designed to execute progressively faster. The loops are normally designed to have limited command excursion. Figure 16.16a shows an arrangement for stator voltage control of induction motors by ac voltage controllers at fixed frequency. The speed controller K_1 processes the speed error and generates the reference current $I_{s(\text{ref})}$. K_2 is the current controller. K_3 generates the delay angle of thyristor converter and the inner current-limit loop sets the torque limit indirectly. The current limiter instead of current clamping has the advantage of feeding back the short-circuit current in case of fault. The speed controller K_1 may be a simple gain (proportional type), proportional-integral type, or a lead-lag compensator. This type of control is characterized by poor dynamic and static performance and is generally used in fans, pumps, and blower drives. A 187-kW, three-phase ac voltage regulator (or controller) for a wire-drawing machine is shown in Figure 16.17, where the control electronics are mounted on the side panel.

The arrangement in Figure 16.16a can be extended to a volt/hertz control with the addition of a controlled rectifier and dc voltage control loop, as shown in Figure 16.16b. After the current limiter, the same signal generates the inverter frequency and provides input to the dc-link gain controller K_3. A small voltage V_0 is added to the dc voltage reference to compensate for the stator resistance drop at low frequency. The dc voltage V_d acts as the reference for the voltage control of the controlled rectifier. In case of PWM inverter, there is no need for the controlled rectifier and the signal V_d controls the inverter voltage directly by varying the modulation index. For current monitoring, it requires a sensor, which introduces a delay in the system response. Shown in Figure 16.18 are 15 forced-commutated inverter cubicles for controlling the undergrate cooler fans of a cement kiln, where each unit is rated at 100 kW.

FIGURE 16.17

A 187-k W, three-phase ac voltage regulator. (Reproduced by permission of Brush Electrical Machines Ltd., England.)

FIGURE 16.18

A cement kiln with 15 forced-commutated inverter cubicles. (Reproduced by permission of Brush Electrical Machines Ltd., England.)

Because the torque of induction motors is proportional to the slip frequency, $\omega_{sl} = \omega_s - \omega_m = s\omega_s$, the slip frequency instead of the stator current can be controlled. The speed error generates the slip frequency command, as shown in Figure 16.16c, where the slip limits set the torque limits. The function generator, which produces the command signal for voltage control in response to the frequency ω_s, is nonlinear and can take also into account the compensating drop V_o at a low frequency. The compensating drop V_o is shown in Figure 16.16c. For a step change in the speed command, the motor accelerates or decelerates within the torque limits to a steady-state slip value corresponding to the load torque. This arrangement controls the torque indirectly within the speed control loop and do not require the current sensor.

A simple arrangement for current control is shown in Figure 16.19. The speed error generates the reference signal for the dc-link current. The slip frequency, $\omega_{sl} = \omega_s - \omega_m$, is fixed. With a step speed command, the machine accelerates with a

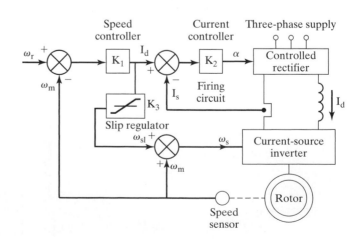

FIGURE 16.19

Current control with constant slip.

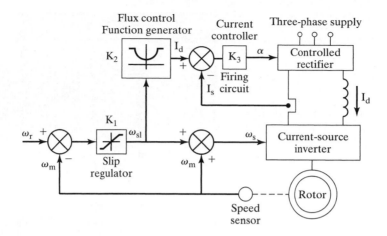

FIGURE 16.20

Current control with constant flux operation.

high current that is proportional to the torque. In the steady state, the motor current is low. However, the air-gap flux fluctuates, and due to varying flux at different operating points, the performance of this drive is poor.

A practical arrangement for current control, where the flux is maintained constant, is shown in Figure 16.20. The speed error generates the slip frequency, which controls the inverter frequency and the dc-link current source. The function generator produces the current command to maintain the air-gap flux constant, normally at the rated value.

The arrangement in Figure 16.16a for speed control with inner current control loop can be applied to a static Kramer drive, as shown in Figure 16.21, where the

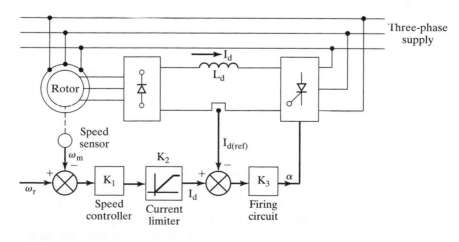

FIGURE 16.21

Speed control of static Kramer drive.

FIGURE 16.22

A 110-kW static Kramer drive. (Reproduced by permission of Brush Electrical Machines Ltd., England.)

torque is proportional to the dc-link current I_d. The speed error generates the dc-link current command. A step increase in speed clamps the current to the maximum value and the motor accelerates at a constant torque that corresponds to the maximum current. A step decrease in the speed sets the current command to zero and the motor decelerates due to the load torque. A 110-kW static Kramer drive for fans in cement works is shown in Figure 16.22.

Key Points of Section 16.3

- The closed-loop is normally used to control the steady-state and transient response of ac drives.
- However, the parameters of induction motors are coupled to each other and the scalar control lacks in producing fast dynamic response.

16.4   VECTOR CONTROLS

The control methods that have been discussed so far provide satisfactory steady-state performance, but their dynamic response is poor. An induction motor exhibits nonlinear multivariables and highly coupled characteristics. The *vector control* technique, which is also known as *field-oriented control (FOC)*, allows a squirrel-cage induction motor to be driven with high dynamic performance that is comparable to the characteristic of a dc motor [11–15]. The FOC technique decouples the two components of stator current: one providing the air-gap flux and the other producing the torque. It provides independent control of flux and torque, and the control characteristic is linearized. The stator currents are converted to a fictitious synchronously rotating reference frame aligned with the flux vector and are transformed back to the stator frame before feeding back to the machine. The two components are d-axis i_{ds} analogous to armature current, and q-axis i_{qs} analogous to the field current of a separately excited dc motor. The rotor flux linkage vector is aligned along the d-axis of the reference frame.

16.4.1   Basic Principle of Vector Control

With a vector control, an induction motor can operate as a separately excited dc motor. In a dc machine, the developed torque is given by

$$T_d = K_t I_a I_f \tag{16.75}$$

where I_a is the armature current and I_f is the field current. The construction of a dc machine is such that the field flux linkage Ψ_f produced by I_f is perpendicular to the armature flux linkage Ψ_a produced by I_a. These flux vectors that are stationary in space are orthogonal or decoupled in nature. As a result, a dc motor has fast transient response. However, an induction motor cannot give such fast response due to its inherent coupling problem. However, an induction motor can exhibit the dc machine characteristic if the machine is controlled in a synchronously rotating frame $(d^e - q^e)$, where the sinusoidal machine variables appear as dc quantities in the steady state.

Figure 16.23a shows an inverter-fed induction motor with two control current inputs: i_{ds}^* and i_{qs}^* are the direct-axis component and quadrature-axis component of the stator current, respectively, in a synchronously rotating reference frame. With vector control, i_{ds}^* is analogous to the field current I_f and i_{ds}^* is analogous to armature current I_a of a dc motor. Therefore, the developed torque of an induction motor is given by

$$T_d = K_m \hat{\Psi}_r I_f = K_t I_{ds} i_{qs_r} \tag{16.76}$$

where $\hat{\Psi}_r$ is the absolute peak value of the sinusoidal space flux linkage vector $\overrightarrow{\Psi}_r$;
   i_{ds} is the field component;
   i_{qs} is the torque component.

Figure 16.23b shows the space-vector diagram for vector control: i_{ds}^* is oriented (or aligned) in the direction of rotor flux $\hat{\lambda}_r$, and i_{qs}^* must be perpendicular to it under all operating conditions. The space vectors rotate synchronously at frequency ω_s. Thus, the vector control must ensure the correct orientation of the space vectors and generate the control input signals.

The implementation of the vector control is shown in Figure 16.23c. The inverter generates currents i_a, i_b, and i_c in response to the corresponding command currents i_a^*, i_b^*, and i_c^* from the controller. The machine terminal currents i_a, i_b, and i_c are converted to i_{ds}^s and i_{qs}^s components by three-phase to two-phase transformation. These are then converted to synchronously rotating frame (into i_{ds} and i_{qs} components) by the unit vector components $\cos\theta_s$ and $\sin\theta_s$ before applying them to the machine. The machine is represented by internal conversions into the $d^e - q^e$ model.

The controller makes two stages of inverse transformation so that the line control currents i_{ds}^* and i_{qs}^* correspond to the machine currents i_{ds} and i_{qs}, respectively. In

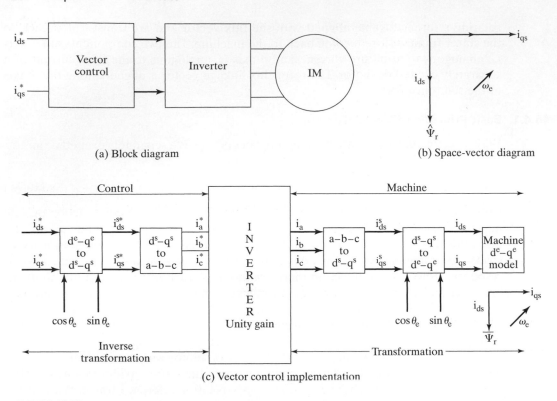

(a) Block diagram   (b) Space-vector diagram

(c) Vector control implementation

FIGURE 16.23

Vector control of induction motor.

addition, the unit vector ($\cos\theta_e$ and $\sin\theta_e$) ensures correct alignment of i_{ds} current with the flux vector Ψ_r and i_{qs} current is perpendicular to it. It is important to note that the transformation and inverse transformation ideally do not incorporate any dynamics. Therefore, the response to i_{ds} and i_{qs} is instantaneous except for any delays due to computational and sampling times.

16.4.2 Direct and Quadrature-Axis Transformation

The vector control technique uses the dynamic equivalent circuit of the induction motor. There are at least three fluxes (rotor, air gap, and stator) and three currents or mmfs (in stator, rotor, and magnetizing) in an induction motor. For fast dynamic response, the interactions between current, fluxes, and speed, must be taken into account in obtaining the dynamic model of the motor and determining appropriate control strategies.

All fluxes rotate at synchronous speed. The three-phase currents create mmfs (stator and rotor), which also rotate at synchronous speed. Vector control aligns axes of an mmf and a flux orthogonally at all times. It is easier to align the stator current mmf orthogonally to the rotor flux.

Any three-phase sinusoidal set of quantities in the stator can be transformed to an orthogonal reference frame by

$$
\begin{bmatrix} f_{\alpha s} \\ f_{\beta s} \\ f_o \end{bmatrix} = \frac{2}{3} \begin{bmatrix} \cos\theta & \cos\left(\theta - \dfrac{2\pi}{3}\right) & \cos\left(\theta - \dfrac{4\pi}{3}\right) \\ \sin\theta & \sin\left(\theta - \dfrac{2\pi}{3}\right) & \sin\left(\theta - \dfrac{4\pi}{3}\right) \\ \dfrac{1}{2} & \dfrac{1}{2} & \dfrac{1}{2} \end{bmatrix} \begin{bmatrix} f_{as} \\ f_{bs} \\ f_{cs} \end{bmatrix} \qquad (16.77)
$$

where θ is the angle of the orthogonal set α-β-0 with respect to any arbitrary reference. If the α-β-0 axes are stationary and the α-axis is aligned with the stator a-axis, then $\theta = 0$ at all times. Thus, we get

$$
\begin{bmatrix} f_{\alpha s} \\ f_{\beta s} \\ f_{os} \end{bmatrix} = \frac{2}{3} \begin{bmatrix} 1 & -\dfrac{1}{2} & -\dfrac{1}{2} \\ 0 & \dfrac{\sqrt{3}}{2} & \dfrac{\sqrt{3}}{2} \\ \dfrac{1}{2} & \dfrac{1}{2} & \dfrac{1}{2} \end{bmatrix} \begin{bmatrix} f_{as} \\ f_{bs} \\ f_{cs} \end{bmatrix} \qquad (16.78)
$$

The orthogonal set of reference that rotates at the synchronous speed ω_s is referred to as d–q–0 axes. Figure 16.24 shows the axis of rotation for various quantities.

The three-phase rotor variables, transformed to the synchronously rotating frame, are given by

$$
\begin{bmatrix} f_{dr} \\ f_{qr} \\ f_o \end{bmatrix} = \frac{2}{3} \begin{bmatrix} \cos(\omega_s - \omega_m)t & \cos\left((\omega_s - \omega_m)t - \dfrac{2\pi}{3}\right) & \cos\left((\omega_s - \omega_m)t - \dfrac{4\pi}{3}\right) \\ \sin(\omega_s - \omega_m)t & \sin\left((\omega_s - \omega_m)t - \dfrac{2\pi}{3}\right) & \sin\left((\omega_s - \omega_m)t - \dfrac{4\pi}{3}\right) \\ \dfrac{1}{2} & \dfrac{1}{2} & \dfrac{1}{2} \end{bmatrix} \begin{bmatrix} f_{ar} \\ f_{br} \\ f_{cr} \end{bmatrix}
$$

$$(16.79)$$

It is important to note that the difference $\omega_s - \omega_m$ is the relative speed between the synchronously rotating reference frame and the frame attached to the rotor. This difference is slip frequency ω_{sl}, which is frequency of the rotor variables. By applying these transformations, voltage equations of the motor in the synchronously rotating frame are reduced to,

$$
\begin{bmatrix} v_{qs} \\ v_{ds} \\ v_{qr} \\ v_{dr} \end{bmatrix} = \begin{bmatrix} R_s + DL_s & \omega_s L_s & DL_m & \omega_s L_m \\ -\omega_s L_s & R_s + DL_s & -\omega_s L_m & DL_m \\ DL_m & (\omega_s - \omega_m)L_m & R_r + DL_r & (\omega_s - \omega_m)L_r \\ -(\omega_s - \omega_m) & DL_m & -(\omega_s - \omega_m)L_r & R_r + DL_r \end{bmatrix} \begin{bmatrix} i_{qs} \\ i_{ds} \\ i_{qr} \\ i_{dr} \end{bmatrix}
$$

$$(16.80)$$

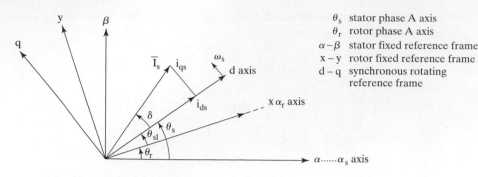

FIGURE 16.24

Axis of rotation for various quantities.

where ω_s is the speed of the reference frame (or synchronous speed), ω_m is rotor speed and

$$L_s = L_{ls} + L_m, L_r = L_{lr} + L_m$$

Subscripts l and m stand for leakage and magnetizing, respectively, and D represents the differential operator d/dt. The dynamic equivalent circuits of the motor in this reference frame are shown in Figure 16.25.

The stator flux linkages are expressed as

$$\Psi_{qs} = L_{ls}i_{qs} + L_m(i_{qs} + i_{qr}) = L_s i_{qs} + L_m i_{qr} \tag{16.81}$$

$$\Psi_{ds} = L_{ls}i_{ds} + L_m(i_{ds} + i_{dr}) = L_s i_{ds} + L_m i_{dr} \tag{16.82}$$

$$\hat{\Psi}_s = \sqrt{(\Psi_{qs}^2 + \Psi_{ds}^2)} \tag{16.83}$$

The rotor flux linkages are given by

$$\Psi_{qr} = L_{lr}i_{qr} + L_m(i_{qs} + i_{qr}) = L_r i_{qr} + L_m i_{qs} \tag{16.84}$$

$$\Psi_{dr} = L_{lr}i_{dr} + L_m(i_{ds} + i_{dr}) = L_r i_{dr} + L_m i_{ds} \tag{16.85}$$

$$\hat{\Psi}_r = \sqrt{(\Psi_{qr}^2 + \Psi_{dr}^2)} \tag{16.86}$$

The air-gap flux linkages are given by

$$\Psi_{mq} = L_m(i_{qs} + i_{qr}) \tag{16.87}$$

$$\Psi_{md} = L_m(i_{ds} + i_{dr}) \tag{16.88}$$

$$\hat{\Psi}_m = \sqrt{(\Psi_{mqs}^2 + \Psi_{mds}^2)} \tag{16.89}$$

Therefore, the torque developed by the motor is given by

$$T_d = \frac{3p}{2}[\Psi_{ds}i_{qs} - \Psi_{qs}i_{ds}] \tag{16.90}$$

where p is the number of poles. Equation (16.80) gives the rotor voltages in d- and q-axis as

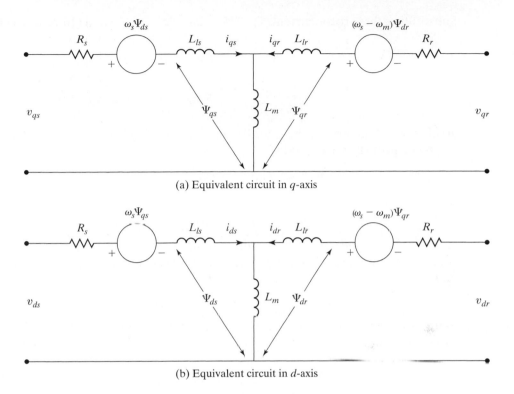

(a) Equivalent circuit in q-axis

(b) Equivalent circuit in d-axis

FIGURE 16.25

Dynamic equivalent circuits in the synchronously rotating frame.

$$v_{qr} = 0 = L_m \frac{di_{qs}}{dt} + (\omega_s - \omega_m)L_m i_{ds} + (R_r + L_r)\frac{di_{qr}}{dt} + (\omega_e - \omega_r)L_r i_{dr} \quad (16.91)$$

$$v_{dr} = 0 = L_m \frac{di_{ds}}{dt} + (\omega_s - \omega_m)L_m i_{qs} + (R_r + L_r)\frac{di_{dr}}{dt} + (\omega_e - \omega_r)L_r i_{qr} \quad (16.92)$$

which, after substituting Ψ_{qr} from Eq. (16.84) and Ψ_{dr} from Eq. (16.85), give

$$\frac{d\Psi_{qr}}{dt} + R_r i_{qr} + (\omega_s - \omega_m)\Psi_{dr} = 0 \quad (16.93)$$

$$\frac{d\Psi_{dr}}{dt} + R_r i_{dr} + (\omega_s - \omega_m)\Psi_{qr} = 0 \quad (16.94)$$

Solving for i_{qr} from Eq. (16.84) and i_{dr} from Eq. (16.85), we get

$$i_{qr} = \frac{1}{L_r}\Psi_{qr} - \frac{L_m}{L_r}i_{qs} \quad (16.95)$$

$$i_{dr} = \frac{1}{L_r}\Psi_{dr} - \frac{L_m}{L_r}i_{ds} \quad (16.96)$$

Substituting these rotor currents i_{qr} and i_{dr} into Eqs. (16.93) and (16.94), we get

$$\frac{d\Psi_{qr}}{dt} + \frac{L_r}{R_r}\Psi_{qr} - \frac{L_m}{L_r}R_r i_{qs} + (\omega_s - \omega_m)\Psi_{dr} = 0 \tag{16.97}$$

$$\frac{d\Psi_{qr}}{dt} + \frac{L_r}{R_r}\Psi_{qr} - \frac{L_m}{L_r}R_r i_{qs} + (\omega_s - \omega_m)\Psi_{dr} = 0 \tag{16.98}$$

To eliminate transients in the rotor flux and the coupling between the two axes, the following conditions must be satisfied.

$$\Psi_{qr} = 0 \text{ and } \hat{\Psi}_r = \sqrt{\Psi_{dr}^2 + \Psi_{qr}^2} = \Psi_{dr} \tag{16.99}$$

Also, the rotor flux should remain constant so that

$$\frac{d\Psi_{dr}}{dt} = \frac{d\Psi_{qr}}{dt} = 0 \tag{16.100}$$

With conditions in Eqs. (16.99) and (16.100), the rotor flux $\hat{\Psi}_r$ is aligned on the d^e-axis, and we get

$$\omega_s - \omega_m = \omega_{sl} = \frac{L_m}{\hat{\Psi}_r}\frac{R_r}{L_r}i_{qs} \tag{16.101}$$

and

$$\frac{L_r}{R_r}\frac{d\hat{\Psi}_r}{dt} + \hat{\Psi}_r = L_m i_{ds} \tag{16.102}$$

Substituting the expressions for i_{qr} from Eq. (16.95) into Eq. (16.84) and i_{dr} from Eq. (16.96) into Eq. (16.85), we get

$$\Psi_{qs} = \left(L_s - \frac{L_m^2}{L_r}\right)i_{qs} + \frac{L_m}{L_r}\Psi_{qr} \tag{16.103}$$

$$\Psi_{ds} = \left(L_s - \frac{L_m^2}{L_r}\right)i_{ds} + \frac{L_m}{L_r}\Psi_{dr} \tag{16.104}$$

Substituting Ψ_{qs} Eq. (16.103) and Ψ_{ds} from Eq. (16.104) into Eq. (16.90) gives the developed torque as

$$T_d = \frac{3p}{2}\frac{L_m}{L_r}\left(\Psi_{dr}i_{qs} - \Psi_{qr}i_{ds}\right) = \frac{3p}{2}\frac{L_m}{L_r}\hat{\Psi}_r i_{qs} \tag{16.105}$$

If the rotor flux $\hat{\Psi}_r$ remains constant, Eq. (16.102) becomes

$$\hat{\Psi}_r = L_m i_{ds} \tag{16.106}$$

which indicates that the rotor flux is directly proportional to current i_{ds}. Thus, T_d becomes

$$T_d = \frac{3p}{2}\frac{L_m^2}{L_r}i_{ds}i_{qs} = K_m i_{ds}i_{qs} \tag{16.107}$$

where $K_m = 3pL_m^2/2L_r$.

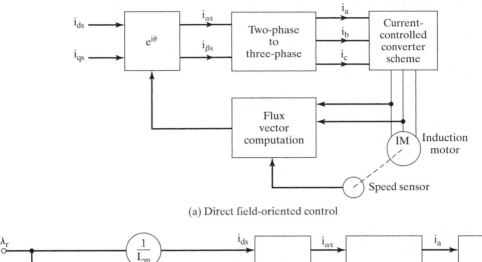

(a) Direct field-oriented control

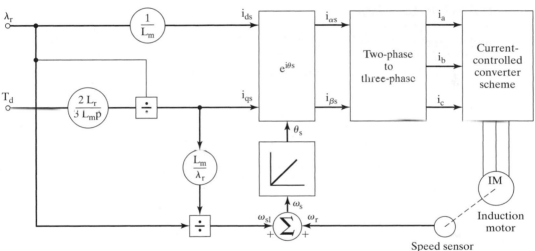

(b) Indirect field-oriented control

FIGURE 16.26

Block diagrams of vector control.

The vector control can be implemented by either direct method or indirect method [4]. The methods are different essentially by how the unit vector ($\cos \theta_s$ and $\sin \theta_s$) is generated for the control. In the direct method, the flux vector is computed from the terminal quantities of the motor, as shown in Figure 16.26a. The indirect method uses the motor slip frequency ω_{sl} to compute the desired flux vector, as shown in Figure 16.26b. It is simpler to implement than the direct method and is used increasingly in induction motor control. T_d is the desired motor torque, Ψ_r is the rotor flux linkage, T_r is the rotor time constant, and L_m is the mutual inductance. The amount of decoupling is dependent on the motor parameters unless the flux is measured directly. Without the exact knowledge of the motor parameters, an ideal decoupling is not possible.

Notes:

1. According to Eq. (16.102), the rotor flux $\hat{\Psi}_{dr}$ is determined by i_{dr}, which is subject to a time delay T_r due to the rotor time constant (L_r/R_r).

2. According to Eq. (16.107), the current i_{qs} controls the developed torque T_d without delay.

3. Currents i_{ds} and i_{qs} are orthogonal to each other and are called the *flux-* and *torque*-producing currents, respectively. This correspondence between flux- and torque-producing currents is subject to maintaining the conditions in Eqs. (16.101) and (16.102). Normally, i_{ds} would remain fixed for operation up to the base speed. Thereafter, it is reduced to weaken the rotor flux so that the motor may be driven with a constant power like characteristic.

16.4.3 Indirect Vector Control

Figure 16.27 shows the block diagram for control implementation of the indirect field-oriented control (IFOC). The flux component of current i_{ds}^* for the desired rotor flux $\hat{\Psi}_r$ is determined from Eq. (16.106), and is maintained constant. The variation of

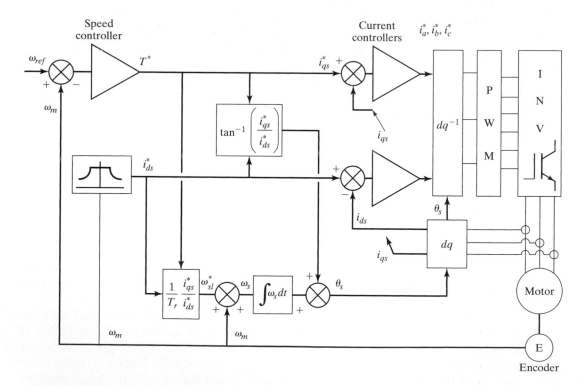

FIGURE 16.27

Indirect rotor flux oriented control scheme.

magnetizing inductance L_m can, however, cause some drift in the flux. Equation (16.101) relates the angular speed error $(\omega_{ref} - \omega_m)$ to i_{qs}^* by

$$i_{qs}^* = \frac{\hat{\Psi}_r L_r}{L_m R_r}(\omega_{ref} - \omega_m) \tag{16.108}$$

which, in turn, generates the torque component of current i_{qs}^* from the speed control loop. The slip frequency ω_{sl}^* is generated from i_{qs}^* in feed-forward manner from Eq. (16.101). The corresponding expression of slip gain K_{sl} is given by

$$K_{sl} = \frac{\omega_{sl}^*}{i_{qs}^*} = \frac{L_m}{\hat{\Psi}_r} \times \frac{R_r}{L_r} = \frac{R_r}{L_r} \times \frac{1}{i_{ds}^*} \tag{16.109}$$

The slip speed ω_{sl}^* is added to the rotor speed ω_m to obtain the stator frequency ω_s. This frequency is integrated with respect to time to produce the required angle θ_s of the stator mmf relative to the rotor flux vector. This angle is used to generate the unit vector signals ($\cos\theta_s$ and $\sin\theta_s$) and to transform the stator currents (i_{ds} and i_{qs}) to the dq reference frame. Two independent current controllers are used to regulate the i_q and i_d currents to their reference values. The compensated i_q and i_d errors are then inverse transformed into the stator a-b-c reference frame for obtaining switching signals for the inverter via PWM or hysteresis comparators.

The various components of the space vectors are shown in Figure 16.28. The d^s–q^s axes are fixed on the stator, but the d^r–q^r axes, which are fixed on the rotor, are moving at speed ω_m. Synchronously rotating axes d^e–q^e are rotating ahead of the

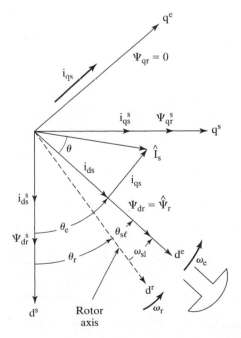

FIGURE 16.28

Phasor diagram showing space vector components for indirect vector control.

d^r-q^r axes by the positive slip angle θ_{sl} corresponding to slip frequency ω_{sl}. Because the rotor pole is directed on the d^e-axis and $\omega_s = \omega_m + \omega_{sl}$, we can write

$$\theta_s = \int \omega_s\, dt = \int (\omega_m + \omega_{sl})dt = \theta_m + \theta_{sl} \tag{16.110}$$

The rotor position θ_s is not absolute, but is slipping with respect to the rotor at frequency ω_{sl}. For decoupling control, the stator flux component of current i_{ds} should be aligned on the d^e-axis, and the torque component of current i_{qs} should be on the q^e-axis.

This method uses a feed forward scheme to generate ω_{sl}^* from i_{ds}^*, i_{qs}^*, and T_r. The rotor time constant T_r may not remain constant for all conditions of operation. Thus, with operating conditions, the slip speed ω_{sl}, which directly affects the developed torque and the rotor flux vector position, may vary widely. The indirect method requires the controller to be matched with the motor being driven. This is because the controller needs also to know some rotor parameter or parameters, which may vary according to the conditions of operation, continuously. Many rotor time constant identification schemes may be adopted to overcome this problem.

16.4.4 Director Vector Control

The air-gap flux linkages in the stator d- and q-axes are used for respective leakage fluxes to determine the rotor flux linkages in the stator reference frame. The air-gap flux linkages are measured by installing quadrature flux sensors in the air gap, as shown in Figure 16.29.

Equations (16.81) to (16.89) in the stator reference frame can be simplified to get the flux linkages as given by

$$\Psi_{qr}^s = \frac{L_r}{L_m}\Psi_{qm}^s - L_{lr}i_{qs}^s \tag{16.111}$$

$$\Psi_{dr}^s = \frac{L_r}{L_m}\Psi_{dm}^s - L_{lr}i_{ds}^s \tag{16.112}$$

where the superscript^s stands for the stator reference frame. Figure 16.30 shows the phasor diagram for space vector components. The current i_{ds} must be aligned in the

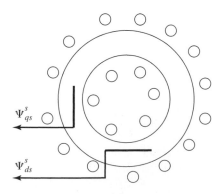

FIGURE 16.29

Quadrature sensors for air-gap flux for direct vector control.

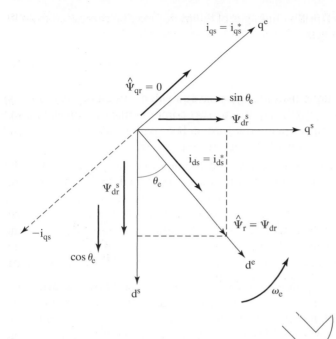

FIGURE 16.30

Phasors d^e–q^e and d^s–q^s for direct vector control.

direction of flux $\hat{\Psi}_r$ and i_{qs} must be perpendicular to $\hat{\Psi}_r$. The d^e–q^e frame is rotating at synchronous speed ω_s with respect to stationary frame d^s–q^s, and at any instant, the angular position of the d^e-axis with respect to the d^s-axis is θ_s in Eq. (16.80).

The stationary frame rotor flux vectors are given by

$$\Psi_{qr}^s = \hat{\Psi}_r \sin \theta_s \tag{16.113}$$

$$\Psi_{dr}^s = \hat{\Psi}_r \cos \theta_s \tag{16.114}$$

which give the unit vector components as

$$\cos \theta_s = \frac{\Psi_{dr}^s}{\hat{\Psi}_r} \tag{16.115}$$

$$\sin \theta_s = \frac{\Psi_{qr}^s}{\hat{\Psi}_r} \tag{16.116}$$

where

$$|\hat{\Psi}_r| = \sqrt{(\Psi_{dr}^2)^2 + (\Psi_{qr}^s)^2} \tag{16.117}$$

The flux signals Ψ_{dr}^s and Ψ_{qr}^s are generated from the machine terminal voltages and currents by using a voltage model estimator. The motor torque is controlled via the current i_{qs} and the rotor flux via the current i_{ds}. This method of control offers better low-speed performance than the IFOC. However, at high speed, the air-gap flux sensors reduce the reliability. The IFOC is normally preferred in practical applications.

However, the d- and q-axes stator flux linkages of the motor may be computed from integrating the stator input voltages.

Key Points of Section 16.4

- The vector control technique uses the dynamic equivalent circuit of the induction motor. It decouples the stator current into two components: one providing the air-gap flux and the other producing the torque. It provides independent control of flux and torque, and the control characteristic is linearized.
- The stator currents are converted to a fictitious synchronously rotating reference frame aligned with the flux vector and are transformed back to the stator frame before feeding back to the machine. The indirect method is normally preferred over the direct method.

16.5   SYNCHRONOUS MOTOR DRIVES

Synchronous motors have a polyphase winding on the stator, also known as armature, and a field winding carrying a dc current on the rotor. There are two mmfs involved: one due to the field current and the other due to the armature current. The resultant mmf produces the torque. The armature is identical to the stator of induction motors, but there is no induction in the rotor. A synchronous motor is a constant-speed machine and always rotates with zero slip at the synchronous speed, which depends on the frequency and the number of poles, as given by Eq. (16.1). A synchronous motor can be operated as a motor or generator. The PF can be controlled by varying the field current. With cycloconverters and inverters the applications of synchronous motors in variable-speed drives are widening. The synchronous motors can be classified into six types:

1. Cylindrical rotor motors
2. Salient-pole motors
3. Reluctance motors
4. Permanent-magnet motors
5. Switched reluctance motors
6. Brushless dc and ac motors

16.5.1   Cylindrical Rotor Motors

The field winding is wound on the rotor, which is cylindrical, and these motors have a uniform air gap. The reactances are independent on the rotor position. The equivalent circuit per phase, neglecting the no-load loss, is shown in Figure 16.31a, where R_a is the armature resistance per phase and X_s is the *synchronous reactance* per phase. V_f, which is dependent on the field current, is known as *excitation*, or *field* voltage.

The PF depends on the field current. The V-curves, which show the typical variations of the armature current against the excitation current, are shown in Figure 16.32. For the same armature current, the PF could be lagging or leading, depending on the excitation current I_f.

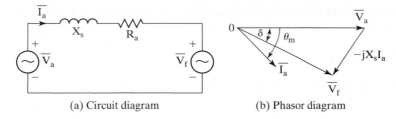

(a) Circuit diagram   (b) Phasor diagram

FIGURE 16.31

Equivalent circuit of synchronous motors.

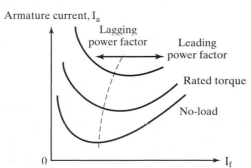

FIGURE 16.32

Typical V-curves of synchronous motors.

If θ_m is the lagging PF angle of the motor, Figure 16.31a gives

$$\overline{V}_f = V_a \underline{/0} - \overline{I}_a(R_a + jX_s)$$

$$= V_a \underline{/0} - I_a(\cos \theta_m - j \sin \theta_m)(R_a + jX_s) \tag{16.118}$$

$$= V_a - I_a X_s \sin \theta_m - I_a R_a \cos \theta_m - j I_a(X_s \cos \theta_m - R_a \sin \theta_m) \tag{16.119a}$$

$$= V_f \underline{/\delta} \tag{16.119b}$$

where

$$\delta = \tan^{-1} \frac{-(I_a X_s \cos \theta_m - I_a R_a \sin \theta_m)}{V_a - I_a X_s \sin \theta_m - I_a R_a \cos \theta_m} \tag{16.120}$$

and

$$V_f = [(V_a - I_a X_s \sin \theta_m - I_a R_a \cos \theta_m)^2 + (I_a X_s \cos \theta_m - I_a R_a \sin \theta_m)^2]^{1/2} \tag{16.121}$$

The phasor diagram in Figure 16.31b yields

$$\overline{V}_f = V_f(\cos \delta + j \sin \delta) \tag{16.122}$$

$$\overline{I}_a = \frac{\overline{V}_a - \overline{V}_f}{R_a + jX_s} = \frac{[V_a - V_f(\cos \delta + j \sin \delta)](R_a - jX_s)}{R_a^2 + X_s^2} \tag{16.123}$$

The real part of Eq. (16.123) becomes

$$I_a \cos \theta_m = \frac{R_a(V_a - V_f \cos \delta) - V_f X_s \sin \delta}{R_a^2 + X_s^2} \qquad (16.124)$$

The input power can be determined from Eq. (16.124),

$$
\begin{aligned}
P_i &= 3 V_a I_a \cos \theta_m \\
&= \frac{3[R_a(V_a^2 - V_a V_f \cos \delta) - V_a V_f X_s \sin \delta]}{R_a^2 + X_s^2}
\end{aligned}
\qquad (16.125)
$$

The stator (or armature) copper loss is

$$P_{su} = 3 I_a^2 R_a \qquad (16.126)$$

The gap power, which is the same as the developed power, is

$$P_d = P_g = P_i - P_{su} \qquad (16.127)$$

If ω_s is the synchronous speed, which is the same as the rotor speed, the developed torque becomes

$$T_d = \frac{P_d}{\omega_s} \qquad (16.128)$$

If the armature resistance is negligible, T_d in Eq. (16.128) becomes

$$T_d = -\frac{3 V_a V_f \sin \delta}{X_s \omega_s} \qquad (16.129)$$

and Eq. (16.120) becomes

$$\delta = -\tan^{-1} \frac{I_a X_s \cos \theta_m}{V_a - I_a X_s \sin \theta_m} \qquad (16.130)$$

For motoring δ is negative and torque in Eq. (16.129) becomes positive. In the case of generating, δ is positive and the power (and torque) becomes negative. The angle δ is called the *torque angle*. For a fixed voltage and frequency, the torque depends on the angle δ and is proportional to the excitation voltage V_f. For fixed values of V_f and δ, the torque depends on the voltage-to-frequency ratio and a constant volts/hertz control can provide speed control at a constant torque. If V_a, V_f, and δ remain fixed, the torque decreases with the speed and the motor operates in the field weakening mode.

If $\delta = 90°$, the torque becomes maximum and the maximum developed torque, which is called the *pull-out torque*, becomes

$$T_p = T_m = -\frac{3 V_a V_f}{X_s \omega_s} \qquad (16.131)$$

The plot of developed torque against the angle δ is shown in Figure 16.33. For stability considerations, the motor is operated in the positive slope of T_d–δ characteristics and this limits the range of torque angle, $-90° \leq \delta \leq 90°$.

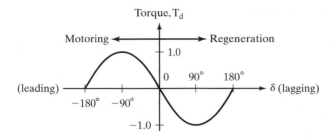

FIGURE 16.33

Torque versus torque angle with cylindrical rotor.

Example 16.8 Finding the Performance Parameters of a Cylindrical Rotor Synchronous Motor

A three-phase, 460-V, 60-Hz, six-pole, Y-connected cylindrical rotor synchronous motor has a synchronous reactance of $X_s = 2.5\ \Omega$ and the armature resistance is negligible. The load torque, which is proportional to the speed squared, is $T_L = 398\ \text{N}\cdot\text{m}$ at 1200 rpm. The PF is maintained at unity by field control and the voltage-to-frequency ratio is kept constant at the rated value. If the inverter frequency is 36 Hz and the motor speed is 720 rpm, calculate (a) the input voltage V_a, (b) the armature current I_a, (c) the excitation voltage V_f, (d) the torque angle δ, and (e) the pull-out torque T_p.

Solution

$PF = \cos\theta_m = 1.0$, $\theta_m = 0$, $V_{a(\text{rated})} = V_b = V_s = 460/\sqrt{3} = 265.58\ \text{V}$, $p = 6$, $\omega = 2\pi \times 60 = 377\ \text{rad/s}$, $\omega_b = \omega_s = \omega_m = 2 \times 377/6 = 125.67\ \text{rad/s}$ or 1200 rpm, and $d = V_b/\omega_b = 265.58/125.67 = 2.1133$. At 720 rpm,

$$T_L = 398 \times \left(\frac{720}{1200}\right)^2 = 143.28\ \text{N}\cdot\text{m} \quad \omega_s = \omega_m = 720 \times \frac{\pi}{30} = 75.4\ \text{rad/s}$$

$$P_0 = 143.28 \times 75.4 = 10{,}803\ \text{W}$$

a. $V_a = d\omega_s = 2.1133 \times 75.4 = 159.34\ \text{V}$.

b. $P_0 = 3V_a I_a\ PF = 10{,}803$ or $I_a = 10{,}803/(3 \times 159.34) = 22.6\ \text{A}$.

c. From Eq. (16.118),

$$\overline{V}_f = 159.34 - 22.6 \times (1 + j0)(j2.5) = 169.1\ \underline{/-19.52°}$$

d. The torque angle, $\delta = -19.52°$.

e. From Eq. (16.131),

$$T_p = \frac{3 \times 159.34 \times 169.1}{2.5 \times 75.4} = 428.82\ \text{N}\cdot\text{m}$$

16.5.2 Salient-Pole Motors

The armature of salient-pole motors is similar to that of cylindrical rotor motors. However, due to saliency, the air gap is not uniform and the flux is dependent on the position of the rotor. The field winding is normally wound on the pole pieces. The armature current and the reactances can be resolved into direct and quadrature axis components.

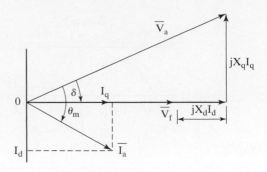

FIGURE 16.34

Phase diagram for salient-pole synchronous motors.

I_d and I_q are the components of the armature current in the direct (or d) axis and quadrature (or q) axis, respectively. X_d and X_q are the d-axis reactance and q-axis reactance, respectively. Using Eq. (16.118), the excitation voltage becomes

$$\overline{\mathbf{V}}_f = \overline{\mathbf{V}}_a - jX_d\overline{\mathbf{I}}_d - jX_q\overline{\mathbf{I}}_q - R_a\overline{\mathbf{I}}_a$$

For negligible armature resistance, the phasor diagram is shown in Figure 16.34. From the phasor diagram,

$$I_d = I_a \sin(\theta_m - \delta) \tag{16.132}$$

$$I_q = I_a \cos(\theta_m - \delta) \tag{16.133}$$

$$I_d X_d = V_a \cos \delta - V_f \tag{16.134}$$

$$I_q X_q = V_a \sin \delta \tag{16.135}$$

Substituting I_q from Eq. (16.133) in Eq. (16.135), we have

$$\begin{aligned} V_a \sin \delta &= X_q I_a \cos(\theta_m - \delta) \\ &= X_q I_a(\cos \delta \cos \theta_m + \sin \delta \sin \theta_m) \end{aligned} \tag{16.136}$$

Dividing both sides by $\cos \delta$ and solving for δ gives

$$\delta = -\tan^{-1} \frac{I_a X_q \cos \theta_m}{V_a - I_a X_q \sin \theta_m} \tag{16.137}$$

where the negative sign signifies that V_f lags V_a. If the terminal voltage is resolved into a d-axis and a q-axis,

$$V_{ad} = -V_a \sin \delta \text{ and } V_{aq} = V_a \cos \delta$$

The input power becomes

$$\begin{aligned} P &= -3(I_d V_{ad} + I_q V_{aq}) \\ &= 3I_d V_a \sin \delta - 3I_q V_a \cos \delta \end{aligned} \tag{16.138}$$

Substituting I_d from Eq. (16.134) and I_q from Eq. (16.135) in Eq. (16.138) yields

$$P_d = -\frac{3V_a V_f}{X_d} \sin \delta - \frac{3V_a^2}{2} \left[\frac{X_d - X_q}{X_d X_q} \sin 2\delta \right] \tag{16.139}$$

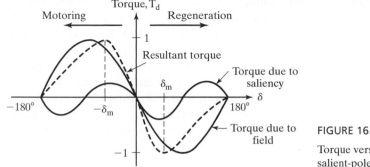

FIGURE 16.35

Torque versus torque angle with salient-pole rotor.

Dividing Eq. (16.139) by speed gives the developed torque as

$$T_d = -\frac{3 V_a V_f}{X_d \omega_s} \sin \delta - \frac{3 V_a^2}{2\omega_s} \left[\frac{X_d - X_q}{X_d X_q} \sin 2\delta \right] \tag{16.140}$$

The torque in Eq. (16.140) has two components. The first component is the same as that of cylindrical rotor if X_d is replaced by X_s and the second component is due to the rotor saliency. The typical plot of T_d against torque angle is shown in Figure 16.35, where the torque has a maximum value at $\delta = \pm \delta_m$. For stability the torque angle is limited in the range of $-\delta_m \leq \delta \leq \delta_m$ and in this stable range, the slope of T_d–δ characteristic is higher than that of cylindrical rotor motor.

16.5.3 Reluctance Motors

The reluctance motors are similar to the salient-pole motors, except that there is no field winding on the rotor. The armature circuit, which produces rotating magnetic field in the air gap, induces a field in the rotor that has a tendency to align with the armature field. The reluctance motors are very simple and are used in applications where a number of motors are required to rotate in synchronism. These motors have low-lagging PF, typically in the range 0.65 to 0.75.

With $V_f = 0$, Eq. (16.140) can be applied to determine the reluctance torque,

$$T_d = -\frac{3 V_a^2}{2\omega_s} \left[\frac{X_d - X_q}{X_d X_q} \sin 2\delta \right] \tag{16.141}$$

where

$$\delta = -\tan^{-1} \frac{I_a X_q \cos \theta_m}{V_a - I_a X_q \sin \theta_m} \tag{16.142}$$

The pull-out torque for $\delta = -45°$ is

$$T_p = \frac{3 V_a^2}{2\omega_s} \left[\frac{X_d - X_q}{X_d X_q} \right] \tag{16.143}$$

16.5.4 Permanent-Magnet Motors

The permanent-magnet motors are similar to salient-pole motors, except that there is no field winding on the rotor and the field is provided by mounting permanent magnets at the rotor. The excitation voltage cannot be varied. For the same frame size, perma-nent-magnet motors have higher pull-out torque. The equations for the salient-pole motors may be applied to permanent-magnet motors if the excitation voltage V_f is as-sumed constant. The elimination of field coil, dc supply, and slip rings reduces the motor loss and the complexity. These motors are also known as *brushless motors* and finding increased applications in robots and machine tools. A permanent-magnet motor can be fed by either rectangular current or sinusoidal current. The rectangular current-fed motors, which have concentrated windings on the stator inducing a square or trapezoidal voltage, are normally used in low-power drives. The sinusoidal current-fed motors, which have distributed windings on the stator, provide smoother torque and are normally used in high-power drives.

Example 16.9 Finding the Performance Parameters of a Reluctance Motor

A three-phase, 230-V, 60-Hz, four-pole, Y-connected reluctance motor has $X_d = 22.5 \, \Omega$ and $X_q = 3.5 \, \Omega$. The armature resistance is negligible. The load torque is $T_L = 12.5 \, \text{N} \cdot \text{m}$. The volt-age-to-frequency ratio is maintained constant at the rated value. If the supply frequency is 60 Hz, determine (a) the torque angle δ, (b) the line current I_a, and (c) the input PF.

Solution

$T_L = 12.5 \, \text{N} \cdot \text{m}$, $V_{a(\text{rated})} = V_b = 230/\sqrt{3} = 132.79 \, \text{V}$, $p = 4$, $\omega = 2\pi \times 60 = 377 \, \text{rad/s}$, $\omega_b = \omega_s = \omega_m = 2 \times 377/4 = 188.5 \, \text{rad/s}$ or 1800 rpm, and $V_a = 132.79 \, \text{V}$.

a. $\omega_s = 188.5 \, \text{rad/s}$. From Eq. (16.141),

$$\sin 2\delta = -\frac{12.5 \times 2 \times 188.5 \times 22.5 \times 3.5}{3 \times 132.79^2 \times (22.5 - 3.5)}$$

and $\delta = -10.84°$.

b. $P_0 = 12.5 \times 188.5 = 2356 \, \text{W}$. From Eq. (16.142),

$$\tan(10.84°) = \frac{3.5 I_a \cos \theta_m}{132.79 - 3.5 I_a \sin \theta_m}$$

and $P_0 = 2356 = 3 \times 132.79 I_a \cos \theta_m$. From these two equations, I_a and θ_m can be de-termined by an iterative method of solution, which yields $I_a = 9.2 \, \text{A}$ and $\theta_m = 49.98°$.

c. PF $= \cos(49.98°) = 0.643$.

16.5.5 Switched Reluctance Motors

A switched reluctance motor (SRM) is a variable reluctance step motor. A cross-sectional view is shown in Figure 16.36a. Three phases ($q = 3$) are shown with six stator teeth, $N_s = 6$, and four rotor teeth, $N_r = 4$. N_r is related to N_s and q by $N_r = N_s \pm N_s/q$. Each phase winding is placed on two diametrically opposite teeth. If phase A is excited by a current i_a, a torque is developed and it causes a pair of rotor poles to be magnetically aligned with the poles of phase A. If the subsequent phase B and phase C are excited in sequence, a further rotation can take place. The motor speed can be varied by exciting in sequence phases A, B, and C. A commonly used circuit to drive an SRM is shown in Figure 16.36b. An absolute position sensor is usually required

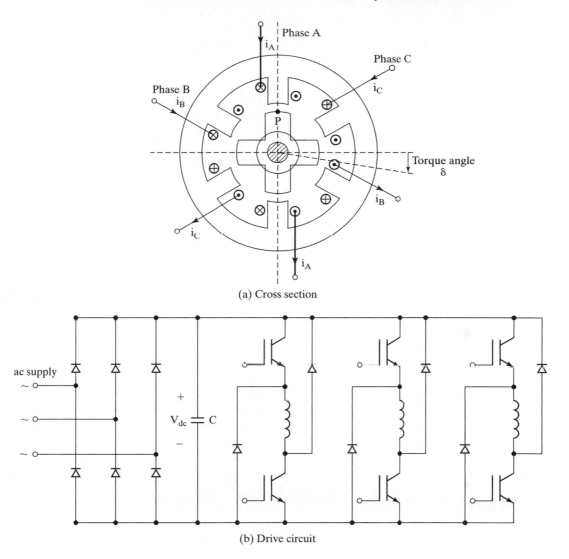

(a) Cross section

(b) Drive circuit

FIGURE 16.36

Switched reluctance motor.

to directly control the angles of stator excitation with respect to the rotor position. A position feedback control is provided in generating the gating signals. If the switching takes place at a fixed rotor position relative to the rotor poles, an SRM would exhibit the characteristics of a dc series motor. By varying the rotor position, a range of operating characteristics can be obtained [16, 17].

16.5.6 Closed-Loop Control of Synchronous Motors

The typical characteristics of torque, current, and excitation voltage against frequency ratio β are shown in Figure 16.37a. There are two regions of operation: constant torque

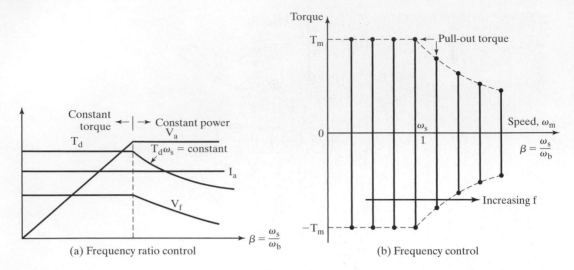

FIGURE 16.37

Torque–speed characteristics of synchronous motors.

and constant power. In the constant-torque region, the volts/hertz is maintained constant, and in the constant-power region, the torque decreases with frequency. Speed–torque characteristics for different frequencies are shown in Figure 16.37b. Similar to induction motors, the speed of synchronous motors can be controlled by varying the voltage, frequency, and current. There are various configurations for closed loop control of synchronous motors. A basic arrangement for constant volts/hertz control of synchronous motors is shown in Figure 16.38, where the speed error generates the frequency and voltage command for the PWM inverter. Because the speed of synchronous motors depends on the supply frequency only, they are employed in multimotor

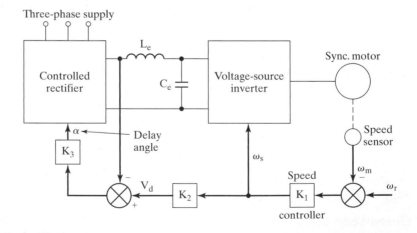

FIGURE 16.38

Volts/hertz control of synchronous motors.

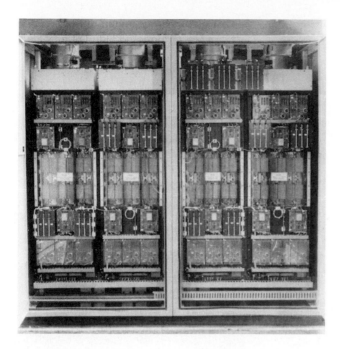

FIGURE 16.39

A 3.3-MW synchronous motor compressor drive. (Reproduced by permission of Brush Electrical Machines Ltd., England.)

drives requiring accurate speed tracking between motors, as in fiber spinning mills, paper mills, textile mills, and machine tools. A 3.3-kW soft-start system for synchronous motor compressor drives is shown in Figure 16.39, where two three-phase stacks are connected in parallel to give 12-pulse operation. The control cubicle for the drive in Figure 16.39 is shown in Figure 16.40.

16.5.7   Brushless Dc and Ac Motor Drives

Brushless drives are basically synchronous motor drives in self-control mode. The armature supply frequency is changed in proportion to the rotor speed changes so that the armature field always moves at the same speed as the rotor. The self-control ensures that for all operating points the armature and rotor fields move exactly at the same speed. This prevents the motor from pulling out of step, hunting oscillations, and instability due to a step change in torque or frequency. The accurate tracking of the speed is normally realized with a rotor position sensor. The PF can be maintained at unity by varying the field current. The block diagrams of a self-controlled synchronous motor fed from a three-phase inverter or a cycloconverter are shown in Figure 16.41.

For an inverter-fed drive, as shown in Figure 16.41a, the input source is dc. Depending on the type of inverter, the dc source could be current source, a constant current, or a controllable voltage source. The inverter's frequency is changed in proportion to the speed so that the armature and rotor mmf waves revolve at the same speed, thereby producing a steady torque at all speeds, as in a dc motor. The rotor position and inverter perform the same function as the brushes and commutator in a dc motor. Due to the similarity in operation with a dc motor, an inverter-fed self-controlled synchronous motor is known as a *commutatorless dc motor*. If the synchronous motor

FIGURE 16.40

Control cubicle for the drive in
Figure 16.39. (Reproduced by permission of
Brush Electrical Machines Ltd., England.)

is a permanent-magnet motor, a reluctance motor, or a wound field motor with a
brushless excitation, it is known as a *brushless* and *commutatorless dc motor* or simply
a *brushless dc motor*. Connecting the field in series with the dc supply gives the characteristics of a dc series motor. The brushless dc motors offer the characteristics of dc motors and do not have limitations such as frequent maintenance and inability to operate
in explosive environments. They are finding increasing applications in servo drives [18].

If the synchronous motor is fed from an ac source shown in Figure 16.41b, it is
called a *brushless* and *commutatorless ac motor* or simply *a brushless ac motor*. These
ac motors are used for high-power applications (up to megawatt range) such as com-
pressors, blowers, fans, conveyers, steel rolling mills, large ship steering, and cement
plants. The self-control is also used for starting large synchronous motors in gas turbine
and pump storage power plants.

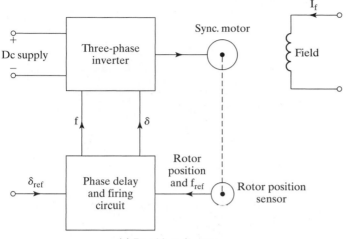

(a) Brushless dc motor

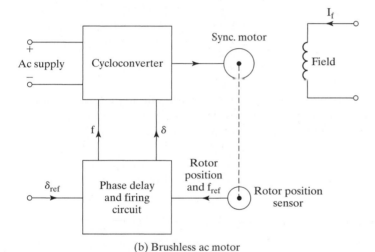

(b) Brushless ac motor

FIGURE 16.41

Self-controlled synchronous motors.

Key Points of Section 16.5

- A synchronous motor is a constant-speed machine and always rotates with zero slip at the synchronous speed.
- The torque is produced by the armature current in the stator winding and the field current in the rotor winding.
- The cycloconverters and inverters are used for applications of synchronous motors in variable-speed drives.

16.6   STEPPER MOTOR CONTROL

Stepper motors are electromechanical motion devices, which are used primarily to convert information in digital form to mechanical motion [19, 20]. These motors rotate at a

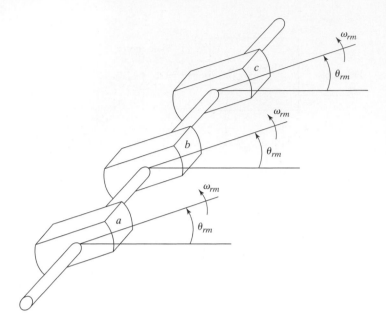

FIGURE 16.42

Rotor of two-pole, three-stack
variable-reluctance stepper motors.

predetermined angular displacement in response to a logic input. Whenever stepping
from one position to another is required, the stepper motors are generally used. They
are found as the drivers for the paper in line printers and in other computer peripheral
equipment such as in positioning of the magnetic-disk head.

Stepper motors fall into two types: (1) the variable-reluctance stepper motor and (2)
the permanent-magnet stepper motor. The operating principle of the variable-reluctance
stepper motor is much the same as that of the synchronous reluctance machine, and the
permanent-magnet stepper motor is similar in principle to the permanent-magnet synchronous machine.

16.6.1 Variable-Reluctance Stepper Motors

These motors can be used as a single unit or as a multistack. In the multistack operation, three or more single-phase reluctance motors are mounted on a common shaft
with their stator magnetic axes displaced from each other. The rotor of a three-stack is
shown in Figure 16.42. It has three cascaded two-pole rotors with a minimum-reluctance
path of each aligned at the angular displacement θ_{rm}. Each of these rotors has a separate, single-phase stator with the magnetic axes of the stators displaced from each
other. The corresponding stators are shown in Figure 16.43.

Each stator has two poles with stator-winding wound around both poles. Positive
current flows into as_1 and out as_1', which is connected to as_2 so that positive current
flows into as_2 and out as_2'. Each winding can have several turns; and the number of
turns from as_1 to as_1' is $N N_s/2$, which is the same as that from as_2 to as_2'. θ_{rm} is referenced from the minimum-reluctance path from the as-axis.

If the windings bs and cs are open circuited and the winding as is excited with
a dc voltage, a constant current i_{as} can be immediately established. The rotor a can
be aligned to the as-axis and $\theta_{rm} = 0$ or 180°. If the as winding is instantaneously de-
energized and the bs winding is energized with a direct current, rotor b aligns itself

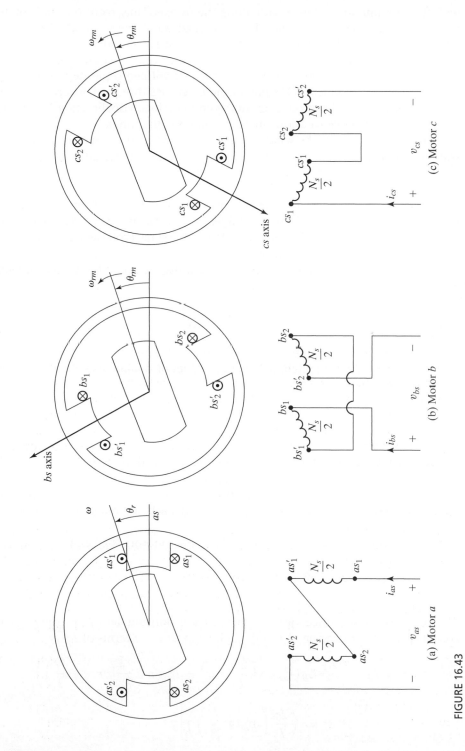

FIGURE 16.43

Stator configurations of two-pole, three-stack, variable-reluctance stepper motors.

with the minimum-reluctance path along the *bs*-axis. Thus, rotor *b* would rotate clockwise from $\theta_{rm} = 0$ to $\theta_{rm} = -60°$. However, instead of energizing the *bs* winding, if we energize the *cs* winding with a direct current, the rotor *c* aligns itself with the minimum-reluctance path along the *cs*-axis. Thus, the rotor *c* would rotate anticlockwise from $\theta_{rm} = 0$ to $\theta_{rm} = 60°$. Thus, applying a dc voltage separately in the sequences *as*, *bs*, *cs*, *as*, ... produces 60° steps in the clockwise direction, whereas the sequence *as*, *cs*, *bs*, *as*, ... produces 60° steps in the counterclockwise direction. We need at least three stacks to achieve rotation (stepping) in both directions.

If both the *as* and *bs* windings are energized at the same time, initially the *as* winding is energized with $\theta_{rm} = 0$ and the *bs* winding is energized without de-energizing the *as* winding. The rotor rotates clockwise from $\theta_{rm} = 0$ to $\theta_{rm} = -30°$. The step length is reduced by one-half. This is referred to as *half-step* operation. A stepper motor is a discrete device, operated by switching a dc voltage from one stator winding to the other. Each stack is often called a *phase*. In other words, a three-stack machine is a three-phase machine. Although more than three stacks (phases) as high as seven may be used, three-stack stepper motors are quite common.

The tooth pitch (T_p), which is the angular displacement between root teeth, is related to the rotor teeth per stack R_T by

$$T_p = \frac{2\pi}{R_T} \tag{16.144}$$

If we energize each stack separately, then going from *as* to *bs* to *cs* back to *as* causes the rotor to rotate one tooth pitch. If N is the number of stacks (phases), the step length S_L is related to T_p by

$$S_L = \frac{T_P}{N} = \frac{2\pi}{N R_T} \tag{16.145}$$

For $N = 3, R_T = 4, T_P = 2\pi/4 = 90°, S_L = 90°/3 = 30°$; and an *as*, *bs*, *cs*, *as*, ... sequence produces 30° steps in the clockwise direction. For $N = 3, R_T = 8, T_P = 2\pi/8 = 45°, S_L = 45°/3 = 15°$; and an *as*, *cs*, *bs*, *as*, ... sequence produces 15° steps in the counterclockwise direction. Therefore, by increasing the number of rotor teeth reduces the step length. The step lengths of multistack stepping motors typically range from 2° to 15°.

The torque developed by multistack stepper motors is given by

$$T_d = -\frac{R_T}{2}L_B[i_{as}^2 \sin(R_T\theta_{rm}) + i_{bs}^2 \sin(R_T(\theta_{rm} \pm S_L)) + i_{cs}^2 \sin(R_T(\theta_{rm} \pm S_L))] \tag{16.146}$$

The stator self-inductance varies with the rotor position, and L_B is the peak value at $\cos(p\theta_{rm}) = \pm 1$. Equation (16.146) can be expressed in terms of T_p as

$$T_d = -\frac{R_T}{2}L_B\left[i_{as}^2 \sin\left(\frac{2\pi}{T_p}\theta_{rm}\right) + i_{bs}^2 \sin\left(\frac{2\pi}{T_p}\left(\theta_{rm} \pm \frac{T_p}{3}\right)\right)\right.$$

$$\left. + i_{cs}^2 \sin\left(\frac{2\pi}{T_p}\left(\theta_{rm} \pm \frac{T_p}{3}\right)\right)\right] \tag{16.147}$$

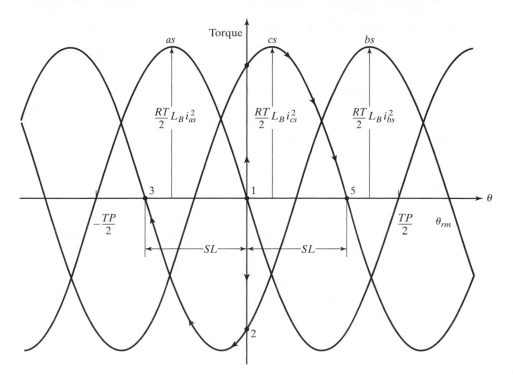

FIGURE 16.44

Steady-state torque components against the rotor angle θ_{rm} for a three-stack motor.

which indicates that the magnitude of the torque is proportional to the number of rotor teeth per stack R_T. The steady-state torque components in Eq. (116.146) against θ_{rm} are shown in Figure 16.44.

16.6.2 Permanent-Magnet Stepper Motors

The permanent-magnet stepper motor is also quite common. It is a permanent-magnet synchronous machine and it may be operated either as a stepping motor or as a continuous-speed device. However, we concern ourselves only with its application as a stepping motor.

The cross section of a two-pole, two-phase permanent-magnet stepper motor is shown in Figure 16.45. For explaining the stepping action, let us assume that the *bs* winding is open-circuited and apply a constant positive current through the *as* winding. As a result, this current establishes a stator south pole at the stator tooth on which the as_1 winding is wound, and stator north pole is established at the stator tooth on which the as_2 winding is wound. The rotor would be positioned at $\theta_{rm} = 0$. Now let us simultaneously de-energize the *as* winding while energizing the *bs* winding with a positive current. The rotor moves one step length in the counterclockwise direction. To continue stepping in the counterclockwise direction, the *bs* winding is de-energized and the *as* winding is energized with a negative current. That is, counterclockwise stepping occurs

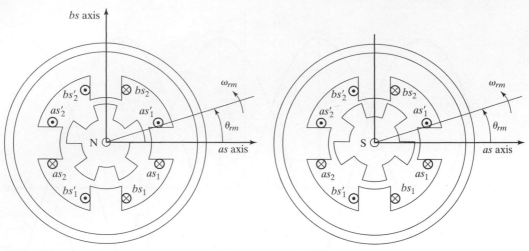

(a) Axial cross section with north–pole   (b) Axial cross section with south north–pole

FIGURE 16.45

Cross section of a two-pole, two-phase permanent-magnet stepper motor.

with a current sequence of i_{as}, $i_{bs} - i_{as}$, $-i_{bs}i_{as}$, i_{bs}.... Clockwise rotation is achieved with a current sequence of i_{as}, $-i_{bs} - i_{as}$, $i_{bs}i_{as}$, $-i_{bs}$....

A counterclockwise rotation is achieved by a sequence of i_{as}, $i_{bs} - i_{as}$, $-i_{bs}i_{as}$, i_{bs}.... Thus, it takes four switching (steps) to advance the rotor one tooth pitch. If N is the number of phases, the step length S_L is related to T_p by

$$S_L = \frac{T_P}{2N} = \frac{\pi}{N\,R_T} \qquad (16.148)$$

For $N = 2$, $R_T = 5$, $T_P = 2\pi/5 = 72°$, $S_L = 72°/(2 \times 2) = 18°$. Therefore, increasing the number of phases and rotor teeth reduces the step length. The step lengths typically range from 2° to 15°. Most permanent-magnet stepper motors have more than two poles and more than 5 rotor teeth; some may have as many as eight poles and as many as 50 rotor teeth.

The torque developed by a permanent-magnet stepper motor is given by

$$T_d = -R_T\lambda'_m[i_{as} \sin(R_T\theta_{rm}) - i_{bs} \sin(R_T\theta_{rm})] \qquad (16.149)$$

where λ'_m is the amplitude of the flux linkages established by the permanent magnet as viewed from the stator phase windings. It is the constant inductance times a constant current. In other words, the magnitude of λ'_m is proportional to the magnitude of the open-circuit sinusoidal voltage induced in each stator phase winding.

The plots of the torque components in Eq. (16.149) are shown in Figure 16.46. The term $\pm T_{d(am)}$ is the torque due to the interaction of the permanent magnet and $\pm i_{as}$, and the term $\pm T_{d(bm)}$ is the torque due to the interaction of the permanent magnet and $\pm i_{bs}$. The reluctance of the permanent magnet is large, approaching that of air. Because the flux established by the phase currents flows through the magnet, the reluctance of the

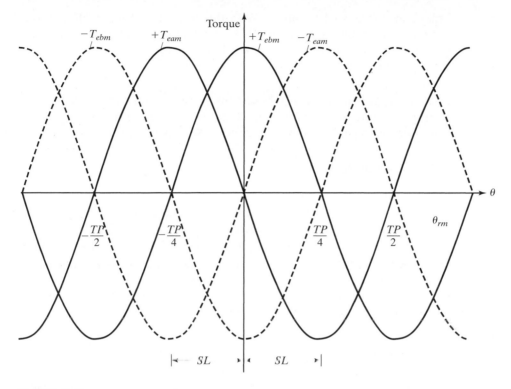

FIGURE 16.46

Steady-state torque components against the rotor angle θ_{rm} for a permanent-magnet stepper motor with constant phase currents.

flux path is relatively large. Hence, the variation in the reluctance due to rotation of the rotor is small and, consequently, the amplitudes of the reluctance torques are small relative to the torque developed by the interaction between the magnet and the phase currents. For these reasons, the reluctance torques are generally neglected.

Notes:

1. For a permanent-magnet stepper motor, it is necessary for the phase currents to flow in both directions to achieve rotation. For a variable-reluctance stepper motor, it is not necessary to reverse the direction of the current in the stator windings to achieve rotation and, therefore, the stator voltage source need only be unidirectional.

2. Generally, stepper motors are supplied from a dc voltage source; hence, the power converter between the phase windings and the dc source must be bidirectional; that is, it must have the capability of applying a positive and negative voltage to each phase winding.

3. Permanent-magnet stepper motors are often equipped with what is referred to as *bifilar* windings. Instead of only one winding on each stator tooth, there are two identical windings with one wound opposite to the other, each having separate

independent external terminals. With this type of winding configuration the direction of the magnetic field established by the stator windings is reversed, not by changing the direction of the current but by reversing the sense of the winding through which current is flowing. This will, however, increase the size and weight of the stepper motor.

4. Hybrid stepper motors [21] whose construction is a hybrid between permanent magnet and reluctance motor topologies broaden the range of applications and offer enhanced performance with simpler and lower cost of power converters.

Key Points of Section 16.6

- Stepper motors are electromechanical motion devices, which are used primarily to convert information in digital form to mechanical motion. Stepper motors are synchronous machines that are operated as stepping motors.

- Stepper motors fall into two types: the variable-reluctance stepper motor and the permanent-magnet stepper motor. A variable-reluctance stepper motor requires only unidirectional current flow. Whereas, a permanent-magnet stepper motor requires bidirectional flow unless the motor has *bifilar* stator windings.

SUMMARY

Although ac drives require advanced control techniques for control of voltage, frequency, and current, they have advantages over dc drives. The voltage and frequency can be controlled by voltage-source inverters. The current and frequency can be controlled by current-source inverters. The slip power recovery schemes use controlled rectifiers to recover the slip power of induction motors. The most common method of closed-loop control of induction motors is volts/hertz, flux, or slip control. Both squirrel-cage and wound-rotor motors are used in variable-speed drives. A voltage-source inverter can supply a number of motors connected in parallel, whereas a current-source inverter can supply only one motor.

Synchronous motors are constant-speed machines and their speeds can be controlled by voltage, frequency, or current. Synchronous motors are of six types: cylindrical rotors, salient poles, reluctance, permanent magnets, switched reluctance motors, and brushless dc and ac motors. Whenever stepping from one position to another is required, the stepper motors are generally used. Synchronous motors can be operated as stepper motors. Stepper motors fall into two types: the variable-reluctance stepper motor and the permanent-magnet stepper motor. Due to the pulsating nature of the converter voltages and currents, it requires special specifications and design of motors for variable-speed applications [22]. There is abundant literature on ac drives, so only the fundamentals are covered in this chapter.

REFERENCES

[1] H. S. Rajamani and R. A. McMahon, "Induction motor drives for domestic appliances," *IEEE Industry Applications Magazine*, Vol. 3, No. 3, May/June 1997, pp. 21–26.

[2] D. G. Kokalj, "Variable frequency drives for commercial laundry machines," *IEEE Industry Applications Magazine*, Vol. 3, No. 3, May/June 1997, pp. 27–36.

[3]   M. F. Rahman, D. Patterson, A. Cheok, and R. Betts, *Power Electronics Handbook*, edited by M. H. Rashid. San Diego, CA: Academic Press. 2001, Chapter 27—Motor Drives.

[4]   B. K. Bose, *Modern Power Electronics and AC Drives*. Upper Saddle River, NJ: Prentice-Hall. 2002, Chapter 8—Control and Estimation of Induction Motor Drives.

[5]   R. Krishnan, *Electric Motor Drives: Modeling, Analysis, and Control*. Upper Saddle River, NJ: Prentice-Hall. 1998, Chapter 8—Stepper Motors.

[6]   I. Boldea and S. A. Nasar, *Electric Drives*. Boca Raton, FL: CRC Press, 1999.

[7]   M. A. El-Sharkawi, *Fundamentals of Electric Drives*. Pacific Grove, CA: Brooks/Cole. 2000.

[8]   S. B. Dewan, G. B. Slemon, and A. Straughen, *Power Semiconductor Drives*. New York: John Wiley & Sons. 1984.

[9]   A. von Jouanne, P. Enjeiti, and W. Gray, "Application issues for PWM adjustable speed ac motors," *IEEE Industry Applications Magazine*, Vol. 2, No. 5, September/October 1996, pp. 10–18.

[10]   S. Shashank and V. Agarwal, "Simple control for wind-driven induction generator," *IEEE Industry Applications Magazine*, Vol. 7, No. 2, March/April 2001, pp. 44–53.

[11]   W. Leonard, *Control of Electrical Drives*. New York: Springer-Verlag. 1985.

[12]   D. W. Novotny and T. A. Lipo, *Vector Control and Dynamics of Drives*. Oxford, UK: Oxford Science Publications, 1996.

[13]   P. Vas, *Electrical Machines and Drives: A Space Vector Theory Approach*. London, UK: Clarendon Press, 1992.

[14]   N. Mohan, *Electric Drives: An Integrative Approach*. Minneapolis, MN: MNPERE, 2000.

[15]   E. Y. Y. Ho and P. C. Sen, "Decoupling control of induction motors," *IEEE Transactions on Industrial Electronics*, Vol. 35, No. 2, May 1988, pp. 253–262.

[16]   T. J. E. Miller, *Switched Reluctance Motors*. London, UK: Oxford Science. 1992.

[17]   C. Pollock and A. Michaelides, "Switched reluctance drives: A comprehensive evaluation," *Power Engineering Journal*, December 1995, pp. 257–266.

[18]   N. Matsui, "Sensorless PM brushless DC motor drives," *IEEE Transactions on Industrial Electronics*, Vol. 43, No. 2, April 1996, pp. 300–308.

[19]   H.-D. Chai, *Electromechanical Motion Devices*. Upper Saddle River, NJ: Prentice Hall. 1998, Chapter 8—Stepper Motors.

[20]   P. C. Krause and O. Wasynczukm, *Electromechanical Motion Devices*. New York: McGraw-Hill. 1989.

[21]   J. D. Wale and C. Pollack, "Hybrid stepping motors," *Power Engineering Journal*, Vol. 15, No. 1, February 2001, pp. 5–12.

[22]   J. A. Kilburn and R. G. Daugherty, "NEMA design E motors and controls—What's it all about," *IEEE Industry Applications Magazine*, Vol. 5, No. 4, July/August 1999, pp. 26–36.

REVIEW QUESTIONS

16.1   What are the types of induction motors?

16.2   What is a synchronous speed?

16.3   What is a slip of induction motors?

16.4   What is a slip frequency of induction motors?

16.5   What is the slip at starting of induction motors?

16.6   What are the torque–speed characteristics of induction motors?

16.7   What are various means for speed control of induction motors?

16.8   What are the advantages of volts/hertz control?

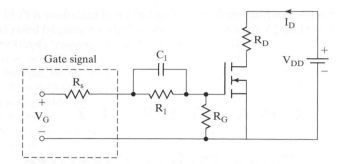

FIGURE 17.1

Fast-turn-on gate circuit.

shown in Figure 17.1 to charge the gate capacitance faster. When the gate voltage is turned on, the initial charging current of the capacitance is

$$I_G = \frac{V_G}{R_S} \tag{17.1}$$

and the steady-state value of gate voltage is

$$V_{GS} = \frac{R_G V_G}{R_S + R_1 + R_G} \tag{17.2}$$

where R_s is the internal resistance of gate drive source.

To achieve switching speeds on the order of 100 ns or less, the gate drive circuit should have a low-output impedance and the ability to sink and source relatively large currents. A totem pole arrangement that is capable of sourcing and sinking a large current is shown in Figure 17.2. The *PNP*- and *NPN*-transistors act as emitter followers and offer a low-output impedance. These transistors operate in the linear region instead of the saturation mode, thereby minimizing the delay time. The gate signal for the power MOSFET may be generated by an op-amp. Feedback via the capacitor C regulates the rate of rise and fall of the gate voltage, thereby controlling the rate of rise and fall of the MOSFET drain current. A diode across the capacitor C allows the gate voltage

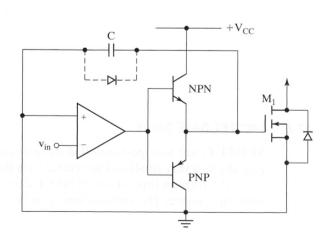

FIGURE 17.2

Totem pole arrangement gate drive with pulse-edge shaping.

to change rapidly in one direction only. There are a number of integrated drive circuits on the market that are designed to drive transistors and are capable of sourcing and sinking large currents for most converters. The totem pole arrangement in gate drive ICs is normally made of two MOSFET devices.

Key Points of Section 17.2

- An MOSFET is a voltage-controlled device.
- Applying a gate voltage turns it on and it draws negligible gate current.
- The gate drive circuit should have low impedance for fast turn-on.

17.3 BJT BASE DRIVE

The switching speed can be increased by reducing turn-on time t_{on} and turn-off time t_{off}. The t_{on} can be reduced by allowing base current peaking during turn-on, resulting in low forced $\beta(\beta_F)$ at the beginning. After turn-on, β_f can be increased to a sufficiently high value to maintain the transistor in the quasi-saturation region. The t_{off} can be reduced by reversing base current and allowing base current peaking during turn-off. Increasing the value of reverse base current I_{B2} decreases the storage time. A typical waveform for base current is shown in Figure 17.3.

Apart from a fixed shape of base current, as in Figure 17.3, the forced β may be controlled continuously to match the collector current variations. The commonly used techniques for optimizing the base drive of a transistor are:

1. Turn-on control
2. Turn-off control
3. Proportional base control
4. Antisaturation control

Turn-on control. The base current peaking can be provided by the circuit of Figure 17.4. When the input voltage is turned on, the base current is limited by resistor R_1 and the initial value of base current is

$$I_B = \frac{V_1 - V_{BE}}{R_1} \tag{17.3}$$

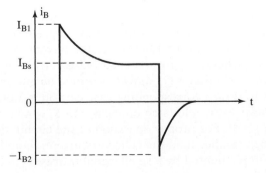

FIGURE 17.3

Base drive current waveform.

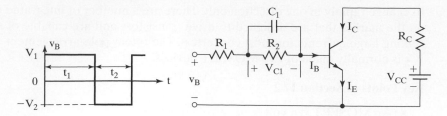

FIGURE 17.4

Base current peaking during turn-on.

and the final value of the base current is

$$I_{BS} = \frac{V_1 - V_{BE}}{R_1 + R_2} \tag{17.4}$$

The capacitor C_1 charges up to a final value of

$$V_c \cong V_1 \frac{R_2}{R_1 + R_2} \tag{17.5}$$

The charging time constant of the capacitor is approximately

$$\tau_1 = \frac{R_1 R_2 C_1}{R_1 + R_2} \tag{17.6}$$

Once the input voltage v_B becomes zero, the base–emitter junction is reverse biased and C_1 discharges through R_2. The discharging time constant is $\tau_2 = R_2 C_1$. To allow sufficient charging and discharging times, the width of the base pulse must be $t_1 \geq 5\tau_1$ and the off-period of the pulse must be $t_2 \geq 5\tau_2$. The maximum switching frequency is $f_s = 1/T = 1/(t_1 + t_2) = 0.2/(\tau_1 + \tau_2)$.

Turn-off control. If the input voltage in Figure 17.4 is changed to $-V_2$ during turn-off, the capacitor voltage V_c in Eq. (17.5) is added to V_2 as a reverse voltage across the transistor. There will be base current peaking during turn-off. As the capacitor C_1 discharges, the reverse voltage can be reduced to a steady-state value V_2. If different turn-on and turn-off characteristics are required, a turn-off circuit (using C_2, R_3, and R_4), as shown in Figure 17.5 may be added. The diode D_1 isolates the forward base drive circuit from the reverse base drive circuit during turn-off.

Proportional base control. This type of control has advantages over the constant drive circuit. If the collector current changes due to change in load demand, the base drive current is changed in proportion to the collector current. An arrangement is shown in Figure 17.6. When switch S_1 is turned on, a pulse current of short duration would flow through the base of transistor Q_1; and Q_1 is turned on into saturation. Once the collector current starts to flow, a corresponding base current is induced due to the transformer action. The transistor would latch on itself, and S_1 can be turned off. The turns ratio is $N_2/N_1 = I_C/I_B = \beta$. For proper operation of the circuit, the magnetizing current, which must be much smaller than the collector current, should be as small as possible. Switch S_1 can be implemented by a small-signal transistor, and an additional

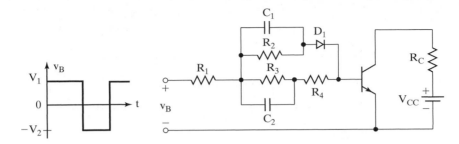

FIGURE 17.5

Base current peaking during turn-on and turn-off.

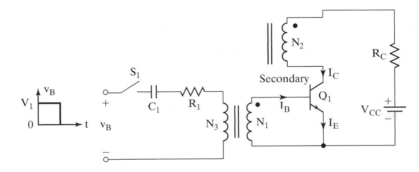

FIGURE 17.6

Proportional base drive circuit.

circuitry is necessary to discharge capacitor C_1 and to reset the transformer core during turn-off of the power transistor

Antisaturation control.   If the transistor is driven hard, the storage time, which is proportional to the base current, increases and the switching speed is reduced. The storage time can be reduced by operating the transistor in soft saturation instead of hard saturation. This can be accomplished by clamping the collector–emitter voltage to a predetermined level and the collector current is given by

$$I_C = \frac{V_{CC} - V_{cm}}{R_C} \tag{17.7}$$

where V_{cm} is the clamping voltage and $V_{cm} > V_{CE(\text{sat})}$. A circuit with clamping action (also known as Baker's clamp) is shown in Figure 17.7.

The base current without clamping, which is adequate to drive the transistor hard, can be found from

$$I_B = I_1 = \frac{V_B - V_{d1} - V_{BE}}{R_B} \tag{17.8}$$

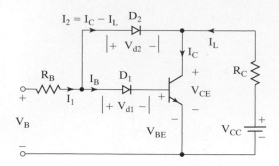

FIGURE 17.7

Collector clamping circuit.

and the corresponding collector current is

$$I_C = \beta I_B \tag{17.9}$$

After the collector current rises, the transistor is turned on, and the clamping takes place (due to the fact that D_2 gets forward biased and conducts). Then

$$V_{CE} = V_{BE} + V_{d1} - V_{d2} \tag{17.10}$$

The load current is

$$I_L = \frac{V_{CC} - V_{CE}}{R_C} = \frac{V_{CC} - V_{BE} - V_{d1} + V_{d2}}{R_C} \tag{17.11}$$

and the collector current with clamping is

$$I_C = \beta I_B = \beta(I_1 - I_C + I_L)$$
$$= \frac{\beta}{1 + \beta}(I_1 + I_L) \tag{17.12}$$

For clamping, $V_{d1} > V_{d2}$ and this can be accomplished by connecting two or more diodes in place of D_1. The load resistance R_C should satisfy the condition

$$\beta I_B > I_L$$

From Eq. (17.11),

$$\beta I_B R_C > (V_{CC} - V_{BE} - V_{d1} + V_{d2}) \tag{17.13}$$

The clamping action results in a reduced collector current and almost elimination of the storage time. At the same time, a fast turn-on is accomplished. However, due to increased V_{CE}, the on-state power dissipation in the transistor is increased, whereas the switching power loss is decreased.

Example 17.1 Finding the Transistor Voltage and Current with Clamping

The base drive circuit in Figure 17.6 has $V_{CC} = 100$ V, $R_C = 1.5\ \Omega$, $V_{d1} = 2.1$ V, $V_{d2} = 0.9$ V, $V_{BE} = 0.7$ V, $V_B = 15$ V, $R_B = 2.5\ \Omega$, and $\beta = 16$. Calculate (a) the collector current without clamping, (b) the collector–emitter clamping voltage V_{CE}, and (c) the collector current with clamping.

Solution

a. From Eq. (17.8), $I_1 = (15 - 2.1 - 0.7)/2.5 = 4.88$ A. Without clamping, $I_C = 16 \times 4.88 = 78.08$ A.

b. From Eq. (17.10), the clamping voltage is

$$V_{CE} = 0.7 + 2.1 - 0.9 = 1.9 \text{ V}$$

c. From Eq. (17.11), $I_L = (100 - 1.9)/1.5 = 65.4$ A. Equation (17.12) gives the collector current with clamping:

$$I_C = 16 \times \frac{4.88 + 65.4}{16 + 1} = 66.15 \text{ A}$$

Key Points of Section 17.3

- A BJT is a current-controlled device.
- Base current peaking can reduce the turn-on time and reversing the base current can reduce the turn-off time.
- The storage time of a BJT increases with the amount of base drive current, and overdrive should be avoided.

17.4   ISOLATION OF GATE AND BASE DRIVES

For operating power transistors as switches, an appropriate gate voltage or base current must be applied to drive the transistors into the saturation mode for low on-state voltage. The control voltage should be applied between the gate and source terminals or between the base and emitter terminals. The power converters generally require multiple transistors and each transistor must be gated individually. Figure 17.8a shows the topology of a single-phase bridge inverter. The main dc voltage is V_s with ground terminal G.

The logic circuit in Figure 17.8b generates four pulses. These pulses, as shown in Figure 17.8c, are shifted in time to perform the required logic sequence for power conversion from dc to ac. However, all four logic pulses have a common terminal C. The common terminal of the logic circuit may be connected to the ground terminal G of the main dc supply, as shown by dashed lines.

The terminal g_1, which has a voltage of V_{g1} with respect to terminal C, cannot be connected directly to gate terminal G_1. The signal V_{g1} should be applied between the gate terminal G_1 and source terminal S_1 of transistor M_1. There is a need for isolation and interfacing circuits between the logic circuit and power transistors. However, transistors M_2 and M_4 can be gated directly without isolation or interfacing circuits if the logic signals are compatible with the gate drive requirements of the transistors.

The importance of gating a transistor between its gate and source instead of applying gating voltage between the gate and common ground can be demonstrated with

17.5 THYRISTOR FIRING CIRCUITS

In thyristor converters, different potentials exist at various terminals. The power circuit is subjected to a high voltage, usually greater than 100 V, and the gate circuit is held at a low voltage, typically 12 to 30 V. An isolation circuit is required between an individual thyristor and its gate-pulse generating circuit. The isolation can be accomplished by either pulse transformers or optocouplers. An optocoupler could be a phototransistor or photo-silicon-controlled rectifier (SCR), as shown in Figure 17.12. A short pulse to the input of an ILED, D_1, turns on the photo-SCR T_1; and the power thyristor T_L is triggered. This type of isolation requires a separate power supply V_{cc} and increases the cost and weight of the firing circuit.

A simple isolation arrangement [1] with pulse transformers is shown in Figure 17.13a. When a pulse of adequate voltage is applied to the base of switching transistor Q_1, the transistor saturates and the dc voltage V_{cc} appears across the transformer primary, inducing a pulsed voltage on the transformer secondary, which is applied between the thyristor gate and cathode terminals. When the pulse is removed from the base of transistor Q_1, the transistor turns off and a voltage of opposite polarity is induced across the primary and the freewheeling diode D_m conducts. The current due to the transformer magnetic energy decays through D_m to zero. During this transient decay a corresponding reverse voltage is induced in the secondary. The pulse width can be made longer by connecting a capacitor C cross the resistor R, as shown in Figure 17.13b. The transformer carries unidirectional current and the magnetic core can saturate, thereby limiting the pulse width. This type of pulse isolation is suitable for pulses of typically 50 to 100 µs.

In many power converters with inductive loads, the conduction period of a thyristor depends on the load power factor (PF); therefore, the beginning of thyristor conduction is not well defined. In this situation, it is often necessary to trigger the thyristors continuously. However, a continuous gating increases thyristor losses. A pulse train that is preferable can be obtained with an auxiliary winding, as shown in Figure 17.13c. When transistor Q_1 is turned on, a voltage is also induced in the auxiliary winding N_3 at the base of transistor Q_1, such that diode D_1 is reverse biased and Q_1 turns off. In the meantime, capacitor C_1 charges up through R_1 and turns on Q_1 again. This process of

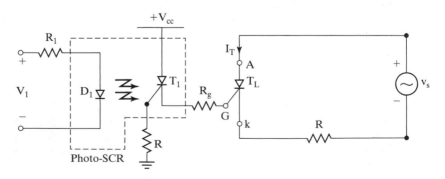

FIGURE 17.12

Photo-SCR coupled isolator.

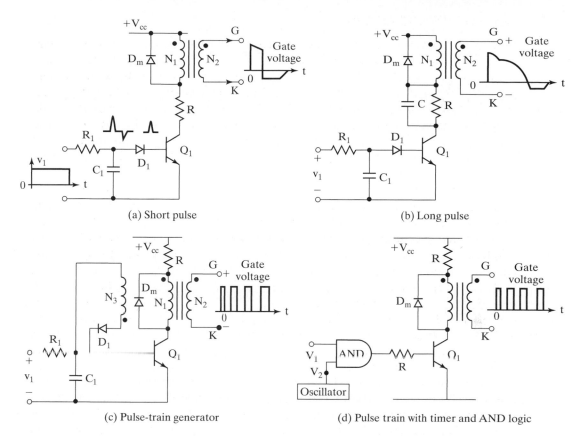

FIGURE 17.13

Pulse transformer isolation.

turn-on and turn-off continues as long as there is an input signal v_1 to the isolator. Instead of using the auxiliary winding as a blocking oscillator, an AND-logic gate with an oscillator (or a timer) could generate a pulse train, as shown in Figure 17.13d. In practice, the AND gate cannot drive transistor Q_1 directly, and a buffer stage is normally connected before the transistor.

The output of gate circuits in Figure 17.12 or Figure 17.13 is normally connected between the gate and cathode along with other gate-protecting components, as shown in Figure 17.14. The resistor R_g in Figure 17.14a increases the dv/dt capability of the thyristor, reduces the turn-off time, and increases the holding and latching currents. The capacitor C_g in Figure 17.14b removes high-frequency noise components and increases dv/dt capability and gate delay time. The diode D_g in Figure 17.14c protects the gate from negative voltage. However, for asymmetric silicon-controlled rectifiers, SCRs, it is desirable to have some amount of negative gate voltage to improve the dv/dt capability and also to reduce the turn-off time. All these features can be combined as shown in Figure 17.14d, where diode D_1 allows only the positive pulses and R_1 damps out any transient oscillation and limits the gate current.

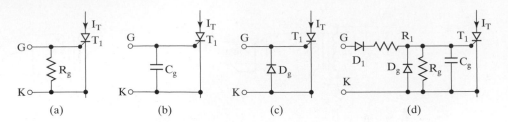

FIGURE 17.14

Gate protection circuits.

Key Points of Section 17.5

- Applying a pulse signal turns on a thyristor.
- The low-level gate circuit must be isolated from the high-level power circuit through isolation techniques.
- The gate should be protected from triggering by a high frequency or an interference signal.

17.6   UNIJUNCTION TRANSISTOR

The unijunction transistor (UJT) is commonly used for generating triggering signals for SCRs. A basic UJT-triggering circuit is shown in Figure 17.15a. A UJT has three terminals, called the emitter E, base-one B_1, and base-two B_2. Between B_1 and B_2 the unijunction has the characteristics of an ordinary resistance. This resistance is the interbase resistance R_{BB} and has values in the range 4.7 to 9.1 kΩ. The static characteristics of a UJT are shown in Figure 17.15b.

When the dc supply voltage V_s is applied, the capacitor C is charged through resistor R because the emitter circuit of the UJT is in the open state. The time constant of the charging circuit is $\tau_1 = RC$. When the emitter voltage V_E, which is the same as the capacitor voltage v_C, reaches the *peak voltage* V_p, the UJT turns on and capacitor C discharges through R_{B1} at a rate determined by the time constant $\tau_2 = R_{B1}C$. τ_2 is much smaller than τ_1. When the emitter voltage V_E decays to the valley point V_v, the emitter ceases to conduct, the UJT turns off, and the charging cycle is repeated. The waveforms of the emitter and triggering voltages are shown in Figure 17.15c.

The waveform of the triggering voltage V_{B1} is identical to the discharging current of capacitor C. The triggering voltage V_{B1} should be designed to be sufficiently large to turn on the SCR. The period of oscillation, T, is fairly independent of the dc supply voltage V_s, and is given by

$$T = \frac{1}{f} \approx RC \ln \frac{1}{1 - \eta} \tag{17.14}$$

where the parameter η is called the *intrinsic stand-off ratio*. The value of η lies between 0.51 and 0.82.

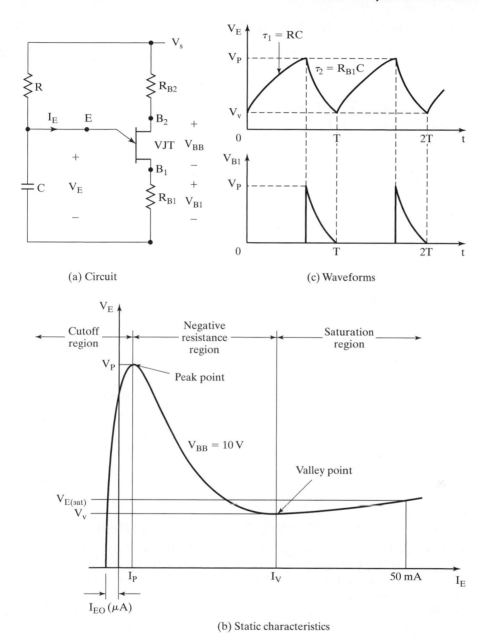

(a) Circuit

(c) Waveforms

(b) Static characteristics

FIGURE 17.15

UJT triggering circuit.

Resistor R is limited to a value between 3 kΩ and 3 MΩ. The upper limit on R is set by the requirement that the load line formed by R and V_s intersects the device characteristics to the right of the peak point but to the left of the valley point. If the load line fails to pass to the right of the peak point, the UJT cannot turn on. This condition can be satisfied if $V_s - I_p R > V_p$. That is,

$$R < \frac{V_s - V_p}{I_p} \tag{17.15}$$

At the valley point $I_E = I_V$ and $V_E = V_v$ so that the condition for the lower limit on R to ensure turning off is $V_s - I_v R < V_v$. That is,

$$R > \frac{V_s - V_v}{I_v} \tag{17.16}$$

The recommended range of supply voltage V_s is from 10 to 35 V. For fixed values of η, the peak voltage V_p varies with the voltage between the two bases, V_{BB}. V_p is given by

$$V_p = \eta V_{BB} + V_D(= 0.5 \text{ V}) \approx \eta V_s + V_D(= 0.5 \text{ V}) \tag{17.17}$$

where V_D is the one-diode forward voltage drop. The width t_g of triggering pulse is

$$t_g = R_{B1} C \tag{17.18}$$

In general, R_{B1} is limited to a value below 100 Ω, although values up to 2 or 3 kΩ are possible in some applications. A resistor R_{B2} is generally connected in series with base-two to compensate for the decrease in V_p due to temperature rise and to protect the UJT from possible thermal runaway. Resistor R_{B2} has a value of 100 Ω or greater and can be determined approximately from

$$R_{B2} = \frac{10^4}{\eta V_s} \tag{17.19}$$

Example 17.2 Finding the Circuit Values of a UJT Triggering Circuit

Design the triggering circuit of Figure 17.15a. The parameters of the UJT are $V_s = 30$ V, $\eta = 0.51$, $I_p = 10 \ \mu\text{A}$, $V_v = 3.5$ V, and $I_v = 10$ mA. The frequency of oscillation is $f = 60$ Hz, and the width of the triggering pulse is $t_g = 50 \ \mu\text{s}$. Assume $V_D = 0.5$.

Solution
$T = 1/f = 1/60 \text{ Hz} = 16.67$ ms. From Eq. (17.17), $V_p = 0.51 \times 30 + 0.5 = 15.8$ V. Let $C = 0.5 \ \mu\text{F}$. From Eqs. (17.15) and (17.16), the limiting values of R are

$$R < \frac{30 - 15.8}{10 \ \mu\text{A}} = 1.42 \text{ M}\Omega$$

$$R > \frac{30 - 3.5}{10 \ \text{mA}} = 2.65 \text{ k}\Omega$$

From Eq. (17.14), 16.67 ms $= R \times 0.5 \ \mu\text{F} \times \ln[1/(1 - 0.51)]$, which gives $R = 46.7 \ \text{k}\Omega$, which falls within the limiting values. The peak gate voltage $V_{B1} = V_p = 15.8 \ \text{V}$. From Eq. (17.18),

$$R_{B1} = \frac{t_g}{C} = \frac{50 \ \mu\text{s}}{0.5 \ \mu\text{F}} = 100 \ \Omega$$

From Eq. (17.19),

$$R_{B2} = \frac{10^4}{0.51 \times 30} = 654 \ \Omega$$

Key Points of Section 17.6

- The PUT can generate a triggering signal for thyristors.
- When the emitter voltage reaches the peak point voltage, the UJT turns on; when the emitter voltages falls to the decay point, it turns off.

17.7 PROGRAMMABLE UNIJUNCTION TRANSISTOR

The programmable unijunction transistor (PUT) is a small thyristor shown in Figure 17.16a. A PUT can be used as a relaxation oscillator, as shown in Figure 17.16b. The gate voltage V_G is maintained from the supply by the resistor dividers R_1 and R_2, and determines the peak point voltage V_p. In the case of the UJT, V_p is fixed for a device by the dc supply voltage. However, V_p of a PUT can be varied by varying the resistor dividers R_1 and R_2. If the anode voltage V_A is less than the gate voltage V_G, the device can remain in its off-state. If V_A exceeds the gate voltage by one diode forward voltage V_D, the peak point is reached and the device turns on. The peak current I_p and the valley point current I_v both depend on the equivalent impedance on the gate $R_G = R_1 R_2/(R_1 + R_2)$ and the dc supply voltage V_s. In general, R_k is limited to a value below 100 Ω.

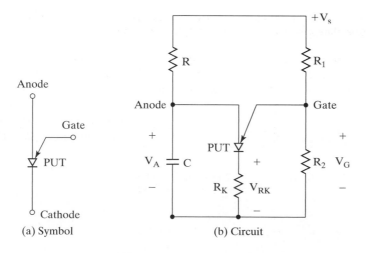

(a) Symbol (b) Circuit

FIGURE 17.16
PUT triggering circuit.

V_p is given by

$$V_p = \frac{R_2}{R_1 + R_2} V_s \qquad (17.20)$$

which gives the intrinsic ratio as

$$\eta = \frac{V_p}{V_s} = \frac{R_2}{R_1 + R_2} \qquad (17.21)$$

R and C control the frequency along with R_1 and R_2. The period of oscillation T is given approximately by

$$T = \frac{1}{f} \approx RC \ln \frac{V_s}{V_s - V_p} = RC \ln \left(1 + \frac{R_2}{R_1}\right) \qquad (17.22)$$

The gate current I_G at the valley point is given by

$$I_G = (1 - \eta) \frac{V_s}{R_G} \qquad (17.23)$$

where $R_G = R_1 R_2/(R_1 + R_2)$. R_1 and R_2 can be found from

$$R_1 = \frac{R_G}{\eta} \qquad (17.24)$$

$$R_2 = \frac{R_G}{1 - \eta} \qquad (17.25)$$

Example 17.3 Finding the Circuit Values of a Programmable UJT Triggering Circuit

Design the triggering circuit of Figure 17.16b. The parameters of the PUT are $V_s = 30$ V and $I_G = 1$ mA. The frequency of oscillation is $f = 60$ Hz. The pulse width is $t_g = 50$ μs, and the peak triggering voltage is $V_{Rk} = 10$ V.

Solution

$T = 1/f = 1/60$ Hz $= 16.67$ ms. The peak triggering voltage $V_{Rk} = V_p = 10$ V. Let $C = 0.5$ μF. From Eq. (17.18), $R_k = t_g/C = 50$ μs/0.5 μF $= 100$ Ω. From Eq. (17.21), $\eta = V_p/V_s = 10/30 = 1/3$. From Eq. (17.22), 16.67 ms $= R \times 0.5$ μF $\times \ln[30/(30 - 10)]$, which gives $R = 82.2$ kΩ. For $I_G = 1$ mA, Eq. (17.23) gives $R_G = (1 - \frac{1}{3}) \times 30/1$ mA $= 20$ kΩ. From Eq. (17.24),

$$R_1 = \frac{R_G}{\eta} = 20 \text{ k}\Omega \times \frac{3}{1} = 60 \text{ k}\Omega$$

From Eq. (17.25),

$$R_2 = \frac{R_G}{1 - \eta} = 20 \text{ k}\Omega \times \frac{3}{2} = 30 \text{ k}\Omega$$

Key Points of Section 17.7

- The UJT can generate triggering signal for thyristors.
- The peak point voltage can be adjusted from an external circuit usually by two resistors forming a potential divider. Thus, the frequency of the triggering pulses can be varied.

Example 17.4 Finding the Minimum Width of the Gate Pulse for a Thyristor Converter

The holding current of thyristors in the single-phase full converter of Figure 10.2a is $I_H = 500$ mA and the delay time is $t_d = 1.5$ μs. The converter is supplied from a 120-V, 60-Hz supply and has a load of $L = 10$ mH and $R = 10$ Ω. The converter is operated with a delay angle of $\alpha = 30°$. Determine the minimum value of gate pulse width, t_G.

Solution

$I_H = 500$ mA $= 0.5$ A, $t_d = 1.5$ μs, $\alpha = 30° = \pi/6$, $L = 10$ mH, and $R = 10$ Ω. The instantaneous value of the input voltage is $v_s(t) = V_m \sin \omega t$, where $V_m = \sqrt{2} \times 120 = 169.7$ V.

At $\omega t = \alpha$,

$$V_1 = v_s(\omega t = \alpha) = 169.7 \times \sin \frac{\pi}{6} = 84.85 \text{ V}$$

The rate of rise of anode current di/dt at the instant of triggering is approximately

$$\frac{di}{dt} = \frac{V_1}{L} = \frac{84.85}{10 \times 10^{-3}} = 8485 \text{ A/s}$$

If di/dt is assumed constant for a short time after the gate triggering, the time t_1 required for the anode current to rise to the level of holding current is calculated from $t_1 \times (di/dt) = I_H$ or $t_1 \times 8485 = 0.5$ and this gives $t_1 = 0.5/8485 = 58.93$ μs. Therefore, the minimum width of the gate pulse is

$$t_G = t_1 + t_d = 58.93 + 1.5 = 60.43 \text{ μs}$$

17.8 THYRISTOR CONVERTER GATING CIRCUITS

The generation of gating signals for thyristors of dc–ac converters requires (1) detecting zero crossing of the input voltage, (2) appropriate phase shifting of signals, (3) pulse shaping to generate pulses of short duration, and (4) pulse isolation through pulse transformers or optocouplers. The block diagram for gating circuit of a single-phase full converter is shown in Figure 17.17.

17.9 GATE DRIVE ICs

The gate drive requirements for an MOSFET or an IGBT switch, as shown in Figure 17.18, are to satisfy as follows:

- Gate voltage must be 10 to 15 V higher than the source or emitter voltage. Because the power drive is connected to the main high voltage rail $+V_S$, the gate voltage must be higher than the rail voltage.
- The gate voltage that is normally referenced to ground must be controllable from the logic circuit. Thus, the control signals have to be level shifted to the source terminal of the power device, which in most applications swings between the two rails V^+.

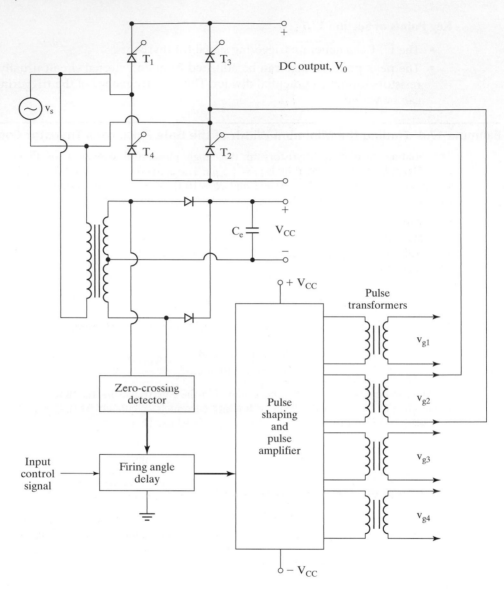

FIGURE 17.17

Block diagram for a thyristor gating circuit.

- A low-side power device generally drives the high-side power device that is connected to the high voltage. Thus, there is one high-side and one low-side power device. The power absorbed by the gate drive circuitry should be low and it should not significantly affect the overall efficiency of the power converter.

There are several techniques, as shown in Table 17.1, that can be used to meet the gate drive requirements. Each basic circuit can be implemented in a wide variety of configurations. An IC gate drive integrates most of the functions required to drive one

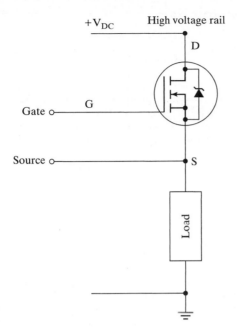

FIGURE 17.18

Power MOSFET connect to the high voltage rail side.

TABLE 17.1   Gate Drive Techniques, Ref. 2. (Courtesy of Siemens Group, Germany)

Method	Basic Circuit	Key Features
Floating gate drive supply		Full gate control for indefinite periods of time, cost impact of isolated supply is significant (one required for each high side MOSFET), level shifting a ground referenced signal can be tricky: level shifter must sustain full voltage, switch fast with minimal propagation delays and low-power consumption, optoisolators tend to be relatively expensive, limited in bandwidth and noise sensitive
Pulse transformer		Simple and cost effective but limited in many respects, operation over wide duty cycles requires complex techniques, transformer size increases significantly as frequency decreases; significant parasitics create less than ideal operation with fast-switching waveforms

TABLE 17.1   (*Continued*)

Method	Basic Circuit	Key Features
Charge pump		Can be used to generate an "over-rail" voltage controlled by a level shifter or to "pump" the gate when MOSFET is turned on, in the first case the problems of a level shifter have to be tackled, in the second case turn-on times tend to be too long for switching applications, in either case, gate can be kept on for an indefinite period of time, inefficiencies in the voltage multiplication circuit may require more than two stages of pumping
Bootstrap		Simple and inexpensive with some of the limitations of the pulse transformer: duty cycle and on time are both constrained by the need to refresh the bootstrap capacitor, if the capacitor is charged from a high voltage rail, power dissipation can be significant, requires level shifter, with its associated difficulties
Carrier drive		Gives full gate control for an indefinite period of time but is somewhat limited in switching performance, this can be improved with added complexity

high-side and one low-side power device in a compact, high-performance package with low-power dissipation. The IC must also have some protection functions to operate under overload and fault conditions.

Three types of circuits can perform the gate drive and protection functions. The first one is the output buffer that is needed to provide sufficient gate voltage or charge to the power device. The second is the level shifters that are needed to interface between the control signals to the high-side and low-side output buffers. The third is the sensing of overload conditions in the power device and the appropriate countermeasure taken in the output buffer as well as fault status feedback.

17.9.1   Drive IC for Converters

There are numerous IC gate drives that are commercially available for gating power converters. These include pulse-width modulation (PWM) control [3], power factor correction (PFC) control [2], combined PWM and PFC control, current mode control [4], bridge driver, servo driver, hall-bridge drivers, stepper motor driver, and thyristor gate driver. These ICs can be used for applications such as buck converters for battery chargers, dual forward converter for switched reluctance motor drives, full-bridge inverter with current-mode control, three-phase inverter for brushless and induction motor drives, push–pull bridge converter for power supplies, and synchronous PWM control of switched-mode power supplies (SMPSs). The block diagram of a typical general-purpose VH floating MOS-gate driver (MGD) is shown in Figure 17.19 [2].

The input logic channels are controlled by TTL/CMOS compatible inputs. The transition thresholds are different from device to device. Some MGDs have the transition threshold proportional to the logic supply V_{DD} (3 to 20 V) and Schmitt trigger buffers with hysteresis equal to 10% of V_{DD} to accept inputs with long risetime, whereas other MGDs have a fixed transition from logic 0 to logic 1 between 1.5 and 2 V. Some MGDs can drive only one high-side power device, whereas others can drive one high-side and one low-side power device. Others can drive a full three-phase bridge. Any high-side driver can also drive a low-side device. Those MGDs with two gate drive channels can have dual, hence independent, input commands or a single input command with complementary drive and predetermined dead time.

The low-side output stage is implemented either with two N-channel MOSFETs in totem pole configuration or with an N-channel and a P-channel CMOS inverter stage. The source follower acts as a current source and common source for current sinking. The source of the lower driver is independently brought out to pin 2 so that a direct connection can be made to the source of the power device for the return of the gate drive current. This can prevent either channel from operating undervoltage lockout if V_{CC} is below a specified value (typically 8.2 V).

The high-side channel has been built into an "isolation tub" capable of floating with respect to common ground (COM). The tub "floats" at the potential of V_S, which is established by the voltage applied to V_{CC} (typically 15 V) and swings between the two rails. The gate charge for the high-side MOSFET is provided by the bootstrap capacitor C_B, which is charged by the V_{CC} supply through the bootstrap diode during the time when the device is off. Because the capacitor is charged from a low-voltage source, the power consumed to drive the gate is small. Therefore, the MOS-gated transistors exhibit capacitive input characteristic, that is, supplying a charge to the gate instead of a continuous current can turn on the device.

A typical application of a PWM current-mode controller is shown in Figure 17.20. Its features include low-standby power, soft start, peak current detection, input undervoltage lockout, thermal shutdown and overvoltage protection, and high-switching frequency of 100 kHz.

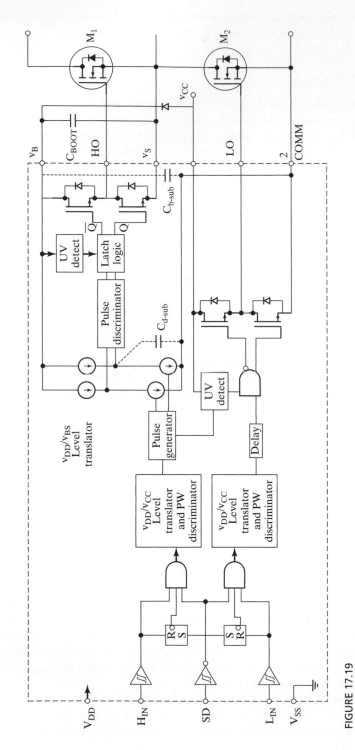

FIGURE 17.19

Block diagram of an MOS-gated driver, Ref. 2. (Courtesy of International Rectifier, Inc.)

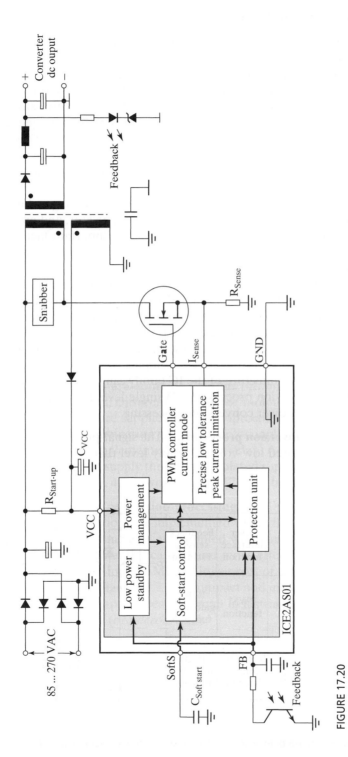

FIGURE 17.20

Typical application of current-model control IC for switched-mode power supply Ref. 4. (Courtesy of Siemens Group, Germany)

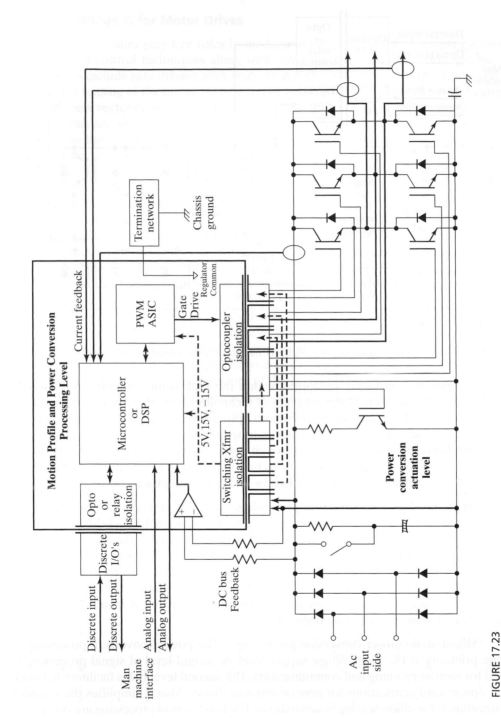

FIGURE 17.23

Single-level power conversion-processing architecture, Ref. 5. (Courtesy of International Rectifier, Inc.)

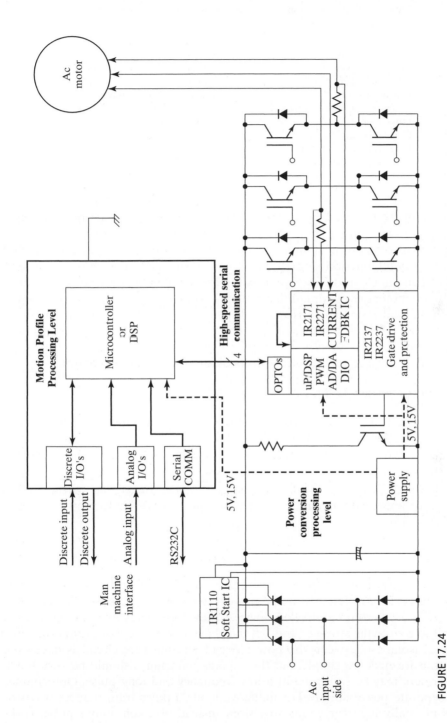

FIGURE 17.24

Single-level power conversion-processing architecture, Ref. 5. (Courtesy of International Rectifier, Inc.)

TABLE 17.2   Comparison of Two-Level versus Single-Level Power
Conversion Architecture

Two-Level Architecture	Single-Level Architecture
Motion and power conversion processed together	Motion and power conversion processed separately
Isolation by optodrivers (sensitive high-speed signals)	Isolation by digital interface (high noise-margin signals)
Large dead time	Small dead time
Complex switching supply	Simple flyback supply
Large hall current sensors	Small HVIC current sensors
Protection at signal level	Protection at power level
Larger size and more EMI	Smaller size and less EMI

One of the key features of the single-level and mixed-mode power conversion-processing architecture is the integration of the gate drive, protection, and sensing functions. The integration is implemented in a high-voltage integrated circuit (HVIC) technology. Multifunction sensing chips that integrate current and voltage feedback with both amplitude and phase information can simplify the designs of ac or brushless dc (BLDC) motor drives. Monolithic integration of the gate drive, protection, linear current sensing, and more functions in a single piece of silicon using HVIC technology is the ultimate goal. Thus, all power conversion functions for robust, efficient cost-effective, compact motor drives should ideally be integrated in modular fashion with appropriately defined serial communication protocol for local or remote control.

Key Points of Section 17-9

- An IC gate drive integrates most of the control functions including some protection functions to operate under overload and fault conditions. There are numerous IC gate drives that are commercially available for gating power converters.
- The especial purpose ICs for motor drives include many features such as gate driving with protection, soft start charging of DC bus, linear current sensing of motor phase current, and control algorithms from V/Hz to sensor-less vector or servo control.

SUMMARY

MOSFETs are voltage-controlled devices and they require very low gating power. The gate signals can be isolated from the power circuit by pulse transformers or optocouplers. BJTs are current-controlled devices. They require reverse base current during turn-off to reduce the storage time, but they have low on-state or saturation voltage. A means of isolation between the power circuit and the gate circuit is necessary. The pulse transformers are simple, but the leakage inductance should be very small. The transformers may be saturated at a low frequency and long pulse. Optocouplers require separate power supply. For inductive loads, a pulse train reduces thyristor loss and is normally used for gating thyristors, instead of a continuous pulse. UJTs and PUTs are used for generating triggering pulses.

There are numerous gate drive ICs for drives that are commercially available for gating power converters. These ICs integrate logic, gate isolation, protection, and control functions. As a result, discrete gate circuits have become obsolete.

Dedicated power conversion processors can integrate many functions such as protection and soft shutdown for the inverter stage, current sensing, analog-to-digital conversion for use in the algorithm for closed loop current control, soft charge of the dc bus capacitor, and an almost bulletproof input converter stage. An ideal PCP for a robust, efficient cost-effective, compact motor drive should ideally be integrated in modular fashion with appropriately defined serial communication protocol for local or remote control.

REFERENCES

[1]   M. S. J. Asghar, *Power Electronics Handbook*, edited by M. H. Rashid. San Diego, CA: Academic Press. 2001, Chapter 18—Gate Drive Circuits.

[2]   "HV Floating MOS-Gate Driver ICs," Application Note AN978, International Rectifier, Inc., El Segunda, CA, July 2001. www.irf.com

[3]   "Enhanced Generation of PWM Controllers," Unitrode Application Note U-128, Texas Instruments, Dallas, Texas, 2000.

[4]   "Off-Line SMPS Current Mode Controller," Application Note ICE2AS01, Infineon Technologies, Munich, Germany, February 2001. www.infineon.com

[5]   "Power Conversion Processor Architecture and HVIC Products for Motor Drives," International Rectifier, Inc., El Segunda, CA, 2001, pp. 1–21. www.irf.com

REVIEW QUESTIONS

17.1 What is the switching model of MOSFETs?
17.2 Why is it necessary to reverse bias BJTs during turn-off?
17.3 What are the base drive techniques for increasing switching speeds of BJTs?
17.4 What is an antisaturation control of BJTs?
17.5 What are the advantages and disadvantages of transformer gate isolation?
17.6 What are the advantages and disadvantages of optocoupled gate isolation?
17.7 What is a UJT?
17.8 What is the peak voltage of a UJT?
17.9 What is the valley point voltage of a UJT?
17.10 What is the intrinsic stand-off ratio of a UJT?
17.11 What is a PUT?
17.12 What are the advantages of a PUT over a UJT?
17.13 What are functional requirements of gate drive ICs?
17.14 What are basic techniques for implementing the functional requirements of gate drivers?
17.15 What is a totem pole arrangement?
17.16 What is the function of the boost capacitor in gate drive ICs?
17.17 What are the three types of architectures for motor drive ICs?
17.18 What is two-level power conversion processing?
17.19 What is single-level power conversion processing?
17.20 What is mixed-mode power conversion processing?
17.21 What are the differences between single-level and two-level power processing?

PROBLEMS

17.1 The base drive voltage to the circuit, as shown in Figure 17.3, is a square wave of 10 V. The peak base current is $I_{BO} \geq 1.5$ mA and the steady-state base current is $I_{BS} \geq 1$ mA. Find **(a)** the values of C_1, R_1, and R_2, and **(b)** the maximum permissible switching frequency f_{max}.

17.2 The base drive circuit in Figure 17.7 has $V_{CC} = 400$ V, $R_C = 4$ Ω, $V_{d1} = 3.6$ V, $V_{d2} = 0.9$ V, $V_{BE(sat)} = 0.7$ V, $V_B = 15$ V, $R_B = 1.1$ Ω, and $\beta = 12$. Calculate **(a)** the collector current without clamping, **(b)** the collector clamping voltage V_{CE}, and **(c)** the collector current with clamping.

17.3 Design the triggering circuit of Figure 17.15a. The parameters of the UJT are $V_s = 20$ V, $\eta = 0.66$, $I_p = 10$ μA, $V_v = 2.5$ V, and $I_v = 10$ mA. The frequency of oscillation is $f = 1$ kHz, and the width of the gate pulse is $t_g = 40$ μs.

17.4 Design the triggering circuit of Figure 17.16b. The parameters of the PUT are $V_s = 20$ V and $I_G = 1.5$ mA. The frequency of oscillation is $f = 1$ kHz. The pulse width is $t_g = 40$ μs, and the peak triggering pulse is $V_{Rk} = 8$ V.

17.5 A 240-V, 50-Hz supply is connected to an RC trigger circuit in Figure P17.5. If R is variable from 1.5 to 24 kΩ, $V_{GT} = 2.5$ V, and $C = 0.47$ μF, find the minimum and maximum values of the firing angle α.

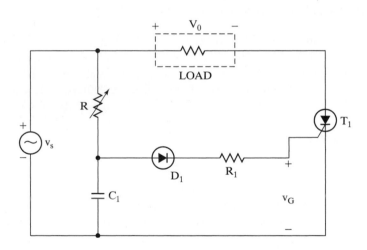

FIGURE P17.5

17.6 If the RC-trigger circuit in Figure P17.5 is used for a full-wave ac–dc thyristor converter, find the minimum and maximum values of the firing angle α.

17.7 The parameters and ratings of a UJT at $T_j = 25°C$ are as follows:

Maximum interbase voltage $(V_{BB}) = 35$ V;
Maximum average power dissipation $= 300$ mW;
Range of interbase resistance $(R_{BB}) = 4.7$ to 9.1 kΩ (typical 5.6 kΩ at $V_{BB} = 12$ V);
Valley point current, $I_V = 4$ mA at $V_{BB} = 20$ V;
Intrinsic stand-off ratio, $\eta = 0.56$ to 0.75 (0.63 typical);
Valley point voltage, $V_V = 2$ V at $V_{BB} = 20$ V;
Peak point current, $I_P = 5$ mA (maximum) at $V_{BB} = 25$ V;
The maximum gate voltage (V_{GD}) that cannot trigger on the SCR $= 0.18$ V;
Design of a suitable UJT base trigger circuit for a single-phase thyristor-controlled converter (rectifier) operating with 60-Hz ac main supply.

CHAPTER 18

Protection of Devices and Circuits

The learning objectives of this chapter are as follows:

- To understand the electrical analog of thermal models and the methods for cooling power devices
- To learn the methods for protecting the devices from excessive di/dt and dv/dt, and transient voltages due to load and supply disconnection
- To learn how to select fast-acting fuses for protecting power devices
- To learn about sources of electromagnetic interference (EMI) and the methods of minimizing EMI effects on receptor circuits

18.1 INTRODUCTION

Due to the reverse recovery process of power devices and switching actions in the presence of circuit inductances, voltage transients occur in the converter circuits. Even in carefully designed circuits, short-circuit fault conditions may exist, resulting in an excessive current flow through the devices. The heat produced by losses in a semiconductor device must be dissipated sufficiently and effectively to operate the device within its upper temperature limit. The reliable operation of a converter would require ensuring that at all times the circuit conditions do not exceed the ratings of the power devices, by providing protection against overvoltage, overcurrent, and overheating. In practice, the power devices are protected from (1) thermal runaway by heat sinks, (2) high dv/dt and di/dt by snubbers, (3) reverse recovery transients, (4) supply and load-side transients, and (5) fault conditions by fuses.

18.2 COOLING AND HEAT SINKS

Due to on-state and switching losses, heat is generated within the power device. This heat must be transferred from the device to a cooling medium to maintain the operating

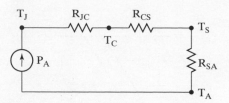

FIGURE 18.1

Electrical analog of heat transfer.

junction temperature within the specified range. Although this heat transfer can be accomplished by conduction, convection, radiation, or natural or forced-air, convection cooling is commonly used in industrial applications.

The heat must flow from the device to the case and then to the heat sink in the cooling medium. If P_A is the average power loss in the device, the electrical analog of a device, which is mounted on a heat sink, is shown in Figure 18.1. The junction temperature of a device T_J is given by

$$T_J = P_A(R_{JC} + R_{CS} + R_{SA}) \qquad (18.1)$$

where R_{JC} = thermal resistance from junction to case, °C/W
       R_{CS} = thermal resistance from case to sink, °C/W
       R_{SA} = thermal resistance from sink to ambient, °C/W
       T_A = ambient temperature, °C

R_{JC} and R_{CS} are normally specified by the power device manufacturers. Once the device power loss P_A is known, the required thermal resistance of the heat sink can be calculated for a known ambient temperature T_A. The next step is to choose a heat sink and its size that would meet the thermal resistance requirement.

A wide variety of extruded aluminum heat sinks are commercially available and they use cooling fins to increase the heat transfer capability. The thermal resistance characteristics of a typical heat sink with natural and forced cooling are shown in Figure 18.2, where the power dissipation against the sink temperature rise is depicted for natural cooling. In forced cooling, the thermal resistance decreases with the air velocity. However, above a certain velocity, the reduction in thermal resistance is not significant. Heat sinks of various types are shown in Figure 18.3.

The contact area between the device and heat sink is extremely important to minimize the thermal resistance between the case and sink. The surfaces should be flat, smooth, and free of dirt, corrosion, and surface oxides. Silicone greases are normally applied to improve the heat transfer capability and to minimize the formation of oxides and corrosion.

The device must be mounted properly on the heat sink to obtain the correct mounting pressure between the mating surfaces. The proper installation procedures are usually recommended by the device manufacturers. In case of stud-mounted devices,

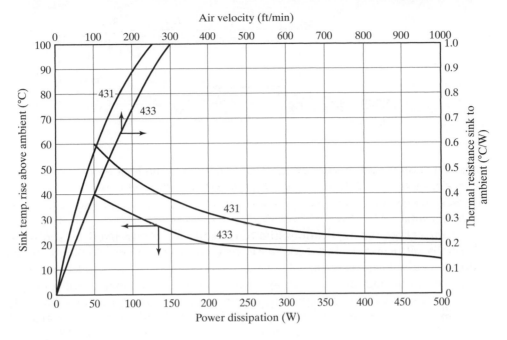

FIGURE 18.2

Thermal resistance characteristics. (Courtesy of EG&G Wakefield Engineering.)

excessive mounting torques may cause mechanical damage of the silicon wafer; and the stud and nut should not be greased or lubricated because the lubrication increases the tension on the stud.

The device may be cooled by heat pipes partially filled with low-vapor-pressure liquid. The device is mounted on one side of the pipe and a condensing mechanism (or

FIGURE 18.3

Heat sinks. (Courtesy of EG&G Wakefield Engineering.)

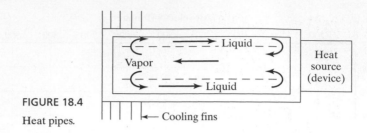

FIGURE 18.4

Heat pipes.

heat sink) on the other side, as shown in Figure 18.4. The heat produced by the device vaporizes the liquid and the vapor, and then flows to the condensing end, where it condenses and the liquid returns to the heat source. The device may be some distance from the heat sink.

In high-power applications, the devices are more effectively cooled by liquids, normally oil or water. Water cooling is very efficient and approximately three times more effective than oil cooling. However, it is necessary to use distilled water to minimize corrosion and antifreeze to avoid freezing. Oil is flammable. Oil cooling, which may be restricted to some applications, provides good insulation and eliminates the problems of corrosion and freezing. Heat pipes and liquid-cooled heat sinks are commercially available. Two water-cooled ac switches are shown in Figure 18.5. Power converters are available in assembly units, as shown in Figure 18.6.

The thermal impedance of a power device is very small, and as a result the junction temperature of the device varies with the instantaneous power loss. The instantaneous junction temperature must always be maintained lower than the acceptable value. A plot of the transient thermal impedance versus square-wave pulse duration is supplied by the device manufacturers as a part of the data sheet. From the knowledge of current waveform through a device, a plot of power loss against time can be determined and then the transient impedance characteristics can be used to calculate the temperature variations with time. If the cooling medium fails in practical systems, the temperature rise of the heat sinks is normally served to switch off the power converters, especially in high-power applications.

The step response of a first-order system can be applied to express the transient thermal impedance. If Z_0 is the steady-state junction case thermal impedance,

FIGURE 18.5

Water-cooled ac switches. (Courtesy of Powerex, Inc.)

FIGURE 18.6

Assembly units. (Courtesy of Powerex, Inc.)

the instantaneous thermal impedance can be expressed as

$$Z(t) = Z_0(1 - e^{-t/\tau_{th}})$$ (18.2)

where τ_{th} is the thermal time constant of the device. If the power loss is P_d, the instantaneous junction temperature rise above the case is

$$T_J = P_d Z(t)$$ (18.3)

If the power loss is a pulsed type, as shown in Figure 18.7, Eq. (18.3) can be applied to plot the step responses of the junction temperature $T_J(t)$. If t_n is the duration of nth power pulse, the corresponding thermal impedances at the beginning and end of nth pulse are $Z_0 = Z(t = 0) = 0$ and $Z_n = Z(t = t_n)$, respectively. The thermal impedance $Z_n = Z(t = t_n)$ corresponding to the duration of t_n can be found from the transient thermal impedance characteristics. If $P_1, P_2, P_3, \ldots$ are the power pulses with $P_2 = P_4 = \cdots = 0$, the junction temperature at the end of mth pulse can be expressed as

$$T_J(t) = T_{J0} + P_1(Z_1 - Z_2) + P_3(Z_3 - Z_4) + P_5(Z_5 - Z_6) + \cdots$$
$$= T_{J0} + \sum_{n=1,3,\ldots}^{m} P_n(Z_n - Z_{n+1})$$ (18.4)

where T_{J0} is the initial junction temperature. The negative signs of $Z_2, Z_4, \ldots$ signify that the junction temperature falls during the intervals $t_2, t_4, t_6, \ldots$.

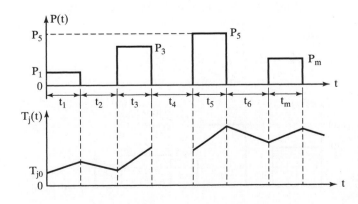

FIGURE 18.7

Junction temperature with rectangular power pulses.

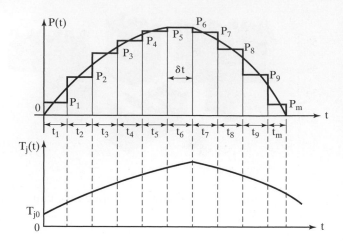

FIGURE 18.8

Approximation of a power pulse by rectangular pulses.

The step response concept of junction temperature can be extended to other power waveforms [13]. A waveform of any shape can be represented approximately by rectangular pulses of equal or unequal duration, with the amplitude of each pulse equal to the average amplitude of the actual pulse over the same period. The accuracy of such approximations can be improved by increasing the number of pulses and reducing the duration of each pulse. This is shown in Figure 18.8.

The junction temperature at the end of mth pulse can be found from

$$T_J(t) = T_{J0} + Z_1 P_1 + Z_2(P_2 - P_1) + Z_3(P_3 - P_2) + \cdots$$

$$= T_{J0} + \sum_{n=1,2\ldots}^{m} Z_n(P_n - P_{n-1}) \tag{18.5}$$

where Z_n is the impedance at the end of nth pulse of duration $t_n = \delta t$. P_n is the power loss for the nth pulse and $P_0 = 0$; t is the time interval.

Example 18.1 Plot the Instantaneous Junction Temperature

The power loss of a device is shown in Figure 18.9. Plot the instantaneous junction temperature rise above the case. $P_2 = P_4 = P_6 = 0$, $P_1 = 800$ W, $P_3 = 1200$ W, and $P_5 = 600$ W. For $t_1 = t_3 = t_5 = 1$ ms, the data sheet gives

$$Z(t = t_1) = Z_1 = Z_3 = Z_5 = 0.035°\text{C/W}$$

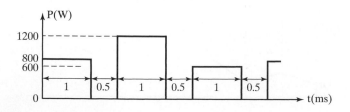

FIGURE 18.9

Device power loss.

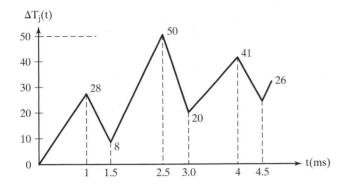

FIGURE 18.10

Junction temperature rise for Example 18.1.

For $t_2 = t_4 = t_6 = 0.5$ ms,

$$Z(t = t_2) = Z_2 = Z_4 = Z_6 = 0.025°\text{C/W}$$

Solution
Equation (18.4) can be applied directly to calculate the junction temperature rise.

$$\Delta T_J(t = 1 \text{ ms}) = T_J(t = 1 \text{ ms}) - T_{J0} = Z_1 P_1 = 0.035 \times 800 = 28°\text{C}$$

$$\Delta T_J(t = 1.5 \text{ ms}) = 28 - Z_2 P_1 = 28 - 0.025 \times 800 = 8°\text{C}$$

$$\Delta T_J(t = 2.5 \text{ ms}) = 8 + Z_3 P_3 = 8 + 0.035 \times 1200 = 50°\text{C}$$

$$\Delta T_J(t = 3 \text{ ms}) = 50 - Z_4 P_3 = 50 - 0.025 \times 1200 = 20°\text{C}$$

$$\Delta T_J(t = 4 \text{ ms}) = 20 + Z_5 P_5 = 20 + 0.035 \times 600 = 41°\text{C}$$

$$\Delta T_J(t = 4.5 \text{ ms}) = 41 - Z_6 P_5 = 41 - 0.025 \times 600 = 26°\text{C}$$

The junction temperature rise above case is shown in Figure 18.10.

Key Points of Section 18.2

- Power devices must be protected by heat sinks from excessive heat generated due to the dissipated power.
- The instantaneous junction temperature must not exceed the maximum manufacturer's specified temperature.

18.3 THERMAL MODELING OF POWER SWITCHING DEVICES

The power generated within a power device increases the device temperature, which, in turn, significantly affects its characteristics. For example, the mobility (both bulk and surface values), threshold voltage, drain resistance, and various oxide capacitances of a metal oxide semiconductor (MOS) transistor are temperature dependent. The temperature dependence of the bulk mobility causes an increasing resistance with a rise in temperature, thereby the power dissipation. These device parameters can affect the accuracy of the transistor model. Therefore, the instantaneous device heating should be coupled directly with the thermal model of the device and heat

TABLE 18.1   Equivalencies between the Electrical and Thermal Variables

Thermal	Electrical
Temperature, T in K	Voltage V in volts
Heat flow, P in watts	Current I in amperes
Thermal resistance, R_{th} in K/W	Resistance R in V/A (Ω)
Thermal capacitance, C_{th} in W.s/K	Capacitance C in A.s/V

sink. That is, the instantaneous power dissipation in the transistor is determined at all times and a current proportional to the dissipated power should be fed into the thermal equivalent network [13]. Table 18.1 shows the equivalence between the electrical and thermal variables.

18.3.1   Electrical Equivalent Thermal Model

The heat path from the chip to the heat sink can be modeled by one analogous to the electrical transmission line shown in Figure 18.11. Thermal resistance and thermal capacitance per unit length are needed for exact characterization of the thermal properties. The electrical power source $P(t)$ represents the power dissipation (heat flow) occurring in the chip in the thermal equivalent.

R_{th} and C_{th} are the lumped equivalent parameters of the elements within a device. These can be derived directly from the structure of the element when it basically exhibits one-dimensional heat flow. Figure 18.12 shows the thermal equivalent elements of a typical transistor in a package with solid cooling tab (e.g., TO-220 or D-Pak). The thermal equivalent elements can be determined directly from the physical structure. The structure is segmented into partial volumes (usually by a factor of 2 to 8) with progressively larger thermal time constants ($R_{th,i}, C_{th,i}$) in the direction of heat propagation.

If the heat-inducing area is smaller than the heat-conducting material cross section, a "heat-spreading" effect occurs, as shown in Figure 18.12. This effect can be taken into account by enlarging the heat-conducting cross-section A [1]. Thermal capacitance

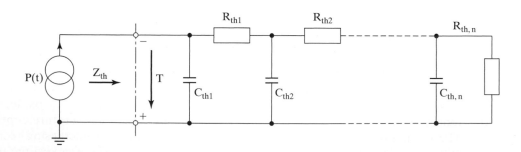

FIGURE 18.11

Electrical transmission line equivalent circuit or modeling heat conduction.

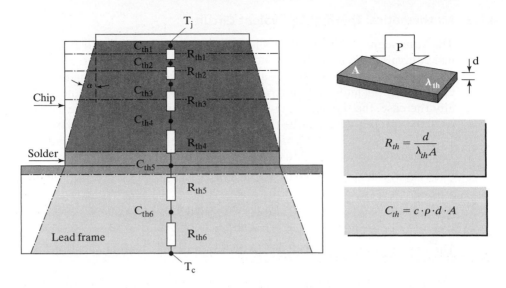

FIGURE 18.12

Thermal equivalent elements for modeling heat conduction. [Ref. 1, M. März]

C_{th} depends on the specific heat c and the mass density ρ. For heat propagation in homogeneous media, it is assumed that the spreading angle is about 40° and subsequent layers do not obstruct the heat propagation with low-heat conductivity. The size of every volume element must be determined exactly because its thermal capacitance has a decisive influence on the thermal impedance of the system when power dissipation pulses, with a very short duration, occur. Table 18.2 shows the thermal data for common materials.

One can also use the finite element analysis (FEA) method to calculate the heat flow. The FEA method divides the entire structure, which sometimes covers several tens or hundred thousand finite elements, into suitable substructures to determine lumped equivalent elements. Unless this process is supported by standard FEA software tools, this solution is too complex for most applications.

TABLE 18.2   Thermal Data for Common Materials [Ref. 1]

	ρ [g/cm³]	λ_{th} [W/(mK)]	c [J/(gK)]
Silicon	2.4	140	0.7
Solder (Sn-Pb)	9	60	0.2
Cu	7.6 to 8.9	310 to 390	0.385 to 0.42
Al	2.7	170 to 230	0.9 to 0.95
Al_2O_3	3.8	24	0.8
FR4	—	0.3	—
Heat conductive paste	—	0.4 to 2.6	—
Insulating foil	—	0.9 to 2.7	—

18.3.2 Mathematical Thermal Equivalent Circuit

The equivalent circuit shown in Figure 18.11 is often referred to as the natural or physical equivalent circuit of heat conduction and describes the internal temperature distribution correctly. It enables a clear correlation of equivalent elements to real structural elements. If the internal temperature distribution is not normally needed, which is usually the case, the thermal equivalent network, as shown in Figure 18.13, is frequently used to correctly describe the thermal behavior at the input terminals of the black box.

The individual RC elements represent the terms of a partial fractional division of the thermal transfer function of the system. Using the partial fractional representation, the step response of the thermal impedance can be expressed as given by

$$Z_{th}(t) = \sum_{i=1}^{n} R_i \left(1 - e^{\frac{t}{R_i C_i}} \right) \qquad (18.6)$$

The equivalent input impedance at the input terminals can be expressed as

$$Z_{th} = \cfrac{1}{sC_{th,1} + \cfrac{1}{sR_{th,1} + \cfrac{1}{sC_{th,2} + \dots + \cfrac{1}{R_{th,n}}}}} \qquad (18.7)$$

Standard curve-fitting algorithms of the computer software tools such as Mathcad can use the data from the transient thermal impedance curve to determine the individual R_{th} and C_{th} elements. The transient thermal impedance curve is normally provided in the device data sheet.

This simple model is based on the parameterization of the elements of the equivalent circuit by using data from measurements and related curve fitting. The usual procedure for the cooling curve, in practice, is to first heat the component with specific power dissipation P_k until it assumes a stationary temperature T_{jk}. If we know the

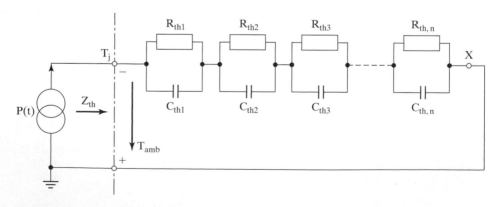

FIGURE 18.13

Simple mathematical equivalent circuit, Ref. 1.

exact temperature dependence of a parameter of the chip such as the forward voltage drop, the plot of $T_j(t)$ known as the *cooling curve* can be determined by reducing the power dissipation P_k progressively to zero. This cooling curve can be used to find the transient thermal impedance of the device.

$$Z_{th} = \frac{T_{jk} - T_j(t)}{P_k} \qquad (18.8)$$

18.3.3   Coupling of Electrical and Thermal Components

Coupling the thermal equivalent circuit with the device model, as shown in Figure 18.14 for a metal oxide semiconductor field-effect transistor (MOSFET), can simulate the instantaneous junction temperature. The instantaneous power dissipation in the device ($I_D V_{DS}$) is determined at all times, and a current proportional to the dissipated power is fed into the thermal equivalent network. The voltage at node T_j then gives the instantaneous junction temperature, which affects directly the temperature-dependent parameters of the MOSFET. The coupled circuit model can simulate the instantaneous junction temperature under dynamic conditions such as short circuit and overload.

The MOS channel can be described with a level-three MOS model (X1) in SPICE. The temperature is defined by global SPICE variable "Temp". The threshold voltage, the drain current, and the drain resistance are scaled according to the instantaneous junction temperature T_j. The drain current I_{Di} (Temp) is scaled with a temperature-dependent factor as given by

$$I_D(T_j) = I_{Di}(\text{Temp})\left(\frac{T_j}{\text{Temp}}\right)^{-3/2} \qquad (18.9)$$

The threshold voltage has a temperature coefficient of $-2.5\,\text{mV/K}$ and the effective gate voltage to the MOS device can be made temperature dependent by using the analog

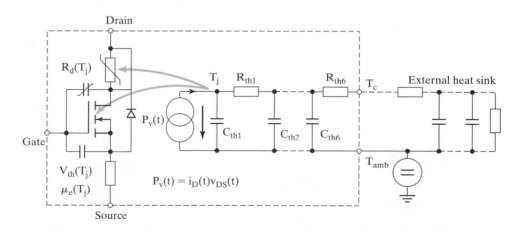

FIGURE 18.14

Coupling of electrical and thermal components. [Ref. 1, M. März]

behavior model of SPICE. Due to the importance of device thermal modeling, some device manufacturers (Infineon Technologies) offer temperature-dependent SPICE and SABER models for power devices.

Example 18.2 Calculating the Parameters of the Thermal Equivalent Circuit

A device in a TO-220 package is mounted with a 0.33-mm thick insulating foil on a small aluminum heat sink. This is shown in Figure 18.15a. The thermal resistance of the heat sink is $R_{th\_KK} = 25$ K/W and its mass is $m_{sk} = 2$ g. The surface area of the TO-220 package is $A_{sk} = 1$ cm². The surface area of the device chip is $A_{cu} = 10$ mm², the amount of copper around the pyramid stump is $m_{cu} = 1$ g, and the thickness of the copper is $d_{cu} = 0.8$ mm. Find the parameters of the thermal equivalent circuit.

Solution

Because the heat sink is small and compact, there is no need to divide the structure into several *RC* elements. The first-order thermal equivalent circuit is shown in Figure 18.15b. $m_{sk} = 2$ g, $R_{th\_KK} = 25$ K/W, $d_{foil} = 0.3$ mm, $A_{foil} = 1$ cm², $A_{cu} = 10$ mm², $m_{cu} = 1$ g, and $d_{cu} = 0.8$ mm. From Table 18.2, the specific heat of the aluminum material is $c_{sk} = 0.95$ J/(gK). Thus, the thermal capacitance of the heat sink becomes

$$C_{th\_KK} = c_{sk}m_{sk} = 0.95 \frac{\text{J}}{\text{gK}} \; 0.2 \frac{J}{K} = 1.9 \text{ g}$$

For insulating foil, Table 18.2 gives $\lambda_{th-foil} = 1.1$ W/mK. Thus, the thermal resistance of the foil is

$$R_{th\_foil} = \frac{d_{foil}}{\lambda_{th-foil}A_{foil}} = \frac{0.3 \text{ mm}}{1.1 \frac{\text{W}}{\text{mK}} \times 1 \text{ cm}^2} = 2.7 \frac{\text{K}}{\text{W}}$$

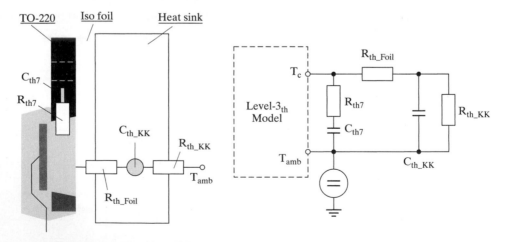

(a) Device mounted on a heat sink   (b) Thermal equivalent circuit

FIGURE 18.15

Device mounted on a heat sink and its thermal equivalent circuit. [Ref. 1, M. März]

For the copper chip, Table 18.2 gives $c_{cu} = 0.39$ J/(gK).

The thermal capacitance of the chip is $C_{th7} = c_{cu}m_{cu} = 0.39 \dfrac{J}{gK} \times 1\,g = 0.39 \dfrac{J}{K}$.

$$R_{th7} = \frac{d_{cu}}{\lambda_{th}A_{cu}} = \frac{0.8\ mm}{390\ \dfrac{W}{mK} \times 10\ mm^2} = 0.205\ \frac{K}{W}$$

Key Point of Section 18.3

- The parameters of a mathematical thermal model can be determined from the cooling curve of the device.

18.4 SNUBBER CIRCUITS

An *RC* snubber is normally connected across a semiconductor device to limit the *dv/dt* within the maximum allowable rating [2, 3]. The snubber could be polarized or unpolarized. A forward-polarized snubber is suitable when a thyristor or transistor is connected with an antiparallel diode, as shown in Figure 18.16a. The resistor, *R* limits the forward *dv/dt*; and R_1 limits the discharge current of the capacitor when the device is turned on.

A reverse-polarized snubber that limits the reverse *dv/dt* is shown in Figure 18.16b, where R_1 limits the discharge current of the capacitor. The capacitor does not discharge through the device, resulting in reduced losses in the device.

When a pair of thyristors is connected in inverse parallel, the snubber must be effective in either direction. An unpolarized snubber is shown in Figure 18.16c.

Key Point of Section 18.4

- Power devices must be protected from excessive *di/dt* and *dv/dt* by adding snubber circuits.

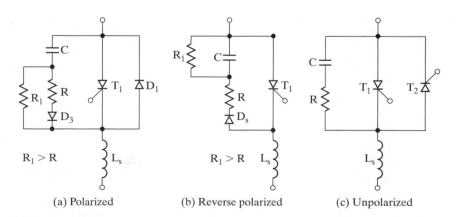

(a) Polarized (b) Reverse polarized (c) Unpolarized

FIGURE 18.16

Snubber networks.

18.5   REVERSE RECOVERY TRANSIENTS

Due to the reverse recovery time t_{rr} and recovery current I_R, an amount of energy is trapped in the circuit inductances and as a result transient voltage appears across the device. In addition to dv/dt protection, the snubber limits the peak transient voltage across the device. The equivalent circuit for a circuit arrangement is shown in Figure 18.17, where the initial capacitor voltage is zero and the inductor carries an initial current of I_R. The values of snubber RC are selected so that the circuit is slightly underdamped and Figure 18.18 shows the recovery current and transient voltage. Critical damping usually results in a large value of initial reverse voltage RI_R; and insufficient damping causes large overshoot of the transient voltage. In the following analysis, it is assumed that the recovery is abrupt and the recovery current is suddenly switched to zero.

The snubber current is expressed as

$$L\frac{di}{dt} + Ri + \frac{1}{C}\int i\, dt + v_c(t = 0) = V_s \tag{18.10}$$

$$v = V_s - L\frac{di}{dt} \tag{18.11}$$

with initial conditions $i(t = 0) = I_R$ and $v_c(t = 0) = 0$. We have seen in Section 2.11 that the form of the solution for Eq. (18.10) depends on the values of RLC. For an underdamped case, the solutions of Eqs. (18.10) and (18.11) yield the reverse voltage across the device as

$$v(t) = V_s - (V_s - RI_R)\left(\cos \omega t - \frac{\alpha}{\omega}\sin \omega t\right)e^{-\alpha t} + \frac{I_R}{\omega C}e^{-\alpha t}\sin \omega t \tag{18.12}$$

where

$$\alpha = \frac{R}{2L} \tag{18.13}$$

The undamped natural frequency is

$$\omega_0 = \frac{1}{\sqrt{LC}} \tag{18.14}$$

The damping ratio is

$$\delta = \frac{\alpha}{\omega_0} = \frac{R}{2}\sqrt{\frac{C}{L}} \tag{18.15}$$

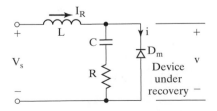

FIGURE 18.17

Equivalent circuit during recovery.

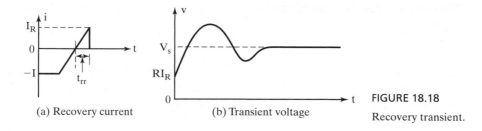

(a) Recovery current	(b) Transient voltage

FIGURE 18.18

Recovery transient.

and the damped natural frequency is

$$\omega = \sqrt{\omega_0^2 - \alpha^2} = \omega_0\sqrt{1 - \delta^2} \tag{18.16}$$

Differentiating Eq. (18.12) yields

$$\frac{dv}{dt} = (V_s - RI_R)\left(2\alpha \cos \omega t + \frac{\omega^2 - \alpha^2}{\omega} \sin \omega t\right)e^{-\alpha t}$$

$$+ \frac{I_R}{C}\left(\cos \omega t - \frac{\alpha}{\omega} \sin \omega t\right)e^{-\alpha t} \tag{18.17}$$

The initial reverse voltage and dv/dt can be found from Eqs. (18.12) and (18.17) by setting $t = 0$:

$$v(t = 0) = RI_R \tag{18.18}$$

$$\frac{dv}{dt}\bigg|_{t=0} = (V_s - RI_R)2\alpha + \frac{I_R}{C} = \frac{(V_s - RI_R)R}{L} + \frac{I_R}{C}$$

$$= V_s\omega_0(2\delta - 4d\delta^2 + d) \tag{18.19}$$

where the current factor (or ratio) d is given by

$$d = \frac{I_R}{V_s}\sqrt{\frac{L}{C}} = \frac{I_R}{I_p} \tag{18.20}$$

If the initial dv/dt in Eq. (18.19) is negative, the initial inverse voltage RI_R is the maximum and this may produce a destructive dv/dt. For a positive dv/dt, $V_s\omega_0(2\delta - 4d\delta^2 + d) > 0$ or

$$\delta < \frac{1 + \sqrt{1 + 4d^2}}{4d} \tag{18.21}$$

and the reverse voltage is maximum at $t = t_1$. The time t_1, which can be obtained by setting Eq. (18.17) equal to zero, is found as

$$\tan \omega t_1 = \frac{\omega[(V_s - RI_R)2\alpha + IR/C]}{(V_s - RI_R)(\omega^2 - \alpha^2) - \alpha I_R/C} \tag{18.22}$$

and the peak voltage can be found from Eq. (18.12):

$$V_p = v(t = t_1) \tag{18.23}$$

The peak reverse voltage depends on the damping ratio δ, and the current factor d. For a given value of d, there is an optimum value of damping ratio δ_o, which minimizes the peak voltage. However, the dv/dt varies with d and minimizing the peak voltage may not minimize the dv/dt. It is necessary to make a compromise between the peak voltage V_p, and dv/dt. McMurray [4] proposed to minimize the product $V_p(dv/dt)$ and the optimum design curves are shown in Figure 18.19, where dv/dt is the average value over time t_1 and d_o is the optimum value of current factor.

The energy stored in the inductor L, which is transferred to the snubber capacitor C, is dissipated mostly in the snubber resistor. This power loss is dependent on the switching frequency and load current. For high-power converters, where the snubber loss is significant, a nondissipative snubber that uses an energy recovery transformer, as shown in Figure 18.20, can improve the circuit efficiency. When the primary current

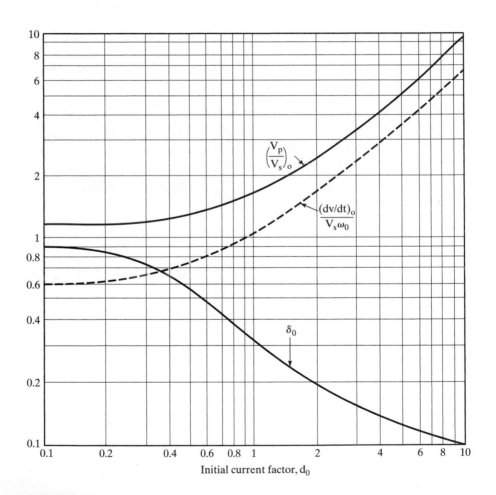

FIGURE 18.19

Optimum snubber parameters for compromise design. (Reproduced from W. McMurray, "Optimum snubbers for power semiconductors," *IEEE Transactions on Industry Applications*, Vol. 1A8, No. 5, 1972, pp. 503–510, Fig. 7, © 1972 by IEEE.)

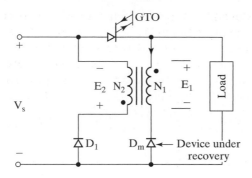

FIGURE 18.20

Nondissipative snubber.

rises, the induced voltage E_2 is positive and diode D_1 is reverse biased. If the recovery current of diode D_m starts to fall, the induced voltage E_2 becomes negative and diode D_1 conducts, returning energy to the dc supply.

Example 18.3 Finding the Values of the Snubber Circuit

The recovery current of a diode, as shown in Figure 18.17, is $I_R = 20$ A and the circuit inductance is $L = 50$ μH. The input voltage is $V_s = 220$ V. If it is necessary to limit the peak transient voltage to 1.5 times the input voltage, determine (a) the optimum value of current factor α_o, (b) the optimum damping factor δ_o, (c) the snubber capacitance C, (d) the snubber resistance R, (e) the average dv/dt, and (f) the initial reverse voltage.

Solution

$I_R = 20$ A, $L = 50$ μH, $V_s = 220$ V, and $V_p = 1.5 \times 220 = 330$ V. For $V_p/V_s = 1.5$, Figure 18.19 gives:

a. The optimum current factor $d_o = 0.75$.
b. The optimum damping factor $\delta_o = 0.4$.
c. From Eq. (18.20), the snubber capacitance (with $d = d_o$) is

$$C = L\left[\frac{I_R}{dV_s}\right]^2$$

$$= 50\left[\frac{20}{0.75 \times 220}\right]^2 = 0.735 \text{ μF} \qquad (18.24)$$

d. From Eq. (18.15), the snubber resistance is

$$R = 2\delta\sqrt{\frac{L}{C}}$$

$$= 2 \times 0.4\sqrt{\frac{50}{0.735}} = 6.6 \text{ Ω} \qquad (18.25)$$

e. From Eq. (18.14),

$$\omega_0 = \frac{10^6}{\sqrt{50 \times 0.735}} = 164{,}957 \text{ rad/s}$$

From Figure 18.19,

$$\frac{dv/dt}{V_s\omega_0} = 0.88$$

or

$$\frac{dv}{dt} = 0.88V_s\omega_0 = 0.88 \times 220 \times 164{,}957 = 31.9 \text{ V/}\mu\text{s}$$

f. From Eq. (18.18), the initial reverse voltage is

$$v(t = 0) = 6.6 \times 20 = 132 \text{ V}$$

Example 18.4   Finding the Peak Value, the *di/dt* and the *dv/dt* Values of the Snubber Circuit

An *RC*-snubber circuit, as shown in Figure 18.16c, has $C = 0.75\ \mu\text{F}$, $R = 6.6\ \Omega$, and input voltage $V_s = 220$ V. The circuit inductance is $L = 50\ \mu\text{H}$. Determine (a) the peak forward voltage V_p, (b) the initial *dv/dt*, and (c) the maximum *dv/dt*.

Solution
$R = 6.6\ \Omega$, $C = 0.75\ \mu\text{F}$, $L = 50\ \mu\text{H}$, and $V_s = 220$ V. By setting $I_R = 0$, the forward voltage across the device can be determined from Eq. (18.12),

$$v(t) = V_s - V_s\left(\cos \omega t - \frac{\alpha}{\omega}\sin \omega t\right)e^{-\alpha t} \tag{18.26}$$

From Eq. (18.17) for $I_R = 0$,

$$\frac{dv}{dt} = V_s\left(2\alpha \cos \omega t + \frac{\omega^2 - \alpha^2}{\omega}\sin \omega t\right)e^{-\alpha t} \tag{18.27}$$

The initial *dv/dt* can be found either from Eq. (18.27) by setting $t = 0$, or from Eq. (18.19) by setting $I_R = 0$

$$\left.\frac{dv}{dt}\right|_{t=0} = V_s 2\alpha = \frac{V_s R}{L} \tag{18.28}$$

The forward voltage is maximum at $t = t_1$. The time t_1, which can be obtained either by setting Eq. (18.27) equal to zero, or by setting $I_R = 0$ in Eq. (18.22), is given by

$$\tan \omega t_1 = -\frac{2\alpha\omega}{\omega^2 - \alpha^2} \tag{18.29}$$

$$\cos \omega t_1 = -\frac{\omega^2 - \alpha^2}{\omega^2 + \alpha^2} \tag{18.30}$$

$$\sin \omega t_1 = -\frac{2\alpha\omega}{\omega^2 + \alpha^2} \tag{18.31}$$

Substituting Eqs. (18.30) and (18.31) into Eq. (18.26), the peak voltage is found as

$$V_p = v(t = t_1) = V_s(1 + e^{-\alpha t_1}) \qquad (18.32)$$

where

$$\omega t_1 = \left(\pi - \tan^{-1}\frac{-2\delta\sqrt{1 - \delta^2}}{1 - 2\delta^2}\right) \qquad (18.33)$$

Differentiating Eq. (18.27) with respect to t, and setting equal to zero, dv/dt is maximum at $t = t_m$ when

$$-\frac{\alpha(3\omega^2 - \alpha^2)}{\omega}\sin \omega t_m + (\omega^2 - 3\alpha^2)\cos \omega t_m = 0$$

or

$$\tan \omega t_m = \frac{\omega(\omega^2 - 3\alpha^2)}{\alpha(3\omega^2 - \alpha^2)} \qquad (18.34)$$

Substituting the value of t_m in Eq. (18.27) and simplifying the sine and cosine terms yield the maximum value of dv/dt,

$$\left.\frac{dv}{dt}\right|_{max} = V_s\sqrt{\omega^2 + \alpha^2}\, e^{-\alpha t_m} \qquad \text{for } \delta \le 0.5 \qquad (18.35)$$

For a maximum to occur, $d(dv/dt)/dt$ must be positive if $t \le t_m$; and Eq. (18.34) gives the necessary condition as

$$\omega^2 - 3\alpha^2 \ge 0 \qquad \text{or} \qquad \frac{\alpha}{\omega} \le \frac{1}{\sqrt{3}} \qquad \text{or} \qquad \delta \le 0.5$$

Equation (18.35) is valid for $\delta \le 0.5$. For $\delta > 0.5$, dv/dt becomes maximum when $t = 0$ is obtained from Eq. (18.27),

$$\left.\frac{dv}{dt}\right|_{max} = \left.\frac{dv}{dt}\right|_{t=0} = V_s2\alpha = \frac{V_sR}{L} \qquad \text{for } \delta > 0.5 \qquad (18.36)$$

a. From Eq. (18.13), $\alpha = 6.6/(2 \times 50 \times 10^{-6}) = 66{,}000$ and from Eq. (18.14),

$$\omega_0 = \frac{10^6}{\sqrt{50 \times 0.75}} = 163{,}299 \text{ rad/s}$$

From Eq. (18.15), $\delta = (6.6/2)\sqrt{0.75/50} = 0.404$, and from Eq. (18.16),

$$\omega = 163{,}299\sqrt{1 - 0.404^2} = 149{,}379 \text{ rad/s}$$

From Eq. (18.33), $t_1 = 15.46$ μs; therefore, Eq. (18.32) yields the peak voltage, $V_p = 220(1 + 0.36) = 299.3$ V.

b. Equation (18.28) gives the initial dv/dt of $(220 \times 6.6/50) = 29$ V/μs.

c. Because $\delta < 0.5$, Eq. (18.35) should be used to calculate the maximum dv/dt. From Eq. (18.34), $t_m = 2.16$ μs and Eq. (18.35) yields the maximum dv/dt as 31.2 V/μs.

Note: $V_p = 299.3$ V and the maximum $dv/dt = 31.2$ V/μs. The optimum snubber design in Example 18.2 gives $V_p = 330$ V, and the average $dv/dt = 31.9$ μs.

Key Points of Section 18.5

- When the power device turns off at the end of the reverse recovery time, the energy stored in the di/dt limiting inductor due to the reverse current may cause a high dv/dt.
- The dv/dt snubber must be designed for optimum performance.

18.6   SUPPLY- AND LOAD-SIDE TRANSIENTS

A transformer is normally connected to the input side of the converters. Under steady-state condition, an amount of energy is stored in the magnetizing inductance L_m of the transformer, and switching off the supply produces a transient voltage to the input of the converter. A capacitor may be connected across the primary or secondary of the transformer to limit the transient voltage, as shown in Figure 18.21a, and in practice a resistance is also connected in series with the capacitor to limit the transient voltage oscillation.

Let us assume that the switch has been closed for a sufficiently long time. Under steady-state conditions, $v_s = V_m \sin \omega t$ and the magnetizing current is given by

$$L_m \frac{di}{dt} = V_m \sin \omega t$$

which gives

$$i(t) = -\frac{V_m}{\omega L_m} \cos \omega t$$

If the switch is turned off at $\omega t = \theta$, the capacitor voltage at the beginning of switch-off is

$$V_c = V_m \sin \theta \tag{18.37}$$

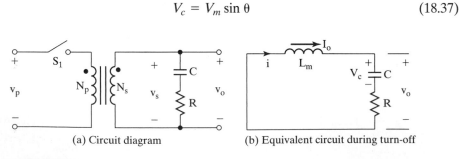

(a) Circuit diagram   (b) Equivalent circuit during turn-off

FIGURE 18.21

Switching off transient.

and the magnetizing current is

$$I_0 = -\frac{V_m}{\omega L_m} \cos \theta \tag{18.38}$$

The equivalent circuit during the transient condition is shown in Figure 18.21b and the capacitor current is expressed as

$$L_m \frac{di}{dt} + Ri + \frac{1}{C} \int i \, dt + v_c(t = 0) = 0 \tag{18.39}$$

and

$$v_0 = -L_m \frac{di}{dt} \tag{18.40}$$

with initial conditions $i(t = 0) = -I_0$ and $v_c(t = 0) = V_c$. The transient voltage $v_0(t)$ can be determined from Eqs. (18.39) and (18.40), for underdamped conditions. A damping ratio of $\delta = 0.5$ is normally satisfactory. The analysis can be simplified by assuming small damping tending to zero (i.e., $\delta = 0$, or $R = 0$). Eq. (D.16), which is similar to Eq. (18.39), can be applied to determine the transient voltage $v_0(t)$. The transient voltage $v_0(t)$ is the same as the capacitor voltage $v_c(t)$.

$$
\begin{aligned}
v_0(t) = v_c(t) &= V_c \cos \omega_0 t + I_0 \sqrt{\frac{L_m}{C}} \sin \omega_0 t \\
&= \left(V_c^2 + I_0^2 \frac{L_m}{C} \right)^{1/2} \sin (\omega_0 t + \phi) \tag{18.41} \\
&= V_m \left(\sin^2 \theta + \frac{1}{\omega^2 L_m C} \cos^2 \theta \right)^{1/2} \sin (\omega_0 t + \phi) \\
&= V_m \left(1 + \frac{\omega_0^2 - \omega^2}{\omega^2} \cos^2 \theta \right)^{1/2} \sin (\omega_0 t + \phi) \tag{18.42}
\end{aligned}
$$

where

$$\phi = \tan^{-1} \frac{V_c}{I_0} \sqrt{\frac{C}{L_m}} \tag{18.43}$$

and

$$\omega_0 = \frac{1}{\sqrt{CL_m}} \tag{18.44}$$

If $\omega_0 < \omega$, the transient voltage in Eq. (18.42) that is maximum when $\cos \theta = 0$ (or $\theta = 90°$) is

$$V_p = V_m \tag{18.45}$$

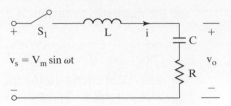

FIGURE 18.22

Equivalent circuit during switching on the supply.

In practice, $\omega_0 > \omega$ and the transient voltage, which is maximum when $\cos \theta = 1$ (or $\theta = 0°$), is

$$V_p = V_m \frac{\omega_0}{\omega} \tag{18.46}$$

which gives the peak transient voltage due to turning off the supply. Using the voltage and current relationship in a capacitor, the required amount of capacitance to limit the transient voltage can be determined from

$$C = \frac{I_0}{V_p \omega_0} \tag{18.47}$$

Substituting ω_0 from Eq. (18.46) into Eq. (18.47) gives us

$$C = \frac{I_0 V_m}{V_p^2 \omega} \tag{18.48}$$

Now with the capacitor connected across the transformer secondary, the maximum instantaneous capacitor voltage depends on the instantaneous ac input voltage at the instant of switching on the input voltage. The equivalent circuit during switch-on is shown in Figure 18.22, where L is the equivalent supply inductance plus the leakage inductance of the transformer.

Under normal operation, an amount of energy is stored in the supply inductance and the leakage inductance of the transformer. When the load is disconnected, transient voltages are produced due to the energy stored in the inductances. The equivalent circuit due to load disconnection is shown in Figure 18.23.

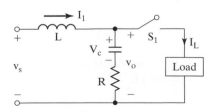

FIGURE 18.23

Equivalent circuit due to load disconnection.

Example 18.5   Finding the Performance Parameters of Switching Transients

A capacitor is connected across the secondary of an input transformer, as shown in Figure 18.21a, with zero damping resistance $R = 0$. The secondary voltage is $V_s = 120$ V, 60 Hz. If the magnetizing inductance referred to the secondary is $L_m = 2$ mH and the input supply to the transformer primary is disconnected at an angle of $\theta = 180°$ of the input ac voltage, determine (a) the initial capacitor value V_0, (b) the magnetizing current I_0, and (c) the capacitor value to limit the maximum transient capacitor voltage to $V_p = 300$ V.

Solution

$V_s = 120$ V, $V_m = \sqrt{2} \times 120 = 169.7$ V, $\theta = 180°$, $f = 60$ Hz, $L_m = 2$ mH, and $\omega = 2\pi \times 60 = 377$ rad/s.

    a. From Eq. (18.37), $V_c = 169.7 \sin \theta = 0$.

    b. From Eq. (18.38),

$$I_0 = -\frac{V_m}{\omega L_m} \cos \theta = \frac{169.7}{377 \times 0.002} = 225 \text{ A}$$

    c. $V_p = 300$ V. From Eq. (18.48), the required capacitance is

$$C = 225 \times \frac{169.7}{300^2 \times 377} = 1125.3 \text{ μF}$$

Key Points of Section 18.6

- Switching transients appear across the converter when the supply to the input transformer of a converter is turned off and also when an inductive load is disconnected from the converter.
- The power devices must be protected from these switching transients.

18.7   VOLTAGE PROTECTION BY SELENIUM DIODES AND METALOXIDE VARISTORS

The selenium diodes may be used for protection against transient overvoltages. These diodes have low forward voltage but well-defined reverse breakdown voltage. The characteristics of selenium diodes are shown in Figure 18.24a. Normally, the operating point lies before the knee of the characteristic curve and draws very small current from

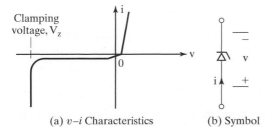

(a) v–i Characteristics   (b) Symbol

FIGURE 18.24

Characteristics of selenium diode.

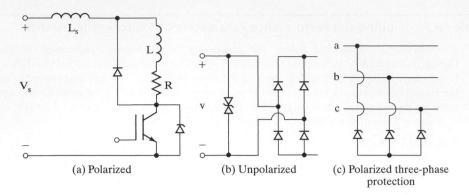

(a) Polarized   (b) Unpolarized   (c) Polarized three-phase protection

FIGURE 18.25

Voltage-suppression diodes.

the circuit. However, when an overvoltage appears, the knee point is crossed and the reverse current flow through the selenium increases suddenly, thereby typically limiting the transient voltage to twice the normal voltage.

A selenium diode (or suppressor) must be capable of dissipating the surge energy without undue temperature rise. Each cell of a selenium diode is normally rated at a root-mean-square (rms) voltage of 25 V, with a clamping voltage of typically 72 V. For the protection of dc circuit, the suppression circuit is polarized, as shown in Figure 18.25a. In ac circuits, as in Figure 18.25b, the suppressors are nonpolarized, so that they can limit overvoltages in both directions. For three-phase circuits, Y-connected-polarized suppressors, as shown in Figure 18.25c, can be used.

If a dc circuit of 240 V is to be protected with 25-V selenium cells, then $240/25 \approx$ 10 cells would be required and the total clamping voltage would be $10 \times 72 = 720$ V. To protect a single-phase ac circuit of 208 V, 60 Hz, with 25-V selenium cells, $208/25 \approx 9$ cells would be required in each direction and total of $2 \times 9 = 18$ cells would be necessary for nonpolarized suppression. Due to low internal capacitance, the selenium diodes do not limit the dv/dt to the same extent as compared to the RC-snubber circuits. However, they limit the transient voltages to well-defined magnitudes. In protecting a device, the reliability of an RC circuit is better than that of selenium diodes.

Varistors are nonlinear variable-impedance devices, consisting of metal oxide particles, separated by an oxide film or insulation. As the applied voltage is increased, the film becomes conductive and the current flow is increased. The current is expressed as

$$I = KV^\alpha \tag{18.49}$$

where K is a constant and V is the applied voltage. The value of α varies between 30 and 40.

Key Points of Section 18.7

- Power devices may be protected against transient overvoltages by selenium diodes or metal oxide varistors.

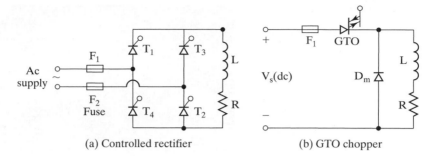

FIGURE 18.26

Protection of power devices.

- These devices draw very small current under normal operating conditions.
- However, when an overvoltage appears, the resistance of these devices decreases with the amount of overvoltage, thereby allowing more current flow and limiting the amount of the transient voltage.

18.8 CURRENT PROTECTIONS

The power converters may develop short circuits or faults and the resultant fault currents must be cleared quickly. Fast-acting fuses are normally used to protect the semiconductor devices. As the fault current increases, the fuse opens and clears the fault current in few millisconds.

18.8.1 Fusing

The semiconductor devices may be protected by carefully choosing the locations of the fuses, as shown in Figure 18.26 [5, 6]. However, the fuse manufacturers recommend placing a fuse in series with each device, as shown in Figure 18.27. The individual

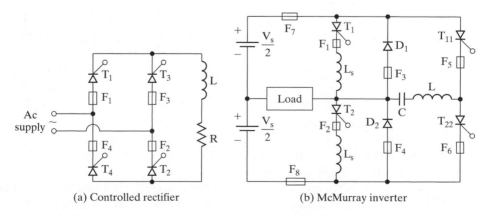

FIGURE 18.27

Individual protection of devices.

FIGURE 18.28

Semiconductor fuses. (Reproduced by
permission of Brush Electrical Machines Ltd.,
England.)

protection that permits better coordination between a device and its fuse, allows superior utilization of the device capabilities, and protects from short through faults (e.g., through T_1 and T_4 in Figure 18.27a). The various sizes of semiconductor fuses are shown in Figure 18.28 [7].

When the fault current rises, the fuse temperature also rises until $t = t_m$, at which time the fuse melts and arcs are developed across the fuse. Due to the arc, the impedance of the fuse is increased, thereby reducing the current. However, an arc voltage is formed across the fuse. The generated heat vaporizes the fuse element, resulting in an increased arc length and further reduction of the current. The cumulative effect is the extinction of the arc in a very short time. When the arcing is complete in time t_a, the fault is clear. The faster the fuse clears, the higher is the arc voltage [8].

The clearing time t_c is the sum of melting time t_m and arc time t_a. t_m is dependent on the load current, whereas t_a is dependent on the power factor or parameters of the fault circuit. The fault is normally cleared before the fault current reaches its first peak, and the fault current, which might have blown if there was no fuse, is called the *prospective fault current*. This is shown in Figure 18.29.

The current–time curves of devices and fuses may be used for the coordination of a fuse for a device. Figure 18.30a shows the current–time characteristics of a device and its fuse, where the device can be protected over the whole range of overloads. This type of protection is normally used for low-power converters. Figure 18.30b shows the more commonly used system in which the fuse is used for short-circuit protection at the beginning of the fault; and the normal overload protection is provided by circuit breaker or other current-limiting system.

If R is the resistance of the fault circuit and i is the instantaneous fault current between the instant of fault occurring and the instant of arc extinction, the energy fed to the circuit can be expressed as

$$W_e = \int Ri^2 \, dt \qquad (18.50)$$

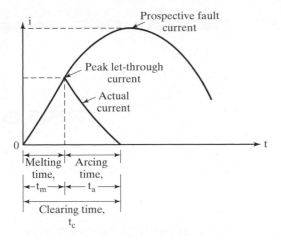

FIGURE 18.29

Fuse current.

If the resistance R remains constant, the value of i^2t is proportional to the energy fed to the circuit. The i^2t value is termed as the *let-through energy* and is responsible for melting the fuse. The fuse manufacturers specify the i^2t characteristic of the fuse and Figure 18.31 shows the typical characteristics of IR fuses, type TT350.

In selecting a fuse it is necessary to estimate the fault current and then to satisfy the following requirements:

1. The fuse must carry continuously the device rated current.
2. The i^2t let-through value of the fuse before the fault current is cleared must be less than the rated i^2t of the device to be protected.
3. The fuse must be able to withstand the voltage, after the arc extinction.
4. The peak arc voltage must be less than the peak voltage rating of the device.

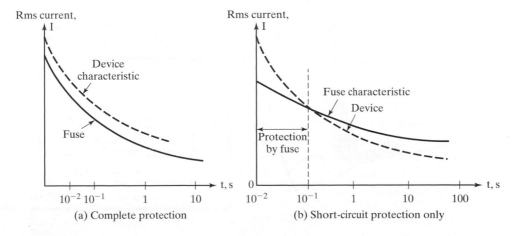

FIGURE 18.30

Current–time characteristics of device and fuse.

INTERNATIONAL RECTIFIER

T350 SERIES

290V/175-450A r.m.s. Semiconductor Fuses

Suitable for protecting High Power Semiconductor Devices

Conforms to BS88: Part 4: 1976 and IEC 269-4. ASTA certificate of short circuit ratings and verification of I^2t cut-off and arc voltage characteristics are available.

IMPORTANT

Note 1: Thyristors/diodes are rated in average current while fuses are rated in r.m.s. current. During steady state operation the fuse must not be operated in excess of its maximum r.m.s. rating.

Note 2: The maximum cap temperature and cap temperature rise above ambient of a fuse are critical design parameters. Caution should be taken during installation to ensure that the specified ratings are not exceeded. Some form of heatsink may be necessary.

The T350 Series of semiconductor fuses are available with I700 indicator fuses already fitted, for dimensional details refer to page E-12. For electrical, thermal and mechanical specifications on I700 refer to page E-5.
To complete part number add prefix "I" e.g. IT350-450.

ELECTRICAL SPECIFICATIONS

Maximum r.m.s. voltage rating:	290V
Maximum tested peak voltage	450V
Maximum d.c. voltage rating (L/R⩽15ms)	160V
Maximum arcing voltage for AC Supply Voltage = 240V	490V

For variation in arcing voltage with AC Supply Voltage

$$V_A = 100 + 1.63\ V_S$$

where V_A = Peak arc voltage, V_S = AC Supply Voltage

Fusing Factor	1.25
Force cooling Current uprating factor at 5 m/s	1.2

THERMAL AND MECHANICAL SPECIFICATIONS

Maximum cap temperature:	100°C
Maximum cap temperature rise above ambient	75°C
Weight:	170g (5.95 oz.)

Part number	RMS CURRENT (1) $T_{amb} = 25°C$	RMS CURRENT (1) $T_{amb} = 25°C$	MAX. POWER LOSS	PRE-ARCING (2) I^2t	TOTAL I^2t (2) at 120 V_{RMS}	TOTAL I^2t (2) at 240 V_{RMS}	NOTES
	A	A	W	A^2s	A^2s	A^2s	
T350–150	175	155	17	1600	7000	16000	1) Maximum current carrying ability, natural convection cooling using test arrangement as BS88: Part 4: 1976 conductors 1.0 to 1.6 A/mm² attachment.
T350–200	210	190	28	2100	10000	20000	
T350–250	250	230	28	4800	20000	40000	
T350–300	315	290	35	9000	34000	70000	
T350–350	355	320	35	13000	50000	100000	2) Typical values of I^2t at 20 times rated RMS current
T350–400	400	350	40	20000	75000	160000	
T350–450	450	400	42	30000	110000	220000	

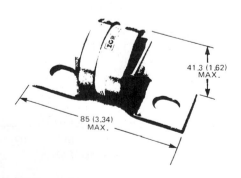

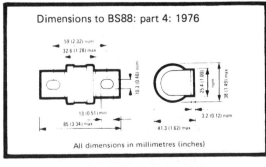

Dimensions to BS88: part 4: 1976

All dimensions in millimetres (inches)

FIGURE 18.31
Data sheet of IR fuse, type T350. (Courtesy of International Rectifier)

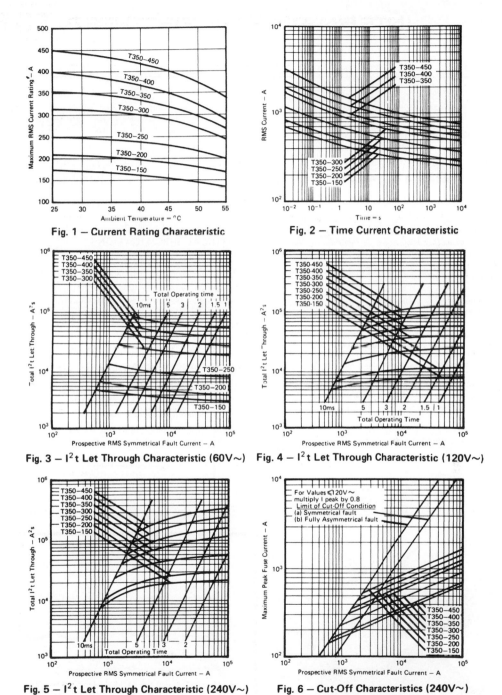

Fig. 1 – Current Rating Characteristic

Fig. 2 – Time Current Characteristic

Fig. 3 – I^2t Let Through Characteristic (60V~)

Fig. 4 – I^2t Let Through Characteristic (120V~)

Fig. 5 – I^2t Let Through Characteristic (240V~)

Fig. 6 – Cut-Off Characteristics (240V~)

FIGURE 18.31   (*Continued*)

INTERNATIONAL RECTIFIER

TT350 SERIES

290V/400-900A r.m.s. Semiconductor Fuses

Suitable for protecting High Power Semiconductor Devices

Conforms to BS88: Part 4: 1976 and IEC 269-4.

IMPORTANT:
Note 1: Thyristors/diodes are calibrated in average current ratings while fuses are calibrated in r.m.s. current ratings. During steady state operation the fuse must not be operated in excess of its maximum r.m.s. rating.

Note 2: The maximum cap temperature and cap temperature rise above ambient of a fuse are critical design parameters. Caution should be taken during installation to ensure that the specified ratings are not exceeded.

The TT350 Series of semiconductor fuses are available with I700 indicator fuses already fitted, for dimensional details refer to page E-67. For electrical, thermal and mechanical specifications refer to page E-66.
To complete part number add prefix ''I'' e.g. ITT350-900.

ELECTRICAL SPECIFICATIONS

Maximum r.m.s. voltage rating:	290V
Maximum tested peak voltage:	450V
Maximum d.c. voltage rating (L/R≤15ms)	160V
Maximum arcing voltage for AC Supply Voltage = 240V	490V

For variation in arcing voltage with AC Supply Voltage

$$V_A = 100 + 1.63\ V_s$$

Where V_A = Peak arc voltage, V_s = AC Supply Voltage

Fusing Factor:	1.25
Force cooling Current uprating factor at 5 m/s	1.2

THERMAL AND MECHANICAL SPECIFICATIONS

Maximum cap temperature:	100°C
Maximum cap temperature rise above ambient	75°C
Maximum gravitational withstand capability:	1500g (52.5 oz.)
(for device mounted radially to rotation.)	

Part number	RMS CURRENT (1) $T_{amb} = 25°C$	RMS CURRENT (1) $T_{amb} = 45°C$	MAX. POWER LOSS	PRE-ARCING (2) I^2t	TOTAL I^2t (2) at 120 V_{RMS}	TOTAL I^2t (L) at 240 V_{RMS}	NOTES
	A	A	W	kA²s	kA²s	kA²s	
TT350−400	400	350	60	8	35	80	1) Maximum current carrying ability, natural convection cooling using test arrangement as BS88: Part 4: 1976.
TT350−500	500	430	64	19	80	170	
TT350−600	630	540	75	35	150	300	
TT350−700	710	580	77	50	200	420	2) Typical values of I^2t at 20 times rated RMS Current.
TT350−800	800	660	82	70	300	650	
TT350−900	900	740	97	100	400	850	

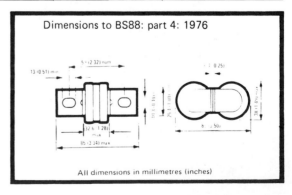

Dimensions to BS88: part 4: 1976

All dimensions in millimetres (inches)

FIGURE 18.31   *(Continued)*

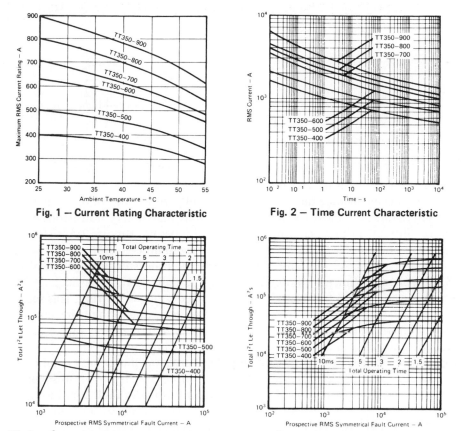

Fig. 1 – **Current Rating Characteristic**

Fig. 2 – **Time Current Characteristic**

Fig. 3 – I^2t Let Through Characteristic (60V~) Fig. 4 – I^2t Let Through Characteristic (120V~)

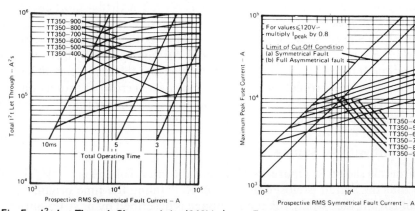

Fig. 5 – I^2t Let Through Characteristics (240V~) Fig. 6 – **Cut-Off Characteristics (240V~)**

FIGURE 18.31   (*Continued*)

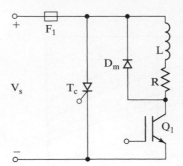

FIGURE 18.32

Protection by crowbar circuit.

In some applications it may be necessary to add a series inductance to limit the *di/dt* of the fault current and to avoid excessive *di/dt* stress on the device and fuse. However, this inductance may affect the normal performance of the converter.

Thyristors have more overcurrent capability than that of transistors. As a result, it is more difficult to protect transistors than thyristors. Bipolar transistors are gain-dependent and current-control devices. The maximum collector current is dependent on its base current. As the fault current rises, the transistor may go out of saturation and the collector–emitter voltage rises with the fault current, particularly if the base current is not changed to cope with the increased collector current. This secondary effect may cause higher power loss within the transistor due to the rising collector–emitter voltage and may damage the transistor, even though the fault current is not sufficient enough to melt the fuse and clear the fault current. Thus, fast-acting fuses may not be appropriate in protecting bipolar transistors under fault conditions.

Transistors can be protected by a crowbar circuit, as shown in Figure 18.32. A crowbar is used for protecting circuits or equipment under fault conditions, where the amount of energy involved is too high and the normal protection circuits cannot be used. A crowbar consists of a thyristor with a voltage- or current-sensitive firing circuit. The crowbar thyristor is placed across the converter circuit to be protected. If the fault conditions are sensed and crowbar thyristor T_c is fired, a virtual short circuit is created and the fuse link F_1 is blown, thereby relieving the converter from overcurrent.

MOSFETS are voltage control devices; and as the fault current rises, the gate voltage does not need to be changed. The peak current is typically three times the continuous rating. If the peak current is not exceeded and the fuse clears quickly enough, a fast-acting fuse may protect an MOSFET. However, a crowbar protection is also recommended. The fusing characteristics of insulated-gate bipolar transistors (IGBTs) are similar to that of bipolar junction transistors (BJTs).

18.8.2 Fault Current with Ac Source

An ac circuit is shown in Figure 18.33, where the input voltage is $v = V_m \sin \omega t$. Let us assume that the switch is closed at $\omega t = \theta$. Redefining the time origin, $t = 0$, at the instant of closing the switch, the input voltage is described by $v_s = V_m \sin(\omega t + \theta)$ for $t \geq 0$. Eq. (11.13) gives the current as

$$ i = \frac{V_m}{|Z_x|} \sin(\omega t + \theta - \phi_x) - \frac{V_m}{|Z_x|} \sin(\theta - \phi_x)e^{-Rt/L} \qquad (18.51) $$

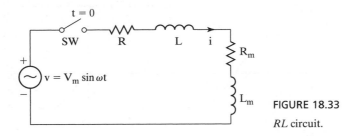

FIGURE 18.33

RL circuit.

where $|Z_x| = \sqrt{R_m^2 + (\omega L_x)^2}$, $\phi_x = \tan^{-1}(\omega L_x/R_x)$, $R_x = R + R_m$, and $L_x = L + L_m$. Figure 18.33 describes the initial current at the beginning of the fault. If there is a fault across the load, as shown in Figure 18.34, Eq. (18.51), which can be applied with an initial current of I_0 at the beginning of the fault, gives the fault current as

$$i = \frac{V_m}{|Z|} \sin(\omega t + \theta - \phi) + \left(I_0 - \frac{V_m}{|Z|}\right)\sin(\theta - \phi)e^{-Rt/L} \qquad (18.52)$$

where $|Z| = \sqrt{R^2 + (\omega L)^2}$ and $\phi = \tan^{-1}(\omega L/R)$. The fault current depends on the initial current I_0, power factor angle of the short-circuit path ϕ, and the angle of fault occurring θ. Figure 18.35 shows the current and voltage waveforms during the fault conditions in an ac circuit. For a highly inductive fault path, $\phi = 90°$ and $e^{-Rt/L} = 1$, and Eq. (18.52) becomes

$$i = -I_0 \cos \theta + \frac{V_m}{|Z|}\left[\cos \theta - \cos(\omega t + \theta)\right] \qquad (18.53)$$

If the fault occurs at $\theta = 0$, that is, at the zero crossing of the ac input voltage, $\omega t = 2n\pi$. Eq. (18.53) becomes

$$i = -I_0 + \frac{V_m}{Z}(1 - \cos \omega t) \qquad (18.54)$$

and Eq. (18.54) gives the maximum peak fault current, $-I_0 + 2V_m/Z$, which occurs $\omega t = \pi$. However, in practice, due to damping the peak current may be less than this.

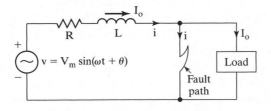

FIGURE 18.34

Fault in ac circuit.

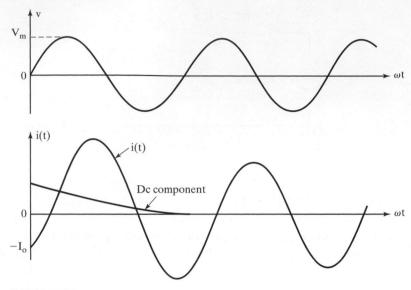

FIGURE 18.35
Transient voltage and current waveforms.

18.8.3 Fault Current with Dc Source

The current of a dc circuit in Figure 18.36 is given by

$$i = \frac{V_s}{R_x}(1 - e^{-R_x t/L_x}) \tag{18.55}$$

With an initial current of I_0 at the beginning of the fault current, as shown in Figure 18.37, the fault current is expressed as

$$i = I_0 e^{-Rt/L} + \frac{V_s}{R}(1 - e^{-Rt/L}) \tag{18.56}$$

The fault current and the fuse clearing time may be dependent on the time constant of the fault circuit. If the prospective current is low, the fuse may not clear the fault and a slowly rising fault current may produce arcs continuously without breaking the fault current. The fuse manufacturers specify the current–time characteristics for ac circuits and there are no equivalent curves for dc circuits. Because the dc fault currents do not have natural periodic zeros, the extinction of the arc is more difficult. For circuits

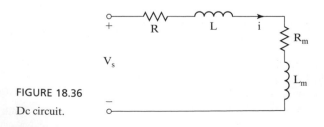

FIGURE 18.36
Dc circuit.

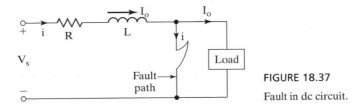

FIGURE 18.37

Fault in dc circuit.

operating from dc voltage, the voltage rating of the fuse should be typically 1.5 times the equivalent ac rms voltage. The fuse protection of dc circuits requires more careful design than that of ac circuits.

Example 18.6 Selecting a Fast-Acting Fuse for Protecting a Thyristor

A fuse is connected in series with each IR thyristor of type S30EF in the single-phase full converter, as shown in Figure 10.2a. The input voltage is 240 V, 60 Hz, and the average current of each thyristor is $I_a = 400$ A. The ratings of thyristors are $I_{T(AV)} = 540$ A, $I_{T(RMS)} = 850$ A, $I^2 t = 300$ kA^{2}s at 8.33 ms, $i^2\sqrt{t} = 4650$ kA$^2\sqrt{s}$, and $I_{TSM} = 10$ kA with reapplied $V_{RRM} = 0$, which may be the case if the fuse opens within a half-cycle. If the resistance of fault circuit is negligible and inductance is $L = 0.07$ mH, select the rating of a suitable fuse from Figure 18.31.

Solution
$V_s = 240$ V, $f_s = 60$ Hz. Let us try an IR fuse, type TT350-600. The short circuit current, that is also known as the prospective rms symmetric fault current is

$$I_{sc} = \frac{V_s}{Z} = 240 \times \frac{1000}{2\pi \times 60 \times 0.07} = 9094 \text{ A}$$

For the 540-A fuse, type TT350-600, and $I_{sc} = 9094$ A, the maximum peak fuse current is 8500 A, which is less than the peak thyristor current of $I_{TSM} = 10$ kA. The fuse $i^2 t$ is 280 kA^{2}s and total clearing time is $t_c = 8$ ms. Because t_c is less than 8.33 ms, the $i^2\sqrt{t}$ rating of the thyristor must be used. If thyristor $i^2\sqrt{t} = 4650 \times 10^3$ kA$^2\sqrt{s}$, then at $t_c = 8$ ms, thyristor $i^2 t = 4650 \times 10^3\sqrt{0.008} = 416$ kA^{2}s, which is 48.6% higher than the $i^2 t$ rating of the fuse (280 kA^{2}s). The $i^2 t$ and peak surge current ratings of the thyristor are higher than those of the fuse. Thus, the thyristor should be protected by the fuse.

Note: As a general rule of thumb, a fast-acting fuse with an rms current rating equal or less than the average current rating of the thyristor or diode normally can provide adequate protection under fault conditions.

Example 18.7 PSpice Simulation of the Instantaneous Fault Current

The ac circuit shown in Figure 18.38a has $R = 1.5$ Ω and $L = 15$ mH. The load parameters are $R_m = 5$ Ω and $L_m = 15$ mH. The input voltage is 208 V (rms), 60 Hz. The circuit has reached a steady-state condition. The fault across the load occurs at $\omega t + \theta = 2\pi$; that is, $\theta = 0$. Use PSpice to plot the instantaneous fault current.

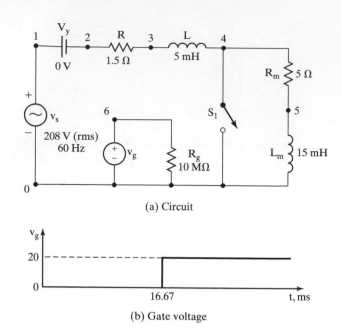

(a) Circuit

FIGURE 18.38

Fault in ac circuit for PSpice simulation.

(b) Gate voltage

Solution

$V_m = \sqrt{2} \times 208 = 294.16 \text{ V}$, $f = 60 \text{ Hz}$. The fault is simulated by a voltage control switch, whose control voltage is shown in Figure 18.38b. The list of the circuit file is as follows:

```
Example 18.7          Fault Current in AC Circuit
VS      1    0    SIN (0      294.16V  60HZ)
VY      1    2    DC    OV  ; Voltage source to measure input   current
Vg      6    0    PWL (16666.67US  OV 16666.68US  20V 60MS 20V)
Rg      6    0    10MEG     ; A very high resistance for control voltage
R       2    3    1.5
L       3    4    5MH
RM      4    5    5
LM      5    0    15MH
S1      4    0    6    0    SMOD     ; Voltage-controlled switch
.MODEL  SMOD  VSWITCH    (RON=0.01    ROFF=10E+5   VON=0.2V    VOFF=OV)
.TRAN   10US  40MS  0  50US           ; Transient analysis
.PROBE                                ; Graphics postprocessor
.options  abstol = 1.00n reltol = 0.01 vntol = 0.1 ITL5=50000  ;
convergence
.END
```

The PSpice plot is shown in Figure 18.39, where $I(VY)$ = fault current. Using the PSpice cursor in Figure 18.39 gives the initial current $I_o = -22.28 \text{ A}$ and the prospective fault current $I_p = 132.132 \text{ A}$.

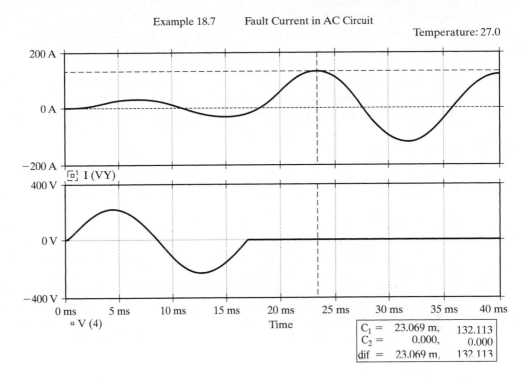

Example 18.7   Fault Current in AC Circuit

Temperature: 27.0

FIGURE 18.39

PSpice plot for Example 18.7.

Key Points of Section 18.8

- Power devices must the protected from the instantaneous fault conditions.
- The device peak, average, and rms currents must be greater than those under fault conditions.
- To protect the power device, the i^2t and $i^2\sqrt{t}$ values of the fuse must be less than that of the device under fault conditions.

18.9 ELECTROMAGNETIC INTERFERENCE

Power electronic circuits switch on and off large amounts of current at high voltages and thus can generate unwanted electrical signals, which affect other electronic systems. These unwanted signals occur at higher frequencies and give rise to EMI, also known as radio frequency interference (RFI). The signals can be transmitted to the other electronic systems by radiation through space or by conduction along cable. The low-level gate control circuit of the power converter can also be affected by EMI generated by its own high-power circuitry. When this occurs, the system is said to possess susceptibility to EMI. Any system that does not emit EMI above a given level, and is not affected by EMI, is said to have electromagnetic compatibility (EMC) [9].

There are three elements to any EMC system: (1) the source of the EMI, (2) the media through which it is transmitted, and (3) the receptor, which is any system that suffers adversely due to the received EMI. Therefore, EMC can be achieved (1) by reducing the EMI levels from the source, (2) blocking the propagation path of the EMI signals, or (3) by making the receiver less susceptible to the received EMI signals.

18.9.1 Sources of EMI

There are numerous sources of EMI, such as atmospheric noise, lightning, radar, radio, television, pagers, and mobile radio. EMI is also caused by sources such as switches, relays, motors, and fluorescent lights [10, 11]. The inrush current of transformers during turn-on is another source of interference, as is the rapid collapse of current in inductive elements, resulting in transient voltages. Integrated circuits also generate EMI due to their high-operating speeds and the close proximity of circuit elements on a silicon die, resulting in stray capacitive coupling.

Any power converter is a primary source of the EMI. The currents or voltages of a converter change very rapidly due to high-frequency switching such as rapid turn-on and turn-off of the power devices, nonsinusoidal voltages and currents through inductive loads, stored energy in stray inductors, and breaking of current by relay contacts and circuit breakers. Stray capacitances and inductances can also create oscillations, which can produce a wide spectrum of unwanted frequencies. The magnitude of the EMI depends on the peak energy stored in the capacitors at the instant of closing any static or power semiconductor switches.

18.9.2 Minimizing EMI Generation

Introducing resistance can damp oscillations. Using high-permeability material for the core can minimize harmonics generated by transformers, although this would cause the device to operate at high flux densities and result in large inrush current. Electrostatic shielding is often used in transformers to minimize coupling between primary and secondary. EMI signals can often be bypassed by high-frequency capacitors, or metal screens around circuitry to protect them from these signals. Twisted signal leads, or leads that are shielded, can be used to reduce coupling of EMI signals.

The collapse of flux in inductive circuits due to the saturation of the magnetic core often results in high-voltage transients, which can be prevented by providing a path for the inductive current to flow such as a freewheeling diode, a zener diode, or a voltage-dependent resistor. Emission from an electronic circuit and its susceptibility to these signals is significantly affected by the layout of the circuit, usually on a printed circuit board operating at high frequencies.

Power converter generated EMI can be reduced by advanced control techniques for minimizing input and output harmonics, operating at unity input power factor, lower total harmonic distribution (THD), and soft switching of power devices. The signal ground should have a low impedance to handle large signal currents, and making the ground plane large usually does this.

18.9.3 EMI Shielding

EMI can be radiated through space as electromagnetic waves, or it can be conducted as a current along a cable. A shield is a conducting material, which is placed in the path of the field to impede it. The effectiveness of the shield is determined by the distance between the source of EMI and the receiver, the type of field, and the characteristic of the material used in the shield. Shielding is effective in attenuating the interfering fields by absorption within its body, or by reflection off its surface.

EMI can also conduct as a current along a cable. If two adjacent cables carry currents i_1 and i_2, we can decompose them two components i_c and i_d such that

$$i_1 = i_c + i_d \qquad (18.57)$$

$$i_2 = i_c - i_d \qquad (18.58)$$

where

$$i_c = (i_1 + i_2)/2 \qquad (18.59)$$

is the common-mode current and

$$i_d = (i_1 - i_2)/2 \qquad (18.60)$$

is the differential-mode current.

Conduction can take the form of common-mode or differential-mode currents. For the differential mode, the currents are equal and opposite on the two wires and are caused primarily by other users on the same lines. Common-mode currents are almost equal in amplitude on the two lines, but travel in the same direction. These currents are mainly caused by coupling of radiated EMI to the power lines and by stray capacitive coupling to the body of the equipment. Transmitted EMI along a cable can be minimized by means of suppression filters, which consist basically of inductive and capacitive elements. There is a variety of such filters. The location of the filter is also important, and it should generally be placed directly at the source of EMI.

18.9.4 EMI Standards

Most countries have their own standards organizations [11, 12] that regulate the EMC, for example, the FCC in the United States, the BSI in the United Kingdom, and the VDE in Germany. The needs of commercial and military equipment are different. The commercial standards specify the requirements to protect radio, telecommunication,

TABLE 18.3   FCC EMI Limits [Ref. 10]

	Conducted EMI Limits			Radiated EMI Limits @ 30 m		
	Maximum R.F. Line Voltage (μV)				Field Strength (μV/m)	
Frequency Range (MHz)	Class A	Class B	Frequency Range (MHz)	Class A	Class B
0.45–1.6	1000	250	30–88	30	100
1.6–30	3000	250	88–216	50	150
			216–1000	70	200

TABLE 18.4   European EN 55022 EMI Limits [Refs. 10, 11]

Conducted EMI Limits			Radiated EMI Limits		
Frequency Range (MHz)	Quasi-Peak Limits dB (μV)		Frequency Range (MHz)	Quasi-Peak Limits dB (μV/m)	
	Class A	Class B		Class A@ 30 m	Class B@ 10 m
0.15–0.5	79	66–56	30–230	30	30
0.50–5	73	56	230–1000	37	37
5–30	73	60			

television, domestic, and industrial systems. The military has its own requirements, referred to as MIL-STD and DEF standards. The military requires that its equipment continue to perform in battlefield conditions.

The FCC administers the use of the frequency spectrum in the United States and its rules span many areas. The FCC classifies approvals into two classes, as shown in Table 18.3. FCC Class A is for commercial users, and FCC Class B has a more stringent requirement and is intended for domestic equipment. The European Standard EN 55022 that covers EMI requirements for information technology equipment grants two classes of approval: Class A and Class B, as shown in Table 18.4. Class A is less stringent and is intended for commercial users. Class B is used for domestic equipment.

Key Points of Section 18.9

- Power electronic circuits, by switching large amounts of current at high voltages, can generate unwanted electrical signals.
- These unwanted signals can give rise to electromagnetic interference and can affect the low-level gate control circuit.
- The sources of the EMI, and their propagation and effects on the receptor circuits should be minimized.

SUMMARY

The power converters must be protected from overcurrents and overvoltages. The junction temperature of power semiconductor devices must be maintained within their maximum permissible values. The instantaneous junction temperature under short-circuit and overload conditions can be simulated in SPICE by using a temperature-dependent device model in conjunction with the thermal equivalent circuit of the heat sink. The heat produced by the device may be transferred to the heat sinks by air and liquid cooling. Heat pipes can also be used. The reverse recovery currents and disconnection of load (and supply line) cause voltage transients due to energy stored in the line inductances.

The voltage transients are normally suppressed by the same *RC* snubber circuit, which is used for *dv/dt* protection. The snubber design is very important to limit the *dv/dt* and peak voltage transients within the maximum ratings. Selenium diodes and varistors can be used for transient voltage suppression.

A fast-acting fuse is normally connected in series with each device for over-current protection under fault conditions. However, fuses may not be adequate for protecting transistors and other protection means, (e.g., a crowbar) may be required.

Power electronic circuits, by switching large amounts of current at high voltages, can generate unwanted electrical signals, which can give rise to EMI. The signals can be transmitted to the other electronic systems by radiation through space or by conduction along cable.

REFERENCES

[1] M. März and P. Nance, "Thermal modeling of power electronic systems," *Infineon Technologies*, 1998, pp. 1–20. www.infenion.com

[2] W. McMurray, "Selection of snubber and clamps to optimize the design of transistor switching converters," *IEEE Transactions on Industry Applications*, Vol. IAI6, No. 4, 1980, pp. 513–523.

[3] T. Undeland, "A snubber configuration for both power transistors and GTO PWM inverter," *IEEE Power Electronics Specialist Conference*, 1984, pp. 42–53.

[4] W. McMurray, "Optimum snubbers for power semiconductors," *IEEE Transactions on Industry Applications*, Vol. IA8, No. 5, 1972, pp. 503–510.

[5] A. F. Howe, P. G. Newbery, and N. P. Nurse, "Dc fusing in semiconductor circuits," *IEEE Transactions on Industry Applications*, Vol. IA22, No. 3, 1986, pp. 483–489.

[6] L. O. Erickson, D. E. Piccone, L. J. Willinger, and W. H. Tobin, "Selecting fuses for power semiconductor devices," *IEEE Industry Applications Magazine*, September/October 1996, pp. 19–23.

[7] International Rectifiers, *Semiconductor Fuse Applications Handbook* (No. HB50), El Segundo, CA: International Rectifiers, 1972.

[8] A. Wright and P. G. Newbery, *Electric Fuses*. London: Peter Peregrinus Ltd. 1994.

[9] T. Tihanyi, *Electromagnetic Compatibility in Power Electronics*. New York: Butterworth–Heinemann. 1995.

[10] F. Mazda, *Power Electronics Handbook*. Oxford, UK: Newnes, Butterworth–Heinemann. 1997, Chapter 4—Electromagnetic Compatibility, pp. 99–120.

[11] G. L. Skibinski, R. J. Kerman, and D. Schlegel, "EMI emissions of modern PWM ac drives," *IEEE Industry Applications Magazine*, November/December 1999, pp. 47–80.

[12] ANSI/IEEE Standard—*518: Guide for the Installation of Electrical Equipment to Minimize Electrical Noise Inputs to Controllers from External Sources*. IEEE Press. 1982.

[13] DYNEX Semiconductor, *Calculation of Junction Temperature*, Application note: AN4506, January 2000. www.dynexsemi.com

REVIEW QUESTIONS

18.1 What is a heat sink?

18.2 What is the electrical analog of heat transfer from a power semiconductor device?

18.3 What are the precautions to be taken in mounting a device on a heat sink?

18.4 What is a heat pipe?

18.5 What are the advantages and disadvantages of heat pipes?

18.6 What are the advantages and disadvantages of water cooling?

18.7 What are the advantages and disadvantages of oil cooling?

18.8 Why is it necessary to determine the instantaneous junction temperature of a device?

18.9 Why it is important to use the temperature-dependent device model for simulating the instantaneous junction temperature of the device?

18.10 What is the physical thermal equivalent circuit model?

18.11 What is the mathematical thermal equivalent circuit model?

18.12 What are the differences between the physical and mathematical thermal equivalent models?

18.13 What is a polarized snubber?

18.14 What is a nonpolarized snubber?

18.15 What is the cause of reverse recovery transient voltage?

18.16 What is the typical value of damping factor for an *RC* snubber?

18.17 What are the considerations for the design of optimum *RC* snubber components?

18.18 What is the cause of load-side transient voltages?

18.19 What is the cause of supply-side transient voltages?

18.20 What are the characteristics of selenium diodes?

18.21 What are the advantages and disadvantages of selenium voltage suppressors?

18.22 What are the characteristics of varistors?

18.23 What are the advantages and disadvantages of varistors in voltage suppressions?

18.24 What is a melting time of a fuse?

18.25 What is an arcing time of a fuse?

18.26 What is a clearing time of a fuse?

18.27 What is a prospective fault current?

18.28 What are the considerations in selecting a fuse for a semiconductor device?

18.29 What is a crowbar?

18.30 What are the problems of protecting bipolar transistors by fuses?

18.31 What are the problems of fusing dc circuits?

18.32 How is EMI transmitted to the receptor circuit?

18.33 What are the sources of EMI?

18.34 How can the EMI generation be minimized?

18.35 How can an electronic or electric circuit be protected from EMI?

PROBLEMS

18.1 The power loss in a device is shown in Figure P18.1. Plot the instantaneous junction temperature rise above case. For $t_1 = t_3 = t_5 = t_7 = 0.5$ ms, $Z_1 = Z_3 = Z_5 = Z_7 = 0.025°C/W$.

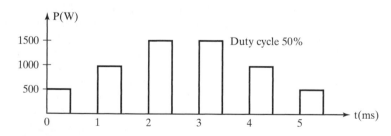

FIGURE P18.1

18.2 The power loss in a device is shown in Figure P18.2. Plot the instantaneous junction temperature rise above the case temperature. For $t_1 = t_2 = \cdots = t_9 = t_{10} = 1$ ms, $Z_1 = Z_2 = \cdots = Z_9 = Z_{10} = 0.035°C/W$. (*Hint*: Approximate by five rectangular pulses of equal duration.)

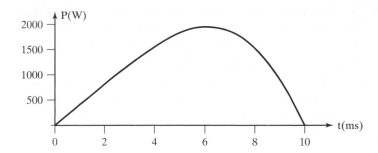

FIGURE P18.2

18.3 The current waveform through an IR thyristor, type S30EF, is shown in Figure 7.26. Plot **(a)** the power loss against time, and **(b)** the instantaneous junction temperature rise above case. (*Hint*: Assume power loss during turn-on and turn-off as rectangles.)

18.4 The recovery current of a device, as shown in Figure 18.17, is $I_R = 30$ A and the circuit inductance is $L = 20$ μH. The input voltage is $V_s = 200$ V. If it is necessary to limit the peak transient voltage to 1.8 times the input voltage, determine **(a)** the optimum value of current ratio d_o; **(b)** the optimum damping factor δ_o; **(c)** the snubber capacitance C; **(d)** the snubber resistance R; **(e)** the average dv/dt; and **(f)** the initial reverse voltage.

18.5 The recovery current of a device, as shown in Figure 18.17, is $I_R = 10$ A and the circuit inductance is $L = 80$ μH. The input voltage is $V_s = 200$ V. The snubber resistance is $R = 2\ \Omega$ and snubber capacitance is $C = 50$ μF. Determine **(a)** the damping ratio δ, **(b)** the peak transient voltage V_p, **(c)** the current ratio d, **(d)** the average dv/dt, and **(e)** the initial reverse voltage.

18.6 An RC snubber circuit, as shown in Figure 18.16c, has $C = 1.5$ μF, $R = 4.5\ \Omega$, and the input voltage is $V_s = 220$ V. The circuit inductance is $L = 20$ μH. Determine **(a)** the peak forward voltage V_p, **(b)** the initial dv/dt, and **(c)** the maximum dv/dt.

18.7 An RC snubber circuit, as shown in Figure 18.16c, has a circuit inductance of $L = 20$ μH. The input voltage is $V_s = 200$ V. If it is necessary to limit the maximum dv/dt to 20 V/μs and damping factor is $\delta = 0.4$, determine **(a)** the snubber capacitance C, and **(b)** the snubber resistance R.

18.8 An RC snubber circuit, as shown in Figure 18.16c, has circuit inductance of $L = 50$ μH. The input voltage $V_s = 220$ V. If it is necessary to limit the peak voltage to 1.5 times the input voltage, and the damping factor, $\alpha = 9500$, determine **(a)** the snubber capacitance C, and **(b)** the snubber resistance R.

18.9 A capacitor is connected to the secondary of an input transformer, as shown in Figure 18.21a, with zero damping resistance $R = 0$. The secondary voltage is $V_s = 208$ V, 60 Hz, and the magnetizing inductance referred to the secondary is $L_m = 3.5$ mH. If the input supply to the transformer primary is disconnected at an angle of $\theta = 120°$ of the input ac voltage, determine **(a)** the initial capacitor value V_0, **(b)** the magnetizing current I_0, and **(c)** the capacitor value to limit the maximum transient capacitor voltage to $V_p = 350$ V.

18.10 The circuit in Figure 18.23 has a load current of $I_L = 10$ A and the circuit inductance is $L = 50$ μH. The input voltage is dc with $V_s = 200$ V. The snubber resistance is $R = 1.5\ \Omega$ and snubber capacitance is $C = 50$ μF. If the load is disconnected, determine **(a)** the damping factor δ, and **(b)** the peak transient voltage V_p.

18.11 Selenium diodes are used to protect a three-phase circuit, as shown in Figure 18.25c. The three-phase voltage is 208 V, 60 Hz. If the voltage of each cell is 25 V, determine the number of diodes.

18.12 The load current at the beginning of a fault in Figure 18.34 is $I_0 = 10$ A. The ac voltage is 208 V, 60 Hz. The resistance and inductance of the fault circuit are $L = 5$ mH and $R = 1.5$ Ω, respectively. If the fault occurs at an angle of $\theta = 45°$, determine the peak value of prospective current in the first half-cycle.

18.13 Repeat Problem 18.12 if $R = 0$.

18.14 The current through a fuse is shown in Figure P18.14. The total i^2t of the fuse is 5400 A^{2}s. If the arcing time $t_a = 0.1$ s and the melting time $t_m = 0.05$ s, determine the peak let-through current I_p.

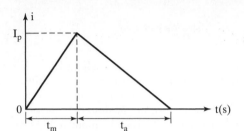

FIGURE P18.14

18.15 The load current in Figure 18.37 is $I_0 = 0$ A. The dc input voltage is $V_s = 220$ V. The fault circuit has an inductance of $L = 2$ mH and negligible resistance. The total i^2t of the fuse is 4500 A^{2}s. The arcing time is 1.5 times the melting time. Determine **(a)** the melting time t_m, **(b)** the clearing time t_c, and **(c)** the peak let-through current I_p.

18.16 Use PSpice to verify the design in Problem 18.7.

18.17 Use PSpice to verify the results of Problem 18.9.

18.18 Use PSpice to verify the results of Problem 18.10.

APPENDIX A

Three-Phase Circuits

In a single-phase circuit, as shown in Figure A.1a, the current is expressed as

$$\bar{I} = \frac{V \angle \alpha}{R + jX} = \frac{V \angle \alpha - \theta}{Z} \tag{A.1}$$

where $Z = (R^2 + X^2)^{1/2}$ and $\theta = \tan^{-1}(X/R)$. The power can be found from

$$P = VI \cos \theta \tag{A.2}$$

where $\cos \theta$ is called the *power factor*, and θ, which is the angle of the load impedance, is known as the *power factor angle*. It is shown in Fig. A.1b.

A three-phase circuit consists of three sinusoidal voltages of equal magnitudes and the phase angles between the individual voltages are 120°. A Y-connected load connected to a three-phase source is shown in Figure A.2a. If the three-phase voltages are

$$\bar{V}_a = V_p \angle 0$$

$$\bar{V}_b = V_p \angle -120°$$

$$\bar{V}_c = V_p \angle -240°$$

the line-to-line voltages as shown in Fig. A.2b are

$$\bar{V}_{ab} = \bar{V}_a - \bar{V}_b = \sqrt{3}\, V_p \angle 30° = V_L \angle 30°$$

$$\bar{V}_{bc} = \bar{V}_b - \bar{V}_c = \sqrt{3}\, V_p \angle -90° = V_L \angle -90°$$

$$\bar{V}_{ca} = \bar{V}_c - \bar{V}_a = \sqrt{3}\, V_p \angle -210° = V_L \angle -210°$$

Thus, a line-to-line voltage, V_L, is $\sqrt{3}$ times of a phase voltage, V_p. The three line currents, which are the same as phase currents, are

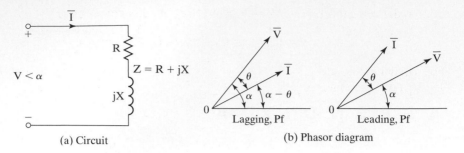

(a) Circuit

(b) Phasor diagram

FIGURE A.1

Single-phase circuit.

$$\bar{\mathbf{I}}_a = \frac{\overline{\mathbf{V}}_a}{Z_a \angle \theta_a} = \frac{V_p}{Z_a} \angle -\theta_a$$

$$\bar{\mathbf{I}}_b = \frac{\overline{\mathbf{V}}_b}{Z_b \angle \theta_b} = \frac{V_p}{Z_b} \angle -120° - \theta_b$$

$$\bar{\mathbf{I}}_c = \frac{\overline{\mathbf{V}}_c}{Z_c \angle \theta_c} = \frac{V_p}{Z_c} \angle -240° - \theta_c$$

The input power to the load is

$$P = V_a I_a \cos \theta_a + V_b I_b \cos \theta_b + V_c I_c \cos \theta_c \qquad (A.3)$$

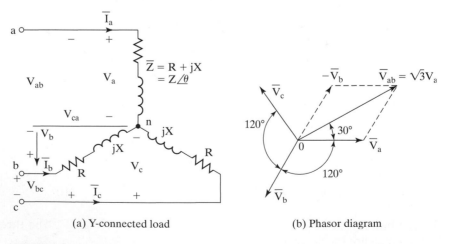

(a) Y-connected load

(b) Phasor diagram

FIGURE A.2

Y-connected three-phase circuit.

For a balanced supply, $V_a = V_b = V_c = V_p$. Eq. (A.3) becomes

$$P = V_p(I_a \cos \theta_a + I_b \cos \theta_b + I_c \cos \theta_c) \tag{A.4}$$

For a balanced load, $Z_a = Z_b = Z_c = Z$, $\theta_a = \theta_b = \theta_c = \theta$, and $I_a = I_b = I_c = I_p = I_L$, Eq. (A.4) becomes

$$P = 3V_p I_p \cos \theta$$
$$= 3\,\frac{V_L}{\sqrt{3}}\,I_L \cos \theta = \sqrt{3}\,V_L I_L \cos \theta \tag{A.5}$$

A delta-connected load is shown in Figure A.3a, where the line voltages are the same as the phase voltages. If the three phase voltages are

$$\overline{\mathbf{V}}_a = \overline{\mathbf{V}}_{ab} = V_L \underline{/0} = V_p \underline{/0}$$
$$\overline{\mathbf{V}}_b = \overline{\mathbf{V}}_{bc} = V_L \underline{/-120°} = V_p \underline{/-120°}$$
$$\overline{\mathbf{V}}_c = \overline{\mathbf{V}}_{ca} = V_L \underline{/-240°} = V_p \underline{/-240°}$$

the three-phase currents as shown in Figure A.3b are

$$\overline{\mathbf{I}}_{ab} = \frac{\overline{\mathbf{V}}_a}{Z_a \underline{/\theta_a}} = \frac{V_L}{Z_a} \underline{/-\theta_a} = I_p \underline{/-\theta_a}$$

$$\overline{\mathbf{I}}_{bc} = \frac{\overline{\mathbf{V}}_b}{Z_b \underline{/\theta_b}} = \frac{V_L}{Z_b} \underline{/-120° - \theta_b} = I_p \underline{/-120° - \theta_b}$$

$$\overline{\mathbf{I}}_{ca} = \frac{\overline{\mathbf{V}}_c}{Z_c \underline{/\theta_c}} = \frac{V_L}{Z_c} \underline{/-240° - \theta_c} = I_p \underline{/-240° - \theta_c}$$

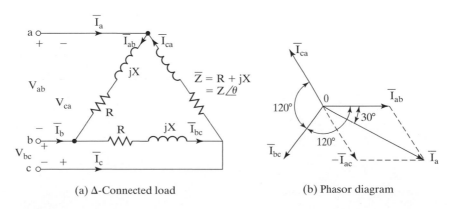

(a) Δ-Connected load   (b) Phasor diagram

FIGURE A.3

Delta-connected load.

and the three line currents are

$$\mathbf{\bar{I}}_a = \mathbf{\bar{I}}_{ab} - \mathbf{\bar{I}}_{ca} = \sqrt{3}\, I_p \angle{-30° - \theta_a} = I_L \angle{-30° - \theta_a}$$

$$\mathbf{\bar{I}}_b = \mathbf{\bar{I}}_{bc} - \mathbf{\bar{I}}_{ab} = \sqrt{3}\, I_p \angle{-150° - \theta_b} = I_L \angle{-150° - \theta_b}$$

$$\mathbf{\bar{I}}_c = \mathbf{\bar{I}}_{ca} - \mathbf{\bar{I}}_{bc} = \sqrt{3}\, I_p \angle{-270° - \theta_c} = I_L \angle{-270° - \theta_c}$$

Therefore, in a delta-connected load, a line current is $\sqrt{3}$ times of a phase current.
   The input power to the load is

$$P = V_{ab}I_{ab} \cos\theta_a + V_{bc}I_{bc} \cos\theta_b + V_{ca}I_{ca} \cos\theta_c \tag{A.6}$$

For a balanced supply, $V_{ab} = V_{bc} = V_{ca} = V_L$, Eq. (A.6) becomes

$$P = V_L(I_{ab} \cos\theta_a + I_{bc} \cos\theta_b + I_{ca} \cos\theta_c) \tag{A.7}$$

For a balanced load, $Z_a = Z_b = Z_c = Z$, $\theta_a = \theta_b = \theta_c = \theta$, and $I_{ab} = I_{bc} = I_{ca} = I_p$, Eq. (A.7) becomes

$$P = 3V_p I_p \cos\theta$$

$$= 3V_L \frac{I_L}{\sqrt{3}} \cos\theta = \sqrt{3} V_L I_L \cos\theta \tag{A.8}$$

Note: Equations (A.5) and (A.8), which express power in a three-phase circuit, are the same. For the same phase voltages, the line currents in a delta-connected load are $\sqrt{3}$ times that of a Y-connected load.

APPENDIX B

Magnetic Circuits

A magnetic ring is shown in Figure B.1. If the magnetic field is uniform and normal to the area under consideration, a magnetic circuit is characterized by the following equations:

$$\phi = BA \tag{B.1}$$

$$B - \mu H \tag{B.2}$$

$$\mu = \mu_r \mu_0 \tag{B.3}$$

$$\mathcal{F} = NI = Hl \tag{B.4}$$

Where ϕ = flux, webers

   B = flux density, webers/m^2 (or teslas)

   H = magnetizing force, ampere-turns/meter

   μ = permeability of the magnetic material

   μ_0 = permeability of the air ($= 4\pi \times 10^{-7}$)

   μ_r = relative permeability of the material

   $\mathcal{F}$ = magnetomotive force, ampere-turns (At)

   N = number of turns in the winding

   I = current through the winding, amperes

   l = length of the magnetic circuit, meters

If the magnetic circuit consists of different sections, Eq. (B.4) becomes

$$\mathcal{F} = NI = \Sigma \, H_i l_i \tag{B.5}$$

where H_i and l_i are the magnetizing force and length of ith section, respectively.

The reluctance of a magnetic circuit is related to the magnetomotive force and flux by

$$\mathcal{R} = \frac{\mathcal{F}}{\phi} = \frac{NI}{\phi} \tag{B.6}$$

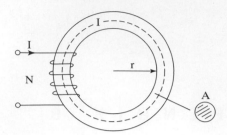

FIGURE B.1

Magnetic ring.

and $\mathcal{R}$ depends on the type and dimensions of the core,

$$\mathcal{R} = \frac{l}{\mu_r \theta_0 A} \tag{B.7}$$

The permeability depends on the B–H characteristic and is normally much larger than that of the air. A typical B–H characteristic that is nonlinear is shown in Figure B.2. For a large value of μ, $\mathcal{R}$ becomes very small, resulting in high value of flux. An air gap is normally introduced to limit the amount of flux flow.

A magnetic circuit with an air gap is shown in Figure B.3a and the analogous electric circuit is shown in Figure B.3b. The reluctance of the air gap is

$$\mathcal{R}_g = \frac{l_g}{\mu_0 A_g} \tag{B.8}$$

and the reluctance of the core is

$$\mathcal{R}_c = \frac{l_c}{\mu_r \mu_0 A_c} \tag{B.9}$$

Where l_g = length of the air gap
l_c = length of the core
A_g = area of the cross section of the air gap
A_c = area of the cross section of the core

The total reluctance of the magnetic circuit is

$$\mathcal{R} = \mathcal{R}_g + \mathcal{R}_c$$

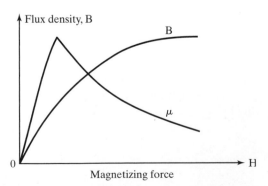

FIGURE B.2

Typical B–H characteristic.

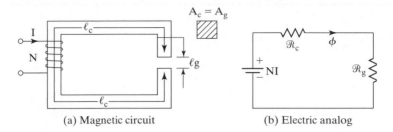

(a) Magnetic circuit   (b) Electric analog

FIGURE B.3

Magnetic circuit with air gap.

Inductance is defined as the flux linkage (λ) per ampere,

$$L = \frac{\lambda}{I} = \frac{N\phi}{I} \tag{B.10}$$

$$= \frac{N^2\phi}{NI} = \frac{N^2}{\mathcal{R}} \tag{B.11}$$

The flux density for various magnetic materials is shown in Table B.1.

TABLE B.1   Flux Density for Various Magnetic Materials

Trade Names	Composition	Saturated Flux Density* (tesla)	DC Coercive Force (amp-turn/ cm)	Squareness Ratio	Material Density (g/cm**)	Loss Factor at 3 kHz and 0.5 T (W/kg)
Magnesil Silectron Microsil Supersil	3% Si 97% Fe	1.5–1.8	0.5–0.75	0.85–1.0	7.63	33.1
Deltamax Orthonol 49 Sq. Mu	50% Ni 50% Fe	1.4–1.6	0.125–0.25	0.94–1.0	8.24	17.66
Allegheny 4750 48 Alloy Carpenter 49	48% Ni 52% Fe	1.15–1.4	0.062–0.187	0.80–0.92	8.19	11.03
4–79 Permalloy Sq. Permalloy 80 Sq. Mu 79	79% Ni 17% Fe 4% Mo	0.66–0.82	0.025–0.05	0.80–1.0	8.73	5.51
Supermalloy	78% Ni 17% Fe 5% Mo	0.65–0.82	0.0037–0.01	0.40–0.70	8.76	3.75

*1 T = 10^4 gauss.

**1 g/cm^3 = 0.036 lb/in.^{3}

Source: Arnold Engineering Company, Magnetics Technology Center, Marengo, IL http://www.grouparnold/com/mtc/index.htm

Example B.1

The parameters of the core in Figure B.3a are $l_g = 1$ mm, $l_c = 30$ cm, $A_g = A_c = 5 \times 10^{-3}$ m^2, $N = 350$, and $I = 2$ A. Calculate the inductance if (a) $\mu_r = 3500$, and (b) the core is ideal, namely, μ_r is very large tending to infinite.

Solution

$\mu_0 = 4\pi \times 10^{-7}$ and $N = 350$.

a. From Eq. (B.8),

$$\mathcal{R}_g = \frac{1 \times 10^{-3}}{4\pi \times 10^{-7} \times 5 \times 10^{-3}} = 159{,}155$$

From Eq. (B.9),

$$\mathcal{R}_c = \frac{30 \times 10^{-2}}{3500 \times 4\pi \times 10^{-7} \times 5 \times 10^{-3}} = 13{,}641$$

$$\mathcal{R} = 159{,}155 + 13{,}641 = 172{,}796$$

From Eq. (B.11), $L = 350^2/172{,}796 = 0.71$ H.

b. If $\mu_r \approx \infty$, $\mathcal{R}_c = 0$, $\mathcal{R} = \mathcal{R}_g = 159{,}155$, and $L = 350^2/159{,}155 = 0.77$ H.

B.1   SINUSOIDAL EXCITATION

If a sinusoidal voltage of $v_s = V_m \sin \omega t = \sqrt{2}\,V_s \sin \omega t$ is applied to the core in Figure B.3a, the flux can be found from

$$V_m \sin \omega t = -N \frac{d\phi}{dt} \tag{B.12}$$

which after integrating gives

$$\phi = \phi_m \cos \omega t = \frac{V_m}{N\omega} \cos \omega t \tag{B.13}$$

Thus

$$\phi_m = \frac{V_m}{2\pi f N} = \frac{\sqrt{2}\,V_s}{2\pi f N} = \frac{V_s}{4.44 f N} \tag{B.14}$$

The peak flux ϕ_m depends on the voltage, frequency, and number of turns. Equation (B.14) is valid if the core is not saturated. If peak flux is high, the core may saturate and the flux cannot be sinusoidal. If the ratio of voltage to frequency is maintained constant, the flux remains constant, provided that the number of turns is unchanged.

B.2   TRANSFORMER

If a second winding, called the *secondary winding*, is added to the core in Figure B.3a and the core is excited from sinusoidal voltage, a voltage is induced in the secondary winding. This is shown in Figure B.4. If N_p and N_s are turns on the primary and secondary

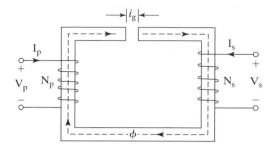

FIGURE B.4

Transformer core.

windings, respectively, the primary voltage V_p and secondary voltage V_s are related to each other as

$$\frac{V_p}{V_s} = \frac{I_s}{I_p} = \frac{N_p}{N_s} = a \qquad (B.15)$$

where a is the turns ratio.

The equivalent circuit of a transformer is shown in Figure B.5, where all the parameters are referred to the primary. To refer a secondary parameter to the primary side, the parameter is multiplied by a^2. The equivalent circuit can be referred to the secondary side by dividing all parameters of the circuit in Figure B.5 by a^2. X_1 and X_2 are the leakage reactances of the primary and secondary windings, respectively. R_1 and R_2 are the resistances of the primary and secondary winding, X_m is the magnetizing reactance, and R_m represents the core loss.

The wire size for a specific bare area is shown in Table B.2.

The variations of the flux due to ac excitation cause two types of losses in the core: (1) hysteresis loss, and (2) eddy-current loss. The hysteresis loss is expressed empirically as

$$P_h = K_h f B_{max}^z \qquad (B.16)$$

where K_h is a hysteresis constant that depends on the material and B_{max} is the peak flux density. z is the Steinmetz constant, which has a value of 1.6 to 2. The eddy-current loss is expressed empirically as

$$P_e = K_e f^2 B_{max}^2 \qquad (B.17)$$

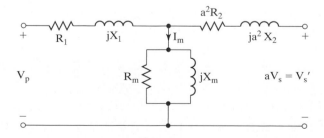

FIGURE B.5

Equivalent circuit of transformer.

TABLE B.2   Wire Size

AWG Wire Size	Bare Area		Resistance	Area		Diameter		Heavy Synthetics				Weight
	cm² 10⁻³*	Cir-Mil**	10^{-6} Ω cm at 20°C	cm² 10⁻³	Cir-Mil**	cm	in.**	Turns-Per: cm	Turns-Per: in.**	Turns-Per: cm²	Turns-Per: in.²	gm/cm
10	52.61	10384	32.70	55.9	11046	0.267	0.1051	3.87	9.5	10.73	69.20	0.468
11	41.68	8226	41.37	44.5	8798	0.238	0.0938	4.36	10.7	13.48	89.95	0.3750
12	33.08	6529	52.09	35.64	7022	0.213	0.0838	4.85	11.9	16.81	108.4	0.2977
13	26.26	5184	65.64	28.36	5610	0.190	0.0749	5.47	13.4	21.15	136.4	0.2367
14	20.82	4109	82.80	22.95	4556	0.171	0.0675	6.04	14.8	26.14	168.6	0.1879
15	16.51	3260	104.3	18.37	3624	0.153	0.0602	6.77	16.6	32.66	210.6	0.1492
16	13.07	2581	131.8	14.73	2905	0.137	0.0539	7.32	18.6	40.73	262.7	0.1184
17	10.39	2052	165.8	11.68	2323	0.122	0.0482	8.18	20.8	51.36	331.2	0.0943
18	8.228	1624	209.5	9.326	1857	0.109	0.0431	9.13	23.2	64.33	414.9	0.07472
19	6.531	1289	263.9	7.539	1490	0.0980	0.0386	10.19	25.9	79.85	515.0	0.05940
20	5.188	1024	332.3	6.065	1197	0.0879	0.0346	11.37	28.9	98.93	638.1	0.04726
21	4.116	812.3	418.9	4.837	954.8	0.0785	0.0309	12.75	32.4	124.0	799.8	0.03757
22	3.243	640.1	531.4	3.857	761.7	0.0701	0.0276	14.25	36.2	155.5	1003	0.02965
23	2.588	510.8	666.0	3.135	620.0	0.0632	0.0249	15.82	40.2	191.3	1234	0.02372
24	2.047	404.0	842.1	2.514	497.3	0.0566	0.0223	17.63	44.8	238.6	1539	0.01884
25	1.623	320.4	1062.0	2.002	396.0	0.0505	0.0199	19.80	50.3	299.7	1933	0.01498
26	1.280	252.8	1345.0	1.603	316.8	0.0452	0.0178	22.12	56.2	374.2	2414	0.01185
27	1.021	201.6	1687.6	1.313	259.2	0.0409	0.0161	24.44	62.1	456.9	2947	0.00945
28	0.8046	158.8	2142.7	1.0515	207.3	0.0366	0.0144	27.32	69.4	570.6	3680	0.00747
29	0.6470	127.7	2664.3	0.8548	169.0	0.0330	0.0130	30.27	76.9	701.9	4527	0.00602
30	0.5067	100.0	3402.2	0.6785	134.5	0.0294	0.0116	33.93	86.2	884.3	5703	0.00472
31	0.4013	79.21	4294.6	0.5596	110.2	0.0267	0.0105	37.48	95.2	1072	6914	0.00372
32	0.3242	64.00	5314.9	0.4559	90.25	0.0241	0.0095	41.45	105.3	1316	8488	0.00305
33	0.2554	50.41	6748.6	0.3662	72.25	0.0216	0.0085	46.33	117.7	1638	10565	0.00241
34	0.2011	39.69	8572.8	0.2863	56.25	0.0191	0.0075	52.48	133.3	2095	13512	0.00189

	A	B	C	D	E	F	G	H	I	J	K	L
35	0.1589	31.36	10849	0.2268	44.89	0.0170	0.0067	58.77	149.3	2645	17060	0.00150
36	0.1266	25.00	13608	0.1813	36.00	0.0152	0.0060	65.62	166.7	3309	21343	0.00119
37	0.1026	20.25	16801	0.1538	30.25	0.0140	0.0055	71.57	181.8	3901	25161	0.000977
38	0.08107	16.00	21266	0.1207	24.01	0.0124	0.0049	80.35	204.1	4971	32062	0.000773
39	0.06207	12.25	27775	0.0932	18.49	0.0109	0.0043	91.57	232.6	6437	41518	0.000593
40	0.04869	9.61	35400	0.0723	14.44	0.0096	0.0038	103.6	263.2	8298	53522	0.000464
41	0.03972	7.84	43405	0.0584	11.56	0.00863	0.0034	115.7	294.1	10273	66260	0.000379
42	0.03166	6.25	54429	0.04558	9.00	0.00762	0.0030	131.2	333.3	13163	84901	0.000299
43	0.02452	4.84	70308	0.03683	7.29	0.00685	0.0027	145.8	370.4	16291	105076	0.000233
44	0.0202	4.00	85072	0.03165	6.25	0.00635	0.0025	157.4	400.0	18957	122272	0.000195

* This notation means the entry in the column must be multiplied by 10^{-3}.

** This data from REA Magnetic Wire Datalator.

Source: Arnold Engineering Company, Magnetics Technology Center, Marengo, IL. www.grouparnold.com/mtc/index.htm

where K_e is the eddy-current constant and depends on the material. The total core loss is

$$P_c = K_h f B_{\max}^2 + K_e f^2 B_{\max}^2 \tag{B.18}$$

The typical magnetic loss against the flux density is shown in Figure B.6.

Note: If a transformer is designed to operate at 60 Hz and it is operated at a higher frequency, the core loss can increase significantly.

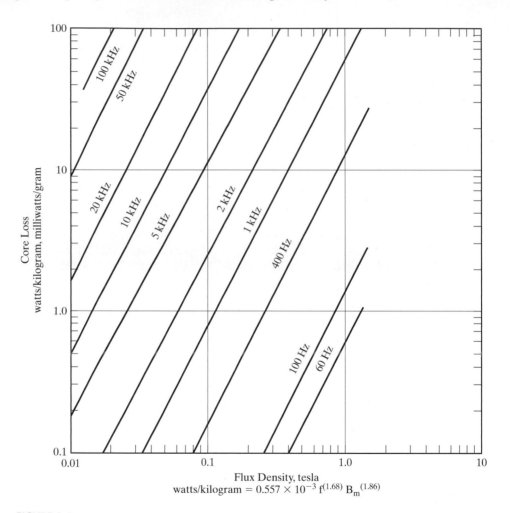

watts/kilogram = $0.557 \times 10^{-3} \, f^{(1.68)} \, B_m^{(1.86)}$

FIGURE B.6

Core loss against the flux density.

APPENDIX C

Switching Functions of Converters

The output of a converter depends on the switching pattern of the converter switches and the input voltage (or current). Similar to a linear system, the output quantities of a converter can be expressed in terms of the input quantities, by spectrum multiplication. The arrangement of a single-phase converter is shown in Figure C.1a. If $V_i(\theta)$ and $I_i(\theta)$ are the input voltage and current, respectively, the corresponding output voltage and current are $V_o(\theta)$ and $I_o(\theta)$, respectively. The input could be either a voltage source or a current source.

Voltage source. For a voltage source, the output voltage $V_o(\theta)$ can be related to input voltage $V_i(\theta)$ by

$$V_o(\theta) = S(\theta)\, V_i(\theta) \tag{C.1}$$

where $S(\theta)$ is the switching function of the converter, as shown in Figure C.1b. $S(\theta)$ depends on the type of converter and the gating pattern of the switches. If g_1, g_2, g_3, and g_4 are the gating signals for switches Q_1, Q_2, Q_3, and Q_4, respectively, the switching function is

$$S(\theta) = g_1 - g_4 = g^2 - g^3$$

Neglecting the losses in the converter switches and using power balance gives us

$$V_i(\theta)I_i(\theta) = V_o(\theta)I_o(\theta)$$

$$S(\theta) = \frac{V_o(\theta)}{V_i(\theta)} = \frac{I_i(\theta)}{I_o(\theta)} \tag{C.2}$$

$$I_i(\theta) = S(\theta)I_o(\theta) \tag{C.3}$$

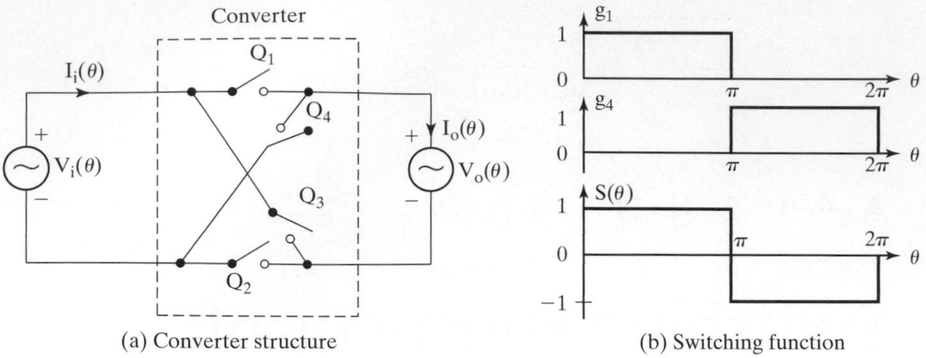

(a) Converter structure   (b) Switching function

FIGURE C.1

Single-phase converter structure.

Once $S(\theta)$ is known, $V_o(\theta)$ can be determined. $V_o(\theta)$ divided by the load impedance gives $I_o(\theta)$; and then $I_i(\theta)$ can be found from Eq. (C.3).

Current source.   In the case of current source, the input current remains constant, $I_i(\theta) = I_i$, and the output current $I_o(\theta)$ can be related to input current I_i,

$$I_o(\theta) = S(\theta)I_i$$
$$V_o(\theta)I_o(\theta) = V_i(\theta)I_i(\theta)$$

(C.4)

which gives

$$V_i(\theta) = S(\theta)V_o(\theta) \tag{C.5}$$

$$S(\theta) = \frac{V_i(\theta)}{V_o(\theta)} = \frac{I_o(\theta)}{I_i(\theta)} \tag{C.6}$$

C.1   SINGLE-PHASE FULL-BRIDGE INVERTERS

The switching function of a single-phase full-bridge inverter in Figure 6.2a, is shown in Figure C.2. If g_1 and g_4 are the gating signals for switches Q_1 and Q_4, respectively, the switching function is

$$
\begin{aligned}
S(\theta) &= g_1 - g_4 \\
&= 1 \qquad \text{for } 0 \le \theta \le \pi \\
&= -1 \qquad \text{for } \pi \le \theta \le 2\pi
\end{aligned}
$$

If f_0 is the fundamental frequency of the inverter,

$$\theta = \omega t = 2\pi f_0 t \tag{C.7}$$

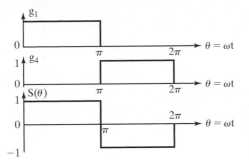

FIGURE C.2

Switching function of single-phase full-bridge inverter.

$S(\theta)$ can be expressed in a Fourier series as

$$S(\theta) = \frac{A_0}{2} + \sum_{n=1,2,\ldots}^{\infty} (A_n \cos n\theta + B_n \sin n\theta)$$

(C.8)

$$B_n = \frac{2}{\pi} \int_0^{\pi} S(\theta) \sin n\theta \, d\theta = \frac{4}{n\pi} \qquad \text{for } n = 1,3, \ldots$$

Due to half-wave symmetry, $A_0 = A_n = 0$.
Substituting A_0, A_n, and B_n in Eq. (C.8) yields

$$S(\theta) = \frac{4}{\pi} \sum_{n=1,3,5,\ldots}^{\infty} \frac{\sin n\theta}{n}$$

(C.9)

If the input voltage, which is dc, is $V_i(\theta) = V_s$, Eq. (C.1) gives the output voltage as

$$V_o(\theta) = S(\theta)V_i(\theta) = \frac{4V_s}{\pi} \sum_{n=1,3,5,\ldots}^{\infty} \frac{\sin n\theta}{n}$$

(C.10)

which is the same as Eq. (6.12). For a three-phase voltage-source inverter in Figure 6.5, there are three switching functions: $S_1(\theta) = g_1 - g_4$, $S_2(\theta) = g_3 - g_6$, and $S_3(\theta) = g_5 - g_2$. There are three line-to-line output voltages corresponding to three switching voltages, namely, $V_{ab}(\theta) = S_1(\theta)V_i(\theta)$, $V_{bc}(\theta) = S_2(\theta)V_i(\theta)$, and $V_{ca}(\theta) = S_3(\theta)V_i(\theta)$.

C.2 SINGLE-PHASE BRIDGE RECTIFIERS

The switching function of a single-phase bridge rectifier is the same as that of the single-phase full-bridge inverter. If the input voltage is $V_i(\theta) = V_m \sin \theta$, Eqs. (C.1) and (C.9) give the output voltage as

$$V_o(\theta) = S(\theta)V_i(\theta) = \frac{4V_m}{\pi} \sum_{n=1,3,5,\ldots}^{\infty} \frac{\sin \theta \sin n\theta}{n}$$

(C.11)

$$= \frac{4V_m}{\pi} \sum_{n=1,3,5,\ldots}^{\infty} \frac{\cos(n-1)\theta - \cos(n+1)\theta}{2n}$$

(C.12)

$$= \frac{2V_m}{\pi}\left[1 - \cos 2\theta + \frac{1}{3}\cos 2\theta - \frac{1}{3}\cos 4\theta \right.$$

$$\left. + \frac{1}{5}\cos 4\theta - \frac{1}{5}\cos 6\theta + \frac{1}{7}\cos 6\theta - \frac{1}{7}\cos 8\theta + \cdots\right]$$

$$= \frac{2V_m}{\pi}\left[1 - \frac{2}{3}\cos 2\theta - \frac{2}{15}\cos 4\theta - \frac{2}{35}\cos 6\theta - \cdots\right]$$

$$= \frac{2V_m}{\pi} - \frac{4V_m}{\pi}\sum_{m=1}^{\infty}\frac{\cos 2m\theta}{4m^2 - 1} \tag{C.13}$$

Equation (C.13) is the same as Eq. (3.22). The first part of Eq. (C.13) is the average output voltage and the second part is the ripple content on the output voltage.

For a three-phase rectifier in Figs. 3.13a and 10.5a, the switching functions are $S_1(\theta) = g_1 - g_4$, $S_2(\theta) = g_3 - g_6$, and $S_3(\theta) = g_5 - g_2$. If the three input phase voltages are $V_{an}(\theta)$, $V_{bn}(\theta)$, and $V_{cn}(\theta)$, the output voltage becomes

$$V_o(\theta) = S_1(\theta)V_{an}(\theta) + S_2(\theta)V_{bn}(\theta) + S_3(\theta)V_{cn}(\theta) \tag{C.14}$$

C.3 SINGLE-PHASE FULL-BRIDGE INVERTERS WITH SINUSOIDAL PULSE-WIDTH MODULATION

The switching function of a single-phase full-bridge inverter with sinusoidal pulse-width modulation (SPWM) is shown in Figure C.3. The gating pulses are generated by

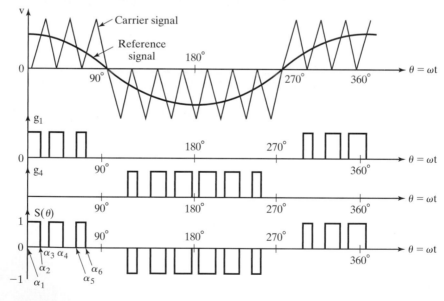

FIGURE C.3

Switching function with SPWM.

comparing a cosine wave with triangular pulses. If g_1 and g_4 are the gating signals for switches Q_1 and Q_4, respectively, the switching function is

$$S(\theta) = g_1 - g_4$$

$S(\theta)$ can be expressed in a Fourier series as

$$S(\theta) = \frac{A_0}{2} + \sum_{n=1,2...}^{\infty} (A_n \cos n\theta + B_n \sin n\theta) \tag{C.15}$$

If there are p pulses per quarter cycle and p is an even number,

$$
\begin{aligned}
A_n &= \frac{2}{\pi} \int_0^\pi S(\theta) \cos n\theta \, d\theta \\
&= \frac{4}{\pi} \int_0^{\pi/2} S(\theta) \cos n\theta \, d\theta \\
&= \frac{4}{\pi} \left[\int_{\alpha 1}^{\alpha 2} \cos n\theta \, d\theta + \int_{\alpha 3}^{\alpha 4} \cos n\theta \, d\theta + \int_{\alpha 5}^{\alpha 6} \cos n\theta \, d\theta + \cdots \right] \tag{C.16} \\
&= \frac{4}{n\pi} \sum_{m=1,2,3,...}^{p} [(-1)^m \sin n\alpha_m]
\end{aligned}
$$

Due to quarter-wave symmetry, $B_n - A_0 = 0$. Substituting A_0, A_n, and B_n in Eq. (C.15) yields

$$
\begin{aligned}
S(\theta) &= \sum_{n=1,3,5...}^{\infty} A_n \cos n\theta \\
&= \frac{4}{n\pi} \sum_{n=1,3,5...}^{\infty} \left[\sum_{m=1,2,3,...}^{p} (-1)^m \sin n\alpha_m \cos n\theta \right]
\end{aligned} \tag{C.17}
$$

If the input voltage is $V_i(\theta) = V_s$, Eqs. (C.1) and (C.17) give the output voltage as

$$V_o(\theta) = V_s \sum_{n=1,3,5,...}^{\infty} A_n \cos n\theta \tag{C.18}$$

C.4 SINGLE-PHASE CONTROLLED RECTIFIERS WITH SPWM

If the input voltage is $V_i(\theta) = V_m \cos \theta$, Eqs. (C.1) and (C.17) give the output voltage as

$$
\begin{aligned}
V_o(\theta) &= V_m \sum_{n=1,3,5,...}^{\infty} A_n \cos n\theta \cos \theta \tag{C.19} \\
&= \frac{V_m}{2} \sum_{n=1,3,5,...}^{\infty} A_n[\cos(n-1)\theta + \cos(n+1)\theta]
\end{aligned}
$$

$$= 0.5V_m[A_1(\cos 0 + \cos 2\theta) + A_3(\cos 2\theta + \cos 4\theta)$$

$$+ A_5(\cos 4\theta + \cos 6\theta) + \cdots]$$

$$= \frac{V_m A_1}{2} + V_m \sum_{n=2,4,6,\ldots}^{\infty} \frac{A_{n-1} + A_{n+1}}{2} \cos n\theta \qquad (C.20)$$

The first part of Eq. (C.20) is the average output voltage and the second part is the ripple voltage. Equation (C.20) is valid, provided that the input voltage and the switching function are cosine waveforms.

In the case of sine waves, the input voltage is $V_i(\theta) = V_m \sin \theta$ and the switching function is

$$S(\theta) = \sum_{n=1,3,5,\ldots}^{\infty} A_n \sin n\theta \qquad (C.21)$$

Equations (C.1) and (C.21) give the output voltage as

$$V_o(\theta) = V_m \sum_{n=1,3,5,\ldots}^{\infty} A_n \sin \theta \sin n\theta \qquad (C.22)$$

$$= \frac{V_m}{2} \sum_{n=1,3,5,\ldots}^{\infty} A_n[\cos(n-1)\theta - \cos(n+1)\theta]$$

$$= 0.5V_m[A_1(\cos 0 - \cos 2\theta) + A_3(\cos 2\theta - \cos 4\theta)$$

$$+ A_5(\cos 4\theta - \cos 6\theta) + \cdots]$$

$$= \frac{V_m A_1}{2} - V_m \sum_{n=2,4,6,\ldots}^{\infty} \frac{A_{n-1} - A_{n+1}}{2} \cos n\theta \qquad (C.23)$$

A P P E N D I X   D

Dc Transient Analysis

D.1   *RC* CIRCUIT WITH STEP INPUT

When the switch S_1 in Figure 2.12a is closed at $t = 0$, the charging current of the capacitor can be found from

$$V_s = v_R + v_c = R\,i + \frac{1}{C}\int i\,dt + v_c(t = 0) \tag{D.1}$$

with initial condition: $v_c(t = 0) = 0$. Using Table D.1, Eq. (D.1) can be transformed into the Laplace's domain of s:

$$\frac{V_s}{s} = RI(s) + \frac{1}{Cs}\,I(s)$$

TABLE D.1   Some Laplace Transformations

$f(t)$	$F(s)$
1	$\dfrac{1}{s}$
t	$\dfrac{1}{s^2}$
$e^{-\alpha t}$	$\dfrac{1}{s + \alpha}$
$\sin \alpha t$	$\dfrac{\alpha}{s^2 + \alpha^2}$
$\cos \alpha t$	$\dfrac{s}{s^2 + \alpha^2}$
$f'(t)$	$sF(s) - F(0)$
$f''(t)$	$s^2F(s) - sF(s) - F'(0)$

which after solving for the current $I(s)$ gives

$$I(s) = \frac{V_s}{R(s + \alpha)} \tag{D.2}$$

where $\alpha = 1/RC$. Inverse transform of Eq. (D.2) in the time domain yields

$$i(t) = \frac{V_s}{R} e^{-\alpha t} \tag{D.3}$$

and the voltage across the capacitor is obtained as

$$v_c(t) = \frac{1}{C} \int_0^t i \, dt = V_s(1 - e^{-\alpha t}) \tag{D.4}$$

In the steady state (at $t = \infty$),

$$I_s = i(t = \infty) = 0$$

$$V_c = v_c(t = \infty) = \frac{V_s}{R}$$

D.2   *RL* CIRCUIT WITH STEP INPUT

Two typical *RL* circuits are shown in Figures 2.13a and 5.4a. The transient current through the inductor in Figure 5.4a can be expressed as

$$V_s = v_L + v_R + E = L\frac{di}{dt} + Ri + E \tag{D.5}$$

with initial condition: $i(t = 0) = I_1$. In Laplace's domain of s, Eq. (D.5) becomes

$$\frac{V_s}{s} = L\, sI(s) - LI_1 + RI(s) + \frac{E}{s}$$

and solving for $I(s)$ gives

$$I(s) = \frac{V_s - E}{L\, s(s + \beta)} + \frac{I_1}{s + \beta}$$

$$= \frac{V_s - E}{R}\left(\frac{1}{s} - \frac{1}{s + \beta}\right) + \frac{I_1}{s + \beta} \tag{D.6}$$

where $\beta = R/L$. Taking inverse transform of Eq. (D.6) yields

$$i(t) = \frac{V_s}{R}\left(1 - e^{-\beta t}\right) + I_1 e^{-\beta t} \tag{D.7}$$

If there is no initial current on the inductor (i.e., $I_1 = 0$), Eq. (D.7) becomes

$$i(t) = \frac{V_s}{R}\left(1 - e^{-\beta t}\right) \tag{D.8}$$

In the steady state (at $t = \infty$), $I_s = i(t = \infty) = V_s/R$.

D.3   *LC* CIRCUIT WITH STEP INPUT

The transient current through the capacitor in Figure 2.15a is expressed as

$$V_s = v_L + v_c = L\frac{di}{dt} + \frac{1}{C}\int i\, dt + v_c(t = 0) \tag{D.9}$$

with initial conditions: $v_c(t = 0) = 0$ and $i(t = 0) = 0$. In the Laplace transform, Eq. (D.9) becomes

$$\frac{V_s}{s} = L\, sI(s) + \frac{1}{Cs}I(s)$$

and solving for $I(s)$ gives

$$I(s) = \frac{V_s}{L(s^2 + \omega_m^2)} \tag{D.10}$$

where $\omega_m = 1/\sqrt{LC}$. The inverse transform of Eq. (D.10) yields the charging current as

$$i(t) = V_s \sqrt{\frac{C}{L}} \sin(\omega_m t) \tag{D.11}$$

and the capacitor voltage is

$$v_c(t) = \frac{1}{C}\int_0^t i(t)\, dt = V_s[1 - \cos(\omega_m t)] \tag{D.12}$$

An *LC* circuit with an initial inductor current of I_m and an initial capacitor voltage of V_o is shown in Figure D.1. The capacitor current is expressed as

$$V_s = L\frac{di}{dt} + \frac{1}{C}\int i\, dt + v_c(t = 0) \tag{D.13}$$

with initial condition $i(t = 0) = Im$ and $vc(t = 0) = V_o$. Note: In Figure D.1, V_o is shown as equal to $-2V_s$. In the Laplace domain of s, Eq. (D.13) becomes

$$\frac{V_s}{s} = L\, sI(s) - LI_m + \frac{1}{Cs} + \frac{V_o}{s}$$

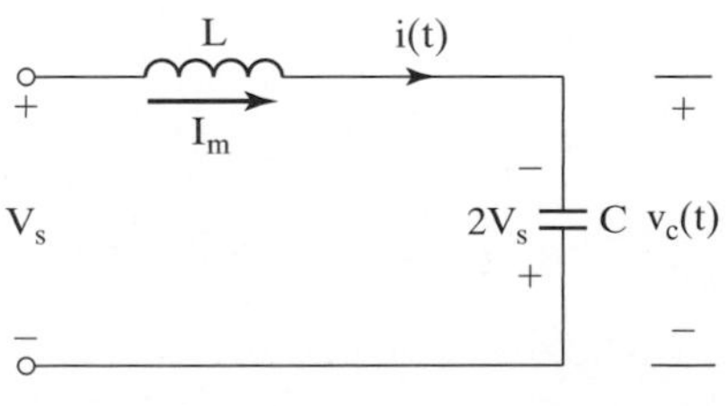

FIGURE D.1
LC-circuit

and solving for the current $I(s)$ gives

$$I(s) = \frac{V_s - V_o}{L(s^2 + \omega_m^2)} + \frac{sI_m}{s^2 + \omega_m^2} \tag{D.14}$$

where $\omega_m = 1/\sqrt{LC}$. The inverse transform of Eq. (D.14) yields

$$i(t) = (V_s - V_o)\sqrt{\frac{C}{L}}\sin(\omega_m t) + I_m \cos(\omega_m t) \tag{D.15}$$

and the capacitor voltage as

$$v_c(t) = \frac{1}{C}\int_0^t i(t)\, dt + V_o$$

$$= I_m \sqrt{\frac{L}{C}}\sin(\omega_m t) - (V_s - V_o)\cos(\omega_m t) + V_s \tag{D.16}$$

APPENDIX E

Fourier Analysis

Under steady-state conditions, the output voltage of power converters is, generally, a periodic function of time defined by

$$v_o(t) = v_o(t + T) \tag{E.1}$$

where T is the periodic time. If f is the frequency of the output voltage in hertz, the angular frequency is

$$\omega = \frac{2\pi}{T} = 2\pi f \tag{E.2}$$

and Eq. (E.1) can be rewritten as

$$v_o(\omega t) = v_o(\omega t + 2\pi) \tag{E.3}$$

The Fourier theorem states that a periodic function $v_o(t)$ can be described by a constant term plus an infinite series of sine and cosine terms of frequency $n\omega$, where n is an integer. Therefore, $v_o(t)$ can be expressed as

$$v_o(t) = \frac{a_o}{2} + \sum_{n=1,2,\ldots}^{\infty} (a_n \cos n\omega t + b_n \sin n\omega t) \tag{E.4}$$

where $a_o/2$ is the average value of the output voltage $v_o(t)$. The constants a_o, a_n, and b_n can be determined from the following expressions:

$$a_o = \frac{2}{T} \int_0^T v_o(t)\, dt = \frac{1}{\pi} \int_0^{2\pi} v_o(\omega t) d(\omega t) \tag{E.5}$$

$$a_n = \frac{2}{T} \int_0^T v_o(t) \cos n\omega t\, dt = \frac{1}{\pi} \int_0^{2\pi} v_o(\omega t) \cos n\omega t\, d(\omega t) \tag{E.6}$$

$$b_n = \frac{2}{T} \int_0^T v_o(t) \sin n\omega t\, dt = \frac{1}{\pi} \int_0^{2\pi} v_o(\omega t) \sin n\omega t\, d(\omega t) \tag{E.7}$$

If $v_o(t)$ can be expressed as an analytical function, these constants can be determined by a single integration. If $v_o(t)$ is discontinuous, which is usually the case for the output of converters, several integrations (over the whole period of the output voltage) must be performed to determine the constants, a_o, a_n, and b_n.

$$a_n \cos n\omega t + b_n \sin n\omega t$$

$$= (a_n^2 + b_n^2)^{1/2} \left(\frac{a_n}{\sqrt{a_n^2 + b_n^2}} \cos n\omega t + \frac{b_n}{\sqrt{a_n^2 + b_n^2}} \sin n\omega t \right) \tag{E.8}$$

Let us define an angle ϕ_n, whose adjacent side is b_n, opposite side is a_n, and the hypotenuse is $(a_n^2 + b_n^2)^{1/2}$. As a result, Eq. (E.8) becomes

$$a_n \cos n\omega t + b_n \sin n\omega t = (a_n^2 + b_n^2)^{1/2}(\sin \phi_n \cos n\omega t + \cos \phi_n \sin n\omega t)$$

$$= (a_n^2 + b_n^2)^{1/2} \sin(n\omega t + \phi_n) \tag{E.9}$$

where

$$\phi_n = \tan^{-1} \frac{a_n}{b_n} \tag{E.10}$$

Substituting Eq. (E.9) into Eq. (E.4), the series may also be written as

$$v_o(t) = \frac{a_o}{2} + \sum_{n=1,2,\dots}^{\infty} C_n \sin(n\omega t + \phi_n) \tag{E.11}$$

where

$$C_n = (a_n^2 + b_n^2)^{1/2} \tag{E.12}$$

C_n and ϕ_n are the peak magnitude and the delay angle of the nth harmonic component of the output voltage $v_o(t)$, respectively.

   If the output voltage has a *half-wave symmetry*, the number of integrations within the entire period can be reduced significantly. A waveform has the property of a half-wave symmetry if the waveform satisfies the following conditions:

$$v_o(t) = -v_o\left(t + \frac{T}{2}\right) \tag{E.13}$$

or

$$v_o(\omega t) = -v_o(\omega t + \pi) \tag{E.14}$$

   In a waveform with a half-wave symmetry, the negative half-wave is the mirror image of the positive half-wave, but phase shifted by $T/2$ s(or π rad) from the positive half-wave. A waveform with a half-wave symmetry does not have the even harmonics (i.e., $n = 2, 4, 6, \dots$) and possess only the odd harmonics (i.e., $n = 1, 3, 5, \dots$). Due to the half-wave symmetry, the average value is zero (i.e., $a_o = 0$). Equations (E.6), (E.7), and (E.11) become

$$a_n = \frac{2}{T} \int_0^T v_o(t) \cos n\omega t \, dt = \frac{1}{\pi} \int_0^{2\pi} v_o(\omega t) \cos n\omega t \, d(\omega t), \quad n = 1, 3, 5, \ldots$$

$$b_n = \frac{2}{T} \int_0^T v_o(t) \sin n\omega t \, dt = \frac{1}{\pi} \int_0^{2\pi} v_o(\omega t) \sin n\omega t \, d(\omega t), \quad n = 1, 3, 5, \ldots$$

$$v_o(t) = \sum_{n=1,3,5,\ldots}^{\infty} C_n \sin(n\omega t + \phi_n)$$

In general, with a half-wave symmetry, $a_o = a_n = 0$, and with a quarter-wave symmetry, $a_o - b_n = 0$.

A waveform has the property of quarter-wave symmetry if the waveform satisfies the following conditions:

$$v_o(t) = -v_o\left(t + \frac{T}{4}\right) \tag{E.15}$$

or

$$v_o(\omega t) = -v_o\left(\omega t + \frac{\pi}{2}\right) \tag{E.16}$$

Bibliography

Power Electronics Books: http://www.smpstech.com/books/booklist.htm

BEDFORD, F. E., and R. G. HOFT, *Principles of Inverter Circuits*. New York: John Wiley & Sons, Inc., 1964.

BILLINGS, K., *Switch Mode Power Supply Handbook*. New York: McGraw-Hill Inc., 1989.

BIRD, B. M., and K. G. KING, *An Introduction to Power Electronics*. Chichester, West Sussex, England: John Wiley & Sons Ltd., 1983.

CSAKI, F., K. GANSZKY, I. IPSITS, and S. MARTI, *Power Electronics*. Budapest: Akademiai Kiadó, 1980.

DATTA, S. M., *Power Electronics & Control*. Reston, VA: Reston Publishing Co., Inc., 1985.

DAVIS, R. M., *Power Diode and Thyristor Circuits*. Stevenage, Herts, England: Institution of Electrical Engineers, 1979.

DEWAN, S. B., and A. STRAUGHEN, *Power Semiconductor Circuits*. New York: John Wiley & Sons, Inc., 1984.

DEWAN, S. B., G. R. SLEMON, and A. STRAUGHEN, *Power Semiconductor Drives*. New York: John Wiley & Sons, Inc., 1975.

DUBEY, G. K., *Power Semiconductor Controlled Drives*. Englewood Cliffs, NJ: Prentice Hall, 1989.

General Electric, GRAFHAN, D. R., and F. B. GOLDEN, eds., *SCR Manual*, 6th ed., Englewood Cliffs, NJ: Prentice Hall, 1982.

GOTTLIEB, I. M., *Power Control with Solid State Devices*. Reston, VA: Reston Publishing Co., Inc., 1985.

HEUMANN, K., *Basic Principles of Power Electronics*. New York: Springer-Verlag, 1986.

HNATEK, E. R., *Design of Solid State Power Supplies*. New York: Van Nostrand Reinhold Company, Inc., 1981.

FISHER, M. J., *Power Electronics*. Boston, MA: PWS-KENT Publishing, 1991.

HINGORANI, N. G., and L. GYUGI, *Understanding FACTS*. Piscataway, NJ: IEEE Press, 2000.

HOFT, R. G., *SCR Applications Handbook*. El Segundo, CA: International Rectifier Corporation, 1974.

HOFT, R. G., *Semiconductor Power Electronics*. New York: Van Nostrand Reinhold Company, Inc., 1986.

KASSAKIAN, J. G., M. SCHLECHT, and G. C. VERGHESE, *Principles of Power Electronics*. Reading, MA: Addison-Wesley Publishing Co., Inc., 1991.

KAZIMIERCZUK, M. K., and D. CZARKOWSKI, *Resonant Power Converters*. New York: John Wiley and Sons Ltd., 1995.

KAZIMIERCZUK, M. K., and D. CZARKOWSKI, *Solutions Manual for Resonant Power Converters*. New York: John Wiley and Sons Ltd., 1995.

KILGENSTEIN, O., *Switch-Mode Power Supplies in Practice*. New York: John Wiley and Sons Ltd., 1989.

KLOSS, A., *A Basic Guide to Power Electronics*. New York: John Wiley & Sons, Inc., 1984.

KUSKO, A., *Solid State DC Motor Drives*. Cambridge, MA: The MIT Press, 1969.

LANDER, C. W., *Power Electronics*. Maidenhead, Berkshire, England: McGraw-Hill Book Company Ltd., 1981.

LENK, R., *Practical Design of Power Supply*. Piscataway, NJ: IEEE Press Inc., 1998.

LEONARD, W., *Control of Electrical Drives*. New York: Springer-Verlag, 1985.

LINDSAY, J. F., and M. H. RASHID, *Electromechanics and Electrical Machinery*. Englewood Cliffs, NJ: Prentice Hall, 1986.

LYE, R. W., *Power Converter Handbook*. Peterborough, Ont.: Canadian General Electric Company Ltd., 1976.

MAZDA, F. F., *Thyristor Control*. Chichester, West Sussex, England: John Wiley & Sons Ltd., 1973.

MAZDA, F., *Power Electronics Handbook*. London: Newnes, 3rd ed., 1997.

McMURRY, W., *The Theory and Design of Cycloconverters*. Cambridge, MA: The MIT Press, 1972.

MITCHELL, D. M., *Switching Regulator Analysis*. New York: McGraw-Hill Inc., 1988.

MOHAN, M., T. M. UNDELAND, and W. P. ROBBINS, *Power Electronics: Converters, Applications and Design*. New York: John Wiley & Sons, Inc., 1989.

MURPHY, I. M. D., *Thyristor Control of AC Motors*. Oxford: Pergamon Press Ltd., 1973.

NOVOTHNY, D. W., and T. A. LIPO, Vector Control and Dynamics of AC Drives. New York: Oxford University Publishing, 1998.

PEARMAN, R. A., *Power Electronics: Solid State Motor Control*. Reston, VA: Reston Publishing Co., Inc., 1980.

PELLY, B. R., *Thyristor Phase Controlled Converters and Cycloconverters*. New York: John Wiley & Sons, Inc., 1971.

RAMAMOORTY, M., *An Introduction to Thyristors and Their Applications*. London: Macmillan Publishers Ltd., 1978.

RAMSHAW, R. S., *Power Electronics: Thyristor Controlled Power for Electric Motors*. London: Chapman & Hall Ltd., 1982.

RASHID, M. H., *SPICE for Power Electronics and Electric Power*. Englewood Cliffs, NJ: Prentice Hall, 1993.

RASHID, M. H., *Power Electronics—Circuits, Devices, and Applications*. Upper Saddle River, NJ: Prentice-Hall, Inc., 1st ed. 1988, 2nd ed., 1993.

RICE, L. R., *SCR Designers Handbook*. Pittsburgh, PA: Westinghouse Electric Corporation, 1970.

ROSE, M. I., *Power Engineering Using Thyristors*, Vol. 1. London: Mullard Ltd., 1970.

SCHAEFER, J., *Rectifier Circuits; Theory and Design*. New York: John Wiley & Sons, Inc., 1965.

SEN, P. C., *Thyristor DC Drives*. New York: John Wiley & Sons, Inc., 1981.

SEVERNS, R. P., and G. BLOOM, *Modern DC-to-DC Switchmode Power Converter Circuits*. New York: Van Nostrand Reinhold Company, Inc., 1985.

SHEPHERD, W., and L. N. HULLEY, *Power Electronics and Motor Drives*. Cambridge, UK: Cambridge University Press, 1987.

SONG, Y. H., and A. T. JOHNS, *Flexible ac Transmission Systems (FACTS)*. London: The Institution of Electrical Engineers, 1999.

STEVEN, R. E., *Electrical Machines and Power Electronics*. Wakingham, Berkshire, UK: Van Nostrand Reinhold Ltd., 1983.

SUBRAHMANYAM, V., *Electric Drives: Concepts and Applications*. New York: McGraw-Hill Inc., 1996.

SUGANDHI, R. K., and K. K. SUGANDHI, *Thyristors: Theory and Applications*. New York: Halsted Press, 1984.

SUM, K. KIT, *Switch Mode Power Conversion: Basic Theory and Design*. New York: Marcel Dekker, Inc., 1984.

TARTER, R. E., *Principles of Solid-State Power Conversion*. Indianapolis, IN: Howard W. Sams & Company, Publishers, Inc., 1985.

VALENTINE, R., *Motor Control Electronics Handbook*. New York: McGraw-Hill Inc., 1998.

WELLS, R., *Static Power Converters*. New York: John Wiley & Sons, Inc., 1962.

WILLIAMS, B. W., *Power Electronics: Devices. Drivers and Applications*. New York: Halsted Press, 1987.

WOOD, P., *Switching Power Converters*. New York: Van Nostrand Reinhold Company, Inc., 1981.

Answers to Selected Problems

3.5 $V_{dc} = 378.15$ V

3.6 $V_{dc} = 371.34$ V

3.7   Diodes: $I_p = 62.834$ A, $I_d = 20$ A, $I_R = 31.417$ A, Transformer: $V_s = 444.3$ V, $I_s = 44.43$ A, TUF = 0.8105

3.8   Diodes: $I_p = 9000$ A, $I_d = 4500$ A, $I_R = 6363.96$ A, Transformer: $V_s = 320.59$ V, $I_s = I_p = 9000$ A, TUF = 0.7798

3.9 $L = 238.4$ mH

3.10 $L = 11.64$ mH

3.11   (a) $\delta = 149.7°$, (b) $R = 1.793$ Ω, (c) $P_R = 488.8$ W, (d) $h = 1$ hr, (e) $\eta = 29\%$, PIV = 104.85 V

3.12   (a) $I_1 = 6.33$ A, (b) $I_d = 8.8$ A, (c) $I_r = 13.83$ A, (d) $I_{rms} = 19.56$ A

3.13   (a) $I_1 = 50.6$ A, (b) $I_d = 17.46$ A, (c) $I_r = 30.2$ A, (d) $I_{rms} = 52.31$ A

3.14   (a) $C_e = 315.46$ μF, (b) $V_{dc} = 158.49$ V

3.15   (a) $C_e = 630.92$ μF, (b) $V_{dc} = 158.49$ V

3.16   (a) $L_{cr} = 12.51$ mH, $\alpha = 37.84°$, $I_{rms} = 29.47$ A, (b) $\alpha = 37.84°$, $I_{rms} = 20.69$ A

3.17 $V_{ac} = V_m/(4\sqrt{2}\,f\,RC)$

3.18 $L_1 = 1.837$ mH

3.19   (b) $L = 11.64$ mH

3.20   (b) PF = 0.9, HF = 0.4834, (c) PF = 0.6366, HF = 1.211

3.21   (b) PF = 0.9, HF = 0.4834, (c) PF = 0.6366, HF = 1.211

3.22   (c) PF = 0.827, HF = 0.68

3.23   (b) $I_1 = \sqrt{3}I_a/(\pi\sqrt{2})$, $\varphi_1 = -\pi/6$, $I_s = \sqrt{2}I_a/3$, (c) PF = 0.827, HF = 0.68

3.24   (b) $I_1 = \sqrt{2}\sqrt{3}I_a/\pi$, $I_s = I_a\sqrt{(2/3)}$, (c) PF = 0.78, HF = 0.803

3.25   (c) PF = 0.9549, HF = 0.3108

3.26   (c) PF = 0.9549, HF = 0.3108

3.27   (c) HF = 0.3108

CHAPTER 4

4.1   (b) $\beta_f = 0.59$, (c) $P_T = 106.5$ W

4.2   (b) $\beta_f = 4.104$, (c) $P_T = 41.104$ W

4.3   (e) $P_T = 482.258$ W

4.4 $R_{SA} = 0.0681°$C/W

4.5 $P_B = 13.2$ W

4.6   (e) $P_T = 125.87$ W

4.7   (e) $P_T = 20.466$ W

4.8 $R_{SA} = 3.863°$K/W

4.9   (b) $I = 10\%$

4.10   (f) $P_s = 1200$ W

4.11   (f) $P_s = 0.75$ W

CHAPTER 5

5.1   (c) $\eta = 99.32\%$

5.2   (e) $I_o = 9.002$ A, (g) $I_R = 15.63$ A

5.3 $L = 27.5$ mH

5.4   (b) $V_{ch} = 0.4033$ Ω

5.6 $k = 0.5$, $I_2 = 615$ A, $I_1 = 585$ A

5.7 $I_{max} = 30$ A

5.8 $I_1 = 1.16$ A, $I_2 = 1.93$ A, $\Delta I = 0.77$ A

5.9   (b) $L = 333.3$ μH, (c) $C = 312.5$ μF, (d) $L_c = 166.75$ μH, $C_c = 0.63$ μF

5.10   (c) $I_2 = 1.61$ A, (d) $V_c = 34.1$ mV, (e) $L_c = 180$ μH, $C_c = 0.5$ μF

5.11   (d) $I_p = 4.326$ A, (e) $L_c = 96$ μH, $C_c = 1$ μF

5.12   (f) $\Delta V_{c2} = 15.58$ mV, (g) $\Delta I_{L2} = 6.25$ A, $I_p = 7.91$ A

5.13 $L_{c1} = 4.32$ mH, $L_{c2} = 0.14$ μH, $C_{c1} = 0.83$ μF, $C_{c1} = 0.42$ μF

5.14 $C_e = 6.25$ μF, $L = 0.582$ mH

5.15   (a) $G(k = 0.5) = 0.5$, (b) $G(k = 0.5) = 1.95$, (c) $G(k = 0.5) = -0.97$

CHAPTER 6

6.1   (e) THD = 48.34%, (f) DF = 3.804%, (g) $V_3 = V_1/3$

6.2 (e) THD = 48.34%,
(f) DF = 3.804%, (g) HF_3 = 33.33%

6.3 (c) HF = 5.61%, (e) I_p = 25.44 A,
I_A = 3.675 A

6.4 (d) P_o = 134.48 W, I_s = 0.612 A,
(e) I_p = 8.2 A, I_A = 0.306 A

6.5 (c) HF = 14.2%, (e) I_p = 31.27 A,
I_A = 5.555 A

6.6 $V_{a1(peak)}$ = 140 V, $V_{ab(peak)}$ =
242.58 V, $I_{a1(peak)}$ = 28 A

6.7 $V_{a1(peak)}$ = 140 V, $V_{ab(peak)}$ =
242.58 V, $I_{a1(peak)}$ = 28 A

6.8 $V_{ab(peak)}$ = 242.58 V, $I_{ab1(peak)}$ =
48.52 A

6.9 $V_{ab(peak)}$ = 242.58 V,
$I_{ab1(peak)}$ = 48.52 A,
$I_{a1(peak)}$ = 84.04 A

6.10 I_S = 0.345 A, I_A = 0.345/2 =
0.1725 A

6.11 δ = 102.07°

6.12 For M = 0.5, $V(1)$ = 48.72%,
$V(3)$ = 39.31%, DF = 10.83%

6.13 For M = 0.8, δ = 72°, α_1 = 9°,
α_2 = 99°, $V_{o(peak)}$ = 232.84 V,
$I_{o1(peak)}$ = 2.315 A

6.14 For M = 0.5, V_{o1} = 70.71%, $V(1)$ =
32.45%, THD = 111.98%,
DF = 1.58%

6.15 For M = 0.5, $V(1)$ = 45.3%,
$V(3)$ = 15.9%, DF = 4.11%

6.16 For M = 0.5, V_{o1} = 55%,
$V(1)$ = 25%, THD = 111.94%,
DF = 0.27%

6.17 For M = 0.5, $V(1)$ = 35.36%,
$V(3)$ = 0, DF = 11.06%

6.18 For M = 0.5, V(1) = 58.67%,
V(3) = 14.65%, DF = 15.39%

6.19 For M = 0.5, V_{o1} = 84.1%,
$V(1)$ = 56.7%, THD = 26.9%,
DF = 0.52%

6.20 δ = 23.04°

6.21 β = 102.07°

6.22 For M = 0.5, α_1 = 7.5°, α_2 = 31.5°,
α_3 = 40.5°, α_4 = 67.5°, α_4 = 76.5°,
$V(1)$ = 95.49%, THD = 70.96%,
DF = 3.97%

6.23 For M = 0.5, α_1 = 4.82°,
α_2 = 20.93°, α_3 = 30.54°,
α_4 = 48.21°, α_5 = 54.64°,
$V(1)$ = 92.51%, THD = 74.56%,
DF = 3.96%

6.24 For M = 0.5, α_1 = 9°, α_2 = 28.13°,
α_3 = 42.75°, α_4 = 66.38°,
α_5 = 77.63°, $V(1)$ = 83.23%,
THD = 81.94%, DF = 3.44%

6.25 For M = 0.5, α_1 = 201.3°,
α_2 = 33.47°, α_3 = 63.73°,
α_4 = 79.10°, α_5 = 93.21°,
$V(1)$ = 72.05%, THD = 95.51%,
DF = 1.478%

6.26 α_1 = 12.53°, α_2 = 21.10°,
α_3 = 41.99°, α_4 = 46.04°

6.27 α_1 = 15.46°, α_2 = 24.33°,
α_3 = 46.12°, α_4 = 49.40°

6.28 α_1 = 10.084°, α_2 = 29.221°,
α_3 = 45.665°, α_4 = 51.681°

6.29 α_1 = 23.663°, α_2 = 33.346°

6.30 $T_1(\theta) = MT_s \sin(\pi/3 - \theta)$

6.31 V_r = 0.8∠29.994°

6.32 (a) G_{dc} = 0.833, G_{ac} = 2.5,
V_m = 1.667

6.33 $V_{o1(peak)}$ = 235.177 V,
$I_{o1(peak)}$ = 9.42 A, I_p = 9.44 A

6.34 f_e = 2799.7 Hz

6.35 $V_{o1(peak)}$ = 229.77 V,
$I_{o1(peak)}$ = 4.04 A, I_p = 4.04 A

6.36 C_e = 38.04 μF

CHAPTER 7

7.1 C_{J2} = 15 pF

7.2 dv/dt = 1.497 V/μs

7.3 $C_s = 10$ pF

7.4 **(a)** $C_s = 0.0392$ μF,
(b) $dv/dt = 375.5$ V/μs

7.5 **(a)** $dv/dt = 66009$ V/s,
$I_o = \pm3.15$ mA

7.6 $I_T = 999.5$ A

7.7 **(a)** $R = 222.22$ kΩ,
(b) $C_1 = 0.675$ μF

7.8 $R_1 = 4.545$ mΩ, $R_2 = 3.7037$ mΩ

CHAPTER 8

8.1 **(b)** $f_{\max} = = 10{,}772$ Hz,
(c) $V_{pp} = 239.86$ V, **(d)** $I_p = 29.65$ A,
(i) $I_{pk} = 29.65$ A, $I_R = 8.28$ A

8.2 **(a)** $I_{ps} = 35.34$ A, **(b)** $I_A = 15.57$ A,
(c) $I_R = 29.26$ A

8.3 **(a)** $I_p = 852$ A, **(b)** $I_A = 198$ A,
(c) $I_R = 364$ A,
(d) $V_{pp} = V_{c1} - V_c = 440$ V

8.4 **(a)** $I_p = 94.36$ A, **(b)** $I_A = 6.07$ A,
(c) $I_R = 21.1$ A, **(e)** $I_s = 12.14$ A

8.5 **(a)** $I_p = 114.6$ A, **(b)** $I_A = 17.12$ A,
(c) $I_R = 39.2$ A, **(e)** $I_s = 23.22$ A

8.6 **(a)** $I_p = 187.7$ A, **(b)** $I_A = 8.42$ A,
(c) $I_R = 35.14$ A, **(e)** $I_s = 33.69$ A

8.7 **(b)** $Q_s = 3.85$, **(c)** $L = 245.1$ μH,
(d) $C = 0.1654$ μF

8.8 **(a)** $V_{i(pk)} = 112.3$ V,
(b) $L = 21.66$ μH, **(d)** $C = 1.871$ μF

8.9 **(a)** $I_s = 15.7$ A, **(b)** $Q_s = 3.85$,
(c) $C = 2.45$ μF, **(d)** $L = 16.54$ μH

8.10 **(a)** $L_e = 12.74$ μH, $C_e = 0.69$ μF,
$L = 222.82$ μH, $C = 0.0479$ μF

8.11 **(a)** $C_f = 14.29$ μF, **(b)** $I_{L(rms)} =$
394 mA, $I_{L(dc)} = 200$ mA,
$I_{C(rms)} = 339.4$ mA, $I_{C(dc)} = 0$

8.12 $C = 0.1018$ μF, $L = 162.88$ μH

8.13 **(a)** $V_p = 32.32$ V, $I_p = 200$ mA,
(b) $I_{L3} = -100$ mA, $t_5 = 13.24$ μs

8.14 $k = 1.5$, $f_o/f_k = 7{,}653$

CHAPTER 9

9.1 THD $= 20.981\%$ for $m = 5$

9.2 $V_{sw} = 1.25$ kV, $V_{D1} = 3.75$ kV,
$V_{D2} = 2.5$ kV, $V_{D3} = 1.25$ kV

9.3 $I_{a(rms)} = 8.157$ A, $I_{b(rms)} = 12.466$ A,
$I_{C1(av)} = 4.918$ A, $I_{C2(av)} = 9.453$ A,
$I_{C1(rms)} = 11.535$ A,
$I_{C2(rms)} = 17.629$ A

9.4 THD $= 20.981\%$ for $m = 5$

9.5 $V_{sw} = 1.25$ kV, $V_{D1} = 1.25$ kV,
$V_{D2} = 1.25$ kV

9.6 For $m = 5$, 4 capacitors for diode
clamp, 10 for flying, 2 for cascaded.

9.7 $I_{rms} = 75$ A, $I_{rms} = 47.746$ A

9.8 $I_{S1(av)} = 7.377$ A, $I_{S2(av)} = 14.032$ A,
$I_{S2(av)} = 19.314$ A

9.9 $\alpha_1 = 12.834°$, $\alpha_2 = 29.908°$,
$\alpha_3 = 50.993°$, and $\alpha_4 = 64.229°$

9.10 $\alpha_1 = 30.653°$, $\alpha_2 = 47.097°$,
$\alpha_3 = 68.041°$, and $\alpha_4 = 59.874°$,
THD $= 38.5\%$,DF $= 4.1\%$

CHAPTER 10

10.1 **(a)** $\cap = 28.32\%$, **(d)** TUF $= 0.1797$

10.2 **(c)** $I_{av} = I_{dc} = 1.35$ A,
$I_R = I_{rms} = 3.75$ A, **(d)** PF $= 0.3126$

10.3 $I_1 = 5.745$ A

10.4 **(b)** DF $= 0.866$, **(c)** PF $= 0.827$

10.5 **(c)** $I_{av} = 1.35$ A, $I_R = 3.75$ A,
(d) PF $= 0.442$

10.6 **(c)** $I_2 = 5.5$ A

10.7 **(b)** DF $= 0.5$, **(c)** PF $= 0.45$

10.8 **(c)** $I_{av} = 1.35$ A, $I_R = 8.49$ A,
(d) PF $= 1.0$

10.9 $I_2 = 6.3$ A

10.10 $I_p = 20$ A, 44.12 A

10.11 **(a)** $\alpha_1 = 0°$, $\alpha_2 = 90°$,
(b) $I_{rms} = 18.97$ A, **(d)** PF $= 0.79$

10.12 **(a)** HF $= 37.26\%$, **(b)** DF $= 0.9707$,
(c) PF $= 0.9096$

10.13 (b) $I_{rms} = 21.53$ A, (d) $P_o = 4635$ W, $PF = 0.897$

10.14 (a) HF $= 31.08\%$, PF $= 0.827$ (lagging)

10.15 (a) HF $= 109.2\%$, (b) DF $= 0.5$, (c) PF $= 0.3377$

10.16 (d) $\eta = 35.75\%$, (e) TUF $= 10.09\%$, (f) PF $= 0.2822$

10.17 $I_3 = 1.16$ A

10.18 (a) HF $= 109.2\%$, (b) DF $= 0.5$, (c) PF $= 0.3377$

10.19 (e) TUF $= 0.1488$, (f) PF $= 0.3829$

10.20 (d) $\eta = 103.7\%$, (e) TUF $= 0.876$, (f) PF $= 0.8430$

10.21 $I_3 = 2.29$ A

10.22 (a) HF $= 31.08\%$, (b) PF $= 0.478$

10.23 (d) $\eta = 41.23\%$, (e) TUF $= 0.1533$, (f) PF $= 0.3717$

10.24 $I_6 = 2.48$ A

10.25 $I_p = 31.11$ A

10.26 (iv) $I_{rms} = 74.4$ A, (v) $I_{dc} = 67.25$ A

10.27 (iv) $I_{rms} = 58.71$ A, (v) $I_{dc} = 55.74$ A

10.28 (iv) $I_{rms} = 93.88$ A, (v) $I_{dc} = 54.2$ A

10.29 HF $= 31.08\%$, PF $= 0.827$

10.30 HF $= 31.08\%$, PF $= 0.827$

10.31 (a) HF $= 80.3\%$, (c) PF $= 0.7797$

10.32 (a) HF $= 102.3\%$, (c) PF $= 0.6753$

10.33 (b) HF $= 53.25\%$, (d) PF $= 0.8827$

10.34 (b) HF $= 58.61\%$, (d) PF $= 0.8627$

10.35 (a) $\mu = 19.33°$, (b) $\mu = 18.35°$

10.36 $t_G = 7.939$ μs

10.37 $t_G = 2.5$ μs

CHAPTER 11

11.1 $I_A = 6.74$ A, $I_R = 13.42$ A

11.2 (a) $k = 0.8333$, (b) PF $= 0.91$

11.3 (b) PF $= 0.9498$, (c) $I_D = 2.7$ A

11.4 $\alpha = 100.3°$

11.5 (b) PF $= 0.442$, (c) $I_A = 2.7$ A, (d) $I_R = 7.5$ A

11.6 (a) $\alpha = 62.74°$, (c) PF $= 0.88$, (d) $I_A = 26.25$ A, (e) $I_R = 49.99$ A

11.8 (e) $I_A = 7.76$ A, (f) PF $= 0.8045$

11.10 (b) PF $= 0.9892$

11.11 (c) PF $= 0.8333$

11.12 (b) PF $= 0.63$

11.13 (b) PF $= 0.978$

11.14 (b) $\alpha = 62°$, (c) PF $= 0.8333$

11.15 (b) PF $= 0.208$

11.18 (d) PF $= 0.894$, (e) $I_R = 26.31$ A

11.19 (b) $I_{R1} = 23.18$ A, (c) $I_{R3} = 10.45$ A, (d) PF $= 0.9$

11.20 (c) PF $= 0.315$

11.21 (c) PF $= 0.707$

11.23 (c) PF $= 0.147$

11.24 (c) PF $= 0.2816$

11.26 (b) $I_{AM} = 43.25$ A, (c) $V_P = 294.1$ V

11.27 (c) $I_{AM} = 11.3$ A, (d) $V_P = 3252.7$ V

11.28 (b) $C = 2070.5$ μF

CHAPTER 12

12.1 $I_R = 98.21$ A, $I_p = 196.42$ A, $I_{av} = 62.52$ A

12.2 $\phi = 25.84°$

12.3 $I_R = 98.21$ A, $I_p = 196.42$ A, $I_{av} = 62.52$ A

12.4 Diodes: $I_p = 196.42$ A, $I_{av} = 62.52$ A, $I_R = 98.21$ A, Thyristors: $I_R = 138.89$ A, $I_p = 196.42$ A, $I_{av} = 125.04$ A

12.5 $180n + 25.84°$, for $n = 0,1,2$

12.6 $I_p = 43.16$ A, $I_{av} = 13.74$ A, $I_R = 21.58$ A

12.7 $\phi = 30.68°$

12.8 $I_p = 24.92$ A, $I_{av} = 7.93$ A, $I_R = 24.92/2 = 12.46$ A

12.9 $I_p = 43.16$ A, $I_{av} = 13.74$ A, $I_R = 21.58$ A

12.10 (a) $I_p = 220$ A, (b) $t_1 = 8.05$ μs

12.11 $t_f = 950$ μs at $i(t) = 0.992$ A

12.12 (a) For thyristor T_1, $I_p = 311.23$ A, $I_{av} = 55$ A, $I_{rms} = 77.78$ A.
For thyristor T_3, $I_p = 311.23$ A, $I_{av} = 2.12$ A, $I_{rms} = 22.718$ A.
For thyristor T_2, $I_p = 311.23$ A, $I_{av} = 4.728$ A, $I_{rms} = 22.718$ A.
(b) $I_{rms} = 22.718$ A

12.13 $t_1 = 20.442$ μs, $t_f = 2100$ μs at $i(t) = 0.946$ A, $t_3 = t_1 + t_f = 2120$ μs

CHAPTER 13

13.1 (a) $I = 140.208$ A,
(b) $P = 25.267$ kW,
(c) $Q = 17.692$ kW

13.2 (a) $I = 140.012$ A,
(b) $P = 30.846$ kW,
(c) $Q = 19.451$ kW

13.3 (a) $\delta = 44.05°$, (b) $I = 203.75$ A,
(c) $Q_p = 37.363$ kW,
(e) $L = 8.48$ mH, (f) $\alpha = 18.64°$

13.4 (a) $V_{co} = 28.47$ kV,
(b) $V_{cpp} = 56.94$ kV, (c) $I_c = 2.96$ A,
(d) $I_{cp} = 5.121$ A.

13.5 (a) $I = 46.74$ A, (b) $P_p = 8.42$ kW,
(c) $Q_p = 8.27$ kW

13.6 (a) $r = 0.852$,
(b) Xcomp $= 13.636$ Ω,
(c) $I = 261.081$ A, (d) $\alpha = 79.7°$

13.7 (a) $r = 0.869$,
(b) Xcomp $= 15.636$ Ω,
(c) $C = 169.64$ μF, (d) $\alpha = 80.04°$

CHAPTER 14

14.1 (a) $I_s = 30$ A, (b) $\eta = 94.87\%$,
(e) $I_R = 21.21$ A, (f) $V_{oc} = 101.2$ V

14.2 (a) $I_s = 30$ A, (b) $\eta = 94.87\%$,
(e) $I_R = 42.43$ A, (f) $V_{oc} = 101.2$ V

14.3 (a) $I_s = 15$ A, (b) $\eta = 94.87\%$,
(e) $I_R = 21.21$ A, (f) $V_{oc} = 101.2$ V

14.4 (a) $I_s = 30$ A, (b) $\eta = 92.66\%$,
(e) $I_R = 21.21$ A, (f) $V_{oc} = 51.8$ V

14.5 (a) $I_s = 14.4$ A, (c) $I_p = 45.24$ A,
(e) $V_{oc} = V_s = 50$ V

14.6 (a) $I_s = 14.4$ A, (d) $I_R = 11.31$ A,
(e) $V_{oc} = 50$ V

14.7 $I_L = 60.32$ A

14.8 $I_L = 14.18$ A

14.9 $N_p = 132$, $N_s = 44$, $I_p = 3.178$ A

14.11 $N = 87$, $A_p = 12.202$ cm^2

14.12 $A_C = 1.32$ cm^2, $N = 32$

CHAPTER 15

15.1 (a) $I_a = 143.99$ A,
(b) $T_L = 294.64$ N·m

15.2 (a) $I_a = 148.4$ A,
(b) $T_L = 294.64$ N·m

15.3 (a) $E_g = 132.28$ V,
(b) $V_a = 141.97$ V,
(c) $I_{rated} = 50.86$ A, (d) speed regulation $= 7.32\%$

15.4 (a) $E_g = 280.28$ V,
(b) $V_a = 287.88$ V,
(c) $I_{rated} = 149.2$ A

15.5 (a) $I_f = 0.851$ A, (b) $\alpha_a = 100.96°$,
(c) PF $= 0.55$

15.6 (a) $T_d = 21.2$ N·m,
(b) $\omega = 2599.75$ rpm,
(c) PF $= 0.6455$

15.7 (a) $\alpha_f = 180°$, (b) $\alpha_a = 145.78°$,
(c) $P_a = 4258$ W

15.8 (a) $\alpha_a = 21.69°$,
(b) $\omega_o = 2063.8$ rpm,
(c) regulation $= 14.66\%$

15.9 (a) $\alpha_a = 30.87°$,
(b) $\omega_o = 2063.8$ rpm,
(c) regulation $= 14.66\%$

15.10 (a) $\alpha_a = 49.58°$, (b) $\omega = 1172.7$ rpm,
(c) $\alpha_f = 49.34°$

15.11 (a) $\alpha_a = 72.73°$, (b) $\omega = 1172.7$ rpm,
(c) $\alpha_f = 72.35°$

15.13 $= 157.17\,k - 1.8768$

15.14 (b) $R_{eq} = 1.778$ Ω, (c) $\omega = 575.35$ rpm, (d) $T_d = 3092.18$ N·m

15.15 **(c)** $R_{eq} = 0.937 \ \Omega$,
(d) $\omega_{min} = 63.67$ rpm,
(e) $\omega_{max} = 1427.84$ rpm,
(f) $\omega = 745.75$ rpm

15.16 **(c)** $R_{eq} = 2.08 \ \Omega$,
(d) $\omega = 1418.75$ rpm,
(e) $V_p = 1250$ V

15.17 $I_{max} = 17.68$ A

15.18 $I_{max} = (440/8) \tanh [2/(3u)]$

15.19 $I_{1s} = 5.53$ A

15.20 $I_{1s} = 112.54/[1 + (u \times 250/113.68)^2]$

15.21 **(b)** $\omega = 2990$ rpm

15.22 **(b)** $\omega = 1528.6$ rpm

15.23 $\omega = 2326.3$ rpm

15.24 **(b)** $V_r = 18.818$ V, **(c)** $= 1747$ rpm

15.25 **(b)** $\omega = 1575.2$ rpm,
(c) $\omega = 1397.9$ rpm,
(d) regulation $= 2.67\%$,
(e) regulation $= 0.172\%$

15.26 **(b)** $\omega = 2808$ rpm

15.27 **(b)** $\omega = 2887.2$ rpm

CHAPTER 16

16.1 **(b)** $s = 0.056$,
(c) $I_i = 137.86/-52.2°$ A,
(d) $P_i = 67,357$ W, **(e)** $PF_s = 0.613$,
(l) $s_m = \pm 0.0648$,
(m) $T_{mm} = 692.06$ N $\cdot$ m,
(n) $T_{mr} = -767.79$ N $\cdot$ m

16.2 **(a)** $\omega_s = 94.25$ rad/s,
(c) $I_i = 141.13/-52.9°$ A,
(d) $P_i = 67,874$ W, **(e)** $PF_s = 0.604$,
(k) $I_{rs} = 172.9$ A,
(m) $T_{mm} = 728.94$ N $\cdot$ m,
(n) $T_{mr} = -728.94$ N $\cdot$ m

16.3 **(c)** $I_i = 28.85/-52.2°$ A,
(j) $P_o = 18,450$ W, **(k)** $I_{rs} = 117.4$ A,
(l) $s_m = \pm 0.1692$,
(m) $T_{mm} = 90.85$ N $\cdot$ m

16.4 **(c)** $V_a = 217.6$ V,
(d) $I_i = 85.91/-69.01°$ A,

(g) $I_{r(max)} = 80.11$ A, **(i)** $T_a = 82.73$ N $\cdot$ m

16.5 **(d)** $I_i = 85.86/-74.19°$ A,
(g) $I_{r(max)} = 80.11$ A,
(i) $T_a = 82.73$ N $\cdot$ m

16.6 **(d)** $I_i = 78.71/-59.3°$ A,
(g) $I_{r(max)} = 91.47$ A,
(i) $T_a = 55.93$ N $\cdot$ m

16.7 **(a)** $R = 4.94 \ \Omega$, **(c)** $k = 0.6505$,
(e) $\eta = 75.5\%$, **(f)** $PF_s = 0.9$

16.8 **(a)** $R = 7.959 \ \Omega$, **(e)** $\eta = 75.5\%$,
(f) $PF_s = 0.9$

16.9 **(a)** $R = 2.791 \ \Omega$, **(c)** $k = 0.547$,
(d) $\eta = 79.5\%$, **(f)** $PF_s = 0.794$

16.10 **(c)** $\alpha = 112°$, **(d)** $\eta = 94.4\%$,
(e) $PF_s = 0.621$

16.11 **(c)** $\alpha = 102°$, **(d)** $\eta = 94.4\%$,
(e) $PF_s = 0.476$

16.12 **(b)** $I_d = 67.46$ A

16.13 **(a)** $= 58.137$ Hz,
(b) $\omega_m = 1524.5$ rpm

16.14 **(a)** $T_d = 114.25$ N $\cdot$ m,
(b) Torque change $= 117.75$ N $\cdot$ m

16.15 **(a)** $T_m = 850.43$ N $\cdot$ m,
(b) $T_m = 850.43$ N $\cdot$ m

16.16 **(a)** $T_m = 479$ N $\cdot$ m,
(b) $T_m = 305.53$ N $\cdot$ m

16.17 **(b)** $s = 0.08065$, **(c)** $\omega_m = 551.6$ rpm,
(d) $V_a = 138.62$ V,
(e) $PF_m = 0.6464$

16.18 **(b)** $s = 0.04053$,
(c) $\omega_m = 1151.3$ rpm,
(d) $V_a = 270.94$ V, **(e)** $PF_m = 0.64$

16.19 **(b)** $I_a = 83.17$ A, **(d)** $\delta = -18.48°$,
(e) $T_p = 2218$ N $\cdot$ m

16.20 **(a)** $\delta = -9.65°$, **(b)** $V_f = 65.44$ V,
(c) $T_d = 265.25$ N $\cdot$ m

16.21 **(a)** $\delta = -3.67°$, **(b)**
$I_a = 7.76$ A, $\Theta_m = 72.25°$,
(c) $PF = 0.3049$

16.22 **(a)** $T_p = 60°$, **(b)** $S_L = 10°$

16.23 **(a)** $T_p = 60°$, **(b)** $S_L = 10°$

16.24 **(a)** $T_L = 72°$, **(b)** $S_L = 7.2°$

CHAPTER 17

17.1 $R_1 = 6.2 \text{ k}\Omega$, $R_2 = 6.2 \text{ k}\Omega$,
$f_{\text{max}} = 3.871 \text{ kHz}$

17.2 $I_C = 100.5 \text{ A}$

17.3 $R_{B1} = 80 \ \Omega$, $R_{B2} = 758 \ \Omega$

17.4 $R_1 = 13.33 \text{ k}\Omega$, $R_2 = 13.33 \text{ k}\Omega$

17.5 $\alpha_{\text{min}} = 15.32°$, $\alpha_{\text{max}} = 78.81°$

17.6 **(a)** $\omega t = 3.325°$, $V_C = 2.001 \text{ V}$,
(b) $\omega t = 12.97°$

17.7 $R_1 = 105.42 \text{ k}\Omega$, $R_2 = 1.323 \text{ k}\Omega$

CHAPTER 18

18.1 $T_J(t = 5.5 \text{ ms}) = 125°C$

18.2 $T_J(t = 10 \text{ ms}) = 3°C$

18.3 $T_J(t = 20.01 \text{ ms}) = 35.13°C$

18.4 **(b)** $\delta_o = 0.27$, **(c)** $C = 0.2296 \ \mu\text{F}$,

(d) $R = 5.04 \ \Omega$,
(e) $dv/dt = 112 \text{ V}/\mu\text{s}$

18.5 **(b)** $V_p = 226.52 \text{ V}$, **(c)** $d = 0.0632$,
(d) $dv/dt = 4.5 \text{ V}/\mu\text{s}$

18.6 **(a)** $V_p = 273.2 \text{ V}$, **(b)**
$dv/dt = 49.5 \text{ V}/\mu\text{s}$,
(c) maximum $dv/dt = 0.23 \text{ V}/\mu\text{s}$

18.7 **(a)** $C = 3.872 \ \mu\text{F}$, **(b)** $R = 1.82 \ \Omega$

18.8 **(a)** $C = 14.5 \ \mu\text{F}$, **(b)** $R = 0.95 \ \Omega$

18.9 **(b)** $I_o = 111.47 \text{ A}$,
(c) $C = 754.94 \ \mu\text{F}$

18.10 $\delta = 0.75$, **(b)** $V_p = 235.34$

18.11 $N = 15$ diodes

18.12 $I_p = 125.45 \text{ A}$

18.13 $I_p = 259.33 \text{ A}$

18.14 $I_p = 328.63 \text{ A}$

18.15 **(c)** $I_p = 840.6 \text{ A}$, **(a)** $t_m = 7.64 \text{ s}$,
(b) $t_c = 11.46 \text{ s}$

Power Electronics Circuits, Devices, And Applications

Third Edition

Index